Practical Optimization Methods
With *Mathematica*® Applications

M. Asghar Bhatti

Practical Optimization Methods
With *Mathematica*® Applications

CD-ROM INCLUDED

M. Asghar Bhatti
Department of Civil and
  Environmental Engineering
University of Iowa
Iowa City, IA 52242
USA
mabhatti@uiowa.edu

Library of Congress Cataloging-in-Publication Data
Bhatti, M. Asghar
    Practical optimization methods : with Mathematica applications /
  M. Asghar Bhatti.
        p. cm.
    Includes bibliographical references and index.
    ISBN 0-387-98631-6 (alk. paper)
    1. Mathematical optimization—Data processing. 2. Mathematica
  (Computer file) I. Title
  QA402.5.B49 1998
  519.3—dc21   98-31038

Printed on acid-free paper.

Mathematica is a registered trademark of Wolfram Research, Inc.

© 2000 Springer-Verlag New York, Inc.
TELOS®, The Electronic Library of Science, is an imprint of Springer-Verlag New York, Inc.

This Work consists of a printed book and a CD-ROM packaged with the book, both of which are protected by federal copyright law and international treaty. The book may not be translated or copied in whole or in part without the written permission of the publisher (Springer-Verlag New York, Inc., 175 Fifth Avenue, New York, NY 10010, USA), except for brief excerpts in connection with reviews or scholarly analysis. For copyright information regarding the CD-ROM, please consult the printed information packaged with the CD-ROM in the back of this publication, and which is also stored as a "readme" file on the CD-ROM. Use of the printed version of this Work in connection with any form of information storage and retrieval, electronic adaptation, computer software, or by similar or dissimilar methodology now known, or hereafter developed, other than those uses expressly granted in the CD-ROM copyright notice and disclaimer information, is forbidden.

The use of general descriptive names, trade names, trademarks, etc., in this publication, even if the former are not especially identified, is not to be taken as a sign that such names, as understood by the Trade Marks and Merchandise Marks Act, may accordingly be used freely by anyone. Where those designations appear in the book and Springer-Verlag was aware of a trademark claim, the designations follow the capitalization style used by the manufacturer.

Production managed by Steven Pisano; manufacturing supervised by Jerome Basma.
Typeset by Integre Technical Publishing Co., Inc., Albuquerque, NM.
Printed and bound by Hamilton Printing Co., Rensselaer, NY.
Printed in the United States of America.

9 8 7 6 5 4 3 2 1

ISBN 0-387-98631-6 Springer-Verlag New York Berlin Heidelberg   SPIN 10693839

Preface

The goal of this book is to present basic optimization theory and modern computational algorithms in a concise manner. The book is suitable for undergraduate and graduate students in all branches of engineering, operations research, and management information systems. The book should also be useful for practitioners who are interested in learning optimization and using these techniques on their own.

Most available books in the field tend to be either too theoretical or present computational algorithms in a cookbook style. An approach that falls somewhere in between these two extremes is adopted in this book. Theory is presented in an informal style to make sense to most undergraduate and graduate students in engineering and business. Computational algorithms are also developed in an informal style by appealing to readers' intuition rather than mathematical rigor.

The available, computationally oriented books generally present algorithms alone and expect readers to perform computations by hand or implement these algorithms by themselves. This obviously is unrealistic for a usual introductory optimization course in which a wide variety of optimization algorithms are discussed. There are some books that present programs written in traditional computer languages such as Basic, FORTRAN, or Pascal. These programs help with computations, but are of limited value in developing understanding of the algorithms because very little information about the intermediate steps is presented. The user interface for these programs is primitive at best and typically requires creating user-defined functions and data files defining a specific problem.

This book fully exploits *Mathematica*'s symbolic, numerical, and graphical capabilities to develop thorough understanding of optimization algorithms. The problems are defined in the form of usual algebraic expressions. By using *Mathematica*'s symbolic manipulation capabilities, appropriate expressions for intermediate results are generated and displayed in customary style. *Mathematica*'s graphical capabilities are used in presenting the progress of numerical algorithms towards the optimum, thus creating a powerful environment for visual comparison of different optimization algorithms.

However, a knowledge of *Mathematica* is not a prerequisite for benefiting from this book. With its numerous examples and graphical illustrations, the book should be useful to anyone interested in the subject. Those familiar with *Mathematica* can further benefit by using the software on the accompanying CD to solve a variety of problems on their own to further develop their understanding. An appendix in included is the book for those interested in a concise introduction to *Mathematica*.

The arrangement of the book is fairly typical of other books covering similar topics. Methods applicable to special classes of problems are presented before those that are more general but computationally expensive. The methods presented in earlier chapters are more developed than those in the later chapters.

General guidelines for formulating problems in the form suitable for an optimization algorithm are presented in Chapter 1. The ideas are illustrated by several examples. Since cost plays an important role in many optimum design formulations, the chapter also contains a brief introduction to the time value of money. Optimization problems involving two variables can be solved very effectively using a graphics method. This topic is covered in detail in Chapter 2. A unique *Mathematica* implementation for a graphics solution is presented in this chapter also. Some of the basic mathematical concepts useful in developing optimization theory are presented in Chapter 3. Chapter 4 presents necessary and sufficient conditions for optimality of unconstrained and constrained problems. A *Mathematica* function is developed that uses necessary conditions to solve small-scale optimization problems. Chapter 5 presents numerical methods for solving unconstrained optimization problems. Several one-dimensional line-search methods are also discussed. All methods are implemented in the form of *Mathematica* functions. Chapter 6 presents a well-known simplex method for solving linear programming problems. Both the tableau form and the revised forms are discussed in detail. These problems can be solved using built-in *Mathematica* functions; however, equivalent new functions are implemented to show all the computational details. The chapter also includes a discussion of post-optimality analysis and presents *Mathematica* functions implementing these ideas. Interior point methods have gained popularity for solving linear and quadratic programming problems. Chapter 7 presents two simple but fairly successful methods belonging to this class for solving linear programming problems. Chapter 8 presents extensions of interior point methods for solving convex quadratic programming problems. This chapter also includes solution methods for special forms of quadratic programming problems that are generated as intermediate problems when solving more general constrained optimization problems. Chapter 9 presents numerical algorithms for solving general nonlinearly constrained problems. Two methods, generally considered among the best in their class, the augmented Lagrangian penalty function method, and the sequential quadratic programming method, are implemented as *Mathematica* functions and several numerical examples are presented.

The accompanying CD contains the implementation of all *Mathematica* functions discussed in the book. Together, these functions constitute what is referred to as the OptimizationToolbox in the text. The functions are written to follow the general steps presented in the text. Computational efficiency was not a major goal in creating these functions. Options are built in many functions to show intermediate calculations to enhance their educational value. The CD also contains additional examples showing the use of these *Mathematica* functions.

All chapters contain problems for homework assignments. Solutions to most homework problems can be generated using the functions defined in the OptimizationToolbox.

The author has been involved in teaching optimization methods to undergraduate engineering students for the past fifteen years. The book grew out of this experience. The book includes more material than what can be covered in a one-semester course. By selecting the most suitable material from this book, and possibly supplementing it with other material, a wide variety of courses can be taught. The first two chapters of the book can be incorporated into any introductory undergraduate engineering design course to give students a quick introduction to optimum design concepts. A one-semester undergraduate course on optimum design can be developed to cover Chapters 1, 2, 3, and 4 (excluding the last section), and parts from Chapters 5 through 9. A graduate course on optimum design can cover the entire book in one semester. A course suitable for management information and operations research students can cover material in Chapters 1 through 4 and 6 through 8 of the book.

Practicing engineers can become proficient users of optimization techniques in everyday design by studying the first two chapters followed by Chapters 5 and 9.

Installing OptimizationToolbox

The OptimizationToolbox consists of the following *Mathematica* packages.

1. CommonFunctions.m
2. EconomicFactors.m
3. GraphicalSolution.m
4. Chap3Tools.m
5. OptimalityConditions.m
6. Unconstrained.m
7. LPSimplex.m
8. InteriorPoint.m
9. QuadraticProgramming.m
10. ConstrainedNLP.m

The CommonFunctions.m package contains functions that are common to several other packages. The other nine packages implement functions described in each of the nine chapters of the book.

All these files must be placed in a folder (directory) called OptimizationToolbox. The best place to put this folder is inside the Mathematica 3.0/AddOns/Applications folder. These packages can then be imported (loaded) into any *Mathematica* notebook by simply executing a Needs command. There is no need to explicitly load the CommonFunctions.m package. It is loaded automatically when any of the other packages are loaded. For example, to import all functions defined in the OptimalityConditions.m package, one needs to execute the following command:

```
Needs["OptimizationToolbox`OptimalityConditions`"];
```

Using OptimizationToolbox

The CD includes *Mathematica* notebooks containing additional examples for each chapter. These notebooks are identified as Chap1CD.nb, Chap2CD.nb, and so forth. Each notebook contains a line for loading appropriate packages for that chapter. The rest of the notebook is divided into sections in the same way they appear in the book. If applicable, for each section, solutions of sample problems are provided using the appropriate OptimizationToolbox functions. Homework problems or any other problem can be solved by simply copying/pasting these examples and making appropriate modifications to define new problems.

M. Asghar Bhatti

Contents

Preface		v
1	**Optimization Problem Formulation**	**1**
	1.1 Optimization Problem Formulation	2
	1.2 The Standard Form of an Optimization Problem	13
	1.3 Solution of Optimization Problems	16
	1.4 Time Value of Money	18
	1.5 Concluding Remarks	31
	1.6 Problems	32
2	**Graphical Optimization**	**47**
	2.1 Procedure for Graphical Optimization	48
	2.2 GraphicalSolution Function	57
	2.3 Graphical Optimization Examples	59
	2.4 Problems	70
3	**Mathematical Preliminaries**	**75**
	3.1 Vectors and Matrices	75
	3.2 Approximation Using the Taylor Series	89
	3.3 Solution of Nonlinear Equations	100

3.4	Quadratic Forms	106
3.5	Convex Functions and Convex Optimization Problems	112
3.6	Problems	125

4 Optimality Conditions — 131

4.1	Optimality Conditions for Unconstrained Problems	132
4.2	The Additive Property of Constraints	144
4.3	Karush-Kuhn-Tucker (KT) Conditions	147
4.4	Geometric Interpretation of KT Conditions	165
4.5	Sensitivity Analysis	175
4.6	Optimality Conditions for Convex Problems	181
4.7	Second-Order Sufficient Conditions	187
4.8	Lagrangian Duality	199
4.9	Problems	208

5 Unconstrained Problems — 227

5.1	Descent Direction	229
5.2	Line Search Techniques—Step Length Calculations	231
5.3	Unconstrained Minimization Techniques	253
5.4	Concluding Remarks	302
5.5	Problems	303

6 Linear Programming — 315

6.1	The Standard LP Problem	316
6.2	Solving a Linear System of Equations	319
6.3	Basic Solutions of an LP Problem	334
6.4	The Simplex Method	339
6.5	Unusual Situations Arising During the Simplex Solution	365
6.6	Post-Optimality Analysis	376
6.7	The Revised Simplex Method	387
6.8	Sensitivity Analysis Using the Revised Simplex Method	402
6.9	Concluding Remarks	420
6.10	Problems	421

7 Interior Point Methods — 437

7.1	Optimality Conditions for Standard LP	438
7.2	The Primal Affine Scaling Method	445
7.3	The Primal-Dual Interior Point Method	464
7.4	Concluding Remarks	481

7.5	Appendix—Null and Range Spaces	481
7.6	Problems	486

8 Quadratic Programming   495
8.1	KT Conditions for Standard QP	495
8.2	The Primal Affine Scaling Method for Convex QP	502
8.3	The Primal-Dual Method for Convex QP	520
8.4	Active Set Method	535
8.5	Active Set Method for the Dual QP Problem	552
8.6	Appendix—Derivation of the Descent Direction Formula for the PAS Method	563
8.7	Problems	573

9 Constrained Nonlinear Problems   581
9.1	Normalization	582
9.2	Penalty Methods	585
9.3	Linearization of a Nonlinear Problem	608
9.4	Sequential Linear Programming—SLP	614
9.5	Basic Sequential Quadratic Programming—SQP	620
9.6	Refined SQP Methods	645
9.7	Problems	660

Appendix An Introduction to *Mathematica*   677
A.1	Basic Manipulations in *Mathematica*	678
A.2	Lists and Matrices	682
A.3	Solving Equations	689
A.4	Plotting in *Mathematica*	691
A.5	Programming in *Mathematica*	695
A.6	Packages in *Mathematica*	702
A.7	Online Help	703

Bibliography   705

Index   709

CHAPTER ONE

Optimization Problem Formulation

Optimization problems arise naturally in many different disciplines. A structural engineer designing a multistory building must choose materials and proportions for different structural components in the building in order to have a safe structure that is as economical as possible. A portfolio manager for a large mutual fund company must choose investments that generate the largest possible rate of return for its investors while keeping the risk of major losses to acceptably low levels. A plant manager in a manufacturing facility must schedule the plant operations such that the plant produces products that maximize company's revenues while meeting customer demands for different products and staying within the available resource limitations. A scientist in a research laboratory may be interested in finding a mathematical function that best describes an observed physical phenomenon.

All these situations have the following three things in common.

1. There is an overall goal, or objective, for the activity. For the structural engineer, the goal may be to minimize the cost of the building, for the portfolio manager it is to maximize the rate of return, for the plant manager it is to maximize the revenue, and for the scientist, the goal is to minimize the difference between the prediction from the mathematical model and the physical observation.

2. In addition to the overall goal, there usually are other requirements, or constraints, that must be met. The structural engineer must meet safety requirements dictated by applicable building standards. The portfolio manager must keep the risk of major losses below levels determined by the company's management. The plant manager must meet customer demands and work within available work force and raw material limitations. For the laboratory scientist, there are no other significant requirements.

3. Implicit in all situations is the notion that there are choices available that, when made properly, will meet the goals and requirements. The choices are known as optimization or design variables. The variables that do not affect the goals are clearly not important. For example, from a structural safety point of view, it does not matter whether a building is painted purple or pink, and therefore the color of a building would not represent a good optimization variable. On the other hand, the height of one story could be a possible design variable because it will determine overall height of the building, which is an important factor in ascertaining structural safety.

1.1 Optimization Problem

Formulation of an optimization problem involves taking statements, defining general goals and requirements of a given activity, and transcribing them into a series of well-defined mathematical statements. More precisely, the formulation of an optimization problem involves:

1. Selecting one or more optimization variables,
2. Choosing an objective function, and
3. Identifying a set of constraints.

The objective function and the constraints must all be functions of one or more optimization variables. The following examples illustrate the process.

1.1.1 Building Design

To save energy costs for heating and cooling, an architect is considering designing a partially buried rectangular building. The total floor space needed is 20,000 m^2. Lot size limits the building plan dimension to 50 m. It has already been decided that the ratio between the plan dimensions must be equal to the *golden ratio* (1.618) and that each story must be 3.5 m high. The heating and cooling costs are estimated at $100 per m² of the exposed surface area of the building. The owner has specified that the annual energy costs should not exceed $225,000. Formulate the problem of determining building dimensions to minimize cost of excavation.

Optimization Variables

From the given data and Figure 1.1, it is easy to identify the following variables associated with the problem:

n = Number of stories
d = Depth of building below ground
h = Height of building above ground
ℓ = Length of building in plan
w = Width of building in plan

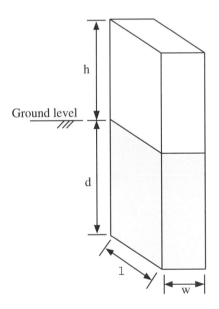

FIGURE 1.1 Partially buried building.

Objective Function

The stated design objective is to minimize excavation cost. Assuming the cost of excavation to be proportional to the volume of excavation, the objective function can be stated as follows:

$$\text{Minimize } d\ell w$$

Constraints

All optimization variables are not independent. Since the height of each story is given, the number of stories and the total height are related to each other as follows:

$$\frac{d+h}{n} = 3.5$$

Also, the requirement that the ratio between the plan dimensions must be equal to the golden ratio makes the two plan dimensions dependent on each other as follows:

$$\ell = 1.618w$$

The total floor space is equal to the area per floor multiplied by the number of stories. Thus, the floor space requirement can be expressed as follows:

$$n\ell w \geq 20{,}000$$

The lot size places the following limits on the plan dimensions:

$$\ell \leq 50 \qquad w \leq 50$$

The energy cost is proportional to the exposed building area, which includes the areas of the exposed sides and the roof. Thus, the energy budget places the following restriction on the design:

$$100(2h\ell + 2hw + \ell w) \leq 225{,}000$$

To make the problem mathematically precise, it is also necessary to explicitly state that the design variables cannot be negative.

$$\ell, w, h, d \geq 0 \qquad n \geq 1, \text{ must be an integer}$$

The complete optimization problem can be stated as follows:

Find (n, ℓ, w, h, d) in order to

Minimize $d\ell w$

Subject to
$$\left(\begin{array}{c} \frac{d+h}{n} = 3.5 \\ \ell = 1.618w \\ n\ell w \geq 20{,}000 \\ \ell \leq 50 \\ w \leq 50 \\ 100(2h\ell + 2hw + \ell w) \leq 225{,}000 \\ n \geq 1 \\ \ell, w, h, d \geq 0 \end{array}\right)$$

1.1.2 Plant Operation

A tire manufacturing plant has the ability to produce both radial and bias-ply automobile tires. During the upcoming summer months, they have contracts to deliver tires as follows.

Date	Radial tires	Bias-ply tires
June 30	5,000	3,000
July 31	6,000	3,000
August 31	4,000	5,000
Total	15,000	11,000

The plant has two types of machines, gold machines and black machines, with appropriate molds to produce these tires. The following production hours are available during the summer months:

Month	Gold machines	Black machine
June	700	1,500
July	300	400
August	1,000	300

The production rates for each machine type and tire combination, in terms of hours per tire, are as follows:

Type	Gold machines	Black machines
Radial	0.15	0.16
Bias-ply	0.12	0.14

The labor costs of producing tires are $10.00 per operating hour, regardless of which machine type is being used or which tire is being produced. The material costs for radial tires are $5.25 per tire, and those for bias-ply tires are $4.15 per tire. Finishing, packing, and shipping costs are $0.40 per tire. The excess tires are carried over into the next month but are subjected to an inventory carrying charge of $0.15 per tire. Wholesale prices have been set at $20 per tire for radials and $15 per tire for bias-ply.

How should the production be scheduled in order to meet the delivery requirements while maximizing profit for the company during the three-month period?

Optimization Variables

From the problem statement, it is clear that the only variables that the production manager has control over are the number and type of tires produced on each machine type during a given month. Thus, the optimization variables are as follows:

x_1	Number of radial tires produced in June on the gold machines
x_2	Number of radial tires produced in July on the gold machines
x_3	Number of radial tires produced in August on the gold machines
x_4	Number of bias-ply tires produced in June on the gold machines
x_5	Number of bias-ply tires produced in July on the gold machines
x_6	Number of bias-ply tires produced in August on the gold machines
x_7	Number of radial tires produced in June on the black machines
x_8	Number of radial tires produced in July on the black machines
x_9	Number of radial tires produced in August on the black machines
x_{10}	Number of bias-ply tires produced in June on the black machines
x_{11}	Number of bias-ply tires produced in July on the black machines
x_{12}	Number of bias-ply tires produced in August on the black machines

Objective Function

The objective of the company is to maximize profit. The profit is equal to the total revenue from sales minus all costs associated with the production, storing, and shipping.

Revenue from sales

$$= \$20(x_1 + x_2 + x_3 + x_7 + x_8 + x_9) + \$15(x_4 + x_5 + x_6 + x_{10} + x_{11} + x_{12})$$

Material costs

$$= \$5.25(x_1 + x_2 + x_3 + x_7 + x_8 + x_9)$$
$$+ \$4.15(x_4 + x_5 + x_6 + x_{10} + x_{11} + x_{12})$$

Labor costs

$$= \$10(0.15(x_1 + x_2 + x_3) + 0.16(x_7 + x_8 + x_9) + 0.12(x_4 + x_5 + x_6)$$
$$+ 0.14(x_{10} + x_{11} + x_{12}))$$

Finishing, packing, and shipping costs

$$= \$0.40(x_1 + x_2 + x_3 + x_4 + x_5 + x_6 + x_7 + x_8 + x_9 + x_{10} + x_{11} + x_{12})$$

The inventory-carrying charges are a little difficult to formulate. Assuming no inventory is carried into or out of the three summer months, we can determine the excess tires produced as follows:

Excess tires produced by June 30

$$= (x_1 + x_7 - 5{,}000) + (x_4 + x_{10} - 3{,}000)$$

Excess tires produced by July 31

$$= (x_1 + x_2 + x_7 + x_8 - 11{,}000) + (x_4 + x_5 + x_{10} + x_{11} - 6{,}000)$$

By assumption, there are no excess tires left by the end of August 31. At $0.15 per tire, the total inventory-carrying charges are as follows:

Inventory cost

$$= \$0.15((x_1 + x_7 - 5{,}000) + (x_4 + x_{10} - 3{,}000)$$
$$+ (x_1 + x_2 + x_7 + x_8 - 11{,}000) + (x_4 + x_5 + x_{10} + x_{11} - 6{,}000))$$

By adding all costs and subtracting from the revenue, the objective function to be maximized can be written as follows:

Maximize $3{,}750 + 12.55x_1 + 12.7x_2 + 12.85x_3 + 8.95x_4 + 9.1x_5 + 9.25x_6$
$+ 12.45x_7 + 12.6x_8 + 12.75x_9 + 8.75x_{10} + 8.9x_{11} + 9.05x_{12}$

Constraints

In a given month, the company must meet the delivery contracts. During July and August, the company can also use the excess inventory to meet the demand. Thus, the delivery contract constraints can be written as follows:

$$x_1 + x_7 \geq 5{,}000$$
$$x_4 + x_{10} \geq 3{,}000$$
$$x_1 + x_2 + x_7 + x_8 \geq 11{,}000$$
$$x_4 + x_5 + x_{10} + x_{11} \geq 6{,}000$$
$$x_1 + x_2 + x_3 + x_7 + x_8 + x_9 = 15{,}000$$
$$x_4 + x_5 + x_6 + x_{10} + x_{11} + x_{12} = 11{,}000$$

Note that the last two constraints are expressed as equalities to stay consistent with the assumption that no inventory is carried into September.

The production hours for each machine are limited. Using the time that it takes to produce a given tire on a given machine type, these limitations can be expressed as follows:

$$0.15x_1 + 0.12x_4 \leq 700$$
$$0.15x_2 + 0.12x_5 \leq 300$$
$$0.15x_3 + 0.12x_6 \leq 1{,}000$$
$$0.16x_7 + 0.14x_{10} \leq 1{,}500$$
$$0.16x_8 + 0.14x_{11} \leq 400$$
$$0.16x_9 + 0.14x_{12} \leq 300$$

The only other requirement is that all optimization variables must be positive.
The complete optimization problem can be stated as follows:

Find $(x_1, x_2, \ldots, x_{12})$ in order to

Maximize $3{,}750 + 12.55x_1 + 12.7x_2 + 12.85x_3 + 8.95x_4 + 9.1x_5 + 9.25x_6 + 12.45x_7 + 12.6x_8 + 12.75x_9 + 8.75x_{10} + 8.9x_{11} + 9.05x_{12}$

$$\text{Subject to} \begin{cases} x_1 + x_7 \geq 5{,}000 \\ x_4 + x_{10} \geq 3{,}000 \\ x_1 + x_2 + x_7 + x_8 \geq 11{,}000 \\ x_4 + x_5 + x_{10} + x_{11} \geq 6{,}000 \\ x_1 + x_2 + x_3 + x_7 + x_8 + x_9 = 15{,}000 \\ x_4 + x_5 + x_6 + x_{10} + x_{11} + x_{12} = 11{,}000 \\ 0.15x_1 + 0.12x_4 \leq 700 \\ 0.15x_2 + 0.12x_5 \leq 300 \\ 0.15x_3 + 0.12x_6 \leq 1{,}000 \\ 0.16x_7 + 0.14x_{10} \leq 1{,}500 \\ 0.16x_8 + 0.14x_{11} \leq 400 \\ 0.16x_9 + 0.14x_{12} \leq 300 \\ x_1, x_2, \ldots, x_{12} \geq 0 \end{cases}$$

1.1.3 Portfolio Management

A portfolio manager for an investment company is looking to make investment decisions such that investors will get at least a 10 percent rate of return while minimizing the risk of major losses. For the past six years the rates of return in four major investment types are as follows:

Type	Annual rates of return						
Year	1	2	3	4	5	6	Average
Blue chip stocks	18.24	12.12	15.23	5.26	2.62	10.42	10.6483
Technology stocks	12.24	19.16	35.07	23.46	−10.62	−7.43	11.98
Real estate	8.23	8.96	8.35	9.16	8.05	7.29	8.34
Bonds	8.12	8.26	8.34	9.01	9.11	8.95	8.6317

Optimization Variables

The portfolio manager must decide what percentage of the total capital to invest in each investment type. Thus, the optimization variables are as follows:

Chapter 1 Optimization Problem Formulation

x_1	Portion of capital invested in blue-chip stocks
x_2	Portion of capital invested in technology stocks
x_3	Portion of capital invested in real estate
x_4	Portion of capital invested in bonds

Objective Function

The objective is to minimize risk of losses. A measure of this risk is the amount of fluctuation in the rate of return from its average value. The *variance* of investment j is defined as follows:

$$v_{jj} = \frac{1}{n} \sum_{k=1}^{n} (r_{jk} - \mu_j)^2$$

where n = total number of observations, r_{jk} = rate of return of investment j for the kth observation (year in the example), and μ_j is the average value of the investment j. Using the numerical data given for this example, the variances are computed as follows:

$$v_{11} = \frac{1}{6}[(18.24 - 10.6483)^2 + \cdots + (10.42 - 10.6483)^2] = 29.0552$$

Similarly,

$$v_{22} = 267.344 \quad v_{33} = 0.3759 \quad v_{44} = 0.1597$$

From the definition, it is clear that the variance measures the risk within one investment type. To measure risk among different investment types, we define *covariance* between two investments i and j as follows:

$$v_{ij} = \frac{1}{n} \sum_{k=1}^{n} (r_{ik} - \mu_i)(r_{jk} - \mu_j)$$

Using the numerical data given for this example, the covariances are computed as follows:

$$v_{12} = \frac{1}{6}[(18.24 - 10.6483)(12.24 - 11.98) + \cdots$$
$$+ (10.42 - 10.6483)(-7.43 - 11.98)]$$
$$= 40.3909$$

Similarly, other covariances can easily be computed. All variances and covariances can be written in a *covariance matrix* as follows:

$$V = \begin{pmatrix} v_{11} & v_{12} & v_{13} & v_{14} \\ v_{21} & v_{22} & v_{23} & v_{24} \\ v_{31} & v_{32} & v_{33} & v_{34} \\ v_{41} & v_{42} & v_{43} & v_{44} \end{pmatrix} = \begin{pmatrix} 29.0552 & 40.3909 & -0.287883 & -1.9532 \\ 40.3909 & 267.344 & 6.83367 & -3.69702 \\ -0.287883 & 6.83367 & 0.375933 & -0.0566333 \\ -1.9532 & -3.69702 & -0.0566333 & 0.159714 \end{pmatrix}$$

Using the covariance matrix, the investment risk is written as the following function:

$$\text{Risk} = \begin{pmatrix} x_1 & x_2 & x_3 & x_4 \end{pmatrix} \begin{pmatrix} 29.0552 & 40.3909 & -0.287883 & -1.9532 \\ 40.3909 & 267.344 & 6.83367 & -3.69702 \\ -0.287883 & 6.83367 & 0.375933 & -0.0566333 \\ -1.9532 & -3.69702 & -0.0566333 & 0.159714 \end{pmatrix} \begin{pmatrix} x_1 \\ x_2 \\ x_3 \\ x_4 \end{pmatrix} \equiv x^T V x$$

Carrying out the matrix products, the objective function can be written explicitly as follows.

$$\text{Minimize } f = 29.0552 x_1^2 + 80.7818 x_2 x_1 - 0.575767 x_3 x_1 - 3.90639 x_4 x_1$$
$$+ 267.344 x_2^2 + 0.375933 x_3^2 + 0.159714 x_4^2 + 13.6673 x_2 x_3$$
$$- 7.39403 x_2 x_4 - 0.113267 x_3 x_4$$

Constraints

Since the selected optimization variables represent portions of total investment, their sum must equal the entire investment. Thus,

$$x_1 + x_2 + x_3 + x_4 = 1$$

The second constraint represents the requirement that the desired average rate of return must be at least 10 percent. It can be written as follows:

$$10.6483 x_1 + 11.98 x_2 + 8.34 x_3 + 8.6317 x_4 \geq 10$$

All optimization variables must also be positive.

$$x_i \geq 0 \quad i = 1, \ldots, 4$$

The complete optimization problem can be stated as follows:

Find $(x_1, x_2, \ldots, x_4)$ in order to

Minimize $29.0552x_1^2 + 80.7818x_2x_1 - 0.575767x_3x_1 - 3.90639x_4x_1 + 267.344x_2^2 + 0.375933x_3^2 + 0.159714x_4^2 + 13.6673x_2x_3 - 7.39403x_2x_4 - 0.113267x_3x_4$

Subject to $\left(\begin{array}{c} x_1 + x_2 + x_3 + x_4 = 1 \\ 10.6483x_1 + 11.98x_2 + 8.34x_3 + 8.6317x_4 \geq 10 \\ x_1, x_2, \ldots, x_4 \geq 0 \end{array} \right)$

1.1.4 Data Fitting

Finding the best function that fits a given set of data can be formulated as an optimization problem. As an example, consider fitting a surface to the data given in the following table:

Point	x	y	$z_{observed}$
1	0	1	1.26
2	0.25	1	2.19
3	0.5	1	0.76
4	0.75	1	1.26
5	1	2	1.86
6	1.25	2	1.43
7	1.5	2	1.29
8	1.75	2	0.65
9	2	2	1.6

The form of the function is first chosen based on prior knowledge of the overall shape of the data surface. For the example data, consider the following general form

$$z_{computed} = c_1 x^2 + c_2 y^2 + c_3 xy$$

The goal now is to determine the best values of coefficients c_1, c_2, and c_3 in order to minimize the sum of squares of error between the computed z values and the observed ones.

Optimization Variables

Values of coefficients c_1, c_2, and c_3

Objective Function

$$\text{Minimize } f = \sum_{i=1}^{9} [z_{\text{observed}}(x_i, y_i) - z_{\text{computed}}(x_i, y_i)]^2$$

Using the given numerical data, the objective function can be written as follows:

$$f = (1.26 - c_2)^2 + (2.19 - 0.0625c_1 - c_2 - 0.25c_3)^2 + \cdots$$
$$+ (1.6 - 4c_1 - 4c_2 - 4c_3)^2$$

or

$$f = 18.7 - 32.8462c_1 + 34.2656c_1^2 - 65.58c_2 + 96.75c_1c_2 + 84c_2^2 - 43.425c_3$$
$$+ 79.875c_1c_3 + 123.c_2c_3 + 48.375c_3^2$$

The complete optimization problem can be stated as follows:

Find $(c_1, c_2,$ and $c_3)$ in order to
Minimize $18.7 - 32.8462c_1 + 34.2656c_1^2 - 65.58c_2 + 96.75c_1c_2 + 84c_2^2 - 43.425c_3 + 79.875c_1c_3 + 123.c_2c_3 + 48.375c_3^2$

This example represents a simple application from a wide field known as Regression Analysis. For more details refer to many excellent books on the subject, e.g., Bates and Watts [1988].

1.2 The Standard Form of an Optimization Problem

As seen in the last section, a large class of situations involving optimization can be expressed in the following form:

Find a vector of optimization variables, $x = (x_1, x_2, \ldots, x_n)^T$ in order to
Minimize an objective (or cost) function, $f(x)$
Subject to

$g_i(x) \le 0$	$i = 1, 2, \ldots, m$	Less than type inequality constraints (LE)
$h_i(x) = 0$	$i = 1, 2, \ldots, p$	Equality constraints (EQ)
$x_{iL} \le x_i \le x_{iU}$	$i = 1, 2, \ldots, n$	Bounds on optimization variables

The bounds constraints are in fact inequality constraints. They are sometimes kept separate from the others because of their simple form. Certain numerical optimization algorithms take advantage of their special form in making the computational procedure efficient.

An equality constraint represents a given relationship between optimization variables. If the constraint expression is simple, it may be possible to solve for one of the variables from the equality constraint equation in terms of the remaining variables. After this value is substituted in all remaining constraints and the objective function, we have a problem with one less optimization variable. Thus, if the number of equality constraints is equal to the number of optimization variables, i.e., $p = n$, the problem is not of an optimization form. We just need to solve constraint equations to get the desired solution. If $p > n$, the problem formulation is not correct because some of the constraints must be dependent on others and therefore are redundant. Thus, for a general optimization problem, p must be less than n. Since the inequality constraints do not represent a specific relationship among optimization variables, there is no limit on the number of inequality constraints in a problem.

The problem as stated above will be referred to as the standard form of an optimization problem in the following chapters. Note that the standard form considers only minimization problems. It will be shown later that a maximization problem can be converted to a minimization problem by simply multiplying the objective function by a negative sign. Furthermore, the standard form considers only the less than type of inequality constraints. If a constraint is actually of the greater than type (GE), it can be converted to a less than type by multiplying both sides by a negative sign. Thus, the standard form is more general than it appears at first glance. In fact, it can handle all situations considered in this text.

1.2.1 Multiple Objective Functions

Sometimes there is more than one objective to be minimized or maximized. For example, we may want to maximize profit from an automobile that we are designing and at the same time minimize the possibility of damage to the car during a collision. These types of problems are difficult to handle because the objective functions are often contradictory.

One possible strategy is to assign weights to each objective function depending on their relative importance and then define a composite objective function as a weighted sum of all these functions, as follows:

$$f(x) = w_1 f_1(x) + w_2 f_2(x) + \cdots$$

where $w_1, w_2, \ldots$ are suitable weighting factors. The success of the method clearly depends on a clever choice of these weighting factors.

Another possibility is to select the most important goal as the single objective function and treat others as constraints with reasonable limiting values.

1.2.2 Classification of Optimization Problems

The methods for solving the general form of the optimization problem tend to be complex and require considerable numerical effort. Special, more efficient methods are available for certain specials forms of the general problem. For this purpose, the optimization problems are usually classified into the following types.

Unconstrained Problems

These problems have an objective function but no constraints. The data-fitting problem, presented in the first section, is an example of an unconstrained optimization problem. The objective function must be nonlinear (because the minimum of an unconstrained linear objective function is obviously $-\infty$). Problems with simple bounds on optimization variables can often be solved first as unconstrained. After examining different options, one can pick a solution that satisfies the bounds on the variables.

Linear Programming (LP) Problems

If the objective function and all the constraints are linear functions of optimization variables, the problem is called a linear programming problem. The tire plant management problem, presented in the first section, is an example of a linear optimization problem. An efficient and robust algorithm, called the Simplex method, is available for solving these problems.

Quadratic Programming (QP) Problems

If the objective function is a quadratic function and all constraint functions are linear functions of optimization variables, the problem is called a quadratic programming problem. The portfolio management problem, presented in the first section, is an example of a quadratic optimization problem. It is possible to solve QP problems using extensions of the methods for LP problems.

Nonlinear Programming (NLP) Problems

The general constrained optimization problems, in which one or more functions are nonlinear, are called nonlinear programming problems. The building design problem, presented in the first section, is an example of a general nonlinear optimization problem.

1.3 Solution of Optimization Problems

Different solutions discussed in this text are as follows:

1.3.1 Graphical Optimization

For problems involving two optimization variables, it is possible to obtain a solution by drawing contours of constraint functions and the objective function. This is a general and powerful method and also gives a great deal of insight into the space of all feasible choices. However, since the results are to be read from graphs, the method cannot give very precise answers. Furthermore, it is obviously not possible to use a graphical method for problems involving more than two optimization variables. Chapter 2 presents details of the graphical optimization procedure.

1.3.2 Optimality Criteria Methods

By extending the ideas of minimization of a function of a single variable—treated in elementary calculus textbooks—it is possible to develop necessary conditions for unconstrained and constrained optimization problems. When these necessary conditions are applied to a given problem, the result is a system of equations. A solution of this system of equations is a candidate for being minimum. Sufficient conditions are also available that can be used to select the optimum solution from all those that satisfy the necessary conditions. The method is suitable for problems with only a few optimization variables and constraints. For large problems, setting up and solving the system of equations becomes impractical. Chapter 3 presents the mathematical background necessary to understand optimality conditions and other methods discussed in later chapters in the book. Solutions based on the optimality conditions are discussed in Chapter 4.

1.3.3 Numerical Methods for Unconstrained Problems

There are several numerical methods that are available for solving unconstrained optimization problems. Some of the methods are Steepest descent, Conjugate gradient, and Newton's method. Different methods have their strengths and weaknesses. Chapter 5 presents several methods for unconstrained problems.

1.3.4 The Simplex Method

This is an efficient and robust algorithm for solving linear programming problems. The method can easily handle problems with thousands of variables and constraints. An extension of the basic algorithm can also handle quadratic programming problems. Chapter 6 presents a detailed treatment of the Simplex method. Both the standard tableau form and the revised matrix form are discussed.

1.3.5 Interior Point Methods for LP and QP Problems

These relatively recent methods are appealing because of their superior theoretical convergence characteristics. In practice, the Simplex method is still more widely used, however. Several methods for solving general nonlinear programming problems generate QP problems as intermediate problems. Interior point methods are becoming important for solving these intermediate QP problems. Chapters 7 and 8 are devoted to interior point methods for LP and QP problems, respectively.

1.3.6 Numerical Methods for General Nonlinear Programming Problems

Numerical methods for solving general nonlinear programming problems are discussed in Chapter 9. The so-called direct methods are based on the idea of finding a search direction by linearizing the objective and constraint functions and then taking a suitable step in this direction. Currently one of the most popular methods, known as the Sequential Quadratic Programming method, is based on this idea.

A fundamentally different approach is adopted by the so-called Penalty methods. In these methods, constraint functions are multiplied by suitable penalty functions and are added to the objective function. The result is an unconstrained optimization problem that can be solved by methods suitable for such problems. A penalty function is defined such that near constraint boundaries, a large positive value is added to the objective function. Since we are trying to minimize this function, the process leads to satisfaction of constraints and eventually to the minimum of the constrained problem.

1.4 Time Value of Money

As seen from the examples in section 1, many optimization problems involve minimizing cost or maximizing profit associated with an activity. Therefore, an understanding of economic factors—such as interest rates, depreciation, inflation, and taxes—is important for proper formulation of these problems. Long-term projects, such as the design of a large dam that may take several years to plan, design, and construct, obviously are directly influenced by these economic factors. Even for smaller projects, one may have to consider economic factors in order to properly assess life cycle cost (which includes initial cost plus cost of any repairs over its useful service life) of such projects.

Several compound interest formulas are derived in this section to gain an understanding of the time value of money. Simple examples explaining the use of these factors are also presented. Using these formulas, it is possible to compare between different alternatives, based on economic factors alone. The second subsection presents a few such examples. For additional details refer to books on engineering economy, e.g., Degarmo, Sullivan, Bontadelli, and Wicks [1997].

1.4.1 Interest Functions

Interest rates are the main reason why the value of money changes over time. This section reviews some of the basic relationships that are useful in computing changes in the value of money over time, assuming that the interest rates are known.

Single Payment Compound Amount Factor—spcaf

Consider the simplest case of determining the future value of a fixed amount of money invested:

$$P = \text{Amount invested} \qquad i = \text{Interest rate per period}$$

The appropriate period depends on the investment type. Most commercial banks base their computations assuming daily compounding. Some of the large capital projects may be based on monthly or annual compounding.

$$\text{Total investment at the end of first period} = P + iP = (1+i)P$$
$$\text{Total investment at the end of second period} = (1+i)P + i(1+i)P$$
$$= (1+i)^2 P$$

Continuing this process for n periods, it is easy to see that

$$\text{Total investment after } n \text{ periods, } S_n = (1+i)^n P \equiv \text{spcaf}[i, n]P$$

Example 1.1 A father deposits $2,000 in his daughter's account on her sixth birthday. If the bank pays an annual interest rate of 8%, compounded monthly, how much money will be there when his daughter reaches her sixteenth birthday?

$$i = 0.08/12 \qquad n = 120 \qquad \text{spcaf}[0.08/12, 120] = (1 + 0.08/12)^{120} = 2.21964$$
$$S_{120} = 2000 \times 2.21964 = \$4,439.28$$

Single Payment Present Worth Factor—sppwf

The inverse of the spcaf will give the present value of a future sum of money. That is,

$$P = (1+i)^{-n} S_n \equiv \text{sppwf}[i, n] S_n$$

Example 1.2 A person will need $10,000 in exactly five years from today. How much money should she deposit in a bank account that pays an annual interest rate of 9%, compounded monthly?

$$i = 0.09/12 \quad n = 60 \quad \text{sppwf}[0.09/12, 60] = (1 + 0.09/12)^{-60} = 0.6387$$

$$P = 10{,}000 \times 0.6387 = \$6{,}387$$

Uniform Series Compound Amount Factor—uscaf

Now consider a more complicated situation in which a series of payments (or investments) are made at regular intervals:

$$R = \text{Uniform series of amounts invested per period}$$
$$i = \text{Interest rate per period}$$

The first payment earns interest over $n - 1$ periods, and thus is equal to $(1 + i)^{n-1}R$.

The second payment earns interest over $n - 2$ periods, and thus is equal to $(1 + i)^{n-2}R$.

Continuing this process, the total investment after n periods is as follows:

$$S_n = (1+i)^{n-1}R + (1+i)^{n-2}R + \cdots + (1+i)R + R$$

This represents a geometric series whose sum, derived in most calculus textbooks, is as follows:

$$S_n = \frac{(1+i)^n - 1}{i} R \equiv \text{uscaf}[i, n] R$$

Example 1.3 A mother deposits $200 every month in her daughter's account starting on her sixth birthday. If the bank pays an annual interest rate of 8%, compounded monthly, how much money will be there when her daughter reaches her sixteenth birthday?

$$i = 0.08/12 \quad n = 120$$

$$\text{uscaf}[0.08/12, 120] = \frac{(1 + 0.08/12)^{120} - 1}{0.08/12} = 182.946$$

$$S_{120} = 200 \times 182.946 = \$36{,}589.2$$

1.4 Time Value of Money

Sinking Fund Deposit Factor—sfdf

The inverse of the uscaf can be used to compute uniform series of payments from a given future amount:

$$R = \frac{i}{(1+i)^n - 1} S_n \equiv \text{sfdf}[i, n] S_n$$

Example 1.4 A person will need $10,000 in exactly five years from today. How much money should he save every month in a bank account that pays an annual interest rate of 9%, compounded monthly?

$$i = 0.09/12 \quad n = 60 \quad \text{sfdf}[0.09/12, 60] = \frac{0.08/12}{(1 + 0.08/12)^{60} - 1} = 0.0136097$$

$$R = 10,000 \times 0.0136097 = \$136.1$$

Capital Recovery Factor—crf

The above formulas can be combined to give a simple formula to convert a given present amount to a series of uniform payments:

$$R = \frac{i}{(1+i)^n - 1} S_n = \frac{i}{(1+i)^n - 1}(1+i)^n P = \frac{i}{1 - (1+i)^{-n}} P$$

or

$$R = \frac{i}{1 - (1+i)^{-n}} P \equiv \text{crf}[i, n] P$$

Example 1.5 A car costs $15,000. You put $5,000 down and finance the rest for three years at 10% annual interest rate compounded monthly. What are your monthly car payments?

Cost = $15,000   Down payment = $5,000   Amount financed = $10,000

$$i = 0.1/12 \quad n = 36 \quad \text{crf}[0.1/12, 36] = \frac{0.1/12}{(1 + 0.1/12)^{36} - 1} = 0.0322672$$

$$R = 10,000 \times 0.0322672 = \$322.67$$

Uniform Series Present Worth Factor—uspwf

The inverse of the crf can be used to compute the present worth of a uniform series of payments:

$$P = \frac{1-(1+i)^{-n}}{i} R \equiv \text{uspwf}[i, n] R$$

Note that when n is large $(1+i)^{-n}$ goes to zero. Thus,

$$\text{uspwf}[i, \infty] = \frac{1}{i} \quad \text{and} \quad \text{crf}[i, \infty] = i$$

Example 1.6 A car costs $15,000. You put $5,000 down and finance the rest for three years at 10% annual interest rate compounded monthly. Your monthly payments are $322.67. Compute the amount needed to pay off the loan after making twelve payments.

One way to compute this is to figure out the principal and interest portions of each of the payments. Obviously, the principal left after the twelve payments will be the required amount. The computations are tedious by hand but can be easily handled with a loop structure in *Mathematica*.

We start by initializing variables for *i*, *monthlyPayment*, and *remainingLoan* to their appropriate values. From each payment, the amount of interest is equal to the interest rate times the remaining loan. This is identified as *interest* in the following computation. The remaining portion of the payment (identified as *principal*) goes to reduce the remaining loan. By repeating this calculation twelve times, we get the following table:

```
i = .1/12; monthlyPayment = 322.67;
remainingLoan = 10000;
tbl = {{"n", "Payment", "Interest", "Principal", "Remainingloan"}};

Do[interest = i * remainingLoan;
  principal = monthlyPayment - interest;
  remainingLoan = remainingLoan - principal;
  AppendTo[tbl, {n, monthlyPayment, interest, principal, remainingLoan}],
  {n, 1, 12}];

TableForm[tbl]
```

n	Payment	Interest	Principal	Remaining loan
1	322.67	83.3333	239.337	9760.66
2	322.67	81.3389	241.331	9519.33
3	322.67	79.3278	243.342	9275.99
4	322.67	77.2999	245.37	9030.62
5	322.67	75.2552	247.415	8783.21
6	322.67	73.1934	249.477	8533.73
7	322.67	71.1144	251.556	8282.17
8	322.67	69.0181	253.652	8028.52
9	322.67	66.9043	255.766	7772.76
10	322.67	64.773	257.897	7514.86
11	322.67	62.6238	260.046	7254.81
12	322.67	60.4568	262.213	6992.6

If we are not interested in all the details, and would only like to know how much we owe the bank after making twelve payments, a much easier way is to convert the remaining twenty-four payments to an equivalent single amount using uspwf as follows:

$$i = 0.1/12 \quad n = 24 \quad \text{uspwf}[0.1/12, 24] = \frac{1 - (1 + 0.1/12)^{-24}}{0.1/12} = 21.6709$$

Remaining loan after 12 payments $= P = 322.67 \times 21.6709 = \$6{,}992.53$

Summary and *Mathematica* Implementation

The interest formulas are summarized in the following table for easy reference.

Given	To find	Multiply given value by
P	S_n	$\text{spcaf}[i, n] = (1 + i)^n$
S_n	P	$\text{sppwf}[i, n] = (1 + i)^{-n}$
R	S_n	$\text{uscaf}[i, n] = \frac{(1+i)^n - 1}{i}$
S_n	R	$\text{sfdf}[i, n] = \frac{i}{(1+i)^n - 1}$
P	R	$\text{crf}[i, n] = \frac{i}{1-(1+i)^{-n}}$, $\text{crf}[i, \infty] = i$
R	P	$\text{uspwf}[i, n] = \frac{1-(1+i)^{-n}}{i}$, $\text{uspwf}[i, \infty] = \frac{1}{i}$

All these compound interest formulas have been implemented into simple *Mathematica* functions and are included in the OptimizationToolbox 'Economic-Factors' package. All functions take interest per period i and number of periods n as arguments and return the appropriate interest factor. The following line computes all these factors with $i = 0.08/12$ and $n = 12$.

```
i = 0.08/12; n = 12;
{uspwf[i,n], crf[i,n], sfdf[i,n], uscaf[i,n], sppwf[i,n], spcaf[i,n]}
```

{11.4958, 0.0869884, 0.0803218, 12.4499, 0.923361, 1.083}

1.4.2 Comparison of Alternatives Based on Economic Factors

The compound interest formulas derived in the previous section can be used to compare different alternatives based on economic factors. The main idea is to bring all costs/revenues associated with different options to a common reference point. The two commonly used approaches are as follows:

Annual-cost comparisons—for each alternative, all costs/revenues are expressed in terms of equivalent annual amounts.

Present-worth comparisons—for each alternative, all costs/revenues are expressed in terms of equivalent present amounts.

It becomes easy to make these comparisons if all financial transactions related to a particular alternative are represented on a time line showing the amount of the transaction and the time when the transaction takes place. Such diagrams are called *cash flow diagrams* and are used in the following examples. The following function, included in the OptimizationToolbox 'EconomicFactors' package, is useful in drawing these cash flow diagrams.

```
Needs["OptimizationToolbox`EconomicFactors`"];
?CashFlowDiagram
```

CashFlowDiagram[data, opts]. Draws a cash flow diagram. The
   data is expected in a two dimensional list. Each sublist
   consists of either 2 or 3 entries. The 2 entries are
   interpreted as {value, period}. The three entries are
   assumed as a uniform series {value, startPeriod,
   endPeriod}. The value may be either a numerical value or
   in the form of 'label'->value. The numerical value or the
   label, if specified, is placed on the diagram. The opts
   may be any valid option for the graphics Show function.

Example 1.7 *Purchase versus rent*   A company is considering the purchase of a new piece of testing equipment that is expected to produce $8,000 additional profit at the end of the first year of operation; this amount will decrease by $500 per year for each additional year of ownership. The equipment costs $18,000 and will have a salvage value of $3,000 after four years of use.

The same equipment could also be rented for $6,500 per year, payable in advance on the first day of each rental year. Use a present worth analysis to compare the alternatives. What is your recommendation to the company—purchase, rent, neither? The interest rate is 25% compounded annually.

(a) Option A—Purchase   If we choose this option, we need to spend $18,000 right away. The additional profits per year are $8,000, $7,500, $7,000, and $6,500. The salvage value of the equipment after four years produces an income of $3,000. Thus, the cash flow diagram for this option is as follows:

```
cost = 18000; p1 = 8000; p2 = 7500; p3 = 7000; p4 = 6500; salvage =
3000; i = .25;
```

```
CashFlowDiagram[{{-cost, 0}, {p1, 1}, {p2, 2}, {p3, 3}, {p4, 4}, {salvage, 4}},
   PlotRange -> All]
```

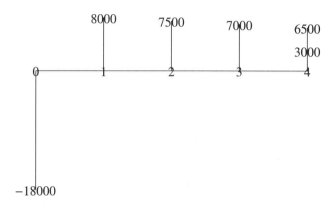

FIGURE 1.2 Cash flow diagram for purchase option.

The −$18,000 is already a present amount. The first income of $8,000 comes after one year (one period because annual compounding is assumed). To convert this amount to the present, we simply need to multiply it with sppwf with $n = 1$. Similarly, all other amounts can be brought to the present. Thus, the present worth of this option is as follows:

```
PWA = -cost + p1*sppwf[i,1] + p2*sppwf[i,2] + p3*sppwf[i,3]
      + p4*sppwf[i,4] + salvage*sppwf[i,4]
675.2
```

Thus, in terms of present dollars, this option results in additional profit of $675.20.

(b) Option B—Rent The costs associated with this option are annual rents. The rent for the following year is paid in advance. Since the company does not own the equipment, there is no salvage value. Thus, the cash flow diagram for this option is as follows:

```
rent = 6500;
```

```
CashFlowDiagram[{{-rent, 0, 3}, {p1, 1}, {p2, 2}, {p3, 3}, {p4, 4}},
 PlotRange -> All]
```

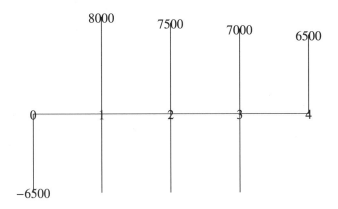

FIGURE 1.3 Cash flow diagram for purchase option.

The profits are converted to the present amount by multiplying them by appropriate sppwf, as in the case of option A. The rent is a uniform series of payments and thus can be converted to the present amount by using the uspwf. There is one subtle point to note, however. The derivation of uspwf assumes that the first payment is at the end of the first period. Thus, we can use the uspwf on the three rent payment, excluding the first one. The present worth of this option is as follows:

```
PWB =  - rent - rent*uspwf[i,3] + p1*sppwf[i,1] + p2*sppwf[i,2] +
 p3*sppwf[i,3] + p4*sppwf[i,4]
-1741.6
```

Thus in terms of present dollars, this option results in a loss of $1741.60.

(c) Conclusion The rent option results in a net loss. The company should purchase the equipment.

Example 1.8 *Water treatment facilities* A city is considering following two proposals for its water supply.

Proposal 1:

First dam and treatment plant—Construction cost = $500,000, annual operations cost = $36,000. The capacity of this facility will be enough for the community for twelve years.

Second dam and treatment plant—Construction cost = $480,000, additional annual operations cost = $28,000. The capacity of the expanded facility will be enough for the foreseeable future.

Proposal 2:

Large dam and first treatment plant—Construction cost = $700,000, annual operations cost = $37,000. The capacity of this facility will be enough for the community for fifteen years.

Second treatment plant—Construction cost = $100,000, additional annual operations cost = $26,000. The capacity of the expanded facility will be enough for the foreseeable future.

Use a present worth analysis to compare the alternatives. Assume annual interest rate of 10% and annual compounding.

(a) Proposal 1 The cash flow diagram for this option is as follows.

```
CashFlowDiagram[{{"Ist dam + plant" -> -1, 0}, {"Operations" -> -1/4, 1, 12},
  {"2nd dam + plant" -> -1, 12}, {"Operations" -> -1/2, 13, 16},
  {" → ∞" -> -1/2, 17}}, PlotRange -> All]
```

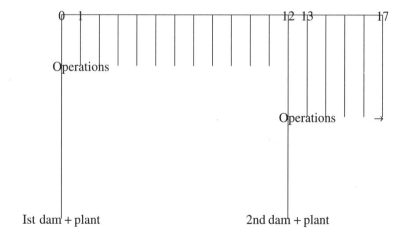

FIGURE 1.4 Cash flow diagram for the water supply problem—proposal 1.

The cost of the first dam and the treatment plant is already in terms of present dollars. Annual operations costs for twelve years can be brought to the present by multiplying by uspwf[i,12]. The cost of the second dam and the treatment plant can be brought to the present by multiplying by sppwf[i,12]. Starting from year thirteen, the operation costs increase and stay the same for the foreseeable future. If we multiply these costs by uspwf[i,∞], this will bring all these costs to a single amount at the year twelve. To bring it to the present, this single amount needs to be multiplied by sppwf[i,12]. Thus, the complete calculations for the present worth are as follows:

```
c1 = 500000; a1 = 36000; a2 = a1+28000; c2 = 480000; i = 0.1;

PW = c1 + a1 * uspwf[i, 12] + c2 * sppwf[i, 12]
    + a2 * uspwf[i, ∞] * sppwf[i, 12]
```

1.10216×10^6

(b) Proposal 2   The cash flow diagram for this option is as follows:

```
CashFlowDiagram[{{"Dam + 1st plant" -> -1, 0},
  {"Operations" -> -1/4, 1, 15}, {"2nd plant" -> -1, 15},
  {"Operations" -> 1/2, 16, 20}, {" → ∞" -> -1/2, 21}}, PlotRange -> All]
```

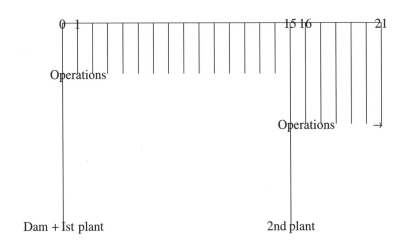

FIGURE 1.5   Cash flow diagram for the water supply problem—proposal 2.

The cost of the dam and the treatment plant is already in terms of present dollars. Annual operation costs for fifteen years can be brought to the present by multiplying by uspwf[i,15]. The cost of the second treatment plant can be brought to the present by multiplying by sppwf[i,15]. Starting from year sixteen, the operation costs increase and stay the same for the foreseeable future. If we multiply these costs by uspwf[i,∞], this will bring all these costs to a single amount at year fifteen. To bring it to the present, this single amount needs to be multiplied by sppwf[i,15]. Thus, the complete calculations for the present worth are as follows:

`c1 = 700000; a1 = 37000; a2 = a1+26000; c2 = 100000; i = 0.1;`

`PW = c1 + a1 * uspwf[i, 15] + c2 * sppwf[i, 15] + a2 * uspwf[i, ∞] * sppwf[i, 15]`

`1.15618 × 10⁶`

(c) Conclusion In terms of present dollars, the cost of the first proposal is 1.10216×10^6 and that of the second proposal is 1.15618×10^6. The first proposal is slightly cheaper for the city and should be adopted.

Example 1.9 *Optimization problem involving annual cost* A company requires open-top rectangular containers to transport material. Using the following data, formulate an optimum design problem to determine the container dimensions for minimum annual cost.

Construction costs	Sides = $65/m² Ends = $80/m² Bottom = $120/m²
Useful life	10 years
Salvage value	20% of the initial construction cost
Yearly maintenance cost	$12/m² of the outside surface area
Minimum required volume of the container	1200 m³
Nominal interest rate	10% (Annual compounding)

As seen from Figure 1.6, the design variables are the three dimensions of the container.

b = Width of container ℓ = Length of container h = Height of container

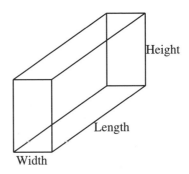

FIGURE 1.6 Open-top rectangular container.

In terms of these design variables, the different costs and container volume are as follows:

Construction cost $= 120\ell b + 2 \times 65\ell h + 2 \times 80 bh = 120\ell b + 130\ell h + 160bh$

Yearly maintenance cost $= 12(\ell b + 2\ell h + 2bh)$

Salvage value $= 0.2(120\ell b + 130\ell h + 160bh)$

Container volume $= bh\ell \geq 1200$

To determine the annual cost, it is useful to draw the cash flow diagram.

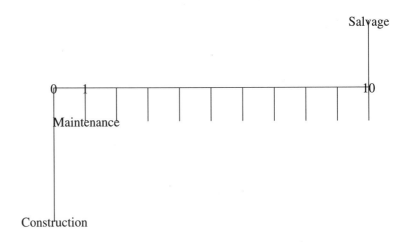

FIGURE 1.7 Cash flow diagram for the container design problem.

```
CashFlowDiagram[{{"Construction" -> -1,0},
  {"Maintenance" -> -1/4,1,10}, {"Salvage" > 1/2,10}},
  PlotRange -> All]
```

Using this cash flow diagram, it is easy to see that to convert all transactions to annual cost, we need to multiply the construction cost by crf[i,10] and the salvage value by sfdf[i,10]. Thus, the annual cost of the project is expressed as follows:

$$\text{Annual cost} = \text{crf}[0.1, 10](120\ell b + 130\ell h + 160bh) + 12(\ell b + 2\ell h + 2bh)$$
$$- \text{sfdf}[0.1, 10] \times 0.2(120\ell b + 130\ell h + 160bh)$$

$$\text{Annual cost} = 48.0314bh + 30.0236b\ell + 43.5255h\ell$$

The optimization problem can now be stated as follows:

Find b, h, and ℓ to

Minimize annual cost $= 48.0314bh + 30.0236b\ell + 43.5255h\ell$

Subject to $bh\ell \geq 1{,}200$ and b, h, and $\ell \geq 0$

1.5 Concluding Remarks

The importance of careful problem formulation cannot be overemphasized. A sophisticated optimization method is useless if the formulation does not capture all relevant variables and constraints. Unfortunately, it is difficult, if not impossible, to teach problem formulation in a textbook on optimization. The examples presented in textbooks such as this must introduce a variety of assumptions to make complex real-world situations easy to understand to readers from different backgrounds. Real-world problems can only be understood and properly formulated by those who are experts in the relevant fields.

The examples presented in this book are intended to provide motivation for studying and applying the optimization techniques. Little attempt has been made to use actual field data for the examples. Readers interested in seeing more engineering design examples formulated as optimization problems should consult Arora [1989], Ertas and Jones [1996], Hayhurst [1987], Hymann [1998], Papalambros and Wilde [1988], Pike [1986], Rao [1996], Stark and Nicholls [1972], Starkey [1988], and Suh [1990]. Excellent industrial engineering examples can be found in Ozan [1986] and Rardin [1998]. For a large collection of linear programming examples and discussion of their solution using commercially available software, see the book by Pannell [1997].

1.6 Problems

Optimization Problem Formulation

1.1. Hawkeye foods owns two types of trucks. Truck type I has a refrigerated capacity of 15 m³ and a nonrefrigerated capacity of 25 m³. Truck type II has a refrigerated capacity of 15 m³ and nonrefrigerated capacity of 10 m³. One of their stores in Gofer City needs products that require 150 m³ of refrigerated capacity and 130 m³ of nonrefrigerated capacity. For the round trip from the distribution center to Gofer City, truck type I uses 300 liters of fuel, while truck type II uses 200 liters. Formulate the problem of determining the number of trucks of each type that the company must use in order to meet the store's needs while minimizing the fuel consumption.

1.2. A manufacturer requires an alloy consisting of 50% tin, 30% lead, and 20% zinc. This alloy can be made by mixing a number of available alloys, the properties and costs of which are tabulated. The goal is to find the cheapest blend. Formulate the problem as an optimization problem.

	Available alloys				
Properties	A	B	C	D	E
Lead (%)	10	10	40	60	30
Zinc (%)	10	30	50	30	30
Tin (%)	80	60	10	10	40
Cost: ($/lb alloy)	8.2	9.3	11.2	13	17

1.3. Dust from an older cement manufacturing plant is a major source of dust pollution in a small community. The plant currently emits two pounds of dust per barrel of cement produced. The Environmental Protection Agency (EPA) has asked the plant to reduce this pollution by 85% (1.7 lbs/barrel). There are two models of electrostatic dust collectors that the plant can install to control dust emission. The higher efficiency model would reduce emissions by 1.8 lbs/barrel and would cost $0.70/barrel to operate. The lower efficiency model would reduce emissions by 1.5 lbs/barrel and would cost $0.50/barrel to operate. Since the higher efficiency model reduces more than the EPA required amount and the lower efficiency less than the required amount, the plant has decided to install one of each. If the plant has a capacity to produce 3 million barrels of cement per year, how many barrels of cement should be produced using each dust control model to meet the EPA requirements at a minimum cost? Formulate the situation as an optimization problem.

1.4. A small firm is capable of manufacturing two different products. The cost of making each product decreases as the number of units produced increases and is given by the following empirical relationships:

$$c_1 = 5 + \frac{1{,}500}{n_1} \qquad c_2 = 7 + \frac{2{,}500}{n_2}$$

where n_1 and n_2 are the number of units of each of the two products produced. The cost of repair and maintenance of equipment used to produce these products depends on the total number of products produced, regardless of its type, and is given by the following quadratic equation:

$$(n_1 + n_2)[0.2 + 2.3 \times 10^{-5}(n_1 + n_2) + 5.3 \times 10^{-9}(n_1 + n_2)^2]$$

The wholesale selling price of the products drops as more units are produced, according to the following relationships:

$$p_1 = 15 - 0.001 n_1 \qquad p_2 = 25 - 0.0015 n_2$$

Formulate the problem of determining how many units of each product the firm should produce to maximize its profit.

1.5. A company can produce three different types of concrete blocks, identified as A, B, and C. The production process is constrained by facilities available for mixing, vibration, and inspection/drying. Using the data given in the following table, formulate the production problem in order to maximize the profit.

	Blocks			Available
	A	B	C	
Mixing (hours/ batch)	1	3	9	900
Vibration (hours/batch)	2	3	6	1,200
Inspection/drying (hours/batch)	0.7	0.8	1	400
Profit: ($/batch)	7	17	30	

1.6. A thin steel plate, 10 in. wide and 1/2 in. thick and carrying a tensile force of $T = 150{,}000$ lbs, is to be connected to a structure through high strength bolts, as shown in Figure 1.8. The plate must extend by a fixed distance $L = 120$ in. from the face of the structure. The material for a steel plate can resist stress up to $F_u = 58{,}000$ lbs/in^2. The bolts are chosen as A325 type (standard bolts available from a number of steel manufacturers). The bolts are arranged in horizontal and vertical rows. The spacing between horizontal rows is $s = 3d$ inches, where d is the diameter of the bolt, and that between vertical rows is $g \geq 3d$ inches. The bolts must be at least $1.5d$ inches from the edges of the plate or the structure.

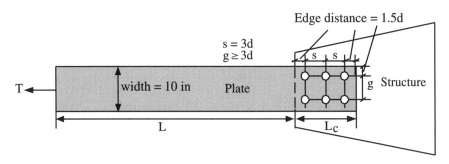

FIGURE 1.8 Steel plate connected to a structure using steel bolts.

The design problem is to determine the diameter of bolts, the total number of bolts required, and the arrangement of bolts (i.e., the number of horizontal and vertical rows) in order to minimize the total cost of the plate and the bolts. Assuming the cost of a steel plate is $2 per inch and that of a bolt is $5 per bolt, formulate the connection design problem as an optimization problem. The pertinent design requirements are as follows:

(a) The tensile force that one A325 bolt can support is given by the smaller of the following two values, each based on a different failure criteria.

Based on failure of bolt $= 12{,}000\pi d^2$ lbs

Based on failure of plate around bolt hole $= 2.4 dt F_u$ lbs

where $d =$ diameter of the bolt, $t =$ thickness of the plate, and $F_u =$ ultimate tensile strength of the plate material $= 58{,}000$ lbs/in^2.

(b) Vertical rows of bolts (those along the width of the plate) make the plate weaker. Taking these holes into consideration, the maximum tensile force that the plate can support is given by the following equation:

$$\text{Tensile load capacity of plate} = 0.75 F_u (w - n_v d_h)$$

where w = width of plate, n_v = number of bolts in one vertical row across the plate width, and d_h = bolt hole diameter ($= d + 1/8$ in).

(c) Practical requirements. The number of bolts in each row is the same. Each row must have at least two bolts, and there must be at least two rows of bolts. The smallest bolt diameter allowed is 1/2 in.

1.7. A mining company operates two mines, identified as A and B. Each mine can produce high-, medium-, and low-grade iron ores. The weekly demand for different ores and the daily production rates and operating costs are given in the following table. Formulate an optimization problem to determine the production schedule for the two mines in order to meet the weekly demand at the lowest cost to the company.

	Weekly Demand	Daily Production	
Ore grade	(tons)	Mine A (tons)	Mine B (tons)
High	12,000	2,000	1,000
Medium	8,000	1,000	1,000
Low	24,000	5,000	2,000
Operations cost ($/day)		210,000	170,000

1.8. A company manufactures fragile gift items and sells them directly to its customers through the mail. An average product weighs 12 kg, has a volume of 0.85 m^3, and costs $60 to produce. The average shipping distance is 120 miles. The shipping costs per mile based on total weight and volume are $0.006/kg plus $0.025/m^3. The products are shipped in cartons that are estimated to cost $2.5/m^3 and weigh 3.2 kg/m^3. The empty space in the carton is completely filled with a packing material to protect the item during shipping. This packing material has negligible weight but costs $0.95/m^3. Based on past experience, the company has developed the following empirical relationship between breakage and the amount of packing material:

$$\% \text{ breakage} = 85\left(1 - \frac{\text{Volume of packing material}}{\text{Volume of the shipping carton}}\right)$$

The manufacturer guarantees delivery in good condition which means that any damaged item must be replaced at the company's expense. Formulate an optimization problem to determine the shipping carton volume and volume of packing material that will result in the minimum overall cost of packing, shipping, and delivery.

1.9. Assignment of parking spaces for its employees has become an issue for an automobile company located in an area with a harsh climate. There are enough parking spaces available for all employees; however, some employees must be assigned spaces in lots that are not adjacent to the buildings in which they work. The following table shows the distances in meters between parking lots (identified as 1, 2, and 3) and office buildings (identified as A, B, C, and D). The number of spaces in each lot and the number of employees who need spaces are also tabulated. Formulate the parking assignment problem to minimize the distances walked by the employees from their parking spaces to their offices.

Parking lot	Distances from parking lot (m)				Spaces Available
	Building A	Building B	Building C	Building D	
1	290	410	260	410	80
2	430	350	330	370	100
3	310	260	290	380	40
# of employees	40	40	60	60	

1.10. An investor is looking to make investment decisions such that she will get at least a 10% rate of return while minimizing the risk of major losses. For the past six years, the rates of return in three major investment types that she is considering are as follows.

Type	Annual rates of return					
Stocks	18.24	17.12	22.23	15.26	12.62	15.42
Mutual funds	12.24	11.16	10.07	8.46	6.62	8.43
Bonds	5.12	6.26	6.34	7.01	6.11	5.95

Formulate the problem as an optimization problem.

1.11. Hawkeye Pharmaceuticals can manufacture a new drug using any one of the three processes identified as A, B, & C. The costs and quantities of ingredients used in *one batch* of these processes are given in the following table. The quantity of a new drug produced during each batch of different processes is also given in the table.

		Ingredients used per batch		
Process	Cost ($ per batch)	Ingredient I (tons)	Ingredient II (tons)	Quantity of drug produced (tons)
A	$12,000	3	2	2
B	$25,000	2	6	5
C	$9,000	7	2	1

The company has a supply of 80 tons of ingredient I and 70 tons of ingredient II at hand and would like to produce 60 tons of the new drug at a minimum cost.

Formulate the problem as an optimization problem. Clearly identify the design variables, constraints, and objective function.

1.12. For a chemical process, pressure measured at different temperatures is given in the following table. Formulate an optimization problem to determine the best values of coefficients in the following exponential model for the data.

$$\text{Pressure} = \alpha e^{\beta T}$$

Temperature ($T°C$)	Pressure (mm of Mercury)
20	15.45
25	19.23
30	26.54
35	34.52
40	48.32
50	68.11
60	98.34
70	120.45

1.13. Consider the cantilever beam-mass system shown in Figure 1.9. The beam cross-section is rectangular. The goal is to select cross-sectional dimensions (b and h) to minimize weight of the beam while keeping the fundamental vibration frequency (ω) larger than 8 rad/sec.

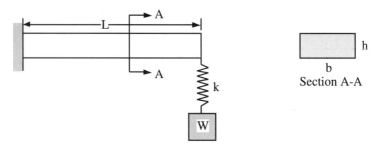

FIGURE 1.9 Rectangular cross-section cantilever beam with a suspended mass.

The numerical data and various equations for the problem are as follows.

Fundamental vibration frequency	$\omega = \sqrt{k_e/m}$ radians/sec
Equivalent spring constant, k_e	$\frac{1}{k_e} = \frac{1}{k} + \frac{L^3}{3EI}$
Mass attached to the spring	$m = W/g$
Gravitational constant	$g = 386$ in/sec^2
Weight attached to the spring	$W = 60$ lbs
Length of beam	$L = 15$ in
Modulus of elasticity	$E = 30 \times 10^6$ lbs/in^2
Spring constant	$k = 10$ lbs/in^2
Moment of inertia	$I = \frac{bh^3}{12}$ in^4
Width of beam cross-section	0.5 in $\leq b \leq 1$ in
Height of beam cross-section	0.2 in $\leq h \leq 2$ in
Unit weight of beam material	0.286 lbs/in^3

1.14. Modern aircraft structures and other sheet metal construction require stiffeners that are normally of I-, Z-, or C-shaped cross sections. An I

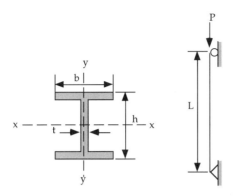

FIGURE 1.10 I-shaped steel stiffener.

shaped cross section, shown in Figure 1.10, has been selected for the present situation. Since a large number of these stiffeners are employed in a typical aircraft, it is important to optimally proportion the dimensions, b, t, and h.

For a preliminary investigation, a stiffener is treated as an axially loaded column, as shown in the figure. The goal is to have a least volume design while meeting the following yield stress and buckling load limits.

Yield stress limit	$P/A \leq \sigma_y$
Overall buckling load	$\frac{\pi^2 E I_{min}}{L^2} \geq P$
Flange buckling load	$0.43 \frac{\pi^2 E}{12(1-\nu^2)} \left(\frac{2t}{b}\right)^2 A \geq P$
Web buckling load	$4 \frac{\pi^2 E}{12(1-\nu^2)} \left(\frac{t}{h}\right)^2 A \geq P$
Area of cross-section, A	$A = 2bt + (h-2t)t$
Minimum moment of inertia, I_{min}	Min$[I_x, I_y]$ $I_x = \frac{1}{6}bt^3 + \frac{1}{2}bt(h-t)^2 + \frac{1}{12}t(h-2t)^3$ $I_y = \frac{1}{6}tb^3 + \frac{1}{12}t^3(h-2t)$
Applied load, P	2000 lbs
Yield stress, σ_y	25,000 lbs/in^2
Young's modulus, E	10^7 lbs/in^2
Poisson's ratio, ν	0.3
Length of stiffener, L	25 in

Formulate the problem in the standard optimization problem format. Clearly indicate your design variables and constraints. Use the given numerical data to express your functions in as simple a form as possible.

1.15. A major auto manufacturer in Detroit, Michigan needs two types of seat assemblies during 1998 on the following quarterly schedule:

	Type 1	Type 2
First quarter	25,000	25,000
Second quarter	35,000	30,000
Third quarter	35,000	25,000
Fourth quarter	25,000	30,000
Total	120,000	110,000

The excess seats from each quarter are carried over to the next quarter but are subjected to an inventory-carrying charge of $20 per thousand seats. However, assume no inventory is carried over to 1999.

The company has contracted with an auto-seat manufacturer that has two plants: one in Detroit and the other in Waterloo, Iowa. Each plant can manufacture either type of seat; however, their maximum capacities and production costs are different. The production costs per seat and the annual capacity at each of the two plants in terms of number of seat assemblies is given as follows:

	Quarterly capacity (either type)	Production cost for type 1	Production cost for type 2
Detroit plant	30,000	$225	$240
Waterloo plant	35,000	$165	$180

The packing and shipping costs from the two plants to the auto manufacturer are as follows:

	Cost/100 seats
Detroit plant	$10
Waterloo plant	$80

Formulate the problem to determine the seat acquisition schedule from the two plants to minimize the overall cost of this operation to the auto manufacturer for the year.

1.16. Consider the optimum design of the rectangular reinforced concrete beam shown in Figure 1.11. There is steel reinforcement near the bottom. Formwork is required on three sides during construction. The beam must support a given bending moment. A least-cost design is required.

The bending strength of the beam is calculated from the following formula:

$$M_u = 0.9 A_s F_y d \left(1 - 0.59 \left(\frac{A_s}{bd}\right)\left(\frac{F_y}{f'_c}\right)\right)$$

where F_y is the specified yield strength of the steel, and f'_c is the specified compressive strength of the concrete. The ductility requirements dictate minimum and maximum limits on the steel ratio $\rho = A_s/bd$.

$$\rho_{\min} \leq \rho \leq \rho_{\max}$$

Use the following numerical data.

Maximum steel ratio	$\rho_{\max} = 0.025$
Minimum steel ratio	$\rho_{\min} = 0.0033$
Required moment capacity	$M_u \geq 400 \times 10^3$ N-m
Minimum beam width	$b \geq 300$ mm
Concrete cover	$c = 65$ mm
Maximum beam depth	$h \leq 1200$ mm
Concrete cost	$100/m^3$
Formwork cost	$2/m^2$
Steel reinforcement cost	$610/ton (1 ton = 907.18 kg)
Density of steel	7850 kg/m^3
Yield stress of steel, F_y	420 MPa
Ultimate concrete strength, f'_c	35 MPa

Formulate the problem of determining the cross-section variables and the amount of steel reinforcement to meet all the design requirements at a minimum cost. Assume a unit beam length for cost computations.

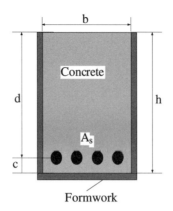

FIGURE 1.11 Reinforced concrete beam.

Clearly indicate your design variables and constraints. Use the given numerical data to express your functions in as simple a form as possible.

Time Value of Money

1.17. You deposit $5,000 in your savings account that pays an annual interest rate of 9%, compounded daily. How much money will be there in five years?

1.18. Collene plans to put enough money into fixed interest CDs to purchase a car for her son in three years. The car is expected to cost $15,000. How much money should she put into a CD that pays an annual interest rate of 8.5%, compounded daily?

1.19. Nancy is thinking about redecorating her bedroom, which is estimated to cost $9,000. If she can save $200 per month, how long will she have to wait till she has enough money for this project? The annual interest rate is 9.5%, compounded monthly.

1.20. A house costs $150,000. You put down 20% down payment and take out a 15-year mortgage for the rest at a fixed annual interest rate of 10%. Assuming monthly compounding, what will be your monthly payments?

1.21. A house costs $200,000. You put 20% down and take out a 20-year mortgage for the rest at a fixed annual interest rate of 10%. After five years (60 payments), you discover that the interest rates have fallen to 7% and you would like to refinance. If you refinance exactly the balance that is left after five years, what will be your new monthly payment? Assume monthly compounding.

1.22. A company is considering acquiring a new piece of equipment that is expected to produce $12,000 in additional revenue per year for the company. The equipment costs $50,000 and will require $1000 per year maintenance. It has a salvage value of $9,000 after seven years of use. The same equipment could also be leased for $8,500 per year, payable in advance on the first day of each year. The leasing company is responsible for maintenance costs.

Use a present worth analysis to compare the alternatives. What is your recommendation to the company—purchase, lease, neither? Assume that the interest rate is 12% compounded annually.

1.23. A company has designs for two facilities, A and B. Initial cost of the facilities, maintenance and operation costs, and additional profits from the facilities are estimated as follows:

Facility	Initial cost	Maintenance & operation costs per year	Profit/year
A	$1,000,000	$100,000	$250,000
B	$800,000	$80,000	$200,000

In addition to the above expenses, it is anticipated that the facilities will need renovation at the end of the tenth year of operation, costing $150,000 for A and $100,000 for B. Using the present worth method, which facility is more profitable? Assume a 20-year life for each facility with a salvage value of $200,000 for A and $160,000 for B. The prevailing interest rate is 9% per year.

1.24. A salesman is trying to sell a $55,000 piece of equipment to the Turner Construction Co. of Iowa City. Using this equipment, the company can generate net annual revenue before taxes of $15,000 over the next five years. The equipment has a salvage value of $5,000 at the end of five years. The tax laws allow straight-line depreciation of equipment, which means that the company can deduct $10,000 from its yearly taxable income. The income tax rate for the company is 34%. The company likes to make at least 8% annually on their investments after all taxes are paid. Would you advise the company to go ahead and make the necessary investment from its internal funds?

1.25. The Hawk City public library needs additional space to accommodate its rapidly expanding multimedia acquisitions. The proposal is to purchase an adjacent property at a cost of $1.2 million to be paid on January 1, 1998. The cost for the new building, including demolition of the existing building on the property, is estimated at $10 million to be paid on January 1, 2002. In addition, semi-annual maintenance is estimated at $50,000 starting on July 1, 2002.

Because of budget constraints, the city is unable to fund the proposal. The library board is considering charging a fee for its multimedia patrons to fund the project. Assuming that there are 18,000 patrons and that fees are collected each January 1 and July 1, with the first payment to start on July 1, 2002, what semiannual fee per user would be needed to support this proposal? Assume a nominal annual interest rate of 8% compounded semiannually and a 15-year period for cost recovery.

Optimization Problems Involving Economic Considerations

1.26. A chemical manufacturer requires an automatic reactor-mixer. The mixing time required is related to the size of the mixer and the stirring power as follows:

$$T = 1{,}000 \frac{\sqrt{S}}{P^2}$$

where S = capacity of the reactor-mixer, kg, P = power of the stirrer, k-Watts and T is the time taken in hours per batch. The cost of building the reactor-mixer is proportional to its capacity and is given by the following empirical relationship:

$$\text{Cost} = \$60{,}000\sqrt{S}$$

The cost of electricity to operate the stirrer is $0.05/k-W-hr and the overhead costs are $137.2 P$ per year. The total reactor to be processed by the mixer per year is 10^7 kg. Time for loading and unloading the mixer is negligible. Using present worth analysis, formulate the problem of determining the capacity of the mixer and the stirrer power in order to minimize cost. Assume a 5-year useful life, 9% annual interest rate compounded monthly, and a salvage value of 10% of the initial cost of the mixer.

1.27. Use the annual cost method in Problem 1.26.

1.28. A multicell evaporator is to be installed to evaporate water from a salt water solution in order to increase the salt concentration in the solution. The initial concentration of the solution is 5% salt by weight. The desired concentration is 10%, which means that half of the water from the solution must be evaporated. The system utilizes steam as the heat source. The evaporator uses 1 lb of steam to evaporate $0.8n$ lb of water, where n is the number of cells. The goal is to determine the number of cells to minimize cost. The other data are as follows:

The facility will be used to process 500,000 lbs of saline solution per day.

The unit will operate for 340 days per year.

Initial cost of the evaporator, including installation = $18,000 per cell.

Additional cost of auxiliary equipment, regardless of the number of cells = $9,000.

Annual maintenance cost = 5% of initial cost.

Cost of steam = $1.55 per 1,000 lbs.

Estimated life of the unit = 10 years.

Salvage value at the end of 10 years = $2,500 per cell.

Annual interest rate = 11%.

Formulate the optimization problem to minimize annual cost.

1.29. Use the present worth method in problem 1.28.

1.30. A small electronics company is planning to expand two of its manufacturing plants. The additional annual revenue expected from the two plants is as follows:

From plant 1: $0.00002x_1^2 - x_2$ From plant 2: $0.00001x_2^2$

where x_1 and x_2 are the investments made to upgrade the facilities. Each plant requires a minimum investment of $30,000. The company can borrow a maximum of $100,000 for this upgrade to be paid back in yearly installments in ten years at an annual interest rate of 12%. The revenue that the company generates can earn interest at an annual rate of 10%. After the 10-year period, the salvage value of the upgrades is expected to be as follows:

For plant 1: $0.1x_1$ For plant 2: $0.15x_2$

Formulate an optimization problem to maximize the net present worth of these upgrades.

CHAPTER TWO

Graphical Optimization

Graphical optimization is a simple method for solving optimization problems involving one or two variables. For problems involving only one optimization variable, the minimum (or maximum) can be read simply from a graph of the objective function. For problems with two optimization variables, it is possible to obtain a solution by drawing contours of constraint functions and the objective function. The procedure is discussed in detail in the first section. After developing the procedure in detail with *hand* calculations, a *Mathematica* function called GraphicalSolution is presented that automates the process of generating the complete contour plots. The second section presents solutions of several optimization problems using this function.

Practical optimization problems most likely will involve more than two variables, and thus, a direct graphical solution may not be feasible. For small-scale problems, it may still be possible to at least get an idea of the optimum by fixing some of the variables to reasonable values based on prior experience and obtaining a graphical solution for two of the most important variables. This is demonstrated in Section 2.1.1 by solving the building design problem formulated in Chapter 1, using the GraphicalSolution function. Perhaps the most important use of graphical optimization is that it enables a clearer understanding of the optimality conditions and the performance of numerical optimization algorithms presented in later chapters. Several options, such as GradientVectors and PlotHistory, are built into the GraphicalSolution function to make this process easier. Later chapters make extensive use of GraphicalSolution and relevant options to demonstrate several fundamental concepts. The function is used in generating *all* contour plots that appear in the textbook. To improve readability, the *Mathematica* code is not shown in all cases. However, the accompanying CD contains complete function calls for a large number of examples.

2.1 Procedure for Graphical Solution

Consider an optimization problem involving two optimization variables and written in the standard form as follows:

Find a vector of optimization variables, (x_1, x_2) in order to

Minimize an objective (or cost) function, $f(x_1, x_2)$

Subject to

$g_i(x_1, x_2) \leq 0 \quad i = 1, 2, \ldots, m \quad$ Less than type inequality constraints
$h_i(x_1, x_2) = 0 \quad i = 1, 2, \ldots, p \quad$ Equality constraints

A graphical solution of this problem involves the following three steps:

1. Choose an appropriate range for optimization variables.
2. Draw a contour for each constraint function to represent its boundary.
3. Draw several contours of the objective function.

The complete procedure is discussed in detail in this section. To make the process clear to all readers, the steps are presented as if the contour plots are to be *hand* drawn. Experienced *Mathematica* users undoubtedly are aware of the built-in ContourPlot function that makes most of the computations shown in the section unnecessary. However, it is instructive to go through at least one detailed example by hand before using the more automated tools. In fact, the GraphicalSolution function presented in the following section goes even a step further than the ContourPlot and completely automates the process of generating complete graphical optimization plots.

2.1.1 Choice of an Appropriate Range for Optimization Variables

The first task is to select a suitable range of values for the two optimization variables. A certain trial-and-error period is usually necessary before a suitable range is determined. The range must obviously include the optimum point. A large range may result in very small contour graphs making it difficult to differentiate between different constraints. On the other hand, a vary small range of variable values may not show the region that represents the optimum solution.

For practical problems, a suitable range can be determined based on prior experience with similar designs. For problems for which one has no prior knowledge of the optimum solution at all, the following procedure may be used.

1. Pick lower and upper limits for one of the variables, say for x_1. Arbitrary values can be chosen if nothing is known about the solution.
2. Solve for the corresponding x_2 values from each of the constraint equations. Set the minimum and maximum values of x_2 based on these values.

As an illustration, consider a graphical solution of the following problem:

Minimize $f(x_1, x_2) = 4x_1^2 - 5x_1x_2 + x_2^2$

Subject to $\begin{pmatrix} g(x_1, x_2) = x_1^2 - x_2 + 2 \leq 0 \\ h(x_1, x_2) = x_1 + x_2 - 6 = 0 \end{pmatrix}$

Assuming we have no idea of the optimum point, we arbitrarily pick lower and upper limits for x_1 such as 0 and 10, respectively. Next, we solve for the corresponding x_2 values from each of the constraint equations, treating all constraints as equalities.

$g:$ with $x_1 = 0$, $x_1^2 - x_2 + 2 = 0$ gives $x_2 = 2$

with $x_1 = 10$, $x_1^2 - x_2 + 2 = 0$ gives $x_2 = 102$

$h:$ with $x_1 = 0$, $x_1 + x_2 - 6 = 0$ gives $x_2 = 6$

with $x_1 = 10$, $x_1 + x_2 - 6 = 0$ gives $x_2 = -4$

Thus for our first trial, we select the range for x_1 as (0, 10) and that for x_2 as (−4, 102).

2.1.2 Contours of Constraint Functions — Feasible Region

The second task is to draw contours for constraint functions and determine the region over which all constraints are satisfied. This region is called the *feasible region*. Assuming that the constraints are written in the standard form (less than or equal to type with the right-hand side being 0), we need to draw lines representing equations $g_i(x_1, x_2) = 0$ and $h_i(x_1, x_2) = 0$. These contour lines can be drawn by selecting several values for one of the variables and solving for the other variable from the constraint function. That is, to plot the contour for constraint g_i, select a few different values for x_1 over the chosen range. For each selected value of x_1, compute the corresponding x_2 by solving the equation $g_i(x_1, x_2) = 0$. A line passing through pairs of such (x_1, x_2) values represents a contour for the function $g_i(x_1, x_2) = 0$. If a function is linear, obviously one needs only two points to draw such a contour. For nonlinear functions, several points are needed so that a reasonably smooth contour can be drawn.

Continuing with the example from the previous section, our task is to draw lines passing through points that satisfy the following equations:

$$\text{Boundary of constraint 1:} \quad x_1^2 - x_2 + 2 = 0$$

$$\text{Boundary of constraint 2:} \quad x_1 + x_2 - 6 = 0$$

To generate the data to draw these contours, we take several points along the x_1 axis and compute the corresponding x_2 values by solving the corresponding equation. The computed values are shown in the following table:

x_1	x_2 from $x_1^2 - x_2 + 2 = 0$	x_2 from $x_1 + x_2 - 6 = 0$
0	2.	6.
2.5	8.25	3.5
5.	27.	1.
7.5	58.25	−1.5
10.	102.	−4.

The graph in Figure 2.1 shows lines that pass through pairs of these points.

For an equality constraint, the contour $h_i(x_1, x_2) = 0$ represents the *feasible* line since any point on this line satisfies the given constraint. For an inequality constraint, the contour $g_i(x_1, x_2) = 0$ represents the boundary of the feasible side of the constraint. On one side of this contour, $g_i > 0$ and on the other side,

$g_i < 0$. Obviously, the side with $g_i < 0$ is the feasible side of the constraint since any point on this side will satisfy the constraint. The infeasible side of each inequality constraint is shown on the graph by shading that side of the constraint contour.

For the example problem, the graph in Figure 2.1 shows that the region below the g line is infeasible. The second constraint being an equality, the feasible region for the problem is that segment of the h line that is above the intersection with the g line.

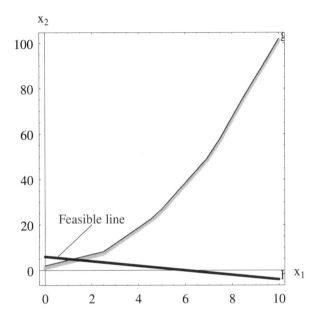

FIGURE 2.1 Initial graph for the example problem showing two constraint lines.

After all constraint contours are drawn, the *intersection* of feasible sides of all constraints represents the feasible region for the optimization problem. The points inside the feasible region satisfy all constraints. Depending on the problem formulation, the feasible region may be disjoint, bounded, unbounded, empty, or just a single point. An empty feasible region obviously means that the problem as formulated does not have a solution. One must go back and reconsider the formulation. A single feasible point also represents a special situation where no further consideration is necessary. There is only one suitable solution that meets all constraints, and thus there is no opportunity to minimize the objective function. In other cases, we need to go to the next step of drawing objective function contours to determine an optimum, if it exists.

2.1.3 Contours of the Objective Function

For a well-posed optimization problem—one with a nonempty feasible region—the next step is to select a point in the feasible domain that has the lowest value of the objective function. This is achieved by drawing a few (at least two) contours for the objective function to determine the direction in which the objective function is decreasing. Visually, one can then select a point that is in the feasible domain as well as has the lowest value of the objective function.

Drawing objective function contours poses one difficulty that was not there in the case of constraint function contours. In the case of constraint functions, it was natural to draw contours for zero function values because such contours represented constraint boundaries. However, in the case of objective functions, there is no standard set of values that can always be used. One must pick a few values of the objective function for which to draw the contours. Values selected totally randomly usually will not work because the corresponding contours may lie outside the range chosen for the graph. A reasonable strategy is to pick two distinct points on the graph and evaluate the objective function f at these two points. For most problems, using f values at the third point of the solution domain works well for the initial objective function contours. In this case, the two values are computed as follows:

$$c_1 = f\left(\frac{1}{3}(x_{1\max} - x_{1\min}), \frac{1}{3}(x_{2\max} - x_{2\min})\right)$$

$$c_2 = f\left(\frac{2}{3}(x_{1\max} - x_{1\min}), \frac{2}{3}(x_{2\max} - x_{2\min})\right)$$

Most likely, the values of f at these two points will be different, say c_1 and c_2. If not, different points are chosen until two distinct values of the objective function are found.

After selecting the contour values, the objective function contours can be drawn in a manner similar to that for the constraint function contours. That is, we pick a few arbitrary values for one of the variables, say x_1, and solve for corresponding x_2 values from $f(x_1, x_2) = c_1$. The line passing through these pairs of points represents a contour for the objective function value of c_1. Similarly, a contour corresponding to $f(x_1, x_2) = c_2$ is drawn. From these two contours, the general shape of the objective function and the direction in which

2.1 Procedure for Graphical Solution

it decreases should become apparent. One can now guess at the optimum point from the graph. To confirm this guess, one needs to draw two more contours, one corresponding to the minimum value of the objective function, and the other with a slightly lower value of f. This last contour should clearly be outside the feasible region.

Continuing with the example from the previous section, the third points of the chosen optimization space are computed as follows:

$$\text{Point 1:} \quad x_1 = \frac{1}{3}(10 - 0) = \frac{10}{3} \quad x_2 = \frac{1}{3}(102 - (-4)) = \frac{106}{3}$$

$$\text{Point 2:} \quad x_1 = \frac{2}{3}(10 - 0) = \frac{20}{3} \quad x_2 = \frac{2}{3}(102 - (-4)) = \frac{212}{3}$$

Evaluating the objective function values at these two points, we select the following two values for the objective function contours:

$$c_1 = f\left(\frac{10}{3}, \frac{106}{3}\right) = 700 \quad c_2 = f\left(\frac{20}{3}, \frac{212}{3}\right) = 2{,}810$$

Our task now is to draw lines passing through points that satisfy the following equations:

$$\text{Objective function contour 1:} \quad 4x_1^2 - 5x_1x_2 + x_2^2 = 700$$

$$\text{Objective function contour 2:} \quad 4x_1^2 - 5x_1x_2 + x_2^2 = 2{,}810$$

To generate the data to draw these contours, we take several points along the x_1 axis and compute the corresponding x_2 values by solving the corresponding equation. The computed values are shown in the following table.

x_1	x_2 from $4x_1^2 - 5x_1x_2 + x_2^2 = 700$	x_2 from $4x_1^2 - 5x_1x_2 + x_2^2 = 2{,}810$
0	26.4575	53.0094
2.5	32.9719	59.3919
5.	40.	66.0374
7.5	47.5	72.9401
10.	55.4138	80.0908

Plotting these contours together with the constraint contours, we get the graph shown in Figure 2.2.

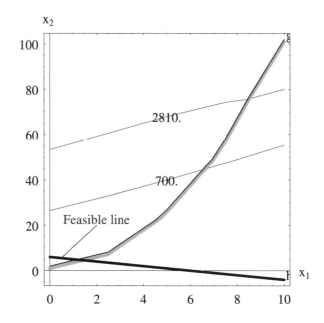

FIGURE 2.2 First complete graph of the example problem.

2.1.4 Solution of the Example Problem

In this section, we continue with the example problem. From Figure 2.2, we see that the feasible region (area where all constraints are satisfied) is the small line near the origin. It is clear that we should adjust ranges for both x_1 and x_2 to get an enlarged view of the feasible domain. Also, the contour lines for nonlinear functions are not very smooth. Thus, we should use more points in plotting these contours.

Second Attempt

For the second attempt, we repeat the steps used in the previous sections but with the following range:

$$x_{1min} = 0 \quad x_{1max} = 3 \quad x_{2min} = 0 \quad x_{2max} = 15$$

The third points of this domain, and the corresponding objective function values, are as follows:

Point 1: $x_1 = 1$ $x_2 = 5$ $f = 4$
Point 2: $x_1 = 2$ $x_2 = 10$ $f = 16$

The data points to draw the four functions are shown in the following table:

	x_2 from			
x_1	$f = 4$	$f = 16$	$g = 0$	$h = 0$
0	2. & −2	4. & −4	2.	6.
0.75	4.1697 & −0.4197	6.0302 & −2.28	2.5625	5.25
1.5	6.7604 & 0.7396	8.3394 & −0.8394	4.25	4.5
2.25	9.5481 & 1.7019	10.8586 & 0.3914	7.0625	3.75
3.	12.4244 & 2.5756	13.5208 & 1.4792	11.	3.

The resulting contours are shown in Figure 2.3.

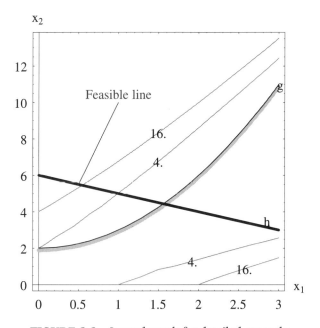

FIGURE 2.3 Second graph for detailed example.

Third Attempt

Figure 2.3 is a big improvement over Figure 2.2 but still shows the feasible region that is a little smaller than the overall graph. To get even a better graph, the limits are adjusted again and the process repeated. The resulting graph is shown in Figure 2.4.

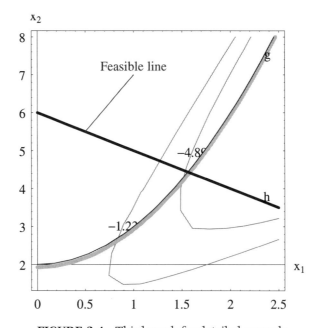

FIGURE 2.4 Third graph for detailed example.

Final Graph

Figure 2.4 shows the feasible region very clearly. The objective function contours show that the minimum point is at the intersection of the two constraints. Solving the two constraint equations simultaneously, we get the following coordinates of the intersection point:

$$\text{Optimum:} \quad x_1^* = 1.56 \quad x_1^* = 4.44 \quad f^* = -5.2$$

The final graph shown in Figure 2.5 shows several different objective function contours. Also, more points are used in drawing these contours for a smoother appearance. The contour with $f = -5.2$ just touches the feasible region. As the contour for $f = -10$ demonstrates, all contours with smaller objective function values will lie outside of the feasible region. Thus, the minimum point is as shown in the figure.

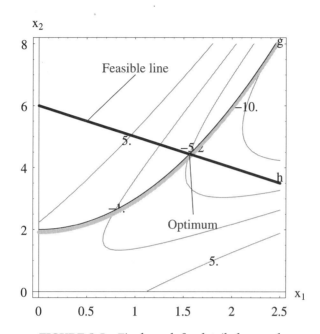

FIGURE 2.5 Final graph for detailed example.

2.2 GraphicalSolution Function

The procedure for graphical optimization described in section 2.1 is implemented in a *Mathematica* function called GraphicalSolution. The function internally calls the built-in function ContourPlot with appropriate contour values to be drawn. The function is included in the OptimizationToolbox 'GraphicalSolution'

```
Needs["OptimizationToolbox`GraphicalSolution`"];
?GraphicalSolution
```

GraphicalSolution[f,{x,xmin,xmax}, {y,ymin,ymax}, options]. The function
  draws contours of the objective function (f) and optionally those for

constraint functions. All contours are labelled. The function returns
a graphics object. See Options[GraphicalSolution] for a description of
different options for this function.

The function accepts the following options:

OptionsUsage[GraphicalSolution]

{ObjectiveContours → 5, ObjectiveContourLabels → True, Constraints → { },
ConstraintLabels → { }, ShadingOffset → 0.1, ShadingThickness → 0.015,
EQConstraintThickness → 0.01, PrintPoints → False, GradientVectors → { },
GradientVectorScale → 0.25, PlotHistory → { }}

ObjectiveContours: Number of objective function contours to be drawn or
   a list of specific values for which the contours are desired. Default
   is ObjectiveContours->5.

ObjectiveContourLabels: Option for GraphicalSolution to put labels
   on the objective function contours. Default is
   ObjectiveContourLabels->True.

Constraints: List of constraint expressions. A constraint can be
   specified as a simple expression in which case it is assumed as a '≤'
   constraint with a zero right hand side. A constraint can also be
   specified as left-hand-side '< , ≤ (both assumed as ≤)', '>, ≥ (both
   assumed as ≥)', or '==' right-hand-side. Such constraints are first
   converted to standard form ('≤' or '==' with right hand side zero).
   Because of these re-arrangements, the order of the constraints may be
   different than that used in the input list of constraints. Default is
   Constraints->{}.

ConstraintLabels: List of labels to be used to identify the constraints
   on the graph. Default is gi for LE constraints and hi for EQ
   constraints.

ShadingOffset: This factor is used to determine the offset for shading
   to show the infeasible side for inequalities. Default is
   ShadingOffset->0.1.

ShadingThickness: defines the thickness of shading to show the
   infeasible side for inequalities. Default is ShadingThickness->0.015.

EQConstraintThickness: defines the thickness of line to show the
   equality constraint. Default is EQConstraintThickness->0.01.

PrintPoints: Option for GraphicalSolution to print values of few data
   points used to draw each contour. Default is PrintPoints->False.

GradientVectors->{{f1,...},{f1_Label,...},{{x1,y1},...}}. Draw gradient
   vectors at given points. A list of functions, labels, and points
   where vectors are drawn is expected. By default no gradient vectors
   are drawn.

```
GradientVectorScale: Scale factor by which the gradient vectors are
   multiplied before plotting. Default is GradientVectorScale->0.25.

PlotHistory->{{x1,y1},...}. Shows a joined list of points on the plot.
   One use is to show history of search in numerical methods.
```

The last three options are useful in later chapters. The gradient vectors at selected points are useful to show physical interpretation of optimality conditions discussed in Chapter 4. The PlotHistory option superimposes a line joining the given points on the graph. This option is useful in later chapters to show the progress of numerical algorithms towards the optimum.

In addition to the above options, the function accepts all relevant options from standard *Mathematica* Graphics and ContourPlot functions. Particularly useful are the options to change the text format (using TextStyle options) and the Epilog option used to place labels and text on the graphs. The Epilog is used frequently in the following examples.

2.3 Graphical Optimization Examples

Several examples are presented in this section to illustrate the graphical optimization procedure and the use of the GraphicalSolution function.

2.3.1 Graphical Optimization Example

In this section, we consider the graphical solution of the following two variable optimization problems involving an inequality and an equality constraint:

Minimize $f = e^{x_1} - x_1 x_2 + x^2$

Subject to $\left(\begin{array}{l} g = 2x_1 + x_2 - 2 \leq 0 \\ h = x_1^2 + x_2^2 - 4 = 0 \end{array} \right)$

The following *Mathematica* expressions define the objective and the constraint functions:

```
f = e^x1 - x1 x2 + x2^2 ;
g = 2x1 + x2 - 2 ≤ 0;
h = x1^2 + x2^2 - 4 == 0;
```

Using GraphicalSolution, we get the left graph shown in Figure 2.6. Since the objective function values for which contours are to be drawn are not specified, five values are automatically selected internally. The feasible domain for this problem is that portion of the dark line (representing the equality constraint) which is on the feasible side of constraint g. From the contours of the objective function, it appears that the optimum will be somewhere in the third quadrant.

We redraw the graph with objective function contours drawn for $\{-1, 1, 3, 10\}$. The number of plot points is increased to 30 to get smoother contours. Also, the ShadingOffset is reduced to 0.06. This brings the line showing the infeasible side closer to the constraint line. Note that the individual graphs (gr1, gr2) are first generated but are not shown. Using the GraphicsArray function, the two graphs are displayed side-by-side.

```
gr1 = GraphicalSolution[f, {x₁, -4, 4}, {x₂, -4, 4}, Constraints → {g, h}];

gr2 = GraphicalSolution[f, {x₁, -4, 4}, {x₂, -4, 4}, Constraints → {g, h},
   ObjectiveContours → {-1, 1, 3, 10}, PlotPoints → 30, ShadingOffset → 0.06];

Show[GraphicsArray[{{gr1, gr2}}]];
```

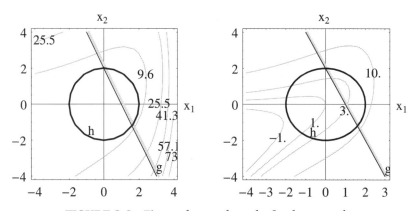

FIGURE 2.6 First and second graphs for the example.

From Figure 2.6, one can easily see that the minimum is near $f = -1$. Precise location of the minimum is difficult to determine from the graph. The following values are read directly from the graph and represent an approximate optimum.

```
xopt = {x₁ → -1.8, x₂ → -0.8};
f/.xopt
```
-0.634701

2.3 Graphical Optimization Examples

To confirm this solution, the final graph is drawn with this value included in the list of ObjectiveContours. Using the Epilog option of *Mathematica*'s built-in graphics functions, we can label the optimum point as well.

```
GraphicalSolution[f, {x₁, -4, 4}, {x₂, -4, 4}, Constraints → {g, h},
  ObjectiveContours → {-2, -0.64, 2}, PlotPoints → 30, ShadingOffset → 0.06,
  Epilog → {RGBColor[1, 0, 0], Line[{{-1.8, -0.8}, {-0.8, 0.2}}],
    Text["≈Optimum", {-0.8, 0.45}]}];
```

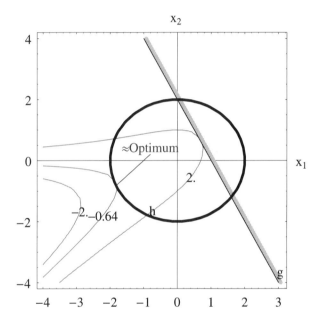

FIGURE 2.7 Final graph for the example.

2.3.2 Disjoint Feasible Region

Consider the solution of the following optimization problem involving two variables and one equality constraint.

Minimize $f = (x + 2)^2 + (y - 3)^2$

Subject to $3x^2 + 4xy + 6y - 140 = 0$

The following *Mathematica* expressions define the objective and the constraint function.

```
f = (x + 2)^2 + (y - 3)^2;
h = 3x^2 + 4xy + 6y - 140 == 0;
```

Using GraphicalSolution, we get the graph shown in Figure 2.8. The feasible region is disjoint and consists of points that lie on the two dark lines representing the equality constraint. The optimum is approximately where the $f = 32$ contour just touches the lower line representing the equality constraint. Note that this contour is still slightly below the upper dark line representing the equality constraint. The optimum solution is as follows:

$$\text{Approximate optimum:} \quad x^* = -7 \quad y^* = 0.3 \quad f^* = 32.2$$

```
GraphicalSolution[f, {x, -10, 10}, {y, -10, 10},
   Constraints → {h}, ObjectiveContours → {20, 32, 40, 50, 80}, PlotPoints → 50,
   Epilog → {RGBColor[1, 0, 0], Line[{{-7, 0.3}, {-2, 3}}],
      Text[Optimum, {-2, 3.5}]}];
```

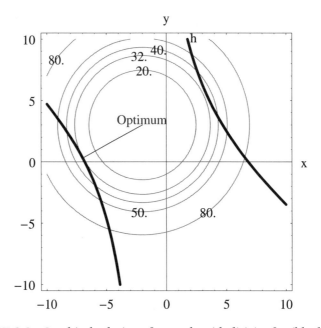

FIGURE 2.8 Graphical solution of example with disjoint feasible domain.

2.3 Graphical Optimization Examples

For a problem with a disjoint feasible region, the optimum solution can change drastically even with a small change in the constraints or the objective function. This example demonstrates this behavior. If the constant 3 in the objective function is changed to 3.5, the solution for the modified problem is as shown in Figure 2.9. The optimum is approximately where the $f = 32$ contour just touches the upper line representing the equality constraint. The optimum solution is as follows:

$$\text{Approximate optimum:} \quad x^* = 3.2 \quad y^* = 5.8 \quad f^* = 32.4$$

Comparing this solution to that of the original problem, we see that the objective function values are close. However, the optimum points have changed drastically.

```
f = (x + 2)^2 + (y - 3.5)^2;
h = 3x^2 + 4xy + 6y - 140 == 0;
GraphicalSolution[f, {x, -10, 10}, {y, -10, 10},
  Constraints → {h}, ObjectiveContours → {20, 32, 40, 50, 80}, PlotPoints → 50,
  Epilog → {RGBColor[1, 0, 0], Line[{{3.2, 5.8}, {-2, 3}}],
         Text["Optimum", {-2, 2.5}]}];
```

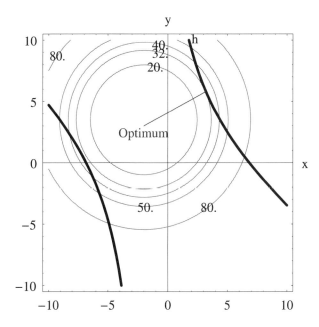

FIGURE 2.9 Graphical solution of example with slightly modified constraint.

2.3.3 Building Design

Consider the building design problem presented in Chapter 1. The problem statement is as follows.

To save energy costs for heating and cooling, an architect is considering designing a partially buried rectangular building. The total floor space needed is 20,000 m^2. Lot size limits the longer building dimensions to 50 m in plan. The ratio between the plan dimensions must be equal to the *golden ratio* (1.618) and each story must be 3.5 m high. The heating and cooling costs are estimated at $100 m^2, of the exposed surface area of the building. The owner has specified that the annual energy costs should not exceed $225,000. The objective is to determine building dimensions such that the cost of excavation is minimized.

A formulation for this problem involving five design variables is discussed in Chapter 1.

$$n = \text{Number of stories}$$
$$d = \text{Depth of building below ground}$$
$$h = \text{Height of building above ground}$$
$$\ell = \text{Length of building in plan}$$
$$w = \text{Width of building in plan}$$

Find (n, d, h, ℓ and w) in order to

Minimize $f = d\ell w$

Subject to
$$\begin{pmatrix} \frac{d+h}{n} = 3.5 \\ \ell = 1.618w \\ 100(2h\ell + 2hw + \ell w) \leq 22,5000 \\ \ell \leq 50 \\ w \leq 50 \\ n\ell w \geq 20000 \\ n \geq 1, d \geq 0, h \geq 0, \ell \geq 0 \text{ and } w \geq 0 \end{pmatrix}$$

For a graphical solution, we need to reduce the number of variables to two. Reducing the formulation to three variables is easy because of the two simple relationships between the optimization variables (equality constraints). Substituting $n = (d+h)/3.5$ and $\ell = 1.618w$, we get the following formulation in terms of three variables.

Find $(d, h,$ and $w)$ in order to

Minimize $f = 1.618dw^2$

Subject to
$$\begin{pmatrix} 100(5.236hw + 1.618w^2) \leq 225{,}000 \\ 1.618w \leq 50 \\ w \leq 50 \\ 0.462286(d+h)w^2 \geq 20{,}000 \\ d \geq 0, h \geq 0, \text{ and } w \geq 0 \end{pmatrix}$$

The third constraint is now redundant and can be dropped. In order to further reduce the number of variables to two, we must fix one of the variables to a reasonable value. For example, if we fix the total number of stories in the building to 25, the total building height is

$$d + h = 25 \times 3.5 = 87.5 \text{ m, giving } h = 87.5 - d$$

Thus in terms of two variables, the problem is as follows:

Find $(d$ and $w)$ in order to

Minimize $f = 1.618dw^2$

Subject to
$$\begin{pmatrix} 100(5.236(87.5 - d)w + 1.618w^2) \leq 225{,}000 \\ 1.618w \leq 50 \\ 40.45w^2 \geq 20{,}000 \\ d \geq 0 \text{ and } w \geq 0 \end{pmatrix}$$

The following *Mathematica* expressions define the objective and the constraint functions:

```
Clear[d, w];
f = 1.618dw²;
g = {100(5.236(87.5 - d)w + 1.618w²) ≤ 225000, 1.618w ≤ 50, 40.45w² ≥ 20000,
     d ≥ 0, w ≥ 0};
```

Using GraphicalSolution, we get the graph shown in Figure 2.10. Note that the objective function is divided by a factor of 10,000 to avoid a large number of digits on the labels of the objective function contours. The variable range and the objective function contours drawn obviously have been determined after several trials. The placement of labels has involved a few trials as well.

```
GraphicalSolution[f/10000, {w, 10, 50}, {d, 50, 100}, Constraints → g,
  AspectRatio → 1, PlotPoints → 25, ObjectiveContours → {3, 6, 9, 12, 18},
  Epilog → {RGBColor[1, 0, 0], Line[{{22.24, 75.05}, {15, 80}}],
    Text["Optimum", {15, 81}], Text["Feasible", {26, 90}]}];
```

Objectivefunction → 0.0001618dw²

$$\text{LEConstraints} \to \begin{pmatrix} g_1 \to -225000 + 100\,(5.236\,(87.5 - d)\,w + 1.618 w^2) \leq 0 \\ g_2 \to -50 + 1.618 w \leq 0 \\ g_3 \to 20000 - 40.45 w^2 \leq 0 \\ g_4 \to -d \leq 0 \\ g_5 \to -w \leq 0 \end{pmatrix}$$

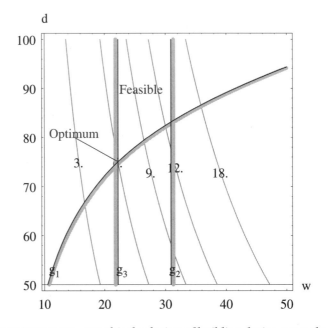

FIGURE 2.10 Graphical solution of building design example.

The optimum is clearly at the intersection of the first and the third constraint. The precise optimum can be located by computing the intersection of these two constraints as follows:

2.3 Graphical Optimization Examples

```
sol = Solve[{100 (5.236 (87.5 - d) w + 1.618w^2) == 225000, 40.45w^2 == 20000},
  {w, d}]
```
$\{\{d \to 75.0459, w \to 22.236\}, \{d \to 99.9541, w \to -22.236\}\}$

f/.sol[[1]]

60036.7

The first solution corresponds to the desired intersection point. Thus, the optimum solution is as follows:

Optimum: $\quad w^* = 22.24$ m $\quad d^* = 75.05$ m $\quad f^* = 60{,}036.7$ m^3

As you probably expected, this optimum solution indicates that we keep most of the building below ground. The height above ground is $h = 87.5 - 75.05 = 12.45$ m. The length in plan is $\ell = 1.618 \times 22.24 = 35.98$ m.

2.3.4 Portfolio Management

Consider the graphical solution of a simple two-variable portfolio management problem. An investor would like to invest in two different stock types to get at least a 15% rate of return while minimizing the risk of losses. For the past six years, the rates of return in the two stock types are as follows:

Type	Annual rate of return					
Blue chip stocks	15.24	17.12	12.23	10.26	12.62	10.42
Technology stocks	12.24	19.16	26.07	23.46	5.62	7.43

The optimization variables are as follows:

x	Portion of capital invested in Blue chip stocks
y	Portion of capital invested in Technology stocks

The given rates of return are used to define a two-dimensional list called *returns* as follows:

```
blueChipStocks = {15.24, 17.12, 12.23, 10.26, 12.62, 10.42};
techStocks = {12.24, 19.16, 26.07, 23.46, 5.62, 7.43};
returns = {blueChipStocks, techStocks}
```

```
{{15.24, 17.12, 12.23, 10.26, 12.62, 10.42},
 {12.24, 19.16, 26.07, 23.46, 5.62, 7.43}}
```

Using *Mathematica*'s Apply and Map functions (for a description of these functions, see the chapter on introduction to *Mathematica*), the average returns for each investment type are computed as follows:

```
averageReturns = Map[Apply[Plus,#]/Length[#]&,returns]
```

{12.9817, 15.6633}

In order to define the objective function, we need to compute the covariance matrix, *V*. Using the formula given in Chapter 1, the following function is defined to generate the covariance coefficient between any two investment types.

```
coVariance[x_, y_] := Module[{xb, yb, n = Length[x]},
    xb = Apply[Plus, x]/n;
    yb = Apply[Plus, y]/n;
    Apply[Plus, (x - xb) (y - yb)]/n
];
```

Using the coVariance function and *Mathematica*'s built-in Outer function, the complete covariance matrix *V* is generated by the following expression:

```
Vmat = Outer[coVariance, returns, returns, 1]; MatrixForm[Vmat]
```

$$\begin{pmatrix} 6.14855 & 0.403411 \\ 0.403411 & 60.2815 \end{pmatrix}$$

The objective function and the constraints can now be generated by the following expressions:

```
vars = {x, y};
f = Expand[vars.Vmat.vars]
```

$6.14855x^2 + 0.806822xy + 60.2815y^2$

```
g = {Apply[Plus, vars] == 1, averageReturns.vars ≥ 15}
```

{x + y == 1, 12.9817x + 15.6633y ≥ 15}

The graphical solution is obtained by using GraphicalSolution as follows:

```
GraphicalSolution[f, {x, 0., 0.5}, {y, 0.5, 1},
  Constraints → g, PlotPoints → 30,
  Epilog → {RGBColor[1, 0, 0], Line[{{.247, .753}, {.4, .8}}],
          Text[Optimum, {.4, .82}]}];
```

2.3 Graphical Optimization Examples

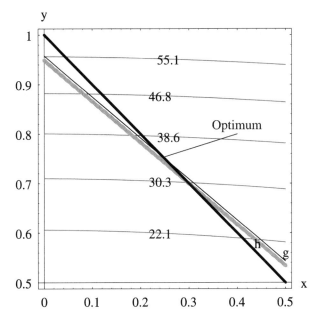

FIGURE 2.11 Graphical solution of portfolio management example.

The optimum is clearly at the intersection of the two constraints. The precise optimum can be located by computing the intersection of these two constraints as follows:

```
sol = Solve[{x + y == 1, 12.982x + 15.663y == 15}, {x, y}]
```

{{x → 0.247296, y → 0.752704}}

```
f/.sol[[1]]
```

34.6795

Thus, the optimum solution is as follows:

Optimum strategy: Blue chip stocks = 25%   Technology stocks = 75%
                                      Risk = 35

2.4 Problems

Use graphical methods to solve the following two variable optimization problems.

2.1. Minimize $f(x_1, x_2) = 3x_1 - x_2$

$$\text{Subject to} \begin{pmatrix} x_1 + x_2 \geq 3 \\ 2x_1 + 4x_2 \leq 2 \end{pmatrix}$$

2.2. Minimize $f(x_1, x_2) = 6x_1 + x_2$

$$\text{Subject to} \begin{pmatrix} 2x_1 + 7x_2 \geq 3 \\ 2x_1 - x_2 \geq 2 \end{pmatrix}$$

2.3. Maximize $f(x_1, x_2) = -6x + 9y$

$$\text{Subject to} \begin{pmatrix} x - y \geq 2 \\ 3x + y \geq 1 \\ 2x - 3y \geq 3 \end{pmatrix}$$

2.4. Minimize $f(x_1, x_2) = x^2 + 2y^2$

$$\text{Subject to} \begin{pmatrix} x + y \geq 1 \\ x, y \geq 0 \end{pmatrix}$$

2.5. Minimize $f(x_1, x_2) = x^2 + 2y^2 - 24x - 20y$

$$\text{Subject to} \begin{pmatrix} x + 2y \geq 0 \\ x + 2y \leq 9 \\ x + y \leq 8 \\ x + y \geq 0 \end{pmatrix}$$

2.6. Minimize $f(x_1, x_2) = x_1 + \frac{x_1}{x_2^2} + \frac{x_2}{x_1}$

$$\text{Subject to} \begin{pmatrix} x_1 + x_2 \geq 2 \\ x_1, x_2 > 0 \end{pmatrix}$$

2.7. Maximize $f(x_1, x_2) = (x_1 - 2)^2 + (x_2 - 10)^2$

$$\text{Subject to} \begin{pmatrix} x_1^2 + x_2^2 \leq 50 \\ x_1^2 + x_2^2 + 2x_1 x_2 - x_1 - x_2 + 20 \geq 0 \\ x_1, x_2 \geq 0 \end{pmatrix}$$

2.8. Minimize $f(x, y) = x^2 + 2yx + y^2 - 15x - 20y$

Subject to $\begin{pmatrix} x^2 + y^2 \leq 20 \\ x^2 - y^2 \leq 10 \end{pmatrix}$

2.9. Minimize $f(x, y) = \frac{1}{xy}$

Subject to $\begin{pmatrix} x + y \leq 5 \\ x, y \geq 1 \end{pmatrix}$

2.10. Minimize $f(x, y) = \frac{1000}{e^{2x+y}}$

Subject to $\begin{pmatrix} e^x + e^y \leq 20 \\ x, y \geq 1 \end{pmatrix}$

2.11. Hawkeye foods owns two types of trucks. Truck type I has a refrigerated capacity of 15 m³ and a nonrefrigerated capacity of 25 m³. Truck type II has a refrigerated capacity of 15 m³ and a nonrefrigerated capacity of 10 m³. One of their stores in Gofer City needs products that require 150 m³ of refrigerated capacity and 130 m³ of nonrefrigerated capacity. For the round trip from the distribution center to Gofer City, truck type I uses 300 liters of fuel, while truck type II uses 200 liters. Formulate the problem of determining the number of trucks of each type that the company must use in order to meet the store's needs while minimizing the fuel consumption. Use graphical methods to determine the optimum solution.

2.12. Dust from an older cement manufacturing plant is a major source of dust pollution in a small community. The plant currently emits 2 pounds of dust per barrel of cement produced. The Environmental Protection Agency (EPA) has asked the plant to reduce this pollution by 85% (1.7 lbs/barrel). There are two models of electrostatic dust collectors that the plant can install to control dust emission. The higher efficiency model would reduce emissions by 1.8 lbs/barrel and would cost $0.70/barrel to operate. The lower efficiency model would reduce emissions by 1.5 lbs/barrel and would cost $0.50/barrel to operate. Since the higher efficiency model reduces more than the EPA required amount and the lower efficiency less than the required amount, the plant has decided to install one of each. If the plant has the capacity to produce 3 million barrels of cement per year, how many barrels of cement should be produced using each dust control model to meet the EPA requirements at a minimum cost? Formulate the situation as an optimization problem. Use graphical methods to determine the optimum solution.

2.13. For a chemical process, pressure measured at different temperatures is given in the following table. Formulate an optimization problem to determine the best values of coefficients in the following exponential model for the data.

$$\text{Pressure} = \alpha e^{\beta T}$$

Temperature ($T°C$)	Pressure (mm of Mercury)
20	15.45
25	19.23
30	26.54
35	34.52
40	48.32
50	68.11
60	98.34
70	120.45

Use graphical methods to determine the optimum solution.

2.14. Consider the cantilever beam-mass system shown in Figure 2.0. The beam cross-section is rectangular. The goal is to select cross-sectional dimensions (b and h) to minimize weight of the beam while keeping the fundamental vibration frequency (ω) larger than 8 rad/sec. Use graphical methods to determine the optimum solution.

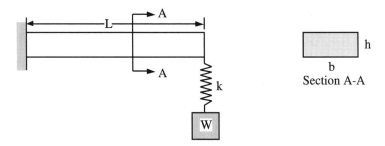

FIGURE 2.12 Rectangular cross-section cantilever beam with a suspended mass.

The numerical data and various equations for the problem are as follows:

Fundamental vibration frequency	$\omega = \sqrt{k_e/m}$ radians/sec
Equivalent spring constant, k_e	$\frac{1}{k_e} = \frac{1}{k} + \frac{L^3}{3EI}$
Mass attached to the spring	$m = W/g$
Gravitational constant	$g = 386$ in/sec^2
Weight attached to the spring	$W = 60$ lbs
Length of beam	$L = 15$ in
Modulus of elasticity	$E = 30 \times 10^6$ lbs/in^2
Spring constant	$k = 10$ lbs/in^2
Moment of inertia	$I = \frac{bh^3}{12}$ in^4
Width of beam cross-section	0.5 in $\leq b \leq 1$ in
Height of beam cross-section	0.2 in $\leq h \leq 2$ in
Unit weight of beam material	0.286 lbs/in^3

2.15. A small electronics company is planning to expand two of its manufacturing plants. The additional annual revenue expected from the two plants is as follows.

From plant 1: $0.00002x_1^2 - x_2$ From plant 2: $0.00001x_2^2$

where x_1 and x_2 are the investments made into upgrading the facilities. Each plant requires a minimum investment of $30,000. The company can borrow a maximum of $100,000 for this upgrade to be paid back in yearly installments in ten years at an annual interest rate of 12%. The revenue that the company generates can earn interest at an annual rate of 10%. After the 10-year period, the salvage value of the upgrades is expected to be as follows:

For plant 1: $0.1x_1$ For plant 2: $0.15x_2$

Formulate an optimization problem to maximize the net present worth of these upgrades. Use graphical methods to determine the optimum solution.

CHAPTER THREE

Mathematical Preliminaries

This chapter presents some of the basic mathematical concepts that are used frequently in the later chapters. A review of matrix notation and some of the basic concepts from linear algebra are presented in the first section. Taylor series approximation, which plays an important role in developing various optimization techniques, is presented in section 2. Definitions of the gradient vector and the Hessian matrix of functions of n variables are introduced in this section also. As an application of the Taylor series, and also since nonlinear equations are encountered frequently in solving optimization problems, the Newton-Raphson method for the numerical solution of nonlinear equations is presented in the third section. A discussion of quadratic functions and convex functions and sets is included in the last two sections.

3.1 Vectors and Matrices

3.1.1 Notation and Basic Operations

Matrix notation is a convenient way of organizing computations when dealing with a function of more than one variable. A vector with n components is written as an $n \times 1$ column vector, as follows:

$$x = \begin{pmatrix} x_1 \\ x_2 \\ \vdots \\ x_n \end{pmatrix}$$

Vectors and matrices are denoted by bold letters in this text. The components of a vector are denoted by using subscripts. A vector is always defined as a column vector. Since the matrix *transpose* operation moves columns into rows, if a computation requires a row vector, it will be defined with a transpose symbol, as follows.

$$x^T = \begin{pmatrix} x_1 & x_2 & \cdots & x_n \end{pmatrix}$$

An $m \times n$ matrix A and its transpose are written as follows:

$$\underset{(m \times n)}{A} = \begin{pmatrix} a_{11} & a_{12} & \cdots & a_{1n} \\ a_{21} & a_{22} & \cdots & a_{2n} \\ \vdots & \vdots & \vdots & \vdots \\ a_{m1} & a_{m2} & \cdots & a_{mn} \end{pmatrix} \qquad \underset{(n \times m)}{A^T} = \begin{pmatrix} a_{11} & a_{21} & \cdots & a_{m1} \\ a_{12} & a_{22} & \cdots & a_{m2} \\ \vdots & \vdots & \vdots & \vdots \\ a_{1n} & a_{2n} & \cdots & a_{mn} \end{pmatrix}$$

The entries $a_{11}, a_{22}, \ldots, a_{mn}$ are called the main diagonal of a matrix. If the number of rows and columns are the same, the matrix is called a *square matrix*; otherwise, it is called a *rectangular matrix*. If the matrix after a transpose operation is the same as the original matrix, then the matrix is known as a *symmetric matrix*.

An *identity matrix* is a square matrix in which all entries on the main diagonal are equal to 1, and the rest of the entries are all 0. An identity matrix is usually denoted by I.

$$\underset{(n \times n)}{I} = \begin{pmatrix} 1 & 0 & \cdots & 0 \\ 0 & 1 & \cdots & 0 \\ \vdots & \vdots & \vdots & \vdots \\ 0 & 0 & \cdots & 1 \end{pmatrix}$$

Multiplication of a matrix by a scalar means all entries in the matrix are multiplied by the scalar. Similarly, matrix addition and subtraction operations apply to each element. Clearly, matrix sizes must be the same for these operations to make sense. The product of two matrices is generated by taking rows of the first matrix and multiplying by the columns of the second matrix.

Therefore, for a matrix product to make sense, the number of columns in the first matrix must be the same as the number of rows in the second matrix. The result of multiplying an $m \times n$ matrix with an $n \times p$ matrix is a matrix of size $m \times p$.

The following relationships, involving matrices of compatible sizes, are useful to remember. Their proofs are given in most elementary books on linear algebra.

1. $IA = AI = A$
2. $AB \neq BA$
3. $ABC = A(BC) = (AB)C$
4. $\left(A^T\right)^T = A$
5. $(ABC)^T = C^T B^T A^T$
6. Given A as a rectangular matrix, $A^T A$ and AA^T are square symmetric matrices.

Example 3.1 In this example, some of the matrix operations are illustrated through numerical examples involving the following matrices. Notice that the matrix A is a rectangular matrix, matrix B is a square symmetric matrix, and c and d are column vectors.

$$A = \begin{pmatrix} 1 & 2 & 3 & 4 \\ 5 & 6 & 7 & 8 \\ 9 & 10 & 11 & 12 \end{pmatrix} \quad B = \begin{pmatrix} 1 & 2 & 3 \\ 2 & 4 & 5 \\ 3 & 5 & 6 \end{pmatrix} \quad c = \begin{pmatrix} 1 \\ 2 \\ 3 \end{pmatrix} \quad d = \begin{pmatrix} 4 \\ 5 \\ 6 \end{pmatrix}$$

(a) Addition, subtraction, and scalar multiplication

$$c + d = \begin{pmatrix} 5 \\ 7 \\ 9 \end{pmatrix} \quad c - d = \begin{pmatrix} -3 \\ -3 \\ -3 \end{pmatrix} \quad 2A = \begin{pmatrix} 2 & 4 & 6 & 8 \\ 10 & 12 & 14 & 16 \\ 18 & 20 & 22 & 24 \end{pmatrix}$$

Obviously, operations such as $A + B$ and $B - c$ do not make sense because of incompatible sizes of the matrices involved.

(b) Transpose of A

$$A^T = \begin{pmatrix} 1 & 5 & 9 \\ 2 & 6 & 10 \\ 3 & 7 & 11 \\ 4 & 8 & 12 \end{pmatrix}$$

(c) Matrix product AB The matrix product AB is not possible since the number of columns of A ($= 4$) is not the same as the number of rows in B ($= 3$)

(d) Matrix product BA

$$BA = \begin{pmatrix} 38 & 44 & 50 & 56 \\ 67 & 78 & 89 & 100 \\ 82 & 96 & 110 & 124 \end{pmatrix}$$

Note that the first entry is obtained by multiplying the first row of matrix B with the first column of matrix A.

$$(1, 2, 3) \times (1, 5, 9) = 1 + 10 + 27 = 38$$

All other entries are computed in a similar way.

(e) Matrix square For square matrices, the following operation makes sense:

$$B^2 \equiv BB = \begin{pmatrix} 14 & 25 & 31 \\ 25 & 45 & 56 \\ 31 & 56 & 70 \end{pmatrix}$$

(f) Matrix products $A^T A$ and AA^T

$$A^T A = \begin{pmatrix} 107 & 122 & 137 & 152 \\ 122 & 140 & 158 & 176 \\ 137 & 158 & 179 & 200 \\ 152 & 176 & 200 & 224 \end{pmatrix} \qquad AA^T = \begin{pmatrix} 30 & 70 & 110 \\ 70 & 174 & 278 \\ 110 & 278 & 446 \end{pmatrix}$$

Notice that even though A is not a symmetric matrix, the products $A^T A$ and AA^T result in symmetric matrices.

(g) Demonstration of $(BA)^T = A^T B^T$

$$BA = \begin{pmatrix} 38 & 44 & 50 & 56 \\ 67 & 78 & 89 & 100 \\ 82 & 96 & 110 & 124 \end{pmatrix} \qquad (BA)^T = \begin{pmatrix} 38 & 67 & 82 \\ 44 & 78 & 96 \\ 50 & 89 & 110 \\ 56 & 100 & 124 \end{pmatrix}$$

$$A^T B^T = \begin{pmatrix} 1 & 5 & 9 \\ 2 & 6 & 10 \\ 3 & 7 & 11 \\ 4 & 8 & 12 \end{pmatrix} \begin{pmatrix} 1 & 2 & 3 \\ 2 & 4 & 5 \\ 3 & 5 & 6 \end{pmatrix} = \begin{pmatrix} 38 & 67 & 82 \\ 44 & 78 & 96 \\ 50 & 89 & 110 \\ 56 & 100 & 124 \end{pmatrix}$$

3.1.2 Vector Norm

The *norm* or *length* of a vector is defined as follows:

$$\|x\| = \sqrt{x_1^2 + x_2^2 + \cdots + x_n^2}$$

The *dot product* or *inner product* of two $n \times 1$ vectors x and y is a scalar given as follows:

$$x^T y = x_1 y_1 + x_2 y_2 + \cdots + x_n y_n$$

The dot product of two vectors gives the cosine of the angle between the two vectors.

$$\cos \theta = \frac{x^T y}{\|x\| \|y\|}$$

The two vectors are *orthogonal* if their dot product is zero.

Using the dot product, the norm of a vector can be written as follows:

$$\|x\| = \sqrt{x_1^2 + x_2^2 + \cdots + x_n^2} \equiv \sqrt{x^T x}$$

Example 3.2 Compute the length of vector c and the cosine of the angle between vectors c and d.

$$c = \begin{pmatrix} 1 \\ 2 \\ 3 \end{pmatrix} \quad d = \begin{pmatrix} 4 \\ 5 \\ 6 \end{pmatrix}$$

(a) Length of vector c

$$\|c\| = \sqrt{c^T c} = \sqrt{14} = 3.74166$$

(b) Cosine of angle between vectors c and d

$$\|c\| = 3.74166 \qquad \|d\| = 8.77496$$

$$c^T d = 32 \qquad \cos \theta = \frac{32}{3.74166 \times 8.77496} = 0.974632$$

3.1.3 Determinant of a Square Matrix

The determinant of a 2 × 2 matrix is computed as follows:

$$A = \begin{pmatrix} a_{11} & a_{12} \\ a_{21} & a_{22} \end{pmatrix}$$

$$\text{Det}[A] \equiv |A| = \begin{vmatrix} a_{11} & a_{12} \\ a_{21} & a_{22} \end{vmatrix} = a_{11}a_{22} - a_{12}a_{21}$$

For larger matrices, the determinant is computed by reducing to a series of smaller matrices, eventually ending up with 2 × 2 matrices.

For a 3 × 3 matrix:

$$A = \begin{pmatrix} a_{11} & a_{12} & a_{13} \\ a_{21} & a_{22} & a_{23} \\ a_{31} & a_{32} & a_{33} \end{pmatrix}$$

$$|A| = \begin{vmatrix} a_{11} & a_{12} & a_{13} \\ a_{21} & a_{22} & a_{23} \\ a_{31} & a_{32} & a_{33} \end{vmatrix} = a_{11} \begin{vmatrix} a_{22} & a_{23} \\ a_{32} & a_{33} \end{vmatrix} - a_{12} \begin{vmatrix} a_{21} & a_{23} \\ a_{31} & a_{33} \end{vmatrix} + a_{13} \begin{vmatrix} a_{21} & a_{22} \\ a_{31} & a_{32} \end{vmatrix}$$

For a 4 × 4 matrix:

$$A = \begin{pmatrix} a_{11} & a_{12} & a_{13} & a_{14} \\ a_{21} & a_{22} & a_{23} & a_{24} \\ a_{31} & a_{32} & a_{33} & a_{34} \\ a_{41} & a_{42} & a_{43} & a_{44} \end{pmatrix}$$

$$|A| = a_{11} \begin{vmatrix} a_{22} & a_{23} & a_{24} \\ a_{32} & a_{33} & a_{34} \\ a_{42} & a_{43} & a_{44} \end{vmatrix} - a_{12} \begin{vmatrix} a_{21} & a_{23} & a_{24} \\ a_{31} & a_{33} & a_{34} \\ a_{41} & a_{43} & a_{44} \end{vmatrix} + a_{13} \begin{vmatrix} a_{21} & a_{22} & a_{24} \\ a_{31} & a_{32} & a_{34} \\ a_{41} & a_{42} & a_{44} \end{vmatrix}$$

$$- a_{14} \begin{vmatrix} a_{21} & a_{22} & a_{23} \\ a_{31} & a_{32} & a_{33} \\ a_{41} & a_{42} & a_{43} \end{vmatrix}$$

The determinants of each of the 3 × 3 matrices are evaluated as in the previous example. Notice that the sign of odd terms is positive and that of even terms is negative.

Example 3.3 Compute the determinant of the following 4 × 4 matrix:

$$A = \begin{pmatrix} 13 & 1 & 2 & 3 \\ 4 & 14 & 5 & 6 \\ 7 & 8 & 15 & 9 \\ 10 & 11 & 12 & 16 \end{pmatrix}$$

$$|A| = 13 \begin{vmatrix} 14 & 5 & 6 \\ 8 & 15 & 9 \\ 11 & 12 & 16 \end{vmatrix} - \begin{vmatrix} 4 & 5 & 6 \\ 7 & 15 & 9 \\ 10 & 12 & 16 \end{vmatrix} + 2 \begin{vmatrix} 4 & 14 & 6 \\ 7 & 8 & 9 \\ 10 & 11 & 16 \end{vmatrix}$$

$$- 3 \begin{vmatrix} 4 & 14 & 5 \\ 7 & 8 & 15 \\ 10 & 11 & 12 \end{vmatrix}$$

$$\begin{vmatrix} 14 & 5 & 6 \\ 8 & 15 & 9 \\ 11 & 12 & 16 \end{vmatrix} = 14 \begin{vmatrix} 15 & 9 \\ 12 & 16 \end{vmatrix} - 5 \begin{vmatrix} 8 & 9 \\ 11 & 16 \end{vmatrix} + 6 \begin{vmatrix} 8 & 15 \\ 11 & 12 \end{vmatrix} = 1289$$

Similarly,

$$\begin{vmatrix} 4 & 5 & 6 \\ 7 & 15 & 9 \\ 10 & 12 & 16 \end{vmatrix} = 22 \qquad \begin{vmatrix} 4 & 14 & 6 \\ 7 & 8 & 9 \\ 10 & 11 & 16 \end{vmatrix} = -210 \qquad \begin{vmatrix} 4 & 14 & 5 \\ 7 & 8 & 15 \\ 10 & 11 & 12 \end{vmatrix} = 633$$

Thus,

$$\text{Det}[A] \equiv |A| = 13 \times 1289 - 1 \times 22 + 2 \times -210 - 3 \times 633 = 14416$$

3.1.4 Inverse of a Square Matrix

Given a square matrix A, its *inverse* is another square matrix, denoted by A^{-1}, such that the following relation holds:

$$AA^{-1} = A^{-1}A = I$$

Efficient computation of the inverse of large matrices involves using the so-called Gauss-Jordan form, and will be discussed in chapter 6. For small matrices and hand computations, the inverse can be computed by using the following formula:

$$A^{-1} = \frac{1}{\text{Det}[A]} (\text{Co}[A])^T$$

where Det[A] is the determinant of matrix A and Co[A] refers to the matrix of cofactors of A. Each term in a cofactor matrix is computed using the following formula:

$$(\text{Co}[A])_{i,j} = (-1)^{i+j} \text{Det}[\bar{A}_{ij}]$$

where $\bar{A}_{ij}$ is the matrix determined by removing the ith row and the jth column from the A matrix. The following relationships, involving matrices of compatible sizes, are useful to remember. Their proofs are given in most elementary books on linear algebra.

1. $(ABC)^{-1} = C^{-1}B^{-1}A^{-1}$
2. $\left(A^T\right)^{-1} = \left(A^{-1}\right)^T$
3. $(A+B)^{-1} \neq A^{-1} + B^{-1}$

Example 3.4 In this example, some of the matrix operations are illustrated through numerical examples involving the following matrices:

$$A = \begin{pmatrix} 1 & 2 & 3 & 4 \\ 5 & 6 & 7 & 8 \\ 9 & 10 & 11 & 12 \end{pmatrix} \qquad B = \begin{pmatrix} 1 & 2 & 3 \\ 2 & 4 & 5 \\ 3 & 5 & 6 \end{pmatrix}$$

(a) Inverse of matrix B The cofactor matrix can be computed as follows:

$$\text{Co}[B] = \begin{pmatrix} \text{Det}\begin{bmatrix}\begin{pmatrix}4 & 5\\5 & 6\end{pmatrix}\end{bmatrix} & -\text{Det}\begin{bmatrix}\begin{pmatrix}2 & 5\\3 & 6\end{pmatrix}\end{bmatrix} & \text{Det}\begin{bmatrix}\begin{pmatrix}2 & 4\\3 & 5\end{pmatrix}\end{bmatrix} \\ -\text{Det}\begin{bmatrix}\begin{pmatrix}2 & 3\\5 & 6\end{pmatrix}\end{bmatrix} & \text{Det}\begin{bmatrix}\begin{pmatrix}1 & 3\\3 & 6\end{pmatrix}\end{bmatrix} & -\text{Det}\begin{bmatrix}\begin{pmatrix}1 & 2\\3 & 5\end{pmatrix}\end{bmatrix} \\ \text{Det}\begin{bmatrix}\begin{pmatrix}2 & 3\\4 & 5\end{pmatrix}\end{bmatrix} & -\text{Det}\begin{bmatrix}\begin{pmatrix}1 & 3\\2 & 5\end{pmatrix}\end{bmatrix} & \text{Det}\begin{bmatrix}\begin{pmatrix}1 & 2\\2 & 4\end{pmatrix}\end{bmatrix} \end{pmatrix}$$

$$= \begin{pmatrix} -1 & 3 & -2 \\ 3 & -3 & 1 \\ -2 & 1 & 0 \end{pmatrix}$$

The determinant of the matrix B can easily be seen to be Det[B] = -1. Thus, the inverse of matrix B is as follows:

$$B^{-1} = \begin{pmatrix} 1 & -3 & 2 \\ -3 & 3 & -1 \\ 2 & -1 & 0 \end{pmatrix}$$

We can easily verify that this inverse is correct by evaluating the matrix product

$$BB^{-1} = \begin{pmatrix} 1 & 2 & 3 \\ 2 & 4 & 5 \\ 3 & 5 & 6 \end{pmatrix} \begin{pmatrix} 1 & -3 & 2 \\ -3 & 3 & -1 \\ 2 & -1 & 0 \end{pmatrix} = \begin{pmatrix} 1 & 0 & 0 \\ 0 & 1 & 0 \\ 0 & 0 & 1 \end{pmatrix} = I$$

(b) Computation of $[(AA^T + B)B)]^{-1}$

$$AA^T + B = \begin{pmatrix} 30 & 70 & 110 \\ 70 & 174 & 278 \\ 110 & 278 & 446 \end{pmatrix} + \begin{pmatrix} 1 & 2 & 3 \\ 2 & 4 & 5 \\ 3 & 5 & 6 \end{pmatrix} = \begin{pmatrix} 31 & 72 & 113 \\ 72 & 178 & 283 \\ 113 & 283 & 452 \end{pmatrix}$$

$$(AA^T + B)B = \begin{pmatrix} 31 & 72 & 113 \\ 72 & 178 & 283 \\ 113 & 283 & 452 \end{pmatrix} \begin{pmatrix} 1 & 2 & 3 \\ 2 & 4 & 5 \\ 3 & 5 & 6 \end{pmatrix} = \begin{pmatrix} 514 & 915 & 1{,}131 \\ 1{,}277 & 2{,}271 & 2{,}804 \\ 2{,}035 & 3{,}618 & 4{,}466 \end{pmatrix}$$

The cofactors and the determinant of the resulting matrix are as follows:

$$\mathrm{Co}[(AA^T + B)B] = \begin{pmatrix} -2{,}586 & 3{,}058 & -1{,}299 \\ 5{,}568 & -6{,}061 & 2{,}373 \\ -2{,}841 & 3{,}031 & -1{,}161 \end{pmatrix}$$

$$\mathrm{Det}[(AA^T + B)B] = -303$$

Taking the transpose of the cofactors matrix and dividing each term by the determinant, we get the following inverse matrix:

$$[(AA^T + B)B]^{-1} = \begin{pmatrix} 514 & 915 & 1{,}131 \\ 1{,}277 & 2{,}271 & 2{,}804 \\ 2{,}035 & 3{,}618 & 4{,}466 \end{pmatrix}^{-1} = \begin{pmatrix} \frac{862}{101} & -\frac{1{,}856}{101} & \frac{947}{101} \\ -\frac{3{,}058}{303} & \frac{6{,}061}{303} & -\frac{3{,}031}{303} \\ \frac{433}{101} & -\frac{791}{101} & \frac{387}{101} \end{pmatrix}$$

From the relationship of the product of inverses,

$$[(AA^T + B)B]^{-1} = B^{-1}(AA^T + B)^{-1}$$

We can demonstrate the validity of this as follows:

$$(AA^T + B)^{-1} = \begin{pmatrix} 31 & 72 & 113 \\ 72 & 178 & 283 \\ 113 & 283 & 452 \end{pmatrix}^{-1} = \begin{pmatrix} \frac{367}{303} & -\frac{565}{303} & \frac{262}{303} \\ -\frac{565}{303} & \frac{1{,}243}{303} & -\frac{637}{303} \\ \frac{262}{303} & -\frac{637}{303} & \frac{334}{303} \end{pmatrix}$$

$$B^{-1} = \begin{pmatrix} 1 & -3 & 2 \\ -3 & 3 & -1 \\ 2 & -1 & 0 \end{pmatrix}$$

$$B^{-1}(AA^T+B)^{-1} = \begin{pmatrix} 1 & -3 & 2 \\ -3 & 3 & -1 \\ 2 & -1 & 0 \end{pmatrix} \begin{pmatrix} \frac{367}{303} & -\frac{565}{303} & \frac{262}{303} \\ -\frac{565}{303} & \frac{1{,}243}{303} & -\frac{637}{303} \\ \frac{262}{303} & -\frac{637}{303} & \frac{334}{303} \end{pmatrix}$$

$$= \begin{pmatrix} \frac{862}{101} & -\frac{1{,}856}{101} & \frac{947}{101} \\ -\frac{3{,}058}{303} & \frac{6{,}061}{303} & -\frac{3{,}031}{303} \\ \frac{433}{101} & -\frac{791}{101} & \frac{387}{101} \end{pmatrix}$$

which is the same as before.

3.1.5 Eigenvalues of a Square Matrix

If A is an $n \times n$ matrix and x is an $n \times 1$ nonzero vector such that $Ax = \lambda x$, then λ is called an eigenvalue of a square matrix A. Eigenvalues of a matrix play an important role in several key optimization concepts. The eigenvalues are obtained by solving the following equation, known as a characteristic equation of a matrix.

$$\text{Det}[A - \lambda I] = 0$$

where I is an identity matrix. For an $n \times n$ matrix, the characteristic equation is a polynomial of nth degree in terms of λ. Solutions of this equation give n eigenvalues of matrix A.

Example 3.5 Compute eigenvalues of the following 3×3 matrix:

$$A = \begin{pmatrix} 13 & 1 & 2 \\ 4 & 14 & 5 \\ 7 & 8 & 15 \end{pmatrix}$$

$$|A - \lambda I| = \left| \begin{pmatrix} 13 & 1 & 2 \\ 4 & 14 & 5 \\ 7 & 8 & 15 \end{pmatrix} - \lambda \begin{pmatrix} 1 & 0 & 0 \\ 0 & 1 & 0 \\ 0 & 0 & 1 \end{pmatrix} \right| = \begin{vmatrix} 13-\lambda & 1 & 2 \\ 4 & 14-\lambda & 5 \\ 7 & 8 & 15-\lambda \end{vmatrix}$$

Evaluating the determinant, we get the following characteristic equation:

$$(13 - \lambda)[(14 - \lambda)(15 - \lambda) - 40] - [4(15 - \lambda) - 35] + 2[32 - 7(14 - \lambda)] = 0$$

or

$$2{,}053 - 529\lambda + 42\lambda^2 - \lambda^3 = 0$$

This nonlinear equation has the following three solutions:

$$\lambda_1 = 7.9401 \quad \lambda_2 = 11.4212 \quad \lambda_3 = 22.6387$$

Mathematica has a built-in function, called Eigenvalues, that gives the eigenvalues for any square matrix.

Eigenvalues[{{13., 1, 2}, {4, 14, 5}, {7, 8, 15}}]

{22.6387, 11.4212, 7.9401}

3.1.6 Principal Minors of a Square Matrix

Principal minors of a matrix are the determinants of square submatrices of the matrix. The first principal minor is the first diagonal element of the matrix. The second principal minor is the determinant of the 2 × 2 matrix obtained from the first two rows and columns. The third principal minor is the determinant of the 3 × 3 matrix obtained from the first three rows and columns. The process is continued until the last principal minor, which is equal to the determinant of the entire matrix. As a specific example, consider A to be the following 3 × 3 matrix:

$$A = \begin{pmatrix} a_{11} & a_{12} & a_{13} \\ a_{21} & a_{22} & a_{23} \\ a_{31} & a_{32} & a_{33} \end{pmatrix}$$

This matrix has the following three principal minors:

$$A_1 = a_{11}$$

$$A_2 = \begin{vmatrix} a_{11} & a_{12} \\ a_{21} & a_{22} \end{vmatrix}$$

$$A_3 = \begin{vmatrix} a_{11} & a_{12} & a_{13} \\ a_{21} & a_{22} & a_{23} \\ a_{31} & a_{32} & a_{33} \end{vmatrix}$$

Example 3.6 Compute the principal minors of the following 3 × 3 matrix:

$$A = \begin{pmatrix} 13 & 1 & 2 \\ 4 & 14 & 5 \\ 7 & 8 & 15 \end{pmatrix}$$

$$A_1 = 13$$

$$A_2 = \begin{vmatrix} 13 & 1 \\ 4 & 14 \end{vmatrix} = 178$$

$$A_3 = \begin{vmatrix} 13 & 1 & 2 \\ 4 & 14 & 5 \\ 7 & 8 & 15 \end{vmatrix} = 2{,}053$$

3.1.7 Rank of a Matrix

The rank of a matrix is equal to the size (dimension) of the *largest* nonsingular square submatrix that can be found by deleting any rows and columns of the given matrix, if necessary. Since the determinant must be nonzero for a matrix to be nonsingular, to determine the rank of a matrix, one must find a square submatrix with a nonzero determinant. The dimension of the largest such submatrix gives the rank of the matrix.

A matrix of *full rank* is a square matrix whose determinant is nonzero.

Example 3.7 Compute the rank of the following 3 × 3 matrix.

$$A = \begin{pmatrix} 13 & 1 & 2 \\ 4 & 14 & 5 \\ 7 & 8 & 15 \end{pmatrix}$$

Since the matrix is already a square matrix, we try computing the determinant of the entire matrix.

$$\text{Det}[A] = \begin{vmatrix} 13 & 1 & 2 \\ 4 & 14 & 5 \\ 7 & 8 & 15 \end{vmatrix} = 2{,}053$$

Since the determinant is nonzero, the rank of the matrix is 3. Thus, the given matrix is of full rank.

Example 3.8 Compute the rank of the following 3 × 4 matrix:

$$A = \begin{pmatrix} 1 & 2 & 3 & 4 \\ 5 & 6 & 7 & 8 \\ 9 & 10 & 11 & 12 \end{pmatrix}$$

The largest square submatrix that can be found from the given matrix is a 3×3. There are four possible submatrices depending upon which column is deleted:

$$\begin{vmatrix} 2 & 3 & 4 \\ 6 & 7 & 8 \\ 10 & 11 & 12 \end{vmatrix} = 0 \qquad \begin{vmatrix} 1 & 3 & 4 \\ 5 & 7 & 8 \\ 9 & 11 & 12 \end{vmatrix} = 0$$

$$\begin{vmatrix} 1 & 2 & 4 \\ 5 & 6 & 8 \\ 9 & 10 & 12 \end{vmatrix} = 0 \qquad \begin{vmatrix} 1 & 2 & 3 \\ 5 & 6 & 7 \\ 9 & 10 & 11 \end{vmatrix} = 0$$

The determinants of all possible 3×3 submatrices are 0. Therefore, the matrix is not of rank 3. Now we must try all 2×2 submatrices. The number of possibilities is now large, however, we don't need to try all possible combinations. We just need to find one submatrix that has a nonzero determinant, and we are done.

$$\begin{vmatrix} 1 & 2 \\ 5 & 6 \end{vmatrix} = -4$$

We see that the first 2×2 submatrix has a nonzero determinant. Thus, we have found a largest possible submatrix with a nonzero determinant and hence, the rank of given matrix A is equal to 2.

A function called Rank is included in the OptimizationToolbox 'CommonFunctions' package for computing the rank of matrices.

?Rank

Rank[a_], returns rank of matrix a.

Rank[{{1, 2, 3, 4}, {5, 6, 7, 8}, {9, 10, 11, 12}}]

2

3.1.8 Linear Independence of Vectors

A given set of $n \times 1$ vectors is linearly independent if any other $n \times 1$ vector can be represented by a suitable linear combination of that set. As an example, consider the following set of two 2×1 vectors:

$$a = \begin{pmatrix} 1 \\ 0 \end{pmatrix} \qquad b = \begin{pmatrix} 1 \\ 1 \end{pmatrix}$$

These two vectors are linearly independent because any 2 × 1 vector can be represented by a combination of these two. For example,

$$c = \begin{pmatrix} 3 \\ -7 \end{pmatrix} \equiv 10a - 7b$$

A given set of vectors can be tested for linear independence by assembling a matrix A whose *columns* are the given vectors. The rank of this matrix determines the number of linearly independent vectors in the set. If the rank is equal to the number of vectors in the given set, then they are linearly independent.

Since the rank of a matrix can never be greater than the number of rows in the matrix, and since the vectors are arranged as columns of this matrix, it should be obvious that the *maximum* number of linearly independent vectors is equal to the dimension of the vector.

Example 3.9 Check to see if the following vectors are linearly independent:

$$a = \begin{pmatrix} 2 \\ -5 \\ 2 \\ -1 \end{pmatrix} \quad b = \begin{pmatrix} 1 \\ 2 \\ 3 \\ 4 \end{pmatrix} \quad c = \begin{pmatrix} -9 \\ 8 \\ 4 \\ -1 \end{pmatrix}$$

The vectors are arranged as columns of the following 4 × 3 matrix:

$$A = \begin{pmatrix} 2 & 1 & -9 \\ -5 & 2 & 8 \\ 2 & 3 & 4 \\ -1 & 4 & -1 \end{pmatrix};$$

Rank[A]

3

The rank of the matrix A is 3. Therefore, these vectors are linearly independent.

Example 3.10 Check to see if the following vectors are linearly independent:

$$a = \begin{pmatrix} 2 \\ -5 \\ 2 \\ -1 \end{pmatrix} \quad b = \begin{pmatrix} 1 \\ 2 \\ 3 \\ 4 \end{pmatrix} \quad c = \begin{pmatrix} -4 \\ 19 \\ 0 \\ 11 \end{pmatrix}$$

The vectors are arranged as columns of the following 4 × 3 matrix:

$$A = \begin{pmatrix} 2 & 1 & -4 \\ -5 & 2 & 19 \\ 2 & 3 & 0 \\ -1 & 4 & 11 \end{pmatrix};$$

Rank[A]

2

The rank of the matrix **A** is 2. Therefore, only two of the three vectors are linearly independent. It is easy to see that the third vector is a linear combination of the first two and can be obtained as follows:

-3a + 2b

{-4, 19, 0, 11}

3.2 Approximation Using the Taylor Series

Any differentiable function can be approximated by a polynomial using the Taylor series. We consider the function of a single variable first and then generalize expressions for functions of two or more variables.

3.2.1 Functions of a Single Variable

Consider a function $f(x)$ of a single variable x. We can approximate this function about $\bar{x}$ using the Taylor series as follows:

$$f(x) \approx f(\bar{x}) + \frac{df(\bar{x})}{dx}(x - \bar{x}) + \frac{1}{2}\frac{d^2 f(\bar{x})}{dx^2}(x - \bar{x})^2 + \cdots$$

A linear approximation of the function is obtained if we retain the first two terms in the series while a quadratic approximation is obtained by including the second derivative term as well.

A *Mathematica* function **TaylorSeriesApprox** is created to perform the necessary computations to generate linear and quadratic approximations. The function graphically compares linear and quadratic approximations with the original function. In order to stay consistent with the notation used for functions of more than one variable, the first derivative of f is indicated by ∇f, and the second derivative by $\nabla^2 f$.

```
Needs["OptimizationToolbox`Chap3Tools`"];
?TaylorSeriesApprox
```

TaylorSeriesApprox[f, vars_List, pt_List], generates linear and quadratic
  approximations of a function 'f(vars)' around given point 'pt' using
  Taylor series. It returns function value at given point and its linear
  and quadratic approximations. For functions of one and two variables,
  it can also generate graph comparing approximations with the original
  function. To get graphs the arguments must be in the following form.
  For f(x): TaylorSeriesApprox[f, {x,xmin,xmax}, pt]. For f(x,y):
  TaylorSeriesApprox[f, {{x,xmin,xmax},{y,ymin,ymax}}, pt]

Mathematica has a much more powerful built-in function called **Series** that generates a Taylor series approximation of any desired order. The main reason for creating the **TaylorSeriesApprox** function is to show the intermediate calculations and to automate the process of graphical comparison.

```
?Series
```

Series[f, {x, x0, n}] generates a power series expansion for f about the
  point x = x0 to order (x - x0)^n. Series[f, {x, x0, nx}, {y, y0, ny}]
  successively finds series expansions with respect to y, then x.

Example 3.11 Determine linear and quadratic approximations of the following function around the given point using the Taylor series.

$$f(x) = \frac{x^2}{4x+3} \qquad \bar{x} = \frac{1}{4}$$

The function and its derivatives at the given point $\bar{x}$ are as follows:

$f \to \frac{x^2}{3+4x}$ $\quad f\{\frac{1}{4}\} \to 0.015625$

$\nabla f \to \left(\frac{2x(3+2x)}{(3+4x)^2}\right)$ $\quad \nabla f\{\frac{1}{4}\} \to (0.109375)$

$\nabla^2 f \to \left(\frac{18}{(3+4x)^3}\right)$ $\quad \nabla^2 f\{\frac{1}{4}\} \to (0.28125)$

$\Delta x \to \left(-\frac{1}{4} + x\right)$

The linear and quadratic Taylor series approximations are as follows:

Linear approximation $\to -0.0117188 + 0.109375x$

Quadratic approximation $\to -0.00292969 + 0.0390625x + 0.140625x^2$

The original and the approximate functions are compared in the following plot. Note that around $\bar{x}$, the approximation is very good. However, as we move away from $\bar{x}$, the actual function and its linear and quadratic approximations become quite different.

3.2 Approximation Using the Taylor Series

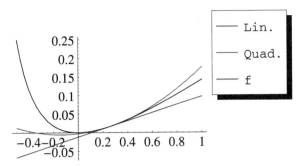

FIGURE 3.1 Comparison of linear and quadratic approximations for a function of a single variable.

Example 3.12 Determine linear and quadratic approximations of the following function around the given point using the Taylor series.

$$f(x) = x^2 \cos(x) \qquad \bar{x} = \frac{\pi}{4}$$

```
TaylorSeriesApprox[x² Cos[x], {x, -2, 2}, π/4];
f → x² Cos[x]    f{π/4} → 0.436179
∇f → (x(2 Cos[x] - x Sin[x]))    ∇f{π/4} → (0.674542)
∇²f → (-(-2 + x²) Cos[x] - 4x Sin[x])    ∇²f{π/4} → (-1.24341)
Δx → (-π/4 + x)
Linear approximation → -0.0936048 + 0.674542x
Quadratic approximation → -0.477103 + 1.65111x - 0.621703x²
```

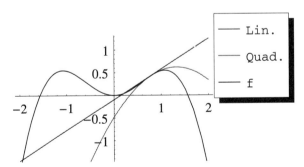

FIGURE 3.2 Comparison of linear and quadratic approximations for a function of a single variable.

3.2.2 Functions of Two Variables

Consider a function $f(x_1, x_2)$ of two variables. We can approximate this function about a known point $\{\bar{x}_1, \bar{x}_2\}$ using the Taylor series as follows:

$$f(x_1, x_2) \approx f(\bar{x}_1 - \bar{x}_2) + \left[\frac{\partial f(\bar{x}_1, \bar{x}_2)}{\partial x_1}(\bar{x}_1 - \bar{x}_1) + \frac{\partial f(\bar{x}, \bar{x})}{\partial x_2}(x_2 - \bar{x}_2)\right]$$

$$+ \frac{1}{2}\left[\frac{\partial^2 f(\bar{x}_1, \bar{x}_2)}{\partial x_1^2}(x_1 - \bar{x}_1)^2 + 2\frac{\partial^2 f(\bar{x}_1, \bar{x}_2)}{\partial x_1 \partial x_2}(x_1 - \bar{x}_1)(x_2 - \bar{x}_2)\right.$$

$$\left. + \frac{\partial^2 f(\bar{x}_1, \bar{x}_2)}{\partial x_2^2}(x_2 - \bar{x}_2)^2\right] + \cdots$$

This expression can be written in a more compact form by using matrix notation, as follows:

$$f(x_1, x_2) \approx f(\bar{x}_1, \bar{x}_2) + \begin{pmatrix} \dfrac{\partial f(\bar{x}_1, \bar{x}_2)}{\partial x_1} & \dfrac{\partial f(\bar{x}_1, \bar{x}_2)}{\partial x_2} \end{pmatrix} \begin{pmatrix} x_1 - \bar{x}_1 \\ x_2 - \bar{x}_2 \end{pmatrix}$$

$$+ \frac{1}{2}\begin{pmatrix} x_1 - \bar{x}_1 & x_2 - \bar{x}_2 \end{pmatrix} \begin{pmatrix} \dfrac{\partial^2 f(\bar{x}_1, \bar{x}_2)}{\partial x_1^2} & \dfrac{\partial^2 f(\bar{x}_1, \bar{x}_2)}{\partial x_1 \partial x_2} \\ \dfrac{\partial^2 f(\bar{x}_1, \bar{x}_2)}{\partial x_1 \partial x_2} & \dfrac{\partial^2 f(\bar{x}_1, \bar{x}_2)}{\partial x_2^2} \end{pmatrix} \begin{pmatrix} x_1 - \bar{x}_1 \\ x_2 - \bar{x}_2 \end{pmatrix} + \cdots$$

Define a 2 × 1 vector called a gradient vector and a 2 × 2 matrix called a Hessian matrix as follows:

$$\text{Gradient vector, } \nabla f(\bar{x}_1, \bar{x}_2) = \begin{pmatrix} \dfrac{\partial f(\bar{x}_1, \bar{x}_2)}{\partial x_1} \\ \dfrac{\partial f(\bar{x}_1, \bar{x}_2)}{\partial x_2} \end{pmatrix}$$

$$\text{Hessian matrix, } \nabla^2 f(\bar{x}_1, \bar{x}_2) = \begin{pmatrix} \dfrac{\partial^2 f(\bar{x}_1, \bar{x}_2)}{\partial x_1^2} & \dfrac{\partial^2 f(\bar{x}_1, \bar{x}_2)}{\partial x_1 \partial x_2} \\ \dfrac{\partial^2 f(\bar{x}_1, \bar{x}_2)}{\partial x_1 \partial x_2} & \dfrac{\partial^2 f(\bar{x}_1, \bar{x}_2)}{\partial x_2^2} \end{pmatrix}$$

Noting that

$$\begin{pmatrix} \dfrac{\partial f(\bar{x}_1, \bar{x}_2)}{\partial x_1} & \dfrac{\partial f(\bar{x}_1, \bar{x}_2)}{\partial x_2} \end{pmatrix} = \nabla f(\bar{x}_1, \bar{x}_2)^T$$

3.2 Approximation Using the Taylor Series

the transpose of the gradient vector and

$$\begin{pmatrix} x_1 - \bar{x}_1 \\ x_2 - \bar{x}_2 \end{pmatrix} = \begin{pmatrix} x_1 \\ x_2 \end{pmatrix} - \begin{pmatrix} \bar{x}_1 \\ \bar{x}_2 \end{pmatrix} = (x - \bar{x}) \equiv \Delta x$$

vector of variables shifted by $\bar{x}$, the Taylor series formula can be written as follows:

$$f(x_1, x_2) \approx f(\bar{x}_1, \bar{x}_2) + \nabla f(\bar{x}_1, \bar{x}_2)^T \Delta x + \frac{1}{2} \Delta x^T \nabla^2 f(\bar{x}_1, \bar{x}_2) \Delta x + \cdots$$

Example 3.13 Determine linear and quadratic approximations of the following function around the given point using the Taylor series.

$$f(x, y) = x^y \qquad (\bar{x}, \bar{y}) = (2, 2)$$

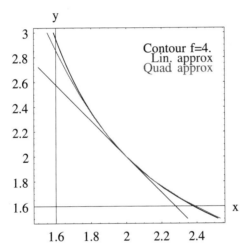

FIGURE 3.3 Comparison of linear and quadratic approximations for a function of two variables.

```
TaylorSeriesApprox[x^y, {{x, 1.5, 3}, {y, 1.5, 3}}, {2, 2}];
```

$f \to x^y \qquad f\{2, 2\} \to 4.$

$\nabla f \to \begin{pmatrix} x^{-1+y} y \\ x^y \text{Log}[x] \end{pmatrix} \qquad \nabla f\{2, 2\} \to \begin{pmatrix} 4. \\ 2.77259 \end{pmatrix}$

$\nabla^2 f \to \begin{pmatrix} x^{-2+y}(-1+y)y & x^{-1+y}(1+y\text{Log}[x]) \\ x^{-1+y}(1+y\text{Log}[x]) & x^y \text{Log}[x]^2 \end{pmatrix}$

$$\nabla^2 f\{2,2\} \to \begin{pmatrix} 2. & 4.77259 \\ 4.77259 & 1.92181 \end{pmatrix}$$

```
Δx → (-2 + x   -2 + y )

Linear approximation → -9.54518 + 4.x + 2.77259y

Quadratic approximation → 17.3888 - 9.54518x + 1.x² - 10.6162y
                          + 4.77259xy + 0.960906y²
```

Example 3.14 Determine linear and quadratic approximations of the following function around the given point using the Taylor series.

$$f(x, y) = 2x^2 + y^2 + 1/(2x^2 + y^2) \qquad (\bar{x}, \bar{y}) = (2, 2)$$

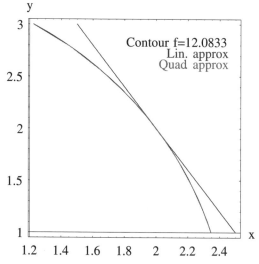

FIGURE 3.4 Comparison of linear and quadratic approximations for a function of two variables.

```
f = 2x² + y² + 1/(2x² + y²);
TaylorSeriesApprox[f, {{x, 1, 3}, {y, 1, 3}}, {2, 2}];
```

$$f \to 2x^2 + y^2 + \frac{1}{2x^2 + y^2} \qquad f\{2, 2\} \to 12.0833$$

$$\nabla f \to \begin{pmatrix} x\left(4 - \frac{4}{\left(2x^2 + y^2\right)^2}\right) \\ y\left(2 - \frac{2}{\left(2x^2 + y^2\right)^2}\right) \end{pmatrix} \qquad \nabla f\{2, 2\} \to \begin{pmatrix} 7.94444 \\ 3.97222 \end{pmatrix}$$

$$\nabla^2 f \to \begin{pmatrix} 4 + \frac{32x^2}{\left(2x^2+y^2\right)^3} - \frac{4}{\left(2x^2+y^2\right)^2} & \frac{16xy}{\left(2x^2+y^2\right)^3} \\ \frac{16xy}{\left(2x^2+y^2\right)^3} & 2 + \frac{8y^2}{\left(2x^2+y^2\right)^3} - \frac{2}{\left(2x^2+y^2\right)^2} \end{pmatrix}$$

$$\nabla^2 f\{2, 2\} \to \begin{pmatrix} 4.0463 & 0.037037 \\ 0.037037 & 2.00463 \end{pmatrix}$$

$$\Delta x \to (-2 + x \quad -2 + y)$$

Linear approximation $\to -11.75 + 7.94444x + 3.97222y$

Quadratic approximation $\to 0.5 - 0.222222x + 2.02315x^2 - 0.111111y$
$\qquad + 0.037037xy + 1.00231y^2$

3.2.3 Functions of *n* Variables

For a function of n variables, $\mathbf{x} = [x_1, x_2, \ldots]^T$, the Taylor series formula can be written by defining an $n \times 1$ gradient vector and an $n \times n$ Hessian matrix as follows:

$$\text{Gradient vector, } \nabla f(x) = \begin{pmatrix} \partial f/\partial x_1 \\ \vdots \\ \partial f/\partial x_n \end{pmatrix}$$

$$\text{Hessian matrix, } \nabla^2 f = \begin{pmatrix} \frac{\partial^2 f}{\partial x_1^2} & \frac{\partial^2 f}{\partial x_1 \partial x_2} & \cdots & \frac{\partial^2 f}{\partial x_1 \partial x_n} \\ \frac{\partial^2 f}{\partial x_2 \partial x_1} & \frac{\partial^2 f}{\partial x_2^2} & \cdots & \frac{\partial^2 f}{\partial x_2 \partial x_n} \\ \vdots & \vdots & \vdots & \vdots \\ \frac{\partial^2 f}{\partial x_n \partial x_2} & \frac{\partial^2 f}{\partial x_n \partial x_2} & \cdots & \frac{\partial^2 f}{\partial x_n^2} \end{pmatrix}$$

The Hessian matrix is also denoted by $\mathbf{H}_f$. For a continuously differentiable function, the mixed second partial derivatives are symmetric, that is

$$\frac{\partial^2 f}{\partial x_i \partial x_j} = \frac{\partial^2 f}{\partial x_j \partial x_i} \qquad i, j = 1, 2, \ldots$$

Hence, the Hessian matrix is always symmetric. The Taylor series approximation around a known point $\bar{x}$ can be written as follows:

$$f(x) \approx f(\bar{x}) + \nabla f(\bar{x})^T \Delta x + \frac{1}{2} \Delta x^T \nabla^2 f(\bar{x}) \Delta x + \cdots$$

The functions Grad and Hessian, included in the OptimizationToolbox 'CommonFunctions' package, compute the gradient vector and Hessian matrix of a function of any number of variables.

```
Needs["OptimizationToolbox`CommonFunctions`"];
?Grad
```

Grad[f, vars], computes gradient of function f (or a list of functions) with respect to variables vars.

```
?Hessian
```

Hessian[f,vars] --- Computes Hessian matrix of function f (or a list of functions).

Example 3.15 Determine linear and quadratic approximations of the following function around the given point using the Taylor series.

$$f(x, y, z) = x^3 + 12xy^2 + 2y^2 + 5z^2 + xz^4 \qquad (\bar{x}, \bar{y}, \bar{z}) = (1, 2, 3)$$

```
f = x^3 + 12xy^2 + 2y^2 + 5z^2 + xz^4;
TaylorSeriesApprox[f, {x, y, z}, {1, 2, 3}];
```

$f \to x^3 + 2y^2 + 12xy^2 + 5z^2 + xz^4 \quad f\{1, 2, 3\} \to 183.$

$\nabla f \to \begin{pmatrix} 3x^2 + 12y^2 + z^4 \\ 4(y + 6xy) \\ 2z(5 + 2xz^2) \end{pmatrix} \quad \nabla f\{1, 2, 3\} \to \begin{pmatrix} 132. \\ 56. \\ 138. \end{pmatrix}$

$\nabla^2 f \to \begin{pmatrix} 6x & 24y & 4z^3 \\ 24y & 4 + 24x & 0 \\ 4z^3 & 0 & 2(5 + 6xz^2) \end{pmatrix} \quad \nabla^2 f\{1, 2, 3\} \to \begin{pmatrix} 6. & 48. & 108. \\ 48. & 28. & 0 \\ 108. & 0 & 118. \end{pmatrix}$

$\Delta x \to (-1 + x \quad -2 + y \quad -3 + z \ )$

Linear approximation $\to -475. + 132.x + 56.y + 138.z$

Quadratic approximation $\to$
  $535. - 294.x + 3.x^2 - 48.y + 48.xy + 14.y^2 - 324.z + 108.xz + 59.z^2$

3.2.4 Surfaces and Their Tangent Planes

The Taylor series approximation gives an important geometrical interpretation to the gradient vector. To see this, consider the equation of a surface in n

3.2 Approximation Using the Taylor Series

dimensions written as follows.

$$f(x_1, \ldots, x_n) = c$$

where c is a given constant. At an arbitrary point, $\bar{x} = (\bar{x}_1, \ldots, \bar{x}_n)$, the linear approximation of f is as follows:

$$f(x_1, \ldots, x_n) \approx f(\bar{x}_1, \ldots, \bar{x}_n) + [\nabla f(\bar{x}_1, \ldots, \bar{x}_n)]^T \begin{pmatrix} x_1 - \bar{x}_1 \\ \vdots \\ x_n - \bar{x}_n \end{pmatrix}$$

If the point $\bar{x}$ is on the surface, then $f(x) - f(\bar{x}) = 0$ and therefore, for any point on the surface

$$\nabla f(\bar{x})^T (x - \bar{x}) = 0 \quad \text{or} \quad \nabla f(\bar{x})^T d = 0$$

where $d = x - \bar{x}$.

This equation says that the dot product of the gradient vector with any vector on the surface is zero. From the definition of the vector dot product, this means that the gradient vector is *normal* to the surface. Furthermore, it can be shown that along the gradient vector locally, the function value is increasing the most rapidly.

The equation of the *tangent plane* to the surface at $\bar{x}$ is given by the equation

$$\text{Tangent plane:} \quad \nabla f(\bar{x})^T d = 0$$

The following two- and three-dimensional examples illustrate these concepts graphically:

Example 3.16 Consider the following function of two variable:

$$f(x, y) = x^2 + y^2$$

For specific function values, the contours are circles centered around the origin. As an example, consider the contour for $f = 2$. The equation for this contour is

$$x^2 + y^2 = 2$$

The point $(1, 1)$ satisfies this equation and is thus on this contour. The gradient of the function at $(1, 1)$ is as follows.

$$\nabla f \to \begin{pmatrix} 2x \\ 2y \end{pmatrix} \qquad \nabla f\{1,1\} \to \begin{pmatrix} 2 \\ 2 \end{pmatrix}$$

The equation of the tangent line to the contour at (1, 1) is

$$(2 \quad 2) \begin{pmatrix} x-1 \\ y-1 \end{pmatrix} = 0 \quad \text{or} \quad 2x + 2y - 4 = 0$$

The contour $f = 2$, the gradient vector, and the tangent line at (1, 1) are illustrated in the following figure. The figure clearly shows the relationship among them.

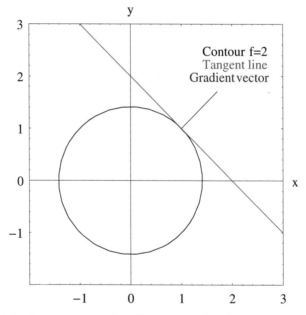

FIGURE 3.5 Tangent line and gradient vector for a function of two variables.

Example 3.17 Consider the following function of three variables:

$$f(x, y) = x^2 + y^2 + z^2$$

For specific function values, the contours are spheres centered around the origin. As an example, consider the contour for $f = 3$. The equation for this contour is

$$x^2 + y^2 + z^2 = 3$$

The point $(1, 1, 1)$ satisfies this equation and is thus on this contour. The gradient of the function at $(1, 1, 1)$ is as follows:

$$\nabla f = \begin{pmatrix} 2x \\ 2y \\ 2z \end{pmatrix} \qquad \nabla f\{1, 1, 1\} = \begin{pmatrix} 2 \\ 2 \\ 2 \end{pmatrix}$$

The equation of the tangent plane to the contour at $(1, 1, 1)$ is

$$(2 \ \ 2 \ \ 2) \begin{pmatrix} x-1 \\ y-1 \\ z-1 \end{pmatrix} = 0 \qquad \text{or} \qquad 2x + 2y + 2z - 6 = 0$$

The contour $f = 3$, the gradient vector, and the tangent line at $(1, 1, 1)$ are illustrated in the following figure. The figure clearly shows the relationship among them.

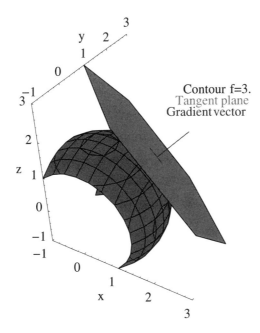

FIGURE 3.6 Tangent plane and gradient vector for a function of three variables.

3.3 Solution of Nonlinear Equations

A practical application of the Taylor series approximation is the development of the so-called Newton-Raphson method for solving the following $n \times n$ system of nonlinear equations:

$$f_1(x_1, x_2, \ldots, x_n) = 0$$
$$f_2(x_1, x_2, \ldots, x_n) = 0$$
$$\vdots$$
$$f_n(x_1, x_2, \ldots, x_n) = 0$$

The method starts from an initial guess and iteratively tries to move it closer to the solution of the equations. The basic iteration is written in the following form:

$$x^{k+1} = x^k + \Delta x^k \qquad k = 0, 1, \ldots$$

with x^0 being the initial guess. The improvement Δx^k is obtained by expanding the equations around x^k using the Taylor series, as follows:

$$f_1(x^{k+1}) \approx f_1(x^k) + \nabla f_1(x^k)^T \Delta x^k = 0$$
$$f_2(x^{k+1}) \approx f_2(x^k) + \nabla f_2(x^k)^T \Delta x^k = 0$$
$$\vdots$$

Writing all n equations in a matrix form, we have

$$\begin{pmatrix} f_1(x^k) \\ f_2(x^k) \\ \vdots \\ f_n(x^k) \end{pmatrix} + \begin{pmatrix} \partial f_1(x^k)/\partial x_1 & \partial f_1(x^k)/\partial x_2 & \cdots & \partial f_1(x^k)/\partial x_n \\ \partial f_2(x^k)/\partial x_1 & \partial f_2(x^k)/\partial x_2 & \cdots & \partial f_2(x^k)/\partial x_n \\ \vdots & \vdots & \vdots & \vdots \\ \partial f_n(x^k)/\partial x_1 & \partial f_n(x^k)/\partial x_2 & \cdots & \partial f_n(x^k)/\partial x_n \end{pmatrix} \Delta x^k = \begin{pmatrix} 0 \\ 0 \\ \vdots \\ 0 \end{pmatrix}$$

or

$$f(x^k) + J(x^k) \Delta x^k = 0$$

The matrix J of partial derivatives of the functions is known as the *Jacobian* matrix. The improvement Δx^k can now be obtained by inverting the Jacobian matrix, giving

$$\Delta x^k = -[J(x^k)]^{-1} f(x^k)$$

3.3 Solution of Nonlinear Equations

With each iteration, the solution in general will improve. That is, taken together, $f_i(x^{k+1})$ should be closer to zero than $f_i(x^k)$. Thus, the convergence of the algorithm can be defined in terms of the norm of the vector $f(x^{k+1})$.

$$\left\| f(x^{k+1}) \right\| \equiv \sqrt{\sum_{i=1}^{n} [f_i(x^{k+1})]^2} \leq \text{tol}$$

where tol is a small tolerance.

The Newton-Raphson method converges fairly rapidly when started from a point that is close to a solution. However, the method is known to diverge in some cases as well because of the Jacobian matrix being nearly singular. Another drawback of the method is that it gives only one solution that is closest to the starting point. Since nonlinear equations, in general, may have several solutions, the only way to get other solutions is to try different starting points. More sophisticated methods for solving nonlinear equations can be found in textbooks on numerical analysis.

The above algorithm is implemented in a function called NewtonRaphson. The main reason for creating this function is to show all intermediate computations. The standard *Mathematica* functions NSolve and FindRoot are much more flexible and efficient for solving general systems of nonlinear equations, and should be used in practice.

```
Needs["OptimizationToolbox`Chap3Tools`"];
?NewtonRaphson
```

NewtonRaphson[fcns, vars, start, maxIterations], computes zero of a
   system of nonlinear functions using Newton-Raphson method. vars = list
   of variables, fcns = list of functions, start = starting values of
   variables, maxIterations = maximum number of iterations allowed
   (optional, Default is 10).

Example 3.18 Determine a solution of the following nonlinear equation using the Newton-Raphson method.

$$x + 2 = e^x$$

(a) Start from $x^0 = 2$ Using NewtonRaphson function calculations for all iterations are as shown below:

```
f = x + 2 - Exp[x];

{sol, hist} = NewtonRaphson[f, x, 2];
```

Jacobian matrix→(1 - E^X)

```
*********** Iteration 1 ***********
xk → (2.)    F(xk) → (-3.38906)    ||F(xk)|| → 3.38906
J(xk) → (-6.38906)
Inverse[J(xk)] → (-0.156518)
Δx → (-0.530447)    New xk → (1.46955)
*********** Iteration 2 ***********
xk → (1.46955)    F(xk) → (-0.877738)    ||F(xk)|| → 0.877738
J(xk) → (-3.34729)
Inverse[J(xk)] → (-0.298749)
Δx → (-0.262223)    New xk → (1.20733)
*********** Iteration 3 ***********
xk → (1.20733)    F(xk) → (-0.137212)    ||F(xk)|| → 0.137212
J(xk) → (-2.34454)
Inverse[J(xk)] → (-0.426523)
Δx → (-0.0585239)    New xk → (1.14881)
*********** Iteration 4 ***********
xk → (1.14881)    F(xk) → (-0.00561748)    ||F(xk)|| → 0.00561748
J(xk) → (-2.15442)
Inverse[J(xk)] → (-0.464161)
Δx → (-0.00260742)    New xk → (1.1462)
*********** Iteration 5 ***********
xk → (1.1462)    F(xk) → (-0.0000107135)    ||F(xk)|| → 0.0000107135
J(xk) → (-2.14621)
Inverse[J(xk)] → (-0.465938)
Δx → (-4.99185 × 10⁻⁶)    New xk → (1.14619)
```

Using the Plot function, the function is plotted as shown in the following figure. The Newton-Raphson iterations are also shown on the plot.

```
Plot[f, {x, 1, 2.1}, PlotRange → All, PlotStyle → {{GrayLevel[0.7]}},
  AxesLabel → {"x", "f"}, TextStyle → {FontFamily → "Times", FontSize → 10},
  Epilog → {RGBColor[1, 0, 0], Line[({#1[[1]], f/.x → #1[[1]]}&) /@hist],
    Text["Start", {2, -3.4}, {-1, 0}], Text["Solution", {1.146, 0.2}]}];
```

3.3 Solution of Nonlinear Equations

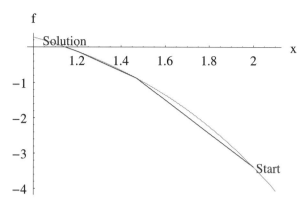

FIGURE 3.7 Graph of function $f(x) = x + 2 - e^x$ and history of Newton-Raphson iterations with $x^0 = 2$.

(b) Start from $x^0 = 0.1$ Using the NewtonRaphson function with this starting point, the method runs into difficulties. The history of points computed by the method are as follows:

x	$\|f(x)\|$
0.1	0.994829
9.55917	0.994829
8.55991	14162.4
7.56174	5207.65
6.5662	1913.63
5.57686	702.096
4.60184	256.663
3.65862	93.066
2.78184	33.149
2.03148	11.3668
1.48904	3.59392
1.21411	0.943797
1.1494	0.15318
1.1462	0.0068998
1.14619	0.0000161414

Near $x = 0.1$, the function is very flat and its first derivative is very close to zero. This causes the method to actually go very far from the solution in the first few iterations. Fortunately, the slope of the function is well behaved in this region and the method eventually converges to the solution. The following graph illustrates the behavior:

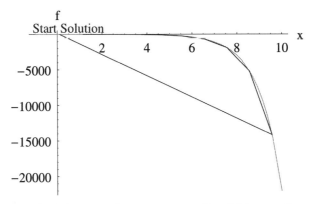

FIGURE 3.8 Graph of function $f(x) = x + 2 - e^x$ and history of Newton-Raphson iterations with $x^0 = 0.1$.

Example 3.19 Starting from $x^0 = (2, 2, 2)$, determine a solution of the following system of nonlinear equations using the Newton-Raphson method.

$$-x_1^2 + x_2^2 + x_1 x_2 x_3 = 1.34 \qquad x_1 x_2 - x_3^2 = 0.09 \qquad e^{x_1} - e^{x_2} + x_3 = 0.41$$

Using the NewtonRaphson function, the following solution is obtained after five iterations.

$$x = \begin{pmatrix} 0.902218 \\ 1.10034 \\ 0.950132 \end{pmatrix}$$

Calculations for all iterations are as shown below.

```
f = {x₁x₂x₃ - x₁² + x₂² - 1.34, x₁x₂ - x₃² - 0.09,
     Exp[x₁] - Exp[x₂] + x₃ - 0.41};

NewtonRaphson[f, {x₁, x₂, x₃}, {2, 2, 2}];
```

$$\text{Jacobian matrix} \to \begin{pmatrix} -2x_1 + x_2 x_3 & 2x_2 + x_1 x_3 & x_1 x_2 \\ x_2 & x_1 & -2x_3 \\ e^{x_1} & -e^{x_2} & 1 \end{pmatrix}$$

3.3 Solution of Nonlinear Equations

*********** Iteration 1 ***********

$$xk \to \begin{pmatrix} 2. \\ 2. \\ 2. \end{pmatrix} \quad F(xk) \to \begin{pmatrix} 6.66 \\ -0.09 \\ 1.59 \end{pmatrix} \quad ||F(xk)|| \to 6.84776$$

$$J(xk) \to \begin{pmatrix} 0. & 8. & 4. \\ 2. & 2. & -4. \\ 7.38906 & -7.38906 & 1. \end{pmatrix}$$

$$\text{Inverse}[J(xk)] \to \begin{pmatrix} 0.0743407 & 0.101319 & 0.107911 \\ 0.0851319 & 0.0797363 & -0.0215823 \\ 0.0797363 & -0.159473 & 0.0431645 \end{pmatrix}$$

$$\Delta x \to \begin{pmatrix} -0.65757 \\ -0.525486 \\ -0.614028 \end{pmatrix} \quad \text{New } xk \to \begin{pmatrix} 1.34243 \\ 1.47451 \\ 1.38597 \end{pmatrix}$$

*********** Iteration 2 ***********

$$xk \to \begin{pmatrix} 1.34243 \\ 1.47451 \\ 1.38597 \end{pmatrix} \quad F(xk) \to \begin{pmatrix} 1.77551 \\ -0.0314865 \\ 0.435397 \end{pmatrix} \quad ||F(xk)|| \to 1.82839$$

$$J(xk) \to \begin{pmatrix} -0.641226 & 4.8096 & 1.97943 \\ 1.47451 & 1.34243 & -2.77194 \\ 3.82834 & -4.36891 & 1. \end{pmatrix}$$

$$\text{Inverse}[J(xk)] \to \begin{pmatrix} 0.145217 & 0.181489 & 0.215631 \\ 0.162998 & 0.110844 & -0.0153911 \\ 0.156186 & -0.210535 & 0.107249 \end{pmatrix}$$

$$\Delta x \to \begin{pmatrix} -0.346005 \\ -0.279214 \\ -0.330635 \end{pmatrix} \quad \text{New } xk \to \begin{pmatrix} 0.996425 \\ 1.1953 \\ 1.05534 \end{pmatrix}$$

*********** Iteration 3 ***********

$$xk \to \begin{pmatrix} 0.996425 \\ 1.1953 \\ 1.05534 \end{pmatrix} \quad F(xk) \to \begin{pmatrix} 0.352813 \\ -0.0127099 \\ 0.0493711 \end{pmatrix} \quad ||F(xk)|| \to 0.356478$$

$$J(xk) \to \begin{pmatrix} -0.731406 & 3.44216 & 1.19103 \\ 1.1953 & 0.996425 & -2.11067 \\ 2.70858 & -3.30455 & 1. \end{pmatrix}$$

$$\text{Inverse}[J(xk)] \to \begin{pmatrix} 0.218674 & 0.269866 & 0.309154 \\ 0.252831 & 0.144751 & 0.00439389 \\ 0.243196 & -0.252618 & 0.177152 \end{pmatrix}$$

$$\Delta x \to \begin{pmatrix} -0.0889844 \\ -0.0875794 \\ -0.0977598 \end{pmatrix} \quad \text{New } xk \to \begin{pmatrix} 0.907441 \\ 1.10772 \\ 0.957577 \end{pmatrix}$$

*********** Iteration 4 ***********

$$xk \to \begin{pmatrix} 0.907441 \\ 1.10772 \\ 0.957577 \end{pmatrix} \quad F(xk) \to \begin{pmatrix} 0.0261437 \\ -0.00176379 \\ -0.00189872 \end{pmatrix} \quad ||F(xk)|| \to 0.0262718$$

$$J(xk) \to \begin{pmatrix} -0.754154 & 3.08439 & 1.00519 \\ 1.10772 & 0.907441 & -1.91515 \\ 2.47797 & -3.02745 & 1. \end{pmatrix}$$

$$\text{Inverse}[J(xk)] \to \begin{pmatrix} 0.244564 & 0.30642 & 0.341009 \\ 0.292712 & 0.162272 & 0.0165448 \\ 0.280148 & -0.26803 & 0.205078 \end{pmatrix}$$

$$\Delta x \to \begin{pmatrix} -0.00520585 \\ -0.00733494 \\ -0.00740746 \end{pmatrix} \quad \text{New } xk \to \begin{pmatrix} 0.902235 \\ 1.10039 \\ 0.95017 \end{pmatrix}$$

*********** Iteration 5 ***********

$$xk \to \begin{pmatrix} 0.902235 \\ 1.10039 \\ 0.95017 \end{pmatrix} \quad F(xk) \to \begin{pmatrix} 0.000155003 \\ -0.0000166859 \\ -0.0000477221 \end{pmatrix} \quad ||F(xk)|| \to 0.000163039$$

$$J(xk) \to \begin{pmatrix} -0.758917 & 3.05805 & 0.992806 \\ 1.10039 & 0.902235 & -1.90034 \\ 2.46511 & -3.00532 & 1. \end{pmatrix}$$

$$\text{Inverse}[J(xk)] \to \begin{pmatrix} 0.246202 & 0.30932 & 0.343383 \\ 0.296171 & 0.164153 & 0.0179052 \\ 0.283177 & -0.269175 & 0.207336 \end{pmatrix}$$

$$\Delta x \to \begin{pmatrix} -0.0000166137 \\ -0.0000423138 \\ -0.0000384902 \end{pmatrix} \quad \text{New } xk \to \begin{pmatrix} 0.902218 \\ 1.10034 \\ 0.950132 \end{pmatrix}$$

3.4 Quadratic Forms

A quadratic form is a function of n variables in which each term is either a square of one variable or is the product of two different variables. Consider the following functions:

1. $f(x_1, x_2, x_3) = x_1^2 + x_1 x_2 - 2x_1 x_3 - 17x_2^2 - x_2 x_3 + 7x_3^2$
2. $f(x_1, x_2, x_3) = x_1^2 + x_1 x_2 x_3 - 2x_1 x_3 - 17x_2^2 - x_2 x_3 + 7x_3^2$
3. $f(x_1, x_2, x_3) = x_1^2 x_2 + x_1 x_2 - 2x_1 x_3 - 17x_2^2 - x_2 x_3 + 7x_3^2$

Out of these three, only the first function is a quadratic form. The second term in the second function and the first term in the third function are the reasons why the other two are not quadratic functions.

3.4 Quadratic Forms

All quadratic forms can be written in the following matrix form:

$$f(x) = \frac{1}{2}x^T A x$$

where x is the vector of variables and A is a symmetric matrix of coefficients organized as follows.

(i) The diagonal entries of matrix A are twice the coefficients of the square terms. The first row diagonal is twice the coefficient of the x_1^2 term, the second row diagonal term is twice the coefficient of the x_2^2 term, and so on.

(ii) The off-diagonal terms contain the coefficients of the cross-product terms. The terms in the first row are coefficients of $x_1 x_2$, $x_1 x_3$, ..., $x_1 x_n$. The second row contains coefficients of $x_2 x_1$, $x_2 x_3$, ..., $x_2 x_n$, and so on.

Example 3.20 Write the following function as a quadratic form:

```
f = -4x₁² - x₁x₂ - 5x₂² + 2x₁x₃ + 4x₂x₃ - 6x₃² - 3x₁x₄ - 5x₂x₄ + 6x₃x₄ - 7x₄²;
vars = {x₁, x₂, x₃, x₄};
```

The symmetric matrix A associated with function $f(x_1, x_2, x_3, x_4)$ is as follows:

$$A = \begin{pmatrix} -8 & -1 & 2 & -3 \\ -1 & -10 & 4 & -5 \\ 2 & 4 & -12 & 6 \\ -3 & -5 & 6 & -14 \end{pmatrix};$$

The following direct computation shows that $f(x) = \frac{1}{2}x^T A x$:

```
1/2 vars.A.vars//Expand
```

$-4x_1^2 - x_1 x_2 - 5x_2^2 + 2x_1 x_3 + 4x_2 x_3 - 6x_3^2 - 3x_1 x_4 - 5x_2 x_4 + 6x_3 x_4 - 7x_4^2$

3.4.1 Differentiation of a Quadratic Form

It can easily be verified by direct computations that the gradient vector and Hessian matrix of a quadratic form $f(x) = \frac{1}{2}x^T A x$ can be written as follows:

$$\nabla f(x) = Ax \qquad \nabla^2 f(x) = A$$

Example 3.21 Verify the above gradient and Hessian expressions for the following quadratic form:

```
f = -4x₁² - x₁x₂ - 5x₂² + 2x₁x₃ + 4x₂x₃ - 6x₃² - 3x₁x₄ - 5x₂x₄ + 6x₃x₄ - 7x₄²;
vars = {x₁, x₂, x₃, x₄};
```

By directly differentiating the given function, the gradient and Hessian are as follows:

Grad[f, vars]//MatrixForm

$$\begin{pmatrix} -8x_1 - x_2 + 2x_3 - 3x_4 \\ -x_1 - 10x_2 + 4x_3 - 5x_4 \\ 2x_1 + 4x_2 - 12x_3 + 6x_4 \\ -3x_1 - 5x_2 + 6x_3 - 14x_4 \end{pmatrix}$$

Hessian[f, vars]//MatrixForm

$$\begin{pmatrix} -8 & -1 & 2 & -3 \\ -1 & -10 & 4 & -5 \\ 2 & 4 & -12 & 6 \\ -3 & -5 & 6 & -14 \end{pmatrix}$$

We can see that the Hessian matrix is the same as the symmetric matrix A associated with the quadratic form:

$$A = \begin{pmatrix} -8 & -1 & 2 & -3 \\ -1 & -10 & 4 & -5 \\ 2 & 4 & -12 & 6 \\ -3 & -5 & 6 & -14 \end{pmatrix};$$

The following direct computation verifies the gradient expression:

A.vars//Expand

```
{-8x₁ - x₂ + 2x₃ - 3x₄, -x₁ - 10x₂ + 4x₃ - 5x₄,
  2x₁ + 4x₂ - 12x₃ + 6x₄, -3x₁ - 5x₂ + 6x₃ - 14x₄}
```

3.4.2 Sign of a Quadratic Form

Depending on the sign of the product $x^T A x$ for all possible vectors x, quadratic forms are classified as follows:

1. Positive definite if $x^T A x > 0$ for all vectors x.
2. Positive semidefinite if $x^T A x \geq 0$ for all vectors x.
3. Negative definite if $x^T A x < 0$ for all vectors x.
4. Negative semidefinite if $x^T A x \leq 0$ for all vectors x.
5. Indefinite if $x^T A x \leq 0$ for some vectors and $x^T A x \geq 0$ for other vectors.

Note that these definitions simply tell us what it means for a matrix to be positive definite, semidefinite, and so on. Using the definitions alone, we cannot determine the sign of a quadratic form because it would involve consideration of an infinite number of vectors x. The following two tests give us a practical way to determine the sign of a quadratic form.

Eigenvalue Test for Determining the Sign of a Quadratic Form

If all eigenvalues, λ_i $i = 1, \ldots, n$, of the $n \times n$ symmetric matrix A in the quadratic form $f(x) = \frac{1}{2} x^T A x$ are known, then the sign of the quadratic form is determined as follows.

1. Positive definite if $\lambda_i > 0$ $i = 1, \ldots, n$.
2. Positive semidefinite if $\lambda_i \geq 0$ $i = 1, \ldots, n$.
3. Negative definite if $\lambda_i < 0$ $i = 1, \ldots, n$.
4. Negative semidefinite if $\lambda_i \leq 0$ $i = 1, \ldots, n$.
5. Indefinite if $\lambda_i \leq 0$ for some i and $\lambda_i \geq 0$ for others.

Principal Minors Test for Determining the Sign of a Quadratic Form

This test usually requires less computational effort than the eigenvalue test. If all principal minors, A_i $i = 1, \ldots, n$, of $n \times n$ symmetric matrix A in the quadratic form $f(x) = \frac{1}{2} x^T A x$ are known, then the sign of the quadratic form is determined as follows:

1. Positive definite if $A_i > 0$ $i = 1, \ldots, n$.
2. Positive semidefinite if $A_i \geq 0$ $i = 1, \ldots, n$.
3. Negative definite if $\begin{cases} A_i < 0 & i = 1, 3, 5, \ldots \text{ (odd indices)} \\ A_i > 0 & i = 2, 4, 6, \ldots \text{ (even indices)} \end{cases}$.
4. Negative semidefinite if $\begin{cases} A_i \leq 0 & i = 1, 3, 5, \ldots \text{ (odd indices)} \\ A_i \geq 0 & i = 2, 4, 6, \ldots \text{ (even indices)} \end{cases}$.
5. Indefinite if none of the above cases applies.

The following two functions, contained in the OptimizationToolbox 'CommonFunctions' package, return principal minors of a matrix and its sign:

```
Needs["OptimizationToolbox`CommonFunctions`"];
?PrincipalMinors
```

?PrincipalMinors

PrincipalMinors[A], returns Principal minors of a matrix A.

?QuadraticFormSign

QuadraticFormSign[A], determines the sign of a quadratic form involving symmetric matrix A using principal minors test. It returns the sign (positive definite, etc.) and the principal minors of A.

Example 3.22 Write the following function as a quadratic form and determine its sign using both the eigenvalue test and the principal minors test:

```
f = -4x₁² - x₁x₂ - 5x₂² + 2x₁x₃ + 4x₂x₃ - 6x₃² - 3x₁x₄ - 5x₂x₄ + 6x₃x₄ - 7x₄²;
xv = {x₁, x₂, x₃, x₄};
```

The symmetric matrix A associated with the function $f(x_1, x_2, x_3, x_4)$ is as follows:

$$A = \begin{pmatrix} -8 & -1 & 2 & -3 \\ -1 & -10 & 4 & -5 \\ 2 & 4 & -12 & 6 \\ -3 & -5 & 6 & -14 \end{pmatrix};$$

The eigenvalues of matrix A are as follows. The Chop function is used to round off results returned by the Eigenvalue function:

Chop[N[Eigenvalues[A]]]

{-7., -6., -23.2621, -7.73791}

The quadratic form is negative definite, since all eigenvalues are negative. The following principal minors test returns the same conclusion (odd principal minors are negative, and even are positive).

QuadraticFormSign[A];

$$\text{Principal minors} \rightarrow \begin{pmatrix} -8 \\ 79 \\ -796 \\ 7560 \end{pmatrix}$$

Quadratic form → NegativeDefinite

3.4 Quadratic Forms

Example 3.23 For the following functions of two variables, determine the quadratic forms and their signs. Compare the contour plots of these functions.

$$f(x, y) = x^2 + xy + 2y^2 \qquad g(x, y) = x^2 + xy - 2y^2$$

The symmetric matrices in the quadratic form for these functions are as follows:

$$A = \begin{pmatrix} 2 & 1 \\ 1 & 4 \end{pmatrix} \qquad B = \begin{pmatrix} 2 & 1 \\ 1 & -4 \end{pmatrix}$$

The principal minors of the two matrices are

$$PM_A = (2, 7) \qquad PM_B = (2, -9)$$

Thus, the function f is positive definite, while g is indefinite. The three-dimensional and contour plots of these functions are shown in the following figure. Note that the positive definite function has a well-defined minimum (at $(0, 0)$ for f), whereas the indefinite function changes sign but does not have a minimum.

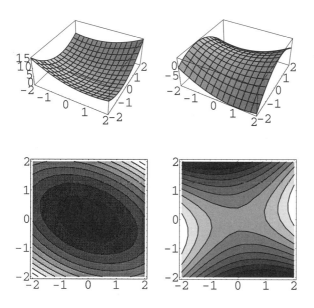

FIGURE 3.9 Three-dimensional and contour plots of functions f and g.

3.5 Convex Functions and Convex Optimization Problems

Optimization problems that are defined in terms of convex functions are very special in optimization literature. There are many theoretical results that are applicable only to such problems. Convergence of many optimization algorithms assumes convexity of objective and constraint functions. This section briefly reviews the basic definitions of convex sets and functions. A test is also described that can be used to determine whether a function is convex or not.

3.5.1 Convex Functions

A function $f(x)$ is convex if for *any* two points $x^{(1)}$ and $x^{(2)}$, the function values satisfy the following inequality:

$$f\left(\alpha x^{(2)} + (1-\alpha)x^{(1)}\right) \leq \alpha f\left(x^{(2)}\right) + (1-\alpha)f\left(x^{(1)}\right) \quad \text{for all } 0 \leq \alpha \leq 1$$

For a function of a single variable, this means that the function is convex if the graph of $f(x)$ lies below the line joining any two points on the graph, as illustrated in Figure 3.10. A so-called *concave* function is defined simply by changing the direction of the inequality in the above expression.

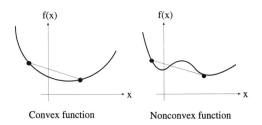

FIGURE 3.10 Convex and nonconvex functions.

Using the definition, it is difficult to check for convexity of a given function because it would require consideration of infinitely many points. However, using the sign of the Hessian matrix of the function, we can determine the convexity of a function as follows:

3.5 Convex Functions and Convex Optimization Problems

(a) A function $f(x)$ is convex if its Hessian $\nabla^2 f(x)$ is at least positive semidefinite.

(b) A function $f(x)$ is concave if its Hessian $\nabla^2 f(x)$ is at least negative semidefinite.

(c) A function $f(x)$ is nonconvex if its Hessian $\nabla^2 f(x)$ is indefinite.

For a linear function, the Hessian matrix consists of all zeros, and hence a linear function is both convex and concave at the same time.

As seen in the previous section, the Hessian of a quadratic function is equal to the coefficient matrix in the quadratic form, and is thus a matrix involving numbers. It is fairly easy to use the principal minor test to determine the sign of the Hessian.

For more complicated functions, the Hessian may be a messy matrix involving variables, and it may not be easy to determine its sign for all possible values of variables. The following results, proved in books on convex analysis, are often useful in dealing with more complex situations:

1. If $f(x)$ is a convex function, then $\alpha f(x)$ is also a convex function for any $\alpha > 0$.
2. The sum of convex functions is also a convex function. That is, if $f_i(x)$, $i = 1, \ldots, k$ are all convex functions, then $f(x) = \sum_{i=1}^{k} f_i(x)$ is a convex function.
3. If $f(x)$ is a convex function, and $g(y)$ is another convex function whose value is continuously increasing then the composite function $g(f(x))$ is also a convex function.

Some functions may not be convex over their entire domain; however, they may be convex over a specified set. These concepts are illustrated through the following examples:

Example 3.24 Determine the convexity of the following function of a single variable:

$$f(x) = x + 1/x \qquad x > 0$$

The Hessian is simply the second derivative of this function:

$$\nabla^2 f = \frac{2}{x^3} > 0 \quad \text{for } x > 0.$$

Therefore, the given function is convex over the specified domain.
The following plot confirms this graphically:

```
Plot[x + 1/x, {x, 0.001, 2},
  TextStyle- > {FontFamily- > "Times", FontSize- > 10}];
```

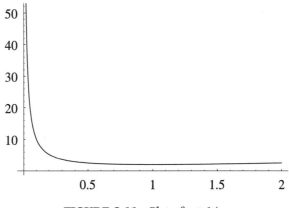

FIGURE 3.11 Plot of $x + 1/x$.

Example 3.25 Determine the convexity of the following function of a single variable:

$$f(x) = x\sin[x]$$

The Hessian is simply the second derivative of this function.

$$\nabla^2 f = 2\cos[x] - x\sin[x]$$

For some values of x, this is positive and for others, it is negative. Therefore, this is a nonconvex function. The following plot confirms this graphically.

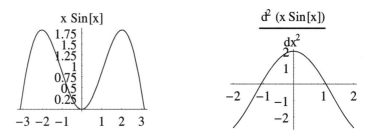

FIGURE 3.12 Plot of $x\sin[x]$ and its second derivative of $x\sin[x]$.

From the graph, one may be tempted to conclude that the function is convex over $-2 \leq x \leq 2$. However, the second derivative actually changes sign at $x = \pm 1.077$, as the graph of the second derivative of the function shows. Thus, between $(-\pi, \pi)$, the function is convex only over $-1.077 \leq x \leq 1.077$.

Example 3.26 Determine the convexity of the following function of two variables:

$$f(x, y) = 5 - 5x - 2y + 2x^2 + 5xy + 6y^2$$

Hessian → $\begin{pmatrix} 4 & 5 \\ 5 & 12 \end{pmatrix}$ PrincipalMinors → $\begin{pmatrix} 4 \\ 23 \end{pmatrix}$

Status → Convex

The following contour plot clearly shows that the function has a well-defined minimum:

```
GraphicalSolution[5 - 5x - 2y + 2x^2 + 5xy + 6y^2, {x, -5, 5}, {y, -5, 5},
  ObjectiveContours → {2, 5, 10, 30}, PlotPoints → 40];
```

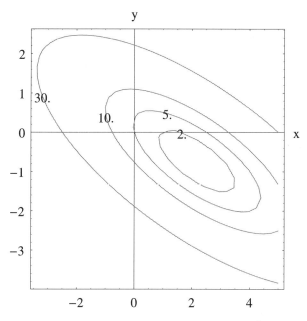

FIGURE 3.13   Contour plot of $f = 5 - 5x - 2y + 2x^2 + 5xy + 6y^2$.

Example 3.27 Determine the convexity of the following function of two variables:

$$f(x, y) = e^{(x^2+y^2)} + e^{(x+2y)}$$

The Hessian matrix for this function is as follows:

$$\nabla^2 f = \begin{pmatrix} 4e^{x^2+y^2}x^2 + e^{x+2y} + 2e^{x^2+y^2} & 4e^{x^2+y^2}xy + 2e^{x+2y} \\ 4e^{x^2+y^2}xy + 2e^{x+2y} & 4e^{x^2+y^2}y^2 + 4e^{x+2y} + 2e^{x^2+y^2} \end{pmatrix}$$

It is obviously hopeless to try to explicitly check the sign of the principal minors of this matrix. However, we can proceed in the following manner to show that the function is indeed convex everywhere.

First, we recognize that the given function is the sum of two functions. Thus, we need to show that the two terms individually are convex. Both terms are of the form e^z, where z is any real number. As the following plot demonstrates, this function is continuously increasing and is convex. Using the composite function rule, therefore, we just need to show that the powers of e are convex functions.

$$f_1(x, y) \equiv x^2 + y^2 \quad \nabla^2 f_1 = \begin{pmatrix} 2 & 0 \\ 0 & 2 \end{pmatrix} \Longrightarrow \text{Convex. Thus, } e^{f_1(x,y)} \text{ is convex.}$$

$$f_2(x, y) = x + 2y \quad \text{Linear} \Longrightarrow \text{Convex. Thus } e^{f_2(x,y)} \text{ is convex.}$$

Thus, each term individually is convex and hence, $f(x, y) = e^{(x^2+y^2)} + e^{(x+2y)}$ is a convex function.

```
Plot[e^z, {z, -5, 1}, TextStyle → {FontFamily → "Times", FontSize → 10}];
```

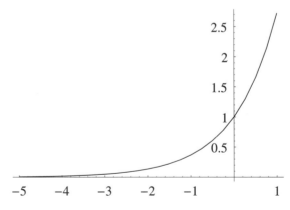

FIGURE 3.14 Plot of e^z showing it is a continuously increasing convex function.

The following contour plot clearly shows that the function has a well-defined minimum:

```
GraphicalSolution[e^(x²+y²) + e^(x+2y), {x, -2, 2}, {y, -2, 2},
  ObjectiveContours → {2, 5, 10, 20, 40}, PlotPoints → 40];
```

Objective function → $e^{x+2y} + e^{x^2+y^2}$

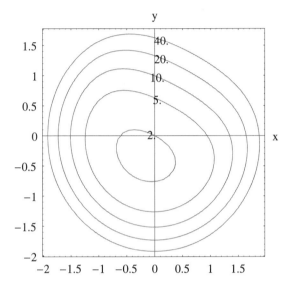

FIGURE 3.15 Contour plot of $f = e^{x+2y} + e^{x^2+y^2}$.

3.5.2 Convex Feasible Sets

A set is convex if a straight line joining any two points in the set lies completely inside the set. Otherwise, it is nonconvex. The following figure shows an example of a convex and a nonconvex set:

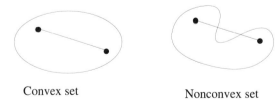

Convex set   Nonconvex set

FIGURE 3.16 Convex and nonconvex sets.

Parametrically, the line segment between two points $x^{(1)}$ and $x^{(2)}$ can be expressed as follows:

$$x = \alpha x^{(2)} + (1-\alpha)x^{(1)} \quad \text{for all } 0 \leq \alpha \leq 1$$

Thus, a set S is convex if all points given by $\alpha x^{(2)} + (1-\alpha)x^{(1)}$ with $0 \leq \alpha \leq 1$, and with any arbitrary pair of points in the set, belong to the set.

Convex sets of importance in optimization are the feasible regions obtained from the intersection of constraint functions. We consider several examples to illustrate these sets.

Example 3.28 *Optimization problem involving linear inequality constraints* As mentioned already, all linear functions are convex. The feasible region for an optimization problem involving linear inequality constraints is always convex. Consider a problem in two variables involving the following four constraints:

$$\begin{pmatrix} g_1 \\ g_2 \\ g_3 \\ g_4 \end{pmatrix} = \begin{pmatrix} -x+y \leq 2 \\ 2x+3y \leq 11 \\ x \geq 0 \\ y \geq 0 \end{pmatrix}$$

The feasible region shown in Figure 3.17 clearly represents a convex set.

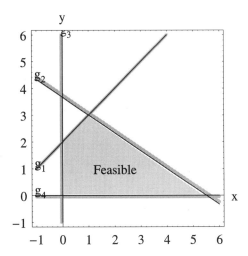

FIGURE 3.17 The feasible region for a problem with linear inequality constraints.

3.5 Convex Functions and Convex Optimization Problems

Example 3.29 *Optimization problem involving linear equality constraints* The feasible region for an optimization problem involving both linear inequality and equality constraints is always convex. Consider a problem in two variables involving the following four constraints:

$$\begin{pmatrix} h \\ g_1 \\ g_2 \\ g_3 \end{pmatrix} = \begin{pmatrix} 2x + 3y = 11 \\ -x + y \leq 2 \\ x \geq 0 \\ y \geq 0 \end{pmatrix}$$

The feasible region now is the straight line between g_3 and g_1, as shown in Figure 3.18. Being a straight line, it trivially satisfies the requirements of being a convex set.

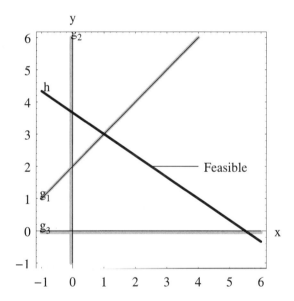

FIGURE 3.18 The feasible region for a problem involving a linear equality constraint.

Example 3.30 *Optimization problem involving convex inequality constraints* The feasible region for an optimization problem involving linear or nonlinear inequality constraints that are all convex functions represents a convex set. Consider a problem in two variables involving the following four constraints:

$$\begin{pmatrix} g_1 \\ g_2 \\ g_3 \\ g_4 \end{pmatrix} = \begin{pmatrix} \left(x - \tfrac{3}{2}\right)^2 + (y-5)^2 \leq 10 \\ 2x^2 + 3y^2 \leq 35 \\ x \geq 0 \\ y \geq 0 \end{pmatrix}$$

It can easily be verified that nonlinear constraints g_1 and g_2 are both convex functions. The feasible region is shown in Figure 3.19 and clearly represents a convex set.

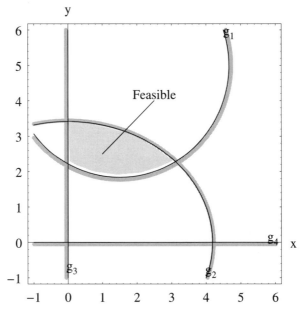

FIGURE 3.19 The feasible region for a problem involving convex nonlinear inequality constraints.

Example 3.31 *Optimization problem involving nonlinear equality constraints* The feasible region for an optimization problem involving nonlinear equality constraints is *not* a convex set regardless of whether the equality constraint is convex or not. Consider a problem in two variables involving the following four constraints.

$$\begin{pmatrix} h \\ g_1 \\ g_2 \\ g_3 \end{pmatrix} = \begin{pmatrix} \left(x - \tfrac{3}{2}\right)^2 + (y-5)^2 = 10 \\ 2x^2 + 3y^2 \leq 35 \\ x \geq 0 \\ y \geq 0 \end{pmatrix}$$

3.5 Convex Functions and Convex Optimization Problems

The functions are the same as those used in the previous example and hence are all convex. However, the feasible region, line segment between constraints g_1 and g_2, is not a convex set, as shown in Figure 3.20. This is because the line joining any two points on this line is not in the feasible line.

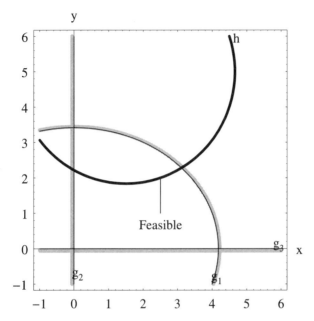

FIGURE 3.20 The feasible region for a problem involving a nonlinear equality constraint.

3.5.3 Convex Optimization Problem

An optimization problem in which the objective function is convex and the feasible region is a convex set is known as a convex programming problem. It is clear from the discussion in the previous section that a linear programming problem is always a convex programming problem. A nonlinear programming problem with a convex objective function and involving linear (equality or inequality) or convex inequality constraints is also a convex optimization problem. A problem involving even a single nonlinear equality constraint is a nonconvex optimization problem.

The most important property of a convex programming problem is that any local minimum point x^* is also a global minimum. We can prove this, by contradiction, as follows:

Suppose x^* is a local minimum. If it is not a global minimum, then there must be another local minimum $\hat{x}$ such that

$$f(\hat{x}) < f(x^*)$$

Consider the line joining the points x^* and $\hat{x}$ as follows:

$$x = \alpha \hat{x} + (1 - \alpha)x^* \qquad 0 \leq \alpha \leq 1$$

Since the feasible region is a convex set, all points x on this line are in the feasible set. Furthermore, since $f(x)$ is a convex function, we have

$$f(x) \leq \alpha f(\hat{x}) + (1 - \alpha)f(x^*) \qquad 0 \leq \alpha \leq 1$$

or

$$f(x) \leq f(x^*) + \alpha[f(\hat{x}) - f(x^*)]$$

From the starting assertion, the term in the square bracket must be less than zero, indicating that

$$f(x) < f(x^*) \qquad \text{for all } 0 \leq \alpha \leq 1$$

This means that even in the small neighborhood of x^*, we have other points where the function value is lower than the minimum point. Clearly, then x^* cannot be a local minimum, which is a contradiction. Hence, there cannot be another local minimum, proving the assertion that x^* must be a global minimum.

Most optimization algorithms discussed in later chapters are designed to find only a local minimum. In general, there is no good way to find a global minimum except for to try several different starting points with the hope that one of these points will lead to the global minimum. However, for convex problems, one can use any optimum solution obtained from these methods with confidence, knowing that there is no other solution that has a better objective function value than this one.

The following function, contained in the OptimizationToolbox 'Common-Function' package, checks convexity of one or more functions of any number of variables:

```
Needs["OptimizationToolbox`CommonFunctions`"];
?ConvexityCheck
```

```
ConvexityCheck[f,vars], checks to see if function f (or a list of
    functions f) is convex.
```

3.5 Convex Functions and Convex Optimization Problems

Example 3.32 Determine if the following optimization problem is a convex programming problem:

Minimize $f(x, y) = (x - 5)^2 + (y - 1)^2 + xy$

Subject to $\begin{pmatrix} x + y \leq 4 \\ 2x + x^2 + y^2 = 16 \\ \frac{x^2}{y} \leq 1 \end{pmatrix}$

The problem involves a nonlinear equality constraint; therefore, it is not a convex programming problem.

Example 3.33 Determine if the following optimization problem is a convex programming problem:

Minimize $f(x, y) = (x - 5)^2 + (y - 1)^2 + xy$

Subject to $\begin{pmatrix} x + y \leq 4 \\ 2x + x^2 + y^2 \leq 16 \\ \frac{x^2}{y} \leq 1 \end{pmatrix}$

The problem does not involve a nonlinear equality constraint; therefore, it is convex if all functions in the problem are convex. We use the Convexity-Check function to determine the convexity status of objective and constraint functions.

```
f = (x - 5)^2 + (y - 1)^2 + xy;
g = {x + y ≤ 4, x^2 + y^2 + 2x ≤ 16, x^2/y ≤ 1};

ConvexityCheck[f, {x, y}];
------ Function → (-5 + x)^2 + (-1 + y)^2 + xy
Hessian → (2  1)   Principal Minors → (2)
          (1  2)                      (3)
Status → Convex

ConvexityCheck[g, {x, y}];
------ Function → -4 + x + y
Hessian → (0  0)   Principal Minors → (0)
          (0  0)                      (0)
Status → Convex
------ Function → -16 + 2x + x^2 + y^2
Hessian → (2  0)   Principal Minors → (2)
          (0  2)                      (4)
```

Status → Convex

- - - - - - Function → $-1 + \dfrac{x^2}{y}$

Hessian → $\begin{pmatrix} \dfrac{2}{y} & -\dfrac{2x}{y^2} \\ -\dfrac{2x}{y^2} & \dfrac{2x^2}{y^3} \end{pmatrix}$ Principal Minors → $\begin{pmatrix} \dfrac{2}{y} \\ 0 \end{pmatrix}$

Status → Undetermined

Note that since the principal minor of the third constraint contains a variable, the function is unable to determine its convexity and returns "Undetermined" status. However, we can see that the first principal minor is positive for $y > 0$. Thus, the problem is convex with the additional constraint that y must be positive. Without this constraint, the problem in nonconvex. The optimum solution in the first quadrant is shown in the following figure.

```
GraphicalSolution[f, {x, 0.1, 5}, {y, 0.1, 5},
  Constraints → g, ObjectiveContours → {5, 10, 15, 22, 30},
  ShadingOffset → 0.05, PlotPoints → 40,
  Epilog → {Line[{{1.56, 2.44}, {3, 3}}], Text["Optimum", {3, 3.2}]}];
```

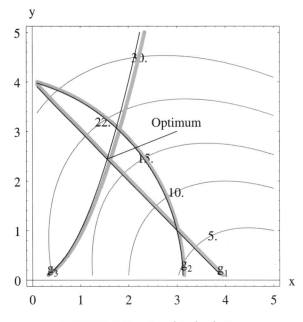

FIGURE 3.21   Graphical solution.

3.6 Problems

Vectors and Matrices

Use the following matrices in Problems 3.1 through 3.8.

$$A = \begin{pmatrix} 1 & -2 & 3 & -4 \\ 5 & 6 & -7 & 8 \\ 9 & -10 & 11 & 12 \end{pmatrix} \quad B = \begin{pmatrix} 1 & 2 & -3 \\ 2 & 4 & -5 \\ -3 & -5 & 6 \end{pmatrix}$$

$$c = \begin{pmatrix} 1 & 2 \\ 2 & 4 \\ -3 & -5 \end{pmatrix} \quad d = \begin{pmatrix} -3 \\ -5 \\ 6 \end{pmatrix}$$

3.1. Perform the following matrix computation:

$$(AA^T - 2B)c$$

3.2. Perform the following matrix computation:

$$A^T B^{-1} c$$

3.3. Compute the determinant of the matrix resulting from the following computation:

$$AA^T - B/2$$

3.4. Compute the cosine of the angle between vector d and the one resulting from the computation Bd.

3.5. Compute the inverse of the matrix resulting from the following computation:

$$AA^T + B$$

3.6. Compute eigenvalues of the matrix resulting from the following computation:

$$AA^T$$

3.7. Compute principal minors of the matrix resulting from the following computation:

$$A^T A$$

3.8. Compute ranks of matrices A, B and c.

3.9. Check to see if the following three vectors are linearly independent:

$$a = \begin{pmatrix} 2 \\ 2 \\ -1 \end{pmatrix} \quad b = \begin{pmatrix} 1 \\ 2 \\ 4 \end{pmatrix} \quad c = \begin{pmatrix} -9 \\ 8 \\ 4 \end{pmatrix}$$

3.10. Show that only two of the following three vectors are linearly independent. Find the linear relationship between the third vector and the two linearly independent ones.

$$a = \begin{pmatrix} 2 \\ 2 \\ -1 \end{pmatrix} \quad b = \begin{pmatrix} -9 \\ -2 \\ 3 \end{pmatrix} \quad c = \begin{pmatrix} -5 \\ 2 \\ 1 \end{pmatrix}$$

3.11. Check to see if the following four vectors are linearly independent:

$$a = \begin{pmatrix} 2 \\ 2 \\ -1 \\ 1 \end{pmatrix} \quad b = \begin{pmatrix} 1 \\ 2 \\ 4 \\ 1 \end{pmatrix} \quad c = \begin{pmatrix} -9 \\ 8 \\ 4 \\ 1 \end{pmatrix} \quad d = \begin{pmatrix} -1 \\ 2 \\ 4 \\ 1 \end{pmatrix}$$

Approximation Using the Taylor Series

Determine linear and quadratic approximations of the following functions around the given points using the Taylor series. For one- and two-variable problems, graphically compare the approximations with the original function.

3.12. $f(x) = x^2 e^{-x^2}$ $x^0 = 1.5$

3.13. $f(x) = x + 1/x - 5$ $x^0 = 2$

3.14. $f(x) = e^{-x^2} \cos[x/\pi] + \sin[x^3] - 1$ $x^0 = \pi/3$

3.15. $f(x, y) = x^2 + 4yx + y^2 - 4$ $(x^0, y^0) = (1, 1)$

3.16. $f(x, y) = 5x^2 - 4yx + 2y^2 - 8$ $(x^0, y^0) = (1, 1)$

3.17. $f(x, y) = e^{(x-y)} - x + y - 5$ $(x^0, y^0) = (1, 2)$

3.18. $f(x, y) = 3 + 2x + 2/y$ $(x^0, y^0) = (1, 2)$

3.19. $f(x, y, z) = 2y/z + x + 2$ $(x^0, y^0, z^0) = (1, 2, 3)$

3.20. $f(x, y, z) = x^2 + y^2 + z^2 - 1$ $(x^0, y^0, z^0) = (1, 2, 3)$

3.21. $f(x_1, x_2, x_3, x_4) = 1/4 + x_1/x_2 - x_2/x_3 + x_3/x_4 - x_4/x_1$
    $(x_1^0, x_2^0, x_3^0, x_4^0) = (1, -1, 1, -1)$

Solution of Nonlinear Equations

Starting from given points, solve the following nonlinear equations using the Newton-Raphson method.

3.22. $x^2 e^{-x^2} = 0 \qquad x^0 = 1.5$

3.23. $x + 1/x = 5 \qquad x^0 = 2$

3.24. $e^{-x^2} \cos[x/\pi] + \sin[x^3] = 1 \qquad x^0 = \pi/3$

3.25. $x^2 + 4yx + y^2 - 4 = 0 \qquad 5x^2 - 4yx + 2y^2 - 8 = 0 \qquad (x^0, y^0) = (1, 1)$

3.26. $e^{(x-y)} - x + y - 5 = 0 \qquad 3 + 2x + 2/y = 0 \qquad (x^0, y^0) = (1, 2)$

3.27. $x^2 + y^2 + z^2 = 1 \qquad 2x + z = 0 \qquad 2y/z + x + 2 = 0 \qquad (x^0, y^0, z^0) = (1, 2, 3)$

Quadratic Forms

For the following problems, write the functions in their quadratic forms and determine their signs using the eigenvalue test. For functions of two variables, draw contour plots of the functions.

3.28. $f(x_1, x_2) = 6x_1^2 + 3x_1 x_2 + 2x_2^2$

3.29. $f(x_1, x_2) = 2x_1^2 - x_1 x_2 + \frac{7x_2^2}{2}$

3.30. $f(x_1, x_2, x_3) = x_1^2 - x_1 x_2 + \frac{3x_2^2}{2} + 3x_1 x_3 - x_2 x_3 + 2x_3^2$

3.31. $f(x_1, x_2, x_3) = 6x_1^2 - 4x_2^2 - 3x_1 x_3 + 2x_3^2$

3.32. $f(x_1, x_2, x_3, x_4) = \frac{x_1^2}{2} + x_2^2 - 3x_1 x_3 + \frac{13x_3^2}{2} + 2x_1 x_4 + x_2 x_4 + 7x_4^2$

3.33. $f(x_1, x_2, x_3, x_4) = -\frac{x_1^2}{2} - x_2^2 + 3x_1 x_3 - 5x_3^2 - 2x_1 x_4 - x_2 x_4 + x_3 x_4 - 15x_4^2$

For the following problems, write the functions in their quadratic forms and determine their signs using the principal minors test. For functions of two variables, draw contour plots of the functions.

3.34. $f(x_1, x_2) = 6x_1^2 + 3x_1 x_2 + 2x_2^2$

3.35. $f(x_1, x_2) = 2x_1^2 - x_1 x_2 + \frac{7x_2^2}{2}$

3.36. $f(x_1, x_2, x_3) = x_1^2 - x_1 x_2 + \frac{3x_2^2}{2} + 3x_1 x_3 - x_2 x_3 + 2x_3^2$

3.37. $f(x_1, x_2, x_3) = 6x_1^2 - 4x_2^2 - 3x_1 x_3 + 2x_3^2$

3.38. $f(x_1, x_2, x_3, x_4) = \frac{x_1^2}{2} + x_2^2 - 3x_1 x_3 + \frac{13x_3^2}{2} + 2x_1 x_4 + x_2 x_4 + 7x_4^2$

3.39. $f(x_1, x_2, x_3, x_4) = -\frac{x_1^2}{2} - x_2^2 + 3x_1 x_3 - 5x_3^2 - 2x_1 x_4 - x_2 x_4 + x_3 x_4 - 15x_4^2$

Convex Functions

For the following problems, determine the convexity status of the given functions. If possible, determine the domain over which a function is convex and plot the function (contour plot for a function of two variables) over this region.

3.40. $f(x) = \sin(x)/(x^2+1)$ $0 \le x \le 2\pi$

3.41. $f(x) = x^7 + x^5 + x^3 + x$

3.42. $f(x) = e^{-x^2}\cos[x/\pi] + \sin[x^3] - 1$ $0 \le x \le 2\pi$

3.43. $f(x, y) = x^2 + 4yx + y^2 - 4$

3.44. $f(x, y) = 5x^2 - 4yx + 2y^2 - 8$

3.45. $f(x, y) = e^{(x-y)} - x + y - 5$

3.46. $f(x, y) = 3 + 2x + 2/y$

3.47. $f(x, y, z) = 2y/z + x + 2$

3.48. $f(x, y, z) = x^2 + y^2 + z^2 - 1$

3.49. $f(x_1, x_2, x_3, x_4) = 1/4 + x_1/x_2 - x_2/x_3 + x_3/x_4 - x_4/x_1$

Convex Optimization Problems

Determine if the following problems are convex optimization problems. Use graphical methods to solve these problems. Verify the global optimality of convex problems.

3.50. Maximize $f(x_1, x_2) = -6x + 9y$

Subject to $\begin{pmatrix} x - y \ge 2 \\ 3x + y \ge 1 \\ 2x - 3y \ge 3 \end{pmatrix}$

3.51. Minimize $f(x_1, x_2) = x^2 + 2y^2$

Subject to $\begin{pmatrix} x + y \ge 1 \\ x, y \ge 0 \end{pmatrix}$

3.52. Minimize $f(x_1, x_2) = x^2 + 2y^2 - 24x - 20y$

Subject to $\begin{pmatrix} x + 2y \ge 0 \\ x + 2y \le 9 \\ x + y \le 8 \\ x + y \ge 0 \end{pmatrix}$

3.53. Maximize $f(x_1, x_2) = x_1 + \frac{x_1}{x_2^2} + \frac{x_2}{x_1}$

Subject to $\begin{pmatrix} x_1 + x_2 = 2 \\ x_1, x_2 \geq 0 \end{pmatrix}$

3.54. Maximize $f(x_1, x_2) = (x_1 - 2)^2 + (x_2 - 10)^2$

Subject to $\begin{pmatrix} x_1^2 + x_2^2 = 50 \\ x_1^2 + x_2^2 + 2x_1 x_2 - x_1 - x_2 + 20 \geq 0 \\ x_1, x_2 \geq 0 \end{pmatrix}$

3.55. Maximize $f(x_1, x_2) = (x_1 - 2)^2 + (x_2 - 10)^2$

Subject to $\begin{pmatrix} x_1^2 + x_2^2 \leq 50 \\ x_1^2 + x_2^2 + 2x_1 x_2 - x_1 - x_2 + 20 \geq 0 \\ x_1, x_2 \geq 0 \end{pmatrix}$

3.56. Minimize $f(x, y) = \frac{1000}{e^{2x+y}}$

Subject to $\begin{pmatrix} e^x + e^y \leq 20 \\ x, y \geq 1 \end{pmatrix}$

3.57. Minimize $f(x, y) = x^2 + y^2$

Subject to $(y - 1)^3 \geq x^2$

3.58. Show that the following is a convex optimization problem.

Minimize $f(x, y) = x^2 + y^2 - \log[x^2 y^2]$

Subject to $x \leq \log[y] \quad x \geq 1 \quad y \geq 1$

Use graphical methods to determine its solution.

3.59. Show that the following is a convex optimization problem.

Minimize $f(x, y, z) = x + y + z$

Subject to $x^{-2} + x^{-2} y^{-2} + x^{-2} y^{-2} z^{-2} \leq 1$

CHAPTER FOUR

Optimality Conditions

This chapter deals with mathematical conditions that must be satisfied by the solution of an optimization problem. A thorough understanding of the concepts presented in this chapter is a prerequisite for understanding the material presented in later chapters. The subject matter of this chapter is quite theoretical and requires use of sophisticated mathematical tools for rigorous treatment. Since the book is not intended to be a theoretical text on mathematical optimization, this chapter is kept simple by appealing to intuition and avoiding precise mathematical statements. Thus, it is implicit in the presentation that all functions are well behaved and have the necessary continuity and differentiability properties. The book by Peressini, Sullivan and Uhl, Jr. [1988] is recommended for those interested in a mathematically rigorous, yet very readable, treatment of most of the concepts presented in this chapter. More details can also be found in Beveridge, Gordon, and Schechter [1970], Jahn [1996], Jeter [1986], McAloon and Tretkoff [1996], Pierre and Lowe [1975], and Shor [1985].

The first section presents necessary and sufficient conditions for optimum of unconstrained problems. Several examples are presented that use these conditions to determine solutions of unconstrained problems. The remaining sections in the chapter consider constrained optimization problems. Section 2 explains the additive property of the less than type inequality ($\leq$) and equality ($=$) constraints that is useful in developing optimality conditions for constrained problems. The first-order necessary conditions for constrained problems, commonly known as Karush-Kuhn-Tucker (KT) conditions, are presented in section 3. Using these conditions, a procedure is developed to find candidate minimum points (KT points) for constrained problems. KT conditions have a simple geometric interpretation that is discussed in section 4. The so-called Lagrange multipliers obtained during a solution using KT conditions are useful in estimating the effect of certain changes in constraints on the optimum solution. This is known as sensitivity analysis and is discussed in section 5. Section 6 shows that for convex problems the KT conditions are necessary and sufficient for the optimum. Second-order sufficient conditions for a general constrained optimization problem are discussed in section 7. The last section briefly considers the topic of duality in optimization. The concept of duality is useful in developing effective computational algorithms for solving linear and quadratic programming problems discussed in later chapters.

4.1 Optimality Conditions for Unconstrained Problems

The unconstrained optimization problem considered in this section is stated as follows:

Find a vector of optimization variables $\mathbf{x} = (x_1 \quad x_2 \quad \ldots \quad x_n)^T$ in order to

Minimize $f(\mathbf{x})$

4.1.1 Necessary Condition for Optimality of Unconstrained Problems

The first-order optimality condition for the minimum of $f(\mathbf{x})$ can be derived by considering linear expansion of the function around the optimum point $\mathbf{x}^*$ using the Taylor series, as follows.

$$f(\mathbf{x}) \approx f(\mathbf{x}^*) + \nabla f(\mathbf{x}^*)^T (\mathbf{x} - \mathbf{x}^*)$$

or

$$f(\mathbf{x}) - f(\mathbf{x}^*) = \nabla f(\mathbf{x}^*)^T \mathbf{d}$$

where $\nabla f(\mathbf{x}^*)$ is a gradient of function $f(\mathbf{x})$ and $\mathbf{d} \equiv \mathbf{x} - \mathbf{x}^*$.

4.1 Optimality Conditions for Unconstrained Problems

If $\mathbf{x}^*$ is a minimum point, then $f(\mathbf{x}) - f(\mathbf{x}^*)$ must be greater than or equal to zero in a small neighborhood of $\mathbf{x}^*$. But since $\mathbf{d}$ can either be negative or positive, the only way to ensure that this condition is always satisfied is if

$$\nabla f(\mathbf{x}^*) = 0$$

Thus, the first-order necessary condition for the minimum of a function is that its gradient is zero at the optimum. Note that all this condition tells us is that the function value in the small neighborhood of $\mathbf{x}^*$ is not increasing. This condition is true even for a maximum point and at any other point where locally the slope is zero (inflection points). Therefore, this is only a necessary condition and is not a sufficient condition for the minimum of f. All points that satisfy this condition are known as *stationary* points.

A plot of a function of a single variable over the interval $\{-2\pi, 2\pi\}$, shown in Figure 4.1, illustrates that it has four points where its first derivative is zero. As shown on the plot, two are local minima, one is a local maximum, and the fourth is simply an inflection point.

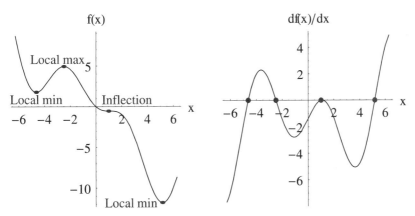

FIGURE 4.1 Graph of a function of single variable $f = x(-\text{Cos}[1] - \text{Sin}[1] + \text{Sin}[x])$ and its first derivative.

4.1.2 Sufficient Condition for Optimality of Unconstrained Problems

The second-order optimality condition for the minimum of $f(\mathbf{x})$ can be derived by considering quadratic expansion of the function around the optimum point

$\mathbf{x}^*$ using the Taylor series, as follows:

$$f(\mathbf{x}) - f(\mathbf{x}^*) = \nabla f(\mathbf{x}^*)^T \mathbf{d} + \frac{1}{2}\mathbf{d}^T \nabla^2 f(\mathbf{x}^*)\mathbf{d}$$

where $\nabla^2 f(\mathbf{x}^*)$ is a Hessian matrix of function $f(\mathbf{x})$ and $\mathbf{d} \equiv \mathbf{x} - \mathbf{x}^*$.

For $\mathbf{x}^*$ to be a local minimum point, $f(\mathbf{x}) - f(\mathbf{x}^*)$ must be greater than or equal to zero in a small neighborhood of $\mathbf{x}^*$. From the first-order necessary condition $\nabla f(\mathbf{x}^*) = 0$ and therefore for $f(\mathbf{x}) - f(\mathbf{x}^*) \geq 0$, we must have

$$\frac{1}{2}\mathbf{d}^T \nabla^2 f(\mathbf{x}^*)\mathbf{d} \geq 0$$

This is a quadratic form, discussed in Chapter 3, and therefore the sign of the product depends on the status of the Hessian matrix $\nabla^2 f(\mathbf{x}^*)$. If the Hessian is positive definite, then $\frac{1}{2}\mathbf{d}^T \nabla^2 f(\mathbf{x}^*)\mathbf{d} > 0$ for any values of $\mathbf{d}$, making $\mathbf{x}^*$ clearly the minimum point. If the Hessian is positive semidefinite, then $\frac{1}{2}\mathbf{d}^T \nabla^2 f(\mathbf{x}^*)\mathbf{d} \geq 0$ and therefore, all we can say is that the function is not increasing in the small neighborhood of $\mathbf{x}^*$, making it just a necessary condition for the minimum.

Thus, the second-order sufficient condition for the *minimum* of a function is that its Hessian matrix is *positive definite* at the optimum.

Proceeding in a similar manner, it can easily be shown that a sufficient condition for the *maximum* of a function is that its Hessian matrix is *negative definite* at the optimum.

4.1.3 Finding the Optimum Using Optimality Conditions

Since all minimum points must satisfy the first-order necessary conditions, we can find these points by solving the following system of equations resulting from setting the gradient of the function to zero:

$$\nabla f(\mathbf{x}) = 0 \implies \begin{pmatrix} \partial f/\partial x_1 = 0 \\ \vdots \\ \partial f/\partial x_n = 0 \end{pmatrix}$$

4.1 Optimality Conditions for Unconstrained Problems

This system of equations is generally nonlinear and hence, may have several possible solutions. These solutions are known as stationary points.

After computing all stationary points, the second-order sufficient conditions are used to further classify these points into a local minimum, maximum, or simply an inflection point as follows:

1. A stationary point at which the Hessian matrix is positive definite is at least a local minimum.

2. A stationary point at which the Hessian matrix is negative definite is at least a local maximum.

3. A stationary point at which the Hessian matrix is indefinite is an inflection point.

4. If the Hessian matrix is semidefinite (either positive or negative) at a stationary point, then even the second-order conditions are inconclusive. One must derive even higher-order conditions or use physical arguments to decide the optimality of such a point.

Example 4.1 Find all stationary points for the following function. Using second-order optimality conditions, classify them as minimum, maximum, or inflection points. Verify the solution using graphical methods.

```
f = x² - xy + 2y² - 2x;
```
Objective function → $-2x + x^2 - xy + 2y^2$

Gradient vector → $\begin{pmatrix} -2 + 2x - y \\ -x + 4y \end{pmatrix}$

Hessian matrix → $\begin{pmatrix} 2 & -1 \\ -1 & 4 \end{pmatrix}$

```
****** First order optimality conditions ******
```
Necessary conditions → $\begin{pmatrix} -2 + 2x - y == 0 \\ -x + 4y == 0 \end{pmatrix}$

Possible solutions (stationary points) → $(x \to 1.14286 \quad y \to 0.285714)$

```
****** Second order optimality conditions ******
------- Point → {x → 1.14286, y → 0.285714}
```
Hessian → $\begin{pmatrix} 2 & -1 \\ -1 & 4 \end{pmatrix}$ Principal minors → $\begin{pmatrix} 2 \\ 7 \end{pmatrix}$

Status → MinimumPoint   Function value → -1.14286

Since the Hessian matrix is positive definite, we know that the function is convex and hence, any minimum is a global minimum. The computations confirm this fact because there is only one possible solution and hence, obviously is a global minimum.

The plots shown in Figure 4.2 confirm this solution as well. The GraphicalSolution function described in Chapter 2 is used first to generate several contours of the objective function. The built-in *Mathematica* function Plot3D is then used to generate a three-dimensional surface plot of the function. Finally, the two plots are shown side-by-side using the GraphicsArray function.

```
gr1 = GraphicalSolution[f, {x, 0, 2}, {y, 0, 1},
    PlotPoints → 20, ObjectiveContours → {-1, -.5, 0, .5, 1},
    Epilog → {RGBColor[1, 0, 0], PointSize[.02],
      Point[{8/7, 2/7}], Text["Minimum", {1.17, 2/7}, {-1, 0}]}];

gr2 = Show[{
    Graphics3D[SurfaceGraphics[Plot3D[f, {x, 0, 2}, {y, 0, 1},
        DisplayFunction → Identity]]],
    Graphics3D[{RGBColor[1, 0, 0], PointSize[.02], Point[{1.17, 2/7, -1.1}]}]
  }, DisplayFunction → $DisplayFunction];

Show[GraphicsArray[{{gr1, gr2}}]];
```

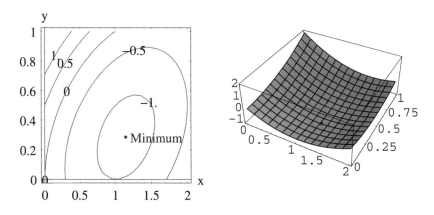

FIGURE 4.2 Graphical solution of $f = x^2 - xy + 2y^2 - 2x$.

Important Observations

The following observations should be noted from the optimality conditions.

1. The minimum point $\mathbf{x}^*$ does not change if we add a constant to the objective function. The reason is that a constant term does not change the gradient

4.1 Optimality Conditions for Unconstrained Problems

and hence, the system of equations based on the necessary conditions remains unchanged.

2. By the same reasoning as above, the minimum point $\mathbf{x}^*$ does not change if we multiply the objective function by a positive constant.

3. The problem changes from a minimization problem to a maximization problem, and vice versa, if we multiply the objective function by a negative sign. This is because if $\mathbf{x}^*$ is a minimum point of $f(\mathbf{x})$, then its Hessian $\nabla^2 f(\mathbf{x}^*)$ is positive definite. The Hessian of $-f(\mathbf{x})$ will be $-\nabla^2 f(\mathbf{x})$, which clearly will be negative definite, making $\mathbf{x}^*$ a maximum point of $-f(x)$. Thus,

$$\mathrm{Min} f(\mathbf{x}) \Longleftrightarrow \mathrm{Max} - f(\mathbf{x})$$

4. Since there are no constraints, the unconstrained problem is a convex programming problem if the objective function is convex. Recall that for a convex case, any local minimum is also a global minimum.

Solutions Using the UnconstrainedOptimality Function

The following function, included in the OptimizationToolbox 'OptimalityConditions' package, implements a solution procedure based on optimality conditions for unconstrained problems. From the necessary conditions, the function generates a system of equations. These equations are then solved using the built-in *Mathematica* functions NSolve (default) or FindRoot. NSolve tries to find all possible solutions but cannot handle certain types of equations, particularly those involving trignometric functions. FindRoot gives only one solution that is closest to the given starting point. See the standard *Mathematica* documentation to find more details on these two functions.

```
Needs["OptimizationToolbox`OptimalityConditions`"];
?UnconstrainedOptimality
```

UnconstrainedOptimality[f, vars, opts], finds all stationary points of
  function f of vars. Using second order sufficient conditions it then
  classifies them into minimum, maximum or inflection points. Options
  can be used to specify a *Mathematica* function to be used for solving
  system of equations. If this function itself has options, they can
  also be specified.

```
OptionsUsage[UnconstrainedOptimality]
```
{SolveEquationsUsing → NSolve, StartingSolution → {}}

SolveEquationsUsing → *Mathematica* function used to solve system of
  equations. Default is NSolve. Options specific to a method are passed
  on to that method.

StartingSolution used with the FindRoot function in *Mathematica*. It is used only if the method specified is FindRoot.

Example 4.2 The following function of two variables has two distinct local minimum points and an inflection point:

`f = (x - 2)^2 + (x - 2y^2)^2;`

`soln = UnconstrainedOptimality[f, {x, y}];`

Objective function $\rightarrow (-2 + x)^2 + (x - 2y^2)^2$

Gradient vector $\rightarrow \begin{pmatrix} -4 + 4x - 4y^2 \\ -8xy + 16y^3 \end{pmatrix}$

Hessian matrix $\rightarrow \begin{pmatrix} 4 & -8y \\ -8y & -8x + 48y^2 \end{pmatrix}$

****** First order optimality conditions ******

Necessary conditions $\rightarrow \begin{pmatrix} -4 + 4x - 4y^2 == 0 \\ -8xy + 16y^3 == 0 \end{pmatrix}$

Possible solutions (stationary points) $\rightarrow \begin{pmatrix} x \rightarrow 1 & y \rightarrow 0 \\ x \rightarrow 2 & y \rightarrow -1 \\ x \rightarrow 2 & y \rightarrow 1 \end{pmatrix}$

****** Second order optimality conditions ******

------- Point $\rightarrow \{x \rightarrow 1, y \rightarrow 0\}$

Hessian $\rightarrow \begin{pmatrix} 4 & 0 \\ 0 & -8 \end{pmatrix}$ Principal minors $\rightarrow \begin{pmatrix} 4 \\ -32 \end{pmatrix}$

Status $\rightarrow$ InflectionPoint   Function value $\rightarrow 2$

------- Point $\rightarrow \{x \rightarrow 2, y \rightarrow -1\}$

Hessian $\rightarrow \begin{pmatrix} 4 & 8 \\ 8 & 32 \end{pmatrix}$ Principal minors $\rightarrow \begin{pmatrix} 4 \\ 64 \end{pmatrix}$

Status $\rightarrow$ MinimumPoint   Function value $\rightarrow 0$

------- Point $\rightarrow \{x \rightarrow 2, y \rightarrow 1\}$

Hessian $\rightarrow \begin{pmatrix} 4 & -8 \\ -8 & 32 \end{pmatrix}$ Principal minors $\rightarrow \begin{pmatrix} 4 \\ 64 \end{pmatrix}$

Status $\rightarrow$ MinimumPoint   Function value $\rightarrow 0$

The following plots, generated using the GraphicalSolution and Plot3D functions as explained in an earlier example, confirm these solutions. In particular, note the inflection point, which is just a flat spot between the two minimums.

4.1 Optimality Conditions for Unconstrained Problems 139

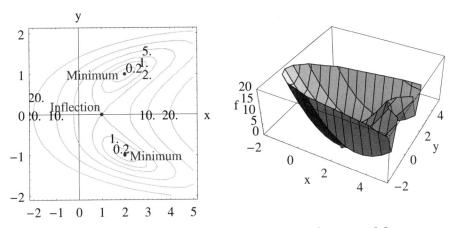

FIGURE 4.3 Graphical solution of $f = (x-2)^2 + (x - 2y^2)^2$.

Example 4.3 The following function of two variables has only one minimum point but two distinct local maximum points and two inflection points.

```
f = 25x^2 - 12x^4 - 6xy + 25y^2 - 24x^2y^2 - 12y^4;

soln = UnconstrainedOptimality[f, {x, y}];
```
Objective function $\to 25x^2 - 12x^4 - 6xy + 25y^2 - 24x^2y^2 - 12y^4$

Gradient vector $\to \begin{pmatrix} 50x - 48x^3 - 6y - 48xy^2 \\ -6x + 50y - 48x^2y - 48y^3 \end{pmatrix}$

Hessian matrix $\to \begin{pmatrix} 50 - 144x^2 - 48y^2 & -6 - 96xy \\ -6 - 96xy & 50 - 48x^2 - 144y^2 \end{pmatrix}$

****** First order optimality conditions ******

Necessary conditions $\to \begin{pmatrix} 50x - 48x^3 - 6y - 48xy^2 == 0 \\ -6x + 50y - 48x^2y - 48y^3 == 0 \end{pmatrix}$

Possible solutions (stationary points) $\to \begin{pmatrix} x \to -0.763763 & y \to 0.763763 \\ x \to -0.677003 & y \to -0.677003 \\ x \to 0. & y \to 0. \\ x \to 0.677003 & y \to 0.677003 \\ x \to 0.763763 & y \to -0.763763 \end{pmatrix}$

****** Second order optimality conditions ******
------- Point $\to \{x \to -0.763763, y \to 0.763763\}$
Hessian $\to \begin{pmatrix} -62. & 50. \\ 50. & -62. \end{pmatrix}$ Principal minors $\to \begin{pmatrix} -62. \\ 1344. \end{pmatrix}$

Status $\to$ MaximumPoint   Function value $\to 16.3333$
------- Point $\to \{x \to -0.677003, y \to -0.677003\}$
Hessian $\to \begin{pmatrix} -38. & -50. \\ -50. & -38. \end{pmatrix}$ Principal minors $\to \begin{pmatrix} -38. \\ -1056. \end{pmatrix}$

140   Chapter 4 Optimality Conditions

```
Status → InflectionPoint  Function value → 10.0833
------- Point → {x → 0., y → 0.}
Hessian → ( 50.  -6. )     Principal minors → ( 50.   )
          ( -6.  50. )                         ( 2464. )
Status → MinimumPoint  Function value → 0.
------- Point → {x → 0.677003, y → 0.677003}
Hessian → ( -38.  -50. )   Principal minors → ( -38.   )
          ( -50.  -38. )                      ( -1056. )
Status → InflectionPoint  Function value → 10.0833
------- Point → {x → 0.763763, y → -0.763763}
Hessian → ( -62.  50.  )   Principal minors → ( -62.  )
          ( 50.   -62. )                      ( 1344. )
Status → MaximumPoint  Function value → 16.3333
```

The following graph confirms these solutions.

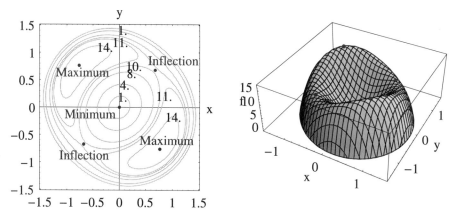

FIGURE 4.4 Graphical solution of $f = 25x^2 - 12x^4 - 6xy + 25y^2 - 24x^2y^2 - 12y^4$.

Example 4.4 The following function of four variables shows three different local minimum points. If this function was from a practical optimization problem, finding a global minimum would be important.

$$f = x_1^4 + x_1^2(1 - 2x_2) - 2x_2 + 2x_2^2 - 2x_3 + x_3^2 + x_3^4 + 2x_1(-2 + x_4) - 4x_4 - 2x_3^2 x_4 + 2x_4^2;$$

soln = UnconstrainedOptimality[f, {x_1, x_2, x_3, x_4}];

Objective function → $x_1^4 + x_1^2(1 - 2x_2) - 2x_2 + 2x_2^2 - 2x_3 + x_3^2 + x_3^4 + 2x_1(-2 + x_4) - 4x_4 - 2x_3^2 x_4 + 2x_4^2$

Gradient vector → $\begin{pmatrix} -4 + 2x_1 + 4x_1^3 - 4x_1 x_2 + 2x_4 \\ -2 - 2x_1^2 + 4x_2 \\ -2 + 2x_3 + 4x_3^3 - 4x_3 x_4 \\ -4 + 2x_1 - 2x_3^2 + 4x_4 \end{pmatrix}$

4.1 Optimality Conditions for Unconstrained Problems

Hessian matrix → $\begin{pmatrix} 2 + 12x_1^2 - 4x_2 & -4x_1 & 0 & 2 \\ -4x_1 & 4 & 0 & 0 \\ 0 & 0 & 2 + 12x_3^2 - 4x_4 & -4x_3 \\ 2 & 0 & -4x_3 & 4 \end{pmatrix}$

****** First order optimality conditions ******

Necessary conditions → $\begin{pmatrix} -4 + 2x_1 + 4x_1^3 - 4x_1x_2 + 2x_4 == 0 \\ -2 - 2x_1^2 + 4x_2 == 0 \\ -2 + 2x_3 + 4x_3^3 - 4x_3x_4 == 0 \\ -4 + 2x_1 - 2x_3^2 + 4x_4 == 0 \end{pmatrix}$

Possible solutions (stationary points) →

$\begin{pmatrix} x_2 \to 0.509352 & x_4 \to 2.00256 & x_3 \to 1.36688 & x_1 \to -0.136761 \\ x_2 \to 0.896053 & x_4 \to 2.70498 & x_3 \to 1.58744 & x_1 \to -0.890003 \\ x_2 \to 1. & x_4 \to 1. & x_3 \to 1. & x_1 \to 1. \end{pmatrix}$

****** Second order optimality conditions ******

------- Point → $\{x_2 \to 0.509352, x_4 \to 2.00256, x_3 \to 1.36688, x_1 \to -0.136761\}$

Hessian → $\begin{pmatrix} 0.187035 & 0.547043 & 0 & 2 \\ 0.547043 & 4 & 0 & 0 \\ 0 & 0 & 16.41 & -5.46751 \\ 2 & 0 & -5.46751 & 4 \end{pmatrix}$

Principal minors → $\begin{pmatrix} 0.187035 \\ 0.448884 \\ 7.3662 \\ -246.514 \end{pmatrix}$

Status → InflectionPoint   Function value → -5.34790

------- Point → $\{x_2 \to 0.896053, x_4 \to 2.70498, x_3 \to 1.58744, x_1 \to -0.890003\}$

Hessian → $\begin{pmatrix} 7.92106 & 3.56001 & 0 & 2 \\ 3.56001 & 4 & 0 & 0 \\ 0 & 0 & 21.4195 & -6.34974 \\ 2 & 0 & -6.34974 & 4 \end{pmatrix}$

Principal minors → $\begin{pmatrix} 7.92106 \\ 19.0106 \\ 407.196 \\ 519.583 \end{pmatrix}$

Status → MinimumPoint   Function value → -5.56484

------- Point → $\{x_2 \to 1., x_4 \to 1., x_3 \to 1., x_1 \to 1.\}$

Hessian → $\begin{pmatrix} 10. & -4. & 0 & 2 \\ -4. & 4 & 0 & 0 \\ 0 & 0 & 10. & -4. \\ 2 & 0 & -4. & 4 \end{pmatrix}$

Principal minors → $\begin{pmatrix} 10. \\ 24. \\ 240. \\ 416. \end{pmatrix}$

Status → MinimumPoint Function value → -6.

Example 4.5 *Data fitting* Consider the data-fitting problem discussed in Chapter 1. The goal is to find a surface of the form $z_{computed} = c_1 x^2 + c_2 y^2 + c_3 xy$ to best approximate the data in the following table:

Point	x	y	$z_{observed}$
1	0	1	1.26
2	0.25	1	2.19
3	0.5	1	0.76
4	0.75	1	1.26
5	1	2	1.86
6	1.25	2	1.43
7	1.5	2	1.29
8	1.75	2	0.65
9	2	2	1.6

The best values of coefficients c_1, c_2, and c_3 are determined to minimize the sum of squares of error between the computed z values and the observed values.

$$\text{Minimize } f = \sum_{i=1}^{9} [z_{observed}(x_i, y_i) - z_{computed}(x_i, y_i)]^2$$

Using the given numerical data, the objective function can be written as follows:

```
xyData = {
  {0,1}, {0.25,1}, {0.5,1}, {0.75,1}, {1,2},
  {1.25,2}, {1.5,2}, {1.75,2}, {2,2}};
zo = {1.26, 2.19, .76, 1.26, 1.86, 1.43, 1.29, .65, 1.6};
zc = Map[(c₁x² + c₂y² + c₃xy)/.{x → #[[1]], y → #[[2]]}&, xyData];
f = Expand[Apply[Plus, (zo - zc)²]]
```

$18.7 - 32.8462 c_1 + 34.2656 c_1^2 - 65.58 c_2 + 96.75 c_1 c_2 + 84 c_2^2 - 43.425 c_3 + 79.875 c_1 c_3 + 123. c_2 c_3 + 48.375 c_3^2$

The minimum of this function is computed as follows:

```
soln = UnconstrainedOptimality[f, {c₁, c₂, c₃}];
```

Objective function → $18.7 - 32.8462 c_1 + 34.2656 c_1^2 - 65.58 c_2 + 96.75 c_1 c_2 + 84 c_2^2 - 43.425 c_3 + 79.875 c_1 c_3 + 123. c_2 c_3 + 48.375 c_3^2$

4.1 Optimality Conditions for Unconstrained Problems

$$\text{Gradient vector} \rightarrow \begin{pmatrix} -32.8462 + 68.5313c_1 + 96.75c_2 + 79.875c_3 \\ -65.58 + 96.75c_1 + 168c_2 + 123.c_3 \\ -43.425 + 79.875c_1 + 123.c_2 + 96.75c_3 \end{pmatrix}$$

$$\text{Hessian matrix} \rightarrow \begin{pmatrix} 68.5313 & 96.75 & 79.875 \\ 96.75 & 168 & 123. \\ 79.875 & 123. & 96.75 \end{pmatrix}$$

****** First order optimality conditions ******

$$\text{Necessary conditions} \rightarrow \begin{pmatrix} -32.8462 + 68.5313c_1 + 96.75c_2 + 79.875c_3 == 0 \\ -65.58 + 96.75c_1 + 168c_2 + 123.c_3 == 0 \\ -43.425 + 79.875c_1 + 123.c_2 + 96.75c_3 == 0 \end{pmatrix}$$

Possible solutions (stationary points) →
$(c_1 \rightarrow 2.09108 \quad c_2 \rightarrow 1.75465 \quad c_3 \rightarrow -3.50823)$

****** Second order optimality conditions ******

------- Point → $\{c_1 \rightarrow 2.09108, c_2 \rightarrow 1.75465, c_3 \rightarrow -3.50823\}$

$$\text{Hessian} \rightarrow \begin{pmatrix} 68.5313 & 96.75 & 79.875 \\ 96.75 & 168 & 123. \\ 79.875 & 123. & 96.75 \end{pmatrix} \quad \text{Principal minors} \rightarrow \begin{pmatrix} 68.5313 \\ 2152.69 \\ 685.547 \end{pmatrix}$$

Status → MinimumPoint   Function value → 2.99556

Thus, the surface of the form $z_{computed} = c_1 x^2 + c_2 y^2 + c_3 xy$ that best fits the given data is as follows:

$$z = 2.09108 x^2 + 1.75465 y^2 - 3.50823 xy$$

The error between the observed z values and those computed from the above formula is shown in the following table. Considering that we used only a few

Point	x	y	$z_{observed}$	$z_{computed}$	$\text{Abs}\left[\dfrac{z_{observed} - z_{computed}}{z_{observed}}\right] 100\%$
1	0	1	1.26	1.75465	39.3
2	0.25	1	2.19	1.00828	54.
3	0.5	1	0.76	0.523305	31.1
4	0.75	1	1.26	0.29971	76.2
5	1	2	1.86	2.09322	12.5
6	1.25	2	1.43	1.51534	5.97
7	1.5	2	1.29	1.19884	7.07
8	1.75	2	0.65	1.14372	76.
9	2	2	1.6	1.35	15.6

data points, the results are reasonable. Using more data points or a different functional form could improve results.

4.2 The Additive Property of Constraints

Consider a general constrained optimization problem with the following inequality and equality constraints:

$$g_i(\mathbf{x}) \leq 0, \quad i = 1, \ldots, m$$
$$h_i(\mathbf{x}) = 0, \quad i = 1, \ldots, p$$

The purpose of this section is to demonstrate that if a point $\bar{\mathbf{x}}$ satisfies all these constraints individually, then it also satisfies the following aggregate constraint (obtained by a linear combination of the constraints)

$$g_a : \sum_{i=1}^{m} u_i g_i(\mathbf{x}) + \sum_{i=1}^{p} v_i h_i(\mathbf{x}) \leq 0$$

where $u_i \geq 0$ and v_i are any arbitrary scalar multipliers. Note that the multipliers for inequality constraints are restricted to be positive, but those for equality constraints have no restrictions. This property of constraints is called the additive property of constraints and is useful in understanding optimality conditions for constrained problems presented in later sections. The following examples graphically illustrate this property of constraints.

Example 4.6 Consider the following example with two inequality constraints:

$$\text{Constraint 1}: x^2 - \frac{1}{4}y^2 \leq 5 \quad \text{Constraint 2}: x - 2y + \frac{1}{4}y^2 \leq 0$$

g₁ = x^2 - $\frac{1}{4}$y^2 - 5 ≤ 0;

g₂ = x - 2y + $\frac{1}{4}$y^2 ≤ 0;

Let's define an aggregate constraint by arbitrarily choosing multipliers 3 and 5.

gₐ = Expand[3First[g₁] + 5First[g₂]] ≤ 0

$$-15 + 5x + 3x^2 - 10y + \frac{y^2}{2} \leq 0$$

4.2 The Additive Property of Constraints

The original and the aggregate constraints are shown in the following graph. It is clear that the feasible domain of the original constraints is entirely on the feasible side of the aggregate constraint. In other words, any feasible point for the original constraints is also feasible for the aggregate constraint.

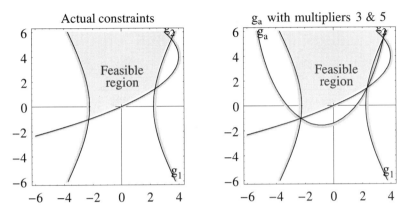

FIGURE 4.5 Graph showing feasible regions for original and aggregate constraints.

As another illustration, a new aggregate constraint is defined using a multiplier 1 for both constraints. Again, the plot clearly shows that the feasible domain of the original constraints is entirely on the feasible side of the aggregate constraint.

g_a = Expand[First[g_1] + First[g_2]] ≤ 0

$-5 + x + x^2 - 2y \leq 0$

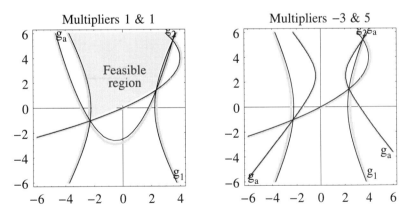

FIGURE 4.6 Graphs showing that positive multipliers work but not negative.

To demonstrate that a negative multiplier for an inequality constraint does not yield a correct aggregate constraint, consider the following case in which the multipliers −3 and 5 are used. The graph shows that the aggregate consists of two disjoint lines labeled g_a. Looking at the infeasible sides of these lines, it is clear that no feasible solution is possible. The feasible region is empty. Thus, using a negative multiplier for an inequality constraint does not work.

g_a = Expand[-3First[g_1] + 5First[g_2]] ≤ 0

$15 + 5x - 3x^2 - 10y + 2y^2 \le 0$

Example 4.7 As a second example, consider a problem with an equality and an inequality constraint:

$$\text{Constraint 1: } x^2 + 2y^2 \le 1 \qquad \text{Constraint 2: } x + y = 0$$

g = -1 + x^2 + 2y^2 ≤ 0;

h = x + y == 0;

Let's define an aggregate constraint by arbitrarily choosing multipliers 3 and 5:

g_a = Expand[3First[g] + 5First[h]] ≤ 0

$-3 + 5x + 3x^2 + 5y + 6y^2 \le 0$

The original and the aggregate constraints are shown in the following graph. The feasible region for the original constraints is the dark line inside the g contour. Clearly, this entire line is still inside the aggregate constraint g_a.

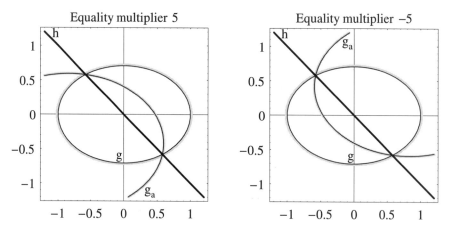

FIGURE 4.7 Graph showing feasible regions for original constraints and aggregate constraint.

To demonstrate that the sign of a multiplier for an equality constraint does not matter, consider the following case in which the multipliers 3 and −5 are used. The original feasible region is still within the feasible region of the aggregate constraint.

```
g_a = Expand[3First[g] - 5First[h]] ≤ 0
```
$-3 - 5x + 3x^2 - 5y + 6y^2 \leq 0$

4.3 Karush-Kuhn-Tucker (KT) Conditions

KT conditions are first-order necessary conditions for a general constrained minimization problem written in the following standard form:

Find **x** in order to

Minimize $f(\mathbf{x})$

Subject to $\begin{pmatrix} g_i(\mathbf{x}) \leq 0, & i = 1, 2, \ldots, m \\ h_i(\mathbf{x}) = 0, & i = 1, 2, \ldots, p \end{pmatrix}$

Note that in this standard form, the objective function is of minimization type, all constraint right-hand sides are zero, and the inequality constraints are of the less than type. Before applying the KT conditions, it is important to make sure that the problem has been converted to this standard form.

4.3.1 Basic Idea

The KT conditions can be derived by considering linear expansion of the objective and the constraint functions around the optimum point $\mathbf{x}^*$ using the Taylor series. The objective function is approximated as follows:

$$f(\mathbf{x}) \approx f(\mathbf{x}^*) + \nabla f(\mathbf{x}^*)^T (\mathbf{x} - \mathbf{x}^*) \quad \text{or} \quad f(\mathbf{x}) - f(\mathbf{x}^*) = \nabla f(\mathbf{x}^*)^T \mathbf{d}$$

For $\mathbf{x}^*$ to be a minimum point, we must have $\nabla f(\mathbf{x}^*)^T \mathbf{d} \geq 0$ where the small changes **d** must be such that the constraints are satisfied. The equality constraints are expanded as follows:

$$h_i(\mathbf{x}) \approx h_i(\mathbf{x}^*) + \nabla h_i(\mathbf{x}^*)^T \mathbf{d} = 0$$

Since the constraints must be satisfied at the optimum point, $h_i(\mathbf{x}^*) = 0$. Thus for sufficiently small $\mathbf{d}$, we have

$$\nabla h_i(\mathbf{x}^*)^T \mathbf{d} = 0, i = 1, 2, \ldots, p$$

For each inequality constraint, we need to consider the following two possibilities:

1. Inactive inequality constraints: constraints for which $g_i(\mathbf{x}^*) < 0$. These constraints do not determine the optimum and hence, are not needed in developing optimality conditions.

2. Active inequality constraints: constraints for which $g_i(\mathbf{x}^*) = 0$. These constraints are called active constraints and are expanded as follows:

$$g_i(\mathbf{x}) \approx g_i(\mathbf{x}^*) + \nabla g_i(\mathbf{x}^*)^T \mathbf{d} \leq 0 \quad \text{or} \quad \nabla g_i(\mathbf{x}^*)^T \mathbf{d} \leq 0, i \in \text{Active}$$

In summary, for $\mathbf{x}^*$ to be a local minimum point, the following conditions must be satisfied simultaneously in a small neighborhood (i.e., with sufficiently small $\mathbf{d}$).

(a) $\nabla f(\mathbf{x}^*)^T \mathbf{d} \geq 0$
(b) $\nabla h_i(\mathbf{x}^*)^T \mathbf{d} = 0, i = 1, 2, \ldots, p$
(c) $\nabla g_i(\mathbf{x}^*)^T \mathbf{d} \leq 0, i \in \text{Active}$

Using the additive property of constraints discussed in the previous section, the constraint conditions can be combined by multiplying the equality constraints by v_i and the active inequality constraints by $u_i \geq 0$. Note that the multipliers for the inequality constraints must be greater than 0; otherwise, the direction of the inequality will be reversed. Thus, we have

$$\left[\sum_{i \in \text{Active}} u_i \nabla g_i(\mathbf{x}^*) + \sum_{i=1}^{p} v_i \nabla h_i(\mathbf{x}^*) \right]^T \mathbf{d} \leq 0$$

or

$$-\left[\sum_{i \in \text{Active}} u_i \nabla g_i(\mathbf{x}^*) + \sum_{i=1}^{p} v_i \nabla h_i(\mathbf{x}^*) \right]^T \mathbf{d} \geq 0$$

Comparing this with (a) above, we can see that all conditions are met if the following equation is satisfied:

$$\nabla f(\mathbf{x}^*) = -\left[\sum_{i \in \text{Active}} u_i \nabla g_i(\mathbf{x}^*) + \sum_{i=1}^{p} v_i \nabla h_i(\mathbf{x}^*)\right]$$

For a detailed mathematical justification of why this condition is necessary for the minimum, see the book by Peressini, Sullivan and Uhl. Thus, the first-order necessary condition for optimality is that there exist multipliers $u_i \geq 0$, $i \in$ Active, and v_i, $i = 1, 2, \ldots, p$ (known as Lagrange multipliers) such that

$$\nabla f(\mathbf{x}^*) + \sum_{i \in \text{Active}} u_i \nabla g_i(\mathbf{x}^*) + \sum_{i=1}^{p} v_i \nabla h_i(\mathbf{x}^*) = 0$$

In other words, a necessary condition for optimality is that the gradient of the objective function is equal and opposite to the linear combination of gradients of equality and active inequality constraints. This physical interpretation is explained in more detail in a later section.

4.3.2 The Regularity Condition

One of the key assumptions in the derivation of the optimality condition is that it is possible to find suitable multipliers such that

$$\nabla f(\mathbf{x}^*) = -\sum_{i \in \text{Active}} u_i \nabla g_i(\mathbf{x}^*) + \sum_{i=1}^{p} v_i \nabla h_i(\mathbf{x}^*)$$

Mathematically this means that the gradient of the objective function can be represented by a linear combination of the gradients of active inequality and equality constraints. A fundamental requirement to make it possible is that the gradients of active inequality and all equality constraints must be linearly independent. This is known as *regularity* condition.

The optimality conditions make sense only at points that are regular, i.e. where the gradients of active inequality and equality constraints are linearly independent. It should be understood that we are not saying that all minimum points must be regular points as well. In fact it is easy to come up with examples of irregular points that can be verified to be minimum points through other means, such as a graphical solution. The requirement of regularity is tied to the use of the optimality conditions derived here. Since regularity assumption is made during the derivation of these conditions, the conclusions drawn from these conditions are valid only at regular points.

Procedure for checking linear independence of a given set of vectors is explained in chapter 3. It involves determining rank of the matrix whose columns are the given vectors. If the rank of this matrix is equal to the number of vectors, then the given set of vectors is linearly independent.

Clearly, if there is only one active constraint, then any point is a regular point, since the question of linear dependence does not even arise. When there are more than one active constraints, only then it is necessary to check for linear independence of gradient vectors of active constraints.

4.3.3 The Lagrangian Function

It is possible to define a single function (called the Lagrangian function) that combines all constraints and the objective function. The function is written in such a way that the necessary conditions for its minimum are the same as those for the constrained problem. In order to define this function, it is convenient to convert all inequalities to equalities. This can be done by adding a suitable positive number (obviously unknown as yet) to the left-hand side of each inequality, as follows:

$$g_i(\mathbf{x}) + s_i^2 = 0, \quad i = 1, 2, \ldots, m$$

The s_i's are known as slack variables. They are squared to make sure that only a positive number is added. The Lagrangian function can now be defined as follows:

$$L(\mathbf{x}, \mathbf{u}, \mathbf{v}, \mathbf{s}) = f(\mathbf{x}) + \sum_{i=1}^{m} u_i \left[g_i(\mathbf{x}) + s_i^2 \right] + \sum_{i=1}^{p} v_i h_i(\mathbf{x})$$

where $\mathbf{x}$ is a vector of optimization variables, $\mathbf{s}$ is a vector of slack variables, $\mathbf{u} \geq 0$ is a vector of Lagrange multipliers for inequality constraints, and $\mathbf{v}$ is a vector of Lagrange multipliers for equality constraints.

Since the Lagrangian function is unconstrained, the necessary conditions for its minimum are that derivatives with respect to all its variables must be equal to zero. That is,

$$\nabla L \equiv \begin{pmatrix} \partial L(\mathbf{x},\mathbf{u},\mathbf{v},\mathbf{s})/\partial \mathbf{x} = \mathbf{0} \\ \partial L(\mathbf{x},\mathbf{u},\mathbf{v},\mathbf{s})/\partial \mathbf{u} = \mathbf{0} \\ \partial L(\mathbf{x},\mathbf{u},\mathbf{v},\mathbf{s})/\partial \mathbf{v} = \mathbf{0} \\ \partial L(\mathbf{x},\mathbf{u},\mathbf{v},\mathbf{s})/\partial \mathbf{s} = \mathbf{0} \end{pmatrix} \implies \begin{pmatrix} \nabla f(\mathbf{x}) + \sum_{i=1}^{m} u_i \nabla g_i(\mathbf{x}) + \sum_{i=1}^{p} v_i \nabla h_i(\mathbf{x}) = \mathbf{0} \\ g_i(\mathbf{x}) + s_i^2 = 0, \quad i = 1, 2, \ldots, m \\ h_i(\mathbf{x}) = 0, \quad i = 1, 2, \ldots, p \\ 2 u_i s_i = 0, \quad i = 1, 2, \ldots, m \end{pmatrix}$$

The first set is known as the gradient condition. All inequality constraints are included in the first sum. However, from the last set of conditions, known as *complementary slackness* or *switching conditions*, we can see that either $u_i = 0$ or $s_i = 0$. If a slack variable is zero, then the corresponding constraint is active with a positive Lagrange multiplier. If a constraint is inactive, then the corresponding Lagrange multiplier must be zero. Hence, in reality the first sum is over active inequality constraints only. Therefore, these conditions are the same as the optimality conditions derived in the previous section. The second and third sets of conditions simply state that the constraints must be satisfied. Thus, the necessary conditions for the minimum of the Lagrangian function are the same as those for the constrained minimization problem.

4.3.4 Summary of KT Conditions

The complete set of necessary conditions for optimality of a constrained minimization problem, known as Karush-Kuhn-Tucker (KT) conditions, are as follows:

For a point $\mathbf{x}^*$ to be minimum of $f(\mathbf{x})$ subject to $h_i(\mathbf{x}) = 0, i = 1, 2, \ldots, p$ and $g_i(\mathbf{x}) \leq 0, i = 1, 2, \ldots, m$ the following conditions must be satisfied.

1. Regularity condition: $\mathbf{x}^*$ must be a regular point
2. Gradient conditions: $\nabla f(\mathbf{x}^*) + \sum_{i=1}^{m} u_i \nabla g_i(\mathbf{x}^*) + \sum_{i=1}^{p} v_i \nabla h_i(\mathbf{x}^*) = \mathbf{0}$
3. Constraints: $\begin{pmatrix} g_i(\mathbf{x}^*) + s_i^2 = 0, \ i = 1, 2, \ldots, m \\ h_i(\mathbf{x}^*) = 0, \ i = 1, 2, \ldots, p \end{pmatrix}$
4. Complementary slackness or switching conditions: $u_i s_i = 0, i = 1, 2, \ldots, m$
5. Feasibility conditions: $s_i^2 \geq 0, i = 1, 2, \ldots, m$
6. Sign of inequality multipliers: $u_i \geq 0, i = 1, 2, \ldots, m$

The KT conditions represent a set of equations that must be satisfied for all local minimum points of a constrained minimization problem. Since they are only necessary conditions, the points that satisfy these conditions are only candidates for being a local minimum, and are usually known as KT points. One must check sufficient conditions (to be discussed later) to determine if a given KT point actually is a local minimum or not.

Example 4.8 Obtain all points satisfying KT conditions for the following optimization problem:

Chapter 4 Optimality Conditions

$$\text{Minimize } f(x, y, z) = 14 - 2x + x^2 - 4y + y^2 - 6z + z^2$$
$$\text{Subject to } x^2 + y^2 + z - 1 = 0$$

```
f = 14 - 2x + x² - 4y + y² - 6z + z²;
h = x² + y² + z - 1 == 0;
vars = {x, y, z};
```

Minimize $f \to 14 - 2x + x^2 - 4y + y^2 - 6z + z^2$

$$\nabla f \to \begin{pmatrix} -2 + 2x \\ -4 + 2y \\ -6 + 2z \end{pmatrix}$$

***** EQ constraints & their gradients

$$h_1 \to -1 + x^2 + y^2 + z == 0 \quad \nabla h_1 \to \begin{pmatrix} 2x \\ 2y \\ 1 \end{pmatrix}$$

***** Lagrangian $\to 14 - 2x + x^2 - 4y + y^2 - 6z + z^2 + (-1 + x^2 + y^2 + z)v_1$

$$\nabla L = 0 \to \begin{pmatrix} -2 + 2x + 2xv_1 == 0 \\ -4 + 2y + 2yv_1 == 0 \\ -6 + 2z + v_1 == 0 \\ -1 + x^2 + y^2 + z == 0 \end{pmatrix}$$

The gradient of Lagrangian yields four nonlinear equations. The following is the only possible solution to these equations:

```
f → 8.0348
x → 0.186935
y → 0.37387
z → 0.825276
v₁ → 4.34945
```

Since there is only one active constraint, the regularity condition is trivially satisfied. Since all requirements are met, this is a valid KT point.

4.3.5 Solution of Optimization Problems Using KT Conditions

Since there are exactly as many equations as the number of unknowns, in principle, it is possible to solve the KT equations simultaneously to obtain candidate minimum points (KT points). However, conventional solution procedures usually get stuck because of many different solution possibilities as a result of switching conditions. As pointed out earlier, the switching conditions say that an inequality constraint is either active ($s_i = 0$) or inactive ($u_i = 0$). A solution can be obtained readily if one can identify the constraints that are active at the optimum. For practical problems, it may be possible to identify the critical constraints based on past experience with similar problems. For a general case, one may consider a trial-and-error approach. A guess is made for the active and inactive constraints, thus setting either u_i or s_i^2 equal to zero for each inequality constraint. The values for unknown u_i and s_i^2 are obtained from the other KT conditions. If all computed u_i and s_i^2 values are greater than or equal to zero, we have identified the correct case of active and inactive constraints. The solution corresponding to this case represents a KT point. Of course, after finding a solution, one must verify that the computed point is a regular point, because as pointed out earlier, the KT conditions are valid only for regular points.

Number of Possible Cases to be Examined

If there are m inequality constraints, then the switching conditions imply that there are a total of 2^m different cases that are possible. For example, a problem with three inequality constraints has the following 8 ($= 2^3$) possibilities:

1. None of the constraints active $\quad u_1 = u_2 = u_3 = 0$
2. g_1 Active $\quad s_1 = u_2 = u_3 = 0$
3. g_2 Active $\quad u_1 = s_2 = u_3 = 0$
4. g_3 Active $\quad u_1 = u_2 = s_3 = 0$
5. g_1 Active, g_2 Active $\quad s_1 = s_2 = u_3 = 0$
6. g_2 Active, g_3 Active $\quad u_1 = s_2 = s_3 = 0$
7. g_1 Active, g_3 Active $\quad s_1 = u_2 = s_3 = 0$
8. All three active $\quad s_1 = s_2 = s_3 = 0$

The actual number of cases to be considered will in general be smaller than this theoretical maximum. This is because, for a well-formulated problem, the number of active constraints must be less than or at most equal to the number of optimization variables. If this is not the case, then the regularity cannot be satisfied. More equations than the number of variables means that either some of the equations are redundant, or there is no solution for the system of equations. In this case, we must reformulate the problem. In the above example, if the problem has two variables then obviously the last case of all three constraints being active is impossible. Hence, the number of cases to be examined reduces to seven.

The presence of equality constraints further reduces the number of possible cases since equality constraints are always active by definition. Thus, if a problem has two variables, three inequality constraints, and one equality constraint, only one of the inequalities can be active. Pairs of g_i cannot be active, and the regularity condition satisfied at the same time, since that would imply more constraints than the variables at the minimum point. Thus, the number of possible cases to be examined reduces to four, as follows.

1. None of the g's active
2. g_1 Active
3. g_2 Active
4. g_3 Active

Even with this reduction in the number of cases, the computational effort increases dramatically as the number of variables and inequality constraints increases. Thus, this method should be used only for problems with few variables and a maximum of three or four inequality constraints. More efficient solution methods for larger problems are presented in later chapters.

Example 4.9 Obtain all points satisfying KT conditions for the following optimization problem:

Minimize $f(x, y, z) = 1/(x^2 + y^2)$

Subject to $\begin{pmatrix} x^2 + y^2 \leq 5 \\ x \leq 10 \\ y \leq 4 \\ x + 2y = 4 \end{pmatrix}$

f = 1/(x² + y²);
cons = {x² + y² ≤ 5, x ≤ 10, y ≤ 4, x + 2y == 4};

Minimize $f \to \dfrac{1}{x^2 + y^2}$

$\nabla f \to \begin{pmatrix} -\dfrac{2x}{(x^2 + y^2)^2} \\ -\dfrac{2y}{(x^2 + y^2)^2} \end{pmatrix}$

***** LE constraints & their gradients

$g_1 \to -5 + x^2 + y^2 \leq 0 \quad g_2 \to -10 + x \leq 0$

$g_3 \to -4 + y \leq 0 \quad \nabla g_1 \to \begin{pmatrix} 2x \\ 2y \end{pmatrix}$

4.3 Karush-Kuhn-Tucker (KT) Conditions

$\nabla g_2 \to \begin{pmatrix} 1 \\ 0 \end{pmatrix}$ $\nabla g_3 \to \begin{pmatrix} 0 \\ 1 \end{pmatrix}$

***** EQ constraints & their gradients

$h_1 \to -4 + x + 2y == 0$ $\nabla h_1 \to \begin{pmatrix} 1 \\ 2 \end{pmatrix}$

***** Lagrangian $\to \dfrac{1}{x^2+y^2} + (-5 + x^2 + y^2 + s_1^2)u_1 + (-10 + x + s_2^2)u_2 + (-4 + y + s_3^2)u_3 + (-4 + x + 2y)v_1$

$\nabla L = 0 \to \begin{pmatrix} -\dfrac{2x}{(x^2+y^2)^2} + 2xu_1 + u_2 + v_1 == 0 \\ -\dfrac{2y}{(x^2+y^2)^2} + 2yu_1 + u_3 + 2v_1 == 0 \\ -5 + x^2 + y^2 + s_1^2 == 0 \\ -10 + x + s_2^2 == 0 \\ -4 + y + s_3^2 == 0 \\ -4 + x + 2y == 0 \\ 2s_1 u_1 == 0 \\ 2s_2 u_2 == 0 \\ 2s_3 u_3 == 0 \end{pmatrix}$

The gradient of Lagrangian yields six nonlinear equations and three switching conditions. Since this is a two-variable problem, and there is one equality constraint already, only one of the inequalities can possibly be active. Thus, there are a total of four cases that must be examined.

************ Case 1 ************
Active inequalities $\to$ None
Known values $\to \{u_1 \to 0, u_2 \to 0, u_3 \to 0\}$

Equations for this case $\to \begin{pmatrix} -\dfrac{2x}{(x^2+y^2)^2} + v_1 == 0 \\ -\dfrac{2y}{(x^2+y^2)^2} + 2v_1 == 0 \\ -5 + x^2 + y^2 + s_1^2 == 0 \\ -10 + x + s_2^2 == 0 \\ -4 + y + s_3^2 == 0 \\ -4 + x + 2y == 0 \end{pmatrix}$

-----Solution 1-----
$\begin{pmatrix} x \\ y \end{pmatrix} \to \begin{pmatrix} 0.8 \\ 1.6 \end{pmatrix}$

$$\begin{pmatrix} u_1 \\ u_2 \\ u_3 \end{pmatrix} \to \begin{pmatrix} 0 \\ 0 \\ 0 \end{pmatrix} \quad \begin{pmatrix} s_1^2 \\ s_2^2 \\ s_3^2 \end{pmatrix} \to \begin{pmatrix} 1.8 \\ 9.2 \\ 2.4 \end{pmatrix}$$

$(v_1) \to (0.15625)$

KT Status → Valid KT Point   Objective function value → 0.3124999

************ Case 2 ************

Active inequalities → {1}

Known values → $\{u_2 \to 0, u_3 \to 0, s_1 \to 0\}$

Equations for this case → $\begin{pmatrix} -\frac{2x}{(x^2+y^2)^2} + 2xu_1 + v_1 == 0 \\ -\frac{2y}{(x^2+y^2)^2} + 2yu_1 + 2v_1 == 0 \\ -5 + x^2 + y^2 == 0 \\ -10 + x + s_2^2 == 0 \\ -4 + y + s_3^2 == 0 \\ -4 + x + 2y == 0 \end{pmatrix}$

-----Solution 1-----

$\begin{pmatrix} x \\ y \end{pmatrix} \to \begin{pmatrix} -0.4 \\ 2.2 \end{pmatrix}$

$\begin{pmatrix} u_1 \\ u_2 \\ u_3 \end{pmatrix} \to \begin{pmatrix} 0.04 \\ 0 \\ 0 \end{pmatrix} \quad \begin{pmatrix} s_1^2 \\ s_2^2 \\ s_3^2 \end{pmatrix} \to \begin{pmatrix} 0 \\ 10.4 \\ 1.8 \end{pmatrix}$

$(v_1) \to (0)$

Constraint gradient matrix → $\begin{pmatrix} -0.8 & 1 \\ 4.4 & 2 \end{pmatrix}$ Rank → 2

Regularity status → RegularPoint

KT Status → Valid KT Point   Objective function value → 0.2

-----Solution 2-----

$\begin{pmatrix} x \\ y \end{pmatrix} \to \begin{pmatrix} 2. \\ 1. \end{pmatrix}$

$\begin{pmatrix} u_1 \\ u_2 \\ u_3 \end{pmatrix} \to \begin{pmatrix} 0.04 \\ 0 \\ 0 \end{pmatrix} \quad \begin{pmatrix} s_1^2 \\ s_2^2 \\ s_3^2 \end{pmatrix} \to \begin{pmatrix} 0 \\ 8. \\ 3 \end{pmatrix}$

$(v_1) \to (0)$

Constraint gradient matrix → $\begin{pmatrix} 4. & 1 \\ 2. & 2 \end{pmatrix}$ Rank → 2

Regularity status → RegularPoint

KT Status → Valid KT Point   Objective function value → 0.2

4.3 Karush-Kuhn-Tucker (KT) Conditions

************ Case 3 ************
Active inequalities → {2}
Known values → {$u_1 \to 0, u_3 \to 0, s_2 \to 0$}

Equations for this case →
$$\begin{pmatrix} -\dfrac{2x}{(x^2+y^2)^2} + u_2 + v_1 == 0 \\ -\dfrac{2y}{(x^2+y^2)^2} + 2v_1 == 0 \\ -5 + x^2 + y^2 + s_1^2 == 0 \\ -10 + x == 0 \\ -4 + y + s_3^2 == 0 \\ -4 + x + 2y == 0 \end{pmatrix}$$

-----Solution 1-----

$\begin{pmatrix} x \\ y \end{pmatrix} \to \begin{pmatrix} 10. \\ -3. \end{pmatrix}$

$\begin{pmatrix} u_1 \\ u_2 \\ u_3 \end{pmatrix} \to \begin{pmatrix} 0 \\ 0.00193586 \\ 0 \end{pmatrix}$ $\begin{pmatrix} s_1^2 \\ s_2^2 \\ s_3^2 \end{pmatrix} \to \begin{pmatrix} -104. \\ 0 \\ 7. \end{pmatrix}$

$(v_1) \to (-0.000252504)$

Constraint gradient matrix → $\begin{pmatrix} 1 & 1 \\ 0 & 2 \end{pmatrix}$ Rank → 2

Regularity status → RegularPoint
KT Status → Invalid KT Point   Objective function value → 0.009174311

The solution is invalid because the slack variable for the first constraint is negative.

************ Case 4 ************
Active inequalities → {3}
Known values → {$u_1 \to 0, u_2 \to 0, s_3 \to 0$}

Equations for this case →
$$\begin{pmatrix} -\dfrac{2x}{(x^2+y^2)^2} + v_1 == 0 \\ -\dfrac{2y}{(x^2+y^2)^2} + u_3 + 2v_1 == 0 \\ -5 + x^2 + y^2 + s_1^2 == 0 \\ -10 + x + s_2^2 == 0 \\ -4 + y == 0 \\ -4 + x + 2y == 0 \end{pmatrix}$$

-----Solution 1-----

$\begin{pmatrix} x \\ y \end{pmatrix} \to \begin{pmatrix} -4. \\ 4. \end{pmatrix}$

$$\begin{pmatrix} u_1 \\ u_2 \\ u_3 \end{pmatrix} \to \begin{pmatrix} 0 \\ 0 \\ 0.0234375 \end{pmatrix} \quad \begin{pmatrix} s_1^2 \\ s_2^2 \\ s_3^2 \end{pmatrix} \to \begin{pmatrix} -27. \\ 14. \\ 0 \end{pmatrix}$$

$(v_1) \to (-0.0078125)$

Constraint gradient matrix $\to \begin{pmatrix} 0 & 1 \\ 1 & 2 \end{pmatrix}$ Rank $\to 2$

Regularity status $\to$ RegularPoint
KT Status $\to$ Invalid KT Point   Objective function value $\to 0.03125$

Again, this solution is invalid because the slack variable for the first constraint is negative. Thus, we get the following three valid KT points:

```
***** Valid KT Point(s) *****
f → 0.2        f → 0.3125     f → 0.2
x → -0.4       x → 0.8        x → 2.
y → 2.2        y → 1.6        y → 1.
u₁ → 0.04      u₁ → 0         u₁ → 0.04
u₂ → 0         u₂ → 0         u₂ → 0
u₃ → 0         u₃ → 0         u₃ → 0
s₁² → 0        s₁² → 1.8      s₁² → 0
s₂² → 10.4     s₂² → 9.2      s₂² → 8.
s₃² → 1.8      s₃² → 2.4      s₃² → 3.
v₁ → 0         v₁ → 0.15625   v₁ → 0
```

4.3.6 The KTSolution Function

The solution procedure based on examining all possible cases is implemented in a function called the KTSolution. This function and other auxiliary functions that it needs are included in the OptimizationToolbox 'OptimalityConditions' package. Similar to the UnconstrainedOptimality function, the equations can be solved either by using the the NSolve (default) or FindRoot functions. NSolve tries to find all possible solutions and should be tried first. If it fails, then FindRoot can be used to get a solution closest to a user-given starting point.

```
Needs["OptimizationToolbox`OptimalityConditions`"];
?KTSolution
```

```
KTSolution[f, con, vars, opts], finds candidate minimum points by
   solving KT conditions. f is the objective function. con is a list
   of constraints. The function automatically converts the constraints
   to standard form before proceeding. vars is a list of problem
   variables. Several options can be used with the function. See
   Options[KTSolution]. By default all possible cases of inequalities
   being active are examined. Active inequalities can be explicitly
   specified through ActiveCases option.
```

```
OptionsUsage[KTSolution]
```
{PrintLevel → 1, ActiveCases → {}, KTVarNames → {u, s, v},
  SolveEquationsUsing → NSolve, StartingSolution → {}}

PrintLevel is an option for most functions in the OptimizationToolbox.
It is specified as an integer. The value of the integer indicates
how much intermediate information is to be printed. A PrintLevel→ 0
suppresses all printing. Default for most functions is set to 1 in
which case they print only the initial problem setup. Higher integers
print more intermediate results.

ActiveCases→ List of active LE constraints to be considered in the
  solution. Default is to consider All possible cases.

KTVarNames→ Variable names used for g ('≤' constraints) multipliers,
  slack variables and h ('=' constraints) multipliers. Default is
  {u, s, v}.

SolveEquationsUsing is an option for Optimality condition based methods.
  Mathematica function used to solve system of equations is specified
  with this option. The choice is between NSolve (default) and FindRoot.

StartingSolution used with the FindRoot function in *Mathematica*. It is
  used only if the method specified is FindRoot.

Example 4.10 Obtain all points satisfying KT conditions for the following optimization problem:

Minimize $f(x, y, z) = 1/(x^2 + y^2 + z^2)$

Subject to $\begin{pmatrix} x^2 + 2y^2 + 3z^2 = 1 \\ x + y + z = 0 \end{pmatrix}$

Using the KTSolution function, the solution is obtained as follows:

```
f = 1/(x² + y² + z²);
h = {x² + 2y² + 3z² == 1, x + y + z == 0};
vars = {x, y, z};

KTSolution[f, h, vars];
```

Minimize $f \to \dfrac{1}{x^2 + y^2 + z^2}$

$\nabla f \to \begin{pmatrix} -\dfrac{2x}{(x^2+y^2+z^2)^2} \\ -\dfrac{2y}{(x^2+y^2+z^2)^2} \\ -\dfrac{2z}{(x^2+y^2+z^2)^2} \end{pmatrix}$

***** EQ constraints & their gradients

$h_1 \to -1 + x^2 + 2y^2 + 3z^2 == 0 \quad h_2 \to x + y + z == 0$

$$\nabla h_1 \rightarrow \begin{pmatrix} 2x \\ 4y \\ 6z \end{pmatrix} \quad \nabla h_2 \rightarrow \begin{pmatrix} 1 \\ 1 \\ 1 \end{pmatrix}$$

***** Lagrangian $\rightarrow \dfrac{1}{x^2+y^2+z^2} + (-1+x^2+2y^2+3z^2)v_1 + (x+y+z)v_2$

$$\nabla L = 0 \rightarrow \begin{pmatrix} -\dfrac{2x}{(x^2+y^2+z^2)^2} + 2xv_1 + v_2 == 0 \\ -\dfrac{2y}{(x^2+y^2+z^2)^2} + 4yv_1 + v_2 == 0 \\ -\dfrac{2z}{(x^2+y^2+z^2)^2} + 6zv_1 + v_2 == 0 \\ -1 + x^2 + 2y^2 + 3z^2 == 0 \\ x + y + z == 0 \end{pmatrix}$$

The gradient of Lagrangian yields five nonlinear equations. There are four possible solutions to these equations.

-----Solution 1-----

$$\begin{pmatrix} x \\ y \\ z \end{pmatrix} \rightarrow \begin{pmatrix} -0.661225 \\ 0.48405 \\ 0.177175 \end{pmatrix}$$

$$\begin{pmatrix} v_1 \\ v_2 \end{pmatrix} \rightarrow \begin{pmatrix} 1.42265 \\ -0.795166 \end{pmatrix}$$

Constraint gradient matrix $\rightarrow \begin{pmatrix} -1.32245 & 1 \\ 1.9362 & 1 \\ 1.06305 & 1 \end{pmatrix}$ Rank $\rightarrow 2$

Regularity status $\rightarrow$ RegularPoint
KT Status $\rightarrow$ Valid KT Point Objective function value $\rightarrow 1.42264$

-----Solution 2-----

$$\begin{pmatrix} x \\ y \\ z \end{pmatrix} \rightarrow \begin{pmatrix} -0.131633 \\ -0.359627 \\ 0.49126 \end{pmatrix}$$

$$\begin{pmatrix} v_1 \\ v_2 \end{pmatrix} \rightarrow \begin{pmatrix} 2.57735 \\ -1.07028 \end{pmatrix}$$

Constraint gradient matrix $\rightarrow \begin{pmatrix} -0.263265 & 1 \\ -1.43851 & 1 \\ 2.94756 & 1 \end{pmatrix}$ Rank $\rightarrow 2$

Regularity status $\rightarrow$ RegularPoint
KT Status $\rightarrow$ Valid KT Point Objective function value $\rightarrow 2.57735$

-----Solution 3-----

$$\begin{pmatrix} x \\ y \\ z \end{pmatrix} \rightarrow \begin{pmatrix} 0.1316323 \\ 0.359627 \\ -0.49126 \end{pmatrix}$$

4.3 Karush-Kuhn-Tucker (KT) Conditions

$$\begin{pmatrix} v_1 \\ v_2 \end{pmatrix} \to \begin{pmatrix} 2.57735 \\ 1.07028 \end{pmatrix}$$

Constraint gradient matrix $\to \begin{pmatrix} 0.263265 & 1 \\ 1.43851 & 1 \\ -2.94756 & 1 \end{pmatrix}$ Rank $\to 2$

Regularity status $\to$ RegularPoint
KT Status $\to$ Valid KT Point   Objective function value $\to 2.57735$

-----Solution 4-----

$$\begin{pmatrix} x \\ y \\ z \end{pmatrix} \to \begin{pmatrix} 0.661225 \\ -0.48405 \\ -0.177175 \end{pmatrix}$$

$$\begin{pmatrix} v_1 \\ v_2 \end{pmatrix} \to \begin{pmatrix} 1.42265 \\ 0.795166 \end{pmatrix}$$

Constraint gradient matrix $\to \begin{pmatrix} 1.32245 & 1 \\ -1.9362 & 1 \\ -1.06305 & 1 \end{pmatrix}$ Rank $\to 2$

Regularity status $\to$ RegularPoint
KT Status $\to$ Valid KT Point   Objective function value $\to 1.42264$

***** Valid KT Point(s) *****

$f \to 1.42265$	$f \to 2.57735$	$f \to 2.57735$	$f \to 1.42265$
$x \to -0.661225$	$x \to -0.131633$	$x \to 0.131633$	$x \to 0.661225$
$y \to 0.48405$	$y \to -0.359627$	$y \to 0.359627$	$y \to -0.48405$
$z \to 0.177175$	$z \to 0.49126$	$z \to -0.49126$	$z \to -0.177175$
$v_1 \to 1.42265$	$v_1 \to 2.57735$	$v_1 \to 2.57735$	$v_1 \to 1.42265$
$v_2 \to -0.795166$	$v_2 \to -1.07028$	$v_2 \to 1.07028$	$v_2 \to 0.795166$

Example 4.11 *Building design*   Consider the building design problem presented in Chapter 1. The problem statement is as follows.

To save energy costs for heating and cooling, an architect is considering designing a partially buried rectangular building. The total floor space needed is 20,000 m². Lot size limits the longer building dimensions to 50*m* in plan. The ratio between the plan dimensions must be equal to the *golden ratio* (1.618), and each story must be 3.5 m high. The heating and cooling costs are estimated at $100 m² of the exposed surface area of the building. The owner has specified that the annual energy costs should not exceed $225,000. The objective is to determine building dimensions such that the cost of excavation is minimized.

A formulation for this problem involving five design variables is discussed in Chapter 1.

$$n = \text{Number of stories}$$
$$d = \text{Depth of building below ground}$$

h = Height of building above ground

ℓ = Length of building in plan

w = Width of building in plan

Find $(n, d, h, \ell, \text{ and } w)$ in order to

Minimize $f = d\ell w$

Subject to
$$\begin{pmatrix} \dfrac{d+h}{n} = 3.5 \\ \ell = 1.618w \\ 100(2h\ell + 2hw + \ell w) \leq 225{,}000 \\ \ell \leq 50 \\ w \leq 50 \\ n\ell w \geq 20{,}000 \\ n \geq 1, d \geq 0, h \geq 0, \ell \geq 0 \text{ and } w \geq 0 \end{pmatrix}$$

Using the equality constraints (substituting $n = (d+h)/3.5$ and $\ell = 1.618w$), we get the following formulation in terms of three variables:

Find $(d, h, \text{ and } w)$ in order to

Minimize $f = 1.61800 dw^2$

Subject to
$$\begin{pmatrix} 100(5.23600hw + 1.61800w^2) \leq 225{,}000 \\ 1.61800w \leq 50 \\ 0.4622857(d+h)w^2 \geq 20{,}000 \\ d \geq 0, h \geq 0, \text{ and } w \geq 0 \end{pmatrix}$$

In order to solve it using KT conditions, there will be $2^6 = 64$ cases to be examined. It is not practical to list the solutions of all these cases. After a few trials, it was determined that the case that produces a valid KT point is when the first and the third constraints are active. The solution for only this case is presented below.

```
Clear[d, h, w];
vars = {d, h, w};
f = 1.618dw^2;
cons = {100(5.236hw + 1.618w^2) ≤ 225,000, 1.618w ≤ 50, 0.4622856(d + h)w^2 ≥
  20,000, d ≥ 0, h ≥ 0, w ≥ 0};

KTSolution[f, cons, vars, PrintLevel → 2, ActiveCases → {{1, 3}}];
```

Minimize $f \to 1.618 dw^2$

$$\nabla f \to \begin{pmatrix} 1.618w^2 \\ 0 \\ 3.236 dw \end{pmatrix}$$

***** LE constraints & their gradients

$g_1 \to -225{,}000 + 100(5.236hw + 1.618w^2) \le 0 \quad g_2 \to -50 + 1.618w \le 0$

$g_3 \to 20{,}000 - 0.462286(d+h)w^2 \le 0 \quad g_4 \to -d \le 0$

$g_5 \to -h \le 0 \quad g_6 \to -w \le 0$

$\nabla g_1 \to \begin{pmatrix} 0 \\ 523.6w \\ 523.6h + 323.6w \end{pmatrix} \quad \nabla g_2 \to \begin{pmatrix} 0 \\ 0 \\ 1.618 \end{pmatrix}$

$\nabla g_3 \to \begin{pmatrix} -0.462286w^2 \\ -0.462286w^2 \\ -0.924571dw - 0.924571hw \end{pmatrix} \quad \nabla g_4 \to \begin{pmatrix} -1 \\ 0 \\ 0 \end{pmatrix}$

$\nabla g_5 \to \begin{pmatrix} 0 \\ -1 \\ 0 \end{pmatrix} \quad \nabla g_6 \to \begin{pmatrix} 0 \\ 0 \\ -1 \end{pmatrix}$

***** Lagrangian $\to 1.618dw^2 + (-225{,}000 + 100(5.236hw + 1.618w^2) + s_1^2)u_1 + (-50 + 1.618w + s_2^2)u_2 + (20{,}000 - 0.462286(d+h)w^2 + s_3^2)u_3 + (-d + s_4^2)u_4 + (-h + s_5^2)u_5 + (-w + s_6^2)u_6$

************ Case 1 ************

Active inequalities $\to \{1, 3\}$

Known values $\to \{u_2 \to 0, u_4 \to 0, u_5 \to 0, u_6 \to 0, s_1 \to 0, s_3 \to 0\}$

Equations for this case

$$\begin{pmatrix} 1.618w^2 - 0.462286w^2 u_3 == 0 \\ 523.6wu_1 - 0.462286w^2 u_3 == 0 \\ 3.236dw + 523.6hu_1 + 323.6wu_1 - 0.924571dwu_3 - 0.924571hwu_3 == 0 \\ -225{,}000 + 100(5.236hw + 1.618w^2) == 0 \\ -50 + 1.618w + s_2^2 == 0 \\ 20{,}000 - 0.462286(d+h)w^2 == 0 \\ -d + s_4^2 == 0 \\ -h + s_5^2 == 0 \\ -w + s_6^2 == 0 \end{pmatrix}$$

-----Solution 1-----

$\begin{pmatrix} d \\ h \\ w \end{pmatrix} \to \begin{pmatrix} 80.0273 \\ 13.3061 \\ 21.5299 \end{pmatrix}$

$\begin{pmatrix} u_1 \\ u_2 \\ u_3 \\ u_4 \\ u_5 \\ u_6 \end{pmatrix} \to \begin{pmatrix} 0.0665304 \\ 0 \\ 3.5 \\ 0 \\ 0 \\ 0 \end{pmatrix} \quad \begin{pmatrix} s_1^2 \\ s_2^2 \\ s_3^2 \\ s_4^2 \\ s_5^2 \\ s_6^2 \end{pmatrix} \to \begin{pmatrix} 0 \\ 15.1647 \\ 0 \\ 80.0273 \\ 13.3061 \\ 21.5299 \end{pmatrix}$

Chapter 4 Optimality Conditions

$$\text{Constraint gradient matrix} \to \begin{pmatrix} 0 & -214.286 \\ 11273. & -214.286 \\ 13934.1 & -1857.88 \end{pmatrix} \quad \text{Rank} \to 2$$

Regularity status → RegularPoint

KT Status → Valid KT Point   Objective function value → 60020.4

-----Solution 2-----

$$\begin{pmatrix} d \\ h \\ w \end{pmatrix} \to \begin{pmatrix} 106.639 \\ -13.3061 \\ -21.5299 \end{pmatrix}$$

$$\begin{pmatrix} u_1 \\ u_2 \\ u_3 \\ u_4 \\ u_5 \\ u_6 \end{pmatrix} \to \begin{pmatrix} -0.0665304 \\ 0 \\ 3.5 \\ 0 \\ 0 \\ 0 \end{pmatrix} \quad \begin{pmatrix} s_1^2 \\ s_2^2 \\ s_3^2 \\ s_4^2 \\ s_5^2 \\ s_6^2 \end{pmatrix} \to \begin{pmatrix} 0 \\ 84.8353 \\ 0 \\ 106.639 \\ -13.3061 \\ -21.5299 \end{pmatrix}$$

$$\text{Constraint gradient matrix} \to \begin{pmatrix} 0 & -214.286 \\ -11273. & -214.286 \\ -13934.1 & 1857.88 \end{pmatrix} \quad \text{Rank} \to 2$$

Regularity status → RegularPoint

KT Status → Invalid KT Point   Objective function value → 79979.6

***** Valid KT Point(s) *****

f → 60020.5
d → 80.0273
h → 13.3061
w → 21.5299
u_1 → 0.0665304
u_2 → 0
u_3 → 3.5
u_4 → 0
u_5 → 0
u_6 → 0
s_1^2 → 0
s_2^2 → 15.1647
s_3^2 → 0
s_4^2 → 80.0273
s_5^2 → 13.3061
s_6^2 → 21.5299

Thus, the KT solution is as follows:

$$d^* = 80.03 \text{ m} \quad h^* = 13.31 \text{ m} \quad w^* = 21.53 \text{ m} \quad f^* = 60020.5 \text{ m}^3$$

This solution is comparable to the one obtained using graphical methods in Chapter 1.

4.4 Geometric Interpretation of KT Conditions

For problems with two variables, it is possible to draw gradient vectors and geometrically interpret the KT conditions. The same concepts apply to the general case of n variables, but obviously pictures cannot be drawn in that case. Considering only the active inequality constraints, the Lagrangian function is

$$L = f(\mathbf{x}) + \sum_{i \in \text{Active}} u_i g_i(\mathbf{x}) + \sum_{i=1}^{p} v_i h_i(\mathbf{x})$$

From the additive property of constraints, the terms $\sum_{i \in \text{Active}} u_i g_i(\mathbf{x})$ $+ \sum_{i=1}^{p} v_i h_i(\mathbf{x})$ represent an aggregate constraint. This is true for any arbitrary multipliers ($u_i \geq 0$ though). The specific multipliers for which the minimum of the unconstrained Lagrangian function is the same as that of the original constrained problem are those that satisfy the following equations:

$$\nabla f(\mathbf{x}^*) = -\left[\sum_{i \in \text{Active}} u_i \nabla g_i(\mathbf{x}^*) + \sum_{i=1}^{p} v_i \nabla h_i(\mathbf{x}^*) \right]$$

If there are no equality constraints and none of the inequalities are active, then the problem is essentially unconstrained, at least in a small neighborhood of $\mathbf{x}^*$, and the above conditions simply state that the gradient of the objective function must be zero. These are obviously the same as the necessary conditions for the minimum of unconstrained problems.

If there is only one constraint (either an equality or an active inequality), then the above equations say that at a KT point, the gradient of the objective function must lie along the same line, but opposite in direction, as the gradient of constraint. The Lagrange multiplier is simply a scale factor that makes the length of the constraint gradient vector the same as that of the objective function gradient.

If there are two constraints (either equality or active inequalities), then the right-hand side represents the vector sum (resultant) of constraint gradients. The Lagrange multipliers are those scale factors that make the length of the resultant vector the same as that of the objective function gradient. In terms of the aggregate constraint, this means that multipliers are defined such that the resultant vector coincides with the gradient of the aggregate constraint. The following examples illustrate these points.

Example 4.12 Consider the solution of the following two variable minimization problems.

```
f = -x - y;
g = {x + y^2 - 5 ≤ 0, x - 2 ≤ 0};
vars = {x, y};
```

The solution using the KT conditions is as follows:

KTSolution[f, g, vars, PrintLevel → 2];
Minimize $f \to -x - y$

$$\nabla f \to \begin{pmatrix} -1 \\ -1 \end{pmatrix}$$

***** LE constraints & their gradients
$g_1 \to -5 + x + y^2 \leq 0 \quad g_2 \to -2 + x \leq 0$

$$\nabla g_1 \to \begin{pmatrix} 1 \\ 2y \end{pmatrix} \quad \nabla g_2 \to \begin{pmatrix} 1 \\ 0 \end{pmatrix}$$

***** Lagrangian $\to -x - y + (-5 + x + y^2 + s_1^2) u_1 + (-2 + x + s_2^2) u_2$

$$\nabla L = 0 \to \begin{pmatrix} -1 + u_1 + u_2 == 0 \\ -1 + 2y u_1 == 0 \\ -5 + x + y^2 + s_1^2 == 0 \\ -2 + x + s_2^2 == 0 \\ 2 s_1 u_1 == 0 \\ 2 s_2 u_2 == 0 \end{pmatrix}$$

Since there are two inequality constraints, we need to examine four possible cases.

```
************ Case 1 ************
Active inequalities → None
Known values → {u_1 → 0, u_2 → 0}
```

$$\text{Equations for this case} \to \begin{pmatrix} \text{False} \\ \text{False} \\ -5 + x + y^2 + s_1^2 == 0 \\ -2 + x + s_2^2 == 0 \end{pmatrix}$$

No solution for this case

Note that the False in the list of equations indicates that by substituting known values for the case, the corresponding equation cannot be satisfied. In

4.4 Geometric Interpretation of KT Conditions

this case, by setting $u_1 = u_2 = 0$, obviously the first two equations become nonsense.

```
************ Case 2 ************
Active inequalities → {1}
Known values → {u₂ → 0, s₁ → 0}
```

$$\text{Equations for this case} \rightarrow \begin{pmatrix} -1 + u_1 == 0 \\ -1 + 2yu_1 == 0 \\ -5 + x + y^2 == 0 \\ -2 + x + s_2^2 == 0 \end{pmatrix}$$

-----Solution 1-----

$$\begin{pmatrix} x \\ y \end{pmatrix} \rightarrow \begin{pmatrix} 4.75 \\ 0.5 \end{pmatrix}$$

$$\begin{pmatrix} u_1 \\ u_2 \end{pmatrix} \rightarrow \begin{pmatrix} 1. \\ 0 \end{pmatrix} \quad \begin{pmatrix} s_1^2 \\ s_2^2 \end{pmatrix} \rightarrow \begin{pmatrix} 0 \\ -2.75 \end{pmatrix}$$

KT Status → Invalid KT Point   Objective function value → -5.25

This solution is invalid because the slack variable for the second constraint is negative.

```
************ Case 3 ************
Active inequalities → {2}
Known values → {u₁ → 0, s₂ → 0}
```

$$\text{Equations for this case} \rightarrow \begin{pmatrix} -1 + u_2 == 0 \\ \text{False} \\ -5 + x + y^2 + s_1^2 == 0 \\ -2 + x == 0 \end{pmatrix}$$

No solution for this case

```
************ Case 4 ************
Active inequalities → {1, 2}
Known values → {s₁ → 0, s₂ → 0}
```

$$\text{Equations for this case} \rightarrow \begin{pmatrix} -1 + u_1 + u_2 == 0 \\ -1 + 2yu_1 == 0 \\ -5 + x + y^2 == 0 \\ -2 + x == 0 \end{pmatrix}$$

-----Solution 1-----

$$\begin{pmatrix} x \\ y \end{pmatrix} \rightarrow \begin{pmatrix} 2. \\ -1.73205 \end{pmatrix}$$

$$\begin{pmatrix} u_1 \\ u_2 \end{pmatrix} \rightarrow \begin{pmatrix} -0.288675 \\ 1.28868 \end{pmatrix} \quad \begin{pmatrix} s_1^2 \\ s_2^2 \end{pmatrix} \rightarrow \begin{pmatrix} 0 \\ 0 \end{pmatrix}$$

Constraint gradient matrix $\to \begin{pmatrix} 1 & 1 \\ -3.4641 & 0 \end{pmatrix}$ Rank $\to 2$

Regularity status $\to$ RegularPoint

KT Status $\to$ Invalid KT Point   Objective function value $\to -0.267949$

This solution is invalid because the Lagrange multiplier for the first constraint is negative.

-----Solution 2-----

$\begin{pmatrix} x \\ y \end{pmatrix} \to \begin{pmatrix} 2. \\ 1.73205 \end{pmatrix}$

$\begin{pmatrix} u_1 \\ u_2 \end{pmatrix} \to \begin{pmatrix} 0.288675 \\ 0.711325 \end{pmatrix}$ $\begin{pmatrix} s_1^2 \\ s_2^2 \end{pmatrix} \to \begin{pmatrix} 0 \\ 0 \end{pmatrix}$

Constraint gradient matrix $\to \begin{pmatrix} 1 & 1 \\ 3.4641 & 0 \end{pmatrix}$ Rank $\to 2$

Regularity status $\to$ RegularPoint

KT Status $\to$ Valid KT Point   Objective function value $\to -3.73205$

There are three points that satisfy the equations resulting from setting the gradient of Lagrangian to zero. However, two of these points are rejected. The first one ($x = 4.75$, $y = 0.5$) was rejected because the slack variable for the second constraint was negative. The second one ($x = 2$, $y = -1.732 = -\sqrt{3}$) was rejected because the Lagrange multiplier for the first constraint was negative at this point. Only one point ($x = 2$, $y = 1.732 = \sqrt{3}$) satisfies all requirements of the KT conditions.

To understand exactly what is going on, all these points are labeled on the graphical solution in Figure 4.8. From the graph, it is clear that the global minimum is at the upper intersection of constraints g_1 and g_2 ($x_1 = 2$ and $x_2 = \sqrt{3}$). This is the same valid KT point computed by the KTSolution. The point labeled Invalid KT-1 clearly violates the second constraint. During the KT solution, this violation was indicated by the negative slack variable for this constraint. The reason why this point satisfied the gradient condition is also clear from the second graph that shows gradients of active constraint (g_1) and the objective function. The gradient of the objective function and the first constraint are pointing exactly in the opposite directions at this point.

The gradient vectors shown in the graphs in Figure 4.9 clarify why the lower intersection of g_1 and g_2 does not satisfy the KT conditions. The resultant vector of gradient vectors ∇g_1 and ∇g_2 will obviously lie in the parallelogram formed by these vectors. The multipliers (with positive values) will simply increase or decrease the size of this parallelogram. From the direction of the ∇f vector at the upper intersection, it is clear that for some values of the multipliers,

4.4 Geometric Interpretation of KT Conditions

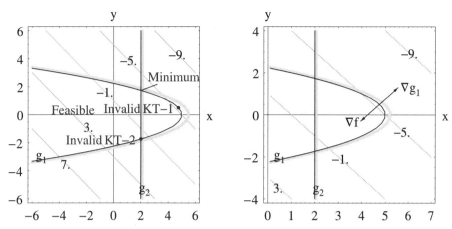

FIGURE 4.8 Graph showing all points computed during the KT solution and gradients at one point.

the resultant vector can be made to lie along the same line (but opposite in direction) to the ∇f vector. However, this is impossible to do at the lower intersection. Thus, even though both points satisfy the same set of constraints, only the upper one is a KT point.

Finally, we can also observe that the resultant of the active constraint gradient vector is exactly the same as that of the aggregate constraint if we use the multipliers obtained from the KT solution. From the KT solution, we get the following values of the Lagrange multipliers for the two active constraints:

$$u_1 = 0.288675 \qquad u_2 = 0.711325$$

With these multipliers, the aggregate constraint is defined as follows:

ga = Expand[0.288675First[g[[1]]] + 0.711325First[g[[2]]]] ≤ 0

$-2.86603 + 1.\text{x} + 0.288675\text{y}^2 \leq 0$

The gradient of this aggregate constraint and that of the objective function are as follows:

grd = Grad[{f, First[ga]}, vars]

{{-1, -1}, {1., 0.57735y}}

At the two intersection points, these gradients are as follows:

{grd/.{x → 2, y → √3}, grd/.{x → 2, y → -√3}}

{{{-1, -1}, {1., 1.}}, {{-1, -1}, {1., -1.}}}

Note that at the upper intersection point, the two are equal and opposite but not at the lower intersection. Therefore, only the upper intersection point satisfies the KT conditions. The same thing is shown graphically by plotting the aggregate constraint, together with the actual constraints.

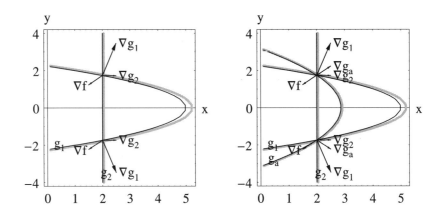

FIGURE 4.9 Graphs showing active constraints, aggregate constraint, and objective function gradients.

Example 4.13 Obtain all points satisfying KT conditions for the following optimization problem:

Minimize $f(x, y) = -x^2 + y^2$

Subject to $x^2 + y^2 - 1 = 0$

Using the KTSolution function, the solution is obtained as follows. The gradient of Lagrangian yields three nonlinear equations. There are four possible solutions to these equations:

```
f = -x^2 + y^2;
h = x^2 + y^2 - 1 == 0;
vars = {x, y};

KTSolution[f, h, vars];
```

Minimize $f \to -x^2 + y^2$

$\nabla f \to \begin{pmatrix} -2x \\ 2y \end{pmatrix}$

***** EQ constraints & their gradients

$h_1 \to -1 + x^2 + y^2 == 0 \quad \nabla h_1 \to \begin{pmatrix} 2x \\ 2y \end{pmatrix}$

4.4 Geometric Interpretation of KT Conditions

```
***** Lagrangian → -x² + y² + (-1 + x² + y²)v₁
```

$$\nabla L = 0 \rightarrow \begin{pmatrix} -2x + 2xv_1 == 0 \\ 2y + 2yv_1 == 0 \\ -1 + x^2 + y^2 == 0 \end{pmatrix}$$

```
***** Valid KT Point(s) *****
f → -1.      f → 1.       f → 1.       f → -1.
x → -1.      x → 0        x → 0        x → 1.
y → 0        y → -1.      y → 1.       y → 0
v₁ → 1.      v₁ → -1.     v₁ → -1.     v₁ → 1.
```

The four KT points are shown on the following graph. The constraint and the objective function gradients are also shown at these four points. Again, we notice that these are the only four points where the objective and the constraint gradients are along the same line. The two points where the gradients point in the opposite directions are the minimum points, while the other two points in fact are the maximum points. The gradient condition is satisfied at the top and the bottom points because of the negative multipliers. Since the sign of Lagrange multipliers for equality constraints is not restricted, KT conditions cannot automatically distinguish a maximum from a minimum.

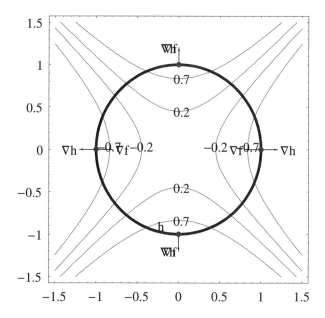

FIGURE 4.10 Graph showing four KT points where the gradient condition is satisfied.

172   Chapter 4   Optimality Conditions

To further demonstrate that the sign of the Lagrange multiplier for an equality constraint has no significance, we multiply the equality constraint by a negative sign (obviously there is nothing wrong in doing this because equality should still hold) and compute the KT solution again.

```
f = -x² + y²;
h = - (x² + y² - 1) == 0;
vars = {x, y};
KTSolution[f, h, vars];

***** Valid KT Point(s) *****

f → -1.      f → 1.       f → 1.       f → -1.
x → -1.      x → 0        x → 0        x → 1.
y → 0        y → -1.      y → 1.       y → 0
v₁ → -1.     v₁ → 1.      v₁ → 1.      v₁ → -1.
```

We get exactly the same solution as before except for the opposite signs for the multipliers. The gradients now are as shown in the following graph. Now the gradients at the minimum point are in the same direction. However, since their multipliers are negative, they satisfy the gradient condition.

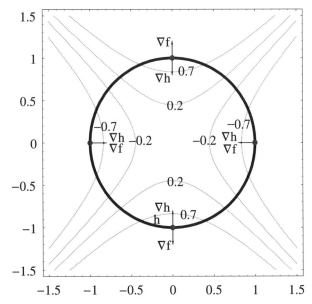

FIGURE 4.11 Graph showing gradients with equality constraint multiplied by a negative sign.

4.4 Geometric Interpretation of KT Conditions

Example 4.14 As a final example, consider a problem that demonstrates an abnormal case because the optimum point is not a regular point:

```
f = -x;
g = {y - (1 - x)^3 ≤ 0, -x ≤ 0, -y ≤ 0};
vars = {x, y};
```

We first consider a graphical solution. From the graph shown in Figure 4.12, it is clear that the point, $x = 1$ and $y = 0$, is the global minimum. Also, constraints g_1 and g_3 are active at this point.

Now consider the KT solution with the known active case.

```
KTSolution[f, g, vars, ActiveCases → {{1, 3}}, PrintLevel → 2];
```

Minimize $f \to -x$

$\nabla f \to \begin{pmatrix} -1 \\ 0 \end{pmatrix}$

***** LE constraints & their gradients

$g_1 \to -(1-x)^3 + y \leq 0 \quad g_2 \to -x \leq 0$

$g_3 \to -y \leq 0 \quad \nabla g_1 \to \begin{pmatrix} 3 - 6x + 3x^2 \\ 1 \end{pmatrix}$

$\nabla g_2 \to \begin{pmatrix} -1 \\ 0 \end{pmatrix} \quad \nabla g_3 \to \begin{pmatrix} 0 \\ -1 \end{pmatrix}$

***** Lagrangian $\to -x + (-(1-x)^3 + y + s_1^2)u_1 + (-x + s_2^2)u_2 + (-y + s_3^2)u_3$

$\nabla L = 0 \to \begin{pmatrix} -1 + 3u_1 - 6xu_1 + 3x^2 u_1 - u_2 == 0 \\ u_1 - u_3 == 0 \\ -(1-x)^3 + y + s_1^2 == 0 \\ -x + s_2^2 == 0 \\ -y + s_3^2 == 0 \\ 2s_1 u_1 == 0 \\ 2s_2 u_2 == 0 \\ 2s_3 u_3 == 0 \end{pmatrix}$

************ Case 1 *************
Active inequalities → {1, 3}
Known values → $\{u_2 \to 0, s_1 \to 0, s_3 \to 0\}$

Equations for this case → $\begin{pmatrix} -1 + 3u_1 - 6xu_1 + 3x^2 u_1 == 0 \\ u_1 - u_3 == 0 \\ -(1-x)^3 + y == 0 \\ -x + s_2^2 == 0 \\ -y == 0 \end{pmatrix}$

No solution for this case
***** No Valid KT Points Found *****

Chapter 4 Optimality Conditions

The first two equations for this case give the solution $x = 1$ and $y = 0$. However, substituting $x = 1$ in the fourth equation gives the nonsense $-1 = 0$. Thus the equations do not have a solution. This is obviously surprising because we know that this case gives the optimum solution. The answer lies in the regularity requirement. The point $(1, 0)$ is not a regular point. To see this, we compute gradients of g_1 and g_3 at $(1, 0)$ as follows:

```
dg = Grad[{g[[1, 1]], g[[3, 1]]}, vars] /. {x → 1, y → 0}
```
{{0, 1}, {0, -1}}

Rank[dg]

1

The rank of this matrix is 1, indicating that the gradient vectors are not linearly independent and thus, the point is not a regular point. By showing the gradients of active constraints at the minimum point, the following graph demonstrates that at an irregular point, it is impossible to express the gradient of the objective function as a linear combination of constraint gradients. Therefore, at such points the gradient condition cannot be satisfied.

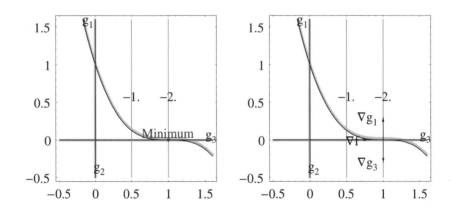

FIGURE 4.12 Graphical solution and gradients at the minimum point showing that it is not a regular point.

This example demonstrates that KT conditions give valid answers only for regular points. An irregular point may or may not be optimum, but we cannot use KT conditions to determine the status of such a point.

4.5 Sensitivity Analysis

After an optimum solution is obtained, the Lagrange multipliers at the optimum point can be used to determine the effect of small changes in the right-hand sides (rhs) of constraints on the optimum solution. Indicating the change in the right-hand side of the ith constraint by Δb_i, it can be shown that

$$f^*_{new} \approx f^* - \sum_{i=1}^{p} v_i \Delta b_i - \sum_{i=1}^{m} u_i \Delta b_i$$

where f^* is the optimum objective function value of the original problem, and u_i and v_i are the associated Lagrange multipliers.

This equation gives a reasonable estimate of the new objective function value as a result of small changes in the right-hand sides of constraints. It should clearly be noted that the equation merely gives an estimate of the new objective function value. It does not tell us the new values of the optimization variables. The only way to find the complete solution is to repeat the solution procedure with the modified constraint.

It is also important to note that this equation is applicable only if the constraints are expressed as the less than or equal to type. In order to determine the appropriate sign for the change, it is best to write the original and modified constraints so that their left-hand sides are exactly the same. The algebraic difference in the right-hand side (rhs) constants (change = new rhs − original rhs) then represents the change.

For example, consider a problem with the original and modified constraints as follows.

$$\text{Original: } x_1 + 2x_2^2 \leq -3 \qquad \text{Modified: } x_1 + 2x_2^2 + 3.1 \leq 0$$

The original constraint is written in the standard form as $x_1 + 2x_2^2 + 3 \leq 0$. Therefore, we write the modified constraint as $x_1 + 2x_2^2 + 3 \leq -0.1$. Thus, the change is $\Delta b = -0.1$.

Derivation of the Sensitivity Equation

For simplicity, consider the original problem with only a single equality constraint written as follows:

Find $\mathbf{x}$ that minimizes $f(\mathbf{x})$ subject to $h(\mathbf{x}) = 0$

The optimum solution of the problem is denoted as follows:

Variables $= \mathbf{x}^*$ $\qquad$ Lagrange multipliers $= v$ $\qquad$ Objective function $= f^*$

The Lagrangian is written as follows:

$$L(\mathbf{x}, v) = f(\mathbf{x}) + vh(\mathbf{x})$$

Now suppose the right-hand side of the constraint is modified as follows:

Modified constraint: $h(x) = \epsilon$ or $h(\mathbf{x}) - \epsilon = 0$

Because of this modification, one would expect the optimum solution to change. Let $\mathbf{x}(\epsilon)$ and $v(\epsilon)$ denote how the solution and the Lagrange multiplier changes as ϵ changes. The Lagrangian for the modified problem is as follows:

$$\tilde{L}(\mathbf{x}, v, \epsilon) = f(\mathbf{x}) + v[h(\mathbf{x}) - \epsilon]$$

At the optimum, the constraint is satisfied and therefore, $f(\mathbf{x}^*(\epsilon)) = \tilde{L}(\mathbf{x}^*(\epsilon), v(\epsilon), \epsilon)$. Using the chain rule

$$\frac{\partial f(\mathbf{x}^*)}{\partial \epsilon} = \nabla_x \tilde{L}(\mathbf{x}^*, v)^T \frac{\partial \mathbf{x}}{\partial \epsilon} + \frac{\partial \tilde{L}}{\partial \mathbf{v}} \frac{\partial \mathbf{v}}{\partial \epsilon} + \frac{\partial \tilde{L}}{\partial \epsilon}$$

Noting that $\frac{\partial \tilde{L}}{\partial \epsilon} = -v$ and from the optimality conditions $\nabla_x \tilde{L}(\mathbf{x}^*, v) = 0$ and $\frac{\partial \tilde{L}}{\partial v} = 0$, we get the following relationship:

$$\frac{\partial f(\mathbf{x}^*)}{\partial \epsilon} = -v$$

Thus, the Lagrange multiplier for a constraint represents the rate of change of the optimum value of the objective function with respect to change in the rhs of that constraint. If the Lagrange multiplier of a constraint has a large magnitude, it means that changing that constraint will have a greater influence on the optimum solution.

Treating the objective function as a function of the change in the right-hand side, we can use the Taylor series to approximate the new objective function value as

$$f^*_{new} \approx f(\mathbf{x}^*) + \frac{\partial f(\mathbf{x}^*)}{\partial \epsilon} \epsilon = f(\mathbf{x}^*) - v\epsilon$$

For a general case, indicating the changes in the right-hand side of the ith constraint by Δb_i (change = new rhs − original rhs), we have

$$f^*_{new} \approx f^* - \sum_{i=1}^{p} v_i \Delta b_i - \sum_{i=1}^{m} u_i \Delta b_i$$

4.5 Sensitivity Analysis

The NewFUsingSensitivity Function

The following function, included in the OptimizationToolbox 'OptimalityConditions' package, computes the new objective function value using the sensitivity theorem.

```
Needs["OptimizationToolbox`OptimalityConditions`"];
?NewFUsingSensitivity
```

NewFUsingSensitivity[vars, f, sol, g, mg] returns the new value of
  the objective function when constants in one or more constraints are
  changed. vars is a list of optimization variables, f is the objective
  function, sol is the list of optimum values of variables followed by
  the Lagrange multipliers, g is a list of original constraints, and mg
  is a list of modified constraints.

Example 4.15 Consider the following optimization problem:

```
f = -x - y^3;
g = {x + y^2 - 10 == 0, -x + 2 ≤ 0, y - 5 ≤ 0};
vars = {x, y};
```

(a) Solution of the original problem All KT points can readily be computed using the KTSolution as follows:

```
soln = KTSolution[f, g, vars];
```

***** Lagrangian $\to -x - y^3 + (2 - x + s_1^2)u_1 + (-5 + y + s_2^2)u_2 + (-10 + x + y^2)v_1$

$$\nabla L = 0 \to \begin{pmatrix} -1 - u_1 + v_1 == 0 \\ -3y^2 + u_2 + 2yv_1 == 0 \\ 2 - x + s_1^2 == 0 \\ -5 + y + s_2^2 == 0 \\ -10 + x + y^2 == 0 \\ 2s_1 u_1 == 0 \\ 2s_2 u_2 == 0 \end{pmatrix}$$

***** Valid KT Point(s) *****

$f \to -24.6274$	$f \to -9.85185$	$f \to -10.$
$x \to 2.$	$x \to 9.55556$	$x \to 10.$
$y \to 2.82843$	$y \to 0.666667$	$y \to 0$
$u_1 \to 3.24264$	$u_1 \to 0$	$u_1 \to 0$
$u_2 \to 0$	$u_2 \to 0$	$u_2 \to 0$
$s_1^2 \to 0$	$s_1^2 \to 7.55556$	$s_1^2 \to 8.$
$s_2^2 \to 2.17157$	$s_2^2 \to 4.33333$	$s_2^2 \to 5.$
$v_1 \to 4.24264$	$v_1 \to 1.$	$v_1 \to 1.$

The objective function at the first KT point is the lowest and therefore is accepted as the optimum solution. This solution is verified graphically as follows:

Chapter 4 Optimality Conditions

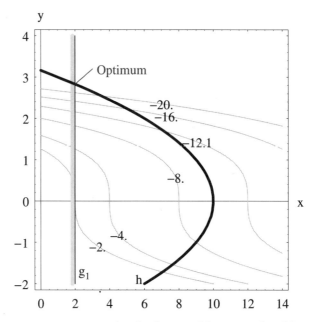

FIGURE 4.13 Graphical solution of the original problem.

(b) Solution of a modified problem Now consider the solution of the problem if the constant 10 in the first constraint is changed to 11. The other constraints remain the same. Thus, the modified list of constraints is as follows:

```
mg = {x + y^2 - 11 == 0, -x + 2 ≤ 0, y - 5 ≤ 0};
sol = {2., 2.8284, 4.24264, 3.24264, 0};
```

Using the sensitivity theorem, the new value of the objective function is computed as follows:

NewFUsingSensitivity[vars, f, sol, g, mg];

$$\text{Original constraints} \rightarrow \begin{pmatrix} -10 + x + y^2 == 0 \\ 2 - x \leq 0 \\ -5 + y \leq 0 \end{pmatrix}$$

$$\text{Modified constraints} \rightarrow \begin{pmatrix} -11 + x + y^2 == 0 \\ 2 - x \leq 0 \\ -5 + y \leq 0 \end{pmatrix}$$

$$\Delta b \rightarrow \begin{pmatrix} 1 \\ 0 \\ 0 \end{pmatrix}$$

Original f → -24.6267   New f → -28.8694

4.5 Sensitivity Analysis

To verify this solution, the modified problem is solved again using the KT-Solution, as follows:

KTSolution[f, mg, vars];

Lagrangian $\to -x - y^3 + (2 - x + s_1^2)u_1 + (-5 + y + s_2^2)u_2 + (-11 + x + y^2)v_1$

$$\nabla L = 0 \to \begin{pmatrix} -1 - u_1 + v_1 == 0 \\ -3y^2 + u_2 + 2yv_1 == 0 \\ 2 - x + s_1^2 == 0 \\ -5 + y + s_2^2 == 0 \\ -11 + x + y^2 == 0 \\ 2s_1 u_1 == 0 \\ 2s_2 u_2 == 0 \end{pmatrix}$$

***** Valid KT Point(s) *****

$f \to -29.$	$f \to -10.8519$	$f \to -11.$
$x \to 2.$	$x \to 10.5556$	$x \to 11.$
$y \to 3.$	$y \to 0.666667$	$y \to 0$
$u_1 \to 3.5$	$u_1 \to 0$	$u_1 \to 0$
$u_2 \to 0$	$u_2 \to 0$	$u_2 \to 0$
$s_1^2 \to 0$	$s_1^2 \to 8.55556$	$s_1^2 \to 9.$
$s_2^2 \to 2.$	$s_2^2 \to 4.33333$	$s_2^2 \to 5.$
$v_1 \to 4.5$	$v_1 \to 1.$	$v_1 \to 1.$

The approximate optimum value of -28.87 compares well with the exact new optimum value of -29.

(c) Solution of another modified problem As another example, consider the solution of the problem if the constants in all three constraints are changed, as follows:

mg = {x + y^2 - 9.5 == 0, -x + 2.5 ≤ 0, y - 4.5 ≤ 0};

Using the sensitivity theorem, the new value of the objective function is computed as follows:

NewFUsingSensitivity[vars, f, sol, g, mg];

Original constraints $\to \begin{pmatrix} -10 + x + y^2 == 0 \\ 2 - x \le 0 \\ -5 + y \le 0 \end{pmatrix}$

Modified constraints $\to \begin{pmatrix} -9.5 + x + y^2 == 0 \\ 2.5 - x \le 0 \\ -4.5 + y \le 0 \end{pmatrix}$

$\Delta b \to \begin{pmatrix} -0.5 \\ -0.5 \\ -0.5 \end{pmatrix}$

Chapter 4 Optimality Conditions

```
Original f → -24.6268   New f → -20.8841
```

To verify this solution, the modified problem is solved again using the KTSolution, as follows:

KTSolution[f, mg, vars];

Lagrangian $\rightarrow -x - y^3 + (2.5 - x + s_1^2)u_1 + (-4.5 + y + s_2^2)u_2 + (-9.5 + x + y^2)v_1$

$$\nabla L = 0 \rightarrow \begin{pmatrix} -1 - u_1 + v_1 == 0 \\ -3y^2 + u_2 + 2yv_1 == 0 \\ 2.5 - x + s_1^2 == 0 \\ -4.5 + y + s_2^2 == 0 \\ -9.5 + x + y^2 == 0 \\ 2s_1 u_1 == 0 \\ 2s_2 u_2 == 0 \end{pmatrix}$$

```
***** Valid KT Point(s) *****
```

f → -21.0203	f → -9.35185	f → 9.5
x → 2.5	x → 9.05556	x → 9.5
y → 2.64575	y → 0.666667	y → 0
u_1 → 2.96863	u_1 → 0	u_1 → 0
u_2 → 0	u_2 → 0	u_2 → 0
s_1^2 → 0	s_1^2 → 6.55556	s_1^2 → 7.
s_2^2 → 1.85424	s_2^2 → 3.83333	s_2^2 → 4.5
v_1 → 3.96863	v_1 → 1.	v_1 → 1.

Again, the approximate optimum value of -20.88 compares well with the exact new optimum value of -21.02.

Example 4.16 *Building design* Consider the building design problem presented in Example 4.11. The following optimum solution was obtained using the KT conditions.

$$d^* = 80.03 \text{ m} \quad h^* = 13.31 \text{ m} \quad w^* = 21.53 \text{ m} \quad f^* = 60{,}020.5 \text{ m}^3$$

The constraints (in standard form) and associated Lagrange multipliers were as follows:

	Constraint	Lagrange multiplier
Energy budget	$100(5.23600hw + 1.61800w^2) - 225{,}000 \leq 0$	0.06653041
Plan dimension	$1.61800w \leq 50$	0
Floor space	$-0.4622857(d+h)w^2 + 20{,}000 \leq 0$	3.50000
Practical dimensions	$d \geq 0, h \geq 0,$ and $w \geq 0$	0

Using sensitivity analysis, we study the effect of the following two changes on the optimum solution.

(a) The energy budget is increased to \$250,000. In this case, the first constraint is modified as follows:

$$100(5.23600hw + 1.61800w^2) - 250{,}000 \le 0$$

Making the left-hand side identical to the original constraint

$$100(5.23600hw + 1.61800w^2) - 225{,}000 \le 25{,}000$$

we see that the change is $\Delta b_1 = 25{,}000$. Assuming this change is small, the new value of the objective function can be estimated as follows:

$$\text{New } f^* = \text{Old } f^* - u_1 \Delta b_1 = 60{,}020.5 - 0.06653041 \times 25{,}000 = 58{,}357.2$$

By solving the problem with the modified first constraint all over again using the KTSolution, we get the exact $f^* = 58{,}311.8$ indicating that the approximate value using sensitivity analysis is fairly good.

(b) The total floor space requirement is reduced to 19,500 m². In this case, the third constraint is modified as follows:

$$-0.4622857(d+h)w^2 + 19{,}500 \le 0$$

Making the left-hand side identical to the original constraint

$$-0.4622857(d+h)w^2 + 20{,}000 \le 500$$

we see that the change is $\Delta b_3 = 500$. Assuming this change is small, the new value of the objective function can be estimated as follows:

$$\text{New } f^* = \text{Old } f^* - u_3 \Delta b_3 = 60{,}020.5 - 3.5 \times 500 = 58{,}270.5$$

By solving the problem with the modified third constraint all over again using the KTSolution, we get the exact $f^* = 58{,}270.5$, which is the same as the one computed by sensitivity analysis.

4.6 Optimality Conditions for Convex Problems

As discussed in Chapter 3, an optimization problem is convex if there are no nonlinear equality constraints and the objective function and all constraint functions are convex (i.e. their Hessian matrices are at least positive semi-definite). For convex problems, any local minimum is also a global minimum. Therefore, a point that satisfies KT conditions is a global minimum point for these problems. Thus, for convex problems, the KT conditions are necessary as well as sufficient for the optimum.

Example 4.17 *Convex case* Consider the following optimization problem:

```
f = (x - 2)^2 + (y - 3)^2;
g = (x - 4)^2 + (y - 5)^2 - 6 ≤ 0;
vars = {x, y};
```

As the following computations show, this is a convex programming problem.

```
ConvexityCheck[{f, g}, vars];
```

------ Function → $(-2 + x)^2 + (-3 + y)^2$

Hessian → $\begin{pmatrix} 2 & 0 \\ 0 & 2 \end{pmatrix}$ Principal Minors → $\begin{pmatrix} 2 \\ 4 \end{pmatrix}$

Status → Convex

------ Function → $-6 + (-4 + x)^2 + (-5 + y)^2$

Hessian → $\begin{pmatrix} 2 & 0 \\ 0 & 2 \end{pmatrix}$ Principal Minors → $\begin{pmatrix} 2 \\ 4 \end{pmatrix}$

Status → Convex

The KTSolution can be obtained as follows:

```
KTSolution[f, g, vars];
```

Minimize f → $(-2 + x)^2 + (-3 + y)^2$

$\nabla f \rightarrow \begin{pmatrix} -4 + 2x \\ -6 + 2y \end{pmatrix}$

***** LE constraints & their gradients

$g_1 \rightarrow -6 + (-4 + x)^2 + (-5 + y)^2 \leq 0$ $\nabla g_1 \rightarrow \begin{pmatrix} -8 + 2x \\ -10 + 2y \end{pmatrix}$

***** Lagrangian → $(-2 + x)^2 + (-3 + y)^2 + (-6 + (-4 + x)^2 + (-5 + y)^2 + s_1^2) u_1$

$\nabla L = 0 \rightarrow \begin{pmatrix} -4 + 2x - 8u_1 + 2xu_1 == 0 \\ -6 + 2y - 10u_1 + 2yu_1 == 0 \\ -6 + (-4 + x)^2 + (-5 + y)^2 + s_1^2 == 0 \\ 2s_1 u_1 == 0 \end{pmatrix}$

***** Valid KT Point(s) *****

f → 0.143594
x → 2.26795
y → 3.26795
u_1 → 0.154701
s_1^2 → 0

4.6 Optimality Conditions for Convex Problems

As expected, there is only one point that satisfies all KT conditions and thus is a global minimum. The following graphical solution confirms this point.

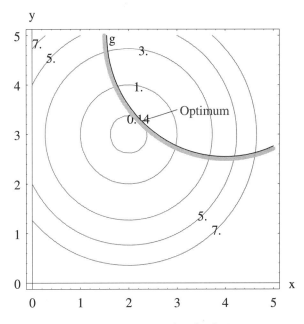

FIGURE 4.14 Graphical solution.

Example 4.18 *Nonconvex case* If the constraint in the previous example is changed to an equality, the problem becomes non-convex (since now we have a nonlinear equality constraint).

```
f = (x - 2)² + (y - 3)²;
h = (x - 4)² + (y - 5)² - 6 == 0;
vars = {x, y};
```

Solving this problem using KT conditions produces two valid KT points.

KTSolution[f, h, vars];

Minimize $f \to (-2+x)^2 + (-3+y)^2$

$\nabla f \to \begin{pmatrix} -4+2x \\ -6+2y \end{pmatrix}$

***** EQ constraints & their gradients

$h_1 \to -6 + (-4+x)^2 + (-5+y)^2 == 0$ $\nabla h_1 \to \begin{pmatrix} -8+2x \\ -10+2y \end{pmatrix}$

***** Lagrangian $\to (-2+x)^2 + (-3+y)^2 + (-6 + (-4+x)^2 + (-5+y)^2) v_1$

$$\nabla L = 0 \to \begin{pmatrix} -4 + 2x - 8v_1 + 2xv_1 == 0 \\ -6 + 2y - 10v_1 + 2yv_1 == 0 \\ -6 + (-4+x)^2 + (-5+y)^2 == 0 \end{pmatrix}$$

```
***** Valid KT Point(s) *****
f → 0.143594        f → 27.8564
x → 2.26795         x → 5.73205
y → 3.26795         y → 6.73205
v₁ → 0.154701       v₁ → -2.1547
```

As seen from the following graphical solution, one of these points corresponds to the global minimum, while the other is in fact the global maximum.

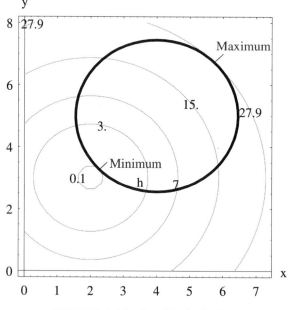

FIGURE 4.15 Graphical solution.

Example 4.19 Consider the following optimization problem:

```
f = x₁⁴ + 2x₂ + 2x₃;
con = {x₁² + x₂ + x₃² ≤ 4, x₁² - x₂ + 2x₃ ≤ 2};
vars = {x₁, x₂, x₃};
```

The problem is convex as seen from the following computations:

```
ConvexityCheck[{f, con}, vars];
------ Function → x₁⁴ + 2x₂ + 2x₃
```

4.6 Optimality Conditions for Convex Problems

$$\text{Hessian} \to \begin{pmatrix} 12x_1^2 & 0 & 0 \\ 0 & 0 & 0 \\ 0 & 0 & 0 \end{pmatrix} \quad \text{Principal Minors} \to \begin{pmatrix} 12x_1^2 \\ 0 \\ 0 \end{pmatrix}$$

Status → Convex

------ Function → $-4 + x_1^2 + x_2 + x_3^2$

$$\text{Hessian} \to \begin{pmatrix} 2 & 0 & 0 \\ 0 & 0 & 0 \\ 0 & 0 & 2 \end{pmatrix} \quad \text{Principal Minors} \to \begin{pmatrix} 2 \\ 0 \\ 0 \end{pmatrix}$$

Status → Convex

------ Function → $-2 + x_1^2 - x_2 + 2x_3$

$$\text{Hessian} \to \begin{pmatrix} 2 & 0 & 0 \\ 0 & 0 & 0 \\ 0 & 0 & 0 \end{pmatrix} \quad \text{Principal Minors} \to \begin{pmatrix} 2 \\ 0 \\ 0 \end{pmatrix}$$

Status → Convex

The KT solution is as follows:

KTSolution[f, con, vars];

Minimize $f \to x_1^4 + 2x_2 + 2x_3$

$$\nabla f \to \begin{pmatrix} 4x_1^3 \\ 2 \\ 2 \end{pmatrix}$$

***** LE constraints & their gradients

$g_1 \to -4 + x_1^2 + x_2 + x_3^2 \leq 0 \quad g_2 \to -2 + x_1^2 - x_2 + 2x_3 \leq 0$

$$\nabla g_1 \to \begin{pmatrix} 2x_1 \\ 1 \\ 2x_3 \end{pmatrix} \quad \nabla g_2 \to \begin{pmatrix} 2x_1 \\ -1 \\ 2 \end{pmatrix}$$

***** Lagrangian → $x_1^4 + 2x_2 + 2x_3 + u_2(-2 + s_2^2 + x_1^2 - x_2 + 2x_3) + u_1(-4 + s_1^2 + x_1^2 + x_2 + x_3^2)$

$$\nabla L = 0 \to \begin{cases} 2u_1 x_1 + 2u_2 x_1 + 4x_1^3 == 0 \\ 2 + u_1 - u_2 == 0 \\ 2 + 2u_2 + 2u_1 x_3 == 0 \\ -4 + s_1^2 + x_1^2 + x_2 + x_3^2 == 0 \\ -2 + s_2^2 + x_1^2 - x_2 + 2x_3 == 0 \\ 2s_1 u_1 == 0 \\ 2s_2 u_2 == 0 \end{cases}$$

***** Valid KT Point(s) *****

$f \to -25.8745$
$x_1 \to 0$
$x_2 \to -9.2915$
$x_3 \to -3.64575$
$u_1 \to 1.13389$
$u_2 \to 3.13389$
$s_1^2 \to 0$
$s_2^2 \to 0$

As expected, there is only one point that satisfies all KT conditions and thus, is a global minimum.

If the objective function were actually the negative of the one given, the problem becomes nonconvex. Several different points satisfy the KT conditions in this case.

ConvexityCheck[-f, vars];

------ Function $\to -x_1^4 - 2x_2 - 2x_3$

Hessian $\to \begin{pmatrix} -12x_1^2 & 0 & 0 \\ 0 & 0 & 0 \\ 0 & 0 & 0 \end{pmatrix}$ Principal Minors $\to \begin{pmatrix} -12x_1^2 \\ 0 \\ 0 \end{pmatrix}$

KTSolution[-f, con, vars];

Minimize $f \to -x_1^4 - 2x_2 - 2x_3$

$\nabla f \to \begin{pmatrix} -4x_1^3 \\ -2 \\ -2 \end{pmatrix}$

***** LE constraints & their gradients

$g_1 \to -4 + x_1^2 + x_2 + x_3^2 \le 0 \quad g_2 \to -2 + x_1^2 - x_2 + 2x_3 \le 0$

$\nabla g_1 \to \begin{pmatrix} 2x_1 \\ 1 \\ 2x_3 \end{pmatrix} \quad \nabla g_2 \to \begin{pmatrix} 2x_1 \\ -1 \\ 2 \end{pmatrix}$

***** Lagrangian $\to -x_1^4 - 2x_2 - 2x_3 + u_2(-2 + s_2^2 + x_1^2 - x_2 + 2x_3) + u_1(-4 + s_1^2 + x_1^2 + x_2 + x_3^2)$

$\nabla L = 0 \to \begin{cases} 2u_1 x_1 + 2u_2 x_1 - 4x_1^3 == 0 \\ -2 + u_1 - u_2 == 0 \\ -2 + 2u_2 + 2u_1 x_3 == 0 \\ -4 + s_1^2 + x_1^2 + x_2 + x_3^2 == 0 \\ -2 + s_2^2 + x_1^2 - x_2 + 2x_3 == 0 \\ 2s_1 u_1 == 0 \\ 2s_2 u_2 == 0 \end{cases}$

***** Valid KT Point(s) *****

$f \to -11.3051$	$f \to -7.5$	$f \to -8.5$	$f \to -7.5$	$f \to -11.3051$
$x_1 \to -1.80305$	$x_1 \to -1.$	$x_1 \to 0$	$x_1 \to 1.$	$x_1 \to 1.80305$
$x_2 \to 0.662419$	$x_2 \to 2.75$	$x_2 \to 3.75$	$x_2 \to 2.75$	$x_2 \to 0.662419$
$x_3 \to -0.29428$	$x_3 \to 0.5$	$x_3 \to 0.5$	$x_3 \to 0.5$	$x_3 \to -0.29428$
$u_1 \to 4.25098$	$u_1 \to 2.$	$u_1 \to 2.$	$u_1 \to 2.$	$u_1 \to 4.25098$
$u_2 \to 2.25098$	$u_2 \to 0$	$u_2 \to 0$	$u_2 \to 0$	$u_2 \to 2.25098$
$s_1^2 \to 0$	$s_1^2 \to 0$	$s_1^2 \to 0$	$s_1^2 \to 0$	$s_1^2 \to 0$
$s_2^2 \to 0$	$s_2^2 \to 2.75$	$s_2^2 \to 4.75$	$s_2^2 \to 2.75$	$s_2^2 \to 0$

4.7 Second-Order Sufficient Conditions

The KT conditions, except for convex problems, are only necessary conditions for the minimum of a constrained optimization problem. Since the Lagrangian function is essentially unconstrained, it is easy to see that the sufficient condition that a given KT point $\{\mathbf{x}^*, \mathbf{u}^*, \mathbf{v}^*\}$ is actually a local minimum is that

$$\mathbf{d}^T \left[\nabla^2 f(\mathbf{x}^*) + \sum_{i \in \text{Active}} u_i^* \nabla^2 g_i(\mathbf{x}^*) + \sum_{i=1}^{p} v_i^* \nabla^2 h_i(\mathbf{x}^*) \right] \mathbf{d} > 0$$

where ∇^2 symbol stands for the Hessian of a function. Note that only active inequalities are considered, and the gradients and Hessians are evaluated at a known KT point.

In the unconstrained case, the changes $\mathbf{d}$ were arbitrary. This resulted in the condition that Hessian must be positive definite. For a constrained problem, we must consider only those changes that do not violate constraints. Thus, the feasible changes $\mathbf{d}$ must satisfy the following linearized versions of the constraints.

$$\nabla g_i(\mathbf{x}^*)^T \mathbf{d} = 0, i \in \text{Active} \qquad \nabla h_i(\mathbf{x}^*)^T \mathbf{d} = 0, i = 1, \ldots, p$$

Because of the need to first determine feasible changes, checking sufficient conditions for a general case is quite tedious. For a special case when

$$\left[\nabla^2 f(\mathbf{x}^*) + \sum_{i \in \text{Active}} u_i \nabla^2 g_i(\mathbf{x}^*) + \sum_{i=1}^{p} v_i \nabla^2 h_i(\mathbf{x}^*) \right]$$

is a positive definite matrix, then the condition is true for any $\mathbf{d}$ and hence, there is no need to compute feasible changes.

Example 4.20 Consider the following optimization problem:

```
f = -x^2 + y;
h = -x^2 - y^2 + 1 == 0;
vars = {x, y};
```

The points satisfying the KT conditions are as follows:

```
soln = KTSolution[f, h, vars];
```

Minimize $f \to -x^2 + y$

$\nabla f \to \begin{pmatrix} -2x \\ 1 \end{pmatrix}$

***** EQ constraints & their gradients

$h_1 \to 1 - x^2 - y^2 == 0 \quad \nabla h_1 \to \begin{pmatrix} -2x \\ -2y \end{pmatrix}$

***** Lagrangian $\to -x^2 + y + (1 - x^2 - y^2) v_1$

$\nabla L = 0 \to \begin{pmatrix} -2x - 2xv_1 == 0 \\ 1 - 2yv_1 == 0 \\ 1 - x^2 - y^2 == 0 \end{pmatrix}$

***** Valid KT Point(s) *****

f → −1.25	f → −1.	f → 1.	f → −1.25
x → −0.8660255	x → 0	x → 0	x → 0.866025
y → −0.5	y → −1.	y → 1.	y → −0.5
v_1 → −1.	v_1 → −0.5	v_1 → 0.5	v_1 → −1.

For this problem, the sufficient condition for $\mathbf{x}^*$ to be minimum is that

$$\mathbf{d}^T \left[\nabla^2 f(\mathbf{x}^*) + v \nabla^2 h(\mathbf{x}^*) \right] \mathbf{d} > 0$$

where the feasible changes $\mathbf{d}$ must satisfy the following equation:

$$\nabla h(\mathbf{x}^*)^T \mathbf{d} = 0$$

The Hessian matrices are as follows:

$$\nabla^2 f = \begin{pmatrix} -2 & 0 \\ 0 & 0 \end{pmatrix} \quad \nabla^2 h = \begin{pmatrix} -2 & 0 \\ 0 & -2 \end{pmatrix}$$

(i) Second-order condition check at point 1

$$x = -0.866025 \quad y = -0.5 \quad v = -1$$

$$\nabla h = \begin{pmatrix} -2x \\ -2y \end{pmatrix} = \begin{pmatrix} 1.73205 \\ 1 \end{pmatrix}$$

Thus, the feasible changes must satisfy the following equation:

$$1.73205 d_1 + d_2 = 0$$

Solving this equation, we get $d_1 = -0.57735 d_2$. The vector of feasible changes is

$$\mathbf{d}^T = (-0.57735 d_2 \quad d_2) \quad \text{with } d_2 \text{ arbitrary}$$

4.7 Second-Order Sufficient Conditions

Substituting into the expression for sufficient conditions, we have

$$(-0.57735d_2 \quad d_2)\left[\begin{pmatrix}-2 & 0\\ 0 & 0\end{pmatrix}+(-1)\begin{pmatrix}-2 & 0\\ 0 & -2\end{pmatrix}\right]\begin{pmatrix}-0.57735d_2\\ d_2\end{pmatrix}$$

This expression evaluates to $2d_2^2$. This is greater than 0 for any $d_2 \neq 0$. Thus, the sufficient condition is satisfied at this point and therefore, it is a local minimum.

(ii) Second-order condition check at point 2

$$x = 0 \quad y = -1 \quad v = -0.5$$

$$\nabla h = \begin{pmatrix}-2x\\ -2y\end{pmatrix} = \begin{pmatrix}0\\ 2\end{pmatrix}$$

Thus, the feasible changes must satisfy the following equation:

$$2d_2 = 0$$

The vector of feasible changes is

$$\mathbf{d}^T = (d_1 \quad 0) \quad \text{with } d_1 \text{ arbitrary}$$

Substituting into the expression for sufficient conditions, we have

$$(d_1 \quad 0)\left[\begin{pmatrix}-2 & 0\\ 0 & 0\end{pmatrix}+(-0.5)\begin{pmatrix}-2 & 0\\ 0 & -2\end{pmatrix}\right]\begin{pmatrix}d_1\\ 0\end{pmatrix}$$

This expression evaluates to $-d_1^2$. This is never greater than 0. Thus, the sufficient condition is not satisfied at this point and therefore, it is not a local minimum.

(iii) Second-order condition check at point 3

$$x = 0 \quad y = -1 \quad v = 0.5$$

$$\nabla h = \begin{pmatrix}-2x\\ -2y\end{pmatrix} = \begin{pmatrix}0\\ 2\end{pmatrix}$$

Thus, the feasible changes must satisfy the following equation:

$$2d_2 = 0$$

The vector of feasible changes is

$$\mathbf{d}^T = (d_1 \quad 0) \quad \text{with } d_1 \text{ arbitrary}$$

Substituting into the expression for sufficient conditions, we have

$$(d_1 \quad 0) \left[\begin{pmatrix} -2 & 0 \\ 0 & 0 \end{pmatrix} + (0.5) \begin{pmatrix} -2 & 0 \\ 0 & -2 \end{pmatrix} \right] \begin{pmatrix} d_1 \\ 0 \end{pmatrix}$$

This expression evaluates to $-3d_1^2$. This is never greater than 0. Thus, the sufficient condition is not satisfied at this point and therefore, it is not a local minimum.

(iv) Second-order condition check at point 4

$$x = 0.866025 \quad y = -0.5 \quad v = -1$$

$$\nabla h = \begin{pmatrix} -2x \\ -2y \end{pmatrix} = \begin{pmatrix} -1.73205 \\ 1 \end{pmatrix}$$

Thus, the feasible changes must satisfy the following equation:

$$-1.73205 d_1 + d_2 = 0$$

Solving this equation, we get $d_1 = 0.57735 d_2$. The vector of feasible changes is

$$\mathbf{d}^T = (0.57735 d_2 \quad d_2) \quad \text{with } d_2 \text{ arbitrary}$$

Substituting into the expression for sufficient conditions, we have

$$(0.57735 d_2 \quad d_2) \left[\begin{pmatrix} -2 & 0 \\ 0 & 0 \end{pmatrix} + (-1) \begin{pmatrix} -2 & 0 \\ 0 & -2 \end{pmatrix} \right] \begin{pmatrix} 0.57735 d_2 \\ d_2 \end{pmatrix}$$

This expression evaluates to $2d_2^2$. This is greater than 0 for any $d_2 \neq 0$. Thus, the sufficient condition is satisfied at this point and therefore, it is a local minimum.

(v) Graphical solution
All four KT points are shown on the following graph. It is clear from the graph that the points 1 and 4 are actually the minimum points, confirming the conclusions drawn from the sufficient conditions.

4.7 Second-Order Sufficient Conditions 191

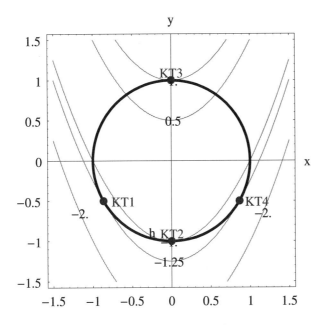

FIGURE 4.16 Graphical solution.

Example 4.21 Consider the following optimization problem:

```
f = x₁³ + 2x₂²x₃ + 2x₃;
h = x₁² + x₂ + x₃² == 4;
g = x₁² - x₂ + 2x₃ ≤ 2;
vars = {x₁, x₂, x₃};
```

The points satisfying the KT conditions are as follows:

soln = KTSolution[f, {g, h}, vars];

Minimize $f \to x_1^3 + 2x_3 + 2x_2^2 x_3$

$\nabla f \to \begin{pmatrix} 3x_1^2 \\ 4x_2 x_3 \\ 2 + 2x_2^2 \end{pmatrix}$

***** LE constraints & their gradients

$g_1 \to -2 + x_1^2 - x_2 + 2x_3 \le 0 \quad \nabla g_1 \to \begin{pmatrix} 2x_1 \\ -1 \\ 2 \end{pmatrix}$

***** EQ constraints & their gradients

Chapter 4 Optimality Conditions

$$h_1 \to -4 + x_1^2 + x_2 + x_3^2 == 0 \quad \nabla h_1 \to \begin{pmatrix} 2x_1 \\ 1 \\ 2x_3 \end{pmatrix}$$

***** Lagrangian $\to x_1^3 + 2x_3 + 2x_2^2 x_3 + u_1 (-2 + s_1^2 + x_1^2 - x_2 + 2x_3) + v_1 (-4 + x_1^2 + x_2 + x_3^2)$

$$\nabla L = 0 \to \begin{pmatrix} 2u_1 x_1 + 2v_1 x_1 + 3x_1^2 == 0 \\ -u_1 + v_1 + 4x_2 x_3 == 0 \\ 2 + 2u_1 + 2x_2^2 + 2v_1 x_3 == 0 \\ -2 + s_1^2 + x_1^2 - x_2 + 2x_3 == 0 \\ -4 + x_1^2 + x_2 + x_3^2 == 0 \\ 2 s_1 u_1 == 0 \end{pmatrix}$$

***** Valid KT Point(s) *****

$f \to -6.74974$	$f \to -3.96447$	$f \to 8.78169$	$f \to -636.782$
$x_1 \to -1.62413$	$x_1 \to -0.342718$	$x_1 \to -0.0372217$	$x_1 \to 0$
$x_2 \to 0.856168$	$x_2 \to 0.065784$	$x_2 \to 1.29184$	$x_2 \to -9.2915$
$x_3 \to -0.711365$	$x_3 \to -1.95365$	$x_3 \to 1.64523$	$x_3 \to -3.64575$
$u_1 \to 0$	$u_1 \to 0$	$u_1 \to 4.27866$	$u_1 \to 219.72$
$s_1^2 \to 1.64111$	$s_1^2 \to 5.85563$	$s_1^2 \to 0$	$s_1^2 \to 0$
$v_1 \to 2.43619$	$v_1 \to 0.514077$	$v_1 \to -4.22283$	$v_1 \to 84.2218$

For this problem, the sufficient condition for $\mathbf{x}^*$ to be minimum is that

$$\mathbf{d}^T \left[\nabla^2 f(\mathbf{x}^*) + v \nabla^2 h(\mathbf{x}^*) + u \nabla^2 g(\mathbf{x}^*) \right] \mathbf{d} > 0$$

where the feasible changes $\mathbf{d}$ must satisfy the following equations:

$$\nabla h(\mathbf{x}^*)^T \mathbf{d} = 0 \qquad \nabla g(\mathbf{x}^*)^T \mathbf{d} = 0$$

The Hessian matrices are as follows:

$$\nabla^2 f = \begin{pmatrix} 6x_1 & 0 & 0 \\ 0 & 4x_3 & 4x_2 \\ 0 & 4x_2 & 0 \end{pmatrix} \quad \nabla^2 h = \begin{pmatrix} 2 & 0 & 0 \\ 0 & 0 & 0 \\ 0 & 0 & 2 \end{pmatrix} \quad \nabla^2 g = \begin{pmatrix} 2 & 0 & 0 \\ 0 & 0 & 0 \\ 0 & 0 & 0 \end{pmatrix}$$

(a) Second-order condition check at point 1

$x_1 = -1.62413 \quad x_2 = 0.856168 \quad x_3 = -0.711365 \quad u = 0 \quad v = 2.43619$

$$\nabla^2 f = \begin{pmatrix} -9.74476 & 0 & 0 \\ 0 & -2.84546 & 3.42467 \\ 0 & 3.42467 & 0 \end{pmatrix}$$

The inequality constraint is inactive; therefore, the feasible changes need to satisfy linearized equality constraint alone.

$$\nabla h = \begin{pmatrix} 2x_1 \\ 1 \\ 2x_3 \end{pmatrix} = \begin{pmatrix} -3.24825 \\ 1 \\ -1.42273 \end{pmatrix}$$

Thus, the feasible changes must satisfy the following equation:
$$-3.24825d_1 + d_2 - 1.42273d_3 = 0$$

Solving this equation, we get $d_1 = 0.3078575(d_2 - 1.42273d_3)$. The vector of feasible changes is

$$\mathbf{d}^T = (0.3078575(d_2 - 1.42273d_3) \quad d_2 \quad d_3) \quad \text{with } d_2 \text{ and } d_3 \text{ arbitrary}$$

Substituting into the expression for sufficient conditions, we have

$$\mathbf{d}^T \left[\begin{pmatrix} -9.74476 & 0 & 0 \\ 0 & -2.84546 & 3.42467 \\ 0 & 3.42467 & 0 \end{pmatrix} + 2.43619 \begin{pmatrix} 2 & 0 & 0 \\ 0 & 0 & 0 \\ 0 & 0 & 2 \end{pmatrix} \right] \mathbf{d}$$

This expression evaluates to

$$-3.30724d_2^2 + 8.16333d_2d_3 + 3.93765d_3^2$$

We can easily see that this expression is not always positive. (For example, choose $d_3 = 0$ and then for any value of d_2, the result is negative.) Therefore, this is not a minimum point.

(b) Second-order condition check at point 2

$$x_1 = -0.3427177 \quad x_2 = 0.06578402 \quad x_3 = -1.95365 \quad u = 0 \quad v = 0.5140766$$

$$\nabla^2 f = \begin{pmatrix} -2.05630 & 0 & 0 \\ 0 & -7.81461 & 0.2631361 \\ 0 & 0.2631361 & 0 \end{pmatrix}$$

The inequality constraint is inactive; therefore, the feasible changes need to satisfy the linearized equality constraint alone.

$$\nabla h = \begin{pmatrix} 2x_1 \\ 1 \\ 2x_3 \end{pmatrix} = \begin{pmatrix} -0.6854355 \\ 1 \\ -3.90730 \end{pmatrix}$$

Thus, the feasible changes must satisfy the following equation:

$$-0.6854355d_1 + d_2 - 3.90730d_3 = 0$$

Solving this equation, we get $d_1 = 1.45892(d_2 - 3.90730d_3)$. The vector of feasible changes is

$$\mathbf{d}^T = (1.45892(d_2 - 3.90730d_3) \quad d_2 \quad d_3) \quad \text{with } d_2 \text{ and } d_3 \text{ arbitrary}$$

Substituting into the expression for sufficient conditions, we have

$$\mathbf{d}^T \left[\begin{pmatrix} -2.05630 & 0 & 0 \\ 0 & -7.81461 & 0.2631361 \\ 0 & 0.2631361 & 0 \end{pmatrix} + 0.5140766 \begin{pmatrix} 2 & 0 & 0 \\ 0 & 0 & 0 \\ 0 & 0 & 2 \end{pmatrix} \right] \mathbf{d}$$

This expression evaluates to

$$-10.0030 d_2^2 + 17.6276 d_2 d_3 - 32.3820 d_3^2$$

It is not always positive. Therefore, this is not a minimum point.

(c) Second-order condition check at point 3

$$x_1 = -0.03722173 \quad x_2 = 1.29184 \quad x_3 = 1.64522 \quad u = 4.27866 \quad v = -4.22282$$

$$\nabla^2 f = \begin{pmatrix} -0.2233304 & 0 & 0 \\ 0 & 6.58091 & 5.16736 \\ 0 & 5.16736 & 0 \end{pmatrix}$$

The inequality constraint is active; therefore, the feasible changes need to satisfy both the linearized equality and inequality constraints.

$$\nabla g = \begin{pmatrix} 2x_1 \\ -1 \\ 2 \end{pmatrix} = \begin{pmatrix} -0.07444346 \\ -1 \\ 2 \end{pmatrix} \quad \nabla h = \begin{pmatrix} 2x_1 \\ 1 \\ 2x_3 \end{pmatrix} = \begin{pmatrix} -0.07444346 \\ 1 \\ 3.29045 \end{pmatrix}$$

Thus, the feasible changes must satisfy the following equations:

$$-0.07444346 d_1 - d_2 + 2 d_3 = 0 \quad -0.07444346 d_1 + d_2 + 3.29045 d_3 = 0$$

Solving these equations, we get $d_1 = 35.5333 d_3 \quad d_2 = -0.6452276 d_3$. The vector of feasible changes is

$$\mathbf{d}^T = (35.5333 d_3 \quad -0.6452276 d_3 \quad d_3) \quad \text{with } d_3 \text{ arbitrary}$$

Substituting into the expression for sufficient conditions, we have

$$\mathbf{d}^T \left[\begin{pmatrix} -0.22333 & 0 & 0 \\ 0 & 6.58091 & 5.16736 \\ 0 & 5.16736 & 0 \end{pmatrix} - 4.22282 \begin{pmatrix} 2 & 0 & 0 \\ 0 & 0 & 0 \\ 0 & 0 & 2 \end{pmatrix} + 4.27866 \begin{pmatrix} 2 & 0 & 0 \\ 0 & 0 & 0 \\ 0 & 0 & 0 \end{pmatrix} \right] \mathbf{d}$$

This expression evaluates to

$$-153.364 d_3^2$$

It is never positive. Therefore, this is not a minimum point.

(d) Second-order condition check at point 4

$$x_1 = 0 \quad x_2 = -9.29150 \quad x_3 = -3.64575 \quad u = 219.719 \quad v = 84.2218$$

$$\nabla^2 f = \begin{pmatrix} 0 & 0 & 0 \\ 0 & -14.5830 & -37.1660 \\ 0 & -37.1660 & 0 \end{pmatrix}$$

The inequality constraint is active; therefore, the feasible changes need to satisfy both the linearized equality and inequality constraints.

$$\nabla g = \begin{pmatrix} 2x_1 \\ -1 \\ 2 \end{pmatrix} = \begin{pmatrix} 0 \\ -1 \\ 2 \end{pmatrix} \quad \nabla h = \begin{pmatrix} 2x_1 \\ 1 \\ 2x_3 \end{pmatrix} = \begin{pmatrix} 0 \\ 1 \\ -7.29150 \end{pmatrix}$$

Thus, the feasible changes must satisfy the following equations:

$$-d_2 + 2d_3 = 0 \quad d_2 - 7.29150 d_3 = 0$$

Solving these equations, we get $d_2 = 0 \quad d_3 = 0$. The vector of feasible changes is

$$\mathbf{d}^T = (d_1 \quad 0 \quad 0) \quad \text{with } d_1 \text{ arbitrary}$$

Substituting into the expression for sufficient conditions, we have

$$\mathbf{d}^T \left[\begin{pmatrix} 0 & 0 & 0 \\ 0 & -14.5830 & -37.1660 \\ 0 & -37.1660 & 0 \end{pmatrix} \right.$$

$$\left. + 84.2218 \begin{pmatrix} 2 & 0 & 0 \\ 0 & 0 & 0 \\ 0 & 0 & 2 \end{pmatrix} + 219.719 \begin{pmatrix} 2 & 0 & 0 \\ 0 & 0 & 0 \\ 0 & 0 & 0 \end{pmatrix} \right] \mathbf{d}$$

This expression evaluates to

$$607.883 d_1^2$$

It is always positive. Therefore, this is a minimum point.

196 Chapter 4 Optimality Conditions

Example 4.22 *Open-top container* Consider the solution of the open-top rectangular container problem formulated in Chapter 1. The problem statement is as follows.

A company requires open-top rectangular containers to transport material. Using the following data, formulate an optimum design problem to determine the container dimensions for the minimum annual cost.

Construction costs	Sides = $65/m^2 Ends = $80/m^2 Bottom = $120/m^2
Useful life	10 years
Salvage value	20% of the initial construction cost
Yearly maintenance cost	$12/m^2 of the outside surface area
Minimum required volume of the container	1200m^3
Nominal interest rate	10% (Annual compounding)

The design variables are the dimensions of the box.

b = Width of container ℓ = Length of container h = height of container

Considering time value of money, the annual cost is written as the following function of design variables (see Chapter 1 for details):

$$\text{Annual cost} = 48.0314bh + 30.0235b\ell + 43.5255h\ell$$

The optimization problem is stated as follows:

Find b, h, and ℓ to

Minimize annual cost = $48.0314bh + 30.0236b\ell + 43.5255h\ell$

Subject to $bh\ell \geq 1200$ and $b, h,$ and $\ell \geq 0$

```
Clear[b, h, ℓ];
vars = {b, h, ℓ};
f = 48.0314bh + 30.0236bℓ + 43.5255hℓ;
g = {-bhℓ + 1200 ≤ 0};
```

A valid KT point is obtained when the volume constraint is active.

```
sol = KTSolution[f, g, vars, ActiveCases → {{1}}];
Minimize f → 48.0314bh + 30.0236bℓ + 43.5255hℓ
```

4.7 Second-Order Sufficient Conditions

$$\nabla f \rightarrow \begin{pmatrix} 48.0314h + 30.0236\ell \\ 48.0314b + 43.5255\ell \\ 30.0236b + 43.5255h \end{pmatrix}$$

***** LE constraints & their gradients

$$g_1 \rightarrow 1200 - bh\ell \leq 0 \quad \nabla g_1 \rightarrow \begin{pmatrix} -h\ell \\ -b\ell \\ -bh \end{pmatrix}$$

***** Lagrangian $\rightarrow 48.0314bh + 30.0236b\ell + 43.5255h\ell + (1200 - bh\ell + s_1^2)u_1$

$$\nabla L = 0 \rightarrow \begin{pmatrix} 48.0314h + 30.0236\ell - h\ell u_1 == 0 \\ 48.0314b + 43.5255\ell - b\ell u_1 == 0 \\ 30.0236b + 43.5255h - bh u_1 == 0 \\ 1200 - bh\ell + s_1^2 == 0 \\ 2s_1 u_1 == 0 \end{pmatrix}$$

***** Valid KT Point(s) *****

$f \rightarrow 13463.3$
$b \rightarrow 11.6384$
$h \rightarrow 8.0281$
$\ell \rightarrow 12.8433$
$u_1 \rightarrow 7.47963$
$s_1^2 \rightarrow 0$

KT point

$$b = 11.6384 \quad h = 8.0281 \quad \ell = 12.8433 \quad u_1 = 7.4796$$

For the sufficient condition, we need to show that

$$\mathbf{d}^T \left[\nabla^2 f(\mathbf{x}^*) + u_1 \nabla^2 g_1(\mathbf{x}^*) \right] \mathbf{d} > 0$$

for all feasible changes $\mathbf{d}$ that satisfy the following equation

$$\nabla g_1(\mathbf{x}^*)^T \mathbf{d} = 0$$

The Hessian matrices are as follows:

$$\nabla^2 f = \begin{pmatrix} 0 & 48.0313 & 30.0235 \\ 48.0313 & 0 & 43.5254 \\ 30.0235 & 43.5254 & 0 \end{pmatrix} \quad \nabla^2 g_1 = \begin{pmatrix} 0 & -\ell & -h \\ -\ell & 0 & -b \\ -h & -b & 0 \end{pmatrix}$$

At the KT point, these matrices are

$$\nabla^2 f = \begin{pmatrix} 0 & 48.0313 & 30.0235 \\ 48.0313 & 0 & 43.5254 \\ 30.0235 & 43.5254 & 0 \end{pmatrix}$$

$$\nabla^2 g_1 = \begin{pmatrix} 0 & -12.8432 & -8.0280 \\ -12.8432 & 0 & -11.6384 \\ -8.0280 & -11.6384 & 0 \end{pmatrix}$$

The feasible changes need to satisfy the linearized inequality constraint.

$$\nabla g_1 = \begin{pmatrix} -h\ell \\ -b\ell \\ -bh \end{pmatrix} = \begin{pmatrix} -103.106 \\ -149.475 \\ -93.4342 \end{pmatrix}$$

Thus, the feasible changes must satisfy the following equation:

$$-103.107 d_1 - 149.475 d_2 - 93.4343 d_3 = 0$$

Solving this equation, we get $d_1 = 0.009698673(-149.475 d_2 - 93.4342 d_3)$. The vector of feasible changes is

$$\mathbf{d}^T = (0.009698673(-149.475 d_2 - 93.4342 d_3) \quad d_2 \quad d_3)$$

with d_2 and d_3 arbitrary

Substituting into the expression for sufficient conditions, we have

$$q = \mathbf{d}^T \left[\begin{pmatrix} 0 & 48.0313 & 30.0235 \\ 48.0313 & 0 & 43.5254 \\ 30.0235 & 43.5254 & 0 \end{pmatrix} \right.$$
$$\left. + 7.47963 \begin{pmatrix} 0 & -12.8432 & -8.02809 \\ -12.8432 & 0 & -11.6384 \\ -8.02809 & -11.6384 & 0 \end{pmatrix} \right] \mathbf{d}$$

This expression evaluates to

$$q = 139.263 d_2^2 + 87.0509 d_2 d_3 + 54.4140 d_3^2$$

This is a quadratic function in (d_2, d_3). To determine the sign of this term, we write it in a quadratic form and determine the principal minors.

$$q = \frac{1}{2} (d_2 \quad d_3) \begin{pmatrix} 278.526 & 87.0509 \\ 87.0509 & 108.828 \end{pmatrix} \begin{pmatrix} d_2 \\ d_3 \end{pmatrix}$$

The principal minors of the matrix are

$$M_1 = 278.526 \qquad M_2 = \text{Det} \left[\begin{pmatrix} 278.526 & 87.0509 \\ 87.0509 & 108.828 \end{pmatrix} \right] = 22733.5$$

Since both principal minors are positive, the matrix is positive definite, and hence the sign of q is always positive. This shows that the computed KT point is a minimum point.

4.8 Lagrangian Duality

For any optimization problem, the Lagrangian function and the KT conditions can be used to define a *dual* optimization problem. The variables in the dual problem are the Lagrange multipliers of the original problem (called the *primal* problem). The dual problems play an important role in theoretical development and implementation of several computational methods that are discussed in later chapters.

It is important to point out that the duality concepts presented in this section apply only to convex problems. This decision is made in order to keep the presentation simple and also recognizing that, in practice, duality is used most often in linear and quadratic programming problems, which are both convex. It is possible to extend the ideas to a general nonconvex case if we assume that locally around an optimum, the functions are convex. In this case, the concept is called *local duality*. For detailed treatment of duality concepts, see books by Bazaraa, Sherali, and Shetty [1995]; Nash and Sofer [1996]; and Bertsekas [1995].

To keep notation simple, the basic concepts of the duality theory are presented by considering problems with inequality constraints alone. The only difference for the equality constraints is that their Lagrange multipliers are free in sign.

The primal problem with inequality constraints alone is stated as follows:

Minimize $f(\mathbf{x})$

Subject to $g_i(\mathbf{x}) \leq 0, i = 1, \ldots, m$

Introducing the Lagrange multiplier vector $\mathbf{u} \geq \mathbf{0}$ and the slack variable vector $\mathbf{s}$, the Lagrangian function is

$$L(\mathbf{x}, \mathbf{u}, \mathbf{s}) = f(\mathbf{x}) + \sum_{i=1}^{m} u_i \left[g_i(\mathbf{x}) + s_i^2\right]$$

As seen from many examples presented in the previous sections, the slack variables give rise to switching conditions that simply state that either a constraint is active (in which case, $u_i > 0$ and $s_i = 0$) or inactive (in which case, $u_i = 0$

and $s_i^2 > 0$). Thus, keeping the switching conditions in mind, we can drop the slack variable term from the Lagrangian function and write it as follows.

$$L(\mathbf{x}, \mathbf{u}) = f(\mathbf{x}) + \sum_{i=1}^{m} u_i g_i(\mathbf{x}) \quad \text{subject to } u_i \geq 0, i = 1, \ldots, m$$

The Lagrangian function is a function of optimization variables and the Lagrange multipliers. If a given point is inside the feasible region, then $u_i = 0$, $i = 1, \ldots, m$, and the second term does not contribute anything to the Lagrangian function. If a constraint is violated, then its multiplier must be increased so that a minimization with respect to $\mathbf{x}$ will force the constraint to be satisfied. Thus, minimization of Lagrangian function can be thought of as a two-step process:

Primal problem: $\text{Minimize}_{\mathbf{x}}[\text{Maximize}_{\mathbf{u} \geq 0} L(\mathbf{x}, \mathbf{u})]$

That is, we first maximize the Lagrangian function with respect to variables $\mathbf{u}$ and then minimize the resulting function with respect to $\mathbf{x}$. This is the primal form.

For *convex problems*, we can write an equivalent dual form by interchanging the order of maximization and minimization as follows.

Dual problem: $\text{Maximize}_{\mathbf{u} \geq 0}[\text{Minimize}_{\mathbf{x}} L(\mathbf{x}, \mathbf{u})]$

Assuming that we can minimize the Lagrangian with respect to optimization variables, a dual function is defined as follows:

$$M(\mathbf{u}) = \text{Minimize}_{\mathbf{x}} L(\mathbf{x}, \mathbf{u}) \equiv \text{Min}_{\mathbf{x}}\left[f(\mathbf{x}) + \sum_{i=1}^{m} u_i g_i(\mathbf{x})\right]$$

The function $M(\mathbf{u})$ is called the dual function. The maximum of the dual function gives Lagrange multipliers that satisfy KT conditions for the primal problem. The dual problem is

Maximize $M(\mathbf{u})$

Subject to $u_i \geq 0, i = 1, \ldots m$

Note that it is implicit in the above discussion that both primal and dual problems have feasible solutions. Only in this case, the maximum of the dual problem is the same as the minimum of the primal problem. Chapters 7 and 8 contain some additional examples to clarify this point.

The following examples illustrate the basic duality concepts.

Example 4.23 Consider the solution of the following two variables minimization problem:

4.8 Lagrangian Duality

```
f = x₁² + x₂² - 2x₁ + 3x₂;
g = {x₁ + x₂ + 5 ≤ 0, x₁ + 2 ≤ 0};
vars = {x₁, x₂};
```

The Lagrangian function for the problem is as follows:

```
L = f + {u₁, u₂}.Map[First, g]
```

$-2x_1 + x_1^2 + u_2(2 + x_1) + 3x_2 + x_2^2 + u_1(5 + x_1 + x_2)$

Note that for each constraint function, only the left-hand side is needed to define the Lagrangian function. In *Mathematica*, it is conveniently done by applying First to each element of list g using the Map function.

We can write the dual function explicitly by minimizing the Lagrangian function with respect to x_1 and x_2, as follows. Since this is an unconstrained situation, the necessary conditions for the minimum are that the gradient of Lagrangian with respect to **x** is zero.

```
eqns = Thread[Grad[L, vars] == 0]
```

$\{-2 + u_1 + u_2 + 2x_1 == 0, 3 + u_1 + 2x_2 == 0\}$

```
sol = Solve[eqns, vars]
```

$\left\{\left\{x_1 \to \frac{1}{2}(2 - u_1 - u_2), x_2 \to \frac{1}{2}(-3 - u_1)\right\}\right\}$

Substituting these values into the Lagrangian, the dual function is written as follows:

```
Mu = Expand[L/.sol[[1]]]
```

$-\frac{13}{4} + \frac{9u_1}{2} - \frac{u_1^2}{2} + 3u_2 - \frac{u_1 u_2}{2} - \frac{u_2^2}{4}$

Since there are only two variables in the dual function, we can graphically determine the maximum of Mu, subject to conditions that u_1 and $u_2 \geq 0$, as follows:

```
GraphicalSolution[Mu, {u₁, 0, 5}, {u₂, 0, 5}, Constraints → {u₁ ≥ 0, u₂ ≥ 0},
   ObjectiveContours → {4, 5, 6, 7, 7.5, 7.9}, PlotPoints → 30,
   Epilog → {RGBColor[1, 0, 0], Disk[{3, 3}, 0.05],
     Text["Maximum", {3, 3.2}]}];
```

The graph shows that the maximum of Mu is at $u_1 = 3$, $u_2 = 3$. Thus, we get the following minimum point for our example.

```
sol/.{u₁ → 3, u₂ → 3}
```

$\{\{x_1 \to -2, x_2 \to -3\}\}$

Chapter 4 Optimality Conditions

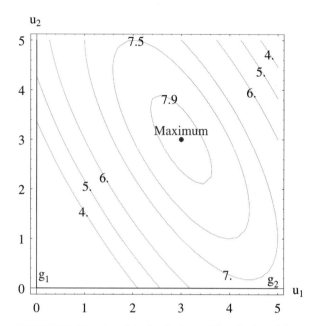

FIGURE 4.17 Graphical solution of the dual problem.

This solution can be verified by solving the problem directly using KT conditions.

KTSolution[f, g, vars];

Minimize $f \to -2x_1 + x_1^2 + 3x_2 + x_2^2$

$\nabla f \to \begin{pmatrix} -2 + 2x_1 \\ 3 + 2x_2 \end{pmatrix}$

***** LE constraints & their gradients

$g_1 \to 5 + x_1 + x_2 \leq 0 \quad g_2 \to 2 + x_1 \leq 0$

$\nabla g_1 \to \begin{pmatrix} 1 \\ 1 \end{pmatrix} \quad \nabla g_2 \to \begin{pmatrix} 1 \\ 0 \end{pmatrix}$

***** Lagrangian $\to -2x_1 + x_1^2 + u_2(2 + s_2^2 + x_1) + 3x_2 + x_2^2 + u_1(5 + s_1^2 + x_1 + x_2)$

$\nabla L = 0 \to \begin{pmatrix} -2 + u_1 + u_2 + 2x_1 == 0 \\ 3 + u_1 + 2x_2 == 0 \\ 5 + s_1^2 + x_1 + x_2 == 0 \\ 2 + s_2^2 + x_1 == 0 \\ 2s_1 u_1 == 0 \\ 2s_2 u_2 == 0 \end{pmatrix}$

***** Valid KT Point(s) *****
$f \to 8.$
$x_1 \to -2.$
$x_2 \to -3.$
$u_1 \to 3.$
$u_2 \to 3.$
$s_1^2 \to 0$
$s_2^2 \to 0$

Example 4.24 *Linear problem* This example illustrates the special form taken by the dual of a linear programming problem.

```
f = 5x + 9y;
g = {-x - y + 3 ≤ 0, x - y - 4 ≤ 0};
vars = {x, y};
```

A solution of this problem using KT conditions can readily be obtained as follows:

```
KTSolution[f, g, vars];
```
$f \to 5x + 9y$

***** Standardized LE (g ≤ 0) constraints
$g_1 \to 3 - x - y \quad g_2 \to -4 + x - y$

$\nabla g_1 \to \begin{pmatrix} -1 \\ -1 \end{pmatrix} \quad \nabla g_2 \to \begin{pmatrix} 1 \\ -1 \end{pmatrix}$

Lagrangian $\to 5x + 9y + (3 - x - y + s_1^2)u_1 + (-4 + x - y + s_2^2)u_2$

$\nabla L = 0 \to \begin{pmatrix} 5 - u_1 + u_2 == 0 \\ 9 - u_1 - u_2 == 0 \\ 3 - x - y + s_1^2 == 0 \\ -4 + x - y + s_2^2 == 0 \\ 2s_1 u_1 == 0 \\ 2s_2 u_2 == 0 \end{pmatrix}$

***** Valid KT Point(s) *****
$f \to 13.$
$x \to 3.5$
$y \to -0.5$
$u_1 \to 7.$
$u_2 \to 2.$
$s_1^2 \to 0$
$s_2^2 \to 0$

The Lagrangian function for the problem is as follows:

```
L = f + {u₁, u₂}.Map[First, g]
```
$5x + 9y + (3 - x - y)u_1 + (-4 + x - y)u_2$

```
Collect[L, {x, y}]
```
$3u_1 + y(9 - u_1 - u_2) - 4u_2 + x(5 - u_1 + u_2)$

Chapter 4 Optimality Conditions

We can write the dual function explicitly by minimizing the Lagrangian with respect to x and y, as follows. Since this is an unconstrained situation, the necessary condition for the minimum is that the gradient of Lagrangian with respect to **x** is zero.

```
eqns = Thread[Grad[L, vars] == 0]
```
$\{5 - u_1 + u_2 == 0, 9 - u_1 - u_2 == 0\}$

These equations do not involve the actual problem variables. However, we notice that these equations are exactly the coefficients of x and y in terms in the Lagrangian function. Thus, using these conditions, we can eliminate primal problem variables from the Lagrangian function.

```
Mu = L/.{x → 0, y → 0}
```
$3u_1 - 4u_2$

The dual problem can now be stated as follows:

Maximize $3u_1 - 4u_2$

Subject to $\begin{pmatrix} 5 - u_1 + u_2 = 0 \\ 9 - u_1 - u_2 = 0 \\ u_1, u_2 \geq 0 \end{pmatrix}$

The solution of this dual problem is obtained using KT conditions, as follows:

```
KTSolution[-Mu, eqns, {u₁, u₂}]
```
Minimize f → $-3u_1 + 4u_2$

$\nabla f \to \begin{pmatrix} -3 \\ 4 \end{pmatrix}$

***** EQ constraints & their gradients
$h_1 \to 5 - u_1 + u_2 == 0 \quad h_2 \to 9 - u_1 - u_2 == 0$

$\nabla h_1 \to \begin{pmatrix} -1 \\ 1 \end{pmatrix} \quad \nabla h_2 \to \begin{pmatrix} -1 \\ -1 \end{pmatrix}$

***** Lagrangian → $-3u_1 + 4u_2 + (5 - u_1 + u_2)v_1 + (9 - u_1 - u_2)v_2$

$\nabla L = 0 \to \begin{pmatrix} -3 - v_1 - v_2 == 0 \\ 4 + v_1 - v_2 == 0 \\ 5 - u_1 + u_2 == 0 \\ 9 - u_1 - u_2 == 0 \end{pmatrix}$

***** Valid KT Point(s) *****

f → -13.
$u_1 \to 7.$
$u_2 \to 2.$
$v_1 \to -3.5$
$v_2 \to 0.5$

4.8 Lagrangian Duality

Comparing this solution with the solution of the primal problem, we see that both problems give the same solution, except that the role of variables is reversed. It is also interesting to note that for linear problems, the primal and dual problems have a very simple relationship. The dual for a general linear programming problem will be presented in Chapter 7.

Example 4.25 *No explicit expression for dual function* In the previous examples, the objective and the constraint function were such that it was possible to get explicit expressions for the dual function. For most nonlinear problems, it is usually difficult to write dual functions explicitly. The dual problem can still be defined, but it remains in terms of both the actual optimization variables and the Lagrange multipliers. This example illustrates this situation:

```
f = x₁² - 4x₁x₂ + 5x₂² - Log[x₁x₂];
g = {-x₁ + 2 ≤ 0, -x₂ + 2 ≤ 0};
vars = {x₁, x₂};
```

It can easily be verified that the problem is convex. The Lagrangian function for the problem is as follows:

```
L = f + {u₁, u₂}.Map[First, g]
```
$-\text{Log}[x_1 x_2] + u_1 (2 - x_1) + x_1^2 + u_2 (2 - x_2) - 4x_1 x_2 + 5x_2^2$

We can write the dual function by minimizing the Lagrangian with respect to x_1 and x_2, as follows:

```
eqns = Thread[Grad[L, vars] == 0]
```
$\left\{-u_1 - \dfrac{1}{x_1} + 2x_1 - 4x_2 == 0,\ -u_2 - 4x_1 - \dfrac{1}{x_2} + 10x_2 == 0\right\}$

An explicit solution of these equations for x_1 and x_2 is difficult. Thus, the dual problem can only be written symbolically, as follows:

Maximize $-\text{Log}[x_1 x_2] + u_1(-2 + x_1) + x_1^2 + u_2(-2 + x_2) - 4x_1 x_2 + 5x_2^2$

Subject to $\begin{pmatrix} -u_1 - \frac{1}{x_1} + 2x_1 - 4x_2 = 0 \\ -u_2 - 4x_1 - \frac{1}{x_2} + 10x_2 = 0 \\ u_1 \geq 0 \\ u_2 \geq 0 \end{pmatrix}$

The solution of this dual problem is obtained by using KT conditions, as follows. Note the use of $-L$ to convert the maximization problem to the minimization form expected by KTSolution. Also note that default KT variables' names must be changed; otherwise, they will conflict with the names of the problem variables.

```
KTSolution[-L, Join[eqns, {u₁ ≥ 0, u₂ ≥ 0}], {x₁, x₂, u₁, u₂},
    KTVarNames → {U, S, V}];
```

Minimize $f \to \text{Log}[x_1 x_2] - u_1(2 - x_1) - x_1^2 - u_2(2 - x_2) + 4x_1 x_2 - 5x_2^2$

$$\nabla f \to \begin{pmatrix} u_1 + \frac{1}{x_1} - 2x_1 + 4x_2 \\ u_2 + 4x_1 + \frac{1}{x_2} - 10x_2 \\ -2 + x_1 \\ -2 + x_2 \end{pmatrix}$$

***** LE constraints & their gradients

$g_1 \to -u_1 \leq 0 \quad g_2 \to -u_2 \leq 0$

$$\nabla g_1 \to \begin{pmatrix} 0 \\ 0 \\ -1 \\ 0 \end{pmatrix} \quad \nabla g_2 \to \begin{pmatrix} 0 \\ 0 \\ 0 \\ -1 \end{pmatrix}$$

***** EQ constraints & their gradients

$h_1 \to -u_1 - \frac{1}{x_1} + 2x_1 - 4x_2 == 0 \quad h_2 \to -u_2 - 4x_1 - \frac{1}{x_2} + 10x_2 == 0$

$$\nabla h_1 \to \begin{pmatrix} 2 + \frac{1}{x_1^2} \\ -4 \\ -1 \\ 0 \end{pmatrix} \quad \nabla h_2 \to \begin{pmatrix} -4 \\ 10 + \frac{1}{x_2^2} \\ 0 \\ -1 \end{pmatrix}$$

***** Lagrangian $\to \text{Log}[x_1 x_2] + (S_1^2 - u_1)U_1 + (S_2^2 - u_2)U_2 - u_1(2 - x_1) - x_1^2 + V_1(-u_1 - \frac{1}{x_1} + 2x_1 - 4x_2) - u_2(2 - x_2) + 4x_1 x_2 - 5x_2^2 + V_2(-u_2 - 4x_1 - \frac{1}{x_2} + 10x_2)$

$$\nabla L = 0 \to \begin{pmatrix} u_1 + 2V_1 - 4V_2 + \frac{V_1}{x_1^2} + \frac{1}{x_1} - 2x_1 + 4x_2 == 0 \\ u_2 - 4V_1 + 10V_2 + 4x_1 + \frac{V_2}{x_2^2} + \frac{1}{x_2} - 10x_2 == 0 \\ -2 - U_1 - V_1 + x_1 == 0 \\ -2 - U_2 - V_2 + x_2 == 0 \\ S_1^2 - u_1 == 0 \\ S_2^2 - u_2 == 0 \\ -u_1 - \frac{1}{x_1} + 2x_1 - 4x_2 == 0 \\ -u_2 - 4x_1 - \frac{1}{x_2} + 10x_2 == 0 \\ 2S_1 U_1 == 0 \\ 2S_2 U_2 == 0 \end{pmatrix}$$

***** Valid KT Point(s) *****

$f \to -1.9054$
$x_1 \to 4.12132$
$x_2 \to 2.$
$u_1 \to 0$
$u_2 \to 3.01472$
$U_1 \to 2.12132$
$U_2 \to 0$
$s_1^2 \to 0$
$s_2^2 \to 3.01472$
$V_1 \to 0$
$V_2 \to 0$

The same solution is obtained by solving the problem directly using KT conditions. Obviously, in this case there is no advantage to defining a dual and then solving it. In fact, since the dual problem has four variables, it is more difficult to solve than the primal problem. The main reason for presenting this example is to show the process of creating a dual problem corresponding to a given primal problem.

KTSolution[f, g, vars];

Minimize $f \to -\text{Log}[x_1 x_2] + x_1^2 - 4x_1 x_2 + 5x_2^2$

$\nabla f \to \begin{pmatrix} -\frac{1}{x_1} + 2x_1 - 4x_2 \\ -4x_1 - \frac{1}{x_2} + 10x_2 \end{pmatrix}$

***** LE constraints & their gradients
$g_1 \to 2 - x_1 \leq 0 \quad g_2 \to 2 - x_2 \leq 0$

$\nabla g_1 \to \begin{pmatrix} -1 \\ 0 \end{pmatrix} \quad \nabla g_2 \to \begin{pmatrix} 0 \\ -1 \end{pmatrix}$

***** Lagrangian $\to -\text{Log}[x_1 x_2] + u_1(2 + s_1^2 - x_1) + x_1^2 + u_2(2 + s_2^2 - x_2) - 4x_1 x_2 + 5x_2^2$

$\nabla L = 0 \to \begin{cases} -u_1 - \frac{1}{x_1} + 2x_1 - 4x_2 == 0 \\ -u_2 - 4x_1 - \frac{1}{x_2} + 10x_2 == 0 \\ 2 + s_1^2 - x_1 == 0 \\ 2 + s_2^2 - x_2 == 0 \\ 2s_1 u_1 == 0 \\ 2s_2 u_2 == 0 \end{cases}$

***** Valid KT Point(s) *****
$f \to 1.9054$
$x_1 \to 4.12132$
$x_2 \to 2.$
$u_1 \to 0$
$u_2 \to 3.01472$
$s_1^2 \to 2.12132$
$s_2^2 \to 0$

4.9 Problems

Optimality Conditions for Unconstrained Problems

Find all stationary points for the following functions. Using second-order optimality conditions, classify them as minimum, maximum, or inflection points. For one and two variable problems, verify solutions graphically.

4.1. $f(x) = x^4 + 1/2x^2 - x$

4.2. $f(x) = 3/4x^6 - 1/3x^3 + 2$

4.3. $f(x, y) = x + 2x^2 + 2y - xy + 2y^2$

4.4. $f(x, y) = 3x - 2x^2 + 2y - 3xy - \frac{7y^2}{2}$

4.5. $f(x, y) = 2x^3 - 3xy + 4y^3$

4.6. $f(x, y) = x^6(y + 2) - xy^3 + 2$

4.7. $f(x, y) = x^3 + xy + \text{Log}[x^5/y^3]$

4.8. $f(x, y, z) = xyz - \frac{1}{2+x^2+y^2+z^2}$

4.9. Show that the following function is convex, and compute its global minimum.

$$f(x, y, z) = 2x^2 - 2xy + y^2 - z - yz + z^2$$

4.10. Show that the following function is convex, and compute its global minimum.

$$f(x, y, z) = 3x + 2x^2 + 2y - 3xy + \frac{7y^2}{2} + z + xz + yz + \frac{5z^2}{2}$$

4.11. Show that the following function of four variables (x_1, x_2, x_3, x_4) is convex, and compute its global minimum.

$$f = x_1 + 4x_1^2 + 2x_2 - x_1x_2 + 4x_2^2 + 3x_3 + 4x_3^2 + 4x_4 + 2x_1x_4 + 2x_2x_4 + 2x_3x_4 + x_4^2$$

4.12. Assume that the power required to propel a barge through a river is proportional to the cube of its speed. Show that, to go upstream, the most economical speed is 1.5 times the river current.

4.13. A small firm is capable of manufacturing two different products. The cost of making each product decreases as the number of units produced increases and is given by the following empirical relationships:

$$c_1 = 5 + \frac{1{,}500}{n_1} \quad c_2 = 7 + \frac{2{,}500}{n_2}$$

where n_1 and n_2 are the number of units of each of the two products produced. The cost of repair and maintenance of equipment used to produce these products depends on the total number of products produced, regardless of its type, and is given by the following quadratic equation:

$$(n_1 + n_2)\left[0.2 + 2.3 \times 10^{-5}(n_1 + n_2) + 5.3 \times 10^{-9}(n_1 + n_2)^2\right]$$

The wholesale selling price of the products drops as more units are produced, according to the following relationships:

$$p_1 = 15 - 0.001 n_1 \quad p_2 = 25 - 0.0015 n_2$$

Formulate the problem of determining how many units of each product the firm should produce to maximize its profit. Find the optimum solution using optimality conditions.

4.14. For a chemical process, pressure measured at different temperatures is given in the following table. Formulate an optimization problem to determine the best values of coefficients in the following exponential model for the data. Find optimum values of these parameters using optimality conditions.

$$\text{Pressure} = \alpha e^{\beta T}$$

Temperature (T^0C)	Pressure (mm of Mercury)
20	15.45
25	19.23
30	26.54
35	34.52
40	48.32
50	68.11
60	98.34
70	120.45

4.15. A chemical manufacturer requires an automatic reactor-mixer. The mixing time required is related to the size of the mixer and the stirring power as follows:

$$T = 1{,}000 \frac{\sqrt{S}}{P^2}$$

where S = capacity of the reactor-mixer, kg, P = power of the stirrer, k-Watts, and T is the time taken in hours per batch. The cost of building the reactor-mixer is proportional to its capacity and is given by the following empirical relationship:

$$\text{Cost} = \$60{,}000\sqrt{S}$$

The cost of electricity to operate the stirrer is $0.05/k-W-hr, and the overhead costs are $137.2 P per year. The total reactor to be processed by the mixer per year is 10^7 kg. Time for loading and unloading the mixer is negligible. Using present worth analysis, formulate the problem of determining the capacity of the mixer and the stirrer power in order to minimize cost. Assume a five-year useful life, 9 percent annual interest rate compounded monthly, and a salvage value of 10 percent of the initial cost of the mixer. Find an optimum solution using optimality conditions.

4.16. Use the annual cost method in problem 4.15.

4.17. A multicell evaporator is to be installed to evaporate water from a salt water solution in order to increase the salt concentration in the solution. The initial concentration of the solution is 5% salt by weight. The desired concentration is 10%, which means that half of the water from the solution must be evaporated. The system utilizes steam as the heat source. The evaporator uses 1 lb of steam to evaporate $0.8n$ lb of water, where n is the number of cells. The goal is to determine the number of cells to minimize cost. The other data are as follows:

The facility will be used to process 500,000 lbs of saline solution per day.

The unit will operate for 340 days per year.

Initial cost of evaporator, including installation = $18,000 per cell.

Additional cost of auxiliary equipment, regardless of the number of cells = $9,000.

Annual maintenance cost = 5% of initial cost.

Cost of steam = $1.55 per 1000 lbs.

Estimated life of the unit = 10 years.

Salvage value at the end of 10 years = $2,500 per cell.

Annual interest rate = 11%.

Formulate the optimization problem to minimize annual cost. Find an optimum solution using optimality conditions.

4.18. Use the present worth method in problem 4.17.

Additive Property of Constraints

Graphically verify the additive property for the following constraints. Try two different multipliers. Also demonstrate that a negative multiplier works for an equality constraint but not for an inequality constraint.

4.19. $2x^2 + y = 5$
$2x - 3y \leq 3$

4.20. $2x + y = 5$
$2x - 3y \leq 3$
$x - y/6 \geq -2$

Karush-Kuhn-Tucker (KT) Conditions and Their Geometric Interpretation

Solve the following problems using KT conditions. For two-variable problems, verify solutions graphically and show gradients of active constraints to illustrate geometric interpretation of KT points.

4.21. Minimize $f(x, y) = -x - 3y$

Subject to $\begin{pmatrix} x + y = 6 \\ -x + y \leq 4 \end{pmatrix}$

4.22. Minimize $f(x_1, x_2, x_3) = 4x_1 + x_2 + x_3$

Subject to $\begin{pmatrix} x_1 + x_2 + 2x_3 \leq 6 \\ 2x_1 + x_2 - x_3 = 4 \\ x_1 \geq 1 \\ x_3 \geq 3 \end{pmatrix}$

4.23. Minimize $f(x_1, x_2, x_3, x_4) = -x_1 + x_2 + x_3 + 4x_4$

Subject to $\begin{pmatrix} x_1 - 5x_2 + x_3 + 3x_4 = 19 \\ x_1 - 4x_2 + 2x_4 = 5 \\ -4x_2 - 5x_3 + 15x_4 = 10 \\ x_1 \geq 2 \end{pmatrix}$

4.24. Minimize $f(x_1, x_2) = 6x_1 + x_2$

Subject to $\begin{pmatrix} 2x_1 + 7x_2 \geq 3 \\ 2x_1 - x_2 \geq 2 \end{pmatrix}$

4.25. Maximize $f(x, y) = -6x + 9y$

Subject to $\begin{pmatrix} x - y \geq 2 \\ 3x + y \geq 1 \\ 2x - 3y \geq 3 \end{pmatrix}$

4.26. Minimize $f(x, y) = x^2 + 2y^2$

Subject to $\begin{pmatrix} x + y \geq 1 \\ x, y \geq 0 \end{pmatrix}$

4.27. Minimize $f(x, y) = x^2 + 2y^2 - 24x - 20y$

Subject to $\begin{pmatrix} x + 2y \geq 0 \\ x + 2y \leq 9 \\ x + y \leq 8 \\ x + y \geq 0 \end{pmatrix}$

4.28. Minimize $f(x, y) = x^2 + y^2 - \text{Log}[x^2 y^2]$

Subject to $x \leq \text{Log}[y]$ $\quad x \geq 1 \quad y \geq 1$

4.29. Minimize $f(x, y, z) = x + y + z$

Subject to $x^{-2} + x^{-2}y^{-2} + x^{-2}y^{-2}z^{-2} \leq 1$

4.30. Minimize $f(x_1, x_2) = x_1 + \frac{x_1}{x_2^2} + \frac{x_2}{x_1}$

Subject to $\begin{pmatrix} x_1 + x_2 \geq 2 \\ x_1, x_2 \geq 0 \end{pmatrix}$

4.31. Maximize $f(x_1, x_2) = x_1 + \frac{x_1}{x_2^2} + \frac{x_2}{x_1}$

Subject to $\begin{pmatrix} x_1 + x_2 = 2 \\ x_1, x_2 \geq 0 \end{pmatrix}$

4.32. Maximize $f(x_1, x_2) = (x_1 - 2)^2 + (x_2 - 10)^2$

Subject to $\begin{pmatrix} x_1^2 + x_2^2 \leq 50 \\ x_1^2 + x_2^2 + 2x_1 x_2 - x_1 - x_2 + 20 \geq 0 \\ x_1, x_2 \geq 0 \end{pmatrix}$

4.33. Minimize $f(x, y) = x^2 + 2yx + y^2 - 15x - 20y$
Subject to $\begin{pmatrix} x^2 + y^2 \leq 20 \\ x^2 - y^2 \leq 10 \end{pmatrix}$

4.34. Minimize $f(x, y) = \frac{1}{xy}$
Subject to $\begin{pmatrix} x + y \leq 5 \\ x, y \geq 1 \end{pmatrix}$

4.35. Minimize $f(x_1, x_2) = \frac{8x_1 + 6x_2 - 5}{-4x_1 + 2x_2 - 40}$
Subject to $\begin{pmatrix} x_1 + x_2 = 10 \\ x_1 \geq 0 \\ 3x_1 - 5x_2 \leq 10 \end{pmatrix}$

4.36. Minimize $f(x_1, x_2, x_3) = x_1^2 + x_2^2/4 + x_3^2/9 - 1$
Subject to $x_1^2 + x_2^2 + x_3^2 = 1$

4.37. Minimize $f(x_1, x_2, x_3) = x_1^2 + x_2^2/4 + x_3^2/9 - 1$
Subject to $x_1^2 + x_2^2 + x_3^2 \leq 1$

4.38. Minimize $f(x_1, x_2, x_3) = 1/(1 + x_1^2 + x_2^2 + x_3^2)$
Subject to $\begin{pmatrix} 2 - 3x_1^2 - 4x_2^2 - 5x_3^2 = 0 \\ 6x_1 + 7x_2 + 8x_3 = 0 \end{pmatrix}$

4.39. Minimize $f(x_1, x_2, x_3) = 1/(1 + x_1^2 + x_2^2 + x_3^2)$

4.40. Subject to $\begin{pmatrix} x_1 + 2x_2 + 3x_3 = 0 \\ 4x_1^2 + 5x_2^2 + 6x_3^2 = 7 \end{pmatrix}$

4.41. Maximize $f(x_1, x_2, x_3) = x_1 x_2 + x_3^2$
Subject to $x_1^2 + 2x_2^2 + 3x_3^2 = 4$

4.42. Minimize $f(x_1, x_2, x_3) = x_1 x_2 + x_3^2$
Subject to $x_1^2 + 2x_2^2 + 3x_3^2 \geq 4$

4.43. Minimize $f(x_1, x_2, x_3) = x_1^2 + 9x_2^2 + x_3^2$
Subject to $x_1 x_2 \geq 1$

4.44. Minimize $f(x_1, x_2, x_3) = x_1^2 + 9x_2^2 + x_3^2$
Subject to $x_1 x_2 x_3 \geq 1$

Chapter 4 Optimality Conditions

4.45. Maximize $f(x_1, x_2, x_3) = 7 \times 10^{-9} x_1^4 x_2 x_3^2$

Subject to $\begin{pmatrix} (x_1^2 x_3^2)/10^7 \leq 0.7 \\ x_1^2 x_2 = 700 \\ x_1 \leq 7 \end{pmatrix}$

4.46. A design problem is formulated in terms of six optimization variables as follows:

Maximize $f = 5x_1 + e^{-2x_2} - e^{-x_2} + x_1 x_3 + 4x_3 + 6x_4 + \frac{5x_5}{x_5+1} + \frac{6x_6}{x_6+1}$

Subject to $\begin{pmatrix} x_1 + x_2 + x_3 + x_4 + x_5 + x_6 \leq 10 \\ x_1 + x_3 + x_4 \leq 5 \\ x_1 - x_2^2 + x_3 + x_5 + x_6^2 \leq 5 \\ x_2 + 2x_4 + x_5 + 0.8 x_6 = 5 \\ x_3^2 + x_5^2 + x_6^2 = 5 \\ x_i \geq 0, \ i = 1, \ldots, 6 \end{pmatrix}$

Assume one is considering solving this problem using KT conditions. What is the total number of cases that must be considered? Find the solution corresponding to the case when the first two inequality constraints are active.

4.47. Consider the following optimization problem:

Maximize $f = x$

Subject to $\begin{pmatrix} x^2 + (y-1)^2 \geq 4 \\ (x-1)^3 - (y-1)^2 = 1 \end{pmatrix}$

Graphically show that the optimum is at (2, 1) and both constraints are active. Write down the KT conditions for the case when both constraints are active and show that they do not support the graphical solution. Explain why.

4.48. A cylindrical vessel, closed at both ends with flat lids, is made of sheet metal. To make a vessel of volume V, show that the least area of sheet metal will be used if the radius is $r = (V/2\pi)^{1/3}$ and height $= 2r$.

4.49. An open-top cylindrical vessel of volume V is made of sheet metal. Show that the least area of sheet metal will be used if the radius is equal to the height.

4.50. Hawkeye foods owns two types of trucks. Truck type I has a refrigerated capacity of 15 m³ and a nonrefrigerated capacity of 25 m³. Truck type II has a refrigerated capacity of 15 m³ and non-refrigerated capacity of 10 m³. One of their stores in Gofer City needs products that require 150 m³ of refrigerated capacity and 130 m³ of nonrefrigerated capacity. For the round trip from the distribution center to Gofer City, truck type I uses 300 liters of fuel, while truck type II uses 200 liters. Use KT conditions to determine the number of trucks of each type that the company must use in order to meet the store's needs while minimizing fuel consumption.

4.51. Dust from an older cement manufacturing plant is a major source of dust pollution in a small community. The plant currently emits 2 pounds of dust per barrel of cement produced. The Environmental Protection Agency (EPA) has asked the plant to reduce this pollution by 85% (1.7 lbs/barrel). There are two models of electrostatic dust collectors that the plant can install to control dust emission. The higher-efficiency model would reduce emissions by 1.8 lbs/barrel and would cost $0.70/barrel to operate. The lower-efficiency model would reduce emissions by 1.5 lbs/barrel and would cost $0.50/barrel to operate. Since the higher-efficiency model reduces more than the EPA required amount and the lower-efficiency less than the required, the plant has decided to install one of each. If the plant has a capacity to produce 3 million barrels of cement per year, how many barrels of cement should be produced using each dust control model to meet the EPA requirements at a minimum cost? Formulate the situation as an optimization problem. Find an optimum solution using KT conditions.

4.52. A small electronics company is planning to expand two of its manufacturing plants. The additional annual revenue expected from the two plants is as follows:

From plant 1: $0.00002x_1^2 - x_2$ From plant 2: $0.00001x_2^2$

where x_1 and x_2 are the investments made into upgrading the facilities. Each plant requires a minimum investment of $30,000. The company can borrow a maximum of $100,000 for this upgrade to be paid back in yearly installments in 10 years at an annual interest rate of 12%. The revenue that the company generates can earn interest at an annual rate of 10%. After the 10-year period, the salvage value of the upgrades is expected to be as follows:

For plant 1: $0.1x_1$ For plant 2: $0.15x_2$

Formulate an optimization problem to maximize the net present worth of these upgrades. Find an optimum solution using KT conditions.

4.53. A company manufactures fragile gift items and sells them directly to its customers through the mail. An average product weighs 12 kg, has a volume of 0.85 m³, and costs $60 to produce. The average shipping distance is 120 miles. The shipping costs per mile based on total weight and volume are $0.006/kg plus $0.025/m³. The products are shipped in cartons that are estimated to cost $2.5/m³ and weigh 3.2 kg/m³. The empty space in the carton is completely filled with a packing material to protect the item during shipping. This packing material has negligible weight but costs $0.95/m³. Based on the past experience, the company has developed the following empirical relationship between breakage and the amount of packing material.

$$\% \text{ breakage} = 85\left(1 - \frac{\text{Volume of packing material}}{\text{Volume of the shipping carton}}\right)$$

The manufacturer guarantees delivery in good condition, which means that any damaged item must be replaced at the company's expense. Formulate an optimization problem to determine the shipping carton volume and volume of packing material that will result in the minimum overall cost of packing, shipping, and delivery. Find an optimum solution using KT conditions.

4.54. An investor is looking to make investment decisions such that she will get at least a 10% rate of return while minimizing the risk of major losses. For the past six years, the rates of return in three major investment types that she is considering are as follows:

Type	Annual rates of return					
Stocks	18.24	17.12	22.23	15.26	12.62	15.42
Mutual funds	12.24	11.16	10.07	8.46	6.62	8.43
Bonds	5.12	6.26	6.34	7.01	6.11	5.95

Formulate the problem as an optimization problem. Find an optimum solution using KT conditions.

4.55. Consider the cantilever beam-mass system shown in Figure 4.18. The beam cross-section is rectangular. The goal is to select cross-sectional dimensions (b and h) to minimize the weight of the beam while keeping the fundamental vibration frequency (ω) larger than 8 rad/sec. Find an optimum solution using KT conditions.

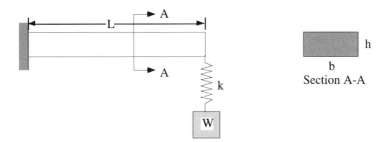

FIGURE 4.18 Rectangular cross-section cantilever beam with a suspended mass.

The numerical data and various equations for the problem are as follows:

Fundamental vibration frequency	$\omega = \sqrt{k_e/m}$ radians/sec
Equivalent spring constant, k_e	$\frac{1}{k_e} = \frac{1}{k} + \frac{L^3}{3EI}$
Mass attached to the spring	$m = W/g$
Gravitational constant	$g = 386$ in/sec^2
Weight attached to the spring	$W = 60$ lbs
Length of beam	$L = 15$ in
Modulus of elasticity	$E = 30 \times 10^6$ lbs/in^2
Spring constant	$k = 10$ lbs/in^2
Moment of inertia	$I = \frac{bh^3}{12}$ in^4
Width of beam cross-section	0.5 in $\leq b \leq 1$ in
Height of beam cross-section	0.2 in $\leq h \leq 2$ in
Unit weight of beam material	0.286 lbs/ in^3

4.56. Consider the optimum design of a rectangular reinforced concrete beam shown in Figure 4.19. There is steel reinforcement near the bottom. Formwork is required on three sides during construction. The beam must support a given bending moment. A least-cost design is required.

The bending strength of the beam is calculated from the following formula:

$$M_u = 0.90 A_s F_y d \left(1 - 0.59 \left(\frac{A_s}{bd}\right)\left(\frac{F_y}{f_c'}\right)\right)$$

where F_y is the specified yield strength of steel, and f'_c is the specified compressive strength of concrete. The ductility requirements dictate minimum and maximum limits on the steel ratio $\rho = A_s/bd$.

$$\rho_{min} \leq \rho \leq \rho_{max}$$

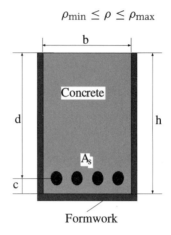

FIGURE 4.19 Reinforced concrete beam.

Use the following numerical data:

Maximum steel ratio	$\rho_{max} = 0.025$
Minimum steel ratio	$\rho_{min} = 0.0033$
Required moment capacity	$M_u \geq 400 \times 10^3$ N-m
Minimum beam width	$b \geq 300$ mm
Concrete cover	$c = 65$ mm
Maximum beam depth	$h \leq 1200$ mm
Concrete cost	$100/m^3$
Formwork cost	$2/m^2$
Steel reinforcement cost	$610/ton (1 ton = 907.18 kg)
Density of steel	7850 kg/m^3
Yield stress of steel, F_y	420 MPa
Ultimate concrete strength, f'_c	35 MPa

Formulate the problem of determining the cross-section variables and amount of steel reinforcement to meet all design requirements at a minimum cost. Assume a unit beam length for cost computations. Find an optimum solution using KT conditions.

Sensitivity Analysis

Solve the following problems using KT conditions. Increase the absolute value of the constants in the constraints by 10%, and use the sensitivity equation to find the effect of this change on the optimum. Verify the results obtained from the sensitivity analysis by solving the modified problem either graphically or by using the KT conditions.

4.57. Minimize $f(x_1, x_2) = x_1 + \frac{x_1}{x_2^2} + \frac{x_2}{x_1}$

Subject to $\begin{pmatrix} x_1 + x_2 \geq 2 \\ x_1, x_2 \geq 0 \end{pmatrix}$

4.58. Maximize $f(x_1, x_2) = x_1 + \frac{x_1}{x_2^2} + \frac{x_2}{x_1}$

Subject to $\begin{pmatrix} x_1 + x_2 = 2 \\ x_1, x_2 \geq 0 \end{pmatrix}$

4.59. Maximize $f(x_1, x_2) = (x_1 - 2)^2 + (x_2 - 10)^2$

Subject to $\begin{pmatrix} x_1^2 + x_2^2 \leq 50 \\ x_1^2 + x_2^2 + 2x_1 x_2 - x_1 - x_2 + 20 \geq 0 \\ x_1, x_2 \geq 0 \end{pmatrix}$

4.60. Minimize $f(x, y) = x^2 + 2yx + y^2 - 15x - 20y$

Subject to $\begin{pmatrix} x^2 + y^2 \leq 20 \\ x^2 - y^2 \leq 10 \end{pmatrix}$

4.61. Minimize $f(x, y) = \frac{1}{xy}$

Subject to $\begin{pmatrix} x + y \leq 5 \\ x, y \geq 1 \end{pmatrix}$

4.62. Minimize $f(x_1, x_2) = \frac{8x_1 + 6x_2 - 5}{-4x_1 + 2x_2 - 40}$

Subject to $\begin{pmatrix} x_1 + x_2 = 10 \\ x_1 \geq 0 \\ 3x_1 - 5x_2 \leq 10 \end{pmatrix}$

Chapter 4 Optimality Conditions

4.63. Minimize $f(x_1, x_2, x_3) = x_1^2 + x_2^2/4 + x_3^2/9 - 1$
Subject to $x_1^2 + x_2^2 + x_3^2 = 1$

4.64. Minimize $f(x_1, x_2, x_3) = x_1^2 + x_2^2/4 + x_3^2/9 - 1$
Subject to $x_1^2 + x_2^2 + x_3^2 \leq 1$

4.65. Minimize $f(x_1, x_2, x_3) = 1/(1 + x_1^2 + x_2^2 + x_3^2)$
Subject to $\begin{pmatrix} 2 - 3x_1^2 - 4x_2^2 - 5x_3^2 = 0 \\ 6x_1 + 7x_2 + 8x_3 = 0 \end{pmatrix}$

4.66. Minimize $f(x_1, x_2, x_3) = 1/(1 + x_1^2 + x_2^2 + x_3^2)$
Subject to $\begin{pmatrix} x_1 + 2x_2 + 3x_3 = 0 \\ 4x_1^2 + 5x_2^2 + 6x_3^2 = 7 \end{pmatrix}$

4.67. Minimize $f(x_1, x_2, x_3) = 1/(1 + x_1^2 + x_2^2 + x_3^2)$
Subject to $\begin{pmatrix} x_1 + 2x_2 + 3x_3 = 0 \\ 4x_1^2 + 5x_2^2 + 6x_3^2 \leq 7 \end{pmatrix}$

4.68. Maximize $f(x_1, x_2, x_3) = x_1 x_2 + x_3^2$
Subject to $x_1^2 + 2x_2^2 + 3x_3^2 = 4$

4.69. Minimize $f(x_1, x_2, x_3) = x_1 x_2 + x_3^2$
Subject to $x_1^2 + 2x_2^2 + 3x_3^2 \geq 4$

4.70. Minimize $f(x_1, x_2, x_3) = x_1^2 + 9x_2^2 + x_3^2$
Subject to $x_1 x_2 \geq 1$

4.71. Minimize $f(x_1, x_2, x_3) = x_1^2 + 9x_2^2 + x_3^2$
Subject to $x_1 x_2 x_3 \geq 1$

4.72. Maximize $f(x_1, x_2, x_3) = 7 * 10^{-9} x_1^4 x_2 x_3^2$
Subject to $\begin{pmatrix} (x_1^2 x_3^2)/10^7 \leq 0.7 \\ x_1^2 x_2 = 700 \\ x_1 \leq 7 \end{pmatrix}$

4.73. Consider the Hawkeye foods Exercise 4.50 again. After solving the original problem, use sensitivity analysis to determine what will be the new optimum objective function value for each of the following changes:

(i) The demand for refrigerated capacity increases to 160 m^3.

(ii) The demand for nonrefrigerated capacity increases to 140 m^3.

(iii) The demand for nonrefrigerated capacity increases to 140 m^3 and that for refrigerated capacity decreases to 140 m^3.

4.74. Consider the investment Exercise 4.54 again. After solving the original problem, use sensitivity analysis to determine what the new optimum objective function value will be for each of the following changes:

(i) The minimum expected rate of return is increased to 12%.

(ii) The minimum expected rate of return is decreased to 9%.

4.75. Consider the cantilever beam Exercise 4.55 again. After solving the original problem, use sensitivity analysis to determine what the new optimum objective function value will be for each of the following changes.

(i) The limiting value of the vibration frequency is increased to 9 rad/sec.

(ii) The limiting value of the vibration frequency is decreased to 7 rad/sec.

Optimality Conditions for Convex Problems

For the following problems, first show that the optimization problem is convex, and then obtain an optimum solution using KT conditions. For problems involving two variables, verify the KT solution by using graphical methods.

4.76. Minimize $f(x, y) = -x - 3y$

Subject to $\begin{pmatrix} x + y = 6 \\ -x + y \leq 4 \end{pmatrix}$

4.77. Minimize $f(x_1, x_2, x_3) = 4x_1 + x_2 + x_3$

Subject to $\begin{pmatrix} x_1 + x_2 + 2x_3 \leq 6 \\ 2x_1 + x_2 - x_3 = 4 \\ x_1 \geq 1 \\ x_3 \geq 3 \end{pmatrix}$

4.78. Minimize $f(x_1, x_2, x_3, x_4) = -x_1 + x_2 + x_3 + 4x_4$

Subject to $\begin{pmatrix} x_1 - 5x_2 + x_3 + 3x_4 = 19 \\ x_1 - 4x_2 + 2x_4 = 5 \\ -4x_2 - 5x_3 + 15x_4 = 10 \\ x_1 \geq 2 \end{pmatrix}$

4.79. Minimize $f(x_1, x_2) = 6x_1 + x_2$

Subject to $\begin{pmatrix} 2x_1 + 7x_2 \geq 3 \\ 2x_1 - x_2 \geq 2 \end{pmatrix}$

4.80. Maximize $f(x, y) = -6x + 9y$

Subject to $\begin{pmatrix} x - y \geq 2 \\ 3x + y \geq 1 \\ 2x - 3y \geq 3 \end{pmatrix}$

4.81. Minimize $f(x, y) = x^2 + 2y^2$

Subject to $\begin{pmatrix} x + y \geq 1 \\ x, y \geq 0 \end{pmatrix}$

4.82. Minimize $f(x, y) = x^2 + 2y^2 - 24x - 20y$

Subject to $\begin{pmatrix} x + 2y \geq 0 \\ x + 2y \leq 9 \\ x + y \leq 8 \\ x + y \geq 0 \end{pmatrix}$

4.83. Minimize $f(x, y) = x^2 + y^2 - \text{Log}[x^2 y^2]$

Subject to $x \leq \text{Log}[y] \quad x \geq 1 \quad y \geq 1$

4.84. Minimize $f(x, y, z) = x + y + z$

Subject to $x^{-2} + x^{-2}y^{-2} + x^{-2}y^{-2}z^{-2} \leq 1$

Second-Order Sufficient Conditions

Use KT conditions and second-order sufficient conditions to solve the following optimization problems.

4.85. Minimize $f(x_1, x_2) = 6x_1 + x_2$

Subject to $\begin{pmatrix} 2x_1 + 7x_2 \geq 3 \\ 2x_1 - x_2 \geq 2 \end{pmatrix}$

4.86. Maximize $f(x, y) = -6x + 9y$

Subject to $\begin{pmatrix} x - y \geq 2 \\ 3x + y \geq 1 \\ 2x - 3y \geq 3 \end{pmatrix}$

4.87. Minimize $f(x, y) = x^2 + 2y^2$

Subject to $\begin{pmatrix} x + y \geq 1 \\ x, y \geq 0 \end{pmatrix}$

4.88. Minimize $f(x, y) = x^2 + 2y^2 - 24x - 20y$

Subject to $\begin{pmatrix} x + 2y \geq 0 \\ x + 2y \leq 9 \\ x + y \leq 8 \\ x + y \geq 0 \end{pmatrix}$

4.89. Minimize $f(x, y) = x^2 + y^2 - \text{Log}[x^2 y^2]$

Subject to $x \leq \text{Log}[y] \quad x \geq 1 \quad y \geq 1$

4.90. Minimize $f(x, y, z) = x + y + z$

Subject to $x^{-2} + x^{-2}y^{-2} + x^{-2}y^{-2}z^{-2} \leq 1$

4.91. Minimize $f(x_1, x_2) = x_1 + \frac{x_1}{x_2^2} + \frac{x_2}{x_1}$

Subject to $\begin{pmatrix} x_1 + x_2 \geq 2 \\ x_1, x_2 \geq 0 \end{pmatrix}$

4.92. Maximize $f(x_1, x_2) = x_1 + \frac{x_1}{x_2^2} + \frac{x_2}{x_1}$

Subject to $\begin{pmatrix} x_1 + x_2 = 2 \\ x_1, x_2 \geq 0 \end{pmatrix}$

4.93. Maximize $f(x_1, x_2) = (x_1 - 2)^2 + (x_2 - 10)^2$

Subject to $\begin{pmatrix} x_1^2 + x_2^2 \leq 50 \\ x_1^2 + x_2^2 + 2x_1 x_2 - x_1 - x_2 + 20 \geq 0 \\ x_1, x_2 > 0 \end{pmatrix}$

4.94. Minimize $f(x, y) = x^2 + 2yx + y^2 - 15x - 20y$

Subject to $\begin{pmatrix} x^2 + y^2 \leq 20 \\ x^2 - y^2 \leq 10 \end{pmatrix}$

4.95. Minimize $f(x, y) = \frac{1}{xy}$

Subject to $\begin{pmatrix} x + y \leq 5 \\ x, y \geq 1 \end{pmatrix}$

4.96. Minimize $f(x_1, x_2) = \frac{8x_1+6x_2-5}{-4x_1+2x_2-40}$

Subject to $\begin{pmatrix} x_1 + x_2 = 10 \\ x_1 \geq 0 \\ 3x_1 - 5x_2 \leq 10 \end{pmatrix}$

4.97. Minimize $f(x_1, x_2, x_3) = x_1^2 + x_2^2/4 + x_3^2/9 - 1$
Subject to $x_1^2 + x_2^2 + x_3^2 = 1$

4.98. Minimize $f(x_1, x_2, x_3) = x_1^2 + x_2^2/4 + x_3^2/9 - 1$
Subject to $x_1^2 + x_2^2 + x_3^2 \leq 1$

4.99. Minimize $f(x_1, x_2, x_3) = 1/(1 + x_1^2 + x_2^2 + x_3^2)$

Subject to $\begin{pmatrix} 2 - 3x_1^2 - 4x_2^2 - 5x_3^2 = 0 \\ 6x_1 + 7x_2 + 8x_3 = 0 \end{pmatrix}$

4.100. Minimize $f(x_1, x_2, x_3) = 1/(1 + x_1^2 + x_2^2 + x_3^2)$

Subject to $\begin{pmatrix} x_1 + 2x_2 + 3x_3 = 0 \\ 4x_1^2 + 5x_2^2 + 6x_3^2 = 7 \end{pmatrix}$

4.101. Minimize $f(x_1, x_2, x_3) = 1/(1 + x_1^2 + x_2^2 + x_3^2)$

Subject to $\begin{pmatrix} x_1 + 2x_2 + 3x_3 = 0 \\ 4x_1^2 + 5x_2^2 + 6x_3^2 \leq 7 \end{pmatrix}$

4.102. Maximize $f(x_1, x_2, x_3) = x_1 x_2 + x_3^2$
Subject to $x_1^2 + 2x_2^2 + 3x_3^2 = 4$

4.103. Minimize $f(x_1, x_2, x_3) = x_1 x_2 + x_3^2$
Subject to $x_1^2 + 2x_2^2 + 3x_3^2 \geq 4$

4.104. Minimize $f(x_1, x_2, x_3) = x_1^2 + 9x_2^2 + x_3^2$
Subject to $x_1 x_2 \geq 1$

4.105. Minimize $f(x_1, x_2, x_3) = x_1^2 + 9x_2^2 + x_3^2$
Subject to $x_1 x_2 x_3 \geq 1$

4.106. Maximize $f(x_1, x_2, x_3) = 7 \times 10^{-9} x_1^4 x_2 x_3^2$

Subject to $\begin{pmatrix} (x_1^2 x_3^2)/10^7 \leq 0.7 \\ x_1^2 x_2 = 700 \\ x_1 \leq 7 \end{pmatrix}$

Lagrangian Duality

Construct dual problems for the following optimization problems. Write explicit dual functions, if possible; otherwise, state the dual optimization problem in terms of both primal and dual variables. Use either KT conditions or graphical methods to verify that both the primal and the dual problems give the same solution.

4.107. Minimize $f(x, y) = -x - 3y$

Subject to $\begin{pmatrix} x + y = 6 \\ -x + y \leq 4 \end{pmatrix}$

4.108. Minimize $f(x_1, x_2, x_3) = 4x_1 + x_2 + x_3$

Subject to $\begin{pmatrix} x_1 + x_2 + 2x_3 \leq 6 \\ 2x_1 + x_2 - x_3 = 4 \\ x_1 \geq 1 \\ x_3 \geq 3 \end{pmatrix}$

4.109. Minimize $f(x_1, x_2, x_3, x_4) = -x_1 + x_2 + x_3 + 4x_4$

Subject to $\begin{pmatrix} x_1 - 5x_2 + x_3 + 3x_4 = 19 \\ x_1 - 4x_2 + 2x_4 = 5 \\ -4x_2 - 5x_3 + 15x_4 = 10 \\ x_1 \geq 2 \end{pmatrix}$

4.110. Minimize $f(x_1, x_2) = 6x_1 + x_2$

Subject to $\begin{pmatrix} 2x_1 + 7x_2 \geq 3 \\ 2x_1 - x_2 \geq 2 \end{pmatrix}$

4.111. Maximize $f(x, y) = -6x + 9y$

Subject to $\begin{pmatrix} x - y \geq 2 \\ 3x + y \geq 1 \\ 2x - 3y \geq 3 \end{pmatrix}$

4.112. Minimize $f(x, y) = x^2 + 2y^2$

Subject to $\begin{pmatrix} x + y \geq 1 \\ x, y \geq 0 \end{pmatrix}$

4.113. Minimize $f(x, y) = x^2 + 2y^2 - 24x - 20y$

Subject to $\begin{pmatrix} x + 2y \geq 0 \\ x + 2y \leq 9 \\ x + y \leq 8 \\ x + y \geq 0 \end{pmatrix}$

4.114. Minimize $f(x, y) = x^2 + y^2 - \text{Log}[x^2 y^2]$
Subject to $x \leq \text{Log}[y] \quad x \geq 1 \quad y \geq 1$

4.115. Minimize $f(x, y, z) = x + y + z$
Subject to $x^{-2} + x^{-2} y^{-2} + x^{-2} y^{-2} z^{-2} \leq 1$

CHAPTER FIVE

Unconstrained Problems

As seen in Chapter 4, it is possible to solve optimization problems by directly using the optimality conditions. However, setting up and solving the resulting nonlinear system of equations becomes very difficult as the problem size increases. Furthermore, at least for constrained problems, checking the second-order sufficient conditions is usually very difficult. In such cases, one must find all stationary points in order to be absolutely sure that a minimum has been found. Otherwise instead of a minimum point, one could actually end up with a maximum point. Clearly, finding all possible solutions for large systems of nonlinear equations is a daunting task, if not impossible.

Starting from this chapter, the remainder of the book is devoted to the presentation of numerically oriented methods that are suitable for practical optimization problems. Methods for solving unconstrained problems are considered in this chapter. Chapters 6 and 7 present methods for large-scale problems involving linear objective and constraint functions. Chapter 8 presents methods for solving an important class of problems known as quadratic programming in which the objective function is a quadratic function but all constraints are linear functions. The last chapter considers the most general case of nonlinearly constrained optimization problems. As you probably expect, the complexity of the methods increases dramatically from unconstrained problems to linearly constrained and finally to nonlinearly constrained problems. The methods presented in this and the next chapter are fairly well developed.

That is not the case for some of the methods presented in Chapters 7, 8, and 9. Keep in mind that numerical methods are designed to find a local minimum point. With the obvious exception of convex problems, there is no guarantee that a solution returned by these methods is a global minimum. The only way to even come close to a global minimum is to try several different starting points and choose the best among the resulting solutions.

This chapter and the remaining chapters in the book start with relatively simple methods suitable for solving the class of problems considered in that chapter. Methods generally become more advanced as we get deeper into each chapter. Unfortunately, the choice of method most suitable for a given problem is usually not that clear cut. Using a more sophisticated method does not automatically guarantee that it is the best method for a given problem. Numerical performance of different methods depends on the type of functions involved and the chosen starting point. Therefore, it is important to understand strengths and weaknesses of different methods. Also, a certain amount of numerical experience with different methods helps in making the right decision.

Numerical methods for solving unconstrained optimization problems are presented in this chapter. The problem is stated as follows:

Find vector of optimization variables $\mathbf{x}$ that minimizes $f(\mathbf{x})$

The basic iteration for all methods presented in this chapter can be written as follows:

$$\mathbf{x}^{k+1} = \mathbf{x}^k + \alpha_k \mathbf{d}^k \qquad k = 0, 1, \ldots$$

where $\mathbf{d}^k$ is known as the descent direction, and α_k is a scalar known as the step length. The starting point $\mathbf{x}^0$ is usually chosen arbitrarily. At each iteration, a step length and a descent direction are chosen such that $f(\mathbf{x}^{k+1}) < f(\mathbf{x}^k)$.

The iteration is stopped when suitable convergence criteria is satisfied. Since the necessary condition for the minimum of an unconstrained problem is that its gradient is zero at the optimum, the convergence criteria is written as follows:

$$\|\nabla f(\mathbf{x}^{k+1})\| \equiv \sqrt{\left(\frac{\partial f(\mathbf{x}^{k+1})}{\partial x_1}\right)^2 + \left(\frac{\partial f(\mathbf{x}^{k+1})}{\partial x_2}\right)^2 + \cdots + \left(\frac{\partial f(\mathbf{x}^{k+1})}{\partial x_n}\right)^2} \leq \text{tol}$$

where tol is a small tolerance (e.g., 10^{-3}).

The first section presents a simple test to determine if a given direction is a descent direction along which the function value decreases. Once a descent direction is known, the problem of computing an appropriate step length is reduced to finding the minimum of a function of a single variable. This process is known as a line search. It is possible to determine the step length using optimality conditions for the minimum of a function of a single variable. However, since this method requires an explicit expression for the derivative of the one-dimensional function, it is useful only for small problems and hand calculations. The method is described in section 2 and is called analytical line search. Following this, several numerical line search methods that do not require derivatives are presented. These methods include interval search, golden section search, and quadratic interpolation. In the third section, line search methods are combined with methods for determining the descent direction to come up with numerical methods for solving unconstrained optimization problems. The methods discussed include Steepest descent, Conjugate gradient, Modified Newton, and Quasi-Newton.

5.1 Descent Direction

A simple test can be derived to determine if a given direction is a direction of descent. For $\mathbf{d}^k$ to be a descent direction, we must have

$$f(\mathbf{x}^{k+1}) < f(\mathbf{x}^k)$$

or

$$f(\mathbf{x}^k + \alpha_k \mathbf{d}^k) < f(\mathbf{x}^k)$$

Using Taylor series expansion, we have

$$f(\mathbf{x}^k) + \alpha_k \nabla f(\mathbf{x}^k)^T \mathbf{d}^k < f(\mathbf{x}^k)$$

or

$$\alpha_k \nabla f(\mathbf{x}^k)^T \mathbf{d}^k < 0$$

If we restrict the step length α_k to positive values, then we get the following criteria for $\mathbf{d}^k$ to be a descent direction at given point $\mathbf{x}^k$:

$$\nabla f(\mathbf{x}^k)^T \mathbf{d}^k < 0$$

Furthermore, the numerical value of the product $\nabla f(\mathbf{x}^k)^T \mathbf{d}^k$ indicates how fast the function is decreasing along this direction.

Chapter 5 Unconstrained Problems

The following function, included in the OptimizationToolbox 'Unconstrained' package, employs this criteria to determine if a given direction is a descent direction:

```
Needs["OptimizationToolbox`Unconstrained`"];
?DescentDirectionCheck
```

DescentDirectionCheck[f, pt, d, vars] Checks to see if a given direction
   is a descent direction for function f. pt = current design point, d =
   direction vector, and vars = list of variables. The function returns
   {status, ∇ f.d} where status is either Descent or NotDescent, ∇ f
   is the gradient at given point and ∇ f.d is the dot product of the
   gradient vector and the given direction vector.

Example 5.1 For the following function of two variables, check if the directions $\mathbf{d}_1$, $\mathbf{d}_2$, and $\mathbf{d}_3$ are directions of descent or not at the given point $\mathbf{x}^k$:

```
f = (x₁² + x₂ - 11)² + (x₁ + x₂² - 7);
vars = {x₁, x₂};
xk = {1, 2};
d₁ = {1, 1}; d₂ = {-1, 1}; d₃ = {31, 12};
```

The DescentDirectionCheck is used to see if d_1 is a descent direction at xk.

DescentDirectionCheck[f, xk, d₁, vars];

$\nabla f \to \begin{pmatrix} 1 - 44x_1 + 4x_1^3 + 4x_1 x_2 \\ -22 + 2x_1^2 + 4x_2 \end{pmatrix}$ $\nabla f \{1., 2.\} \to \begin{pmatrix} -31 \\ -12 \end{pmatrix}$

∇f.d → -43   Status → Descent

Similarly, with directions d_2 and d_3, we get

DescentDirectionCheck[f, xk, d₂, vars];
∇f.d → 19   Status → NotDescent

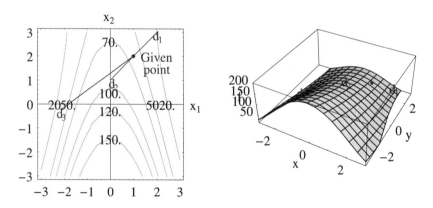

FIGURE 5.1 Graphical illustration of three directions from the given point.

```
DescentDirectionCheck[f, xk, d₃, vars];
```
∇f.d → -1105   Status → Descent

Notice that the direction d_3 is the negative of the gradient at point xk. It is always a descent direction. Furthermore, the product $\nabla f^T \mathbf{d}$ has the largest negative value as well, indicating that the negative gradient direction is the direction of *steepest descent*. The computed results are supported by the contour and surface plots of the function shown in Figure 5.1.

Example 5.2 For the following function of three variables, check if the directions $\mathbf{d}_1$, $\mathbf{d}_2$, and $\mathbf{d}_3$ are directions of descent or not at the given point $\mathbf{x}^k$:

```
f = (x₁ - 1)⁴ + (x₂ - 3)² + 4(x₃ + 5)⁴;
vars = {x₁, x₂, x₃};
xk = {4, 2, -1};
d₁ = {-1, 2, 1}; d₂ = {-1, 10, -1}; d₃ = {-108, 2, -1024};
DescentDirectionCheck[f, xk, d₁, vars];
```

$$\nabla f \to \begin{pmatrix} -4 + 12x_1 - 12x_1^2 + 4x_1^3 \\ -6 + 2x_2 \\ 2000 + 1200x_3 + 240x_3^2 + 16x_3^3 \end{pmatrix} \quad \nabla f\{4., 2., -1.\} \to \begin{pmatrix} 108 \\ -2 \\ 1024 \end{pmatrix}$$

∇f.d → 912   Status → NotDescent

```
DescentDirectionCheck[f, xk, d₂, vars];
```
∇f.d → -1152   Status → Descent

```
DescentDirectionCheck[f, xk, d₃, vars];
```
∇f.d → -1060244   Status → Descent

As before, this example demonstrates that the negative gradient direction is the steepest descent direction.

5.2 Line Search Techniques—Step Length Calculations

At each iteration of a numerical optimization method, we need to determine a descent direction and an appropriate step length. The step length calculations are discussed in this section. Methods for determining descent directions are discussed in the later sections.

At the $(k+1)$ iteration, with a known descent direction, the minimization problem reduces to

Find α in order to minimize $f(\mathbf{x}^{k+1}) = f(\mathbf{x}^k + \alpha_k \mathbf{d}^k) \equiv \phi(\alpha)$

where the subscript k on α is dropped for convenience. The problem therefore reduces to finding minimum of a function of a single variable. To find this minimum, an analytical and several numerical methods are presented in the following subsections. Note that since the intention is to use these methods for computing step length, it will be assumed that we are interested in a positive value of α, usually in the neighborhood of $\alpha = 1$.

5.2.1 Analytical Line Search

If an explicit expression for $\phi(\alpha)$ is available, the optimum step length can easily be found from the necessary and sufficient conditions for the minimum of a function of a single variable, namely

$$\frac{d\phi}{d\alpha} = 0 \quad \text{and} \quad \frac{d^2\phi}{d\alpha^2} > 0$$

Example 5.3 For the following function, compute the optimum step length along the given direction using an analytical line search:

```
f = (x₁² + x₂ - 11)² + (x₁ + x₂² - 7);
vars = {x₁, x₂};
xk = {1, 2}; d = {1, 2};
```

We first check to see if the given direction is a descent direction or not.

DescentDirectionCheck[f, xk, d, vars];

$\nabla f \to \begin{pmatrix} 1 - 44x_1 + 4x_1^3 + 4x_1 x_2 \\ -22 + 2x_1^2 + 4x_2 \end{pmatrix}$ $\nabla f \{1., 2.\} \to \begin{pmatrix} -31 \\ -12 \end{pmatrix}$

$\nabla f \cdot d \to -55$ Status $\to$ Descent

We now construct the function ϕ by setting $\mathbf{x} = \mathbf{x}^k + \alpha \mathbf{d}^k$, as follows:

```
xk1 = xk + αd;
φ = Expand[f/.Thread[vars → xk1]]
```

$62 - 55\alpha + 4\alpha^2 + 8\alpha^3 + \alpha^4$

We compute the step length by solving the equation $d\phi/d\alpha = 0$.

5.2 Line Search Techniques—Step Length Calculations

```
sol = FindRoot[Evaluate[D[ϕ, α] == 0], {α, 1}]
{α → 1.24639}
```

We check the second-order necessary condition to make sure that the computed α is really a minimum.

```
D[ϕ, {α, 2}]/.sol
86.4687
```

A positive value indicates that we have a minimum value of α. Thus, the following is the next point along the given direction:

```
newpt = xk1/.sol
{2.24639, 4.49278}
```

By evaluating f at the given point and this new point, we see that the objective value indeed is reduced.

```
f/.{Thread[vars → xk], Thread[vars → newpt]}
{62, 17.5658}
```

As illustrated in Figure 5.2, the step computed by the line search procedure is at a point where the given direction is tangent to one of the objective function contours.

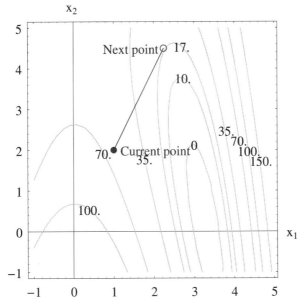

FIGURE 5.2 Optimum step length from the current point along a given descent direction.

Direct Computation of $\frac{d\phi}{d\alpha}$

As seen from the above example, in order to solve for α, we need $\frac{d\phi}{d\alpha}$. Instead of first developing an expression for $\phi(\alpha)$ and then differentiating it, it is possible to develop an expression directly for $\frac{d\phi}{d\alpha}$ as follows:

$$\text{Minimize } f(\mathbf{x}^{k+1}) = f(\mathbf{x}^k + \alpha \mathbf{d}^k) = \phi(\alpha)$$

Using the chain rule of differentiation, and treating f as a function of α and $\mathbf{x} = \mathbf{x}^k + \alpha \mathbf{d}^k$, we have

$$\frac{d\phi}{d\alpha} = \frac{\partial f}{\partial x_1}\frac{\partial x_1}{\partial \alpha} + \frac{\partial f}{\partial x_2}\frac{\partial x_2}{\partial \alpha} + \cdots \equiv \nabla f\left(\mathbf{x}^{k+1}\right)^T \mathbf{d}^k$$

That is, $\frac{d\phi}{d\alpha}$ is written directly by taking the dot product of the gradient of the function at the new point $(\mathbf{x}^k + \alpha \mathbf{d}^k)$ with the direction vector.

The following function, included in the OptimizationToolbox 'Unconstrained' package, implements this procedure to determine the optimum step length.

Needs["OptimizationToolbox`Unconstrained`"];
?AnalyticalLineSearch

AnalyticalLineSearch[gradf, pt, d, vars, prResults:True, opts] computes
    optimum step length using analytical approach. gradf = gradient of
    given function, pt = current design point, d = direction vector,
    vars = list of variables. See Options[AnalyticalLineSearch] for valid
    options and their usage.

OptionsUsage[AnalyticalLineSearch]
{SecantPoints → {0, 0.1}, StepLengthVar → α}

SecantPoints is an option for AnalyticalLineSearch. This option is used
    to specify two initial values to start the Secant method for finding
    root. Default {0, .1}.

StepLengthVar is an option for several unconstrained optimization
    methods. It specifies the symbol used for step length variable.
    Default is StepLengthVar→ α.

Example 5.4 For the following function, compute optimum step length along the given direction using an analytical line search.

```
f = (x₁² + x₂ - 11)² + (x₁ + x₂² - 7);
vars = {x₁, x₂};
xk = {1, 2}; d = {1, 2};
```

5.2 Line Search Techniques—Step Length Calculations

```
df = Grad[f, vars]; MatrixForm[df]
```
$$\begin{pmatrix} 1 - 44x_1 + 4x_1^3 + 4x_1 x_2 \\ -22 + 2x_1^2 + 4x_2 \end{pmatrix}$$

```
DescentDirectionCheck[f, xk, d, vars];
```

$\nabla f \to \begin{pmatrix} 1 - 44x_1 + 4x_1^3 + 4x_1 x_2 \\ -22 + 2x_1^2 + 4x_2 \end{pmatrix} \quad \nabla f\{1., 2.\} \to \begin{pmatrix} -31 \\ -12 \end{pmatrix}$

$\nabla f.d \to -55.$ Status $\to$ Descent

```
step = AnalyticalLineSearch[df, xk, d, vars];
```

$xk1 \to \begin{pmatrix} 1 + \alpha \\ 2 + 2\alpha \end{pmatrix}$

$\nabla f(xk1) \to \begin{pmatrix} -31 - 16\alpha + 20\alpha^2 + 4\alpha^3 \\ 2(-6 + 6\alpha + \alpha^2) \end{pmatrix}$

$d\phi/d\alpha \equiv \nabla f(xk1).d = 0 \to -55 + 8\alpha + 24\alpha^2 + 4\alpha^3 == 0$

$\alpha \to 1.24639$

This step length is the same as that computed in the previous example.

Example 5.5 For the following function, compute optimum step length along the given direction using an analytical line search.

```
f = (x1 - 1)^4 + (x2 - 3)^2 + 4(x3 + 5)^4;
vars = {x1, x2, x3};
xk = {4, 2, -1}; d = {1, 2, -3};
```

```
df = Grad[f, vars]; MatrixForm[df]
```
$$\begin{pmatrix} -4 + 12x_1 - 12x_1^2 + 4x_1^3 \\ -6 + 2x_2 \\ 2000 + 1200x_3 + 240x_3^2 + 16x_3^3 \end{pmatrix}$$

```
DescentDirectionCheck[f, xk, d, vars];
```

$\nabla f \to \begin{pmatrix} -4 + 12x_1 - 12x_1^2 + 4x_1^3 \\ -6 + 2x_2 \\ 2000 + 1200x_3 + 240x_3^2 + 16x_3^3 \end{pmatrix} \quad \nabla f\{4., 2., -1.\} \to \begin{pmatrix} 108 \\ -2 \\ 1024 \end{pmatrix}$

$\nabla f.d \to -2968.$ Status $\to$ Descent

```
step = AnalyticalLineSearch[df, xk, d, vars];
```

$xk1 \to \begin{pmatrix} 4 + \alpha \\ 2 + 2\alpha \\ -1 - 3\alpha \end{pmatrix}$

$\nabla f(xk1) \to \begin{pmatrix} 4(3 + \alpha)^3 \\ -2 + 4\alpha \\ -16(-4 + 3\alpha)^3 \end{pmatrix}$

```
dϕ/dα ≡ ∇f(xk1).d = 0 → 4 (-742 + 1757α - 1287α² + 325α³) == 0
α → 0.78094
```

newpt = xk + stepd
{4.78094, 3.56188, -3.34282}

By evaluating f at the given point and at this new point, we see that the objective value indeed is reduced.

f/.{Thread[vars → xk], Thread[vars → newpt]}
{1106, 234.845}

5.2.2 Equal Interval Search

The analytical method requires an explicit expression for $\phi(\alpha)$ or $d\phi/d\alpha$. For a purely numerical solution, such expressions are not available and hence one must resort to numerical line search techniques.

The simplest line search technique is the equal interval search. In this approach, bounds are found for the minimum of $\phi(\alpha)$. By successively refining these bounds, the minimum is bracketed to any desired degree of precision. Starting from a given lower bound (say $\alpha = 0$) and an interval step parameter $\delta > 0$ (say $\delta = 0.5$), we can compute the bounds as follows:

1. Set $\alpha_1 = $ initial value of α. Compute $\phi(\alpha_1)$.
2. Set $\alpha_2 = \alpha_1 + \delta$. Compute $\phi(\alpha_2)$.

As illustrated in Figure 5.3, one of the following two situations is possible.

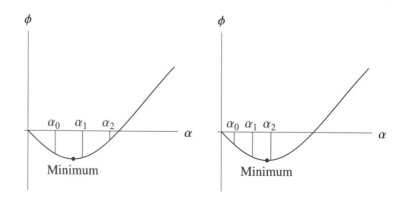

FIGURE 5.3 Illustration of two situations occurring during interval search.

5.2 Line Search Techniques—Step Length Calculations

3. If $\phi(\alpha_2) \leq \phi(\alpha_1)$, then the function is continuing to decrease. The minimum is either between α_1 and α_2 or has not yet been bracketed. Therefore, set $\alpha_1 = \alpha_2$ and go to step 2.

4. If $\phi(\alpha_2) > \phi(\alpha_1)$, we have gone past the minimum. The minimum must be in the interval just tried or in the one before. Thus, we have found the upper and lower limits of the interval in which the minimum lies.

$$\alpha_l \equiv \alpha_1 - \delta \leq \alpha_{min} \leq \alpha_u \equiv \alpha_2$$

The interval step parameter δ is reduced to δ/F, where F is a refinement parameter (say $F = 10$) and the process is repeated. The convergence is achieved when $I = (\alpha_u - \alpha_l)$ is reduced to a specified tolerance. After convergence, the minimum value is set to the average of the upper and lower bounds.

$$\alpha_{min} = \frac{\alpha_u + \alpha_l}{2}$$

The procedure is implemented in the following function that is included in the OptimizationToolbox 'Unconstrained' package.

```
Needs["OptimizationToolbox`Unconstrained`"];
?EqualIntervalSearch
```

```
EqualIntervalSearch[ϕ, α, {αi, δ, F}, tol:10⁻³] --- Determines minimum
   of ϕ(α) using Equal Interval Search. Arguments have the following
   meaning. ϕ = function of single variable α, {αi, δ, F} initial value
   of α, δ = initial delta, and F = refinement factor, respectively, tol
   (optional) = convergence tolerance (default is 10⁻³). The function
   returns {Estimated minimum point, Interval containing minimum}.
```

Example 5.6 Determine a minimum of the following function using equal interval search:

```
ϕ = 1 - 1/(1 - α + 2α²);
```

A plot of the function ϕ is shown in Figure 5.4. The plot shows that a local minimum exists near $\alpha = 0.25$.

We use EqualIntervalSearch function to compute the minimum, starting with $\alpha = 0$, $\delta = 0.5$, $F = 5$, and tol = 0.01.

```
EqualIntervalSearch[ϕ, α, {0, 0.5, 5}, 0.01]
δ → 0.5
     α      ϕ(α)
     0      0
     0.5    0.
     1.     0.5
```

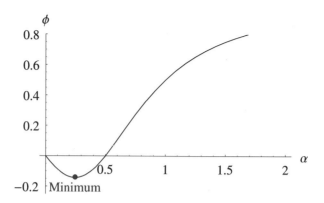

FIGURE 5.4 Plot of $1 - \frac{1}{2\alpha^2 - \alpha + 1}$.

```
Bounds → {0., 1.}
δ → 0.1
    α        φ(α)
    0.       0.
    0.1     -0.0869565
    0.2     -0.136364
    0.3     -0.136364
    0.4     -0.0869565
Bounds → {0.2, 0.4}
δ → 0.02
    α        φ(α)
    0.2     -0.136364
    0.22    -0.140511
    0.24    -0.142596
    0.26    -0.142596
    0.28    -0.140511
Bounds → {0.24, 0.28}
δ → 0.004
    α        φ(α)
    0.24    -0.142596
    0.244   -0.142763
    0.248   -0.142847
    0.252   -0.142847
    0.256   -0.142763
Bounds → {0.248, 0.256}
{0.252, {0.248, 0.256}}
```

Example 5.7 Determine a minimum of the following function using equal interval search:

$\phi = 2 - 4\alpha + \text{Exp}[\alpha]$;

A plot of the function ϕ is shown in Figure 5.5. The plot shows that a local minimum exists near $\alpha = 1.4$.

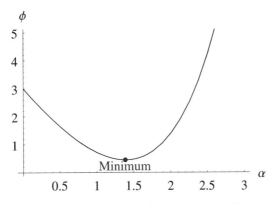

FIGURE 5.5 Plot of $2 - 4\alpha + \text{Exp}[\alpha]$.

We use EqualIntervalSearch function to compute the minimum, starting with $\alpha = 0$, $\delta = 0.5$, $F = 10$, and tol $= 0.01$.

```
EqualIntervalSearch[ϕ, α, {0., 0.5, 10.}, .01]
```

$\delta \to 0.5$

α	$\phi(\alpha)$
0.	3.
0.5	1.64872
1.	0.718282
1.5	0.481689
2.	1.38906

Bounds $\to \{1., 2.\}$

$\delta \to 0.05$

α	$\phi(\alpha)$
1.	0.718282
1.05	0.657651
1.1	0.604166
1.15	0.558193
1.2	0.520117
1.25	0.490343
1.3	0.469297
1.35	0.457426
1.4	0.4552
1.45	0.463115

Bounds → {1.35, 1.45}

δ → 0.005

α	$\phi(\alpha)$
1.35	0.457426
1.355	0.456761
1.36	0.456193
1.365	0.455723
1.37	0.455351
1.375	0.455077
1.38	0.454902
1.385	0.454826
1.39	0.45485

Bounds → {1.38, 1.39}

{1.385, {1.38, 1.39}}

5.2.3 Section Search

The equal interval search is straightforward but requires a large number of function evaluations before it locates a minimum. A considerably more efficient procedure is to use section search in which the initial set of bounds (α_l, α_u) are computed using the same procedure as the equal interval search. However, the bounds are then refined by taking two points, denoted by α_a and α_b, within these bounds and evaluating the function at these two points. If the intermediate points are placed at the third points between the bounds, then

$$\alpha_a = \alpha_l + (\alpha_u - \alpha_l)/3 \quad \alpha_b = \alpha_l + 2(\alpha_u - \alpha_l)/3$$

As illustrated in Figure 5.6, one of the following two situations is possible.

1. If $\phi(\alpha_a) < \phi(\alpha_b)$, then it is clear from the figure that the minimum must be in the interval (α_l, α_b).
2. If $\phi(\alpha_a) \geq \phi(\alpha_b)$, then the minimum must be in the interval (α_a, α_u).

Thus, we have found new bounds for the minimum. The process can now be repeated with these new lower and upper limits. Clearly, at each step the interval in which the minimum lies is reduced by 1/3.

As before, the convergence is achieved when $I = (\alpha_u - \alpha_l)$ is reduced to a specified tolerance. After convergence, the minimum value is set to the

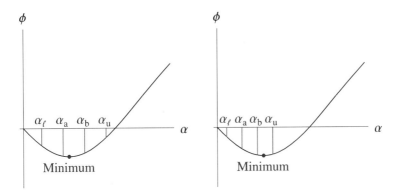

FIGURE 5.6 Illustration of two situations occurring during section search.

average of the upper and lower bounds.

$$\alpha_{\min} = \frac{\alpha_u + \alpha_l}{2}$$

The procedure is implemented in the following function that is included in the OptimizationToolbox 'Unconstrained' package.

```
Needs["OptimizationToolbox`Unconstrained`"];
?SectionSearch
```

SectionSearch[ϕ, α, {αi, δ },tol:10⁻³] --- Determines minimum of ϕ (α) using Section Search. Arguments have the following meaning. ϕ = function of single variable α, {αi, δ } initial value of α, δ = initial delta. tol (optional) = convergence tolerance (default is 10⁻³). The function returns {Estimated minimum point, Interval containing minimum}.

Example 5.8 Determine a minimum of the following function using section search:

ϕ = 1 - 1/(1 - α + 2α²);

We use SectionSearch function to compute the minimum starting with $\alpha = 0$, $\delta = 0.1$, and tol $= 0.01$.

```
SectionSearch[ϕ, α, {0, 0.1}, 0.01]
```

******** Bounding phase ********

δ → 0.1

α	ϕ(α)
0	0
0.1	-0.0869565
0.2	-0.136364
0.3	-0.136364
0.4	-0.0869565

Bounds → {0.2, 0.4}

******** Refinement phase ********

$α_L$	$α_U$	$α_a$	$α_b$	ϕ($α_a$)	ϕ($α_b$)	I
0.2	0.4	0.266667	0.333333	-0.142132	-0.125	0.2
0.2	0.333333	0.244444	0.288889	-0.142777	-0.13892	0.133333
0.2	0.288889	0.22963	0.259259	-0.141774	-0.142633	0.0888889
0.22963	0.288889	0.249383	0.269136	-0.142856	-0.141901	0.0592593
0.22963	0.269136	0.242798	0.255967	-0.142722	-0.142764	0.0395062
0.242798	0.269136	0.251578	0.260357	-0.142851	-0.142577	0.0263374
0.242798	0.260357	0.248651	0.254504	-0.142852	-0.142804	0.0175583
0.242798	0.254504	0.2467	0.250602	-0.142829	-0.142856	0.0117055
0.2467	0.254504	0.249301	0.251903	-0.142856	-0.142848	0.00780369

{0.249301, {0.2467, 0.251903}}

Example 5.9 Determine the minimum of the following function using section search.

```
ϕ = 2 - 4α + Exp[α];
```

We use SectionSearch function to compute the minimum starting with $α = 0$, $δ = 0.5$, and tol $= 0.01$.

```
SectionSearch[ϕ, α, {0, 0.5}, 0.01]
```

******** Bounding phase ********

δ → 0.5

α	ϕ(α)
0	3
0.5	1.64872
1.	0.718282
1.5	0.481689
2.	1.38906

Bounds → {1., 2.}

5.2 Line Search Techniques—Step Length Calculations

******** Refinement phase ********

$$\begin{pmatrix} \alpha_L & \alpha_U & \alpha_a & \alpha_b & \phi(\alpha a) & \phi(\alpha b) & I \\ 1. & 2. & 1.33333 & 1.66667 & 0.460335 & 0.627823 & 1. \\ 1. & 1.66667 & 1.22222 & 1.44444 & 0.505834 & 0.461718 & 0.666667 \\ 1.22222 & 1.66667 & 1.37037 & 1.51852 & 0.455327 & 0.491382 & 0.444444 \\ 1.22222 & 1.51852 & 1.32099 & 1.41975 & 0.46317 & 0.457087 & 0.296296 \\ 1.32099 & 1.51852 & 1.38683 & 1.45267 & 0.454823 & 0.463834 & 0.197531 \\ 1.32099 & 1.45267 & 1.36488 & 1.40878 & 0.455733 & 0.455841 & 0.131687 \\ 1.32099 & 1.40878 & 1.35025 & 1.37952 & 0.45739 & 0.454914 & 0.0877915 \\ 1.35025 & 1.40878 & 1.36976 & 1.38927 & 0.455366 & 0.45484 & 0.0585277 \\ 1.36976 & 1.40878 & 1.38277 & 1.39577 & 0.454847 & 0.455003 & 0.0390184 \\ 1.36976 & 1.39577 & 1.37843 & 1.3871 & 0.454946 & 0.454824 & 0.0260123 \\ 1.37843 & 1.39577 & 1.38421 & 1.38999 & 0.454831 & 0.45485 & 0.0173415 \\ 1.37843 & 1.38999 & 1.38229 & 1.38614 & 0.454855 & 0.454823 & 0.011561 \\ 1.38229 & 1.38999 & 1.38485 & 1.38742 & 0.4548267 & 0.454825 & 0.00770735 \end{pmatrix}$$

{1.38742, {1.38485, 1.38999}}

5.2.4 The Golden Section Search

The golden section search is similar to the basic section search but with two main differences that make it a very clever interval search method. In the bounding phase, instead of adding a fixed δ to the previous α, the α_2 is computed as follows:

$$\alpha_2 = \alpha_1 + \tau^{n-1}\delta$$

where n is an iteration counter ($n = 1, 2, \ldots$) and τ is the golden ratio given by

$$\tau = \frac{1 + \sqrt{5}}{2} \approx 1.618$$

Since τ is greater than 1, the steps get larger and larger as the iteration progresses. Thus, the bounds are located quickly regardless of how small δ value is selected. The complete bounding phase algorithm is as follows:

Given interval step parameter $\delta > 0$ (say $\delta = 0.5$)

1. Set $\alpha_1 =$ initial value of α. Compute $\phi(\alpha_1)$. Set $n = 1$ an iteration counter.
2. Set $\alpha_2 = \alpha_1 + \tau^{n-1}\delta$. Compute $\phi(\alpha_2)$.

Chapter 5 Unconstrained Problems

3. If $\phi(\alpha_2) \leq \phi(\alpha_1)$, the minimum has not been surpassed. Therefore, set $\alpha_1 = \alpha_2$ and go to step 2. Otherwise, the minimum has been surpassed and must lie in the following interval:

$$\alpha_l \equiv \alpha_1 - \delta \leq \alpha_{\min} \leq \alpha_u \equiv \alpha_2$$

In the refinement phase, the two intermediate points are placed as follows:

$$\alpha_a = \alpha_l + \left(1 - \frac{1}{\tau}\right)(\alpha_u - \alpha_l) \quad \text{and} \quad \alpha_b = \alpha_l + (\alpha_u - \alpha_l)/\tau$$

As will be seen from the numerical examples, the advantage of this scheme is that one of the two intermediate points is always the same as the one used in the previous iteration. Thus, each iteration requires only one new function evaluation.

As before, the convergence is achieved when $I = (\alpha_u - \alpha_l)$ is reduced to a specified tolerance. After convergence, the minimum value is set to the average of the upper and lower bounds.

$$\alpha_{\min} = \frac{\alpha_u + \alpha_l}{2}$$

The procedure is implemented in the following function that is included in the OptimizationToolbox 'Unconstrained' package:

```
Needs["OptimizationToolbox`Unconstrained`"];
?GoldenSectionSearch
```

```
GoldenSectionSearch[ϕ, α, {αi, δ }, tol:10⁻³ ]  ---  Determines minimum
of ϕ (α) using Golden Section Search. Arguments have the following
meaning. ϕ = function of single variable α, {αi, δ } initial value of
α, δ = initial delta. tol (optional) = convergence tolerance (default
is 10⁻³). The function returns {Estimated minimum point, Interval
containing minimum}.
```

Example 5.10 Determine a minimum of the following function using the golden section search:

```
ϕ = 1 - 1/(1 - α + 2α²);
```

We use GoldenSectionSearch function to compute the minimum starting with $\alpha = 0$, $\delta = 0.1$, and tol = 0.01.

```
GoldenSectionSearch[ϕ, α, {0, 0.1}, 0.01]
```

******** Bounding phase ********

$\delta \to 0.1$

α	$\phi(\alpha)$
0	0
0.1	-0.0869565
0.261803	-0.142493
0.523607	0.024125

Bounds → {0.1, 0.523607}

******** Refinement phase ********

$$\begin{pmatrix} \alpha_L & \alpha_U & \alpha_a & \alpha_b & \phi(\alpha a) & \phi(\alpha b) & I \\ 0.1 & 0.523607 & 0.261803 & 0.361803 & -0.142493 & -0.111111 & 0.423607 \\ 0.1 & 0.361803 & 0.2 & 0.261803 & -0.136364 & -0.142493 & 0.261803 \\ 0.2 & 0.361803 & 0.261803 & 0.3 & -0.142493 & -0.136364 & 0.161803 \\ 0.2 & 0.3 & 0.238197 & 0.261803 & -0.142493 & -0.142493 & 0.1 \\ 0.238197 & 0.3 & 0.261803 & 0.276393 & -0.142493 & -0.14104 & 0.0618034 \\ 0.238197 & 0.276393 & 0.252786 & 0.261803 & -0.142837 & -0.142493 & 0.0381966 \\ 0.238197 & 0.261803 & 0.247214 & 0.252786 & -0.142837 & -0.142837 & 0.0236068 \\ 0.247214 & 0.261803 & 0.252786 & 0.256231 & -0.142837 & -0.142756 & 0.0145898 \\ 0.247214 & 0.256231 & 0.250658 & 0.252786 & -0.142856 & -0.142837 & 0.00901699 \end{pmatrix}$$

{0.251722, {0.247214, 0.256231}}

Notice that one of the two intermediate points (α_a or α_b) is always the same as the one used in the previous iteration. Thus, each iteration requires only one new function evaluation. Also, notice that the overall number of iterations is smaller than the earlier methods.

Example 5.11 Determine a minimum of the following function using the golden section search.

$\phi = 2 - 4\alpha + \text{Exp}[\alpha];$

We use GoldenSectionSearch function to compute the minimum starting with $\alpha = 0$, $\delta = 0.5$, and tol = 0.01.

GoldenSectionSearch[$\phi, \alpha, \{0, 0.5\}, 0.01$]

******** Bounding phase ********

$\delta \to 0.5$

α	$\phi(\alpha)$
0	3
0.5	1.64872
1.30902	0.466464
2.61803	5.23661

Bounds → {0.5, 2.61803}

******** Refinement phase ********

$$\begin{pmatrix} \alpha_L & \alpha_U & \alpha_a & \alpha_b & \phi(\alpha a) & \phi(\alpha b) & I \\ 0.5 & 2.61803 & 1.30902 & 1.80901 & 0.466464 & 0.868376 & 2.11803 \\ 0.5 & 1.80902 & 1. & 1.30901 & 0.718282 & 0.466464 & 1.30902 \\ 1. & 1.80902 & 1.30902 & 1.49999 & 0.466464 & 0.481689 & 0.809017 \\ 1. & 1.5 & 1.19098 & 1.30901 & 0.526382 & 0.466464 & 0.5 \\ 1.19098 & 1.5 & 1.30902 & 1.38196 & 0.466464 & 0.45486 & 0.309017 \\ 1.30902 & 1.5 & 1.38197 & 1.42705 & 0.45486 & 0.45819 & 0.190983 \\ 1.30902 & 1.42705 & 1.3541 & 1.38196 & 0.456873 & 0.45486 & 0.118034 \\ 1.3541 & 1.42705 & 1.38197 & 1.39918 & 0.45486 & 0.455156 & 0.072949 \\ 1.3541 & 1.39919 & 1.37132 & 1.38196 & 0.455269 & 0.45486 & 0.045085 \\ 1.37132 & 1.39919 & 1.38197 & 1.38854 & 0.45486 & 0.454833 & 0.027864 \\ 1.38197 & 1.39919 & 1.38854 & 1.39260 & 0.454833 & 0.454902 & 0.0172209 \\ 1.38197 & 1.39261 & 1.38603 & 1.38854 & 0.454823 & 0.454833 & 0.0106431 \\ 1.38197 & 1.38854 & 1.38448 & 1.38603 & 0.454829 & 0.454823 & 0.00657781 \end{pmatrix}$$

{1.38525, {1.38197, 1.38854}}

5.2.5 The Quadratic Interpolation Method

All the interval search methods discussed so far compute function values at predetermined locations without explicitly considering the form of the function itself, and thus, could be very slow to converge. Another class of methods known as interpolation methods are based on fitting a polynomial function through a given number of points. As the name indicates, the quadratic interpolation method uses three given points and fits a quadratic function through these points. The minimum of this quadratic function is computed using necessary conditions. A new set of three points is selected by comparing function values at this minimum point with the given three points. The decision process is similar to the one used in the section search. The process is repeated with the three new points until the interval in which the minimum lies becomes fairly small. Similar to the golden section search, this method requires only one new function evaluation at each iteration. As the interval becomes small, the quadratic approximation becomes closer to the actual function, which speeds up convergence.

Main Step

Assuming that we are given three points α_l, α_m, and α_u, a quadratic function ϕ_q passing through the corresponding function values ϕ_l, ϕ_m, and ϕ_u is given by the following equation:

$$\phi_q(\alpha) = \phi_l \frac{(\alpha - \alpha_m)(\alpha - \alpha_u)}{(\alpha_l - \alpha_m)(\alpha_l - \alpha_u)} + \phi_m \frac{(\alpha - \alpha_l)(\alpha - \alpha_u)}{(\alpha_m - \alpha_l)(\alpha_m - \alpha_u)}$$
$$+ \phi_u \frac{(\alpha - \alpha_l)(\alpha - \alpha_m)}{(\alpha_u - \alpha_l)(\alpha_u - \alpha_m)}$$

The necessary condition for the minimum of this quadratic function is

$$\frac{d\phi_q}{d\alpha} = \phi_l \frac{2\alpha - \alpha_m - \alpha_u}{(\alpha_l - \alpha_m)(\alpha_l - \alpha_u)} + \phi_m \frac{2\alpha - \alpha_l - \alpha_u}{(\alpha_m - \alpha_l)(\alpha_m - \alpha_u)} + \phi_u \frac{2\alpha - \alpha_l - \alpha_m}{(\alpha_u - \alpha_l)(\alpha_u - \alpha_m)} = 0$$

Solving this equation for α, we get the following minimum point denoted by α_q:

$$\alpha_q = \frac{1}{2} \left(\frac{\phi_l(\alpha_m^2 - \alpha_u^2) + \phi_m(\alpha_u^2 - \alpha_l^2) + \phi_u(\alpha_l^2 - \alpha_m^2)}{\phi_l(\alpha_m - \alpha_u) + \phi_m(\alpha_u - \alpha_l) + \phi_u(\alpha_l - \alpha_m)} \right)$$

Knowing the minimum, the next task is to determine which one of the given three points to discard before repeating the process. We have one of the following two situations:

1. $\alpha_q \leq \alpha_m$

    If $\phi(\alpha_m) \geq \phi(\alpha_q)$, then the minimum of the actual function is in the interval (α_l, α_m); therefore, we use $(\alpha_l, \alpha_q, \alpha_m)$ as the three points for the next iteration.

    If $\phi(\alpha_m) < \phi(\alpha_q)$, then the minimum of the actual function is in the interval (α_q, α_u); therefore, we use $(\alpha_q, \alpha_m, \alpha_u)$ as the three points for the next iteration.

2. $\alpha_q > \alpha_m$

    If $\phi(\alpha_m) \geq \phi(\alpha_q)$, then the minimum of the actual function is in the interval (α_m, α_u); therefore, we use $(\alpha_m, \alpha_q, \alpha_u)$ as the three points for the next iteration.

    If $\phi(\alpha_m) < \phi(\alpha_q)$, then the minimum of the actual function is in the interval (α_l, α_q); therefore, we use $(\alpha_l, \alpha_m, \alpha_q)$ as the three points for the next iteration.

Two of these four situations are illustrated in Figure 5.7. The interpolated quadratic function is shown in the lighter shade.

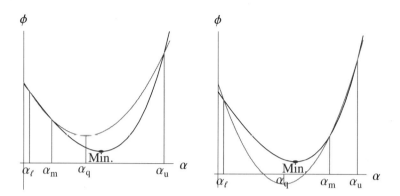

FIGURE 5.7 Illustration of two situations occurring during quadratic interpolation.

Convergence

As the interval in which the minimum lies becomes smaller, the quadratic function becomes closer to the actual function. An indication of this is that at the minimum point of the quadratic function (α_q), both the quadratic function and the actual function have nearly identical values. Using this as convergence criteria, the process is terminated when the following condition is satisfied:

$$\text{Abs}\left[\frac{\phi_q(\alpha_q) - \phi(\alpha_q)}{\phi(\alpha_q)}\right] \leq \text{tol}$$

where tol is a small convergence tolerance.

Choosing Three Initial Points and Establishing Bounds

The procedure is essentially complete, except for the choice of three initial points. Choosing an arbitrary three values of α may cause problems if the denominator of the α_q equation is 0. To see the situation more clearly, assume that the three points are chosen as 0, δ, and 2δ, where δ is a chosen parameter (say $\delta = 1$). In this case, the expression for α_q takes the following form:

$$\alpha_q = \frac{(3\phi_l - 4\phi_m + \phi_u)\delta}{2\phi_l - 4\phi_m + 2\phi_u}$$

For the denominator to be greater than 0, we must have

$$2\phi_l - 4\phi_m + 2\phi_u > 0 \quad \text{or} \quad \frac{\phi_l + \phi_u}{2} > \phi_m$$

Keeping this condition in mind, the following procedure is used to establish three initial points. In addition, the procedure establishes upper and lower bounds for the minimum.

1. Choose initial step δ (say $\delta = 1$). Set $\alpha_l = 0$. Compute $\phi(\alpha_l)$. Set $\alpha_1 = \delta$. Compute $\phi(\alpha_1)$.
2. If $\phi(\alpha_1) > \phi(\alpha_l)$, then the minimum must be at a point less than α_1. Try α_1 as the third point, set $\alpha_u = \alpha_1$, and $\phi(\alpha_u) = \phi(\alpha_1)$. Set the middle point at $\alpha_m = \delta/2$, compute $\phi(\alpha_m)$. Go to step (6).
3. If $\phi(\alpha_1) \leq \phi(\alpha_l)$, then the minimum is not bracketed. We choose α_1 as the middle point, set $\alpha_m = \alpha_1$, and $\phi(\alpha_m) = \phi(\alpha_1)$. For locating the third point, set $\alpha_2 = 2\delta$, compute $\phi(\alpha_2)$, and continue to the next step.
4. If $\phi(\alpha_2) > \phi(\alpha_1)$, then we have located the correct third point. Set $\alpha_u = \alpha_2$, and $\phi(\alpha_u) = \phi(\alpha_2)$, and go to step (6).
5. If $\phi(\alpha_2) < \phi(\alpha_1)$, then the minimum is still not bracketed. Set $\alpha_1 = \alpha_2$, and $\phi(\alpha_1) = \phi(\alpha_2)$, set $\delta = 2\delta$, and go back to step (2).
6. If $\frac{\phi_l + \phi_u}{2} > \phi_m$, we are done. Otherwise, set $\delta = 2\delta$, set $\alpha_1 = \delta$, compute $\phi(\alpha_1)$, and go back to step (2).

The complete solution procedure is implemented in the following function that is included in the OptimizationToolbox 'Unconstrained' package:

Needs["OptimizationToolbox`Unconstrained`"];
?QuadraticSearch

```
QuadraticSearch[ϕ, α, δ:1, maxIter:20, tol:0.001, prResults:True] ---
   Determines minimum of ϕ(α) using Quadratic Interpolation method. ϕ
   = function of single variable α, δ (default = 1) parameter used to
   establish initial three values of α, maxIter (optional) = maximum
   number of iterations allowed (default = 20), tol (optional) =
   convergence tolerance (default is 10^-3). prResults (optional) = If
   True (default) prints all intermediate results. The function returns
   {Estimated minimum point, Interval containing minimum}.
```

Example 5.12 Determine a minimum of the following function using quadratic interpolation.

$\phi = 1 - 1/(1 - \alpha + 2\alpha^2);$

We use QuadraticSearch function to compute the minimum, starting with default $\delta = 1$ and tol = 0.001.

QuadraticSearch[ϕ, α]
**** Iteration 1
$\alpha_l \to 0.$ $\phi_l \to 0.$ $\alpha_1 \to 1.$ $\phi_1 \to 0.5$

Since $\phi_1 > \phi_\ell$, the test in step (2) passes, and we have located suitable bounds and three initial points.

$$\text{Initial three points} \rightarrow \begin{pmatrix} 0. \\ 0.5 \\ 1. \end{pmatrix} \quad \text{Function values} \rightarrow \begin{pmatrix} 0. \\ 0. \\ 0.5 \end{pmatrix}$$

We can now use the logic in the main step to locate the minimum. The computations are summarized as follows:

$$\begin{pmatrix} \alpha_\ell & \alpha_m & \alpha_u & \alpha_q & \phi_\ell & \phi_m & \phi_u & \phi_q & \text{Conv.} \\ 0. & 0.5 & 1. & 0.25 & 0. & 0. & 0.5 & -0.142857 & 0.5625 \\ 0. & 0.25 & 0.5 & 0.25 & 0. & -0.142857 & 0. & -0.142857 & 0. \\ 0.25 & 0.25 & 0.5 & -- & -0.142857 & -0.142857 & 0. & -- & 0. \end{pmatrix}$$

{0.25, {0.25, 0.5}}

As seen from Figure 5.4, the function is fairly flat after $\alpha = 2$ with no minimum; therefore, no bounds can be established with $\delta \geq 2$ using the procedure given in this section. In this situation, one should reduce δ and try again.

QuadraticSearch[$\phi, \alpha, 2, 5$]
```
**** Iteration 1
α_ℓ → 0.  φ_ℓ → 0.  α_1 → 2.   φ_1 → 0.857143
**** Iteration 2
α_ℓ → 0.  φ_ℓ → 0.  α_1 → 4.   φ_1 → 0.965517
**** Iteration 3
α_ℓ → 0.  φ_ℓ → 0.  α_1 → 8.   φ_1 → 0.991736
**** Iteration 4
α_ℓ → 0.  φ_ℓ → 0.  α_1 → 16.  φ_1 → 0.997988
**** Iteration 5
α_ℓ → 0.  φ_ℓ → 0.  α_1 → 32.  φ_1 → 0.999504
Change δ --- No bounds found in 5 tries.
$Aborted
```

The function values at the second point are always larger than the first point. The test in step (2) passes, but that in step (4) fails and a larger value of δ is tried in each successive iteration. The process is finally aborted after five tries fail.

Example 5.13 Determine the minimum of the following function using quadratic interpolation.

$\phi = 2 - 4\alpha + \text{Exp}[\alpha]$;

We use QuadraticSearch function to compute the minimum, starting with the default $\delta = 1$ and tol $= 0.001$.

```
QuadraticSearch[ϕ, α]
**** Iteration 1
α_l → 0.   ϕ_l → 3.   α_1 → 1.   ϕ_1 → 0.718282
α_2 → 2.   ϕ_2 → 1.38906
```

$$\text{Initial three points} \to \begin{pmatrix} 0. \\ 1. \\ 2. \end{pmatrix} \quad \text{Function values} \to \begin{pmatrix} 3. \\ 0.718282 \\ 1.38906 \end{pmatrix}$$

α_l	α_m	α_u	α_q	ϕ_l	ϕ_m	ϕ_u	ϕ_q	Conv.
0.	1.	2.	1.27281	3.	0.718282	1.38906	0.479632	0.268495
1.	1.27281	2.	1.3422	0.718282	0.479632	1.38906	0.458655	0.0234257
1.27281	1.3422	2.	1.37153	0.479632	0.458655	1.38906	0.455256	0.00300283
1.3422	1.37153	2.	1.38066	0.458655	0.455256	1.38906	0.454886	0.000368196
1.37153	1.38066	2.	--	0.455256	0.454886	1.38906	--	0.000368196

{1.38066, {1.37153, 2.}}

5.2.6 Approximate Line Search Based on Armijo's Rule

The line search techniques considered so far attempt to find an exact minimum of function $\phi(\alpha)$. As seen from the numerical examples, all methods require a large number of function evaluations. For use as step length, it is desirable to obtain a reasonable value of α without too many calculations. Thus, the approximate line search methods are popular for use with optimization algorithms. Instead of finding the true minimum, these methods terminate when the step length is within a specified percentage of the exact minimum.

A popular rule, known as Armijo's rule, says to accept α if the following two conditions are satisfied:

$$\phi(\alpha) \leq \phi(0) + \alpha \epsilon \phi'(0) \quad \text{and} \quad \phi(\eta \alpha) \leq \phi(0) + \alpha \eta \epsilon \phi'(0)$$

where $0 < \epsilon < 1$ and $\eta > 1$ are user-specified parameters. Usually $\epsilon = 0.2$ and $\eta = 2$ are selected. As illustrated in Figure 5.8, the upper limit on the step length is where $\phi(0)$ line intersects the graph of ϕ. Assuming the minimum exists, the slope of the function should be negative at $\alpha = 0$. Thus, the conditions define an acceptable step length in the range α_a and α_b.

Based on this criteria, an approximate line search algorithm can be defined as follows:

1. Start with an arbitrary value of α.
2. Compute $\phi(\alpha)$. If $\phi(\alpha) \leq \phi(0) + \alpha \epsilon \phi'(0)$, increase α by setting it to $\eta \alpha$, and repeat until the test fails. Then the step length is equal to the previous value of α.

252　Chapter 5　Unconstrained Problems

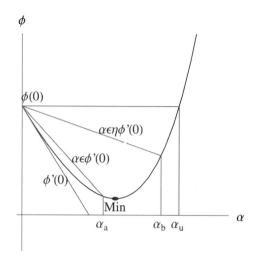

FIGURE 5.8　Illustration of Armijo's approximate line search strategy.

3. If the test $\phi(\alpha) \leq \phi(0) + \alpha\epsilon\eta\phi'(0)$ fails with the initial value of α, decrease α by setting it to α/η and repeat until a value of α is found that makes the test pass.

The procedure is implemented in the following function that is included in the OptimizationToolbox 'Unconstrained' package:

```
Needs["OptimizationToolbox`Unconstrained`"];
?ArmijoLineSearch
```

```
ArmijoLineSearch[ϕ, α, ϵ, η, options] --- performs line search using
   Armijo's rule. Arguments have the following meaning. ϕ = function
   of single variable α. ϵ and η are the two parameters. There are two
   optional parameters. The first is prResults (default True) to control
   printing of intermediate results. The second is MaxIterations allowed.
   Default is MaxIterations set for FindMinimum function. The functions
   returns approximate step length.
```

Example 5.14　Determine an approximate minimum of the following function using Armijo's line search:

```
ϕ = 1 - 1/(1 - α + 2α^2);
```

We use ArmijoLineSearch function to compute the minimum starting with $\epsilon = 0.2$ and $\eta = 2$.

```
ArmijoLineSearch[ϕ,α,0.2,2]
```
$\phi(0) \to 0 \quad \phi'(0) \to -1$

α	ϕ(α)	ϕ(0) + αϕ'(0)ϵ	ϕ(0) + αϕ'(0)ηϵ
1.	0.5	-0.2	--
0.5	0.	--	-0.2
0.25	-0.142857	--	-0.1
0.25			

In this example, the method in fact gives the exact step length.

Example 5.15 Determine an approximate minimum of the following function using Armijo's line search:

```
ϕ = 2 - 4α + Exp[α];
```

We use ArmijoLineSearch to compute the minimum starting with $\epsilon = 0.2$ and $\eta = 2$.

```
ArmijoLineSearch[ϕ,α,.2,2]
```
$\phi(0) \to 3 \quad \phi'(0) \to -3$

α	ϕ(α)	ϕ(0) + αϕ'(0)ϵ	ϕ(0) + αϕ'(0)ηϵ
1.	0.718282	2.4	--
2.	1.38906	1.8	--
4.	40.5982	0.6	--
2.			

The exact step length for this example is 1.386. The approximate step length of 2 is computed in only three function evaluations.

5.3 Unconstrained Minimization Techniques

At each iteration of a numerical optimization method, we need to determine a descent direction and an appropriate step length:

$$\mathbf{x}^{k+1} = \mathbf{x}^k + \alpha_k \mathbf{d}^k \quad k = 0, 1, \ldots$$

The step-length calculations were presented in the last section. These are combined with several different methods for determining descent directions to get methods for solving unconstrained problems. Any of the step-length strategies discussed in the previous section can be used. However, for clarity of presentation, the examples given in this section use the analytical line search method.

5.3.1 The Steepest Descent Method

As mentioned earlier, the gradient vector of a function points towards the direction in which locally the function is increasing the most rapidly. Thus, a natural choice for the descent direction is to use the negative gradient direction. Since locally, the function is changing most rapidly in this direction, it is known as the *steepest descent direction*. Thus, in this method we choose the direction as follows:

$$\mathbf{d}^k = -\nabla f\left(\mathbf{x}^k\right)$$

As seen from the following examples, the method produces successive directions that are perpendicular to each other. Thus, when the point is away from the optimum, the method generally makes good progress towards the optimum. However, near the optimum, because of *zigzagging*, the convergence is very slow.

The procedure is implemented in the following function that is included in the OptimizationToolbox 'Unconstrained' package:

Needs["OptimizationToolbox`Unconstrained`"];
?SteepestDescent

```
SteepestDescent[f, vars, x0, opts]. Computes minimum of f(vars) starting
   from x0 using Steepest Descent method. The step length is computed
   using analytical line search. See Options[SteepestDescent] to see a
   list of options for the function. The function returns {x, hist}. 'x'
   is either the optimum point or the next point after MaxIterations.
   'hist' contains history of values tried at different iterations.
```

Options[SteepestDescent]
{PrintLevel → 1, MaxIterations → 50,
   ConvergenceTolerance → 0.01, StepLengthVar → α}

?PlotSearchPath

```
PlotSearchPath[f, {x1, x1min, x1max}, {x2, x2min, x2max}, hist, opts]
   shows complete search path superimposed on a contour plot of the
   function f over the specified range. 'hist' is assumed to be of the
   form {pt1, pt2, ...}, where pt1 = {x1,x2} is first point, etc. The
   function accepts all relevant options of the standard ContourPlot and
   Graphics functions.
```

5.3 Unconstrained Minimization Techniques

Example 5.16 Use the steepest descent method and the given starting point to find the minimum of the following function:

$f = x^4 + y^4 + 2x^2y^2 - 4x + 3;$
$vars = \{x, y\}; x0 = \{1.25, 1.25\};$

All intermediate calculations are shown for the first two iterations.

$\text{SteepestDescent}[f, vars, x0, \text{PrintLevel} \to 2, \text{MaxIterations} \to 2];$

$f \to 3 - 4x + x^4 + 2x^2y^2 + y^4$

$\nabla f \to \begin{pmatrix} -4 + 4x^3 + 4xy^2 \\ 4x^2y + 4y^3 \end{pmatrix}$

***** Iteration 1 ***** Current point $\to \{1.25, 1.25\}$

Direction finding phase:

$\nabla f(x) \to \begin{pmatrix} 11.625 \\ 15.625 \end{pmatrix} \quad d \to \begin{pmatrix} -11.625 \\ -15.625 \end{pmatrix}$

$||\nabla f(x)|| \to 19.4751 \quad f(x) \to 7.76563$

Step length calculation phase:

$xk1 \to \begin{pmatrix} 1.25 - 11.625\alpha \\ 1.25 - 15.625\alpha \end{pmatrix}$

$\nabla f(xk1) \to \begin{pmatrix} -17636.6(-0.0353314 + \alpha)(0.018656 - 0.251812\alpha + \alpha^2) \\ -23705.0(-0.08 + \alpha)(0.00823927 - 0.179616\alpha + \alpha^2) \end{pmatrix}$

$d\phi/d\alpha \equiv \nabla f(xk1).d = 0 \to$
$575417.(-0.0479137 + \alpha)(0.0137569 - 0.22151\alpha + \alpha^2) == 0$
$\alpha \to 0.0479137$

***** Iteration 2 ***** Current point $\to \{0.693003, 0.501349\}$

Direction finding phase:

$\nabla f(x) \to \begin{pmatrix} -1.97198 \\ 1.46716 \end{pmatrix} \quad d \to \begin{pmatrix} 1.97198 \\ -1.46716 \end{pmatrix}$

$||\nabla f(x)|| \to 2.4579 \quad f(x) \to 0.763231$

Step length calculation phase:

$xk1 \to \begin{pmatrix} 0.693003 + 1.97198\alpha \\ 0.501349 - 1.46716\alpha \end{pmatrix}$

$\nabla f(xk1) \to \begin{pmatrix} 47.6531(-0.141029 + \alpha)(0.293429 + 0.701362\alpha + \alpha^2) \\ -35.4539(-0.341715 + \alpha)(0.121101 + 0.208908\alpha + \alpha^2) \end{pmatrix}$

$d\phi/d\alpha \equiv \nabla f(xk1).d = 0 \to$
$145.987(-0.179066 + \alpha)(0.2311 + 0.492428\alpha + \alpha^2) == 0$
$\alpha \to 0.179066$

New Point (Non-Optimum): $\{1.04612, 0.238631\}$ after 2 iterations

The printout of intermediate results is suppressed (default option), and the method is allowed to continue until convergence. Computation history is saved in *hist*.

```
{opt, hist} = SteepestDescent[f, vars, x0];
```
Optimum: {0.999362, 0.00143171} after 9 iterations

TableForm[hist]

x	d	$\|\|\nabla f(x)\|\|$	f(x)
1.25	-11.625	19.4751	7.76563
1.25	-15.625		
0.693003	1.97198	2.4579	0.763231
0.501349	-1.46716		
1.046112	-0.81762	1.36974	0.141038
0.238631	-1.09895		
0.949342	0.532874	0.664179	0.036264
0.108555	-0.396458		
1.01403	-0.185565	0.310875	0.00871469
0.0604258	-0.249416		
0.988317	0.135914	0.169405	0.00211974
0.0258635	-0.10112		
1.00355	-0.0435646	0.0729841	0.000501114
0.0145326	-0.058556		
0.997266	0.0325665	0.0405914	0.000118541
0.0060905	-0.0242299		
1.00085	-0.0102337	0.0171208	0.0000278277
0.00342556	-0.0137257		
0.999362	0.00764796	0.0095501	6.53879×10^{-6}
0.00143171	-0.00571954		

The first column is extracted from the history, and the heading is removed to get a list of all intermediate points tried by the algorithm.

xhist = Drop[Transpose[hist][[1]], 1]

{{1.25, 1.25}, {0.693003, 0.501344}, {1.04612, 0.238631},
 {0.949342, 0.108555}, {1.01403, 0.0604258}, {0.988317, 0.0258635},
 {1.00355, 0.01453265}, {0.997266, 0.0060905}, {1.00085, 0.00342556},
 {0.999362, 0.00143171}}

Using the PlotSearchPath function, the search path is shown on a contour map of function f in Figure 5.9. The direction is the negative gradient direction, which is always perpendicular to the tangent to contour at the given point. As mentioned in section 5.2, the optimum step length selects the next point such that the direction is tangent to the contour of f at the new point. For the subsequent iteration, then the negative gradient direction, becomes perpendicular to the previous direction. This makes the search path zigzag towards the optimum. As a result, convergence near the optimum becomes extremely slow.

```
PlotSearchPath[f, {x, -.1, 1.3}, {y, -.1, 1.3}, xhist, PlotPoints → 30,
  Epilog → {RGBColor[1, 0, 0], Disk[{1, 0}, 0.015],
    Text["Optimum", {1.01, 0.05}, {-1, 0}]}];
```

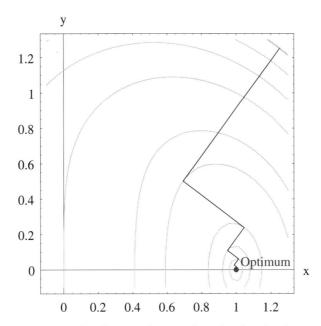

FIGURE 5.9 A contour plot showing the search path taken by the steepest descent method.

The column containing the norm of the function gradient is extracted in preparation for a plot showing how the $\|\nabla f\|$ is decreasing with the number of iterations. The plot shows the rate of convergence of the method.

```
normGradhist = Drop[Transpose[hist][[3]], 1]
```
{19.4751, 2.4579, 1.36974, 0.664179, 0.310875,
 0.169405, 0.0729841, 0.0405914, 0.0171208, 0.0095501}

```
ListPlot[normGradhist,
  PlotJoined → True, AxesLabel → {"Iterations", "||∇f||"}];
```

258   Chapter 5 Unconstrained Problems

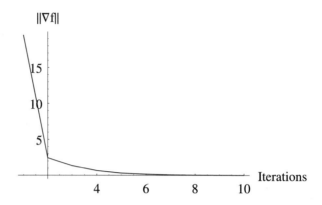

FIGURE 5.10 Plot showing $\|\nabla f\|$ at different iterations.

Example 5.17 Use the steepest descent method and the given starting point to find minimum of the following function:

```
f = (x + y)² + (2 (x² + y² - 1) - 1/3)²;
vars = {x, y}; x0 = {-1.25, 0.25};

SteepestDescent[f, vars, x0, PrintLevel → 2, MaxIterations → 2];
```

$f \to (x+y)^2 + \left(-\frac{1}{3} + 2\left(-1 + x^2 + y^2\right)\right)^2$

$\nabla f \to \begin{pmatrix} -\frac{50x}{3} + 16x^3 + 2y + 16xy^2 \\ 2x - \frac{50y}{3} + 16x^2y + 16y^3 \end{pmatrix}$

***** Iteration 1 ***** Current point → {-1.25, 0.25}

Direction finding phase:

$\nabla f(x) \to \begin{pmatrix} -11.1667 \\ -0.166667 \end{pmatrix} \quad d \to \begin{pmatrix} 11.1667 \\ 0.166667 \end{pmatrix}$

$\|\nabla f(x)\| \to 11.1679 \quad f(x) \to 1.84028$

Step length calculation phase:

$xk1 \to \begin{pmatrix} -1.25 + 11.1667\alpha \\ 0.25 + 0.166667\alpha \end{pmatrix}$

$\nabla f(xk1) \to \begin{pmatrix} 22283.7(-0.198059 + \alpha)(-0.115053 + \alpha)(-0.0219909 + \alpha) \\ 332.593(-0.182098 + \alpha)(0.00188867 + \alpha)(1.45705 + \alpha) \end{pmatrix}$

$d\phi/d\alpha \equiv \nabla f(xk1).d = 0 \to$
$\qquad 248890.(-0.197979 + \alpha)(-0.114697 + \alpha)(-0.0220681 + \alpha) == 0$
$\alpha \to 0.0220681$

***** Iteration 2 ***** Current point → {-1.00357, 0.253678}

Direction finding phase:

5.3 Unconstrained Minimization Techniques

$$\nabla f(x) \to \begin{pmatrix} 0.0281495 \\ -1.88601 \end{pmatrix} \quad d \to \begin{pmatrix} -0.0281495 \\ 1.88601 \end{pmatrix}$$

$||\nabla f(x)|| \to 1.88622 \quad f(x) \to 0.598561$

Step length calculation phase:

$$xk1 \to \begin{pmatrix} -1.00357 - 0.0281495\alpha \\ 0.253678 + 1.88601\alpha \end{pmatrix}$$

$$\nabla f(xk1) \to \begin{pmatrix} -1.60242\,(-0.00222672 + \alpha)\,(0.220873 + \alpha)\,(35.7178 + \alpha) \\ 107.362\,(-0.140625 + \alpha)\,(0.124919 + 0.55996\alpha + \alpha^2) \end{pmatrix}$$

$d\phi/d\alpha \equiv \nabla f(xk1).d = 0 \to$
$$202.532\,(-0.138857 + \alpha)\,(0.126511 + 0.566102\alpha + \alpha^2) == 0$$

$\alpha \to 0.138857$
New Point (Non-Optimum): {-1.00748, 0.515563} after 2 iterations

{opt, hist} = SteepestDescent[f, vars, x0];
Optimum: {-0.764703, 0.762588} after 27 iterations

The following graphs show the path taken by the procedure in moving towards the optimum. Again, notice the zigzag path taken by the method. Also notice that the function has two local minima. As is true of most numerical methods, the algorithm returns the one that is closest to the starting point.

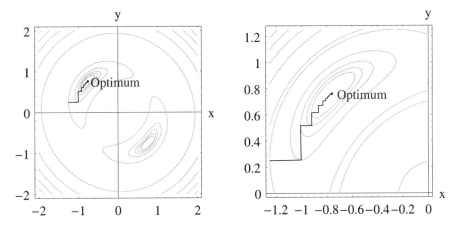

FIGURE 5.11 Graphical solution and the search path taken by the steepest descent method.

The following plot of the norm of the gradient of the objective function shows the convergence rate of the method. The first few iterations show substantial reduction in this norm. However, it takes more than 20 iterations to reduce it from 2 to the convergence tolerance of 10^{-3}.

```
normGradhist = Drop[Transpose[hist][[3]],1];
ListPlot[normGradhist,
  PlotJoined → True, AxesLabel → {"Iterations", "||∇f||"}];
```

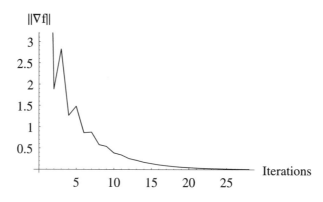

FIGURE 5.12 Plot showing $\|\nabla f\|$ at different iterations.

Example 5.18 Use the steepest descent method and the given starting point to find the minimum of the following function of three variables:

```
f = (x₁ - 1)⁴ + (x₂ - 3)² + 4(x₃ + 5)²;
vars = {x₁, x₂, x₃}; x0 = {-1, -2, 1};

SteepestDescent[f, vars, x0, MaxIterations → 2, PrintLevel → 2];
```

$f \to \left(-1+x_1\right)^4 + \left(-3+x_2\right)^2 + 4\left(5+x_3\right)^2$

$\nabla f \to \begin{pmatrix} -4 + 12x_1 - 12x_1^2 + 4x_1^3 \\ -6 + 2x_2 \\ 40 + 8x_3 \end{pmatrix}$

```
***** Iteration 1 ***** Current point → {-1., -2., 1.}
Direction finding phase:
```

$\nabla f(x) \to \begin{pmatrix} -32 \\ -10 \\ 48 \end{pmatrix} \quad d \to \begin{pmatrix} 32 \\ 10 \\ -48 \end{pmatrix}$

$\|\nabla f(x)\| \to 58.5491 \quad f(x) \to 185$

Step length calculation phase:

$xk1 \to \begin{pmatrix} -1. + 32\alpha \\ -2. + 10\alpha \\ 1. - 48\alpha \end{pmatrix}$

$\nabla f(xk1) \to \begin{pmatrix} 131072. (-0.0625 + \alpha)(-0.0625 + \alpha)^2 \\ -10. + 20\alpha \\ 48 - 384\alpha \end{pmatrix}$

5.3 Unconstrained Minimization Techniques

$d\phi/d\alpha \equiv \nabla f(xk1).d = 0 \to$
$$4.1943 \times 10^6 (-0.107925 + \alpha)(0.00757284 - 0.0795749\alpha + \alpha^2) == 0$$
$\alpha \to 0.107925$
***** Iteration 2 ***** Current point $\to \{2.45360, -0.9207491, -4.18040\}$

Direction finding phase:
$$\nabla f(x) \to \begin{pmatrix} 12.2856 \\ -7.8415 \\ 6.55677 \end{pmatrix} \quad d \to \begin{pmatrix} -12.2856 \\ 7.8415 \\ -6.55677 \end{pmatrix}$$

$||\nabla f(x)|| \to 15.9818 \quad f(x) \to 22.5238$

Step length calculation phase:
$$xk1 \to \begin{pmatrix} 2.4536 - 12.2856\alpha \\ -0.920749 + 7.8415\alpha \\ -4.1804 - 6.55677\alpha \end{pmatrix}$$

$$\nabla f(xk1) \to \begin{pmatrix} -7417.39(-0.118316 + \alpha)(0.0139992 - 0.236636\alpha + \alpha^2) \\ -7.8415 + 15.683\alpha \\ 6.5567 - 52.4541\alpha \end{pmatrix}$$

$d\phi/d\alpha \equiv \nabla f(xk1).d = 0 \to$
$$91127.3(-0.179366 + \alpha)(0.0156265 - 0.175587\alpha + \alpha^2) == 0$$
$\alpha \to 0.179366$
New Point (Non-Optimum):
$\{0.249986, 0.485745, -5.35646\}$ after 2 iterations

{opt, hist} = SteepestDescent[f, vars, x0];

Optimum: $\{0.875831, 3., -5.00063\}$ after 26 iterations

normGradhist = Drop[Transpose[hist][[3]], 1];

ListPlot[normGradhist,
  PlotJoined → True, AxesLabel → {"Iterations", "||∇f||"}];

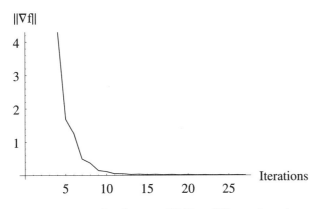

FIGURE 5.13 Plot showing $\|\nabla f\|$ at different iterations.

5.3.2 Conjugate Gradient

The conjugate gradient method attempts to improve the behavior of the steepest descent method by adding a portion of the previous direction to the current negative gradient direction as follows:

$$\mathbf{d}^k = -\nabla f\left(\mathbf{x}^k\right) + \beta \mathbf{d}^{k-1}$$

The scalar multiplier β determines the portion of the previous direction to be added to determine the new direction. As pointed out earlier, when the point is far away from the optimum, moving along the negative gradient direction is a good idea. Near the optimum, however, we need to make changes in this direction. Thus, β must be defined in such a way that it should have small values at points away from the minimum and have relatively large values near it. One of the following two formulas for β is used most often in practice:

Fletcher-Reeves formula: $\quad \beta = \dfrac{[\nabla f(\mathbf{x}^k)]^T \nabla f(\mathbf{x}^k)}{[\nabla f(\mathbf{x}^{k-1})]^T \nabla f(\mathbf{x}^{k-1})}$

Polak-Ribiere formula: $\quad \beta = \dfrac{[\nabla f(\mathbf{x}^{k-1}) - \nabla f(\mathbf{x}^k)]^T \nabla f(\mathbf{x}^k)}{[\nabla f(\mathbf{x}^{k-1})]^T \nabla f(\mathbf{x}^{k-1})}$

The denominator in both formulas is the same and is equal to the square of the norm of the gradient of f at the previous point. In the Fletcher-Reeves formula, the numerator is the square of the norm of the gradient of f at the current point, whereas in the Polak-Ribiere formula, it is slightly modified. The Polak-Ribiere formula usually gives better results than the Fletcher-Reeves formula.

At the beginning, we obviously don't have any previous direction and therefore, the steepest descent direction is chosen at the first iteration. The following numerical examples show the improved performance with this method.

The procedure is implemented in the following function that is included in the OptimizationToolbox 'Unconstrained' package:

```
Needs["OptimizationToolbox`Unconstrained`"];
?ConjugateGradient
```

```
ConjugateGradient[f, vars, x0, opts]. Computes minimum of f(vars)
   starting from x0 using Conjugate Gradient method. User has the option
   of using either 'PolakRibiere' (default) or 'FletcherReeves' method.
   Also step length can be computed using either the 'Exact' analytical
   line search or an 'Approximate' line search using Armijo's rule. See
   Options[ConjugateGradient] to see a list of options for the function.
   The function returns {x, hist}. 'x' is either the optimum point or
   the next point after MaxIterations. 'hist' contains history of values
   tried at different iterations.
```

5.3 Unconstrained Minimization Techniques

Options[ConjugateGradient]
{PrintLevel → 1, MaxIterations → 50, ConvergenceTolerance → 0.01,
 StepLengthVar → α, Method → PolakRibiere, LineSearch → Exact,
 ArmijoParameters → {0.2, 2}}

Example 5.19 Use the conjugate gradient method and the given starting point to find the minimum of the following function:

$f = x^4 + y^4 + 2x^2y^2 - 4x + 3;$
$vars = \{x, y\}; x0 = \{1.25, 1.25\};$

All intermediate calculations are shown for the first two iterations.

**ConjugateGradient[f, vars, x0, PrintLevel → 2, MaxIterations → 2,
 Method → FletcherReeves];**

$f \to 3 - 4x + x^4 + 2x^2y^2 + y^4$

$\nabla f \to \begin{pmatrix} -4 + 4x^3 + 4xy^2 \\ 4x^2y + 4y^3 \end{pmatrix}$

```
Using FletcherReeves method with Exact line search
***** Iteration 1 ***** Current point → {1.25, 1.25}

Direction finding phase:
```

$\nabla f(x) \to \begin{pmatrix} 11.625 \\ 15.625 \end{pmatrix} \quad d \to \begin{pmatrix} -11.625 \\ -15.625 \end{pmatrix}$

$||\nabla f(x)|| \to 19.4751 \quad \beta \to 0.$
$f(x) \to 7.76563$

```
Step length calculation phase:
```

$xk1 \to \begin{pmatrix} 1.25 - 11.625\alpha \\ 1.25 - 15.625\alpha \end{pmatrix}$

$\nabla f(xk1) \to \begin{pmatrix} -17636.6(-0.0353314 + \alpha)(0.018656 - 0.251812\alpha + \alpha^2) \\ -23705.1(-0.08 + \alpha)(0.00823927 - 0.179616\alpha + \alpha^2) \end{pmatrix}$

$d\phi/d\alpha \equiv \nabla f(xk1).d = 0 \to$
$\qquad 575417.(-0.0479137 + \alpha)(0.0137569 - 0.22151\alpha + \alpha^2) == 0$

$\alpha \to 0.0479137$
```
***** Iteration 2 ***** Current point → {0.693003, 0.501349}

Direction finding phase:
```

$\nabla f(x) \to \begin{pmatrix} -1.97198 \\ 1.46716 \end{pmatrix} \quad d \to \begin{pmatrix} 1.78682 \\ -1.71603 \end{pmatrix}$

$||\nabla f(x)|| \to 2.4579 \quad \beta \to 0.0159282$
$f(x) \to 0.763231$

Step length calculation phase:

$$xk1 \to \begin{pmatrix} 0.693003 + 1.7868\alpha \\ 0.501349 - 1.71603\alpha \end{pmatrix}$$

$$\nabla f(xk1) \to \begin{pmatrix} 43.8663(-0.162591 + \alpha)(0.276488 + 0.673591\alpha + \alpha^2) \\ -42.1285(-0.292156 + \alpha)(0.119203 + 0.123158\alpha + \alpha^2) \end{pmatrix}$$

$d\phi/d\alpha \equiv \nabla f(xk1).d = 0 \to$

$$150.675(-0.197956 + \alpha)(0.202543 + 0.382693\alpha + \alpha^2) == 0$$

$\alpha \to 0.197956$
New Point (Non-Optimum): {1.04672, 0.161649} after 2 iterations

The printout of intermediate results is suppressed (default option) and the method is allowed to continue until convergence. Computation history is saved in *hist*.

{opt, hist} = ConjugateGradient[f, vars, x0, Method → FletcherReeves];

Using FletcherReeves method with Exact line search
Optimum: {1.00004, 0.000174657} after 6 iterations

TableForm[hist]

x	d	$||\nabla f(x)||$	f(x)	β
1.25	-11.625	19.4751	7.76563	1.
1.25	-15.625			
0.693003	1.78682	2.4579	0.763231	0.0
0.501349	-1.71603			
1.04672	-0.397468	1.00564	0.0714469	0.1
0.161649	-1.01258			
0.988576	0.126566	0.144823	0.00113522	0.0
0.0135342	-0.0739167			
1.00119	-0.00949338	0.028638	0.0000847527	0.0
0.00616745	-0.0276199			
0.999313	0.0073571	0.00870763	3.8322×10^{-6}	0.0
0.000708579	-0.00538393			
1.00004	--	0.000866728	7.1970×10^{-8}	--
0.000174657				

The first column is extracted from the history, and the heading is removed to get a list of all intermediate points tried by the algorithm.

xhist = Drop[Transpose[hist][[1]], 1]
{{1.25, 1.25}, {0.693003, 0.501349}, {1.04672, 0.161649},
 {0.988576, 0.0135342}, {1.00119, 0.00616745},
 {0.999313, 0.000708579}, {1.00004, 0.000174657}}

Using the PlotSearchPath function, the search path is shown on a contour map of function f. Note that the subsequent directions are not perpendicular to the previous ones. The search path shows considerably less zigzagging near the optimum, as compared to the steepest descent method. The rate of convergence to the optimum is thus improved considerably.

```
PlotSearchPath[f, {x, -.1, 1.3}, {y, -.1, 1.3}, xhist, PlotPoints → 30,
  Epilog → {RGBColor[1, 0, 0], Disk[{1, 0}, 0.015],
    Text["Optimum", {1.01, 0.05}, {-1, 0}]}];
```

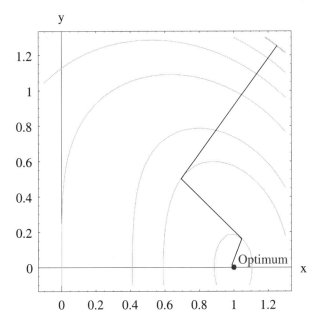

FIGURE 5.14 A contour plot showing the search path taken by the conjugate gradient method.

The column containing the norm of the function gradient is extracted in preparation for a plot showing how the $||\nabla f||$ is decreasing with the number of iterations. The plot shows the rate of convergence of the method:

```
normGradhist = Drop[Transpose[hist][[3]], 1]
```
{19.4751, 2.4579, 1.00564, 0.144823,
  0.028638, 0.00870763, 0.000866728}

```
ListPlot[normGradhist,
  PlotJoined → True, AxesLabel → {"Iterations", "||∇f||"}];
```

Chapter 5 Unconstrained Problems

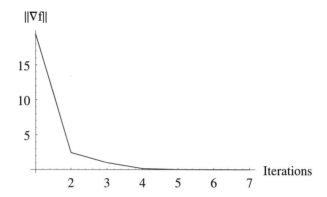

FIGURE 5.15 A plot showing $\|\nabla f\|$ at different iterations.

The approximate line search gives the same results but takes a few more iterations.

```
ConjugateGradient[f, vars, x0, LineSearch → Approximate];
```

```
Using PolakRibiere method with Approximate line search
Optimum: {1.00001, 0.000350558} after 10 iterations
```

The Polak-Ribiere method takes one less iteration as compared to the Fletcher-Reeves method.

```
ConjugateGradient[f, vars, x0];
```

```
Using PolakRibiere method with Exact line search
Optimum: {1., 2.64104 × 10⁻⁸} after 5 iterations
```

Example 5.20 Use the conjugate gradient method and the given starting point to find the minimum of the following function:

```
f = (x + y)² + (2 (x² + y² - 1) - ⅓)²;
vars = {x, y}; x0 = {-1.25, 0.25};
```

All intermediate calculations are shown for the first two iterations.

```
ConjugateGradient[f, vars, x0, PrintLevel → 2, MaxIterations → 2];
```

$$f \to (x+y)^2 + \left(-\frac{1}{3} + 2(-1 + x^2 + y^2)\right)^2$$

$$\nabla f \to \begin{pmatrix} -\frac{50x}{3} + 16x^3 + 2y + 16xy^2 \\ 2x - \frac{50y}{3} + 16x^2y + 16y^3 \end{pmatrix}$$

5.3 Unconstrained Minimization Techniques

Using PolakRibiere method with Exact line search
***** Iteration 1 ***** Current point → {-1.25, 0.25}

Direction finding phase:

$\nabla f(x) \to \begin{pmatrix} -11.1667 \\ -0.166667 \end{pmatrix} \quad d \to \begin{pmatrix} 11.1667 \\ 0.166667 \end{pmatrix}$

$||\nabla f(x)|| \to 11.1679 \quad \beta \to 0.$

$f(x) \to 1.84028$

Step length calculation phase:

$xk1 \to \begin{pmatrix} -1.25 + 11.1667\alpha \\ 0.25 + 0.166667\alpha \end{pmatrix}$

$\nabla f(xk1) \to \begin{pmatrix} 22283.7(-0.198059 + \alpha)(-0.115053 + \alpha)(-0.0219909 + \alpha) \\ 332.593(-0.182098 + \alpha)(0.00188867 + \alpha)(1.45705 + \alpha) \end{pmatrix}$

$d\phi/d\alpha \equiv \nabla f(xk1).d = 0 \to$
$\qquad 248890.(-0.197979 + \alpha)(-0.114697 + \alpha)(-0.0220681 + \alpha) == 0$

$\alpha \to 0.0220681$
***** Iteration 2 ***** Current point → {-1.00357, 0.253678}

Direction finding phase:

$\nabla f(x) \to \begin{pmatrix} 0.0281495 \\ -1.88601 \end{pmatrix} \quad d \to \begin{pmatrix} 0.290392 \\ 1.89077 \end{pmatrix}$

$||\nabla f(x)|| \to 1.88622 \quad \beta \to 0.0285261$

$f(x) \to 0.598561$

Step length calculation phase:

$xk1 \to \begin{pmatrix} -1.00357 + 0.290392\alpha \\ 0.253678 + 1.8907\alpha \end{pmatrix}$

$\nabla f(xk1) \to \begin{pmatrix} 17.0023\left(-3.38977 + \alpha\right)\left(-0.0103728 + \alpha\right)\left(0.0470866 + \alpha\right) \\ 110.703\left(-0.173294 + \alpha\right)\left(0.098311 + 0.41033\alpha + \alpha^2\right) \end{pmatrix}$

$d\phi/d\alpha \equiv \nabla f(xk1).d = 0 \to$
$\qquad 214.251\left(-0.188279 + \alpha\right)\left(0.0881985 + 0.342583\alpha + \alpha^2\right) == 0$

$\alpha \to 0.188279$
New Point (Non-Optimum): {-0.948898, 0.60967} after 2 iterations

{opt, hist} = ConjugateGradient[f, vars, x0];

Using PolakRibiere method with Exact line search
Optimum: {-0.76376, 0.763765} after 6 iterations

TableForm[hist]

x	d	$\|\|\nabla f(x)\|\|$	f(x)	β
-1.25 0.25	11.1667 0.166667	11.1679	1.84028	0.
-1.00357 0.253678	0.290392 1.89077	1.88622	0.598561	0.
-0.948898 0.60967	2.77239 2.86082	2.30597	0.159544	1.
-0.784161 0.779661	0.253614 -1.1650	0.99292	0.0126152	-0
-0.777963 0.751188	0.0782121 0.0687128	0.0910377	0.000749258	-0
-0.763743 0.763681	-0.000580251 0.003031	0.00266003	9.82337×10^{-8}	0.
-0.76376 0.763765	--	0.0000196313	3.46106×10^{-11}	--

The search path shown in the following figure clearly demonstrates the improvement in convergence that this method makes as compared to the steepest descent method.

```
xhist = Drop[Transpose[hist][[1]], 1];

PlotSearchPath[f, {x, -1.25, 0}, {y, 0, 1.25}, xhist, PlotPoints → 50,
  Epilog → {RGBColor[1, 0, 0], Disk[{-.76, .76}, 0.01],
    Text["Optimum", {-.74, .76}, {-1, 0}]}];
```

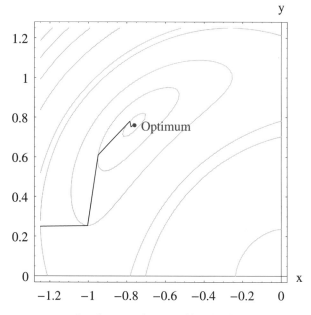

FIGURE 5.16 A contour plot showing the search path taken by the conjugate gradient method.

5.3 Unconstrained Minimization Techniques

```
normGradhist = Drop[Transpose[hist][[3]], 1]
```
{11.1679, 1.88622, 2.30597, 0.99292,
  0.0910377, 0.00266003, 0.0000196313}

```
ListPlot[normGradhist,
  PlotJoined → True, AxesLabel → {"Iterations", "||∇f||"}];
```

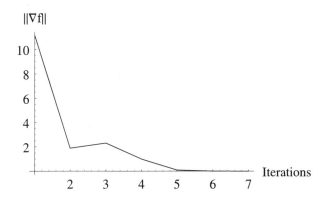

FIGURE 5.17 A plot showing $\|\nabla f\|$ at different iterations.

Example 5.21 Use the conjugate gradient method and the given starting point to find minimum of the following function:

```
f = (x₁ - 1)^4 + (x₂ - 3)^2 + 4(x₃ + 5)^2;
vars = {x₁, x₂, x₃}; x0 = {-1, -2, 1};
```

All intermediate calculations are shown for the first two iterations.

```
ConjugateGradient[f, vars, x0, PrintLevel → 2, MaxIterations → 2];
```

$f \to (-1 + x_1)^4 + (-3 + x_2)^2 + 4(5 + x_3)^2$

$\nabla f \to \begin{pmatrix} -4 + 12x_1 - 12x_1^2 + 4x_1^3 \\ -6 + 2x_2 \\ 40 + 8x_3 \end{pmatrix}$

```
Using PolakRibiere method with Exact line search
***** Iteration 1 ***** Current point → {-1, -2, 1}

Direction finding phase:
```

$\nabla f(x) \to \begin{pmatrix} -32 \\ -10 \\ 48 \end{pmatrix} \quad d \to \begin{pmatrix} 32 \\ 10 \\ -48 \end{pmatrix}$

$||\nabla f(x)|| \to 58.5491 \quad \beta \to 0$

$f(x) \to 185$

Step length calculation phase:

$\text{xk1} \to \begin{pmatrix} -1 + 32\alpha \\ -2 + 10\alpha \\ 1 - 48\alpha \end{pmatrix}$

$\nabla f(\text{xk1}) \to \begin{pmatrix} 32(-1+16\alpha)^3 \\ -10 + 20\alpha \\ 48 - 384\alpha \end{pmatrix}$

$d\phi/d\alpha \equiv \nabla f(\text{xk1}) \cdot d = 0 \to 4(-857 + 16946\alpha - 196608\alpha^2 + 1048576\alpha^3) == 0$

$\alpha \to 0.107925$

***** Iteration 2 ***** Current point $\to \{2.4536, -0.920749, -4.1804\}$

Direction finding phase:

$\nabla f(x) \to \begin{pmatrix} 12.2856 \\ -7.8415 \\ 6.55677 \end{pmatrix} \quad d \to \begin{pmatrix} -9.9013 \\ 8.58659 \\ -10.1332 \end{pmatrix}$

$||\nabla f(x)|| \to 15.9818 \quad \beta \to 0.074509$

$f(x) \to 22.5238$

Step length calculation phase:

$\text{xk1} \to \begin{pmatrix} 2.4536 - 9.90133\alpha \\ -0.920749 + 8.5865\alpha \\ -4.1804 - 10.1332\alpha \end{pmatrix}$

$\nabla f(\text{xk1}) \to \begin{pmatrix} -3882.76(-0.146808 + \alpha)(0.0215529 - 0.293618\alpha + \alpha^2) \\ -7.8415 + 17.1732\alpha \\ 6.55677 - 81.0656\alpha \end{pmatrix}$

$d\phi/d\alpha \equiv \nabla f(\text{xk1}) \cdot d = 0 \to$
$$38444.5 \left(-0.138091 + \alpha\right) \left(0.0481115 - 0.302335\alpha + \alpha^2\right) == 0$$

$\alpha \to 0.138091$

New Point (Non-Optimum): $\{1.08632, 0.264982, -5.1\}$ after 2 iterations

{opt, hist} = ConjugateGradient[f, vars, x0];

5.3 Unconstrained Minimization Techniques

```
Using PolakRibiere method with Exact line search

Optimum: {0.942926, 2.99843, -5.00018} after 11 iterations
```

TableForm[hist]

x	d	$\|\|\nabla f(x)\|\|$	f(x)	β
-1	32	58.5491	185	0
-2	10			
1	-48			
2.4536	-9.90133	15.9818	22.5238	0.0745
-0.920749	8.58659			
-4.1804	-10.1332			
1.08632	-1.51103	7.17142	8.82463	0.1523
0.264982	6.77819			
-5.57971	3.093889			
0.627855	-0.264824	3.18313	0.995014	0.3116
2.32155	3.46962			
-4.64099	-1.9077			
0.577535	0.300057	0.305303	0.03227	0.0058
2.98083	0.0585633			
-5.00348	0.0167521			
0.837298	0.054962	0.109742	0.00218038	0.1257
3.03153	-0.0556896			
-4.98898	-0.0860442			
0.847277	0.0311044	0.0582551	0.00108743	0.3066
3.02142	-0.0599118			
-5.0046	0.0104356			
0.8599967	0.0125769	0.0128732	0.000394154	0.0514
2.99692	0.00308432			
-5.00034	0.00322151			
0.875389	0.075712	0.0299098	0.000293639	5.4045
3.00069	0.0152858			
-4.99639	-0.0114471			
0.904155	0.0325175	0.0147169	0.000128831	0.3829
3.0065	-0.00714454			
-5.00074	0.00155074			
0.939289	0.0212302	0.00790878	0.0000185614	0.6253
2.99878	-0.00202767			
-4.99907	-0.00649969			
0.942926	--	0.00352847	0.0000131975	--
2.99843				
-5.00018				

```
normGradhist = Drop[Transpose[hist][[3]], 1];

ListPlot[normGradhist,
  PlotJoined → True, AxesLabel → {"Iterations", "||∇f||"}];
```

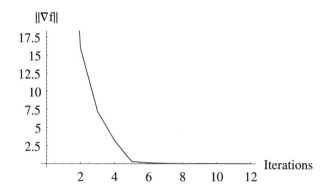

FIGURE 5.18 A plot showing $\|\nabla f\|$ at different iterations.

5.3.3 The Modified Newton Method

In its basic form, the Newton method is derived by considering quadratic approximation of the function using the Taylor series:

$$f(\mathbf{x}^{k+1}) \approx f(\mathbf{x}^k) + \nabla f(\mathbf{x}^k)^T \mathbf{d}^k + \frac{1}{2}(\mathbf{d}^k)^T \mathbf{H}(\mathbf{x}^k)\mathbf{d}^k$$

where $\mathbf{H}(\mathbf{x}^k)$ is the Hessian matrix at $\mathbf{x}^k$. Using the necessary condition for the minimum of this approximate function (i.e., differentiating with respect to $\mathbf{d}^k$), we get

$$\nabla f(\mathbf{x}^k) + \mathbf{H}(\mathbf{x}^k)\mathbf{d}^k = \mathbf{0}$$

The direction can now be obtained by solving this system of equations as follows:

$$\mathbf{d}^k = -[\mathbf{H}(\mathbf{x}^k)]^{-1}\nabla f(\mathbf{x}^k)$$

Note that in actual computations the inverse of Hessian is not computed. It is more efficient to compute the direction by solving the linear system of equations. The above form is used just for symbolic purposes.

In the original form, this method was used without any step-length calculations. Thus, the iterative scheme was as follows:

$$\mathbf{x}^{k+1} = \mathbf{x}^k - [\mathbf{H}(\mathbf{x}^k)]^{-1}\nabla f(\mathbf{x}^k)$$

However, in this form the method has a tendency to diverge when started from a point that is far away from the optimum. The so-called *Modified Newton method* uses the direction given by the Newton method and then computes an appropriate step length along this direction. This makes the method very stable. Thus, the iterations are as follows:

$$\mathbf{x}^{k+1} = \mathbf{x}^k + \alpha_k \mathbf{d}^k \quad k = 0, 1, \ldots$$

with $\mathbf{d}^k = -[\mathbf{H}(\mathbf{x}^k)]^{-1} \nabla f(\mathbf{x}^k)$ and α_k obtained from minimizing $f(\mathbf{x}^k + \alpha_k \mathbf{d}^k)$.

The method has the fastest convergence rate of all the methods considered in this chapter. Each iteration, however, requires more computations because of the need to evaluate the Hessian matrix and then to solve the system of equations for getting the direction. The following numerical examples show the performance of this method.

The procedure is implemented in the following function that is included in the OptimizationToolbox 'Unconstrained' package.

Needs["OptimizationToolbox`Unconstrained`"];
?ModifiedNewton

ModifiedNewton[f, vars, x0, opts]. Computes minimum of f(vars) starting
  from x0 using Modified Newton method. See Options[ModifiedNewton] to
  see a list of options for the function. The function returns {x, hist}.
  'x' is either the optimum point or the next point after MaxIterations.
  'hist' contains history of values tried at different iterations.

Options[ModifiedNewton]
{PrintLevel → 1, MaxIterations → 50,
  ConvergenceTolerance → 0.01, StepLengthVar → α}

Example 5.22 Use the modified Newton method and the given starting point to find the minimum of the following function:

f = x⁴ + y⁴ + 2x²y² - 4x + 3;
vars = {x, y}; x0 = {1.25, 1.25};

All intermediate calculations are shown for the first iteration.

ModifiedNewton[f, vars, x0, PrintLevel → 2, MaxIterations → 1];

$f \to 3 - 4x + x^4 + 2x^2 y^2 + y^4$

$\nabla f \to \begin{pmatrix} -4 + 4x^3 + 4xy^2 \\ 4x^2 y + 4y^3 \end{pmatrix}$

$\nabla^2 f \to \begin{pmatrix} 12x^2 + 4y^2 & 8xy \\ 8xy & 4x^2 + 12y^2 \end{pmatrix}$

***** Iteration 1 ***** Current point → {1.25, 1.25}

Direction finding phase:

$$\nabla^2 f \to \begin{pmatrix} 25. & 12.5 \\ 12.5 & 25. \end{pmatrix} \quad \nabla f(x) \to \begin{pmatrix} 11.625 \\ 15.625 \end{pmatrix}$$

$||\nabla f(x)|| \to 19.4751 \quad f(x) \to 7.76563$

$d \to (-0.203333 \quad -0.523333)$

Step length calculation phase:

$$xk1 \to \begin{pmatrix} 1.25 - 0.203333\alpha \\ 1.25 - 0.523333\alpha \end{pmatrix}$$

$$\nabla f(xk1) \to \begin{pmatrix} -0.256381(-1.547 + \alpha)(29.3101 - 10.3637\alpha + \alpha^2) \\ -0.659865(-2.3885 + \alpha)(9.91364 - 5.76313\alpha + \alpha^2) \end{pmatrix}$$

$d\phi/d\alpha \equiv \nabla f(xk1) \cdot d = 0 \to 0.39746(-2.01494 + \alpha)(13.1619 - 6.62975\alpha + \alpha^2) == 0$

$\alpha \to 2.01494$

New Point (Non-Optimum): {0.840295, 0.195514} after 1 iterations

The printout of intermediate results is suppressed (default option), and the method is allowed to continue until convergence. Computation history is saved in *hist*.

{opt, hist} = ModifiedNewton[f, vars, x0];

Optimum: $\{1., 2.60644 \times 10^{-7}\}$ after 3 iterations

TableForm[hist]

x	d	$\|\|\nabla f(x)\|\|$	f(x)
1.25	-0.203333	19.4751	7.76563
1.25	-0.523333		
0.840295	0.213735	1.60731	0.192834
0.195514	-0.262867		
0.999754	0.000246209	0.00380209	1.08252×10^{-6}
-0.000599898	0.000600194		
1.	--	2.25383×10^{-6}	3.02404×10^{-13}
2.60644×10^{-7}			

The first column is extracted from the history, and the heading is removed to get a list of all intermediate points tried by the algorithm.

xhist = Drop[Transpose[hist][[1]], 1]

$\{\{1.25, 1.25\}, \{0.840295, 0.195514\},$
$\{0.999754, -0.000599898\}, \{1., 2.60644 \times 10^{-7}\}\}$

5.3 Unconstrained Minimization Techniques

Using the PlotSearchPath function, the search path is shown on a contour map of function f. The search path clearly demonstrates that the method eliminates the zigzagging associated with the steepest descent method.

```
PlotSearchPath[f, {x, 0, 1.3}, {y, 0, 1.3}, xhist, PlotPoints → 30,
  Epilog → {RGBColor[1, 0, 0], Disk[{1, 0}, 0.015],
    Text["Optimum", {1.01, 0.05}, {-1, 0}]}];
```

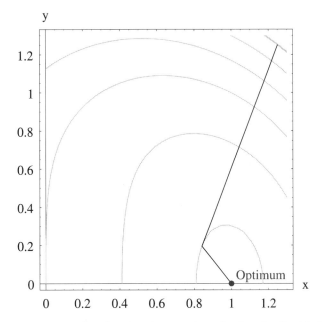

FIGURE 5.19 A contour plot showing the search path taken by the Modified Newton method.

The column containing the norm of the function gradient is extracted in preparation for a plot showing how the $\|\nabla f\|$ is decreasing with the number of iterations. The plot shows the rate of convergence of the method.

```
normGradhist = Drop[Transpose[hist][[3]], 1]
```

$\{19.4751, 1.60731, 0.00380209, 2.25383 \times 10^{-6}\}$

```
ListPlot[normGradhist,
  PlotJoined → True, AxesLabel → {"Iterations", "||∇f||"}];
```

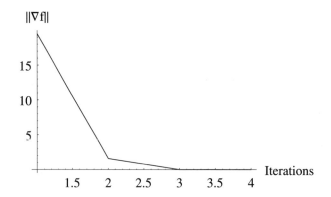

FIGURE 5.20 A plot showing $\|\nabla f\|$ at different iterations.

Example 5.23 Use the modified Newton method and the given starting point to find the minimum of the following function.

```
f = (x + y)² + (2 (x² + y² - 1) - ⅓)²;
vars = {x, y}; x0 = {-1.25, 0.25};
```

All intermediate calculations are shown for the first two iterations.

```
ModifiedNewton[f, vars, x0, PrintLevel → 2, MaxIterations → 2];
```

$$f \to (x+y)^2 + \left(-\frac{1}{3} + 2(-1 + x^2 + y^2)\right)^2$$

$$\nabla f \to \begin{pmatrix} -\frac{50x}{3} + 16x^3 + 2y + 16xy^2 \\ 2x - \frac{50y}{3} + 16x^2y + 16y^3 \end{pmatrix}$$

$$\nabla^2 f \to \begin{pmatrix} -\frac{50}{3} + 48x^2 + 16y^2 & 2 + 32xy \\ 2 + 32xy & -\frac{50}{3} + 16x^2 + 48y^2 \end{pmatrix}$$

***** Iteration 1 ***** Current point → {-1.25, 0.25}

Direction finding phase:

$$\nabla^2 f \to \begin{pmatrix} 59.3333 & -8. \\ -8. & 11.3333 \end{pmatrix} \quad \nabla f(x) \to \begin{pmatrix} -11.1667 \\ -0.166667 \end{pmatrix}$$

$\|\nabla f(x)\| \to 11.1679 \quad f(x) \to 1.84028$

$d \to \begin{pmatrix} 0.21019 & 0.163075 \end{pmatrix}$

Step length calculation phase:

$$xk1 \to \begin{pmatrix} -1.25 + 0.21019\alpha \\ 0.25 + 0.163075\alpha \end{pmatrix}$$

5.3 Unconstrained Minimization Techniques   277

$$\nabla f(xk1) \to \begin{pmatrix} 0.238013\,(-1.53958+\alpha)\,(30.4732-10.6801\alpha+\alpha^2) \\ 0.184662\,(-4.58573+\alpha)\,(0.196817-0.153898\alpha+\alpha^2) \end{pmatrix}$$

$d\phi/d\alpha \equiv \nabla f(xk1).d = 0 \to 0.0801418\,(-2.00245+\alpha)\,(14.795-7.40655\alpha+\alpha^2) == 0$

$\alpha \to 2.00244$

***** Iteration 2 ***** Current point $\to \{-0.829106, 0.57655\}$

Direction finding phase:

$$\nabla^2 f \to \begin{pmatrix} 21.6479 & -13.296 \\ -13.2967 & 10.2876 \end{pmatrix} \quad \nabla f(x) \to \begin{pmatrix} 1.44283 \\ -1.85969 \end{pmatrix}$$

$||\nabla f(x)|| \to 2.35377 \quad f(x) \to 0.150034$

$d \to \begin{pmatrix} 0.215323 & 0.459071 \end{pmatrix}$

Step length calculation phase:

$$xk1 \to \begin{pmatrix} -0.829106 + 0.215323\alpha \\ 0.57655 + 0.459071\alpha \end{pmatrix}$$

$$\nabla f(xk1) \to \begin{pmatrix} 0.885788\,(-3.5121+\alpha)\,(-0.535061+\alpha)\,(0.866795+\alpha) \\ 1.88851\,(-0.467862+\alpha)\,(2.10476+2.39392\alpha+\alpha^2) \end{pmatrix}$$

$d\phi/d\alpha \equiv \nabla f(xk1).d = 0 \to 1.05769\,(-0.443886+\alpha)\,(1.15668+1.44912\alpha+\alpha^2) == 0$

$\alpha \to 0.443886$

New Point (Non-Optimum): $\{-0.733527, 0.780325\}$ after 2 iterations

{opt, hist} = ModifiedNewton[f, vars, x0];

Optimum: $\{-0.7637625, 0.763763\}$ after 4 iterations

TableForm[hist]

x	d	$\|\|\nabla f(x)\|\|$	f(x)
-1.25	0.21019	11.1679	1.84028
0.25	0.163075		
-0.829106	0.215323	2.35377	0.150034
0.57655	0.459071		
-0.733527	-0.0327749	0.358729	0.0037421
0.780325	-0.0180651		
-0.763825	0.0000626412	0.00205812	9.25196×10^{-8}
0.763625	0.000137636		
-0.763763	--	4.88312×10^{-7}	9.63452×10^{-15}
0.763763			

xhist = Drop[Transpose[hist][[1]], 1];

PlotSearchPath[f, {x, -1.25, 0}, {y, 0, 1.25}, xhist, PlotPoints → 50,
  Epilog → {RGBColor[1, 0, 0], Disk[{-.76, .76}, 0.01],
    Text["Optimum", {-.74, .76}, {-1, 0}]}];

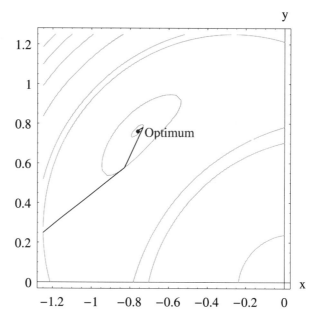

FIGURE 5.21 A contour plot showing the search path taken by the Modified Newton method.

```
normGradhist = Drop[Transpose[hist][[3]],1];

ListPlot[normGradhist,
   PlotJoined → True, AxesLabel → {"Iterations","||∇f||"}];
```

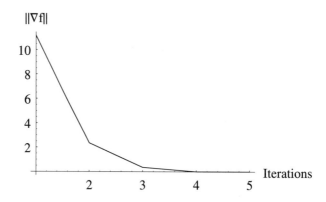

FIGURE 5.22 A plot showing $\|\nabla f\|$ at different iterations.

5.3 Unconstrained Minimization Techniques

Example 5.24 Use the modified Newton method and the given starting point to find the minimum of the following function:

```
f = (x₁ - 1)^4 + (x₂ - 3)^2 + 4(x₃ + 5)^2;
vars = {x₁, x₂, x₃}; x0 = {-1, -2, 1};
```

All intermediate calculations are shown for the first two iterations.

ModifiedNewton[f, vars, x0, PrintLevel → 2, MaxIterations → 2];

$f \to (-1 + x_1)^4 + (-3 + x_2)^2 + 4(5 + x_3)^2$

$\nabla f \to \begin{pmatrix} -4 + 12x_1 - 12x_1^2 + 4x_1^3 \\ -6 + 2x_2 \\ 40 + 8x_3 \end{pmatrix}$

$\nabla^2 f \to \begin{pmatrix} 12 - 24x_1 + 12x_1^2 & 0 & 0 \\ 0 & 2 & 0 \\ 0 & 0 & 8 \end{pmatrix}$

***** Iteration 1 ***** Current point → {-1, -2, 1}

Direction finding phase:

$\nabla^2 f \to \begin{pmatrix} 48 & 0 & 0 \\ 0 & 2 & 0 \\ 0 & 0 & 8 \end{pmatrix} \quad \nabla f(x) \to \begin{pmatrix} -32 \\ -10 \\ 48 \end{pmatrix}$

$||\nabla f(x)|| \to 58.5491 \quad f(x) \to 185$

$d \to \begin{pmatrix} \frac{2}{3} & 5 & -6 \end{pmatrix}$

Step length calculation phase:

$xk1 \to \begin{pmatrix} -1 + \frac{2\alpha}{3} \\ -2 + 5\alpha \\ 1 - 6\alpha \end{pmatrix}$

$\nabla f(xk1) \to \begin{pmatrix} \frac{32}{27}(-3 + \alpha)^3 \\ 10(-1 + \alpha) \\ -48(-1 + \alpha) \end{pmatrix}$

$d\phi/d\alpha \equiv \nabla f(xk1) \cdot d = 0 \to \frac{2}{81}(-14553 + 14553\alpha - 288\alpha^2 + 32\alpha^3) == 0$

$\alpha \to 1.0182$

***** Iteration 2 ***** Current point → {-0.321203, 3.09098, -5.10917}

Direction finding phase:

$\nabla^2 f \to \begin{pmatrix} 20.9469 & 0 & 0 \\ 0 & 2 & 0 \\ 0 & 0 & 8 \end{pmatrix} \quad \nabla f(x) \to \begin{pmatrix} -9.22505 \\ 0.181954 \\ -0.873378 \end{pmatrix}$

$||\nabla f(x)|| \to 9.26809 \quad f(x) \to 3.10299$

$d \to \begin{pmatrix} 0.440401 & -0.0909768 & 0.109172 \end{pmatrix}$

Step length calculation phase:

$$\text{xk1} \to \begin{pmatrix} -0.321203 + 0.440401\alpha \\ 3.09098 - 0.0909768\alpha \\ -5.10917 + 0.109172\alpha \end{pmatrix}$$

$$\nabla f(\text{xk1}) \to \begin{pmatrix} 0.341669\,(-2.99998 + \alpha)\,(9.00005 - 6.00002\alpha + \alpha^2) \\ 0.181954 - 0.181954\alpha \\ -0.873378 + 0.873378\alpha \end{pmatrix}$$

$d\phi/d\alpha \equiv \nabla f(\text{xk1}) \cdot d = 0 \to 0.150471\,(-2.07259 + \alpha)\,(13.386 - 6.92741\alpha + \alpha^2) == 0$

$\alpha \to 2.07259$

New Point (Non-Optimum): {0.591567, 2.90242, -4.8829} after 2 iterations

{opt, hist} = ModifiedNewton[f, vars, x0];

Optimum: {0.96445, 2.99963, -4.99956} after 6 iterations

TableForm[hist]

x	d	$\|\|\nabla f(x)\|\|$	f(x)
-1	2	58.5491	185
-2	3		
1	5		
	-6		
-0.321203	0.440401	9.26809	3.10299
3.09098	-0.0909768		
-5.10917	0.109172		
0.591567	0.136144	0.994942	0.0921967
2.90242	0.0975807		
-4.8829	-0.117097		
0.738062	0.0873128	0.102274	0.00507961
3.00742	-0.00741852		
-5.0089	0.00890222		
0.886003	0.037999	0.0508607	0.000348261
2.99485	0.0051513		
-4.99382	-0.00618156		
0.929607	0.0234643	0.00758056	0.0000284567
3.00076	-0.000759837		
-5.00091	0.000911804		
0.96445	--	0.0036177	2.51501×10^{-6}
2.99963			
-4.99956			

normGradhist = Drop[Transpose[hist][[3]], 1];

ListPlot[normGradhist,
  PlotJoined → True, AxesLabel → {"Iterations", "||∇f||"}];

5.3 Unconstrained Minimization Techniques

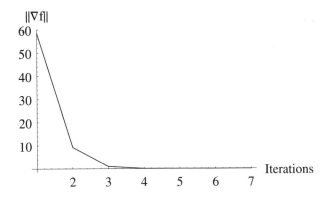

FIGURE 5.23 A plot showing $\|\nabla f\|$ at different iterations.

Example 5.25 To compare the performance of different unconstrained methods, consider the following example:

```
f = Exp[-x - y] + xy + y^2;
vars = {x, y}; x0 = {-1.4, 1.5};
```

Using optimality conditions, we can see that the function does not have a maximum or a minimum. It only has a stationary point at $(-1, 1)$.

```
UnconstrainedOptimality[f, vars,
   SolveEquationsUsing → FindRoot, StartingSolution → x0];
```

Objective function $\to E^{-x-y} + xy + y^2$

Gradient vector $\to \begin{pmatrix} -E^{-x-y} + y \\ -E^{-x-y} + x + 2y \end{pmatrix}$

Hessian matrix $\to \begin{pmatrix} E^{-x-y} & 1 + E^{-x-y} \\ 1 + E^{-x-y} & 2 + E^{-x-y} \end{pmatrix}$

****** First order optimality conditions ******

Necessary conditions $\to \begin{pmatrix} -E^{-x-y} + y == 0 \\ -E^{-x-y} + x + 2y == 0 \end{pmatrix}$

Possible solutions (stationary points) $\to (x \to -1.\ \ y \to 1.)$

****** Second order optimality conditions ******

------- Point $\to \{x \to -1., y \to 1.\}$

Hessian $\to \begin{pmatrix} 1. & 2. \\ 2. & 3. \end{pmatrix}$ Principal minors $\to \begin{pmatrix} 1. \\ -1. \end{pmatrix}$

Status $\to$ InflectionPoint   Function value $\to 1$.

Using the modified Newton method, we readily get this solution in two iterations.

{sol, hist} = ModifiedNewton[f, vars, x0, PrintLevel → 2, MaxIterations → 2];

$f \to E^{-x-y} + xy + y^2$

$\nabla f \to \begin{pmatrix} -E^{-x-y} + y \\ -E^{-x-y} + x + 2y \end{pmatrix}$

$\nabla^2 f \to \begin{pmatrix} E^{-x-y} & 1 + E^{-x-y} \\ 1 + E^{-x-y} & 2 + E^{-x-y} \end{pmatrix}$

***** Iteration 1 ***** Current point → {-1.4, 1.5}

Direction finding phase:

$\nabla^2 f \to \begin{pmatrix} 0.904837 & 1.90484 \\ 1.90484 & 2.90484 \end{pmatrix}$ $\nabla f(x) \to \begin{pmatrix} 0.595163 \\ 0.695163 \end{pmatrix}$

$||\nabla f(x)|| \to 0.915134$ $f(x) \to 1.05484$

$d \to (0.404679 \quad -0.504679)$

Step length calculation phase:

$xk1 \to \begin{pmatrix} -1.4 + 0.404679\alpha \\ 1.5 - 0.504679\alpha \end{pmatrix}$

$\nabla f(xk1) \to \begin{pmatrix} 1.5 - 0.9048374 E^{0.1\alpha} - 0.504679\alpha \\ 1.6 - 0.9048374 E^{0.1\alpha} - 0.604679\alpha \end{pmatrix}$

$d\phi/d\alpha \equiv \nabla f(xk1).d = 0 \to -0.200468 + 0.09048374 E^{0.1\alpha} + 0.100936\alpha == 0$

$\alpha \to 0.995782$

***** Iteration 2 ***** Current point → {-0.997028, 0.99745}

Direction finding phase:

$\nabla^2 f \to \begin{pmatrix} 0.999578 & 1.99958 \\ 1.99958 & 2.99958 \end{pmatrix}$ $\nabla f(x) \to \begin{pmatrix} -0.00212859 \\ -0.00170681 \end{pmatrix}$

$||\nabla f(x)|| \to 0.00272838$ $f(x) \to 0.999999$

$d \to (-0.00297194 \quad 0.00255017)$

Step length calculation phase:

$xk1 \to \begin{pmatrix} -0.997028 - 0.00297194\alpha \\ 0.99745 + 0.00255017\alpha \end{pmatrix}$

$\nabla f(xk1) \to \begin{pmatrix} 0.99745 - 0.999578 E^{0.000421769\alpha} + 0.00255017\alpha \\ 0.997872 - 0.999578 E^{0.000421769\alpha} + 0.0021284\alpha \end{pmatrix}$

$d\phi/d\alpha \equiv \nabla f(xk1).d = 0 \to -0.000419618 + 0.000421591 E^{0.000421769\alpha} - 2.15117 \times 10^{-6}\alpha == 0$

$\alpha \to 1.$

Optimum: {-1., 1.} after 2 iterations

xhist = Drop[Transpose[hist][[1]], 1]

{{-1.4, 1.5}, {-0.997028, 0.99745}, {-1., 1.}}

The search path shown in the following figure shows the direct path that the method takes.

5.3 Unconstrained Minimization Techniques

```
PlotSearchPath[f, {x, -1.5, 0}, {y, 0, 1.5}, xhist, PlotPoints → 30, Epilog → {
   RGBColor[1, 0, 0], Disk[{-.96, .98}, 0.015],
   Line[{{-0.96, 0.98}, {-.75, 1.2}}],
   Text["Stationarypoint", {-.72, 1.2}, {-1, 0}]}];
```

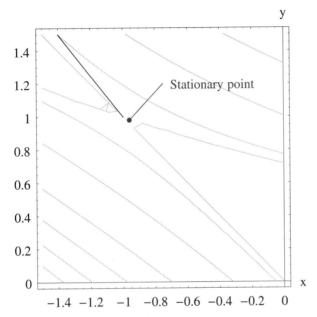

FIGURE 5.24 Convergence to the inflection point of $f(x, y) = e^{-x-y} + xy + y^2$ using the modified Newton method.

The steepest descent method fails to find a positive step length at the second iteration.

```
SteepestDescent[f, vars, x0, PrintLevel → 2, MaxIterations → 2];
```

$f \to E^{-x-y} + xy + y^2$

$\nabla f \to \begin{pmatrix} -E^{-x-y} + y \\ -E^{-x-y} + x + 2y \end{pmatrix}$

***** Iteration 1 ***** Current point → {-1.4, 1.5}

Direction finding phase:

$\nabla f(x) \to \begin{pmatrix} 0.595163 \\ 0.695163 \end{pmatrix} \quad d \to \begin{pmatrix} -0.595163 \\ -0.695163 \end{pmatrix}$

$||\nabla f(x)|| \to 0.915134 \quad f(x) \to 1.05484$

Step length calculation phase:

$$xk1 \to \begin{pmatrix} -1.4 - 0.595163\alpha \\ 1.5 - 0.695163\alpha \end{pmatrix}$$

$$\nabla f(xk1) \to \begin{pmatrix} 1.5 - 0.904837 E^{1.29033\alpha} - 0.695163\alpha \\ 1.6 - 0.904837 E^{1.29033\alpha} - 1.98549\alpha \end{pmatrix}$$

$d\phi/d\alpha \equiv \nabla f(xk1).d = 0 \to -2.005 + 1.16753 E^{1.29033\alpha} + 1.79397\alpha == 0$

$\alpha \to 0.235606$

***** Iteration 2 ***** Current point $\to \{-1.54022, 1.33622\}$

Direction finding phase:

$$\nabla f(x) \to \begin{pmatrix} 0.109909 \\ -0.0940984 \end{pmatrix} \quad d \to \begin{pmatrix} -0.109909 \\ 0.0940984 \end{pmatrix}$$

$||\nabla f(x)|| \to 0.144688 \quad f(x) \to 0.953709$

Step length calculation phase:

$$xk1 \to \begin{pmatrix} -1.54022 - 0.109909\alpha \downarrow \\ 1.33622 + 0.0940984\alpha \end{pmatrix}$$

$$\nabla f(xk1) \to \begin{pmatrix} 1.33622 - 1.22631 E^{0.0158105\alpha} + 0.0940984\alpha \\ 1.13221 - 1.22631 E^{0.0158105\alpha} + 0.0782874\alpha \end{pmatrix}$$

$d\phi/d\alpha \equiv \nabla f(xk1).d = 0 \to -0.0403231 + 0.0193886 E^{0.0158105\alpha} - 0.00297549\alpha == 0$

$\alpha \to -7.79046$

Unconstrained:: step: Negative step length (-7.79046) found. Check direction or change line search parameters.

$$xk1 \to \begin{pmatrix} -1.54022 - 0.109909\alpha \\ 1.33622 + 0.0940984\alpha \end{pmatrix}$$

$$\nabla f(xk1) \to \begin{pmatrix} 1.33622 - 1.22631 E^{0.0158105\alpha} + 0.0940984\alpha \\ 1.13221 - 1.22631 E^{0.0158105\alpha} + 0.0782879\alpha \end{pmatrix}$$

$d\phi/d\alpha \equiv \nabla f(xk1).d = 0 \to -0.0403231 + 0.0193886 E^{0.0158105\alpha} - 0.00297549\alpha == 0$

$\alpha \to -7.79046$

New Point (Non-Optimum): $\{-0.683983, 0.603147\}$ after 2 iterations

In general, a negative step length is not acceptable because it implies an increase in the function value. However, if we ignore this and let the method continue, it does converge to the inflection point in 35 iterations.

{opt, hist} = SteepestDescent[f, vars, x0];
Optimum: $\{-0.964315, 0.978407\}$ after 35 iterations

The following search path dramatically shows the difficulty the method is having with this function.

xhist = Drop[Transpose[hist][[1]], 1];

**PlotSearchPath[f, {x, -1.5, 0}, {y, 0, 1.5}, xhist, PlotPoints → 30,
   Epilog → {RGBColor[1, 0, 0], Disk[{-.96, .98}, 0.015],**

5.3 Unconstrained Minimization Techniques 285

```
Line[{{-0.96,0.98},{-.75,1.2}}],
Text["Stationarypoint",{-.72,1.2},{-1,0}]}];
```

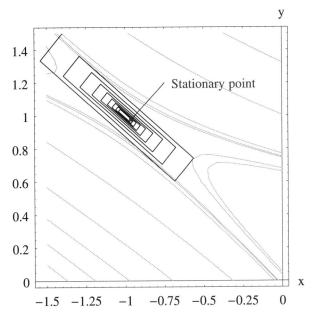

FIGURE 5.25 Convergence to the inflection point of $f(x, y) = e^{-x-y} + xy + y^2$ using the steepest descent method.

The conjugate gradient method encounters a negative step length also; however, if we let it continue, the convergence is quite rapid.

```
ConjugateGradient[f, vars, x0, PrintLevel → 2, MaxIterations → 2];
```

$f \to E^{-x-y} + xy + y^2$

$\nabla f \to \begin{pmatrix} -E^{-x-y} + y \\ -E^{-x-y} + x + 2y \end{pmatrix}$

Using PolakRibiere method with Exact line search
***** Iteration 1 ***** Current point → {-1.4, 1.5}

Direction finding phase:

$\nabla f(x) \to \begin{pmatrix} 0.595163 \\ 0.695163 \end{pmatrix} \quad d \to \begin{pmatrix} -0.595163 \\ -0.695163 \end{pmatrix}$

$||\nabla f(x)|| \to 0.915134 \quad \beta \to 0.$

$f(x) \to 1.05484$

Step length calculation phase:

$$xk1 \to \begin{pmatrix} -1.4 - 0.595163\alpha \\ 1.5 - 0.695163\alpha \end{pmatrix}$$

$$\nabla f(xk1) \to \begin{pmatrix} 1.5 - 0.904837E^{1.29033\alpha} - 0.695163\alpha \\ 1.6 - 0.904837E^{1.29033\alpha} - 1.98549\alpha \end{pmatrix}$$

$d\phi/d\alpha \equiv \nabla f(xk1).d = 0 \to -2.005 + 1.16753E^{1.29033\alpha} + 1.79397\alpha == 0$

$\alpha \to 0.235605$

***** Iteration 2 ***** Current point $\to \{-1.54022, 1.33622\}$

Direction finding phase:

$$\nabla f(x) \to \begin{pmatrix} 0.109909 \\ -0.0940984 \end{pmatrix} \quad d \to \begin{pmatrix} -0.124786 \\ 0.0767212 \end{pmatrix}$$

$||\nabla f(x)|| \to 0.144688 \quad \beta \to 0.0249973$

$f(x) \to 0.953709$

Step length calculation phase:

$$xk1 \to \begin{pmatrix} -1.54022 - 0.124786\alpha \\ 1.33622 + 0.0767212\alpha \end{pmatrix}$$

$$\nabla f(xk1) \to \begin{pmatrix} 1.33622 - 1.22631E^{0.0480652\alpha} + 0.0767212\alpha \\ 1.13221 - 1.22631E^{0.0480652\alpha} + 0.028656\alpha \end{pmatrix}$$

$d\phi/d\alpha \equiv \nabla f(xk1).d = 0 \to -0.0798772 + 0.0589427E^{0.0480652\alpha} - 0.00737524\alpha == 0$

$\alpha \to -4.34465$

Unconstrained:: step: Negative step length (-4.34465) found. Check direction or change line search parameters.

$$xk1 \to \begin{pmatrix} -1.54022 - 0.124786\alpha \\ 1.33622 + 0.0767212\alpha \end{pmatrix}$$

$$\nabla f(xk1) \to \begin{pmatrix} 1.33622 - 1.22631E^{0.0480652\alpha} + 0.0767212\alpha \\ 1.13221 - 1.22631E^{0.0480652\alpha} + 0.028656\alpha \end{pmatrix}$$

$d\phi/d\alpha \equiv \nabla f(xk1).d = 0 \to -0.0798772 + 0.0589427E^{0.0480652\alpha} - 0.00737524\alpha == 0$

$\alpha \to -4.34465$

New Point (Non-Optimum): $\{-0.99807, 1.00289\}$ after 2 iterations

{opt, hist} = ConjugateGradient[f, vars, x0];

$$f \to E^{-x-y} + xy + y^2 \quad \nabla f \to \begin{pmatrix} -E^{-x-y} + y \\ -E^{-x-y} + x + 2y \end{pmatrix}$$

Using PolakRibiere method with Exact line search

Negative step length found.

Negative step length found.

Optimum: $\{-1., 0.999998\}$ after 4 iterations

```
TableForm[hist]
```

x	d	||∇ f(x)||	f(x)	β
-1.4	-0.595163	0.915134	1.05484	0.
1.5	-0.695163			
-1.54022	-0.124786	0.144688	0.953709	0.024997
1.33622	0.0767212			
-0.99807	-0.0109613	0.0146931	1.0000	0.026161
1.00289	-0.0105088			
-1.00067	-0.000171299	0.000192734	1.	0.004173
1.0004	0.000102372			
-1.	--	9.23086×10^{-6}	1.	--
0.999998				

```
xhist = Drop[Transpose[hist][[1]],1];

PlotSearchPath[f, {x,-1.5,0}, {y,0,1.5}, xhist, PlotPoints → 30,
  Epilog → {RGBColor[1,0,0], Disk[{-.96,.98},0.015],
    Line[{{-0.96,0.98},{-.75,1.2}}],
    Text["Stationarypoint",{-.72,1.2},{-1,0}]}];
```

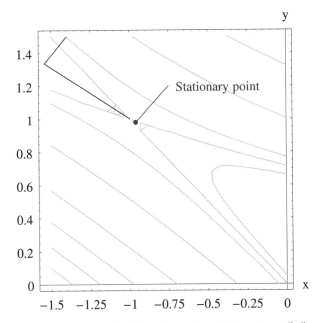

FIGURE 5.26 Convergence to the inflection point of $f(x,y) = e^{-x-y} + xy + y^2$ using the conjugate gradient method.

5.3.4 Quasi-Newton Methods

The Modified Newton method has the fastest convergence rate but has the drawback that it requires computation of the Hessian matrix at each iteration. A class of methods, known as Quasi-Newton methods, has been developed that approaches the convergence rate of the Newton's method but without explicitly requiring the Hessian matrix. In place of the Hessian matrix, these methods start with a positive definite matrix (usually an identity matrix), denoted by $\mathbf{Q}$. Using the gradient information at the current and the previous steps, the matrix $\mathbf{Q}$ is updated so that it approximates the *inverse* of the Hessian matrix. Thus, the basic iteration of all Quasi-Newton methods is as follows:

1. Determine direction: $\quad \mathbf{d}^k = -\mathbf{Q}^k \nabla f(\mathbf{x}^k)$

2. Find step length and update current point $\quad \mathbf{x}^{k+1} = \mathbf{x}^k + \alpha_k \mathbf{d}^k$

3. Update matrix $\mathbf{Q}$ for next iteration $\quad \mathbf{Q}^{k+1}$

Different Quasi-Newton methods differ in the way the matrix $\mathbf{Q}$ is updated.

DFP (Davidon, Fletcher, and Powell) Update

The DFP formula starts with an identity matrix and updates it to approximate the inverse of the Hessian matrix as follows.

$$\mathbf{Q}^{k+1} = \mathbf{Q}^k + \frac{\mathbf{s}^k (\mathbf{s}^k)^T}{(\mathbf{q}^k)^T \mathbf{s}^k} - \frac{(\mathbf{Q}^k \mathbf{q}^k)(\mathbf{Q}^k \mathbf{q}^k)^T}{(\mathbf{q}^k)^T \mathbf{Q}^k \mathbf{q}^k}$$

where

$$\mathbf{q}^k = \nabla f(\mathbf{x}^{k+1}) - \nabla f(\mathbf{x}^k)$$
$$\mathbf{s}^k = \mathbf{x}^{k+1} - \mathbf{x}^k \equiv \alpha_k \mathbf{d}^k$$

It can be shown that $\mathbf{Q}^k$ is positive definite as long as $(\mathbf{q}^k)^T \mathbf{s}^k > 0$. If the step length is computed exactly, this should always be the case. However, with approximate step-length calculations, this is not necessarily the case. Hence, in numerical implementations, the matrix $\mathbf{Q}^k$ is updated only if $(\mathbf{q}^k)^T \mathbf{s}^k > 0$. To avoid problems due to round off, it is desirable to reset the $\mathbf{Q}$ matrix to the identity matrix after a specified number of iterations. Usually the number of iterations after which it is reset is equal to the number of optimization variables.

BFGS (Broyden, Fletcher, Goldfarb, and Shanon) Update

The BFGS formula is similar to DFP in that it also updates the inverse Hessian matrix. However, the numerical experiments show it to be consistently superior to the DFP. The formula is as follows:

$$\mathbf{Q}^{k+1} = \mathbf{Q}^k + \left(1 + \frac{(\mathbf{q}^k)^T \mathbf{Q}^k \mathbf{q}^k}{(\mathbf{q}^k)^T \mathbf{s}^k}\right) \frac{\mathbf{s}^k (\mathbf{s}^k)^T}{(\mathbf{q}^k)^T \mathbf{s}^k}$$

$$- \frac{1}{(\mathbf{q}^k)^T \mathbf{s}^k} \left[\left\{\mathbf{s}^k (\mathbf{q}^k)^T \mathbf{Q}^k\right\}^T + \mathbf{s}^k (\mathbf{q}^k)^T \mathbf{Q}^k\right]$$

The numerical considerations to update and reset $\mathbf{Q}$ are the same as those for DFP.

The procedures are implemented in the following function that is included in the OptimizationToolbox 'Unconstrained' package.

```
Needs["OptimizationToolbox`Unconstrained`"];
?QuasiNewtonMethod
```

QuasiNewtonMethod[f, vars, x0, opts]. Computes minimum of f(vars)
  starting from x0 using a Quasi-Newton method. User has the option
  of using either 'BFGS' (default) or 'DFP' method. Also step length
  can be computed using either the 'Exact' analytical line search
  or an 'Approximate' line search using Armijo's rule. The number of
  iterations after which the inverse hessian approximation is reset
  to identity matrix can be specified using ResetHessian option. The
  default is Automatic in which case the reset interval is set to the
  number of variables. See Options[QuasiNewtonMethod] to see a list
  of options for the function. The function returns {x, hist}. 'x' is
  either the optimum point or the next point after MaxIterations. 'hist'
  contains history of values tried at different iterations.

```
Options[QuasiNewtonMethod]
```
{PrintLevel → 1, MaxIterations → 50, ConvergenceTolerance → 0.01,
  StepLengthVar → α, Method → BFGS, LineSearch → Exact,
  ArmijoParameters → {0.2, 2}, ResetHessian → Automatic}

Example 5.26 Use a Quasi-Newton method and the given starting point to find the minimum of the following function:

```
f = x^4 + y^4 + 2x^2y^2 - 4x + 3;
vars = {x, y}; x0 = {1.25, 1.25};
```

All intermediate calculations are shown for the first two iterations.

```
QuasiNewtonMethod[f, vars, x0, PrintLevel → 2, MaxIterations → 2];
```
$f \to 3 - 4x + x^4 + 2x^2y^2 + y^4$

Chapter 5 Unconstrained Problems

$$\nabla f \to \begin{pmatrix} -4 + 4x^3 + 4xy^2 \\ 4x^2y + 4y^3 \end{pmatrix}$$

Using BFGS method with approximate inverse hessian reset after 2 iterations and Exact line search

***** Iteration 1 ***** Current point → {1.25, 1.25}

Direction finding phase:

Inverse Hessian → $\begin{pmatrix} 1 & 0 \\ 0 & 1 \end{pmatrix}$ $\nabla f(x) \to \begin{pmatrix} 11.625 \\ 15.625 \end{pmatrix}$

$||\nabla f(x)|| \to 19.4751$ $f(x) \to 7.76563$

d → (-11.625 -15.625)

Step length calculation phase:

xk1 → $\begin{pmatrix} 1.25 - 11.625\alpha \\ 1.25 - 15.625\alpha \end{pmatrix}$

$\nabla f(xk1) \to \begin{pmatrix} -17636.5(-0.0353314 + \alpha)(0.018656 - 0.251812\alpha + \alpha^2) \\ -23705.0(-0.08 + \alpha)(0.00823927 - 0.179616\alpha + \alpha^2) \end{pmatrix}$

$d\phi/d\alpha \equiv \nabla f(xk1).d = 0 \to$
$$575417.(-0.04791367 + \alpha)(0.0137569 - 0.22151\alpha + \alpha^2) == 0$$

α → 0.0479137

Updating inverse hessian:

q → $\begin{pmatrix} -13.597 \\ -14.1578 \end{pmatrix}$ qT.s → 18.1728

Q.q → $\begin{pmatrix} -13.597 \\ -14.1578 \end{pmatrix}$ s.sT → $\begin{pmatrix} 0.310245 & 0.416996 \\ 0.416996 & 0.560479 \end{pmatrix}$

Q.q.sT → $\begin{pmatrix} 7.57347 & 10.1794 \\ 7.88587 & 10.5993 \end{pmatrix}$

[Q.q.sT]T + Q.q.sT → $\begin{pmatrix} 15.1469 & 18.0653 \\ 18.0653 & 21.1986 \end{pmatrix}$

Q → $\begin{pmatrix} 0.545557 & -0.484603 \\ -0.484603 & 0.518285 \end{pmatrix}$

***** Iteration 2 ***** Current point → {0.693003, 0.501349}

Direction finding phase:

Inverse Hessian → $\begin{pmatrix} 0.545557 & -0.484603 \\ -0.484603 & 0.518285 \end{pmatrix}$ $\nabla f(x) \to \begin{pmatrix} -1.97198 \\ 1.46715 \end{pmatrix}$

$||\nabla f(x)|| \to 2.4579$ $f(x) \to 0.763231$

d → (1.78682 -1.71603)

5.3 Unconstrained Minimization Techniques

Step length calculation phase:

$$xk1 \to \begin{pmatrix} 0.693003 + 1.78682\alpha \\ 0.501349 - 1.71603\alpha \end{pmatrix}$$

$$\nabla f(xk1) \to \begin{pmatrix} 43.8663 \, (-0.162591 + \alpha) \, (0.276488 + 0.673591\alpha + \alpha^2) \\ -42.1285 \, (-0.292156 + \alpha) \, (0.119203 + 0.123158\alpha + \alpha^2) \end{pmatrix}$$

$d\phi/d\alpha \equiv \nabla f(xk1).d = 0 \to$
$$150.675 \, (-0.197956 + \alpha) \, (0.202543 + 0.382693\alpha + \alpha^2) == 0$$

$\alpha \to 0.197956$

New Point (Non-Optimum): {1.0467, 0.161649} after 2 iterations

The printout of the intermediate results is suppressed (default option), and the method is allowed to continue until convergence. Computation history is saved in *hist*.

{opt, hist} = QuasiNewtonMethod[f, vars, x0, ResetHessian → 10];

Using BFGS method with approximate inverse hessian reset after 10 iterations and Exact line search

Optimum: {1.00004, 0.000106729} after 4 iterations

TableForm[hist]

x	d	$\|\|\nabla f(x)\|\|$	f(x)	Inverse Hessian	
1.25	-11.625	19.4751	7.76563	1	0
1.25	-15.625			0	1
0.693003	1.78682	2.45789	0.763231	0.545557	
0.501349	-1.71603			-0.484603	
1.04672	-0.0144283	1.00564	0.0714469	0.108982	
0.161649	-0.051902			-0.0847718	
1.00144	-0.00231565	0.018013	0.0000154181	0.119708	
-0.00120258	0.0021541			-0.0493572	
1.00004	--	0.000608993	3.06396×10^{-8}	--	
0.000106729					

The first column is extracted from the history, and the heading is removed to get a list of all intermediate points tried by the algorithm.

xhist = Drop[Transpose[hist][[1]], 1]

{{1.25, 1.25}, {0.693003, 0.501349}, {1.04672, 0.161649},
 {1.00144, -0.00120258}, {1.00004, 0.000106729}}

Using the PlotSearchPath function, the search path is shown on a contour map of function f. The search path clearly demonstrates that the method is behaving similar to the ModifiedNewton's method.

```
PlotSearchPath[f, {x, 0, 1.3}, {y, 0, 1.3}, xhist, PlotPoints → 30,
  Epilog → {RGBColor[1, 0, 0], Disk[{1, 0}, 0.015],
    Text["Optimum", {1.01, 0.05}, {-1, 0}]}];
```

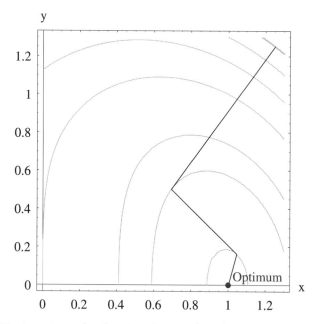

FIGURE 5.27 A contour plot showing the search path taken by the Quasi-Newton's method.

The column containing the norm of the function gradient is extracted in preparation for a plot showing how the $\|\nabla f\|$ is decreasing with the number of iterations. The plot shows the rate of convergence of the method.

```
normGradhist = Drop[Transpose[hist][[3]], 1]
```
{19.4751, 2.4579, 1.00564, 0.018013, 0.000608993}

```
ListPlot[normGradhist,
  PlotJoined → True, AxesLabel → {"Iterations", "||∇f||"}];
```

5.3 Unconstrained Minimization Techniques

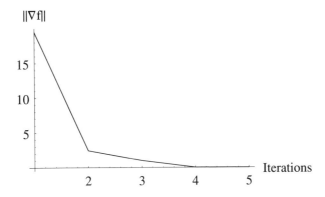

FIGURE 5.28 A plot showing $\|\nabla f\|$ at different iterations.

The same solution is obtained using the DFP updates.

`{opt, hist} = QuasiNewtonMethod[f, vars, x0, Method → DFP, ResetHessian → 10];`

```
Using DFP method with approximate inverse hessian reset after 10
iterations and Exact line search
Optimum: {1.00004, 0.000106729} after 4 iterations
```

Example 5.27 Use a Quasi-Newton method and the given starting point to find the minimum of the following function:

```
f = (x + y)^2 + (2 (x^2 + y^2 - 1) - 1/3)^2;
vars = {x, y}; x0 = {-1.25, 0.25};
```

All intermediate calculations are shown for the first two iterations.

```
QuasiNewtonMethod[f, vars, x0, Method → DFP, PrintLevel → 2,
  MaxIterations → 2, ResetHessian → 10];
```

$$f \to (x+y)^2 + \left(-\frac{1}{3} + 2(-1 + x^2 + y^2)\right)^2$$

$$\nabla f \to \begin{pmatrix} -\frac{50x}{3} + 16x^3 + 2y + 16xy^2 \\ 2x - \frac{50y}{3} + 16x^2y + 16y^3 \end{pmatrix}$$

```
Using DFP method with approximate inverse hessian reset after 10
iterations and Exact line search
***** Iteration 1 ***** Current point → {-1.25, 0.25}

Direction finding phase:
```

$$\text{Inverse Hessian} \to \begin{pmatrix} 1 & 0 \\ 0 & 1 \end{pmatrix} \quad \nabla f(x) \to \begin{pmatrix} -11.1667 \\ -0.166667 \end{pmatrix}$$

$||\nabla f(x)|| \to 11.1679 \quad f(x) \to 1.84028$

$d \to (11.1667 \quad 0.166667)$

Step length calculation phase:

$xk1 \to \begin{pmatrix} -1.25 + 11.1667\alpha \\ 0.25 + 0.166667\alpha \end{pmatrix}$

$\nabla f(xk1) \to \begin{pmatrix} 22283.7 \left(-0.198059 + \alpha\right) \left(-0.115053 + \alpha\right) \left(-0.0219909 + \alpha\right) \\ 332.593 \left(-0.182098 + \alpha\right) \left(0.00188867 + \alpha\right) \left(1.45705 + \alpha\right) \end{pmatrix}$

$d\phi/d\alpha \equiv \nabla f(xk1).d = 0 \to$
$$248890.(-0.197979 + \alpha)(-0.114697 + \alpha)(-0.0220681 + \alpha) == 0$$

$\alpha \to 0.0220681$

Updating inverse hessian:

$q \to \begin{pmatrix} 11.1948 \\ -1.71935 \end{pmatrix} \quad qT.s \to 2.75239$

$Q.q \to \begin{pmatrix} 11.1948 \\ -1.71935 \end{pmatrix} \quad s.sT \to \begin{pmatrix} 0.0607265 & 0.000906365 \\ 0.000906365 & 0.0000135278 \end{pmatrix}$

$Q.q.sT \to \begin{pmatrix} 2.75871 & 0.0411748 \\ -0.423694 & -0.00632379 \end{pmatrix}$

$[Q.q.sT]T + Q.q.sT \to \begin{pmatrix} 5.51742 & -0.382519 \\ -0.382519 & -0.0126476 \end{pmatrix}$

$Q \to \begin{pmatrix} 0.0457651 & 0.154654 \\ 0.154654 & 1.00483 \end{pmatrix}$

***** Iteration 2 ***** Current point $\to \{-1.00357, 0.253678\}$

Direction finding phase:

Inverse Hessian $\to \begin{pmatrix} 0.0457651 & 0.154654 \\ 0.154654 & 1.00483 \end{pmatrix} \quad \nabla f(x) \to \begin{pmatrix} 0.0281495 \\ -1.88601 \end{pmatrix}$

$||\nabla f(x)|| \to 1.88622 \quad f(x) \to 0.598561$

$d \to (0.290392 \quad 1.89077)$

Step length calculation phase:

$xk1 \to \begin{pmatrix} -1.00357 + 0.290392\alpha \\ 0.253678 + 1.8907\alpha \end{pmatrix}$

$\nabla f(xk1) \to \begin{pmatrix} 17.0023 \left(-3.38977 + \alpha\right) \left(-0.0103728 + \alpha\right) \left(0.0470866 + \alpha\right) \\ 110.703 \left(-0.173294 + \alpha\right) \left(0.098311 + 0.41033\alpha + \alpha^2\right) \end{pmatrix}$

$d\phi/d\alpha \equiv \nabla f(xk1).d = 0 \to$
$$214.251(-0.188279 + \alpha)(0.0881986 + 0.342583\alpha + \alpha^2) == 0$$

$\alpha \to 0.188279$

Updating inverse hessian:

$$q \to \begin{pmatrix} -2.3074 \\ 2.23607 \end{pmatrix} \quad qT.s \to 0.669866$$

$$Q.q \to \begin{pmatrix} 0.24022 \\ 1.89002 \end{pmatrix} \quad s.sT \to \begin{pmatrix} 0.00298932 & 0.0194637 \\ 0.0194637 & 0.12673 \end{pmatrix}$$

$$Q.q.sT \to \begin{pmatrix} 0.0131339 & 0.0855162 \\ 0.103336 & 0.672831 \end{pmatrix}$$

$$[Q.q.sT]T + Q.q.sT \to \begin{pmatrix} 0.0262679 & 0.188852 \\ 0.188852 & 1.34566 \end{pmatrix}$$

$$Q \to \begin{pmatrix} 0.035476 & 0.0610589 \\ 0.061059 & 0.222211 \end{pmatrix}$$

New Point (Non-Optimum): {-0.948898, 0.60967} after 2 iterations

```
{opt, hist} = QuasiNewtonMethod[f, vars, x0, ResetHessian → 10];
```

Using BFGS method with approximate inverse hessian reset after 10 iterations and Exact line search

Optimum: {-0.763746, 0.763679} after 5 iterations

TableForm[hist]

x	d	$\|\|\nabla f(x)\|\|$	f(x)	Inverse Hessian	
-1.25	11.1667	11.1679	1.84028	1	0
0.25	0.166667			0	1
-1.00357	0.290392	1.88622	0.598561	0.0457651	0.15
0.253678	1.89077			0.154654	1.00
-0.948898	0.0594846	2.30597	0.159544	0.035476	0.06
0.60967	0.061382			0.0610589	0.22
-0.784161	0.0164554	0.99292	0.0126152	0.0901217	0.06
0.779661	-0.075594			0.0691822	0.18
-0.777963	0.00961856	0.0910383	0.000749257	0.0923642	0.07
0.751188	0.00845028			0.0724209	0.10
-0.763746	--	0.00265284	9.83226×10^{-8}	--	
0.763679					

```
xhist = Drop[Transpose[hist][[1]], 1];

PlotSearchPath[f, {x, -1.25, 0}, {y, 0, 1.25}, xhist, PlotPoints → 50,
   Epilog → {RGBColor[1, 0, 0], Disk[{-.76, .76}, 0.01],
      Text["Optimum", {-.74, .76}, {-1, 0}]}];
```

Chapter 5 Unconstrained Problems

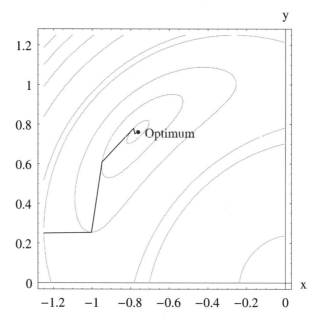

FIGURE 5.29 A contour plot showing the search path taken by the Quasi-Newton method.

```
normGradhist = Drop[Transpose[hist][[3]], 1];

ListPlot[normGradhist,
  PlotJoined → True, AxesLabel → {"Iterations", "||∇f||"}];
```

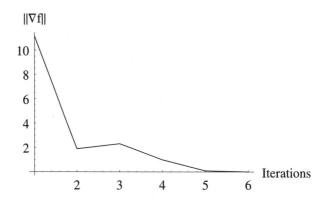

FIGURE 5.30 A plot showing $\|\nabla f\|$ at different iterations.

5.3 Unconstrained Minimization Techniques

Using the DFP updates the solution as follows:

```
{opt, hist} = QuasiNewtonMethod[f, vars, x0, Method → DFP, ResetHessian → 10];
```

Using DFP method with approximate inverse hessian reset after 10
iterations and Exact line search
Optimum: {-0.763746, 0.763679} after 5 iterations

Example 5.28 Use a Quasi-Newton method and the given starting point to find the minimum of the following function:

```
f = (x₁ - 1)⁴ + (x₂ - 3)² + 4(x₃ + 5)²;
vars = {x₁, x₂, x₃}; x0 = {-1, -2, 1};
```

All intermediate calculations are shown for the first two iterations.

```
QuasiNewtonMethod[f, vars, x0, ResetHessian → 10, PrintLevel → 2,
  MaxIterations → 2];
```

$f \to (-1 + x_1)^4 + (-3 + x_2)^2 + 4(5 + x_3)^2$

$\nabla f \to \begin{pmatrix} -4 + 12x_1 - 12x_1^2 + 4x_1^3 \\ -6 + 2x_2 \\ 40 + 8x_3 \end{pmatrix}$

Using BFGS method with approximate inverse hessian reset after 10
iterations and Exact line search
***** Iteration 1 ***** Current point → {-1, -2, 1}

Direction finding phase:

Inverse Hessian → $\begin{pmatrix} 1 & 0 & 0 \\ 0 & 1 & 0 \\ 0 & 0 & 1 \end{pmatrix}$ $\nabla f(x) \to \begin{pmatrix} -32 \\ -10 \\ 48 \end{pmatrix}$

$||\nabla f(x)|| \to 58.5491$ $f(x) \to 185$
$d \to (32 \quad 10 \quad -48)$

Step length calculation phase:

$xk1 \to \begin{pmatrix} -1 + 32\alpha \\ -2 + 10\alpha \\ 1 - 48\alpha \end{pmatrix}$

$\nabla f(xk1) \to \begin{pmatrix} 32(-1 + 16\alpha)^3 \\ -10 + 20\alpha \\ 48 - 384\alpha \end{pmatrix}$

$d\phi/d\alpha \equiv \nabla f(xk1).d = 0 \to 4(-857 + 16946\alpha - 196608\alpha^2 + 1048576\alpha^3) == 0$
$\alpha \to 0.107925$

Updating inverse hessian:

$q \to \begin{pmatrix} 44.2856 \\ 2.1585 \\ -41.4432 \end{pmatrix}$ $qT.s \to 369.967$

Chapter 5 Unconstrained Problems

$$Q.q \to \begin{pmatrix} 44.2856 \\ 2.1585 \\ -41.4432 \end{pmatrix} \quad s.sT \to \begin{pmatrix} 11.9274 & 3.7273 & -17.8911 \\ 3.7273 & 1.16478 & -5.59096 \\ -17.8911 & -5.59096 & 26.8366 \end{pmatrix}$$

$$Q.q.sT \to \begin{pmatrix} 152.945 & 47.7953 & -229.417 \\ 7.45461 & 2.32956 & -11.1819 \\ -143.128 & -44.7276 & 214.693 \end{pmatrix}$$

$$[Q.q.sT]T + Q.q.sT \to \begin{pmatrix} 305.89 & 55.2499 & -372.546 \\ 55.2499 & 4.65913 & -55.9096 \\ -372.546 & -55.9096 & 429.385 \end{pmatrix}$$

$$Q \to \begin{pmatrix} 0.52641 & -0.0389584 & 0.477151 \\ -0.0389584 & 1.0219 & -0.0144481 \\ 0.477151 & -0.0144481 & 0.634124 \end{pmatrix}$$

***** Iteration 2 ***** Current point → {2.4536, -0.920749, -4.1804}

Direction finding phase:

$$\text{Inverse Hessian} \to \begin{pmatrix} 0.52641 & -0.0389584 & 0.477151 \\ -0.0389584 & 1.0219 & -0.0144481 \\ 0.477151 & -0.0144481 & 0.634124 \end{pmatrix} \quad \nabla f(x) \to \begin{pmatrix} 12.285 \\ -7.841 \\ 6.5567 \end{pmatrix}$$

$||\nabla f(x)|| \to 15.9818 \quad f(x) \to 22.5238$

$d \to (-9.90133 \quad 8.58659 \quad -10.1332)$

Step length calculation phase:

$$xk1 \to \begin{pmatrix} 2.45360 - 9.90133\alpha \\ -0.920749 + 8.58659\alpha \\ -4.1804 - 10.1332\alpha \end{pmatrix}$$

$$\nabla f(xk1) \to \begin{pmatrix} -3882.76(-0.146809 + \alpha)(0.0215527 - 0.293617\alpha + \alpha^2) \\ -7.8415 + 17.1732\alpha \\ 6.55677 - 81.0656\alpha \end{pmatrix}$$

$d\phi/d\alpha \equiv \nabla f(xk1).d = 0 \to$

$$38444.5(-0.138091 + \alpha)(0.0481115 - 0.302335\alpha + \alpha^2) == 0$$

$\alpha \to 0.138091$

Updating inverse hessian:

$$q \to \begin{pmatrix} -12.283 \\ 2.37146 \\ -11.1944 \end{pmatrix} \quad qT.s \to 35.2708$$

$$Q.q \to \begin{pmatrix} -11.8997 \\ 3.06366 \\ -12.9938 \end{pmatrix} \quad s.sT \to \begin{pmatrix} 1.86947 & -1.62123 & 1.91325 \\ -1.62123 & 1.40596 & -1.6592 \\ 1.91325 & -1.6592 & 1.95805 \end{pmatrix}$$

$$Q.q.sT \to \begin{pmatrix} 16.2704 & -14.1099 & 16.6514 \\ -4.1889 & 3.63268 & -4.287 \\ 17.7662 & -15.4072 & 18.1823 \end{pmatrix}$$

$$[Q.q.sT]T + Q.q.sT \to \begin{pmatrix} 32.5407 & -18.2988 & 34.4176 \\ -18.2988 & 7.26536 & -19.6942 \\ 34.4176 & -19.6942 & 36.364 \end{pmatrix}$$

$$Q \to \begin{pmatrix} 0.105973 & 0.0443696 & 0.0152603 \\ 0.0443696 & 1.1935 & 0.0982430 \\ 0.0152603 & 0.098243 & 0.129068 \end{pmatrix}$$

New Point (Non-Optimum): {1.08632, 0.264982, -5.57971} after 2 iterations

`{opt, hist} = QuasiNewtonMethod[f, vars, x0, ResetHessian → 10];`

Using BFGS method with approximate inverse hessian reset after 10 iterations and Exact line search
Optimum: {1.03202, 2.99885, -4.99988} after 5 iterations

`normGradhist = Drop[Transpose[hist][[3]], 1]`

{58.5491, 15.9818, 7.17142, 0.990177, 0.0413141, 0.00249873}

`ListPlot[normGradhist,`
`  PlotJoined → True, AxesLabel → {"Iterations", "||∇f||"}];`

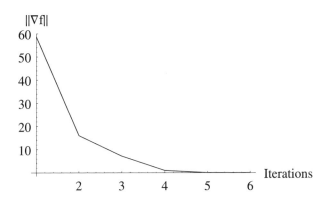

FIGURE 5.31 A plot showing $\|\nabla f\|$ at different iterations.

Using the DFP updates, the solution is as follows:

`{opt, hist} = QuasiNewtonMethod[f, vars, x0, Method → DFP, ResetHessian → 10];`

Using DFP method with approximate inverse hessian reset after 10 iterations and Exact line search
Optimum: {1.03203, 2.99885, -4.99988} after 5 iterations

Example 5.29 *Data fitting* Consider the data-fitting problem first discussed in Chapter 1. The goal is to find a surface of the form $z_{\text{computed}} = c_1 x^2 + c_2 y^2 + c_3 xy$ to best approximate the data in the following table.

Chapter 5 Unconstrained Problems

Point	x	y	z_{observed}
1	0	1	1.26
2	0.25	1	2.19
3	0.5	1	0.76
4	0.75	1	1.26
5	1	2	1.86
6	1.25	2	1.43
7	1.5	2	1.29
8	1.75	2	0.65
9	2	2	1.6

The best values of coefficients c_1, c_2, and c_3 are determined to minimize the sum of the squares of error between the computed z values and the observed values.

Minimize $f = \sum_{i=1}^{9} [z_{\text{observed}}(x_i, y_i) - z_{\text{computed}}(x_i, y_i)]^2$

Using the given numerical data, the objective function can be written as follows:

```
xyData = {{0, 1}, {0.25, 1}, {0.5, 1}, {0.75, 1},
    {1, 2}, {1.25, 2}, {1.5, 2}, {1.75, 2}, {2, 2}};
zo = {1.26, 2.19, .76, 1.26, 1.86, 1.43, 1.29, .65, 1.6};
zc = Map[(c1 x^2 + c2 y^2 + c3 xy) /. {x -> #[[1]], y -> #[[2]]} &, xyData];
f = Expand[Apply[Plus, (zo - zc)^2]]
```

$18.7 - 32.8462 c_1 + 34.2656 c_1^2 - 65.58 c_2 + 96.75 c_1 c_2 + 84 c_2^2 - 43.425 c_3 + 79.875 c_1 c_3 + 123. c_2 c_3 + 48.375 c_3^2$

Using the BFGS method, the minimum of this function is computed as follows. All calculations are shown for the first iteration.

```
QuasiNewtonMethod[f, {c1, c2, c3}, {1, 1, 1},
    ResetHessian -> 10, MaxIterations -> 1, PrintLevel -> 1];
```

```
Using BFGS method with approximate inverse hessian reset after 10
iterations and Exact line search
***** Iteration 1 ***** Current point -> {1, 1, 1}

Direction finding phase:
```

$$\text{Inverse Hessian} \rightarrow \begin{pmatrix} 1 & 0 & 0 \\ 0 & 1 & 0 \\ 0 & 0 & 1 \end{pmatrix} \quad \nabla f(x) \rightarrow \begin{pmatrix} 212.31 \\ 322.17 \\ 256.2 \end{pmatrix}$$

5.3 Unconstrained Minimization Techniques

$||\nabla f(x)|| \to 463.15 \quad f(x) \to 343.114$
$d \to (-212.31 \quad -322.17 \quad -256.2)$

Step length calculation phase:

$xk1 \to \begin{pmatrix} 1 - 212.31\alpha \\ 1 - 322.17\alpha \\ 1 - 256.2\alpha \end{pmatrix}$

$\nabla f(xk1) \to \begin{pmatrix} 212.31 - 66183.8\alpha \\ 322.17 - 106178.\alpha \\ 256.2 - 81372.5\alpha \end{pmatrix}$

$d\phi/d\alpha \equiv \nabla f(xk1).d = 0 \to -214507. + 6.91065 * 10^7 \alpha == 0$
$\alpha \to 0.00310401$

Updating inverse hessian:

$q \to \begin{pmatrix} -205.435 \\ -329.578 \\ -252.581 \end{pmatrix} \quad qT.s \to 665.834$

$Q.q \to \begin{pmatrix} -205.435 \\ -329.578 \\ -252.581 \end{pmatrix} \quad s.sT \to \begin{pmatrix} 0.434298 & 0.659025 & 0.524078 \\ 0.659025 & 1.00004 & 0.795263 \\ 0.524078 & 0.795263 & 0.632419 \end{pmatrix}$

$Q.q.sT \to \begin{pmatrix} 135.384 & 205.439 & 163.372 \\ 217.196 & 329.585 & 262.096 \\ 166.454 & 252.586 & 200.865 \end{pmatrix}$

$[Q.q.sT]T + Q.q.sT \to \begin{pmatrix} 270.769 & 422.635 & 329.826 \\ 422.635 & 659.169 & 514.682 \\ 329.826 & 514.682 & 401.729 \end{pmatrix}$

$Q \to \begin{pmatrix} 0.804239 & -0.314716 & -0.240859 \\ -0.314716 & 0.49564 & -0.3868 \\ -0.240859 & -0.3868 & 0.703762 \end{pmatrix}$

New Point (Non-Optimum):
{0.340987, -0.0000193931, 0.204752} after 1 iterations

{sol, hist} = QuasiNewtonMethod[f, {c_1, c_2, c_3}, {1, 1, 1}, ResetHessian → 10];

Using BFGS method with approximate inverse hessian reset after 10 iterations and Exact line search
Optimum: {2.09108, 1.75465, -3.50823} after 3 iterations

Thus, the surface of the form $z_{computed} = c_1 x^2 + c_2 y^2 + c_3 xy$ that best fits the given data is as follows:

$$z = 2.09108 x^2 + 1.75465 y^2 - 3.50823 xy$$

This is the same solution that was obtained in Chapter 4 using optimality conditions.

5.4 Concluding Remarks

By combining different line-search strategies and direction-finding methods, one can come up with a large number of methods for solving unconstrained problems. For practical problems, one has to base the choice on the relative computational effort required by different methods.

Since the conjugate gradient method has superior convergence properties but requires very little additional computational effort as compared to the steepest descent method, clearly there is no reason to use the steepest method for a practical problem. The modified Newton method has the fastest convergence rate. However, each iteration of this method requires evaluation of the gradient vector and the Hessian matrix. Furthermore, the direction must be computed by solving a linear system of equations. Thus, computationally, each iteration of this method is a lot more expensive than the other methods. The Quasi-Newton methods avoid solving equations and need modest effort to update the approximate inverse Hessian. For most practical problems, therefore, a Quasi-Newton method with BFGS updates is a good choice.

Regarding the step-length calculations, a method that requires the least number of objective function evaluations during line search is most preferred. The analytical line-search method obviously is a very good choice in this regard. However, it is not suitable for a purely numerical solution. The examples presented in this chapter take advantage of *Mathematica*'s symbolic algebra capabilities to implement this procedure. It will be extremely difficult to implement this strategy in a traditional computer programming language, such as C or Fortran. For such implementations, one must choose a numerical line-search method. For exact line search, the choice is usually between the golden section search and the quadratic interpolation method. The quadratic interpolation method usually requires fewer function evaluations and has a slight edge over the golden section search. Most numerical implementations, however, use an approximate line search based on Armijo's rule, since this method frequently needs very few function evaluations. However, because of the approximate nature of the computed step length, the overall convergence rate may be slower as a result. For additional details refer to Dennis and Schnabel [1983], Hestenes [1980], Himmelblau [1972], Polak [1971], Polak and Polak [1997], and Schittkowski [1980, 1987].

All methods presented in this chapter require computation of gradients. Obviously, if an objective function is not differentiable, or if it is very difficult to compute its gradient, none of these methods can be used. For such problems, one must use search methods that require only the function values or generate an approximation of a gradient vector using finite differences. The search methods are essentially of combinatorial nature and are not discussed in this text. A good reference for a study of this subject is the book edited by Colin R. Reeves [1993]. For global optimization refer to Hansen [1992] and Hentenryck, Michel, and Deville [1997]. Recently, genetic algorithms have become popular for solving these problems as well. A recent book by Mitsuo Gen and Runwei Cheng [1997] is a good starting point for a study of this field. Additional references are Goldberg [1989] and Xie and Steven [1997].

5.5 Problems

Descent Direction

For the following problems, check to see if the given direction is a descent direction at the given point. If so, compute the optimum step length using analytical line search. Compute the new point and compare the function values at the initial and the new point.

5.1. $f(x_1, x_2) = 2x_1^2 + x_2^2 - 2x_1x_2 + 2x_1^3 + x_1^4 \quad \mathbf{x}^0 = \{2, -1\} \quad \mathbf{d}^0 = \{-2, 3\}$

5.2. $f(x, y) = 2x^2 + x \operatorname{Sin}[y] \quad \mathbf{x}^0 = \{2, -1\} \quad \mathbf{d}^0 = \{-2, 3\}$

5.3. $f(x, y, z) = 2x^2 + xy + y/z^3 \quad \mathbf{x}^0 = \{2, -1, 1\} \quad \mathbf{d}^0 = \{-2, 3, -4\}$

5.4. $f(x_1, x_2) = x_1^2 + x_2^4 + x_1 x_2^3 + 3 \quad \mathbf{x}^0 = \{2, -1\} \quad \mathbf{d}^0 = \{-2, -3\}$

5.5. $f(x_1, x_2) = (x_1^2 - x_2^3)^3 + x_2^2 + (x_1 x_2^3 + 3)x_2 \quad \mathbf{x}^0 = \{2, -1\} \quad \mathbf{d}^0 = \{-2, 3\}$

Equal Interval Search

For the following problems, use the negative gradient direction and determine the function $\phi(\alpha)$ for computing the step length. Compute the step length using *equal interval search*. Compute the new point and compare the function values at the initial and the new point. Start with the default parameters used in the examples. Adjust if necessary.

5.6. $f(x_1, x_2) = 2x_1^2 + x_2^2 - 2x_1x_2 + 2x_1^3 + x_1^4 \quad \mathbf{x}^0 = \{2, -1\}$

5.7. $f(x, y) = 2x^2 + x \operatorname{Sin}[y] \quad \mathbf{x}^0 = \{2, -1\}$

5.8. $f(x, y, z) = 2x^2 + xy + y/z^3 \quad \mathbf{x}^0 = \{2, -1, 1\}$

5.9. $f(x_1, x_2) = x_1^2 + x_2^4 + x_1 x_2^3 + 3 \quad \mathbf{x}^0 = \{2, -1\}$

5.10. $f(x_1, x_2) = (x_1^2 - x_2^3)^3 + x_2^2 + (x_1 x_2^3 + 3)x_2 \quad \mathbf{x}^0 = \{2, -1\}$

5.11. A multicell evaporator is to be installed to evaporate water from a salt water solution in order to increase the salt concentration in the solution. The initial concentration of the solution is 5% salt by weight. The desired concentration is 10%, which means that half of the water from the solution must be evaporated. The system utilizes steam as the heat source. The evaporator uses 1 lb of steam to evaporate $0.8n$ lb of water, where n is the number of cells. The goal is to determine the number of cells to minimize cost. The other data are as follows:

The facility will be used to process 500,000 lbs of saline solution per day.

The unit will operate for 340 days per year.

The initial cost of the evaporator, including installation, is $18,000 per cell.

The additional cost of auxiliary equipment, regardless of the number of cells, is $9,000.

The annual maintenance cost is 5% of the initial cost.

The cost of steam is $1.55 per 1000 lbs.

The estimated life of the unit is 10 years.

The salvage value at the end of 10 years is $2,500 per cell.

The annual interest rate is 11%.

Formulate the optimization problem to minimize annual cost. Find the optimum number of cells using equal interval search.

5.12. Use the present worth method to solve problem 5.11.

Section Search

For the following problems, use the negative gradient direction and determine the function $\phi(\alpha)$ for computing the step length. Compute the step length using *section search*. Compute the new point and compare the function values at the initial and the new point. Start with the default parameters used in the examples. Adjust if necessary.

5.13. $f(x_1, x_2) = 2x_1^2 + x_2^2 - 2x_1 x_2 + 2x_1^3 + x_1^4 \quad \mathbf{x}^0 = \{2, -1\}$

5.14. $f(x, y) = 2x^2 + x \operatorname{Sin}[y] \quad \mathbf{x}^0 = \{2, -1\}$

5.15. $f(x, y, z) = 2x^2 + xy + y/z^3 \quad \mathbf{x}^0 = \{2, -1, 1\}$

5.16. $f(x_1, x_2) = x_1^2 + x_2^4 + x_1 x_2^3 + 3$ $\mathbf{x}^0 = \{2, -1\}$

5.17. $f(x_1, x_2) = (x_1^2 - x_2^3)^3 + x_2^2 + (x_1 x_2^3 + 3)x_2$ $\mathbf{x}^0 = \{2, -1\}$

5.18. Same as problem 5.11, except use section search to find an optimum.

5.19. Same as problem 5.12, except use section search to find an optimum.

Golden Section Search

For the following problems, use the negative gradient direction and determine the function $\phi(\alpha)$ for computing the step length. Compute the step length using *golden section search*. Compute the new point and compare the function values at the initial and the new point. Start with the default parameters used in the examples. Adjust if necessary.

5.20. $f(x_1, x_2) = 2x_1^2 + x_2^2 - 2x_1 x_2 + 2x_1^3 + x_1^4$ $\mathbf{x}^0 = \{2, -1\}$

5.21. $f(x, y) = 2x^2 + x \operatorname{Sin}[y]$ $\mathbf{x}^0 = \{2, -1\}$

5.22. $f(x, y, z) = 2x^2 + xy + y/z^3$ $\mathbf{x}^0 = \{2, -1, 1\}$

5.23. $f(x_1, x_2) = x_1^2 + x_2^4 + x_1 x_2^3 + 3$ $\mathbf{x}^0 = \{2, -1\}$

5.24. $f(x_1, x_2) = (x_1^2 - x_2^3)^3 + x_2^2 + (x_1 x_2^3 + 3)x_2$ $\mathbf{x}^0 = \{2, -1\}$

5.25. Same as problem 5.11, except use golden section search to find an optimum.

5.26. Same as problem 5.12, except use golden section search to find an optimum.

Quadratic Interpolation

For the following problems, use the negative gradient direction and determine the function $\phi(\alpha)$ for computing the step length. Compute the step length using *quadratic interpolation*. Compute the new point and compare the function values at the initial and the new point. Start with the default parameters used in the examples. Adjust if necessary.

5.27. $f(x_1, x_2) = 2x_1^2 + x_2^2 - 2x_1 x_2 + 2x_1^3 + x_1^4$ $\mathbf{x}^0 = \{2, -1\}$

5.28. $f(x, y) = 2x^2 + x \operatorname{Sin}[y]$ $\mathbf{x}^0 = \{2, -1\}$

5.29. $f(x, y, z) = 2x^2 + xy + y/z^3$ $\mathbf{x}^0 = \{2, -1, 1\}$

5.30. $f(x_1, x_2) = x_1^2 + x_2^4 + x_1 x_2^3 + 3$ $\mathbf{x}^0 = \{2, -1\}$

5.31. $f(x_1, x_2) = (x_1^2 - x_2^3)^3 + x_2^2 + (x_1 x_2^3 + 3)x_2$ $\mathbf{x}^0 = \{2, -1\}$

5.32. Same as problem 5.11, except use quadratic interpolation to find an optimum.

5.33. Same as problem 5.12, except use quadratic interpolation to find an optimum.

Approximate Line Search Armijo's Rule

For the following problems, use the negative gradient direction and determine the function $\phi(\alpha)$ for computing the step length. Compute the step length using *Armijo's approximate line search*. Compute the new point and compare the function values at the initial and the new point. Start with default parameters used in the examples. Adjust if necessary.

5.34. $f(x_1, x_2) = 2x_1^2 + x_2^2 - 2x_1x_2 + 2x_1^3 + x_1^4 \quad \mathbf{x}^0 = \{2, -1\}$

5.35. $f(x, y) = 2x^2 + x\operatorname{Sin}[y] \quad \mathbf{x}^0 = \{2, -1\}$

5.36. $f(x, y, z) = 2x^2 + xy + y/z^3 \quad \mathbf{x}^0 = \{2, -1, 1\}$

5.37. $f(x_1, x_2) = x_1^2 + x_2^4 + x_1x_2^3 + 3 \quad \mathbf{x}^0 = \{2, -1\}$

5.38. $f(x_1, x_2) = (x_1^2 - x_2^3)^3 + x_2^2 + (x_1x_2^3 + 3)x_2 \quad \mathbf{x}^0 = \{2, -1\}$

Steepest Descent Method

For the following problems, use the *steepest descent method*, with analytical line search, to compute the minimum point. Show complete calculations for the first two iterations. Plot a search path for problems with two variables. Plot the history of the norm of the gradient. Verify the solution using optimality conditions.

5.39. $f(x_1, x_2) = 2x_1^2 + x_2^2 - 2x_1x_2 + 2x_1^3 + x_1^4 \quad \mathbf{x}^0 = \{2, -1\}$

5.40. $f(x, y) = 2x^2 + x\operatorname{Sin}[y] \quad \mathbf{x}^0 = \{2, -1\}$

5.41. $f(x_1, x_2) = x_1^2 + x_2^4 + x_1x_2^3 + 3 \quad \mathbf{x}^0 = \{2, -1\}$

5.42. $f(x_1, x_2) = (x_1^2 - x_2^3)^3 + x_2^2 + (x_1x_2^3 + 3)x_2 \quad \mathbf{x}^0 = \{2, -1\}$

5.43. $f(x, y) = x + 2x^2 + 2y - xy + 2y^2 \quad \mathbf{x}^0 = \{2, -1\}$

5.44. $f(x, y) = -3x + 2x^2 - 2y + 3xy + \dfrac{7y^2}{2} \quad \mathbf{x}^0 = \{2, -1\}$

5.45. $f(x, y, z) = xyz - \dfrac{1}{2 + x^2 + y^2 + z^2}$ $\mathbf{x}^0 = \{-0.1, -0.2, 1\}$

5.46. $f(x, y, z) = 2x^2 - 2xy + y^2 - z - yz + z^2$ $\mathbf{x}^0 = \{1, 2, 3\}$

5.47. $f(x, y, z) = 3x + 2x^2 + 2y - 3xy + \dfrac{7y^2}{2} + z + xz + yz + \dfrac{5z^2}{2}$ $\mathbf{x}^0 = \{1, 2, 3\}$

5.48. $f = x_1 + 4x_1^2 + 2x_2 - x_1 x_2 + 4x_2^2 + 3x_3 + 4x_3^2 + 4x_4 + 2x_1 x_4 + 2x_2 x_4 + 2x_3 x_4 + x_4^2$

$\mathbf{x}^0 = \{1, 2, 3, 4\}$

5.49. A small firm is capable of manufacturing two different products. The cost of making each product decreases as the number of units produced increases and is given by the following empirical relationships,

$$c_1 = 5 + \dfrac{1{,}500}{n_1} \quad c_2 = 7 + \dfrac{2{,}500}{n_2}$$

where n_1 and n_2 are the number of units of each of the two products produced. The cost of repair and maintenance of equipment used to produce these products depends on the total number of products produced, regardless of its type, and is given by the following quadratic equation:

$$(n_1 + n_2)\left[0.2 + 2.3 * 10^{-5}(n_1 + n_2) + 5.3 * 10^{-9}(n_1 + n_2)^2\right]$$

The wholesale selling price of the products drops as more units are produced, according to the following relationships:

$$p_1 = 15 - 0.001 n_1 \quad p_2 = 25 - 0.0015 n_2$$

Determine how many units of each product the firm should produce to maximize its profit. Choose any arbitrary starting point and find an optimum solution using the steepest descent method.

5.50. For a chemical process, pressure measured at different temperatures is given in the following table. Formulate an optimization problem to determine the best values of coefficients in the following exponential model for the data. Choose any arbitrary starting point and find optimum values of these parameters using the steepest descent method.

$$\text{Pressure} = \alpha e^{\beta T}$$

Temperature (T^0 C)	Pressure (mm of Mercury)
20	15.45
25	19.23
30	26.54
35	34.52
40	48.32
50	68.11
60	98.34
70	120.45

5.51. A chemical manufacturer requires an automatic reactor-mixer. The mixing time required is related to the size of the mixer and the stirring power, as follows:

$$T = 1,000 \frac{\sqrt{S}}{P^2}$$

where S = capacity of the reactor-mixer, kg, P = power of the stirrer, k-Watts, and T is the time taken in hours per batch. The cost of building the reactor-mixer is proportional to its capacity and is given by the following empirical relationship:

$$\text{Cost} = \$60,000\sqrt{S}$$

The cost of electricity to operate the stirrer is $0.05/k-W-hr, and the overhead costs are $137.2 P per year. The total reactor to be processed by the mixer per year is 10^7 kg. Time for loading and unloading the mixer is negligible. Using present worth analysis, determine the capacity of the mixer and the stirrer power in order to minimize cost. Assume a 5-year useful life, a 9% annual interest rate compounded monthly, and a salvage value of 10% of the initial cost of the mixer. Find an optimum solution using the steepest descent method.

5.52. Use the annual cost formulation in problem 5.51.

Conjugate Gradient Method—Polak-Ribiere

For the following problems, use the *conjugate gradient method—Polak-Ribiere*, with analytical line search, to compute the minimum point. Show complete calculations for the first two iterations. Plot a search path for problems with two variables. Plot the history of the norm of the gradient of functions. Verify the solution using optimality conditions.

5.53. $f(x_1, x_2) = 2x_1^2 + x_2^2 - 2x_1x_2 + 2x_1^3 + x_1^4 \quad \mathbf{x}^0 = \{2, -1\}$

5.54. $f(x, y) = 2x^2 + x\operatorname{Sin}[y] \quad \mathbf{x}^0 = \{2, -1\}$

5.55. $f(x_1, x_2) = x_1^2 + x_2^4 + x_1x_2^3 + 3 \quad \mathbf{x}^0 = \{2, -1\}$

5.56. $f(x_1, x_2) = (x_1^2 - x_2^3)^3 + x_2^2 + (x_1x_2^3 + 3)x_2 \quad \mathbf{x}^0 = \{2, -1\}$

5.57. $f(x, y) = x + 2x^2 + 2y - xy + 2y^2 \quad \mathbf{x}^0 = \{2, -1\}$

5.58. $f(x, y) = -3x + 2x^2 - 2y + 3xy + \dfrac{7y^2}{2} \quad \mathbf{x}^0 = \{2, -1\}$

5.59. $f(x, y, z) = xyz - \dfrac{1}{2 + x^2 + y^2 + z^2} \quad \mathbf{x}^0 = \{-0.1, -0.2, 1\}$

5.60. $f(x, y, z) = 2x^2 - 2xy + y^2 - z - yz + z^2 \quad \mathbf{x}^0 = \{1, 2, 3\}$

5.61. $f(x, y, z) = 3x + 2x^2 + 2y - 3xy + \dfrac{7y^2}{2} + z + xz + yz + \dfrac{5z^2}{2} \quad \mathbf{x}^0 = \{1, 2, 3\}$

5.62. $f = x_1 + 4x_1^2 + 2x_2 - x_1x_2 + 4x_2^2 + 3x_3 + 4x_3^2 + 4x_4 + 2x_1x_4 + 2x_2x_4 + 2x_3x_4 + x_4^2$
$\mathbf{x}^0 = \{1, 2, 3, 4\}$

5.63. Find an optimum solution of the manufacturing problem 5.49, using the conjugate gradient method—Polak-Ribiere.

5.64. Find an optimum solution of the data-fitting problem 5.50, using the conjugate gradient method—Polak-Ribiere.

5.65. Find an optimum solution of the mixer problem 5.51, using the conjugate gradient—Polak-Ribiere method.

5.66. Find an optimum solution of the mixer problem 5.52, using the conjugate gradient method—Polak-Ribiere.

Conjugate Gradient Method—Fletcher-Reeves

For the following problems, use the *conjugate gradient method—Fletcher-Reeves*, with analytical line search, to compute the minimum point. Show complete calculations for the first two iterations. Plot a search path for problems with two variables. Plot the history of the norm of the gradient of functions. Verify the solution using optimality conditions.

5.67. $f(x_1, x_2) = 2x_1^2 + x_2^2 - 2x_1x_2 + 2x_1^3 + x_1^4 \quad \mathbf{x}^0 = \{2, -1\}$

5.68. $f(x, y) = 2x^2 + x\,\text{Sin}[y] \quad \mathbf{x}^0 = \{2, -1\}$

5.69. $f(x_1, x_2) = x_1^2 + x_2^4 + x_1x_2^3 + 3 \quad \mathbf{x}^0 = \{2, -1\}$

5.70. $f(x_1, x_2) = (x_1^2 - x_2^3)^3 + x_2^2 + (x_1x_2^3 + 3)x_2 \quad \mathbf{x}^0 = \{2, -1\}$

5.71. $f(x, y) = x + 2x^2 + 2y - xy + 2y^2 \quad \mathbf{x}^0 = \{2, -1\}$

5.72. $f(x, y) = -3x + 2x^2 - 2y + 3xy + \dfrac{7y^2}{2} \quad \mathbf{x}^0 = \{2, -1\}$

5.73. $f(x, y, z) = xyz - \dfrac{1}{2 + x^2 + y^2 + z^2} \quad \mathbf{x}^0 = \{-0.1, -0.2, 1\}$

5.74. $f(x, y, z) = 2x^2 - 2xy + y^2 - z - yz + z^2 \quad \mathbf{x}^0 = \{1, 2, 3\}$

5.75. $f(x, y, z) = 3x + 2x^2 + 2y - 3xy + \dfrac{7y^2}{2} + z + xz + yz + \dfrac{5z^2}{2} \quad \mathbf{x}^0 = \{1, 2, 3\}$

5.76. $f = x_1 + 4x_1^2 + 2x_2 - x_1x_2 + 4x_2^2 + 3x_3 + 4x_3^2 + 4x_4 + 2x_1x_4 + 2x_2x_4 + 2x_3x_4 + x_4^2$
$\mathbf{x}^0 = \{1, 2, 3, 4\}$

5.77. Find an optimum solution of the manufacturing problem 5.49, using the conjugate gradient method—Fletcher-Reeves.

5.78. Find an optimum solution of the data-fitting problem 5.50, using the conjugate gradient method—Fletcher-Reeves.

5.79. Find an optimum solution of the mixer problem 5.51, using the conjugate gradient method—Fletcher-Reeves.

5.80. Find an optimum solution of the mixer problem 5.52, using the conjugate gradient method—Fletcher-Reeves.

Modified Newton Method

For the following problems, use the *modified Newton method*, with analytical line search, to compute the minimum point. Show complete calculations for the first two iterations. Plot a search path for problems with two variables. Plot the history of the norm of the gradient of functions. Verify the solution using optimality conditions.

5.5 Problems

5.81. $f(x_1, x_2) = 2x_1^2 + x_2^2 - 2x_1x_2 + 2x_1^3 + x_1^4 \quad \mathbf{x}^0 = \{2, -1\}$

5.82. $f(x, y) = 2x^2 + x \operatorname{Sin}[y] \quad \mathbf{x}^0 = \{2, -1\}$

5.83. $f(x_1, x_2) = x_1^2 + x_2^4 + x_1 x_2^3 + 3 \quad \mathbf{x}^0 = \{2, -1\}$

5.84. $f(x_1, x_2) = (x_1^2 - x_2^3)^3 + x_2^2 + (x_1 x_2^3 + 3)x_2 \quad \mathbf{x}^0 = \{2, -1\}$

5.85. $f(x, y) = x + 2x^2 + 2y - xy + 2y^2 \quad \mathbf{x}^0 = \{2, -1\}$

5.86. $f(x, y) = -3x + 2x^2 - 2y + 3xy + \dfrac{7y^2}{2} \quad \mathbf{x}^0 = \{2, -1\}$

5.87. $f(x, y, z) = 2x^2 - 2xy + y^2 - z - yz + z^2 \quad \mathbf{x}^0 = \{1, 2, 3\}$

5.88. $f(x, y, z) = 3x + 2x^2 + 2y - 3xy + \dfrac{7y^2}{2} + z + xz + yz + \dfrac{5z^2}{2} \quad \mathbf{x}^0 = \{1, 2, 3\}$

5.89. $f = x_1 + 4x_1^2 + 2x_2 - x_1x_2 + 4x_2^2 + 3x_3 + 4x_3^2 + 4x_4 + 2x_1x_4 + 2x_2x_4 + 2x_3x_4 + x_4^2$
$\mathbf{x}^0 = \{1, 2, 3, 4\}$

5.90. Find an optimum solution of the manufacturing problem 5.49, using the modified Newton method.

5.91. Find an optimum solution of the data-fitting problem 5.50, using the modified Newton method.

5.92. Find an optimum solution of the mixer problem 5.51, using the modified Newton method.

5.93. Find an optimum solution of the mixer problem 5.52, using the modified Newton method.

BFGS Method

For the following problems, use the *BFGS method*, with analytical line search, to compute the minimum point. Show complete calculations for the first two iterations. Plot a search path for problems with two variables. Plot the history of the norm of the gradient of functions. Verify the solution using optimality conditions.

5.94. $f(x_1, x_2) = 2x_1^2 + x_2^2 - 2x_1x_2 + 2x_1^3 + x_1^4 \quad \mathbf{x}^0 = \{2, -1\}$

5.95. $f(x, y) = 2x^2 + x \operatorname{Sin}[y] \quad \mathbf{x}^0 = \{2, -1\}$

5.96. $f(x_1, x_2) = x_1^2 + x_2^4 + x_1 x_2^3 + 3 \quad \mathbf{x}^0 = \{2, -1\}$

5.97. $f(x_1, x_2) = (x_1^2 - x_2^3)^3 + x_2^2 + (x_1 x_2^3 + 3)x_2 \quad \mathbf{x}^0 = \{2, -1\}$

5.98. $f(x, y) = x + 2x^2 + 2y - xy + 2y^2$ $\mathbf{x}^0 = \{2, -1\}$

5.99. $f(x, y) = -3x + 2x^2 - 2y + 3xy + \dfrac{7y^2}{2}$ $\mathbf{x}^0 = \{2, -1\}$

5.100. $f(x, y, z) = xyz - \dfrac{1}{2 + x^2 + y^2 + z^2}$ $\mathbf{x}^0 = \{-0.1, -0.2, 1\}$

5.101. $f(x, y, z) = 2x^2 - 2xy + y^2 - z - yz + z^2$ $\mathbf{x}^0 = \{1, 2, 3\}$

5.102. $f(x, y, z) = 3x + 2x^2 + 2y - 3xy + \dfrac{7y^2}{2} + z + xz + yz + \dfrac{5z^2}{2}$ $\mathbf{x}^0 = \{1, 2, 3\}$

5.103. $f = x_1 + 4x_1^2 + 2x_2 - x_1x_2 + 4x_2^2 + 3x_3 + 4x_3^2 + 4x_4 + 2x_1x_4 + 2x_2x_4 + 2x_3x_4 + x_4^2$
$\mathbf{x}^0 = \{1, 2, 3, 4\}$

5.104. Find an optimum solution of the manufacturing problem 5.49, using the BFGS Quasi-Newton method.

5.105. Find an optimum solution of the data-fitting problem 5.50, using the BFGS Quasi-Newton method.

5.106. Find an optimum solution of the mixer problem 5.51, using the BFGS Quasi-Newton method.

5.107. Find an optimum solution of the mixer problem 5.52, using the BFGS Quasi-Newton method.

DFP Method

For the following problems, use the *DFP method*, with analytical line search, to compute the minimum point. Show complete calculations for the first two iterations. Plot a search path for problems with two variables. Plot the history of the norm of the gradient of functions. Verify the solution using optimality conditions.

5.108. $f(x_1, x_2) = 2x_1^2 + x_2^2 - 2x_1x_2 + 2x_1^3 + x_1^4$ $\mathbf{x}^0 = \{2, -1\}$

5.109. $f(x, y) = 2x^2 + x \operatorname{Sin}[y]$ $\mathbf{x}^0 = \{2, -1\}$

5.110. $f(x_1, x_2) = x_1^2 + x_2^4 + x_1 x_2^3 + 3$ $\mathbf{x}^0 = \{2, -1\}$

5.111. $f(x_1, x_2) = (x_1^2 - x_2^3)^3 + x_2^2 + (x_1 x_2^3 + 3)x_2$ $\mathbf{x}^0 = \{2, -1\}$

5.112. $f(x, y) = x + 2x^2 + 2y - xy + 2y^2$ $\mathbf{x}^0 = \{2, -1\}$

5.113. $f(x, y) = -3x + 2x^2 - 2y + 3xy + \dfrac{7y^2}{2}$ $\mathbf{x}^0 = \{2, -1\}$

5.114. $f(x, y, z) = xyz - \dfrac{1}{2 + x^2 + y^2 + z^2}$ $\mathbf{x}^0 = \{-0.1, -0.2, 1\}$

5.115. $f(x, y, z) = 2x^2 - 2xy + y^2 - z - yz + z^2$ $\mathbf{x}^0 = \{1, 2, 3\}$

5.116. $f(x, y, z) = 3x + 2x^2 + 2y - 3xy + \dfrac{7y^2}{2} + z + xz + yz + \dfrac{5z^2}{2}$ $\mathbf{x}^0 = \{1, 2, 3\}$

5.117. $f = x_1 + 4x_1^2 + 2x_2 - x_1 x_2 + 4x_2^2 + 3x_3 + 4x_3^2 + 4x_4 + 2x_1 x_4 + 2x_2 x_4 + 2x_3 x_4 + x_4^2$
$\mathbf{x}^0 = \{1, 2, 3, 4\}$

5.118. Find an optimum solution of the manufacturing problem 5.49, using the DFP Quasi-Newton method.

5.119. Find an optimum solution of the data-fitting problem 5.50, using the DFP Quasi-Newton method.

5.120. Find an optimum solution of the mixer problem 5.51, using the DFP Quasi-Newton method.

5.121. Find an optimum solution of the mixer problem 5.52, using the DFP Quasi-Newton method.

CHAPTER SIX

Linear Programming

When the objective function and all constraints are linear functions of optimization variables, the problem is known as a linear programming (LP) problem. A large number of engineering and business applications have been successfully formulated and solved as LP problems. LP problems also arise during the solution of nonlinear problems as a result of linearizing functions around a given point.

It is important to recognize that a problem is of the LP type because of the availability of well-established methods, such as the simplex method, for solving such problems. Problems with thousands of variables and constraints can be handled with the simplex method in a routine manner. In contrast, most methods for solving general nonlinear constrained problems, such as those discussed in chapter 9, do not perform as well for large problems.

This chapter presents the well-known simplex method for solving LP problems. Another class of methods, known as interior point methods, are discussed in the following chapter. The simplex method requires that the LP problem be stated in a standard form that involves only the equality constraints. Conversion of a given LP problem to this form is discussed in the first section. In the standard LP form, since the constraints are linear equalities, the simplex method essentially boils down to solving systems of linear equations. A review of solving linear systems of equations using the Gauss-Jordan form and the LU decomposition is presented in the second section. The basic solutions for an LP problem are presented in section 4. The traditional tableau form of the simplex method is presented in detail in section 5. The procedure for sensitivity analysis based on the simplex method is presented in section 6. The tableau form is convenient for organizing computations for a hand solution; however, it is inefficient for computer implementation. The so-called revised simplex method, in which the computations are organized as a series of matrix operations, is discussed in section 7. The last section considers the sensitivity of the optimum solution as the objective function and the constraint coefficients are changed based on the revised simplex method.

6.1 The Standard LP Problem

This chapter presents methods for solving linear programming (LP) problems expressed in the following standard form:

Find $\mathbf{x}$ in order to

Minimize $f(\mathbf{x}) \equiv \mathbf{c}^T \mathbf{x}$

Subject to $\mathbf{A}\mathbf{x} = \mathbf{b}$ and $\mathbf{x} \geq 0$.

where

$\mathbf{x} = [x_1, x_2, \ldots, x_n]^T$ vector of optimization variables

$\mathbf{c} = [c_1, c_2, \ldots, c_n]^T$ vector of objective or cost coefficients

$\mathbf{A} = \begin{pmatrix} a_{11} & a_{12} & \cdots & a_{1n} \\ a_{21} & a_{22} & \cdots & a_{2n} \\ \vdots & \vdots & \vdots & \vdots \\ a_{m1} & a_{m2} & \cdots & a_{mn} \end{pmatrix}$ $m \times n$ matrix of constraint coefficients

$\mathbf{b} = [b_1, b_2, \ldots, b_m]^T \geq 0$ vector of right-hand sides of constraints

Note that in this standard form, the problem is of minimization type. All constraints are expressed as equalities with the right-hand side greater than or equal to ($\geq$) 0. Furthermore, all optimization variables are restricted to be positive.

6.1.1 Conversion to Standard LP Form

At first glance, it may appear that the standard LP form is very restrictive. However, as shown below, it is possible to convert any LP problem to the above standard form.

Maximization Problem

As already mentioned in previous chapters, a maximization problem can be converted to a minimization problem simply by multiplying the objective function by a negative sign. For example,

Maximize $z(\mathbf{x}) = 3x_1 + 5x_2$ is the same as Minimize $f(\mathbf{x}) = -3x_1 - 5x_2$

Constant Term in the Objective Function

From the optimality conditions, it is easy to see that the optimum solution $\mathbf{x}^*$ does not change if a constant is either added to or subtracted from the objective function. Thus, a constant in the objective function can simply be ignored. After the solution is obtained, the optimum value of the objective function is adjusted to account for this constant.

Alternatively, a new *dummy* optimization variable can be defined to multiply the constant and a constraint added to set the value of this variable to 1. For example, consider the following objective function of two variables:

Minimize $f(\mathbf{x}) = 3x_1 + 5x_2 + 7$

In standard LP form, it can be written as follows:

Minimize $\quad f(\mathbf{x}) = 3x_1 + 5x_2 + 7x_3$
Subject to $\quad x_3 = 1$

Negative Values on the Right-Hand Sides of Constraints

The standard form requires that all constraints must be arranged such that the constant term, if any, is a positive quantity on the right-hand side. If a constant appears as negative on the right-hand side of a given constraint, multiply the constraint by a negative sign. Keep in mind that the direction of inequality changes (that $\leq$ becomes $\geq$, and vice versa) when both sides are multiplied by a negative sign. For example,

$3x_1 + 5x_2 \leq -7$ is the same as $-3x_1 - 5x_2 \geq 7$

Less than Type Constraints

Add a new positive variable (called a *slack* variable) to convert a $\leq$ constraint (LE) to an equality. For example, $3x_1 + 5x_2 \leq 7$ is converted to $3x_1 + 5x_2 + x_3 = 7$, where $x_3 \geq 0$ is a slack variable

Greater than Type Constraints

Subtract a new positive variable (called a *surplus* variable) to convert a $\geq$ constraint (GE) to equality. For example, $3x_1 + 5x_2 \geq 7$ is converted to $3x_1 + 5x_2 - x_3 = 7$, where $x_3 \geq 0$ is a surplus variable. Note that, since the right-hand sides of the constraints are restricted to be positive, we cannot simply multiply both sides of the GE constraints by -1 to convert them into the LE type, as was done for the KT conditions in Chapter 4.

Unrestricted Variables

The standard LP form restricts all variables to be positive. If an actual optimization variable is unrestricted in sign, it can be converted to the standard form by defining it as the difference of two new positive variables. For example, if variable x_1 is unrestricted in sign, it is replaced by two new variables y_1 and y_2 with $x_1 = y_1 - y_2$. Both the new variables are positive. After the solution is obtained, if $y_1 > y_2$, then x_1 will be positive and if $y_1 < y_2$, then x_1 will be negative.

Example 6.1 Convert the following problem to the standard LP form.

Maximize $z = 3x_1 + 8x_2$

Subject to $\begin{pmatrix} 3x_1 + 4x_2 \geq -20 \\ x_1 + 3x_2 \geq 6 \\ x_1 \geq 0 \end{pmatrix}$

Note that x_2 is unrestricted in sign. Define new variables (all ≥ 0)

$$x_1 = y_1 \quad x_2 = y_2 - y_3$$

Substituting these and multiplying the first constraint by a negative sign, the problem is as follows:

Maximize $z = 3y_1 + 8y_2 - 8y_3$

Subject to $\begin{pmatrix} -3y_1 - 4y_2 + 4y_3 \leq 20 \\ y_1 + 3y_2 - 3y_3 \geq 6 \\ y_1, y_2, y_3 \geq 0 \end{pmatrix}$

Multiplying the objective function by a negative sign and introducing slack/surplus variables in the constraints, the problem in the standard LP form is as follows:

Minimize $f = -3y_1 - 8y_2 + 8y_3$

Subject to $\begin{pmatrix} -3y_1 - 4y_2 + 4y_3 + y_4 = 20 \\ y_1 + 3y_2 - 3y_3 - y_5 = 6 \\ y_1, \ldots, y_5 \geq 0 \end{pmatrix}$

6.1.2 The Optimum of LP Problems

Since linear functions are always convex, the LP problem is a convex programming problem. This means that if an optimum solution exists, it is a global optimum.

The optimum solution of an LP problem always lies on the boundary of the feasible domain. We can easily prove this by contradiction. Suppose the solution lies inside the feasible domain; then the optimum is an unconstrained point, and hence, the necessary conditions for optimality would imply that $\partial f/\partial x_i \equiv c_i = 0, i = 1, 2, \ldots, n$, which obviously is not possible (because all $c_i = 0$ means $f \equiv \mathbf{c}^T\mathbf{x} = 0$). Thus, the solution cannot lie on the inside of the feasible domain for LP problems.

Once an LP problem is converted to its standard form, the constraints represent a system of n equations in m unknowns. If $m = n$ (i.e., the number of constraints is equal to the number of optimization variables), then the solution for all variables is obtained from the solution of constraint equations and there is no consideration of the objective function. This situation clearly does not represent an optimization problem. On the other hand, $m > n$ does not make sense because in this case, some of the constraints must be linearly dependent on the others. Thus, from an optimization point of view, the only meaningful case is when the number of constraints is smaller than the number of variables (after the problem has been expressed in the standard LP form).

6.2 Solving Linear Systems of Equations

It should be clear from the previous section that solving LP problems involves solving a system of undetermined linear equations. (The number of equations is less than the number of unknowns.) A review of the Gauss-Jordan procedure for solving a system of linear equations is presented in this section.

Consider the solution of the following system of equations:

$$\mathbf{Ax} = \mathbf{b}$$

where $\mathbf{A}$ is an $m \times n$ coefficient matrix, $\mathbf{x}$ is $n \times 1$ vector of unknowns, and $\mathbf{b}$ is an $m \times 1$ vector of known right-hand sides.

6.2.1 A Solution Using the Gauss-Jordan Form

A solution of the system of equations can be obtained by first writing an $(m \times n + 1)$ augmented matrix as follows:

Augmented matrix: $(\mathbf{A} \mid \mathbf{b})$

A suitable series of row operations (adding appropriate multiples of rows together) is then performed on this matrix in an attempt to convert it to the following form:

Gauss-Jordan form: $\begin{pmatrix} \mathbf{I} & \mathbf{c} \\ \hline \mathbf{0} & \mathbf{0} \end{pmatrix}$

where $\mathbf{I}$ is a $p \times p$ identity matrix ($p \leq m$), $\mathbf{c}$ represents the remaining $p \times (n + 1 - p)$ entries in the augmented matrix, and the $\mathbf{0}$'s are appropriate-sized zero matrices. This form is known as the Gauss-Jordan form. Since the variables corresponding to the identity matrix appear only in one of the equations, a general solution of the system of equations can be written directly in terms of these variables.

To get the Gauss-Jordan form, we first perform row operations to reduce the system to an upper triangular form in which all entries on the diagonal are 1 and those below the diagonal are 0. We do it systematically by starting in the first column, making the diagonal entry 1, and then using the appropriate multiples of the first row to make all other entries in the first column zero. Then we proceed to the second column and perform the same series of steps. The process is continued until we have reached the diagonal element of the last row. This completes what is known as the *forward* pass. Next, we start from the last column in which the forward pass ended and perform another series of row operations to zero out entries above the diagonal. This is known as the *backward* pass.

6.2 Solving Linear Systems of Equations

The following examples illustrate the process of converting a system of equations to an equivalent Gauss-Jordan form and recovering a general solution from this form.

Example 6.2 As an example, consider the following system of equations:

$$2x_1 + 5x_2 - x_3 + x_4 = 1$$
$$-3x_1 - 8x_2 + 2x_3 + 3x_4 = 4$$
$$x_1 + 2x_2 + 5x_4 = 6$$

The augmented matrix for this system of equations is as follows:

$$\begin{pmatrix} x_1 & x_2 & x_3 & x_4 & \text{rhs} \\ 2 & 5 & -1 & 1 & 1 \\ -3 & -8 & 2 & 3 & 4 \\ 1 & 2 & 0 & 5 & 6 \end{pmatrix}$$

To get 1 on the first row diagonal, divide the first row by 2.

$$\begin{pmatrix} x_1 & x_2 & x_3 & x_4 & \text{rhs} \\ 1 & \frac{5}{2} & -\frac{1}{2} & \frac{1}{2} & \frac{1}{2} \\ -3 & -8 & 2 & 3 & 4 \\ 1 & 2 & 0 & 5 & 6 \end{pmatrix}$$

Adding ($3 \times$ row 1) to row 2 and ($-1 \times$ row 1) to row 3 will make the remaining entries in the first column zero, giving

$$\begin{pmatrix} x_1 & x_2 & x_3 & x_4 & \text{rhs} \\ 1 & \frac{5}{2} & -\frac{1}{2} & \frac{1}{2} & \frac{1}{2} \\ 0 & -\frac{1}{2} & \frac{1}{2} & \frac{9}{2} & \frac{11}{2} \\ 0 & -\frac{1}{2} & \frac{1}{2} & \frac{9}{2} & \frac{11}{2} \end{pmatrix}$$

To make the diagonal entry in the second column 1, divide row 2 by $-1/2$.

$$\begin{pmatrix} x_1 & x_2 & x_3 & x_4 & \text{rhs} \\ 1 & \frac{5}{2} & -\frac{1}{2} & \frac{1}{2} & \frac{1}{2} \\ 0 & 1 & -1 & -9 & -11 \\ 0 & -\frac{1}{2} & \frac{1}{2} & \frac{9}{2} & \frac{11}{2} \end{pmatrix}$$

Now we use this modified row 2 to zero out all entries below the diagonal of the second column. Adding $1/2 \times$ row 2 to the third row will make the entry below the diagonal of the second row zero.

$$\begin{pmatrix} x_1 & x_2 & x_3 & x_4 & \text{rhs} \\ 1 & \frac{5}{2} & -\frac{1}{2} & \frac{1}{2} & \frac{1}{2} \\ 0 & 1 & -1 & -9 & -11 \\ 0 & 0 & 0 & 0 & 0 \end{pmatrix}$$

Since the diagonal element of the third column is already zero, we have reached the end of the forward pass. In the backward pass, starting from the last column of the forward pass, we make the entries above the diagonal go to zero. We only have one entry in the second column that needs to be made zero. This is done by adding $(-5/2 \times \text{row 2})$ to row 1, giving the complete Gauss-Jordan form for the problem as follows:

$$\begin{pmatrix} x_1 & x_2 & x_3 & x_4 & \text{rhs} \\ 1 & 0 & 2 & 23 & 28 \\ 0 & 1 & -1 & -9 & -11 \\ 0 & 0 & 0 & 0 & 0 \end{pmatrix}$$

We see that we have a 2×2 identity matrix and a 2×3 matrix **c**. This means that only two of the three equations are independent, and we can only solve for two unknowns in terms of the remaining two. The following general solution can readily be written from the above system of equations:

From the first equation: $x_1 + 2x_3 + 23x_4 = 28$, giving $x_1 = 28 - 2x_3 - 23x_4$

From the second equation: $x_2 - x_3 - 9x_4 = -11$, giving $x_2 = -11 + x_3 + 9x_4$

Row Exchanges

A key step in converting a matrix to its Gauss-Jordan form is division by the diagonal elements. Obviously, if a zero diagonal element is encountered during elimination, this step cannot be carried out. In these cases, it may be necessary to perform row exchanges to bring another row with a nonzero element at the diagonal before proceeding. Since each row represents an equation, a row exchange simply means reordering of equations and hence does not affect the solution. The following example involves a situation requiring a row exchange.

Example 6.3 Find the solution of the following system of equations:

$$x_1 + 2x_2 + 3x_3 + 4x_4 = 5$$

$$x_1 + 2x_2 + 4x_3 - 9x_4 = 9$$

$$-x_1 - x_2 + x_3 + x_4 = 6$$

6.2 Solving Linear Systems of Equations

The augmented matrix for this system of equations is as follows:

$$\begin{pmatrix} x_1 & x_2 & x_3 & x_4 & \text{rhs} \\ 1 & 2 & 3 & 4 & 5 \\ 1 & 2 & 4 & -9 & 9 \\ -1 & -1 & 1 & 1 & 6 \end{pmatrix}$$

The conversion to the Gauss-Jordan form proceeds as follows.

************Forward pass***********

$$(\text{Row 2}) - (1) * \text{Row 1} \rightarrow \begin{pmatrix} 1 & 2 & 3 & 4 & 5 \\ 0 & 0 & 1 & -13 & 4 \\ -1 & -1 & 1 & 1 & 6 \end{pmatrix}$$

$$(\text{Row 3}) - (-1) * \text{Row 1} \rightarrow \begin{pmatrix} 1 & 2 & 3 & 4 & 5 \\ 0 & 0 & 1 & -13 & 4 \\ 0 & 1 & 4 & 5 & 11 \end{pmatrix}$$

We cannot proceed with the second column because of a zero on the diagonal. However, if we interchange rows 2 and 3 (which physically means writing equation 3 before equation 2), we have the following situation:

$$\text{Rows 2 and 3 interchanged} \rightarrow \begin{pmatrix} 1 & 2 & 3 & 4 & 5 \\ 0 & 1 & 4 & 5 & 11 \\ 0 & 0 & 1 & -13 & 4 \end{pmatrix}$$

Since the diagonal entry in the second column is already 1 and those below it are 0, we are done with the second column. Next, we proceed to the third column. The diagonal is again already 1, and therefore, we are done with the forward pass.

**********Backward pass***********

$$(\text{Row 1}) - (3) * \text{Row 3} \rightarrow \begin{pmatrix} 1 & 2 & 0 & 43 & -7 \\ 0 & 1 & 4 & 5 & 11 \\ 0 & 0 & 1 & -13 & 4 \end{pmatrix}$$

$$(\text{Row 2}) - (4) * \text{Row 3} \rightarrow \begin{pmatrix} 1 & 2 & 0 & 43 & -7 \\ 0 & 1 & 0 & 57 & -5 \\ 0 & 0 & 1 & -13 & 4 \end{pmatrix}$$

$$(\text{Row 1}) - (2) * \text{Row 2} \rightarrow \begin{pmatrix} 1 & 0 & 0 & -71 & 3 \\ 0 & 1 & 0 & 57 & -5 \\ 0 & 0 & 1 & -13 & 4 \end{pmatrix}$$

We have a Gauss-Jordan form with a 3×3 identity matrix and a 3×2 matrix **c**. The general solution from each equation can be written as follows. Remember the ith column represents the coefficient of the ith variable, and the last column is the right-hand side of the equations.

$$x_1 = 3 + 71x_4 \qquad x_2 = -5 - 57x_4 \qquad x_3 = 4 + 13x_4$$

The GaussJordanForm Function

Given an augmented matrix, the following *Mathematica* function converts the matrix to a Gauss-Jordan form. The function also considers row exchanges if a zero is encountered on the diagonal. There are built-in *Mathematica* functions that do the job more efficiently. The main reason for writing the following function is to document the intermediate steps. The function reports details of all operations by printing the augmented matrix after each operation.

```
Needs["OptimizationToolbox`LPSimplex`"];
?GaussJordanForm
```

GaussJordanForm[mat]---converts the given matrix into a Gauss-Jordan
    form, printing results at all intermediate steps. The function
    performs row exchanges, if necessary.

Example 6.4 Find the solution of the following system of equations:

$$3x_1 + x_2 + x_3 = 8$$

$$2x_1 - x_2 - x_3 = -3$$

$$x_1 + 2x_2 - x_3 = 2$$

The augmented matrix for this system of equations is as follows:

$$\begin{pmatrix} x_1 & x_2 & x_3 & \text{rhs} \\ 3 & 1 & 1 & 8 \\ 2 & -1 & -1 & -3 \\ 1 & 2 & -1 & 2 \end{pmatrix}$$

The conversion to the Gauss-Jordan form proceeds as follows:

$$a = \begin{pmatrix} 3 & 1 & 1 & 8 \\ 2 & -1 & -1 & -3 \\ 1 & 2 & -1 & 2 \end{pmatrix}; \quad \text{GaussJordanForm[a]};$$

**********Forward pass**********

$$(\text{Row 1})/3 \to \begin{pmatrix} 1 & \frac{1}{3} & \frac{1}{3} & \frac{8}{3} \\ 2 & -1 & -1 & -3 \\ 1 & 2 & -1 & 2 \end{pmatrix}$$

$$(\text{Row 2}) - (2) * \text{Row 1} \to \begin{pmatrix} 1 & \frac{1}{3} & \frac{1}{3} & \frac{8}{3} \\ 0 & -\frac{5}{3} & -\frac{5}{3} & -\frac{25}{3} \\ 1 & 2 & -1 & 2 \end{pmatrix}$$

$$(\text{Row 3}) - (1) * \text{Row 1} \to \begin{pmatrix} 1 & \frac{1}{3} & \frac{1}{3} & \frac{8}{3} \\ 0 & -\frac{5}{3} & -\frac{5}{3} & -\frac{25}{3} \\ 0 & \frac{5}{3} & -\frac{4}{3} & -\frac{2}{3} \end{pmatrix}$$

6.2 Solving Linear Systems of Equations

$$(\text{Row 2})/-\frac{5}{3} \to \begin{pmatrix} 1 & \frac{1}{3} & \frac{1}{3} & \frac{8}{3} \\ 0 & 1 & 1 & 5 \\ 0 & \frac{5}{3} & -\frac{4}{3} & -\frac{2}{3} \end{pmatrix}$$

$$(\text{Row 3}) - (\frac{5}{3}) * \text{Row 2} \to \begin{pmatrix} 1 & \frac{1}{3} & \frac{1}{3} & \frac{8}{3} \\ 0 & 1 & 1 & 5 \\ 0 & 0 & -3 & -9 \end{pmatrix}$$

$$(\text{Row 3})/-3 \to \begin{pmatrix} 1 & \frac{1}{3} & \frac{1}{3} & \frac{8}{3} \\ 0 & 1 & 1 & 5 \\ 0 & 0 & 1 & 3 \end{pmatrix}$$

**********Backward pass**********

$$(\text{Row 1}) - (\frac{1}{3}) * \text{Row 3} \to \begin{pmatrix} 1 & \frac{1}{3} & 0 & \frac{5}{3} \\ 0 & 1 & 1 & 5 \\ 0 & 0 & 1 & 3 \end{pmatrix}$$

$$(\text{Row 2}) - (1) * \text{Row 3} \to \begin{pmatrix} 1 & \frac{1}{3} & 0 & \frac{5}{3} \\ 0 & 1 & 0 & 2 \\ 0 & 0 & 1 & 3 \end{pmatrix}$$

$$(\text{Row 1}) - (\frac{1}{3}) * \text{Row 2} \to \begin{pmatrix} 1 & 0 & 0 & 1 \\ 0 & 1 & 0 & 2 \\ 0 & 0 & 1 & 3 \end{pmatrix}$$

Solution

$$x_1 = 1 \quad x_2 = 2 \quad x_3 = 3$$

Inconsistent Systems of Equations

For some systems of equations, using the previous series of steps, the Gauss-Jordan form comes out as follows:

$$\left(\begin{array}{c|c} \mathbf{I} & \mathbf{c} \\ \hline \mathbf{0} & \mathbf{d} \end{array} \right)$$

The last equation says that a nonzero constant **d** is equal to **0**, which obviously is nonsense. Such a system of equations is inconsistent and has no solution.

Example 6.5 Consider the same system of equations as the previous example, except for the right-hand side of the third equation:

$$2x_1 + 5x_2 - x_3 + x_4 = 1$$
$$-3x_1 - 8x_2 + 2x_3 + 3x_4 = 4$$
$$x_1 + 2x_2 + 5x_4 = 7$$

The augmented matrix for this system of equations is as follows:

$$\begin{pmatrix} x_1 & x_2 & x_3 & x_4 & \text{rhs} \\ 2 & 5 & -1 & 1 & 1 \\ -3 & -8 & 2 & 3 & 4 \\ 1 & 2 & 0 & 5 & 7 \end{pmatrix}$$

$a = \begin{pmatrix} 2 & 5 & -1 & 1 & 1 \\ -3 & -8 & 2 & 3 & 4 \\ 1 & 2 & 0 & 5 & 7 \end{pmatrix}$; GaussJordanForm[a];

**********Forward pass**********

(Row 1) /2 $\rightarrow \begin{pmatrix} 1 & \frac{5}{2} & -\frac{1}{2} & \frac{1}{2} & \frac{1}{2} \\ -3 & -8 & 2 & 3 & 4 \\ 1 & 2 & 0 & 5 & 7 \end{pmatrix}$

(Row 2) $-$ (-3) $*$ Row 1 $\rightarrow \begin{pmatrix} 1 & \frac{5}{2} & -\frac{1}{2} & \frac{1}{2} & \frac{1}{2} \\ 0 & -\frac{1}{2} & \frac{1}{2} & \frac{9}{2} & \frac{11}{2} \\ 1 & 2 & 0 & 5 & 7 \end{pmatrix}$

(Row 3) $-$ (1) $*$ Row 1 $\rightarrow \begin{pmatrix} 1 & \frac{5}{2} & -\frac{1}{2} & \frac{1}{2} & \frac{1}{2} \\ 0 & -\frac{1}{2} & \frac{1}{2} & \frac{9}{2} & \frac{11}{2} \\ 0 & -\frac{1}{2} & \frac{1}{2} & \frac{9}{2} & \frac{13}{2} \end{pmatrix}$

(Row 2) $/-\frac{1}{2} \rightarrow \begin{pmatrix} 1 & \frac{5}{2} & -\frac{1}{2} & \frac{1}{2} & \frac{1}{2} \\ 0 & 1 & -1 & -9 & -11 \\ 0 & -\frac{1}{2} & \frac{1}{2} & \frac{9}{2} & \frac{13}{2} \end{pmatrix}$

(Row 3) $-$ ($-\frac{1}{2}$) $*$ Row 2 $\rightarrow \begin{pmatrix} 1 & \frac{5}{2} & -\frac{1}{2} & \frac{1}{2} & \frac{1}{2} \\ 0 & 1 & -1 & -9 & -11 \\ 0 & 0 & 0 & 0 & 1 \end{pmatrix}$

**********Backward pass**********

(Row 1) $-$ ($\frac{5}{2}$) $*$ Row 2 $\rightarrow \begin{pmatrix} 1 & 0 & 2 & 23 & 28 \\ 0 & 1 & -1 & -9 & -11 \\ 0 & 0 & 0 & 0 & 1 \end{pmatrix}$

The first two equations are fine, but the third equation is nonsense because it says that

$$0x_1 + 0x_2 + 0x_3 + 0x_4 = 1$$

6.2 Solving Linear Systems of Equations

Thus, the Gauss-Jordan form is inconsistent. The system of equations has no solution.

Matrix Inversion Using the Gauss-Jordan Form

It is also possible to find the inverse of a square matrix using the Gauss-Jordan form. For this purpose, the augmented matrix is defined as follows:

Initial augmented matrix: $(\mathbf{A} \mid \mathbf{I})$

where $\mathbf{A}$ is the matrix whose inverse is to be found, and $\mathbf{I}$ is an identity matrix of the same size. After carrying out the operations to convert the matrix to the Gauss-Jordan form, the final matrix is as follows:

Final Gauss-Jordan form: $(\mathbf{I} \mid \mathbf{A}^{-1})$

Example 6.6 As an example, consider finding an inverse of the matrix

$$\mathbf{A} = \begin{pmatrix} 3 & 1 & 1 \\ 2 & -1 & -1 \\ 1 & 2 & -1 \end{pmatrix}$$

For finding the inverse, we define the augmented matrix as follows:

$$a = \begin{pmatrix} 3 & 1 & 1 & 1 & 0 & 0 \\ 2 & -1 & -1 & 0 & 1 & 0 \\ 1 & 2 & -1 & 0 & 0 & 1 \end{pmatrix};$$

Its Gauss-Jordan form is computed as follows:

GaussJordanForm[a];
```
**********Forward pass**********
```

$(\text{Row 1})/3 \rightarrow \begin{pmatrix} 1 & \frac{1}{3} & \frac{1}{3} & \frac{1}{3} & 0 & 0 \\ 2 & -1 & -1 & 0 & 1 & 0 \\ 1 & 2 & -1 & 0 & 0 & 1 \end{pmatrix}$

$(\text{Row 2}) - (2) * \text{Row 1} \rightarrow \begin{pmatrix} 1 & \frac{1}{3} & \frac{1}{3} & \frac{1}{3} & 0 & 0 \\ 0 & -\frac{5}{3} & -\frac{5}{3} & -\frac{2}{3} & 1 & 0 \\ 1 & 2 & -1 & 0 & 0 & 1 \end{pmatrix}$

$(\text{Row 3}) - (1) * \text{Row 1} \rightarrow \begin{pmatrix} 1 & \frac{1}{3} & \frac{1}{3} & \frac{1}{3} & 0 & 0 \\ 0 & -\frac{5}{3} & -\frac{5}{3} & -\frac{2}{3} & 1 & 0 \\ 0 & \frac{5}{3} & -\frac{4}{3} & -\frac{1}{3} & 0 & 1 \end{pmatrix}$

$$(\text{Row 2}) / -\frac{5}{3} \rightarrow \begin{pmatrix} 1 & \frac{1}{3} & \frac{1}{3} & \frac{1}{3} & 0 & 0 \\ 0 & 1 & 1 & \frac{2}{5} & -\frac{3}{5} & 0 \\ 0 & \frac{5}{3} & -\frac{4}{3} & -\frac{1}{3} & 0 & 1 \end{pmatrix}$$

$$(\text{Row 3}) - (\frac{5}{3}) * \text{Row 2} \rightarrow \begin{pmatrix} 1 & \frac{1}{3} & \frac{1}{3} & \frac{1}{3} & 0 & 0 \\ 0 & 1 & 1 & \frac{2}{5} & -\frac{3}{5} & 0 \\ 0 & 0 & -3 & -1 & 1 & 1 \end{pmatrix}$$

$$(\text{Row 3}) / -3 \rightarrow \begin{pmatrix} 1 & \frac{1}{3} & \frac{1}{3} & \frac{1}{3} & 0 & 0 \\ 0 & 1 & 1 & \frac{2}{5} & -\frac{3}{5} & 0 \\ 0 & 0 & 1 & \frac{1}{3} & -\frac{1}{3} & -\frac{1}{3} \end{pmatrix}$$

**********Backward pass**********

$$(\text{Row 1}) - (\frac{1}{3}) * \text{Row 3} \rightarrow \begin{pmatrix} 1 & \frac{1}{3} & 0 & \frac{2}{9} & \frac{1}{9} & \frac{1}{9} \\ 0 & 1 & 1 & \frac{2}{5} & -\frac{3}{5} & 0 \\ 0 & 0 & 1 & \frac{1}{3} & -\frac{1}{3} & -\frac{1}{3} \end{pmatrix}$$

$$(\text{Row 2}) - (1) * \text{Row 3} \rightarrow \begin{pmatrix} 1 & \frac{1}{3} & 0 & \frac{2}{9} & \frac{1}{9} & \frac{1}{9} \\ 0 & 1 & 0 & \frac{1}{15} & -\frac{4}{15} & \frac{1}{3} \\ 0 & 0 & 1 & \frac{1}{3} & -\frac{1}{3} & -\frac{1}{3} \end{pmatrix}$$

$$(\text{Row 1}) - (\frac{1}{3}) * \text{Row 2} \rightarrow \begin{pmatrix} 1 & 0 & 0 & \frac{1}{5} & \frac{1}{5} & 0 \\ 0 & 1 & 0 & \frac{1}{15} & -\frac{4}{15} & \frac{1}{3} \\ 0 & 0 & 1 & \frac{1}{3} & -\frac{1}{3} & -\frac{1}{3} \end{pmatrix}$$

Thus, the inverse of the given matrix is

$$\mathbf{A}^{-1} = \begin{pmatrix} \frac{1}{5} & \frac{1}{5} & 0 \\ \frac{1}{15} & -\frac{4}{15} & \frac{1}{3} \\ \frac{1}{3} & -\frac{1}{3} & -\frac{1}{3} \end{pmatrix}$$

6.2.2 Solution Using LU Decomposition for Square Matrices

Consider a square system of equations (m equations in m unknowns)

$$\mathbf{Ax} = \mathbf{b}$$

where $\mathbf{A}$ is an $m \times m$ matrix. We can obviously solve this system using the Gauss-Jordan form. However, it is more efficient to first decompose the matrix

A into product of a lower triangular matrix **L** and an upper triangular matrix **U**. The matrix **L** has 1's on the diagonal and all of its entries above the diagonal are 0. All entries below the diagonal of the matrix **U** are 0.

$$\mathbf{A} = \mathbf{LU} \quad \mathbf{L} = \begin{pmatrix} 1 & 0 & 0 & \cdots & 0 \\ L_{21} & 1 & 0 & \cdots & 0 \\ L_{31} & L_{32} & 1 & \cdots & 0 \\ \vdots & \vdots & \vdots & \ddots & 0 \\ L_{m1} & L_{m2} & L_{m3} & \cdots & 1 \end{pmatrix} \quad \mathbf{U} = \begin{pmatrix} U_{11} & U_{12} & U_{13} & \cdots & U_{1m} \\ 0 & U_{22} & U_{23} & \cdots & U_{2m} \\ 0 & 0 & U_{33} & \cdots & U_{3m} \\ \vdots & \vdots & \vdots & \ddots & \vdots \\ 0 & 0 & 0 & \cdots & U_{mm} \end{pmatrix}$$

Such a decomposition is always possible for nonsingular square matrices. Assuming this decomposition is known, we can solve the system of equations in two steps as follows:

$$\mathbf{LUx} = \mathbf{b}$$

If we define $\mathbf{Ux} = \mathbf{y}$, then

$$\mathbf{Ly} = \mathbf{b} \quad \text{or} \quad \begin{pmatrix} 1 & 0 & 0 & \cdots & 0 \\ L_{21} & 1 & 0 & \cdots & 0 \\ L_{31} & L_{32} & 1 & \cdots & 0 \\ \vdots & \vdots & \vdots & \ddots & 0 \\ L_{m1} & L_{m2} & L_{m3} & \cdots & 1 \end{pmatrix} \begin{pmatrix} y_1 \\ y_2 \\ y_3 \\ \vdots \\ y_m \end{pmatrix} = \begin{pmatrix} b_1 \\ b_2 \\ b_3 \\ \vdots \\ b_m \end{pmatrix}$$

Since **L** is a lower triangular matrix, the first row gives the solution for the first variable as $y_1 = b_1$. Knowing this, we can solve for $y_2 = b_2 - L_{21}y_1$ from the second equation. Thus, proceeding forward from the first equation, we can solve for all intermediate variables y_i. This is known as *forward elimination*.

Now we can solve for x_i as follows:

$$\mathbf{Ux} = \mathbf{y} \quad \text{or} \quad \begin{pmatrix} U_{11} & U_{12} & U_{13} & \cdots & U_{1m} \\ 0 & U_{22} & U_{23} & \cdots & U_{2m} \\ 0 & 0 & U_{33} & \cdots & U_{3m} \\ \vdots & \vdots & \vdots & \ddots & \vdots \\ 0 & 0 & 0 & \cdots & U_{mm} \end{pmatrix} \begin{pmatrix} x_1 \\ x_2 \\ x_3 \\ \vdots \\ x_m \end{pmatrix} = \begin{pmatrix} y_1 \\ y_2 \\ y_3 \\ \vdots \\ y_m \end{pmatrix}$$

In this system, the last equation has only one unknown, giving $x_m = y_m/U_{mm}$. The second to last equation is used next to get x_{m-1}. Thus, working backward from the last equation, we can solve for all unknowns. This is known as *backward substitution*.

The procedure is especially efficient if solutions for several different right-hand sides are needed. Since in this case, the major computational task of decomposing the matrix into the LU form is not necessary for every right-hand side (rhs), we simply need to perform a forward elimination and a backward substitution for each different rhs.

An Algorithm for LU Decomposition

Given a nonsingular square matrix **A**, the matrices **L** and **U** are determined by applying a series of steps similar to those used for generating the Gauss-Jordan form. The main difference is that each elimination step is expressed as a matrix multiplication operation. At the end, the product of these elimination matrices gives the required decomposition.

For the following discussion, entries in matrix **A** are denoted by $a_{ij}, i, j = 1, 2, \ldots$. Assuming $a_{11} \neq 0$ (called pivot), it is easy to see that if we multiply matrix **A** by the following matrix $\mathbf{M}^{(1)}$, all entries below the diagonal in the first column will become zero.

$$\mathbf{M}^{(1)} = \begin{pmatrix} 1 & 0 & 0 & \ldots & 0 \\ -m_{21} & 1 & 0 & \ldots & 0 \\ -m_{31} & 0 & 1 & \ldots & 0 \\ \vdots & \vdots & \vdots & \ddots & \vdots \\ -m_{m1} & 0 & 0 & \ldots & 1 \end{pmatrix}$$

where $m_{k1} = a_{k1}/a_{11} \quad k = 2, 3, \ldots, m$. The new matrix will have the following form:

$$\mathbf{A}^{(2)} \equiv \mathbf{M}^{(1)}\mathbf{A} = \begin{pmatrix} a_{11}^{(2)} & a_{12}^{(2)} & a_{13}^{(2)} & \ldots & a_{1m}^{(2)} \\ 0 & a_{22}^{(2)} & a_{23}^{(2)} & \ldots & a_{2m}^{(2)} \\ 0 & a_{33}^{(2)} & a_{34}^{(2)} & \ldots & a_{3m}^{(2)} \\ \vdots & \vdots & \vdots & \ddots & \vdots \\ 0 & a_{m2}^{(2)} & a_{m3}^{(2)} & \ldots & a_{mm}^{(2)} \end{pmatrix}$$

The superscript (2) is used to indicate changed entries in **A** as a result of the matrix multiplication. We now define a matrix $\mathbf{M}^{(2)}$ to make entries below the diagonal in the second column to go to zero (again assuming a nonzero pivot, i.e., $a_{22}^{(2)} \neq 0$).

$$\mathbf{M}^{(2)} = \begin{pmatrix} 1 & 0 & 0 & \cdots & 0 \\ 0 & 1 & 0 & \cdots & 0 \\ 0 & -m_{32} & 1 & \cdots & 0 \\ \vdots & \vdots & \vdots & \ddots & \vdots \\ 0 & -m_{m2} & 0 & \cdots & 1 \end{pmatrix}$$

where $m_{k2} = a_{k2}^{(2)}/a_{22}^{(2)}$ $k = 3, \ldots, m$. The new matrix $\mathbf{A}^{(3)}$ will have the following form:

$$\mathbf{A}^{(3)} \equiv \mathbf{M}^{(2)}\mathbf{A}^{(2)} = \begin{pmatrix} a_{11}^{(3)} & a_{12}^{(3)} & a_{13}^{(3)} & \cdots & a_{1m}^{(3)} \\ 0 & a_{22}^{(3)} & a_{23}^{(3)} & \cdots & a_{2m}^{(3)} \\ 0 & 0 & a_{34}^{(3)} & \cdots & a_{3m}^{(3)} \\ \vdots & \vdots & \vdots & \ddots & \vdots \\ 0 & 0 & a_{m3}^{(3)} & \cdots & a_{mm}^{(3)} \end{pmatrix}$$

Repeating these steps for all columns in the matrix, we get an upper triangular matrix. Thus,

$$\mathbf{U} = \mathbf{M}^{(m-1)}\mathbf{M}^{(m-2)} \ldots \mathbf{M}^{(1)}\mathbf{A} \equiv \mathbf{M}\mathbf{A}$$

The lower triangular matrix can be written as follows:

$$\mathbf{L}\mathbf{U} = \mathbf{A} \quad \text{or} \quad \mathbf{U} = \mathbf{L}^{-1}\mathbf{A}$$

Since $\mathbf{U} = \mathbf{M}\mathbf{A}$, we see that

$$\mathbf{L}^{-1} = \mathbf{M} \quad \text{giving} \quad \mathbf{L} = \mathbf{M}^{-1} \equiv (\mathbf{M}^{(1)})^{-1}(\mathbf{M}^{(2)})^{-1} \ldots (\mathbf{M}^{(m-1)})^{-1}$$

Because of the special form of $\mathbf{M}$ matrices, it can easily be verified that the product of their inverses, and hence $\mathbf{L}$, is as follows:

$$\mathbf{L} = \begin{pmatrix} 1 & 0 & 0 & \cdots & 0 \\ m_{21} & 1 & 0 & \cdots & 0 \\ m_{31} & m_{32} & 1 & \cdots & 0 \\ \vdots & \vdots & \vdots & \ddots & \vdots \\ m_{m1} & m_{m2} & m_{m3} & \cdots & 1 \end{pmatrix}$$

When generating M matrices, it was assumed that the diagonal entries are nonzero. Sometimes the diagonal entries do become zero. If the overall system of equations is nonsingular, it is always possible to get a nonzero pivot by reordering the equations, as was done while generating the Gauss-Jordan form.

Chapter 6 Linear Programming

The following *Mathematica* implementation is designed to illustrate these ideas. The function keeps track of row exchanges to avoid zero pivots. *Mathematica* has built-in functions for performing LUDecomposition and LUBackSubstitution. These functions should be used in actual implementations.

```
Needs["OptimizationToolbox`LPSimplex`"];
?LUDecompositionSteps
```

LUDecompositionSteps[A]---decomposes matrix A into LU form, printing all
    intermediate steps. Equations are reordered, if necessary.

Example 6.7 Determine LU factors for the following matrix:

$$A = \begin{pmatrix} 2 & 1 & 1 \\ 4 & 1 & 0 \\ -2 & 2 & 1 \end{pmatrix};$$

`LUDecompositionSteps[A];`

$$M^1 \to \begin{pmatrix} 1 & 0 & 0 \\ -2 & 1 & 0 \\ 1 & 0 & 1 \end{pmatrix} \quad A^2 \to \begin{pmatrix} 2 & 1 & 1 \\ 0 & -1 & -2 \\ 0 & 3 & 2 \end{pmatrix}$$

$$M^2 \to \begin{pmatrix} 1 & 0 & 0 \\ 0 & 1 & 0 \\ 0 & 3 & 1 \end{pmatrix} \quad A^3 \to \begin{pmatrix} 2 & 1 & 1 \\ 0 & -1 & -2 \\ 0 & 0 & -4 \end{pmatrix}$$

$$\text{Product of all M matrices} \to \begin{pmatrix} 1 & 0 & 0 \\ -2 & 1 & 0 \\ -5 & 3 & 1 \end{pmatrix}$$

$$L \to \begin{pmatrix} 1 & 0 & 0 \\ 2 & 1 & 0 \\ -1 & -3 & 1 \end{pmatrix} \quad U \to \begin{pmatrix} 2 & 1 & 1 \\ 0 & -1 & -2 \\ 0 & 0 & -4 \end{pmatrix} \quad \text{Equation order} \to \begin{pmatrix} 1 \\ 2 \\ 3 \end{pmatrix}$$

Example 6.8 Find the solution of the following system of equations using LU decomposition:

$$x_1 + 2x_2 + 3x_3 + 4x_4 = 5$$
$$x_1 + 2x_2 + 4x_3 - 9x_4 = 9$$
$$-2x_1 - x_2 + x_3 + 2x_4 = 6$$
$$-x_1 - x_2 + x_3 + x_4 = 6$$

In matrix form:

$$A = \begin{pmatrix} 1 & 2 & 3 & 4 \\ 1 & 2 & 4 & -9 \\ -2 & -1 & 1 & 2 \\ -1 & -1 & 1 & 1 \end{pmatrix}; b = \{5, 9, 6, 6\};$$

6.2 Solving Linear Systems of Equations

We first generate the LU decomposition of the **A** matrix as follows. Note that a zero pivot is encountered after the first step. The equations are reordered and the process is continued until complete **L** and **U** matrices are obtained.

`{L, U, p} = LUDecompositionSteps[A];`

$$M^1 \to \begin{pmatrix} 1 & 0 & 0 & 0 \\ -1 & 1 & 0 & 0 \\ 2 & 0 & 1 & 0 \\ 1 & 0 & 0 & 1 \end{pmatrix} \quad A^2 \to \begin{pmatrix} 1 & 2 & 3 & 4 \\ 0 & 0 & 1 & -13 \\ 0 & 3 & 7 & 10 \\ 0 & 1 & 4 & 5 \end{pmatrix}$$

Re-ordered equations

$$A \to \begin{pmatrix} 1 & 2 & 3 & 4 \\ -2 & -1 & 1 & 2 \\ 1 & 2 & 4 & -9 \\ -1 & -1 & 1 & 1 \end{pmatrix} \quad \text{Equation order} \to \begin{pmatrix} 1 \\ 3 \\ 2 \\ 4 \end{pmatrix}$$

$$M^1 \to \begin{pmatrix} 1 & 0 & 0 & 0 \\ 2 & 1 & 0 & 0 \\ -1 & 0 & 1 & 0 \\ 1 & 0 & 0 & 1 \end{pmatrix} \quad A^2 \to \begin{pmatrix} 1 & 2 & 3 & 4 \\ 0 & 3 & 7 & 10 \\ 0 & 0 & 1 & -13 \\ 0 & 1 & 4 & 5 \end{pmatrix}$$

$$M^2 \to \begin{pmatrix} 1 & 0 & 0 & 0 \\ 0 & 1 & 0 & 0 \\ 0 & 0 & 1 & 0 \\ 0 & -\frac{1}{3} & 0 & 1 \end{pmatrix} \quad A^3 \to \begin{pmatrix} 1 & 2 & 3 & 4 \\ 0 & 3 & 7 & 10 \\ 0 & 0 & 1 & -13 \\ 0 & 0 & \frac{5}{3} & \frac{5}{3} \end{pmatrix}$$

$$M^3 \to \begin{pmatrix} 1 & 0 & 0 & 0 \\ 0 & 1 & 0 & 0 \\ 0 & 0 & 1 & 0 \\ 0 & 0 & -\frac{5}{3} & 1 \end{pmatrix} \quad A^4 \to \begin{pmatrix} 1 & 2 & 3 & 4 \\ 0 & 3 & 7 & 10 \\ 0 & 0 & 1 & -13 \\ 0 & 0 & 0 & \frac{70}{3} \end{pmatrix}$$

$$\text{Product of all M matrices} \to \begin{pmatrix} 1 & 0 & 0 & 0 \\ 2 & 1 & 0 & 0 \\ -1 & 0 & 1 & 0 \\ 2 & -\frac{1}{3} & -\frac{5}{3} & 1 \end{pmatrix}$$

$$L \to \begin{pmatrix} 1 & 0 & 0 & 0 \\ -2 & 1 & 0 & 0 \\ 1 & 0 & 1 & 0 \\ -1 & \frac{1}{3} & \frac{5}{3} & 1 \end{pmatrix} \quad U \to \begin{pmatrix} 1 & 2 & 3 & 4 \\ 0 & 3 & 7 & 10 \\ 0 & 0 & 1 & -13 \\ 0 & 0 & 0 & \frac{70}{3} \end{pmatrix} \quad \text{Equation order} \to \begin{pmatrix} 1 \\ 3 \\ 2 \\ 4 \end{pmatrix}$$

The next step is the solution for intermediate variables (forward elimination). Note the order of the right-hand side is changed to make it consistent with the order of the equations.

$$\begin{pmatrix} 1 & 0 & 0 & 0 \\ -2 & 1 & 0 & 0 \\ 1 & 0 & 1 & 0 \\ -1 & \frac{1}{3} & \frac{5}{3} & 1 \end{pmatrix} \begin{pmatrix} y_1 \\ y_2 \\ y_3 \\ y_4 \end{pmatrix} = \begin{pmatrix} 5 \\ 6 \\ 9 \\ 6 \end{pmatrix}$$

$$y_1 = 5$$
$$y_2 = 6 + 2y_1 = 16$$
$$y_3 = 9 - y_1 = 4$$
$$y_4 = 6 + y_1 - \frac{1}{3}y_3 - \frac{5}{3}y_2 = -1$$

The following is the solution for the actual variables (backward substitution):

$$\begin{pmatrix} 1 & 2 & 3 & 4 \\ 0 & 3 & 7 & 10 \\ 0 & 0 & 1 & -13 \\ 0 & 0 & 0 & \frac{70}{3} \end{pmatrix} \begin{pmatrix} x_1 \\ x_2 \\ x_3 \\ x_4 \end{pmatrix} = \begin{pmatrix} 5 \\ 16 \\ 4 \\ -1 \end{pmatrix}$$

$$x_4 = -\frac{1}{\frac{70}{3}} = -\frac{3}{70}$$

$$x_3 = 4 + 13x_4 = \frac{241}{70}$$

$$x_2 = \frac{16 - 7x_3 - 10x_4}{3} = -\frac{179}{70}$$

$$x_1 = 5 - 2x_2 - 3x_3 - 4x_4 = -\frac{3}{70}$$

The same solution can directly be obtained by using the built-in function, called LinearSolve.

LinearSolve[A, b]

$$\left\{ -\frac{3}{70}, -\frac{179}{70}, \frac{241}{70}, -\frac{3}{70} \right\}$$

6.3 Basic Solutions of an LP Problem

As mentioned before, the solution of an LP problem reduces to solving a system of *underdetermined* (fewer equations than variables) linear equations. From m equations, at most we can solve for m variables in terms of the remaining $n - m$ variables. The variables that we choose to solve for are called *basic*, and the remaining variables are called *nonbasic*.

Consider the following example:

Minimize $f = -x_1 + x_2$

Subject to $\begin{pmatrix} x_1 - 2x_2 \geq 2 \\ x_1 + x_2 \leq 4 \\ x_1 \leq 3 \\ x_i \geq 0, i = 1, 2 \end{pmatrix}$

In the standard LP form

Minimize $f = -x_1 + x_2$

Subject to $\begin{pmatrix} x_1 - 2x_2 - x_3 = 2 \\ x_1 + x_2 + x_4 = 4 \\ x_1 + x_5 = 3 \\ x_i \geq 0, i = 1, \ldots, 5 \end{pmatrix}$

where x_3 is a surplus variable for the first constraint, and x_4 and x_5 are slack variables for the two less-than type constraints. The total number of variables is $n = 5$, and the number of equations is $m = 3$. Thus, we can have three basic variables and two nonbasic variables. If we arbitrarily choose x_3, x_4, and x_5 as basic variables, a general solution of the constraint equations can readily be written as follows:

$$x_3 = -2 + x_1 - 2x_2 \qquad x_4 = 4 - x_1 - x_2 \qquad x_5 = 3 - x_1$$

The general solution is valid for any values of the nonbasic variables. Since all variables are positive and we are interested in minimizing the objective function, we assign 0 values to nonbasic variables. A solution from the constraint equations obtained by setting nonbasic variables to zero is called a *basic solution*. Therefore, one possible basic solution for the above example is as follows:

$$x_3 = -2 \qquad x_4 = 4 \qquad x_5 = 3$$

Since all variables must be ≥ 0, this basic solution is infeasible because x_3 is negative.

Let's find another basic solution by choosing (again arbitrarily) x_1, x_4, and x_5 as basic variables and x_2 and x_3 as nonbasic. By setting nonbasic variables to zero, we need to solve for the basic variables from the following equations:

$$x_1 = 2 \qquad x_1 + x_4 = 4 \qquad x_1 + x_5 = 3$$

It can easily be verified that the solution is $x_1 = 2$, $x_4 = 2$, and $x_5 = 1$. Since all variables have positive values, this basic solution is feasible as well.

The maximum number of possible basic solutions depends on the number of constraints and the number of variables in the problem, and can be determined from the following equation:

Number of possible basic solutions = Binomial$[n, m] \equiv \frac{n!}{m!\,(n-m)!}$

where "!" stands for *factorial*. For the example problem where $m = 3$ and $n = 5$, therefore, the maximum number of basic solutions is

$$\frac{5!}{3!\,2!} = \frac{5 \times 4 \times 3!}{3! \times 2} = 10$$

All these basic solutions are computed from the constraint equations and are summarized in the following table. The set of basic variables for a particular solution is called a *basis* for that solution.

	Basis	Solution	Status	f
(1)	$\{x_1, x_2, x_3\}$	$\{3, 1, -1, 0, 0\}$	Infeasible	—
(2)	$\{x_1, x_2, x_4\}$	$\{3, \frac{1}{2}, 0, \frac{1}{2}, 0\}$	Feasible	$-\frac{5}{2}$
(3)	$\{x_1, x_2, x_5\}$	$\{\frac{10}{3}, \frac{2}{3}, 0, 0, -\frac{1}{3}\}$	Infeasible	—
(4)	$\{x_1, x_3, x_4\}$	$\{3, 0, 1, 1, 0\}$	Feasible	-3
(5)	$\{x_1, x_3, x_5\}$	$\{4, 0, 2, 0, -1\}$	Infeasible	—
(6)	$\{x_1, x_4, x_5\}$	$\{2, 0, 0, 2, 1\}$	Feasible	-2
(7)	$\{x_2, x_3, x_4\}$	$\{-\}$	NoSolution	—
(8)	$\{x_2, x_3, x_5\}$	$\{0, 4, -10, 0, 3\}$	Infeasible	—
(9)	$\{x_2, x_4, x_5\}$	$\{0, -1, 0, 5, 3\}$	Infeasible	—
(10)	$\{x_3, x_4, x_5\}$	$\{0, 0, -2, 4, 3\}$	Infeasible	—

The case (7) produces an inconsistent system of equations and hence, there is no solution. There are only three cases that give a basic feasible solution. The objective function values are computed for these solutions. Since the original problem was a two-variable problem, we can obtain a graphical solution to gain further insight into the basic feasible solutions. As seen from the graph in Figure 6.1, the three basic feasible solutions correspond to the three vertices of the feasible region. The infeasible basic solutions correspond to constraint intersections that are outside of the feasible region.

A Brute-force Method for Solving an LP Problem

As illustrated in the previous section, the vertices of the feasible region of an LP problem correspond to basic feasible solutions. Furthermore, since the solution of an LP problem must lie on the boundary of the feasible domain,

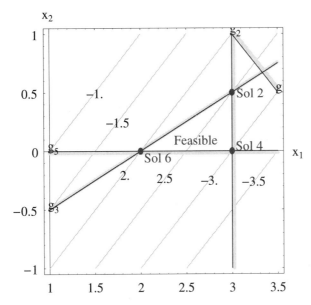

FIGURE 6.1 A graphical solution showing all basic feasible solutions.

one of these basic feasible solutions must be the optimum. Thus, a *brute-force* method to find an optimum is to compute all possible basic solutions. The one that is feasible and has the lowest value of the objective function is the optimum solution.

For the example problem, the fourth basic solution is feasible and has the lowest value of the objective function. Thus, this represents the optimum solution for the problem.

Optimum solution:

$$x_1^* = 3 \quad x_2^* = 0 \quad x_3^* = 1 \quad x_4^* = 1 \quad x_5^* = 0 \quad f^* = -3$$

Mathematica Function to Compute All Basic Solutions

Computation of all basic solutions is quite tedious. The following *Mathematica* function is developed to compute all possible basic solutions for an LP problem written in the standard form. The number of possible combinations increases very rapidly as the number of variables and the constraints increase. Therefore, this function, by default, computes a maximum of 20 solutions. More solutions can be obtained by using a larger number when calling the function.

```
Needs["OptimizationToolbox`LPSimplex`"];
?BasicSolutions
```

BasicSolutions[f,eqns,vars,maxSolutions:20]---determines basic solutions
 for an LP problem written in standard form. f is the objective
 functions, eqns = constraint equations, vars = list of variables.
 maxSolutions = maximum number of solutions computed (default ≤ 20).

Example 6.9 Compute all basic solutions of the following LP problem:

Minimize $f = x_1 + x_2$

Subject to $\begin{pmatrix} 2x_1 + x_2 \leq 8 \\ 3x_1 + 2x_2 \leq 10 \\ x_i \geq 0, i = 1, 2 \end{pmatrix}$

In the standard LP form

Minimize $f = x_1 + x_2$

Subject to $\begin{pmatrix} 2x_1 + x_2 + x_3 = 8 \\ 3x_1 + 2x_2 + x_4 = 10 \\ x_i \geq 0, i = 1, \ldots, 4 \end{pmatrix}$

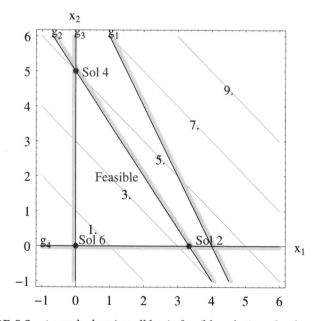

FIGURE 6.2 A graph showing all basic feasible solutions for the example.

```
f = x₁ + x₂;
eqns = {2x₁ + x₂ + x₃ == 8, 3x₁ + 2x₂ + x₄ == 10};
BasicSolutions[f, eqns, {x₁, x₂, x₃, x₄}];
```

$$\begin{pmatrix} \text{Basis} & \text{Solution} & \text{Status} & f \\ \{x_1, x_2\} & \{6, -4, 0, 0\} & \text{Infeasible} & -- \\ \{x_1, x_3\} & \{\frac{10}{3}, 0, \frac{4}{3}, 0\} & \text{Feasible} & \frac{10}{3} \\ \{x_1, x_4\} & \{4, 0, 0, -2\} & \text{Infeasible} & -- \\ \{x_2, x_3\} & \{0, 5, 3, 0\} & \text{Feasible} & 5 \\ \{x_2, x_4\} & \{0, 8, 0, -6\} & \text{Infeasible} & -- \\ \{x_3, x_4\} & \{0, 0, 8, 10\} & \text{Feasible} & 0 \end{pmatrix}$$

The graph of the original problem shown in Figure 6.2 again confirms that the three basic feasible solutions correspond to the three vertices of the feasible region.

6.4 The Simplex Method

The optimum solution of an LP problem corresponds to one of the basic feasible solutions and thus can be found by examining all basic solutions. However, the number of possible basic solutions can be very large for practical problems. Thus, the goal is to develop a procedure that quickly finds the basic feasible solution with the lowest objective function value without examining all possibilities.

6.4.1 Basic Idea

The basic idea of the simplex method is to start with a basic feasible solution and then try to obtain a neighboring basic feasible solution that has the objective function with a value lower than the current basic feasible solution. With each try, one of the current basic variables is made nonbasic and is replaced with a variable from the nonbasic set. An optimum is reached when no other basic feasible solution can be found with a lower objective function value. Rules are established such that for most problems, the method finds an optimum in a lot fewer steps than the total number of possible basic solutions.

The complete algorithm needs procedures for (a) finding a starting basic feasible solution, (b) bringing a currently nonbasic variable into the basic set, and (c) moving a currently basic variable out of the basic set to make room for the new basic variable.

6.4.2 Using the Simplex Method for Problems with LE Constraints

Problems that initially have all less than type (LE, $\leq$) constraints are easiest to deal with in the simplex method. The procedure will be explained with reference to the following example:

Minimize $f = 5x_1 - 3x_2 - 8x_3$

Subject to $\begin{pmatrix} 2x_1 + 5x_2 - x_3 \leq 1 \\ -2x_1 - 12x_2 + 3x_3 \leq 9 \\ -3x_1 - 8x_2 + 2x_3 \leq 4 \\ x_i \geq 0, i = 1, \ldots, 3 \end{pmatrix}$

Starting Basic Feasible Solution

The starting basic feasible solution is easy to obtain for problems that involve only LE type constraints (after making the right-hand side positive, if necessary). A different slack variable must be added to each LE constraint to convert it into equality. We have as many slack variables as the number of constraint equations. If we treat these slack variables as basic variables and the actual variables as nonbasic (set to 0), then the right-hand side of each constraint represents the basic solution. Since the right-hand sides must all be positive (a requirement of the standard LP form), this basic solution is feasible as well.

For the example problem, introducing slack variables x_4, x_5, and x_6, the constraints are written in the standard LP form as follows.

$$2x_1 + 5x_2 - x_3 + x_4 = 1$$

$$-2x_1 - 12x_2 + 3x_3 + x_5 = 9$$

$$-3x_1 - 8x_2 + 2x_3 + x_6 = 4$$

Treating the slack variables as basic and the others as nonbasic, the starting basic feasible solution is

Basic: $x_4 = 1, x_5 = 9, x_6 = 4$

Nonbasic: $x_1 = x_2 = x_3 = 0$

$f = 0$

Note that the objective function is expressed in terms of nonbasic variables. In this first step, we did not have to do anything special to achieve this. However, in subsequent steps, it is necessary to explicitly eliminate basic variables from the objective function. The following rule used to make decisions regarding which variable to bring into the basic set assumes that the objective function is written in terms of nonbasic variables alone.

Bringing a New Variable Into the Basic Set

In order to find a new basic feasible solution, one of the currently nonbasic variables must be made basic. The best candidate for this purpose is the variable that causes the largest decrease in the objective function. The nonbasic variable that has the largest negative coefficient in the objective function is the best choice. This makes sense because the nonbasic variables are set to zero and so the current value of the objective function is not influenced by them. But if one of them is made basic, it will have a positive value and therefore if its coefficient is a large negative number, it has the greatest potential of causing a large reduction in the objective function.

Continuing with the previous example, we have the following situation:

Basic: $x_4 = 1, x_5 = 9, x_6 = 4$

Nonbasic: $x_1 = x_2 = x_3 = 0$

$f = 0 = 5x_1 - 3x_2 - 8x_3$

The largest negative coefficient in the f equation is that of x_3. Thus, our next basic feasible solution should use x_3 as one of its basic variables.

Moving an Existing Basic Variable Out of the Basic Set

Since the number of basic variables is fixed by the number of constraint equations, one of the currently basic variables must be removed in order to make room for a new basic variable. The decision to remove a variable from the basic set is based on the need to keep the solution feasible. A little algebra shows that we should remove the basic variable that corresponds to the smallest ratio of the right-hand side of constraints and the coefficients of the new variable that is being brought into the basic set. Furthermore, if the coefficient is negative, then there is no danger that the associated basic variable will become negative; thus, during this process, we need to look at ratios of right-hand sides and positive constraint coefficients.

To understand the reasoning behind this rule, let's continue with the example problem. The constraints and the corresponding basic variables are as follows:

Constraint 1: x_4 basic: $2x_1 + 5x_2 - x_3 + x_4 = 1$
Constraint 2: x_5 basic: $-2x_1 - 12x_2 + 3x_3 + x_5 = 9$
Constraint 3: x_6 basic: $-3x_1 - 8x_2 + 2x_3 + x_6 = 4$

The variable x_3 is to be brought into the basic set which means that it will have a value greater than or equal to zero in the next basic feasible solution. For constraint 1, the coefficient of x_3 is negative and therefore, the solution from this constraint will remain positive as a result of making x_3 basic. In constraints 2 and 3, the coefficients of x_3 are positive and therefore, these constraints could result in negative values of variables. From the constraint 2, we see that the new basic variable x_3 must have a value less than or equal to $9/3 = 3$; otherwise, this constraint will give a negative solution. Similarly, the third constraint shows that x_3 must have a value less than or equal to $4/2 = 2$. The constraint 3 is more critical and therefore, we should make x_3 a basic variable for this constraint and hence remove x_6 from the basic set.

The Next Basic Feasible Solution

Now we are in a position to compute the next basic feasible solution. We need to solve the system of constraint equations for the new set of basic variables. Also, the objective function must be expressed in terms of new nonbasic variables in order to continue with the subsequent steps of the simplex method.

For the example problem, the constraints are currently written as follows:

Constraint 1: x_4 basic: $2x_1 + 5x_2 - x_3 + x_4 = 1$
Constraint 2: x_5 basic: $-2x_1 - 12x_2 + 3x_3 + x_5 = 9$
Constraint 3: x_6 basic: $-3x_1 - 8x_2 + 2x_3 + x_6 = 4$

The new basic variable set is (x_4, x_5, x_3). We can achieve this by eliminating x_3 from the first and the second constraints. We divide the third constraint by 2 first to make the coefficient of x_3 equal to 1.

Constraint 3: x_3 basic: $-\frac{3}{2}x_1 - 4x_2 + x_3 + \frac{1}{2}x_6 = 2$

This constraint is known as the *pivot row* for computing the new basic feasible solution, and is used to eliminate x_3 from the other constraints and the objective function. Variable x_3 can be eliminated from constraint 1 by adding the pivot row to the first constraint.

Constraint 1: x_4 basic: $\frac{1}{2}x_1 + x_2 + x_4 + \frac{1}{2}x_6 = 3$

From the second constraint, variable x_3 is eliminated by adding (-3) times the pivot row to it.

Constraint 2: x_5 basic: $\frac{5}{2}x_1 + x_5 - \frac{3}{2}x_6 = 3$

The objective function is

$$5x_1 - 3x_2 - 8x_3 = f$$

The variable x_3 is eliminated by adding eight times the pivot row to it.

$$-7x_1 - 35x_2 + 4x_6 = f + 16$$

We now have a new basic feasible solution, as follows:

Basic: $x_3 = 2$, $x_4 = 3$, $x_5 = 3$

Nonbasic: $x_1 = x_2 = x_6 = 0$

$f = -16$

Considering the objective function value, this solution is better than our starting solution.

The Optimum Solution

The series of steps are repeated until all coefficients in the objective function become positive. When this happens, then bringing any of the current nonbasic variables into the basic set will increase the objective function value. This indicates that we have reached the lowest value possible and the current basic feasible solution represents the optimum solution.

The next step for the example problem is summarized as follows:

Constraint 1: x_4 basic: $\frac{1}{2}x_1 + x_2 + x_4 + \frac{1}{2}x_6 = 3$
Constraint 2: x_5 basic: $\frac{5}{2}x_1 + x_5 - \frac{3}{2}x_6 = 3$
Constraint 3: x_3 basic: $-\frac{3}{2}x_1 - 4x_2 + x_3 + \frac{1}{2}x_6 = 2$
Objective: $-7x_1 - 35x_2 + 4x_6 = f + 16$

In the objective row, the variable x_2 has the largest negative coefficient. Thus, the next variable to be made basic is x_2. In the constraint expressions, positive coefficient of x_2 shows up only in the first constraint, which has x_4 as the basic variable. Thus, x_4 should be removed from the basic set. The coefficient of x_2 in the first row is already 1; thus, there is nothing that needs to be done to this equation. For eliminating x_2 from the remaining equation, now we must use the first equation as the pivot row (PR) as follows.

Constraint 1: x_2 basic: $\frac{1}{2}x_1 + x_2 + x_4 + \frac{1}{2}x_6 = 3$ (PR)

Constraint 2: x_5 basic: $\frac{5}{2}x_1 + x_5 - \frac{3}{2}x_6 = 3$ (no change)

Constraint 3: x_3 basic: $\frac{1}{2}x_1 + x_3 + 4x_4 + \frac{5}{2}x_6 = 14$ (Added 4 × PR)

Objective: $\frac{21}{2}x_1 + 35x_4 + \frac{43}{2}x_6 = f + 121$ (Added 35 × PR)

Looking at the objective function row, we see that all coefficients are positive. This means that bringing any of the current nonbasic variables into the basic set will increase the objective function. Thus, we have reached the lowest value possible and hence, the above basic feasible solution represents the optimum solution.

Optimum solution:

     Basic: $x_2 = 3, x_3 = 14, x_5 = 3$

     Nonbasic: $x_1 = x_4 = x_6 = 0$

     $f + 121 = 0$ giving $f^* = -121$

It is interesting to note that the total number of possible basic solutions for this example was Binomial[6, 3] = 20. However, the simplex method found the optimum in only three steps.

6.4.3 Simplex Tableau

A tabular form is convenient to organize the calculations involved in the simplex method. The tableau is essentially the augmented matrix used in the Gauss-Jordan form for computing the solution of equations. Thus, the rows represent coefficients of the constraint equations. The objective function is written in the last row. The first column indicates the basic variable associated with the constraint in that row. Recall that this variable should appear only in one constraint with a coefficient of 1. The last column represents the right-hand sides of the equations. The other columns represent coefficients of variables, usually arranged in ascending order with the actual variables first, followed by the slack variables. The exact form of the simplex tableau is illustrated through the following examples.

Example 6.10   Consider the solution of the problem from the previous section using the tableau form.

Minimize $f = 5x_1 - 3x_2 - 8x_3$

Subject to $\begin{pmatrix} 2x_1 + 5x_2 - x_3 \leq 1 \\ -2x_1 - 12x_2 + 3x_3 \leq 9 \\ -3x_1 - 8x_2 + 2x_3 \leq 4 \\ x_i \geq 0, i = 1, \ldots, 3 \end{pmatrix}$

Introducing slack variables, the constraints are written in the standard LP form as follows:

$$2x_1 + 5x_2 - x_3 + x_4 = 1$$

$$-2x_1 - 12x_2 + 3x_3 + x_5 = 9$$

$$-3x_1 - 8x_2 + 2x_3 + x_6 = 4$$

The starting tableau is as follows:

Initial Tableau:
$\begin{pmatrix} \text{Basis} & x_1 & x_2 & x_3 & x_4 & x_5 & x_6 & \text{RHS} \\ x_4 & 2 & 5 & -1 & 1 & 0 & 0 & 1 \\ x_5 & -2 & -12 & 3 & 0 & 1 & 0 & 9 \\ x_6 & -3 & -8 & 2 & 0 & 0 & 1 & 4 \\ \text{Obj.} & 5 & -3 & -8 & 0 & 0 & 0 & f \end{pmatrix}$

The first three rows are simply the constraint equations. The fourth row is the objective function, expressed in the form of an equation, $5x_1 - 3x_2 - 8x_3 = f$. The right-hand side of the objective function equation is set to f. Since x_4 appears only in the first row, it is the basic variable associated with the first constraint. Similarly, the basic variables associated with the other two constraints are x_5 and x_6. The basic variables for each constraint row are identified in the first column.

From the tableau, we can read the basic feasible solution simply by setting the basis to the rhs (since the nonbasic variables are all set to 0).

Basic: $x_4 = 1$, $x_5 = 9$, $x_6 = 4$

Nonbasic: $x_1 = x_2 = x_3 = 0$

$f = 0$

We now proceed to the first iteration of the simplex method. To bring a new variable into the basic set, we look at the largest negative number in the objective function row. From the simplex tableau, we can readily identify that the coefficient corresponding to x_3 is most negative (-8). Thus, x_3 should be made basic.

The variable that must be removed from the basic set corresponds to the smallest ratio of the entries in the constraint right-hand sides and the positive entries in the column corresponding to the new variable to be made basic. From the simplex tableau, we see that in the column corresponding to x_3, there are two constraint rows that have positive coefficients. Ratios of the right-hand side and these entries are as follows:

Ratios: $\{\frac{9}{3} = 3, \frac{4}{2} = 2\}$

The minimum ratio corresponds to the third constraint for which x_6 is the current basic variable. Thus, we should make x_6 nonbasic.

Based on these decisions, our next tableau must be of the following form.

$$\begin{pmatrix} \text{Basis} & x_1 & x_2 & x_3 & x_4 & x_5 & x_6 & \text{RHS} \\ x_4 & - & - & 0 & 1 & 0 & - & - \\ x_5 & - & - & 0 & 0 & 1 & - & - \\ x_3 & - & - & 1 & 0 & 0 & - & - \\ \text{Obj.} & - & - & 0 & 0 & 0 & - & - \end{pmatrix}$$

That is, we need to eliminate variable x_3 from the first, second, and the objective function row. Since each row represents an equation, this can be done by adding or subtracting appropriate multiples of rows together. However, we must be careful in how we perform these steps because we need to preserve x_4 and x_5 as basic variables for the first and the second constraints. That is, the form of columns x_4 and x_5 must be maintained during these row operations.

The systematic procedure to actually bring the tableau into the desired form is to first divide row 3 (because it involves the new basic variable) by its diagonal element (the coefficient corresponding to the new basic variable). Thus, dividing row 3 by 2, we have the following situation:

$$\begin{pmatrix} \text{Basis} & x_1 & x_2 & x_3 & x_4 & x_5 & x_6 & \text{RHS} \\ x_4 & - & - & 0 & 1 & 0 & - & - \\ x_5 & - & - & 0 & 0 & 1 & - & - \\ x_3 & -\frac{3}{2} & -4 & 1 & 0 & 0 & \frac{1}{2} & 2 \\ \text{Obj.} & 0 & - & - & - & 0 & 0 & - \end{pmatrix}$$

We call this modified row as *pivot row* (PR) and use it to eliminate x_3 from the other rows. The computations are as follows:

Basis	x_1	x_2	x_3	x_4	x_5	x_6	RHS		
x_4	$\frac{1}{2}$	1	0	1	0	$\frac{1}{2}$	3	$\Leftarrow$	PR + Row1
x_5	$\frac{5}{2}$	0	0	0	1	$-\frac{3}{2}$	3	$\Leftarrow$	$-3 \times$ PR + Row2
x_3	$-\frac{3}{2}$	-4	1	0	0	$\frac{1}{2}$	2	$\Leftarrow$	PR
Obj.	-7	-35	0	0	0	4	$16 + f$	$\Leftarrow$	$8 \times$ PR + Obj.Row

6.4 The Simplex Method

This completes one step of the simplex method, and we have a second basic feasible solution.

Second Tableau:
$$\begin{pmatrix} \text{Basis} & x_1 & x_2 & x_3 & x_4 & x_5 & x_6 & \text{RHS} \\ x_4 & \frac{1}{2} & 1 & 0 & 1 & 0 & \frac{1}{2} & 3 \\ x_5 & \frac{5}{2} & 0 & 0 & 0 & 1 & -\frac{3}{2} & 3 \\ x_3 & -\frac{3}{2} & -4 & 1 & 0 & 0 & \frac{1}{2} & 2 \\ \text{Obj.} & -7 & -35 & 0 & 0 & 0 & 4 & 16+f \end{pmatrix}$$

Basic: $x_3 = 2, x_4 = 3, x_5 = 3$ Nonbasic: $x_1 = x_2 = x_6 = 0$ $f = -16$

The same series of steps can now be repeated for additional tableaus. For the third tableau, we should make x_2 be basic (the largest negative coefficient in the obj. row = -35). In the column corresponding to x_2, only the first row has a positive coefficient and thus, we have no choice but to make x_4 (the current basic variable for the first row) nonbasic. Based on these decisions, our next tableau must be of the following form:

$$\begin{pmatrix} \text{Basis} & x_1 & x_2 & x_3 & x_4 & x_5 & x_6 & \text{RHS} \\ x_2 & - & 1 & 0 & - & 0 & - & - \\ x_5 & - & 0 & 0 & - & 1 & - & - \\ x_3 & - & 0 & 1 & - & 0 & - & - \\ \text{Obj.} & - & 0 & 0 & - & 0 & - & - \end{pmatrix}$$

The first row already has a 1 in the x_2 column; therefore, we don't need to do anything and use it as our new pivot row to eliminate x_2 from the other rows. The computations are as follows:

Basis	x_1	x_2	x_3	x_4	x_5	x_6	RHS		
x_2	$\frac{1}{2}$	1	0	1	0	$\frac{1}{2}$	3	$\Longleftarrow$	PR
x_5	$\frac{5}{2}$	0	0	0	1	$-\frac{3}{2}$	3	$\Longleftarrow$	Row2
x_3	$\frac{1}{2}$	0	1	4	0	$\frac{5}{2}$	14	$\Longleftarrow$	$4 \times$ PR + Row3
Obj.	$\frac{21}{2}$	0	0	35	0	$\frac{43}{2}$	$121+f$	$\Longleftarrow$	$35 \times$ PR + Obj.Row

This completes the second iteration of the simplex method, and we have a third basic feasible solution:

Third Tableau:
$$\begin{pmatrix} \text{Basis} & x_1 & x_2 & x_3 & x_4 & x_5 & x_6 & \text{RHS} \\ x_2 & \frac{1}{2} & 1 & 0 & 1 & 0 & \frac{1}{2} & 3 \\ x_5 & \frac{5}{2} & 0 & 0 & 0 & 1 & -\frac{3}{2} & 3 \\ x_3 & \frac{1}{2} & 0 & 1 & 4 & 0 & \frac{5}{2} & 14 \\ \text{Obj.} & \frac{21}{2} & 0 & 0 & 35 & 0 & \frac{43}{2} & 121+f \end{pmatrix}$$

Basic: $x_2 = 3, x_3 = 14, x_5 = 3$ Nonbasic: $x_1 = x_4 = x_6 = 0$ $f = -121$

Since all coefficients in the Obj. row are positive, we cannot reduce the objective function any further and thus have reached the minimum. The optimum solution is

Optimum: $x_1 = 0, x_2 = 3, x_3 = 14, x_4 = 0, x_5 = 3, x_6 = 0$ $f = -121$

Example 6.11 Solve the following LP problem using the tableau form of the simplex method.

Maximize $-7x_1 - 4x_2 + 15x_3$

Subject to $\begin{pmatrix} \frac{x_1}{3} - \frac{32x_2}{9} + \frac{20x_3}{9} \leq 1 \\ \frac{x_1}{6} - \frac{13x_2}{9} + \frac{5x_3}{18} \leq 2 \\ \frac{2x_1}{3} - \frac{16x_2}{9} + \frac{x_3}{9} \geq -3 \\ x_i \geq 0, i = 1, \ldots, 3 \end{pmatrix}$

Note that the problem as stated has a greater than type constraint. However, since the right-hand side of all constraints must be positive, as soon as we multiply the third constraint by a negative sign, all constraints become of LE type and therefore, we can handle this problem with the procedure developed so far.

$$\frac{2x_1}{3} - \frac{16x_2}{9} + \frac{x_3}{9} \geq -3 \quad \text{same as} \quad -\frac{2x_1}{3} + \frac{16x_2}{9} - \frac{x_3}{9} \leq 3$$

In the standard LP form

Minimize $f = 7x_1 + 4x_2 - 15x_3$

Subject to $\begin{pmatrix} \frac{x_1}{3} - \frac{32x_2}{9} + \frac{20x_3}{9} + x_4 = 1 \\ \frac{x_1}{6} - \frac{13x_2}{9} + \frac{5x_3}{18} + x_5 = 2 \\ -\frac{2x_1}{3} + \frac{16x_2}{9} - \frac{x_3}{9} + x_6 = 3 \\ x_i \geq 0, i = 1, \ldots, 6 \end{pmatrix}$

where x_4, x_5, and x_6 are slack variables for the three constraints.

Initial Tableau:
$$\begin{pmatrix} \text{Basis} & x_1 & x_2 & x_3 & x_4 & x_5 & x_6 & \text{RHS} \\ x_4 & \frac{1}{3} & -\frac{32}{9} & \frac{20}{9} & 1 & 0 & 0 & 1 \\ x_5 & \frac{1}{6} & -\frac{13}{9} & \frac{5}{18} & 0 & 1 & 0 & 2 \\ x_6 & -\frac{2}{3} & \frac{16}{9} & -\frac{1}{9} & 0 & 0 & 1 & 3 \\ \text{Obj.} & 7 & 4 & -15 & 0 & 0 & 0 & f \end{pmatrix}$$

New basic variable $= x_3 (-15$ is the largest negative number in the Obj. row)

Ratios: $\{\frac{1}{20/9} = 0.45, \frac{2}{5/18} = 7.2\}$ Minimum $= 0.45 \Longrightarrow x_4$ out of the basic set.

Basis	x_1	x_2	x_3	x_4	x_5	x_6	RHS		
x_3	$\frac{3}{20}$	$-\frac{8}{5}$	1	$\frac{9}{20}$	0	0	$\frac{9}{20}$	$\Longleftarrow$	PR $(= \text{Row } 1/\frac{20}{9})$
x_5	$\frac{1}{8}$	-1	0	$-\frac{1}{8}$	1	0	$\frac{15}{8}$	$\Longleftarrow$	$-\frac{5}{18} \times \text{PR} + \text{Row2}$
x_6	$-\frac{13}{20}$	$\frac{8}{5}$	0	$\frac{1}{20}$	0	1	$\frac{61}{20}$	$\Longleftarrow$	$\frac{1}{9} \times \text{PR} + \text{Row3}$
Obj.	$\frac{37}{4}$	-20	0	$\frac{27}{4}$	0	0	$\frac{27}{4} + f$	$\Longleftarrow$	$15 \times \text{PR} + \text{Obj.Row}$

This completes one step of the simplex method, and we have a second basic feasible solution.

Second Tableau:

$$\begin{pmatrix} \text{Basis} & x_1 & x_2 & x_3 & x_4 & x_5 & x_6 & \text{RHS} \\ x_3 & \frac{3}{20} & -\frac{8}{5} & 1 & \frac{9}{20} & 0 & 0 & \frac{9}{20} \\ x_5 & \frac{1}{8} & -1 & 0 & -\frac{1}{8} & 1 & 0 & \frac{15}{8} \\ x_6 & -\frac{13}{20} & \frac{8}{5} & 0 & \frac{1}{20} & 0 & 1 & \frac{61}{20} \\ \text{Obj.} & \frac{37}{4} & -20 & 0 & \frac{27}{4} & 0 & 0 & \frac{27}{4} + f \end{pmatrix}$$

New basic variable $= x_2$ (-20 is the largest negative number in the Obj. row)

Ratios: $\{\frac{61/20}{8/5}\}$ Minimum (only choice) $\Longrightarrow x_6$ out of the basic set.

Basis	x_1	x_2	x_3	x_4	x_5	x_6	RHS		
x_3	$-\frac{1}{2}$	0	1	$\frac{1}{2}$	0	1	$\frac{7}{2}$	$\Longleftarrow$	$\frac{8}{5} \times \text{PR} + \text{Row1}$
x_5	$-\frac{9}{32}$	0	0	$-\frac{3}{32}$	1	$\frac{5}{8}$	$\frac{121}{32}$	$\Longleftarrow$	$\text{PR} + \text{Row2}$
x_2	$-\frac{13}{32}$	1	0	$\frac{1}{32}$	0	$\frac{5}{8}$	$\frac{61}{32}$	$\Longleftarrow$	PR $(= \text{Row3}/\frac{8}{5})$
Obj.	$\frac{9}{8}$	0	0	$\frac{59}{8}$	0	$\frac{25}{2}$	$\frac{359}{8} + f$	$\Longleftarrow$	$20 \times \text{PR} + \text{Obj.Row}$

Third Tableau:

$$\begin{pmatrix} \text{Basis} & x_1 & x_2 & x_3 & x_4 & x_5 & x_6 & \text{RHS} \\ x_3 & -\frac{1}{2} & 0 & 1 & \frac{1}{2} & 0 & 1 & \frac{7}{2} \\ x_5 & -\frac{9}{32} & 0 & 0 & -\frac{3}{32} & 1 & \frac{5}{8} & \frac{121}{32} \\ x_2 & -\frac{13}{32} & 1 & 0 & \frac{1}{32} & 0 & \frac{5}{8} & \frac{61}{32} \\ \text{Obj.} & \frac{9}{8} & 0 & 0 & \frac{59}{8} & 0 & \frac{25}{2} & \frac{359}{8} + f \end{pmatrix}$$

Since all coefficients in the Obj. row are positive, we cannot reduce the objective function any further, and we have reached the minimum. The optimum solution is

Optimum: $x_1 = 0$, $x_2 = \frac{61}{32}$, $x_3 = \frac{7}{2}$, $x_4 = 0$, $x_5 = \frac{121}{32}$, $x_6 = 0$ $f = -\frac{359}{8}$

6.4.4 Using the Simplex Method for Problems with GE or EQ Constraints

The starting basic feasible solution is more difficult to obtain for problems that involve greater than (GE) type (after making the right-hand side positive, if necessary) or equality (EQ) constraints. The reason is that there is no unique positive variable associated with each constraint. A unique surplus variable is present in each GE constraint, but it is multiplied by a negative sign and thus will give an infeasible solution if treated as a basic variable. An equality constraint does not need a slack/surplus variable, and thus, one cannot assume that there will always be a unique variable for each equality constraint.

The situation is handled by what is known as the *Phase I* simplex method. A unique artificial variable is added to each GE and EQ type constraint. Treating these artificial variables as basic and the actual variables as nonbasic gives a starting basic feasible solution. An artificial objective function, denoted by $\phi(\mathbf{x})$, is defined as the sum of all artificial variables needed in the problem. During this so-called Phase I, this artificial objective function is minimized using the usual simplex procedure. Since there are no real constraints on ϕ, the optimum solution of Phase I is reached when $\phi = 0$. That is when all artificial variables are equal to zero (out of the basis), which is the lowest value possible because all variables are positive in LP. This optimum solution of Phase I is a basic feasible solution for the original problem since when the artificial variables are set to zero, the original constraints are recovered. Using this basic feasible solution, we are then in a position to start solving the original problem with the actual objective function. This is known as *Phase II*, and is the same as that described for LE constraints in the previous section.

Consider the following example with two GE constraints:

Minimize $f = 2x_1 + 4x_2 + 3x_3$

Subject to $\begin{pmatrix} -x_1 + x_2 + x_3 \geq 2 \\ 2x_1 + x_2 \geq 1 \\ x_i \geq 0, i = 1, \ldots, 3 \end{pmatrix}$

Introducing surplus variables x_4 and x_5, the constraints are written in the standard LP form as follows:

$$-x_1 + x_2 + x_3 - x_4 = 2 \quad 2x_1 + x_2 - x_5 = 1$$

Now introducing artificial variables x_6 and x_7, the Phase I objective function and constraints are as follows:

Phase I Problem:

Minimize $\phi = x_6 + x_7$

Subject to
$$\begin{pmatrix} -x_1 + x_2 + x_3 - x_4 + x_6 = 2 \\ 2x_1 + x_2 - x_5 + x_7 = 1 \\ x_i \geq 0, i = 1, \ldots, 7 \end{pmatrix}$$

The starting basic feasible solution for Phase I is as follows:

Basic: $x_6 = 2, x_7 = 1$

Nonbasic: $x_1 = x_2 = x_3 = x_4 = x_5 = 0$

$\phi = 3$

Before proceeding with the simplex method, the artificial objective function must be expressed in terms of nonbasic variables. It can easily be done by solving for the artificial variables from the constraint equations and substituting into the artificial objective function.

From the constraints, we have

$$x_6 = 2 + x_1 - x_2 - x_3 + x_4 \qquad x_7 = 1 - 2x_1 - x_2 + x_5$$

Thus, the artificial objective function is written as

$$\phi = x_6 + x_7 = 3 - x_1 - 2x_2 - x_3 + x_4 + x_5$$

or

$$\phi - 3 = -x_1 - 2x_2 - x_3 + x_4 + x_5$$

Obviously, the actual objective function is not needed during Phase I. However, all reduction operations are performed on it as well so that at the end of Phase I, f is in the correct form (that is, it is expressed in terms on nonbasic variables only) for the simplex method. The complete solution is as follows:

Phase I: Initial Tableau

$$\begin{pmatrix} \text{Basis} & x_1 & x_2 & x_3 & x_4 & x_5 & x_6 & x_7 & \text{RHS} \\ x_6 & -1 & 1 & 1 & -1 & 0 & 1 & 0 & 2 \\ x_7 & 2 & 1 & 0 & 0 & -1 & 0 & 1 & 1 \\ \text{Obj.} & 2 & 4 & 3 & 0 & 0 & 0 & 0 & f \\ \text{ArtObj.} & -1 & -2 & -1 & 1 & 1 & 0 & 0 & -3 + \phi \end{pmatrix}$$

New basic variable = x_2 (-2 is the largest negative number in the ArtObj. row)

Ratios: $\{\frac{2}{1} = 2, \frac{1}{1} = 1\}$ Minimum $= 1 \Longrightarrow x_7$ out of the basic set.

Phase I: Second Tableau

$$\begin{pmatrix} \text{Basis} & x_1 & x_2 & x_3 & x_4 & x_5 & x_6 & x_7 & \text{RHS} \\ x_6 & -3 & 0 & 1 & -1 & 1 & 1 & -1 & 1 \\ x_2 & 2 & 1 & 0 & 0 & -1 & 0 & 1 & 1 \\ \text{Obj.} & -6 & 0 & 3 & 0 & 4 & 0 & -4 & -4+f \\ \text{ArtObj.} & 3 & 0 & -1 & 1 & -1 & 0 & 2 & -1+\phi \end{pmatrix} \begin{matrix} \Longleftarrow -\text{PR} + \text{Row1} \\ \Longleftarrow \text{PR} \\ \Longleftarrow -4 \times \text{PR} + \text{Obj.Row} \\ \Longleftarrow 2 \times \text{PR} + \text{ArtObj.Row} \end{matrix}$$

New basic variable = x_3 (-1 is the first largest negative number in the ArtObj. row)

Ratios: $\{\frac{1}{1}\}$ Minimum (only choice) $\Longrightarrow x_6$ out of the basic set.

Phase I: Third Tableau

$$\begin{pmatrix} \text{Basis} & x_1 & x_2 & x_3 & x_4 & x_5 & x_6 & x_7 & \text{RHS} \\ x_3 & -3 & 0 & 1 & -1 & 1 & 1 & -1 & 1 \\ x_2 & 2 & 1 & 0 & 0 & -1 & 0 & 1 & 1 \\ \text{Obj.} & 3 & 0 & 0 & 3 & 1 & -3 & -1 & -7+f \\ \text{ArtObj.} & 0 & 0 & 0 & 0 & 0 & 1 & 1 & \phi \end{pmatrix} \begin{matrix} \Longleftarrow \text{PR} \\ \Longleftarrow \text{Row2} \\ \Longleftarrow -3 \times \text{PR} + \text{Obj.Row} \\ \Longleftarrow \text{PR} + \text{ArtObj.Row} \end{matrix}$$

All coefficient in the artificial objective function row are now positive, signalling that the optimum of Phase I has been reached. The solution is as follows:

Basic: $x_2 = 1 \quad x_3 = 1$

Nonbasic: $x_1 = x_4 = \cdots = x_7 = 0 \quad \phi = 0$

Since the artificial variables are now zero, the constraint equations now represent the original constraints, and we have a basic feasible solution for our original problem.

Phase II with the actual objective function can now begin. Ignoring the artificial objective function row, we have the following initial simplex tableau for Phase II. Note that the columns associated with artificial variables are really not needed anymore either. However, as will be seen later, the entries in these columns are useful in the sensitivity analysis. Thus, we carry these columns through Phase II as well. However, we don't use these columns for any decision-making.

Phase II: Initial Tableau

$$\begin{pmatrix} \text{Basis} & x_1 & x_2 & x_3 & x_4 & x_5 & x_6 & x_7 & \text{RHS} \\ x_3 & -3 & 0 & 1 & -1 & 1 & 1 & -1 & 1 \\ x_2 & 2 & 1 & 0 & 0 & -1 & 0 & 1 & 1 \\ \text{Obj.} & 3 & 0 & 0 & 3 & 1 & -3 & -1 & -7+f \end{pmatrix}$$

All coefficient in the objective function row (excluding the artificial variables) are positive, meaning that we cannot find another basic feasible solution without increasing the objective function value. Thus, this basic feasible solution is the optimum solution of the problem, and we are done.

Optimum solution:
$$x_1 = 0, x_2 = 1, x_3 = 1, x_4 = 0, x_5 = 0 \quad f = 7$$

6.4.5 The BasicSimplex Function

The above procedure is implemented in a *Mathematica* function called BasicSimplex. The function is intended to be used for educational purposes. Several intermediate results can be printed to gain understanding of the process. The function also implements the post-optimality (sensitivity) analysis discussed in a later section.

Mathematica has built-in functions, ConstrainedMin and LinearProgramming, for solving LP problems. However, these functions do not give intermediate results. They also do not perform any sensitivity analysis. For large problems, and when no sensitivity analysis is required, the built-in functions should be used.

```
Needs["OptimizationToolbox`LPSimplex`"];
?BasicSimplex
```

BasicSimplex[f, g, vars, options]. Solves an LP problem using Phase
 I and II simplex algorithm. f is the objective function, g is
 a list of constraints, and vars is a list of variables. See
 Options[BasicSimplex] to find out about a list of valid options
 for this function.

OptionsUsage[BasicSimplex]
{UnrestrictedVariables → {}, MaxIterations → 10, ProblemType → Min,
 SimplexVariables → {x, s, a}, PrintLevel → 1, SensitivityAnalysis → False}

UnrestrictedVariables is an option for LP and several QP problems.
 A list of variables that are not restricted to be positive can be
 specified with this option. Default is {}.

MaxIterations is an option for several optimization methods. It
 specifies maximum number of iterations allowed.

ProblemType is an option for most optimization methods. It can either be
 Min (default) or Max.

SimplexVariables is an option of the BasicSimplex. t specifies symbols
 to use when creating variable names. Default is {x,s,a}, where 'x'

is used for problem variables, 's' for slack/surplus, and 'a' for artificial.

PrintLevel is an option for most functions in the OptimizationToolbox. It is specified as an integer. The value of the integer indicates how much intermediate information is to be printed. A PrintLevel→ 0 suppresses all printing. Default for most functions is set to 1 in which case they print only the initial problem setup. Higher integers print more intermediate results.

SensitivityAnalysis is an option of simplex method. It controls whether a post-optimality (sensitivity) analysis is performed after obtaining an optimum solution. Default is False.

Example 6.12 Solve the following LP problem using the simplex method:

$f = -\frac{3}{4}x1 + 20x2 - \frac{1}{2}x3 + 6x4;$

$g = \{\frac{1}{4}x1 - 8x2 - x3 + 9x4 \leq 1, \frac{1}{2}x1 - 12x2 - \frac{1}{2}x3 + 3x4 \leq 3, x3 + x4 \leq 1\};$

BasicSimplex[f, g, {x1, x2, x3, x4}, PrintLevel → 2];

Problem variables redefined as: $\{x1 \to x_1, x2 \to x_2, x3 \to x_3, x4 \to x_4\}$

Minimize $\frac{-3x_1}{4} + 20x_2 - \frac{x_3}{2} + 6x_4$

Subject to $\begin{pmatrix} \frac{x_1}{4} - 8x_2 - x_3 + 9x_4 \leq 1 \\ \frac{x_1}{2} - 12x_2 - \frac{x_3}{2} + 3x_4 \leq 3 \\ x_3 + x_4 \leq 1 \end{pmatrix}$

All variables ≥ 0

********** Initial simplex tableau **********

Note that variables 5, 6, and 7 are the slack variables associated with the three constraints. Also, since the problem involves only LE constraints, no Phase I is needed, and we are directly into Phase II of the simplex.

New problem variables: $\{x_1, x_2, x_3, x_4, s_1, s_2, s_3\}$

Basis	1	2	3	4	5	6	7	RHS
5	$\frac{1}{4}$	-8	-1	9	1	0	0	1
6	$\frac{1}{2}$	-12	$-\frac{1}{2}$	3	0	1	0	3
7	0	0	1	1	0	0	1	1
Obj.	$-\frac{3}{4}$	20	$-\frac{1}{2}$	6	0	0	0	f

Variable to be made basic → 1
Ratios: RHS/Column 1 → $(4 \quad 6 \quad \infty)$
Variable out of the basic set → 5

6.4 The Simplex Method

**********Phase II --- Iteration 1**********

$$\begin{pmatrix} \text{Basis} & 1 & 2 & 3 & 4 & 5 & 6 & 7 & \text{RHS} \\ --- & --- & --- & --- & --- & --- & --- & --- & --- \\ 1 & 1 & -32 & -4 & 36 & 4 & 0 & 0 & 4 \\ 6 & 0 & 4 & \frac{3}{2} & -15 & -2 & 1 & 0 & 1 \\ 7 & 0 & 0 & 1 & 1 & 0 & 0 & 1 & 1 \\ \text{Obj.} & 0 & -4 & -\frac{7}{2} & 33 & 3 & 0 & 0 & 3+f \end{pmatrix}$$

Variable to be made basic $\rightarrow$ 2

Ratios: RHS/Column 2 $\rightarrow \begin{pmatrix} \infty & \frac{1}{4} & \infty \end{pmatrix}$

Variable out of the basic set $\rightarrow$ 6

**********Phase II --- Iteration 2**********

$$\begin{pmatrix} \text{Basis} & 1 & 2 & 3 & 4 & 5 & 6 & 7 & \text{RHS} \\ --- & --- & --- & --- & --- & --- & --- & --- & --- \\ 1 & 1 & 0 & 8 & -84 & -12 & 8 & 0 & 12 \\ 2 & 0 & 1 & \frac{3}{8} & -\frac{15}{4} & -\frac{1}{2} & \frac{1}{4} & 0 & \frac{1}{4} \\ 7 & 0 & 0 & 1 & 1 & 0 & 0 & 1 & 1 \\ \text{Obj.} & 0 & 0 & -2 & 18 & 1 & 1 & 0 & 4+f \end{pmatrix}$$

Variable to be made basic $\rightarrow$ 3

Ratios: RHS/Column 3 $\rightarrow \begin{pmatrix} \frac{3}{2} & \frac{2}{3} & 1 \end{pmatrix}$

Variable out of the basic set $\rightarrow$ 2

**********Phase II --- Iteration 3**********

$$\begin{pmatrix} \text{Basis} & 1 & 2 & 3 & 4 & 5 & 6 & 7 & \text{RHS} \\ --- & --- & --- & --- & --- & --- & --- & --- & --- \\ 1 & 1 & -\frac{64}{3} & 0 & -4 & -\frac{4}{3} & \frac{8}{3} & 0 & \frac{20}{3} \\ 3 & 0 & \frac{8}{3} & 1 & -10 & -\frac{4}{3} & \frac{2}{3} & 0 & \frac{2}{3} \\ 7 & 0 & -\frac{8}{3} & 0 & 11 & \frac{4}{3} & -\frac{2}{3} & 1 & \frac{1}{3} \\ \text{Obj.} & 0 & \frac{16}{3} & 0 & -2 & -\frac{5}{3} & \frac{7}{3} & 0 & \frac{16}{3}+f \end{pmatrix}$$

Variable to be made basic $\rightarrow$ 4

Ratios: RHS/Column 4 $\rightarrow \begin{pmatrix} \infty & \infty & \frac{1}{33} \end{pmatrix}$

Variable out of the basic set $\rightarrow$ 7

**********Phase II --- Iteration 4**********

$$\begin{pmatrix} \text{Basis} & 1 & 2 & 3 & 4 & 5 & 6 & 7 & \text{RHS} \\ --- & --- & --- & --- & --- & --- & --- & --- & --- \\ 1 & 1 & -\frac{736}{33} & 0 & 0 & -\frac{28}{33} & \frac{80}{33} & \frac{4}{11} & \frac{224}{33} \\ 3 & 0 & \frac{8}{33} & 1 & 0 & -\frac{4}{33} & \frac{2}{33} & \frac{10}{11} & \frac{32}{33} \\ 4 & 0 & -\frac{8}{33} & 0 & 1 & \frac{4}{33} & -\frac{2}{33} & \frac{1}{11} & \frac{1}{33} \\ \text{Obj.} & 0 & \frac{160}{33} & 0 & 0 & -\frac{47}{33} & \frac{73}{33} & \frac{2}{11} & \frac{178}{33} + f \end{pmatrix}$$

Variable to be made basic $\to$ 5

Ratios: RHS/Column 5 $\to \left(\infty \quad \infty \quad \frac{1}{4} \right)$

Variable out of the basic set $\to$ 4

**********Phase II --- Iteration 5**********

$$\begin{pmatrix} \text{Basis} & 1 & 2 & 3 & 4 & 5 & 6 & 7 & \text{RHS} \\ --- & --- & --- & --- & --- & --- & --- & --- & --- \\ 1 & 1 & -24 & 0 & 7 & 0 & 2 & 1 & 7 \\ 3 & 0 & 0 & 1 & 1 & 0 & 0 & 1 & 1 \\ 5 & 0 & -2 & 0 & \frac{33}{4} & 1 & -\frac{1}{2} & \frac{3}{4} & \frac{1}{4} \\ \text{Obj.} & 0 & 2 & 0 & \frac{47}{4} & 0 & \frac{3}{2} & \frac{5}{4} & \frac{23}{4} + f \end{pmatrix}$$

Optimum solution $\to \{\{\{x1 \to 7, x2 \to 0, x3 \to 1, x4 \to 0\}\}\}$

Optimum objective function value $\to -\frac{23}{4}$

The solution can be verified using the built-in *Mathematica* function, ConstrainedMin, as follows:

ConstrainedMin[f, g, {x1, x2, x3, x4}]

$\left\{ -\frac{23}{4}, \{x1 \to 7, x2 \to 0, x3 \to 1, x4 \to 0\} \right\}$

Example 6.13 Find the maximum of the following LP problem using the simplex method. Note that variable $x2$ is not restricted to be positive.

- ```
f = -3x1 + 2x2 - 4x3 + x4 - x5;
g = {2x1 + 3x2 + x3 + 4x4 + 4x5 == 12,
    4x1 - 5x2 + 3x3 - x4 - 4x5 == 10, 3x1 - x2 + 2x3 + 2x4 + x5 ≥ 8};
BasicSimplex[f, g, {x1, x2, x3, x4, x5},
    ProblemType → Max, UnrestrictedVariables → {x2}, PrintLevel → 2];
```

Problem variables redefined as: $\{x1 \to x_1, x2 \to x_2 - x_3, x3 \to x_4, x4 \to x_5, x5 \to x_6\}$

Minimize $3x_1 - 2x_2 + 2x_3 + 4x_4 - x_5 + x_6$

Subject to $\begin{pmatrix} 2x_1 + 3x_2 - 3x_3 + x_4 + 4x_5 + 4x_6 == 12 \\ 4x_1 - 5x_2 + 5x_3 + 3x_4 - x_5 - 4x_6 == 10 \\ 3x_1 - x_2 + x_3 + 2x_4 + 2x_5 + x_6 \geq 8 \end{pmatrix}$

All variables $\geq 0$

********** Initial simplex tableau **********
```

6.4 The Simplex Method

The third constraint needs a surplus variable. This is placed in the seventh column of the tableau. All three constraints need artificial variables to start the Phase I solution. These are placed in the last three columns. The artificial objective function is the sum of the three artificial variables. As explained earlier, it is then expressed in terms of nonbasic variables to give the form included in the following tableau.

New problem variables: $\{x_1, x_2, x_3, x_4, x_5, x_6, s_3, a_1, a_2, a_3\}$

$$\begin{pmatrix}
\text{Basis} & 1 & 2 & 3 & 4 & 5 & 6 & 7 & 8 & 9 & 10 & \text{RHS} \\
--- & --- & --- & --- & --- & --- & --- & --- & --- & --- & --- & --- \\
8 & 2 & 3 & -3 & 1 & 4 & 4 & 0 & 1 & 0 & 0 & 12 \\
9 & 4 & -5 & 5 & 3 & -1 & -4 & 0 & 0 & 1 & 0 & 10 \\
10 & 3 & -1 & 1 & 2 & 2 & 1 & -1 & 0 & 0 & 1 & 8 \\
\text{Obj.} & 3 & -2 & 2 & 4 & -1 & 1 & 0 & 0 & 0 & 0 & f \\
\text{ArtObj.} & -9 & 3 & -3 & -6 & -5 & -1 & 1 & 0 & 0 & 0 & -30 + \phi
\end{pmatrix}$$

Variable to be made basic $\to 1$

Ratios: RHS/Column $1 \to \left(6 \quad \dfrac{5}{2} \quad \dfrac{8}{3}\right)$

Variable out of the basic set $\to 9$

**********Phase I --- Iteration 1**********

$$\begin{pmatrix}
\text{Basis} & 1 & 2 & 3 & 4 & 5 & 6 & 7 & 8 & 9 & 10 & \text{RHS} \\
--- & --- & --- & --- & --- & --- & --- & --- & --- & --- & --- & --- \\
8 & 0 & \frac{11}{2} & -\frac{11}{2} & -\frac{1}{2} & \frac{9}{2} & 6 & 0 & 1 & -\frac{1}{2} & 0 & 7 \\
1 & 1 & -\frac{5}{4} & \frac{5}{4} & \frac{3}{4} & -\frac{1}{4} & -1 & 0 & 0 & \frac{1}{4} & 0 & \frac{5}{2} \\
10 & 0 & \frac{11}{4} & -\frac{11}{4} & -\frac{1}{4} & \frac{11}{4} & 4 & -1 & 0 & -\frac{3}{4} & 1 & \frac{1}{2} \\
\text{Obj.} & 0 & \frac{7}{4} & -\frac{7}{4} & \frac{7}{4} & -\frac{1}{4} & 4 & 0 & 0 & -\frac{3}{4} & 0 & -\frac{15}{2} + f \\
\text{ArtObj.} & 0 & -\frac{33}{4} & \frac{33}{4} & \frac{3}{4} & -\frac{29}{4} & -10 & 1 & 0 & \frac{9}{4} & 0 & -\frac{15}{2} + \phi
\end{pmatrix}$$

Variable to be made basic $\to 6$

Ratios: RHS/Column $6 \to \left(\dfrac{7}{6} \quad \infty \quad \dfrac{1}{8}\right)$

Variable out of the basic set $\to 10$

**********Phase I --- Iteration 2**********

$$\begin{pmatrix}
\text{Basis} & 1 & 2 & 3 & 4 & 5 & 6 & 7 & 8 & 9 & 10 & \text{RHS} \\
--- & --- & --- & --- & --- & --- & --- & --- & --- & --- & --- & --- \\
8 & 0 & \frac{11}{8} & -\frac{11}{8} & -\frac{1}{8} & \frac{3}{8} & 0 & \frac{3}{2} & 1 & \frac{5}{8} & -\frac{3}{2} & \frac{25}{4} \\
1 & 1 & -\frac{9}{16} & \frac{9}{16} & \frac{11}{16} & \frac{7}{16} & 0 & -\frac{1}{4} & 0 & \frac{1}{16} & \frac{1}{4} & \frac{21}{8} \\
6 & 0 & \frac{11}{16} & -\frac{11}{16} & -\frac{1}{16} & \frac{11}{16} & 1 & -\frac{1}{4} & 0 & -\frac{3}{16} & \frac{1}{4} & \frac{1}{8} \\
\text{Obj.} & 0 & -1 & 1 & 2 & -3 & 0 & 1 & 0 & 0 & -1 & -8 + f \\
\text{ArtObj.} & 0 & -\frac{11}{8} & \frac{11}{8} & \frac{1}{8} & -\frac{3}{8} & 0 & -\frac{3}{2} & 0 & \frac{3}{8} & \frac{5}{2} & -\frac{25}{4} + \phi
\end{pmatrix}$$

Variable to be made basic → 7
Ratios: RHS/Column 7 → $\left(\frac{25}{6} \quad \infty \quad \infty\right)$
Variable out of the basic set → 8

**********Phase I --- Iteration 3**********

$$\begin{pmatrix} \text{Basis} & 1 & 2 & 3 & 4 & 5 & 6 & 7 & 8 & 9 & 10 & \text{RHS} \\ --- & --- & --- & --- & --- & --- & --- & --- & --- & --- & --- & --- \\ 7 & 0 & \frac{11}{12} & -\frac{11}{12} & -\frac{1}{12} & \frac{1}{4} & 0 & 1 & \frac{2}{3} & \frac{5}{12} & -1 & \frac{25}{6} \\ 1 & 1 & -\frac{1}{3} & \frac{1}{3} & \frac{2}{3} & \frac{1}{2} & 0 & 0 & \frac{1}{6} & \frac{1}{6} & 0 & \frac{11}{3} \\ 6 & 0 & \frac{11}{12} & -\frac{11}{12} & -\frac{1}{12} & \frac{3}{4} & 1 & 0 & \frac{1}{6} & -\frac{1}{12} & 0 & \frac{7}{6} \\ \text{Obj.} & 0 & -\frac{23}{12} & \frac{23}{12} & \frac{25}{12} & -\frac{13}{4} & 0 & 0 & -\frac{2}{3} & -\frac{5}{12} & 0 & -\frac{73}{6}+f \\ \text{ArtObj.} & 0 & 0 & 0 & 0 & 0 & 0 & 0 & 1 & 1 & 1 & \phi \end{pmatrix}$$

End of phase I
Variable to be made basic → 5
Ratios: RHS/Column 5 → $\left(\frac{50}{3} \quad \frac{22}{3} \quad \frac{14}{9}\right)$
Variable out of the basic set → 6

**********Phase II --- Iteration 1**********

$$\begin{pmatrix} \text{Basis} & 1 & 2 & 3 & 4 & 5 & 6 & 7 & 8 & 9 & 10 & \text{RHS} \\ --- & --- & --- & --- & --- & --- & --- & --- & --- & --- & --- & --- \\ 7 & 0 & \frac{11}{18} & -\frac{11}{18} & -\frac{1}{18} & 0 & -\frac{1}{3} & 1 & \frac{11}{18} & \frac{4}{9} & -1 & \frac{34}{9} \\ 1 & 1 & -\frac{17}{18} & \frac{17}{18} & \frac{13}{18} & 0 & -\frac{2}{3} & 0 & \frac{1}{18} & \frac{2}{9} & 0 & \frac{26}{9} \\ 5 & 0 & \frac{11}{9} & -\frac{11}{9} & -\frac{1}{9} & 1 & \frac{4}{3} & 0 & \frac{2}{9} & -\frac{1}{9} & 0 & \frac{14}{9} \\ \text{Obj.} & 0 & \frac{37}{18} & -\frac{37}{18} & \frac{31}{18} & 0 & \frac{13}{3} & 0 & \frac{1}{18} & -\frac{7}{9} & 0 & -\frac{64}{9}+f \end{pmatrix}$$

Variable to be made basic → 3
Ratios: RHS/Column 3 → $\left(\infty \quad \frac{52}{17} \quad \infty\right)$
Variable out of the basic set → 1

**********Phase II --- Iteration 2**********

$$\begin{pmatrix} \text{Basis} & 1 & 2 & 3 & 4 & 5 & 6 & 7 & 8 & 9 & 10 & \text{RHS} \\ --- & --- & --- & --- & --- & --- & --- & --- & --- & --- & --- & --- \\ 7 & \frac{11}{17} & 0 & 0 & \frac{7}{17} & 0 & -\frac{13}{17} & 1 & \frac{11}{17} & \frac{10}{17} & -1 & \frac{96}{17} \\ 3 & \frac{18}{17} & -1 & 1 & \frac{13}{17} & 0 & -\frac{12}{17} & 0 & \frac{1}{17} & \frac{4}{17} & 0 & \frac{52}{17} \\ 5 & \frac{22}{17} & 0 & 0 & \frac{14}{17} & 1 & \frac{8}{17} & 0 & \frac{5}{17} & \frac{3}{17} & 0 & \frac{90}{17} \\ \text{Obj.} & \frac{37}{17} & 0 & 0 & \frac{56}{17} & 0 & \frac{49}{17} & 0 & \frac{3}{17} & -\frac{5}{17} & 0 & -\frac{14}{17}+f \end{pmatrix}$$

Optimum solution → $\left\{\left\{\left\{x1 \to 0, x2 \to -\frac{52}{17}, x3 \to 0, x4 \to \frac{90}{17}, x5 \to 0\right\}\right\}\right\}$

Optimum objective function value → $-\frac{14}{17}$

Example 6.14 *Shortest route problem* This example demonstrates formulation and solution of an important class of problems known as *network* problems in the LP literature. In these problems a network of *nodes* and *links* is given. The problem is usually to find the maximum flow or the shortest route. As an example, consider the problem of finding the shortest route between two cities while traveling on a given network of available roads. A typical situation is shown in Figure 6.3. The nodes represent cities and the links are the roads that connect these cities. The distances in kilometers along each road are noted in the figure.

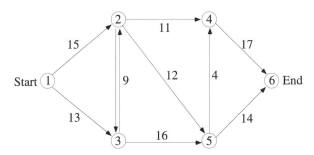

FIGURE 6.3 A network diagram showing distances and direction of travel between cities.

The optimization variables are the roads that one can take to reach the destination. Indicating the roads by the indices of the nodes that they connect, with the order indicating the direction of travel, we have the following set of optimization variables. Note that two separate variables are needed for the roads where travel in either direction is possible.

Variables = $\{x_{12}, x_{13}, x_{23}, x_{32}, x_{24}, x_{25}, x_{35}, x_{54}, x_{46}, x_{56}\}$

The objective function is to minimize the distance travelled and is simply the sum of miles along each route, as follows:

Minimize $f = 15x_{12} + 13x_{13} + 9x_{23} + 9x_{32} + 11x_{24} + 12x_{25} + 16x_{35} + 4x_{54} + 17x_{46} + 14x_{56}$

The constraints express the relationship between the links. The *inflow* at a node must equal the *outflow*.

Node 2: $x_{12} + x_{32} = x_{24} + x_{25} + x_{23}$

Node 3: $x_{13} + x_{23} = x_{32} + x_{35}$

Node 4: $x_{24} + x_{54} = x_{46}$

Node 5: $x_{35} + x_{25} = x_{54} + x_{56}$

The origin and destination nodes are indicated by the fact that the outflow from the origin node and inflow into the destination node are equal to 1.

Node 1: $x_{12} + x_{13} = 1$

Node 6: $x_{46} + x_{56} = 1$

In terms of *Mathematica* expressions, the problem is defined as follows:

```
vars = {x₁₂, x₁₃, x₂₃, x₃₂, x₂₄, x₂₅, x₃₅, x₅₄, x₄₆, x₅₆};
f = 15x₁₂ + 13x₁₃ + 9x₂₃ + 9x₃₂ + 11x₂₄ + 12x₂₅ + 16x₃₅ + 4x₅₄ + 17x₄₆ + 14x₅₆;
g = {x₁₂ + x₃₂ == x₂₄ + x₂₅ + x₂₃, x₁₃ + x₂₃ == x₃₂ + x₃₅, x₂₄ + x₅₄ == x₄₆,
     x₃₅ + x₂₅ == x₅₄ + x₅₆, x₁₂ + x₁₃ == 1, x₄₆ + x₅₆ == 1};
```

The solution is obtained by using the BasicSimplex, as follows:

```
BasicSimplex[f, g, vars, PrintLevel → 2];
Problem variables redefined as: {x₁₂ → x₁, x₁₃ → x₂, x₂₃ → x₃,
  x₃₂ → x₄, x₂₄ → x₅, x₂₅ → x₆, x₃₅ → x₇, x₅₄ → x₈, x₄₆ → x₉, x₅₆ → x₁₀}
Minimize 15x₁ + 13x₂ + 9x₃ + 9x₄ + 11x₅ + 12x₆ + 16x₇ + 4x₈ + 17x₉ + 14x₁₀
```

$$\text{Subject to} \begin{pmatrix} x_1 - x_3 + x_4 - x_5 - x_6 == 0 \\ x_2 + x_3 - x_4 - x_7 == 0 \\ x_5 + x_8 - x_9 == 0 \\ x_6 + x_7 - x_8 - x_{10} == 0 \\ x_1 + x_2 == 1 \\ x_9 + x_{10} == 1 \end{pmatrix}$$

```
All variables ≥ 0

********** Initial simplex tableau **********

New problem variables:
{x₁, x₂, x₃, x₄, x₅, x₆, x₇, x₈, x₉, x₁₀, a₁, a₂, a₃, a₄, a₅, a₆}
```

6.4 The Simplex Method

```
⎛ Basis    1    2    3    4    5    6    7    8
  ---     ---  ---  ---  ---  ---  ---  ---  ---
   11      1    0   -1    1   -1   -1    0    0
   12      0    1    1   -1    0    0   -1    0
   13      0    0    0    0    1    0    0    1
   14      0    0    0    0    0    1    1   -1
   15      1    1    0    0    0    0    0    0
   16      0    0    0    0    0    0    0    0
  Obj.    15   13    9    9   11   12   16    4
⎝ArtObj.  -2   -2    0    0    0    0    0    0

           9   10   11   12   13   14   15   16   RHS ⎞
          ---  ---  ---  ---  ---  ---  ---  ---  ---
           0    0    1    0    0    0    0    0    0
           0    0    0    1    0    0    0    0    0
          -1    0    0    0    1    0    0    0    0
           0   -1    0    0    0    1    0    0    0
           0    0    0    0    0    0    1    0    1
           1    1    0    0    0    0    0    1    1
          17   14    0    0    0    0    0    0    f
           0    0    0    0    0    0    0    0  -2+φ ⎠
```

Variable to be made basic → 1
Ratios: RHS/Column 1 → $\begin{pmatrix} 0 & \infty & \infty & \infty & 1 & \infty \end{pmatrix}$
Variable out of the basic set → 11
**********Phase I --- Iteration 1**********

```
⎛ Basis    1    2    3    4    5    6    7    8
  ---     ---  ---  ---  ---  ---  ---  ---  ---
    1      1    0   -1    1   -1   -1    0    0
   12      0    1    1   -1    0    0   -1    0
   13      0    0    0    0    1    0    0    1
   14      0    0    0    0    0    1    1   -1
   15      0    1    1   -1    1    1    0    0
   16      0    0    0    0    0    0    0    0
  Obj.     0   13   24   -6   26   27   16    4
⎝ArtObj.   0   -2   -2    2   -2   -2    0    0

           9   10   11   12   13   14   15   16   RHS ⎞
          ---  ---  ---  ---  ---  ---  ---  ---  ---
           0    0    1    0    0    0    0    0    0
           0    0    0    1    0    0    0    0    0
          -1    0    0    0    1    0    0    0    0
           0   -1    0    0    0    1    0    0    0
           0    0   -1    0    0    0    1    0    1
           1    1    0    0    0    0    0    1    1
          17   14  -15    0    0    0    0    0    f
           0    0    2    0    0    0    0    0  -2+φ ⎠
```

Variable to be made basic → 2
Ratios: RHS/Column 2 → $\begin{pmatrix} \infty & 0 & \infty & \infty & 1 & \infty \end{pmatrix}$
Variable out of the basic set → 12

**********Phase I --- Iteration 2**********

$$\begin{pmatrix}
\text{Basis} & 1 & 2 & 3 & 4 & 5 & 6 & 7 & 8 \\
--- & --- & --- & --- & --- & --- & --- & --- & --- \\
1 & 1 & 0 & -1 & 1 & -1 & -1 & 0 & 0 \\
2 & 0 & 1 & 1 & -1 & 0 & 0 & -1 & 0 \\
13 & 0 & 0 & 0 & 0 & 1 & 0 & 0 & 1 \\
14 & 0 & 0 & 0 & 0 & 0 & 1 & 1 & -1 \\
15 & 0 & 0 & 0 & 0 & 1 & 1 & 1 & 0 \\
16 & 0 & 0 & 0 & 0 & 0 & 0 & 0 & 0 \\
\text{Obj.} & 0 & 0 & 11 & 7 & 26 & 27 & 29 & 4 \\
\text{ArtObj.} & 0 & 0 & 0 & 0 & -2 & -2 & -2 & 0
\end{pmatrix}$$

$$\begin{pmatrix}
9 & 10 & 11 & 12 & 13 & 14 & 15 & 16 & \text{RHS} \\
--- & --- & --- & --- & --- & --- & --- & --- & --- \\
0 & 0 & 1 & 0 & 0 & 0 & 0 & 0 & 0 \\
0 & 0 & 0 & 1 & 0 & 0 & 0 & 0 & 0 \\
-1 & 0 & 0 & 0 & 1 & 0 & 0 & 0 & 0 \\
0 & -1 & 0 & 0 & 0 & 1 & 0 & 0 & 0 \\
0 & 0 & -1 & -1 & 0 & 0 & 1 & 0 & 1 \\
1 & 1 & 0 & 0 & 0 & 0 & 0 & 1 & 1 \\
17 & 14 & -15 & -13 & 0 & 0 & 0 & 0 & f \\
0 & 0 & 2 & 2 & 0 & 0 & 0 & 0 & -2+\phi
\end{pmatrix}$$

Variable to be made basic → 5
Ratios: RHS/Column 5 → $(\infty \;\; \infty \;\; 0 \;\; \infty \;\; 1 \;\; \infty)$
Variable out of the basic set → 13

**********Phase I --- Iteration 3**********

$$\begin{pmatrix}
\text{Basis} & 1 & 2 & 3 & 4 & 5 & 6 & 7 & 8 \\
--- & --- & --- & --- & --- & --- & --- & --- & --- \\
1 & 1 & 0 & -1 & 1 & 0 & -1 & 0 & 1 \\
2 & 0 & 1 & 1 & -1 & 0 & 0 & -1 & 0 \\
5 & 0 & 0 & 0 & 0 & 1 & 0 & 0 & 1 \\
14 & 0 & 0 & 0 & 0 & 0 & 1 & 1 & -1 \\
15 & 0 & 0 & 0 & 0 & 0 & 1 & 1 & -1 \\
16 & 0 & 0 & 0 & 0 & 0 & 0 & 0 & 0 \\
\text{Obj.} & 0 & 0 & 11 & 7 & 0 & 27 & 29 & -22 \\
\text{ArtObj.} & 0 & 0 & 0 & 0 & 0 & -2 & -2 & 2
\end{pmatrix}$$

$$\begin{pmatrix}
9 & 10 & 11 & 12 & 13 & 14 & 15 & 16 & \text{RHS} \\
--- & --- & --- & --- & --- & --- & --- & --- & --- \\
-1 & 0 & 1 & 0 & 1 & 0 & 0 & 0 & 0 \\
0 & 0 & 0 & 1 & 0 & 0 & 0 & 0 & 0 \\
-1 & 0 & 0 & 0 & 1 & 0 & 0 & 0 & 0 \\
0 & -1 & 0 & 0 & 0 & 1 & 0 & 0 & 0 \\
1 & 0 & -1 & -1 & -1 & 0 & 1 & 0 & 1 \\
1 & 1 & 0 & 0 & 0 & 0 & 0 & 1 & 1 \\
43 & 14 & -15 & -13 & -26 & 0 & 0 & 0 & f \\
-2 & 0 & 2 & 2 & 2 & 0 & 0 & 0 & -2+\phi
\end{pmatrix}$$

Variable to be made basic → 6
Ratios: RHS/Column 6 → $(\infty \;\; \infty \;\; \infty \;\; 0 \;\; 1 \;\; \infty)$

6.4 The Simplex Method

Variable out of the basic set → 14
**********Phase I --- Iteration 4**********

Basis	1	2	3	4	5	6	7	8	9	10	11	12	13	14	15	16	RHS
1	1	0	-1	1	0	0	1	0	-1	-1	1	0	1	1	0	0	0
2	0	1	1	-1	0	0	-1	0	0	0	0	1	0	0	0	0	0
5	0	0	0	0	1	0	0	1	-1	0	0	0	1	0	0	0	0
6	0	0	0	0	0	1	1	-1	0	-1	0	0	0	1	0	0	0
15	0	0	0	0	0	0	0	0	1	1	-1	-1	-1	-1	1	0	1
16	0	0	0	0	0	0	0	0	1	1	0	0	0	0	0	1	1
Obj.	0	0	11	7	0	0	2	5	43	41	-15	-13	-26	-27	0	0	f
ArtObj.	0	0	0	0	0	0	0	0	-2	-2	2	2	2	2	0	0	$-2 + \phi$

Variable to be made basic → 9
Ratios: RHS/Column 9 → $(\infty \ \infty \ \infty \ \infty \ 1 \ 1)$
Variable out of the basic set → 16
**********Phase I --- Iteration 5**********

Basis	1	2	3	4	5	6	7	8	9	10	11	12	13	14	15	16	RHS
1	1	0	-1	1	0	0	1	0	0	0	1	0	1	1	0	1	1
2	0	1	1	-1	0	0	-1	0	0	0	0	1	0	0	0	0	0
5	0	0	0	0	1	0	0	1	0	1	0	0	1	0	0	1	1
6	0	0	0	0	0	1	1	-1	0	-1	0	0	0	1	0	0	0
15	0	0	0	0	0	0	0	0	0	0	-1	-1	-1	-1	1	-1	0
9	0	0	0	0	0	0	0	0	1	1	0	0	0	0	0	1	1
Obj.	0	0	11	7	0	0	2	5	0	-2	-15	-13	-26	-27	0	-43	$-43 + f$
ArtObj.	0	0	0	0	0	0	0	0	0	0	2	2	2	2	0	2	ϕ

Note that the variable 15, which is the artificial variable for the fifth constraint, is still in the basis. However, since it has a zero rhs, the artificial objective function value is reduced to 0 and we are done with Phase I. Furthermore, we also notice that in the same constraint row (fifth), all coefficients corresponding to nonartificial variables (columns 1 through 10) are 0. This indicates that this constraint is redundant and can be removed from the subsequent iterations. For ease of implementation, however, this constraint is kept in the following tableaus.

```
End of phase I

Variable to be made basic → 10

Ratios: RHS/Column 10 → (∞  ∞  1  ∞  ∞  1)

Variable out of the basic set → 9

**********Phase II --- Iteration 1**********
```

$$\begin{pmatrix}
\text{Basis} & 1 & 2 & 3 & 4 & 5 & 6 & 7 & 8 \\
\text{---} & \text{---} & \text{---} & \text{---} & \text{---} & \text{---} & \text{---} & \text{---} & \text{---} \\
1 & 1 & 0 & -1 & 1 & 0 & 0 & 1 & 0 \\
2 & 0 & 1 & 1 & -1 & 0 & 0 & -1 & 0 \\
5 & 0 & 0 & 0 & 0 & 1 & 0 & 0 & 1 \\
6 & 0 & 0 & 0 & 0 & 0 & 1 & 1 & -1 \\
15 & 0 & 0 & 0 & 0 & 0 & 0 & 0 & 0 \\
10 & 0 & 0 & 0 & 0 & 0 & 0 & 0 & 0 \\
\text{Obj.} & 0 & 0 & 11 & 7 & 0 & 0 & 2 & 5
\end{pmatrix}$$

$$\begin{pmatrix}
9 & 10 & 11 & 12 & 13 & 14 & 15 & 16 & \text{RHS} \\
\text{---} & \text{---} & \text{---} & \text{---} & \text{---} & \text{---} & \text{---} & \text{---} & \text{---} \\
0 & 0 & 1 & 0 & 1 & 1 & 0 & 1 & 1 \\
0 & 0 & 0 & 1 & 0 & 0 & 0 & 0 & 0 \\
-1 & 0 & 0 & 0 & 1 & 0 & 0 & 0 & 0 \\
1 & 0 & 0 & 0 & 0 & 1 & 0 & 1 & 1 \\
0 & 0 & -1 & -1 & -1 & -1 & 1 & -1 & 0 \\
1 & 1 & 0 & 0 & 0 & 0 & 0 & 1 & 1 \\
2 & 0 & -15 & -13 & -26 & -27 & 0 & -41 & -41 + f
\end{pmatrix}$$

```
Optimum solution → {{{x₁₂ → 1, x₁₃ → 0, x₂₃ → 0, x₃₂ → 0, x₂₄ → 0,
                     x₂₅ → 1, x₃₅ → 0, x₅₄ → 0, x₄₆ → 0, x₅₆ → 1}}}

Optimum objective function value → 41
```

The solution indicates that the shortest route is $1 \to 2$, $2 \to 5$, and $5 \to 6$, with a total distance of 41 kilometers.

6.5 Unusual Situations Arising During the Simplex Solution

For certain problems, unusual situations may arise during the solution using the simplex method. These situations signal special behavior of the optimization problem being solved.

6.5.1 No Feasible Solution

During Phase I, if the artificial objective function still has a positive value (i.e., an artificial variable still is in the basic set) but all coefficients in the artificial objective row are positive, we obviously cannot proceed any further. The optimum of Phase I has been reached, but we still don't have a basic feasible solution for the actual problem. This situation indicates that the original problem does not have a feasible solution. The following numerical example demonstrates this behavior.

Example 6.15

```
f = x2;
g = {x1 + x2 ≤ -3, x1 - 2x2 ≤ -1};
BasicSimplex[f, g, {x1, x2}, ProblemType → Max, PrintLevel → 2];
```

Problem variables redefined as: $\{x1 \to x_1, x2 \to x_2\}$

Minimize $-x_2$

Subject to $\begin{pmatrix} -x_1 - x_2 \geq 3 \\ -x_1 + 2x_2 \geq 1 \end{pmatrix}$

All variables ≥ 0

********** Initial simplex tableau **********

New problem variables: $\{x_1, x_2, s_1, s_2, a_1, a_2\}$

Basis	1	2	3	4	5	6	RHS
5	-1	-1	-1	0	1	0	3
6	-1	2	0	-1	0	1	1
Obj.	0	-1	0	0	0	0	f
ArtObj.	2	-1	1	1	0	0	-4 + ϕ

Variable to be made basic → 2

Ratios: RHS/Column 2 → $\left(\infty, \frac{1}{2}\right)$

Variable out of the basic set → 6

```
**********Phase I --- Iteration 1**********
Basis      1      2    3      4     5     6    RHS
---       ---    ---  ---    ---   ---   ---   ---
  5      -3/2     0   -1    -1/2    1    1/2   7/2
  2      -1/2     1    0    -1/2    0    1/2   1/2
 Obj.    -1/2     0    0    -1/2    0    1/2   1/2 + f
ArtObj.   3/2     0    1     1/2    0    1/2  -7/2 + φ
End of phase I
Unbounded or infeasible problem
```

The graphical solution, shown in Figure 6.4, clearly shows that there is no feasible solution for this problem.

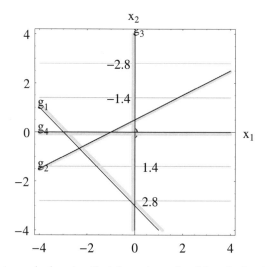

FIGURE 6.4 A graph showing that there is no feasible solution for the problem.

6.5.2 Unbounded Solution

Assume that the objective function row of a given problem indicates that one of the variables can be brought into the basic set. This clearly means that a lower value of the objective function is possible. Now if coefficients of that variable in all the constraint equations are negative, it means that none of the constraints is in danger of being violated. That is, we can lower the value of the objective function without worrying about constraints. Clearly, this means

6.5 Unusual Situations Arising During the Simplex Solution

that the problem is unbounded and has a solution $f^* = -\infty$. The following numerical example illustrates this situation:

Example 6.16

```
f = 2x1 + 3x2;
g = {x1 + 3x2 ≥ 3, x1 - x2 ≤ 2};
BasicSimplex[f, g, {x1, x2}, ProblemType → Max, PrintLevel → 2];
```

Problem variables redefined as: $\{x1 \to x_1, x2 \to x_2\}$

Minimize $-2x_1 - 3x_2$

Subject to $\begin{pmatrix} x_1 + 3x_2 \geq 3 \\ x_1 - x_2 \leq 2 \end{pmatrix}$

All variables ≥ 0

********** Initial simplex tableau **********

New problem variables: $\{x_1, x_2, s_1, s_2, a_1\}$

$$\begin{pmatrix} \text{Basis} & 1 & 2 & 3 & 4 & 5 & \text{RHS} \\ --- & --- & --- & --- & --- & --- & --- \\ 5 & 1 & 3 & -1 & 0 & 1 & 3 \\ 4 & 1 & -1 & 0 & 1 & 0 & 2 \\ \text{Obj.} & -2 & -3 & 0 & 0 & 0 & f \\ \text{ArtObj.} & -1 & -3 & 1 & 0 & 0 & -3+\phi \end{pmatrix}$$

Variable to be made basic $\to 2$

Ratios: RHS/Column 2 $\to (1 \;\; \infty)$

Variable out of the basic set $\to 5$

**********Phase I --- Iteration 1**********

$$\begin{pmatrix} \text{Basis} & 1 & 2 & 3 & 4 & 5 & \text{RHS} \\ --- & --- & --- & --- & --- & --- & --- \\ 2 & \frac{1}{3} & 1 & -\frac{1}{3} & 0 & \frac{1}{3} & 1 \\ 4 & \frac{4}{3} & 0 & -\frac{1}{3} & 1 & \frac{1}{3} & 3 \\ \text{Obj.} & -1 & 0 & -1 & 0 & 1 & 3+f \\ \text{ArtObj.} & 0 & 0 & 0 & 0 & 1 & \phi \end{pmatrix}$$

End of phase I

Variable to be made basic $\to 1$

Ratios: RHS/Column 1 $\to \left(3 \;\; \frac{9}{4}\right)$

Variable out of the basic set $\to 4$

**********Phase II --- Iteration 1**********

$$\begin{pmatrix} \text{Basis} & 1 & 2 & 3 & 4 & 5 & \text{RHS} \\ --- & --- & --- & --- & --- & --- & --- \\ 2 & 0 & 1 & -\frac{1}{4} & -\frac{1}{4} & \frac{1}{4} & \frac{1}{4} \\ 1 & 1 & 0 & -\frac{1}{4} & \frac{3}{4} & \frac{1}{4} & \frac{9}{4} \\ \text{Obj.} & 0 & 0 & -\frac{5}{4} & \frac{3}{4} & \frac{5}{4} & \frac{21}{4}+f \end{pmatrix}$$

Unbounded or infeasible problem

The graphical solution, shown in Figure 6.5, clearly shows that the feasible region is unbounded and the minimum is $f^* = -\infty$.

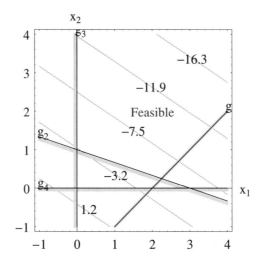

FIGURE 6.5 A graph showing the unbounded feasible region.

6.5.3 Multiple Solutions

We know that at the optimum point, the coefficients of all basic variables in the objective function row are zero. Normally, the coefficients of nonbasic variables in the row are positive, indicating that if one of these variables is brought into the basic set, the objective function value will increase. The same reasoning would suggest that if the coefficient of one of these nonbasic variables is zero, then another basic feasible solution is possible without increasing the objective function value. This obviously means that the problem may have multiple optimum points (different points but all with the same objective function value). This situation occurs when one of the active constraints is parallel to the objective function. The following numerical example illustrates this situation:

Example 6.17

```
f = -x1 - x2;
g = {x1 - x2 ≤ 1, x1 + x2 ≤ 2};
BasicSimplex[f, g, {x1, x2}, PrintLevel → 2];
```

6.5 Unusual Situations Arising During the Simplex Solution

Problem variables redefined as: $\{x1 \rightarrow x_1, x2 \rightarrow x_2\}$

Minimize $-x_1 - x_2$

Subject to $\begin{pmatrix} x_1 - x_2 \leq 1 \\ x_1 + x_2 \leq 2 \end{pmatrix}$

All variables ≥ 0

********** Initial simplex tableau **********

New problem variables: $\{x_1, x_2, s_1, s_2\}$

$$\begin{pmatrix} \text{Basis} & 1 & 2 & 3 & 4 & \text{RHS} \\ --- & --- & --- & --- & --- & --- \\ 3 & 1 & -1 & 1 & 0 & 1 \\ 4 & 1 & 1 & 0 & 1 & 2 \\ \text{Obj.} & -1 & -1 & 0 & 0 & f \end{pmatrix}$$

Variable to be made basic $\rightarrow 1$

Ratios: RHS/Column $1 \rightarrow \begin{pmatrix} 1 & 2 \end{pmatrix}$

Variable out of the basic set $\rightarrow 3$

**********Phase II --- Iteration 1**********

$$\begin{pmatrix} \text{Basis} & 1 & 2 & 3 & 4 & \text{RHS} \\ --- & --- & --- & --- & --- & --- \\ 1 & 1 & -1 & 1 & 0 & 1 \\ 4 & 0 & 2 & -1 & 1 & 1 \\ \text{Obj.} & 0 & -2 & 1 & 0 & 1+f \end{pmatrix}$$

Variable to be made basic $\rightarrow 2$

Ratios: RHS/Column $2 \rightarrow \begin{pmatrix} \infty & \frac{1}{2} \end{pmatrix}$

Variable out of the basic set $\rightarrow 4$

**********Phase II --- Iteration 2**********

$$\begin{pmatrix} \text{Basis} & 1 & 2 & 3 & 4 & \text{RHS} \\ --- & --- & --- & --- & --- & --- \\ 1 & 1 & 0 & \frac{1}{2} & \frac{1}{2} & \frac{3}{2} \\ 2 & 0 & 1 & -\frac{1}{2} & \frac{1}{2} & \frac{1}{2} \\ \text{Obj.} & 0 & 0 & 0 & 1 & 2+f \end{pmatrix}$$

Optimum solution $\rightarrow \left\{\left\{\left\{x1 \rightarrow \frac{3}{2}, x2 \rightarrow \frac{1}{2}\right\}\right\}\right\}$

Optimum objective function value $\rightarrow -2$

** The problem may have multiple solutions.

As demonstrated by the graph shown in Figure 6.6, any point on the boundary of constraint g_2 in the feasible region has the same objective function value as the optimum solution given by the simplex method.

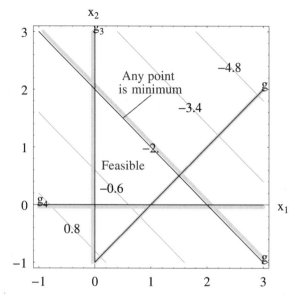

FIGURE 6.6 A graph showing multiple solutions.

6.5.4 Degenerate Solution

In a basic feasible solution, if any of the basic variables has a zero value, then that solution is called a *degenerate* basic feasible solution. An LP problem is called *degenerate* if one or more of its basic feasible solutions is degenerate; otherwise, it is known as *nondegenerate*.

For a nondegenerate LP problem, each iteration of the simplex method reduces the objective function value. Thus, the method cannot go to a previous basic feasible solution because it will have a higher objective function value. For a degenerate LP problem, since the basic variable with a zero value does not have any influence on the objective function value, it is possible to have iterations in which the objective function value is not decreased. Furthermore, at least theoretically, it is possible for the method to keep iterating infinitely between two adjacent basic feasible solutions without decreasing f. This phenomenon is known as *cycling*. As the following examples demonstrate, not every degenerate case will have a cycling problem. However, to avoid any possibility of cycling the commercial implementations of the simplex method, incorporate additional rules to avoid visiting a previous basic feasible solution.

6.5 Unusual Situations Arising During the Simplex Solution

One of the simplest rules to avoid cycling is known as *Bland's rule*. According to this rule, when deciding which variable to bring into the basic set, instead of selecting the variable corresponding to the largest negative number in the objective function row, we simply select the *first negative* coefficient in that row. Using this rule, it is easy to see that a vertex will be visited only once and thus, there is no possibility of cycling.

Example 6.18 *Degenerate case* Consider the following problem with three variables and three equality constraints:

```
f = x1 + x2 + x3;
g = {-x1 + x2 == 0, -x2 + x3 == 0, x1 + x3 == 3};
```

Since there are three equality constraints, it is not really an optimization problem. The solution is simply the one that satisfies the constraint equations. Thus, solving the constraint equations, we get the solution as follows:

```
Solve[g]
```
$$\left\{\left\{x1 \to \frac{3}{2}, x2 \to \frac{3}{2}, x3 \to \frac{3}{2}\right\}\right\}$$

Solving the problem with the simplex method, we notice that the initial basic feasible solution for Phase I is degenerate. The method performs several iterations without any improvement in the objective function. Eventually, for this problem, the method does find the solution. However, in general, there is no such guarantee for problems involving degenerate basic feasible solutions.

```
BasicSimplex[f, g, {x1, x2, x3}, PrintLevel → 2];
```
Problem variables redefined as: $\{x1 \to x_1, x2 \to x_2, x3 \to x_3\}$
Minimize $x_1 + x_2 + x_3$
Subject to $\begin{pmatrix} -x_1 + x_2 == 0 \\ -x_2 + x_3 == 0 \\ x_1 + x_3 == 3 \end{pmatrix}$
All variables ≥ 0
********** Initial simplex tableau **********
New problem variables: $\{x_1, x_2, x_3, a_1, a_2, a_3\}$

Basis	1	2	3	4	5	6	RHS
---	---	---	---	---	---	---	---
4	-1	1	0	1	0	0	0
5	0	-1	1	0	1	0	0
6	1	0	1	0	0	1	3
Obj.	1	1	1	0	0	0	f
ArtObj.	0	0	-2	0	0	0	$-3 + \phi$

Variable to be made basic → 3
Ratios: RHS/Column 3 → $(\infty \quad 0 \quad 3)$
Variable out of the basic set → 5
**********Phase I --- Iteration 1**********

$$\begin{pmatrix} \text{Basis} & 1 & 2 & 3 & 4 & 5 & 6 & \text{RHS} \\ --- & --- & --- & --- & --- & --- & --- & --- \\ 4 & -1 & 1 & 0 & 1 & 0 & 0 & 0 \\ 3 & 0 & -1 & 1 & 0 & 1 & 0 & 0 \\ 6 & 1 & 1 & 0 & 0 & -1 & 1 & 3 \\ \text{Obj.} & 1 & 2 & 0 & 0 & -1 & 0 & f \\ \text{ArtObj.} & 0 & -2 & 0 & 0 & 2 & 0 & -3+\phi \end{pmatrix}$$

Variable to be made basic → 2
Ratios: RHS/Column 2 → $(0 \quad \infty \quad 3)$
Variable out of the basic set → 4
**********Phase I --- Iteration 2**********

$$\begin{pmatrix} \text{Basis} & 1 & 2 & 3 & 4 & 5 & 6 & \text{RHS} \\ --- & --- & --- & --- & --- & --- & --- & --- \\ 2 & -1 & 1 & 0 & 1 & 0 & 0 & 0 \\ 3 & -1 & 0 & 1 & 1 & 1 & 0 & 0 \\ 6 & 2 & 0 & 0 & -1 & -1 & 1 & 3 \\ \text{Obj.} & 3 & 0 & 0 & -2 & -1 & 0 & f \\ \text{ArtObj.} & -2 & 0 & 0 & 2 & 2 & 0 & -3+\phi \end{pmatrix}$$

Variable to be made basic → 1
Ratios: RHS/Column 1 → $\left(\infty \quad \infty \quad \frac{3}{2}\right)$
Variable out of the basic set → 6
**********Phase I --- Iteration 3**********

$$\begin{pmatrix} \text{Basis} & 1 & 2 & 3 & 4 & 5 & 6 & \text{RHS} \\ --- & --- & --- & --- & --- & --- & --- & --- \\ 2 & 0 & 1 & 0 & \frac{1}{2} & -\frac{1}{2} & \frac{1}{2} & \frac{3}{2} \\ 3 & 0 & 0 & 1 & \frac{1}{2} & \frac{1}{2} & \frac{1}{2} & \frac{3}{2} \\ 1 & 1 & 0 & 0 & -\frac{1}{2} & -\frac{1}{2} & \frac{1}{2} & \frac{3}{2} \\ \text{Obj.} & 0 & 0 & 0 & -\frac{1}{2} & \frac{1}{2} & -\frac{3}{2} & -\frac{9}{2}+f \\ \text{ArtObj.} & 0 & 0 & 0 & 1 & 1 & 1 & \phi \end{pmatrix}$$

End of phase I

Optimum solution → $\left\{\left\{\left\{x1 \to \frac{3}{2}, x2 \to \frac{3}{2}, x3 \to \frac{3}{2}\right\}\right\}\right\}$

Optimum objective function value → $\frac{9}{2}$

Example 6.19 *Nondegenerate case* In order to further clarify the difference between a degenerate and a nondegenerate case, consider the following example again with three variables and three equality constraints. In fact, the problem is just a slight modification of the previous example.

6.5 Unusual Situations Arising During the Simplex Solution

```
f = x1 + x2 + x3;
g = {-x1 + x2 == 1, -x2 + x3 == 1, x1 + x3 == 3};
```

Solving this problem using the simplex method, we see that each iteration involves a reduction in the objective function value and thus, there is no chance of cycling.

```
BasicSimplex[f, g, {x1, x2, x3}, PrintLevel → 2];
```

Problem variables redefined as: $\{x1 \to x_1, x2 \to x_2, x3 \to x_3\}$

Minimize $x_1 + x_2 + x_3$

Subject to $\begin{pmatrix} -x_1 + x_2 == 1 \\ -x_2 + x_3 == 1 \\ x_1 + x_3 == 3 \end{pmatrix}$

All variables ≥ 0

********** Initial simplex tableau **********

New problem variables: $\{x_1, x_2, x_3, a_1, a_2, a_3\}$

Basis	1	2	3	4	5	6	RHS
---	---	---	---	---	---	---	---
4	-1	1	0	1	0	0	1
5	0	-1	1	0	1	0	1
6	1	0	1	0	0	1	3
Obj.	1	1	1	0	0	0	f
ArtObj.	0	0	-2	0	0	0	$-5 + \phi$

Variable to be made basic → 3
Ratios: RHS/Column 3 → $(\infty \;\; 1 \;\; 3)$
Variable out of the basic set → 5

**********Phase I --- Iteration 1**********

Basis	1	2	3	4	5	6	RHS
---	---	---	---	---	---	---	---
4	-1	1	0	1	0	0	1
3	0	-1	1	0	1	0	1
6	1	1	0	0	-1	1	2
Obj.	1	2	0	0	-1	0	$-1 + f$
ArtObj.	0	-2	0	0	2	0	$-3 + \phi$

Variable to be made basic → 2
Ratios: RHS/Column 2 → $(1 \;\; \infty \;\; 2)$
Variable out of the basic set → 4

**********Phase I --- Iteration 2**********

Basis	1	2	3	4	5	6	RHS
---	---	---	---	---	---	---	---
2	-1	1	0	1	0	0	1
3	-1	0	1	1	1	0	2
6	2	0	0	-1	-1	1	1
Obj.	3	0	0	-2	-1	0	$-3 + f$
ArtObj.	-2	0	0	2	2	0	$-1 + \phi$

Variable to be made basic → 1

Ratios: RHS/Column 1 → $\left(\infty \quad \infty \quad \dfrac{1}{2}\right)$

Variable out of the basic set → 6

**********Phase I --- Iteration 3**********

$$\begin{pmatrix} \text{Basis} & 1 & 2 & 3 & 4 & 5 & 6 & \text{RHS} \\ \text{---} & \text{---} & \text{---} & \text{---} & \text{---} & \text{---} & \text{---} & \text{---} \\ 2 & 0 & 1 & 0 & \frac{1}{2} & -\frac{1}{2} & \frac{1}{2} & \frac{3}{2} \\ 3 & 0 & 0 & 1 & \frac{1}{2} & \frac{1}{2} & \frac{1}{2} & \frac{5}{2} \\ 1 & 1 & 0 & 0 & -\frac{1}{2} & -\frac{1}{2} & \frac{1}{2} & \frac{1}{2} \\ \text{Obj.} & 0 & 0 & 0 & -\frac{1}{2} & \frac{1}{2} & -\frac{3}{2} & -\frac{9}{2}+f \\ \text{ArtObj.} & 0 & 0 & 0 & 1 & 1 & 1 & \phi \end{pmatrix}$$

End of phase I

Optimum solution → $\left\{\left\{\left\{x1 \to \dfrac{1}{2}, x2 \to \dfrac{3}{2}, x3 \to \dfrac{5}{2}\right\}\right\}\right\}$

Optimum objective function value → $\dfrac{9}{2}$

Example 6.20 *Degenerate case* Example 6.18 started with a degenerate solution. As the following example demonstrates, a degenerate solution can also show up at any stage. For this example, the initial basic feasible solution is not degenerate, but the solution at the next two iterations is degenerate. However, the degeneracy does not cause any difficulty for this problem as well.

The example also demonstrates that degeneracy is related to the solution being *overdetermined*. Two equations are enough to solve for two variables. Thus, degeneracy will occur whenever a basic feasible solution is determined by more constraints than the number of variables.

```
f = -3x1 - 2x2;
g = {x1 + x2 ≤ 4, 2x1 + x2 ≤ 6, x1 ≤ 3};
BasicSimplex[f, g, {x1, x2}, PrintLevel → 2];
```

Problem variables redefined as: $\{x1 \to x_1, x2 \to x_2\}$

Minimize $-3x_1 - 2x_2$

Subject to $\begin{pmatrix} x_1 + x_2 \leq 4 \\ 2x_1 + x_2 \leq 6 \\ x_1 \leq 3 \end{pmatrix}$

All variables ≥ 0

********** Initial simplex tableau **********

New problem variables: $\{x_1, x_2, s_1, s_2, s_3\}$

6.5 Unusual Situations Arising During the Simplex Solution

$$\begin{pmatrix} \text{Basis} & 1 & 2 & 3 & 4 & 5 & \text{RHS} \\ --- & --- & --- & --- & --- & --- & --- \\ 3 & 1 & 1 & 1 & 0 & 0 & 4 \\ 4 & 2 & 1 & 0 & 1 & 0 & 6 \\ 5 & 1 & 0 & 0 & 0 & 1 & 3 \\ \text{Obj.} & -3 & -2 & 0 & 0 & 0 & f \end{pmatrix}$$

Variable to be made basic $\to 1$
Ratios: RHS/Column 1 $\to \begin{pmatrix} 4 & 3 & 3 \end{pmatrix}$
Variable out of the basic set $\to 5$
**********Phase II --- Iteration 1**********

$$\begin{pmatrix} \text{Basis} & 1 & 2 & 3 & 4 & 5 & \text{RHS} \\ --- & --- & --- & --- & --- & --- & --- \\ 3 & 0 & 1 & 1 & 0 & -1 & 1 \\ 4 & 0 & 1 & 0 & 1 & -2 & 0 \\ 1 & 1 & 0 & 0 & 0 & 1 & 3 \\ \text{Obj.} & 0 & -2 & 0 & 0 & 3 & 9+f \end{pmatrix}$$

Variable to be made basic $\to 2$
Ratios: RHS/Column 2 $\to \begin{pmatrix} 1 & 0 & \infty \end{pmatrix}$
Variable out of the basic set $\to 4$
**********Phase II --- Iteration 2**********

$$\begin{pmatrix} \text{Basis} & 1 & 2 & 3 & 4 & 5 & \text{RHS} \\ --- & --- & --- & --- & --- & --- & --- \\ 3 & 0 & 0 & 1 & -1 & 1 & 1 \\ 2 & 0 & 1 & 0 & 1 & -2 & 0 \\ 1 & 1 & 0 & 0 & 0 & 1 & 3 \\ \text{Obj.} & 0 & 0 & 0 & 2 & -1 & 9+f \end{pmatrix}$$

Variable to be made basic $\to 5$
Ratios: RHS/Column 5 $\to \begin{pmatrix} 1 & \infty & 3 \end{pmatrix}$
Variable out of the basic set $\to 3$
**********Phase II --- Iteration 3**********

$$\begin{pmatrix} \text{Basis} & 1 & 2 & 3 & 4 & 5 & \text{RHS} \\ --- & --- & --- & --- & --- & --- & --- \\ 5 & 0 & 0 & 1 & -1 & 1 & 1 \\ 2 & 0 & 1 & 2 & -1 & 0 & 2 \\ 1 & 1 & 0 & -1 & 1 & 0 & 2 \\ \text{Obj.} & 0 & 0 & 1 & 1 & 0 & 10+f \end{pmatrix}$$

Optimum solution $\to \{\{x1 \to 2, x2 \to 2\}\}$
Optimum objective function value $\to -10$

The first and the second iterations show degenerate solutions. Consider the solution at the second iteration with $x_1 = 3$ and $x_2 = 0$. As the graph shown in Figure 6.7 illustrates, at this point there are three active constraints, namely ($g_2 = 0, g_3 = 0$, and $x_2 = 0$). Since the problem has only two variables, one of the conditions must be redundant, resulting in a degenerate case. The optimum would clearly not change if, say, the constraint g_3 is eliminated from the problem. Furthermore, from the KT conditions point of view, it is easy to see that this point is not a regular point.

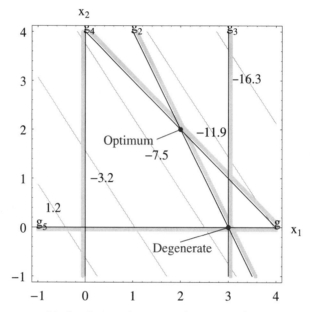

FIGURE 6.7 A graphical solution showing a degenerate basic feasible solution encountered during a simplex solution.

6.6 Post-Optimality Analysis

The final simplex tableau, in addition to the optimum solution, contains information that allows us to determine the active status of constraints, recover Lagrange multipliers, and determine allowable ranges of constraint right-hand sides and objective function coefficients for which the basic variables remain the same. The process is known as sensitivity or post-optimality analysis. This section simply presents the procedures for sensitivity analysis. The derivation of these procedures is easier to see using the revised simplex method that is presented in the next section.

The post-optimality analysis procedures will be explained with reference to the following example:

Minimize $f = -\frac{3}{4}x_1 + 20x_2 - \frac{1}{2}x_3 + 6x_4$

Subject to
$$\begin{pmatrix} \frac{1}{4}x_1 - 8x_2 - x_3 + 9x_4 \geq 1 \\ \frac{1}{2}x_1 - 12x_2 - \frac{1}{2}x_3 + 3x_4 = 3 \\ x_3 + x_4 \leq 1 \\ x_i \geq 0, i = 1, \ldots, 4 \end{pmatrix}$$

To start the simplex solution the constraints are written as follows:

Constraint 1: $\frac{1}{4}x_1 - 8x_2 - x_3 + 9x_4 - x_5 + x_7 = 1$ $x_5 \to$ Surplus $x_7 \to$ Artificial

Constraint 2: $\frac{1}{2}x_1 - 12x_2 - \frac{1}{2}x_3 + 3x_4 + x_8 = 3$ $x_8 \to$ Artificial

Constraint 3: $x_3 + x_4 + x_6 = 1$ $x_6 \to$ Slack

After several iterations of the simplex method, the following final simplex tableau is obtained:

$$\begin{pmatrix} \text{Basis} & x_1 & x_2 & x_3 & x_4 & x_5 & x_6 & x_7 & x_8 & \text{RHS} \\ x_1 & 1 & -\frac{736}{33} & 0 & 0 & \frac{28}{33} & \frac{4}{11} & -\frac{28}{33} & \frac{80}{33} & \frac{224}{33} \\ x_3 & 0 & \frac{8}{33} & 1 & 0 & \frac{4}{33} & \frac{10}{11} & -\frac{4}{33} & \frac{2}{33} & \frac{32}{33} \\ x_4 & 0 & -\frac{8}{33} & 0 & 1 & -\frac{4}{33} & \frac{1}{11} & \frac{4}{33} & -\frac{2}{33} & \frac{1}{33} \\ \text{Obj.} & 0 & \frac{160}{33} & 0 & 0 & \frac{47}{33} & \frac{2}{11} & -\frac{47}{33} & \frac{73}{33} & \frac{178}{33} + f \end{pmatrix}$$

The optimum solution is:

Basic: $x_1 = \frac{224}{33}, x_3 = \frac{32}{33}, x_4 = \frac{1}{33}$

Nonbasic: $x_2 = x_5 = x_6 = x_7 = x_8 = 0$ $f^* = -\frac{178}{33}$

6.6.1 Status of Constraints

The status of active and inactive constraints at optimum is determined from the values of surplus/slack variables. If a slack/surplus variable associated with a given constraint is zero, then obviously that constraint is active.

For the example the optimum solution shows the following values for slack/surplus variables.

$$x_5 = 0, x_6 = 0$$

This means that both constraints 1 and 3 are active. Of course constraint 2, being an equality to begin with, is always active.

6.6.2 Recovery of Lagrange Multipliers

The Lagrange multipliers for the constraints can be read directly from the objective function row of the final simplex tableau. For LE constraints, the Lagrange multiplier values are read from the associated slack variable column. For GE or EQ constraints, they are read from the associated artificial variable column. The sign of Lagrange multipliers for GE constraints should always be negative and that for LE constraints should always be positive. For equality constraints, the sign of a Lagrange multiplier has no special significance.

For the example problem, the first constraint is a GE constraint and therefore, its Lagrange multiplier, $u_1 = -47/33$, is read from the objective function row and the x_7 column (because this is its associated artificial variable). The second constraint is an EQ constraint, and its Lagrange multiplier, $u_2 = 77/33$, is read from x_8 column. The third constraint is an LE constraint, and its Lagrange multiplier, $u_3 = 2/11$, is read from column x_6 associated with its slack variable. Thus, the Lagrange multipliers are as follows:

$$u_1 = -\frac{47}{33}, u_2 = \frac{73}{33}, u_3 = \frac{2}{11}$$

Using KT conditions, it can be shown that the Lagrange multiplier of a GE constraint is the negative of the same constraint written in the LE form. Thus, for the first constraint in the example, we can interpret the multipliers as follows:

For a constraint written as: $\frac{1}{4}x_1 - 8x_2 - x_3 + 9x_4 \geq 1$
the Lagrange multiplier $= -\frac{47}{33}$

For a constraint written as: $-\frac{1}{4}x_1 + 8x_2 + x_3 - 9x_4 \leq -1$
the Lagrange multiplier $= \frac{47}{33}$

6.6.3 Allowable Changes in the Right-Hand Sides of Constraints

If the right-hand side constant of a constraint is changed, the optimum solution obviously may change. However, it is possible to determine the allowable changes of the right-hand sides of constraints for which the basic variable set remains the same. This means that the status of active and inactive constraints will not change and therefore, we can use the Lagrange multipliers to obtain a new value of the objective function based on the sensitivity analysis discussed in Chapter 4 with the KT conditions.

Recall that the decision to remove a variable from the basic set is based on the ratios of right-hand sides and the column corresponding to the variable being brought into the basic set. Thus, intuitively, one would expect to see the allowable changes based on these ratios. The procedure for determining the allowable changes is presented below.

(a) Choice of an appropriate column For determining allowable change in the right-hand side of a constraint, one has to choose a column from the coefficient matrix. For LE constraints, the appropriate column corresponds to the slack variable associated with the constraint. For GE and EQ constraints, the appropriate column corresponds to the artificial variable associated with the constraint.

(b) Computation of ratios The ratios of the negative of the right-hand sides (rhs) and coefficients in the chosen column are then computed.

(c) Determination of allowable range The lower limit of the allowable change is given by the maximum of the negative ratios (negative number with the smallest absolute value), and the upper limit is obtained by the minimum of the positive ratios (smallest positive number).

For the first constraint in the example, the appropriate column is the x_7 column, and the negative of the ratios of the entries in this column and the rhs are as follows:

$$\left\{ -\frac{224/33}{-28/33} = 8, \ -\frac{32/33}{-4/33} = 8, \ -\frac{1/33}{4/33} = -\frac{1}{4} \right\}$$

Denoting the change in the right-hand side of the first constraint by Δb_1, we get the following range:

$$\text{Max}\left[-\frac{1}{4}\right] \leq \Delta b_1 \leq \text{Min}\,[8, 8] \quad \text{or} \quad -\frac{1}{4} \leq \Delta b_1 \leq 8$$

Adding the current value of the right-hand side to both the upper and the lower limit, we can express the allowable range of the right-hand of this constraint as follows:

$$-\frac{1}{4} + 1 \leq b_1 \leq 8 + 1 \quad \text{or} \quad \frac{3}{4} \leq b_1 \leq 9$$

This result implies that the basic variables at the optimum point will remain the same as long as the right-hand side of constraint 1 is between 3/4 and 9.

For the second constraint in the example, the appropriate column is the x_8 column, and the ratios are as follows:

$$\left\{-\frac{224/33}{80/33} = -\frac{14}{5}, -\frac{32/33}{2/33} = -16, -\frac{1/33}{-2/33} = \frac{1}{2}\right\}$$

Denoting the change in the right-hand side of the second constraint by Δb_2, we get the following range:

$$\text{Max}\left[-\frac{14}{5}, -16\right] \leq \Delta b_2 \leq \frac{1}{2} \quad \text{or} \quad -\frac{14}{5} \leq \Delta b_2 \leq \frac{1}{2}$$

Adding the current value of the right-hand side to both the upper and the lower limit, we can express the allowable range of the right-hand of this constraint as follows:

$$-\frac{14}{5} + 3 \leq b_2 \leq \frac{1}{2} + 3 \quad \text{or} \quad \frac{1}{5} \leq b_2 \leq \frac{7}{2}$$

For the third constraint in the example, the appropriate column is the x_6 column and the ratios are as follows:

$$\left\{-\frac{224/33}{4/11} = -\frac{56}{3}, -\frac{32/33}{10/11} = -\frac{16}{15}, -\frac{1/33}{1/11} = -\frac{1}{3}\right\}$$

Denoting the change in the right-hand side of the third constraint by Δb_3, we get the following range:

$$\text{Max}\left[-\frac{56}{3}, -\frac{16}{15}, -\frac{1}{3}\right] \leq \Delta b_3 \leq \infty \quad \text{or} \quad -\frac{1}{3} \leq \Delta b_3 \leq \infty$$

Adding the current value of the right-hand side to both the upper and the lower limit we can express the allowable range of the right-hand of this constraint as follows:

$$-\frac{1}{3} + 1 \leq b_3 \leq \infty + 1 \quad \text{or} \quad \frac{2}{3} \leq b_3 \leq \infty$$

6.6.4 Use of Lagrange Multipliers to Determine the New Optimum

The sensitivity analysis, presented with the KT conditions, shows that

$$\frac{\partial f}{\partial b_i} = -u_i$$

6.6 Post-Optimality Analysis

Thus, if the right-hand side of the ith constraint is changed (and the change is within the allowable limits), the new optimum value of the objective function can be computed using its Lagrange multiplier as follows:

$$\text{New } f^* = \text{Original } f^* - u_i (\text{New } b_i - \text{Original } b_i)$$

It is important to note that this equation gives us a new value of the objective function but does not give us values of the new optimization variables. Modified constraint equations can be solved to get new optimization variables values, if desired.

Also recall that this equation is based on the assumption that all inequality constraints are converted to LE form. In LP problems, we have to deal with GE constraints as well. Therefore, for proper application of this equation, it is best to first convert the modified and the original constraints and their multipliers to LE form and then use the above equation.

Consider the situation when the right-hand side of the first constraint in the example is changed to 5. From the previous section, we know that this change is allowable. The sign of the Lagrange multiplier depends on how we write the constraint.

For a constraint written as: $\frac{1}{4}x_1 - 8x_2 - x_3 + 9x_4 \geq 1$

the Lagrange multiplier $= -\frac{47}{33}$

For a constraint written as: $-\frac{1}{4}x_1 + 8x_2 + x_3 - 9x_4 \leq -1$

the Lagrange multiplier $= \frac{47}{33}$

The modified constraint is written as follows:

$$\frac{1}{4}x_1 - 8x_2 - x_3 + 9x_4 \geq 5 \quad \text{or} \quad -\frac{1}{4}x_1 + 8x_2 + x_3 - 9x_4 \leq -5$$

With the constraints in the LE form, we can get the new objective function value as follows:

$$\text{New } f^* = \text{Original } f^* - u_i (\text{New } b_i - \text{Original } b_i)$$

$$= -\frac{178}{33} - \left(\frac{47}{33}\right)(-5 - (-1)) = \frac{10}{33}$$

As another example, consider the effect of changing the right-hand side of the second constraint to its maximum allowed value. From the previous section, we know that the maximum allowable value for the right-hand side of this constraint is $7/2$.

Original constraint 2: $\frac{1}{2}x_1 - 12x_2 - \frac{1}{2}x_3 + 3x_4 = 3$

$$\text{Lagrange multiplier} = \frac{73}{33}$$

Modified constraint 2: $\frac{1}{2}x_1 - 12x_2 - \frac{1}{2}x_3 + 3x_4 = \frac{7}{2}$

New $f^* = $ Original $f^* - u_i(\text{New } b_i - \text{Original } b_i) = -\frac{178}{33} - (\frac{73}{33})(\frac{7}{2} - 3) = -\frac{13}{2}$

This new optimum value of f can be verified by redoing the problem with the modified constraint.

6.6.5 Allowable Changes in the Objective Function Coefficients

If a coefficient in the objective function is changed, without modifying the constraints, then the feasible domain will remain the same. The only thing that changes is the slope of the objective function contours. Thus, as long as the changes are within certain limits, the optimum point will remain the same. From the entries in the final simplex tableau, it is possible to define the allowable changes in the objective function coefficients for which the basic solution will not change. The procedure for determining the allowable changes is presented below.

For Changes in Coefficients of Variables in the Nonbasic Set

For changes in coefficients of nonbasic variables, the lower limit for the allowable change is negative of the entry in the objective function row and the column that corresponds to the variable. The upper limit for coefficients of all nonbasic variables is ∞.

The variable x_2 is a nonbasic variable for the example problem. Denoting the change by Δc_2, the allowable change in the coefficient of x_2 is given by

$$-\frac{160}{33} \leq \Delta c_2 \leq \infty$$

Adding the current value of the coefficient to both the upper and the lower limit, we can express the allowable range of the coefficient x_2 as follows:

$$-\frac{160}{33} + 20 \leq c_2 \leq \infty + 20 \quad \text{or} \quad 15.15 \leq c_2 \leq \infty$$

Thus, the basic variables and their optimum values will remain the same regardless of how much the coefficient of x_2 in the objective function is increased. However, it only goes down to 15.15 without affecting the optimum solution.

For Changes in Coefficients of Variables in the Basic Set

For changes in coefficients of basic variables, the allowable range is determined by taking the ratios of entries in the objective function row and the constraint row for which the basic variable is the same as the variable whose coefficient is being changed. The entries in the objective function row corresponding to basic and artificial variables are not considered in computing the ratios. The lower limit of the allowable change is given by the maximum of the negative ratios (the negative number with the smallest absolute value) and the upper limit is obtained by the minimum of the positive ratios (the smallest positive number).

The allowable change in the coefficient of variable x_1 in the example problem is determined from the ratios of the objective function row and the first constraint row (because x_1 is the basic variable associated with this constraint). The ratios are as follows:

$$\left\{ \frac{160/33}{-736/33} = -\frac{5}{23}, \frac{47/33}{28/33} = \frac{47}{28}, \frac{2/11}{4/11} = \frac{1}{2} \right\}$$

Denoting the change by Δc_1, the allowable change in the coefficient of x_1 is given by

$$-\frac{5}{23} \leq \Delta c_1 \leq \text{Min}\left[\frac{47}{28}, \frac{1}{2}\right] \quad \text{or} \quad -\frac{5}{23} \leq \Delta c_1 \leq \frac{1}{2}$$

Adding the current value of the coefficient to both the upper and the lower limit, we can express the allowable range of the coefficient x_1 as follows:

$$-\frac{5}{23} - \frac{3}{4} \leq c_1 \leq \frac{1}{2} - \frac{3}{4} \quad \text{or} \quad -\frac{89}{92} \leq c_1 \leq -\frac{1}{4}$$

The allowable changes in the coefficient of variable x_3 in the example problem are determined from the ratios of the objective function row and the second constraint row (because x_3 is the basic variable associated with this constraint). The ratios are as follows:

$$\left\{ \frac{160/33}{8/33} = 20, \frac{47/33}{4/33} = \frac{47}{4}, \frac{2/11}{10/11} = \frac{1}{5} \right\}$$

Denoting the change by Δc_3, the allowable change in the x_3 coefficient in the objective function is given by

$$-\infty \leq \Delta c_3 \leq \text{Min}\left[20, \frac{47}{4}, \frac{1}{5}\right] \quad \text{or} \quad -\infty \leq \Delta c_3 \leq \frac{1}{5}$$

Adding the current value of the coefficient to both the upper and the lower limit, we can express the allowable range of the coefficient x_3 as follows:

$$-\infty - \frac{1}{2} \leq c_3 \leq \frac{1}{5} - \frac{1}{2} \quad \text{or} \quad -\infty \leq c_3 \leq -\frac{3}{10}$$

Finally, the allowable changes in the coefficient of variable x_4 in the example problem are determined from the ratios of the objective function row and the third constraint row (because x_4 is the basic variable associated with this constraint). The ratios are as follows:

$$\left\{\frac{160/33}{-8/33} = -20, \frac{47/33}{-4/33} = -\frac{47}{4}, \frac{2/11}{1/11} = 2\right\}$$

Denoting the change by Δc_4, the allowable change in the x_4 coefficient in the objective function is given by

$$\text{Max}\left[-\frac{47}{4}, -20\right] \leq \Delta c_4 \leq 2 \quad \text{or} \quad -\frac{47}{4} \leq \Delta c_4 \leq 2$$

Adding the current value of the coefficient to both the upper and the lower limit, we can express the allowable range of the coefficient x_4 as follows:

$$-\frac{47}{4} + 6 \leq c_4 \leq 2 + 6 \quad \text{or} \quad -\frac{23}{4} \leq c_4 \leq 8$$

Consider a specific situation when the objective function is modified as follows:

Modified $f = -\frac{3}{4}x_1 + 17x_2 - \frac{1}{2}x_3 - 5x_4$
Original $f = -\frac{3}{4}x_1 + 20x_2 - \frac{1}{2}x_3 + 6x_4$

The coefficients of x_2 and x_4 are changed in the modified f. Both changes are within the allowable range computed above. Thus, the optimum variable values remain the same as that of the original problem.

Optimum variable values: Basic : $x_1 = \frac{224}{33}, x_3 = \frac{32}{33}, x_4 = \frac{1}{33}$
Nonbasic : $x_2 = x_5 = x_6 = x_7 = x_8 = 0$

Substituting these values into the modified objective function, we get New $f^* = -\frac{63}{11}$. This solution can be verified by redoing the problem with the modified objective function.

6.6.6 The SensitivityAnalysis Option of the BasicSimplex Function

The SensitivityAnalysis option of the BasicSimplex function returns Lagrange multipliers and allowable ranges for right-hand sides and cost coefficients.

```
Options[BasicSimplex]
```
{UnrestrictedVariables → {}, MaxIterations → 10, ProblemType → Min, SimplexVariables → {x, s, a}, PrintLevel → 1, SensitivityAnalysis → False}

```
?SensitivityAnalysis
```
SensitivityAnalysis is an option of simplex method. It controls whether a post-optimality (sensitivity) analysis is performed after obtaining an optimum solution. Default is False.

Example 6.21

```
f = 2x1 + x2 - x3;
g = {x1 + 2x2 + x3 ≤ 8, -x1 + x2 - 2x3 ≤ 4};
BasicSimplex[f, g, {x1, x2, x3}, SensitivityAnalysis → True];
```

Problem variables redefined as: $\{x1 \to x_1, x2 \to x_2, x3 \to x_3\}$

Minimize $2x_1 + x_2 - x_3$

Subject to $\begin{pmatrix} x_1 + 2x_2 + x_3 \le 8 \\ -x_1 + x_2 - 2x_3 \le 4 \end{pmatrix}$

All variables ≥ 0
********** Initial simplex tableau **********
New problem variables: $\{x_1, x_2, x_3, s_1, s_2\}$

$\begin{pmatrix} \text{Basis} & 1 & 2 & 3 & 4 & 5 & \text{RHS} \\ \text{---} & \text{---} & \text{---} & \text{---} & \text{---} & \text{---} & \text{---} \\ 4 & 1 & 2 & 1 & 1 & 0 & 8 \\ 5 & -1 & 1 & -2 & 0 & 1 & 4 \\ \text{Obj.} & 2 & 1 & -1 & 0 & 0 & f \end{pmatrix}$

********** Final simplex tableau **********

$\begin{pmatrix} \text{Basis} & 1 & 2 & 3 & 4 & 5 & \text{RHS} \\ \text{---} & \text{---} & \text{---} & \text{---} & \text{---} & \text{---} & \text{---} \\ 3 & 1 & 2 & 1 & 1 & 0 & 8 \\ 5 & 1 & 5 & 0 & 2 & 1 & 20 \\ \text{Obj.} & 3 & 3 & 0 & 1 & 0 & 8+f \end{pmatrix}$

```
Optimum solution → {{{x1 → 0, x2 → 0, x3 → 8}}}
Optimum objective function value → -8
Lagrange multipliers → {1, 0}
Allowable constraint RHS changes and ranges
  -8 ≤ Δb₁ ≤ ∞      0 ≤ b₁ ≤ ∞
  -20 ≤ Δb₂ ≤ ∞    -16 ≤ b₂ ≤ ∞
Allowable cost coefficient changes and ranges
  -3 ≤ Δc₁ ≤ ∞     -1 ≤ c₁ ≤ ∞
  -3 ≤ Δc₂ ≤ ∞     -2 ≤ c₂ ≤ ∞
  -∞ ≤ Δc₃ ≤ 1     -∞ ≤ c₃ ≤ 0
```

The Lagrange multipliers tell that the first constraint is active while the second is inactive.

As an example of the use of Lagrange multipliers in sensitivity analysis, assume that the rhs of the first constraint is changed to 5. The change is $(5 - 8) = -3$. Since the change is allowable, from the Lagrange theorem, the new minimum is

$$f = -8 - 1 \times (-3) = -5.$$

Re-solving the problem with modified constraints confirms this solution.

```
ConstrainedMin[2x1 + x2 - x3, {x1 + 2x2 + x3 ≤ 5, -x1 + x2 - 2x3 ≤ 4}, {x1, x2, x3}]
{-5, {x1 → 0, x2 → 0, x3 → 5}}
```

The allowable range for the change in the coefficient of x_2 in the objective function is $\{-3, \infty\}$. Thus, the solution will remain the same if we increase the coefficient of x_2 in f to any value but will change if the coefficient is made smaller than -2. This is confirmed by re-solving the problem as follows.

As demonstrated by the following solution, if we make the coefficient of x_2 very large, say 600, still the optimum values of variables remain the same.

```
ConstrainedMin[2x1 + 600x2 - x3, {x1 + 2x2 + x3 ≤ 8, -x1 + x2 - 2x3 ≤ 4}, {x1, x2, x3}]
{-8, {x1 → 0, x2 → 0, x3 → 8}}
```

However, reducing the coefficient of x_2 to just outside of the range, say making it -2.1, changes the solution drastically.

```
ConstrainedMin[2x1 - 2.1x2 - x3, {x1 + 2x2 + x3 ≤ 8, -x1 + x2 - 2x3 ≤ 4}, {x1, x2, x3}]
{-8.4, {x1 → 0, x2 → 4., x3 → 0}}
```

6.7 The Revised Simplex Method

The tableau form of the simplex method considered so far is convenient for hand calculations but is inefficient for computer implementation. The method can alternatively be presented in a matrix form that lends itself to efficient numerical implementation. It also gives a clearer picture of the computations involved and makes it straightforward to derive sensitivity analysis procedures.

6.7.1 The Matrix Form of the Simplex Method

Consider an LP problem expressed in the standard form as

Find $\mathbf{x}$ in order to Minimize $f(\mathbf{x}) = \mathbf{c}^T \mathbf{x}$ subject to $\mathbf{A}\mathbf{x} = \mathbf{b}$ and $\mathbf{x} \geq 0$

where

$\mathbf{x} = [x_1, x_2, \ldots, x_n]^T$ vector of optimization variables

$\mathbf{c} = [c_1, c_2, \ldots, c_n]^T$ vector of cost coefficients

$$\mathbf{A} = \begin{pmatrix} a_{11} & a_{12} & \cdots & a_{1n} \\ a_{21} & a_{22} & \cdots & a_{2n} \\ \vdots & \vdots & \vdots & \vdots \\ a_{m1} & a_{m2} & \cdots & a_{mn} \end{pmatrix} \quad m \times n \text{ matrix of constraint coefficients}$$

$\mathbf{b} = [b_1, b_2, \ldots, b_m]^T \geq 0$ vector of right-hand sides of constraints

Identifying columns of the constraint coefficient matrix that multiply a given optimization variable, we write the constraint equations as follows:

$$\mathbf{A}_1 x_1 + \mathbf{A}_2 x_2 + \ldots + \mathbf{A}_n x_n = \mathbf{b}$$

where $\mathbf{A}_i$ is the ith column of the $\mathbf{A}$ matrix. For a basic feasible solution, denoting the vector of m basic variables by $\mathbf{x}_B$ and the $(n-m)$ vector of nonbasic variables by $\mathbf{x}_N$, the problem can be written in the partitioned form as follows:

$$\mathbf{x} = \begin{pmatrix} \mathbf{x}_B \\ \mathbf{x}_N \end{pmatrix} \quad (\mathbf{B} \ \mathbf{N}) \begin{pmatrix} \mathbf{x}_B \\ \mathbf{x}_N \end{pmatrix} = \mathbf{b} \quad f = \begin{pmatrix} \mathbf{c}_B^T & \mathbf{c}_N^T \end{pmatrix} \begin{pmatrix} \mathbf{x}_B \\ \mathbf{x}_N \end{pmatrix}$$

where $m \times m$ matrix $\mathbf{B}$ consists of those columns of matrix $\mathbf{A}$ that correspond to the basic variables and $m \times (n-m)$ matrix $\mathbf{N}$ consists of those that correspond to nonbasic variables. Similarly, vector $\mathbf{c}_B$ contains cost coefficients corresponding to basic variables, and $\mathbf{c}_N$ those that correspond to nonbasic variables.

The general solution of the constraint equations can now be written as follows:

$$\mathbf{B}\mathbf{x}_B + \mathbf{N}\mathbf{x}_N = \mathbf{b}$$

giving $\mathbf{x}_B = \mathbf{B}^{-1}\mathbf{b} - \mathbf{B}^{-1}\mathbf{N}\mathbf{x}_N$

Substituting this into the objective function, we get

$$f = \mathbf{c}_B^T \left(\mathbf{B}^{-1}\mathbf{b} - \mathbf{B}^{-1}\mathbf{N}\mathbf{x}_N\right) + \mathbf{c}_N^T \mathbf{x}_N = \mathbf{c}_B^T \mathbf{B}^{-1}\mathbf{b} + \left(\mathbf{c}_N^T - \mathbf{c}_B^T \mathbf{B}^{-1}\mathbf{N}\right)\mathbf{x}_N$$

or

$$f = \mathbf{w}^T \mathbf{b} + \mathbf{r}^T \mathbf{x}_N$$

where $\mathbf{w}^T = \mathbf{c}_B^T \mathbf{B}^{-1}$ and $\mathbf{r}^T = \mathbf{c}_N^T - \mathbf{c}_B^T \mathbf{B}^{-1}\mathbf{N} = \mathbf{c}_N^T - \mathbf{w}^T \mathbf{N}$.

The vector $\mathbf{w}$ is referred to as the *simplex multipliers* vector. As will be pointed out later, these multipliers are related to the Lagrange multipliers.

From these general expressions, the basic feasible solution for the problem is obtained by setting $\mathbf{x}_N = \mathbf{0}$ and thus

$$\text{Basic feasible solution: } \mathbf{x} = \begin{pmatrix} \mathbf{x}_B \\ \mathbf{0} \end{pmatrix} \quad \mathbf{x}_B = \mathbf{B}^{-1}\mathbf{b} \quad f = \mathbf{w}^T \mathbf{b}$$

Also from the objective function expression, it is clear that if a new variable is brought into the basic set, the change in the objective function will be proportional to $\mathbf{r}$. Thus, this term represents the reduced objective function row of the simplex tableau.

Using this new notation, the problem can be expressed in the following form:

$$\begin{pmatrix} \text{Basic} & \text{Nonbasic} & \text{RHS} \\ \hline \mathbf{B} & \mathbf{N} & \mathbf{b} \\ \hline \mathbf{c}_B^T & \mathbf{c}_N^T & f \end{pmatrix}$$

and a basic solution can be written as follows:

$$\begin{pmatrix} \text{Basic} & \text{Nonbasic} & \text{RHS} \\ \hline \mathbf{I} & \mathbf{B}^{-1}\mathbf{N} & \mathbf{B}^{-1}\mathbf{b} \\ \hline \mathbf{0} & \mathbf{r}^T & f - \mathbf{w}^T \mathbf{b} \end{pmatrix}$$

6.7 The Revised Simplex Method

This form gives us a way to write the simplex tableau corresponding to any basic solution by simply performing a series of matrix operations.

Example 6.22 Consider the LP problem considered in example 6.12.

Minimize $f = -3/4x_1 + 20x_2 - 1/2x_3 + 6x_4$

Subject to
$$\begin{pmatrix} 1/4x_1 - 8x_2 - x_3 + 9x_4 \leq 1 \\ 1/2x_1 - 12x_2 - 1/2x_3 + 3x_4 \leq 3 \\ x_3 + x_4 \leq 1 \\ x_i \geq 0, i = 1, \ldots, 4 \end{pmatrix}$$

Introducing slack variables, we have the standard LP form as follows:

Minimize $f = -3/4x_1 + 20x_2 - 1/2x_3 + 6x_4$

Subject to
$$\begin{pmatrix} 1/4x_1 - 8x_2 - x_3 + 9x_4 + x_5 = 1 \\ 1/2x_1 - 12x_2 - 1/2x_3 + 3x_4 + x_6 = 3 \\ x_3 + x_4 + x_7 = 1 \\ x_i \geq 0, i = 1, \ldots, 7 \end{pmatrix}$$

The problem in the matrix form is as follows:

$$\mathbf{c} = \left\{ -\frac{3}{4}, 20, -\frac{1}{2}, 6, 0, 0, 0 \right\}^T$$

$$\mathbf{A} = \begin{pmatrix} \frac{1}{4} & -8 & -1 & 9 & 1 & 0 & 0 \\ \frac{1}{2} & -12 & -\frac{1}{2} & 3 & 0 & 1 & 0 \\ 0 & 0 & 1 & 1 & 0 & 0 & 1 \end{pmatrix} \quad \mathbf{b} = \begin{pmatrix} 1 \\ 3 \\ 1 \end{pmatrix}$$

Using the matrix form, we can write the simplex tableau corresponding to any set of basic and non-basic variables. As an example, consider writing the simplex tableau corresponding to $\{x_1, x_2, x_7\}$ as basic, and $\{x_3, x_4, x_5, x_6\}$ as nonbasic variables.

The partitioned matrices corresponding to selected basic and nonbasic variables are as follows:

$$\mathbf{c}_B = \left\{ -\frac{3}{4}, 20, 0 \right\}^T \quad \mathbf{c}_N = \left\{ -\frac{1}{2}, 6, 0, 0 \right\}^T$$

$$\mathbf{B} = \begin{pmatrix} \frac{1}{4} & -8 & 0 \\ \frac{1}{2} & -12 & 0 \\ 0 & 0 & 1 \end{pmatrix} \quad \mathbf{N} = \begin{pmatrix} -1 & 9 & 1 & 0 \\ -\frac{1}{2} & 3 & 0 & 1 \\ 1 & 1 & 0 & 0 \end{pmatrix}$$

The corresponding basic solution tableau can be written by simply performing the required matrix operations as follows:

$$\mathbf{B}^{-1} = \begin{pmatrix} -12 & 8 & 0 \\ -\frac{1}{2} & \frac{1}{4} & 0 \\ 0 & 0 & 1 \end{pmatrix} \quad \mathbf{B}^{-1}\mathbf{N} = \begin{pmatrix} 8 & -84 & -12 & 8 \\ \frac{3}{8} & -\frac{15}{4} & -\frac{1}{2} & \frac{1}{4} \\ 1 & 1 & 0 & 0 \end{pmatrix} \quad \mathbf{B}^{-1}\mathbf{b} = \begin{pmatrix} 12 \\ \frac{1}{4} \\ 1 \end{pmatrix}$$

$$\mathbf{w}^T = \mathbf{c}_B^T \mathbf{B}^{-1} = \{-1, -1, 0\} \quad \mathbf{w}^T \mathbf{b} = -4$$

$$\mathbf{r}^T = \mathbf{c}_N^T - \mathbf{w}^T \mathbf{N} = \{-2, 18, 1, 1\}$$

The simplex tableau can now be written by simply placing these values in their appropriate places.

$$\begin{pmatrix} \text{Basic} & \text{Nonbasic} & \text{RHS} \\ \hline \mathbf{I} & \mathbf{B}^{-1}\,\mathbf{N} & \mathbf{B}^{-1}\,\mathbf{b} \\ \mathbf{0} & \mathbf{r}^T & f - \mathbf{w}^T\mathbf{b} \end{pmatrix}$$

$$\begin{pmatrix} \text{Basis} & x_1 & x_2 & x_7 & x_3 & x_4 & x_5 & x_6 & \text{RHS} \\ x_1 & 1 & 0 & 0 & 8 & -84 & -12 & 8 & 12 \\ x_2 & 0 & 1 & 0 & \frac{3}{8} & -\frac{15}{4} & -\frac{1}{2} & \frac{1}{4} & \frac{1}{4} \\ x_7 & 0 & 0 & 1 & 1 & 1 & 0 & 0 & 1 \\ \text{Obj.} & 0 & 0 & 0 & -2 & 18 & 1 & 1 & 4+f \end{pmatrix}$$

Except for a slight rearrangement of columns, this tableau is exactly the same as the second tableau given in example 6.12.

6.7.2 Revised Simplex Algorithm

Using the matrix form, it is possible to present an algorithm for solving an LP problem without need for writing complete tableaus.

In order to decide which variable to bring into the basic set, we need to compute the reduced objective function coefficient given by vector $\mathbf{r}$. In the component form, we have

$$r_j = c_j - \mathbf{w}^T \mathbf{A}_j \quad j \in \text{nonbasic}$$

where $\mathbf{A}_j$ is the jth column of matrix $\mathbf{A}$, and $\mathbf{w}^T = \mathbf{c}_B^T \mathbf{B}^{-1}$.

If all of these reduced cost coefficients are positive, we have reached the optimum. Otherwise, we can bring the variable corresponding to the largest negative coefficient into the basic set. As pointed out before, the selection of largest negative coefficient could possibly cause cycling in case of degenerate problems. Alternatively, following Bland's rule, we can select the variable corresponding to the first negative r_j value. This rule is adopted here. For the following discussion, we will denote the index of the new basic variable by p.

The next key step in the simplex solution is to decide which variable to remove from the basic set. This is done by taking the ratios of the rhs and the pth column (using only those entries that are greater than 0) of the coefficient matrix. In order to perform this computation, we don't need the entire reduced submatrix $\mathbf{B}^{-1}\mathbf{N}$. We simply need the pth column of this matrix, which can be generated by $\mathbf{B}^{-1}\mathbf{A}_p$. This is where computationally, the revised procedure is more efficient. Instead of performing a reduction on the entire matrix $\mathbf{N}$, we are just computing the column that is needed. Using the terminology introduced in the last chapter for numerical optimization methods, we can organize this computation as a descent direction and step-length calculation as follows:

Direction of descent, $\mathbf{d} = -\mathbf{B}^{-1}\mathbf{A}_p$

Current basic solution, $\mathbf{x}_B = \mathbf{B}^{-1}\mathbf{b}$

Step length, $\alpha = \text{Min}\{-\frac{x_i}{d_i}, d_i < 0, i \in \text{basic}\}$

Update values, $x_p = \alpha \quad x_i = x_i + \alpha d_i, i \in \text{basic}$

Since α is set to the minimum ratio, one of the current basic variables will go to zero, which is then removed from the basic set. Also note that if $\mathbf{d} \geq \mathbf{0}$, the problem is unbounded.

The complete revised simplex algorithm is summarized in the following steps:

1. Start with a basic feasible solution. If necessary, the starting basic feasible solution is found using the Phase I procedure of the simplex method. Recall that the objective function for this phase is the sum of artificial variables. The following vectors and matrices are known corresponding to the current basic feasible solution.

    $basic$ = Vector of indices of current basic variables

    $nonBasic$ = Vector of indices of current nonbasic variables

    $\mathbf{x}$ = current solution vector

    $\mathbf{A}$ = Complete constraint coefficient matrix

    $\mathbf{c}$ = complete vector of cost coefficients

    $\mathbf{b}$ = vector of rhs of constraints

2. Form $m \times m$ matrix $\mathbf{B}$ and $m \times 1$ vector $\mathbf{c}_B$, consisting of columns of constraint coefficient matrix and cost coefficients corresponding to current basic

variables. That is,

$$\mathbf{B} = \begin{bmatrix} \mathbf{A}_j \end{bmatrix} \text{ and } \mathbf{c}_B = \begin{bmatrix} c_j \end{bmatrix} \text{ using all } j \in \text{basic}$$

3. Compute the multipliers $\mathbf{w}^T = \mathbf{c}_B^T \mathbf{B}^{-1}$.

   The computation involves the inverse of matrix $\mathbf{B}$. However, instead of actually inverting matrix $\mathbf{B}$, it is more efficient to obtain $\mathbf{w}$ by solving a system of equations obtained as follows:

$$\mathbf{w}^T = \mathbf{c}_B^T \mathbf{B}^{-1} \implies \mathbf{w}^T \mathbf{B} = \mathbf{c}_B^T \mathbf{B}^{-1} \mathbf{B} \implies \mathbf{B}^T \mathbf{w} = \mathbf{c}_B$$

   In order to solve equations $\mathbf{B}^T \mathbf{w} = \mathbf{c}_B$, the matrix $\mathbf{B}$ is first decomposed into lower ($\mathbf{L}$) and upper triangular ($\mathbf{U}$) matrices ($\mathbf{B} = \mathbf{LU}$). The system of equations is then $\mathbf{U}^T \mathbf{L}^T \mathbf{w} = \mathbf{c}_B$. The solution is obtained in two steps. First, a set of intermediate values ($\mathbf{y}$) are computed from $\mathbf{U}^T \mathbf{y} = \mathbf{c}_B$ by forward solution. Finally, vector $\mathbf{w}$ is computed from $\mathbf{L}^T \mathbf{w} = \mathbf{y}$ by back-substitution.

4. Compute the reduced cost coefficients.

   $r_j = c_j - \mathbf{w}^T \mathbf{A}_j \quad j \in \text{nonbasic}$, where $\mathbf{A}_j$ is the jth column of matrix $\mathbf{A}$.

5. Check for optimality.

   If $r_j \geq 0$ for all $j \in$ nonbasic, then stop. The current basic feasible solution is optimum.

6. Choose a variable to bring into the basic set.

   Select the lowest index $p \in$ nonbasic such that $r_p < 0$ (following Bland's rule to avoid cycling in degenerate cases).

7. Compute move direction

   This step corresponds to eliminating basic variables from the column corresponding to a new basic variable. In the basic simplex method, this step is carried out by defining the pivot row and carrying out row operations. Here it is accomplished by solving the system of equations $\mathbf{Bd} = -\mathbf{A}_p$, where $\mathbf{A}_p$ is the column corresponding to the new basic variable. Since the $\mathbf{B}$ matrix is already factored, we simply need to solve $\mathbf{LUd} = -\mathbf{A}_p$. Thus, this step requires a forward solution to get the intermediate solution from $\mathbf{Ly} = -\mathbf{A}_p$ and then a back-substitution to get $\mathbf{d}$ from $\mathbf{Ud} = \mathbf{y}$.

8. Check for unboundedness.

   If $\mathbf{d} \geq \mathbf{0}$, the problem is unbounded. Stop.

9. Compute step length and choose a variable to be made nonbasic.

   Compute the negative of ratios of current variable values and the direction vector.

   $$\alpha = \text{Min}\left\{-\frac{x_i}{d_i}, d_i < 0, i \in \text{basic}\right\}$$

   The step length α is the minimum of these ratios. The index of the minimum ratio corresponds to the variable that is to go out of the basis. Denote this index by q.

10. New basic and nonbasic variables.

    Define a new vector of basic variable indices by replacing the qth entry by p. Set the new values of the variables as follows.

    $$x_p = \alpha \quad x_i = x_i + \alpha d_i, i \in \text{basic}$$

6.7.3 The RevisedSimplex Function

The following RevisedSimplex function implements the revised simplex algorithm for solving LP problems. The function usage and its options are explained first. The function is intended to be used for educational purposes. Several intermediate results can be printed to gain understanding of the process. The procedure for performing sensitivity analysis using the revised simplex procedure is also implemented. The derivation of the sensitivity equations is presented in a later section.

The built-in *Mathematica* functions ConstrainedMin and LinearProgramming do the same thing. For large problems, the built-in functions may be more efficient.

```
Needs["OptimizationToolbox`LPSimplex`"];
?RevisedSimplex
```

RevisedSimplex[f, g, vars, options]. Solves an LP problem using
  Phase I and II simplex algorithm. f is the objective function, g
  is a list of constraints, and vars is a list of variables. See
  Options[RevisedSimplex] to find out about a list of valid options for
  this function.

OptionsUsage[RevisedSimplex]

{UnrestrictedVariables → {}, MaxIterations → 10, ProblemType → Min,
  StandardVariableName → x, PrintLevel → 1, SensitivityAnalysis → False}

UnrestrictedVariables is an option for LP and several QP problems. A list of variables that are not restricted to be positive can be specified with this option. Default is {}.

MaxIterations is an option for several optimization methods. It specifies maximum number of iterations allowed.

ProblemType is an option for most optimization methods. It can either be Min (default) or Max.

StandardVariableName is an option for LP and QP methods. It specifies the symbol to use when creating variable names during conversion to the standard form. Default is x.

PrintLevel is an option for most functions in the OptimizationToolbox. It is specified as an integer. The value of the integer indicates how much intermediate information is to be printed. A PrintLevel→ 0 suppresses all printing. Default for most functions is set to 1 in which case they print only the initial problem setup. Higher integers print more intermediate results.

SensitivityAnalysis is an option of simplex method. It controls whether a post-optimality (sensitivity) analysis is performed after obtaining an optimum solution. Default is False.

Example 6.23 Solve the following LP problem using the revised simplex method.

```
f = - 3x1/4 + 20x2 - x3/2 + 6x4;
g = {x1/4 - 8x2 - x3 + 9x4 ≥ 2, x1/2 - 12x2 - x3/2 + 3x4 == 3, x3 + x4 ≤ 1};
vars = {x1, x2, x3, x4};
RevisedSimplex[f, g, vars, UnrestrictedVariables → {x2}, PrintLevel → 2];
```

**** Problem in Starting Simplex Form ****

New variables are defined as → $\{x1 \to x_1, x2 \to x_2 - x_3, x3 \to x_4, x4 \to x_5\}$

Slack/surplus variables → $\{x_6, x_7\}$

Artificial variables → $\{x_8, x_9\}$

Minimize $\frac{-3x_1}{4} + 20x_2 - 20x_3 - \frac{x_4}{2} + 6x_5$

Subject to $\begin{pmatrix} \frac{x_1}{4} - 8x_2 + 8x_3 - x_4 + 9x_5 - x_6 + x_8 == 2 \\ \frac{x_1}{2} - 12x_2 + 12x_3 - \frac{x_4}{2} + 3x_5 + x_9 == 3 \\ x_4 + x_5 + x_7 == 1 \end{pmatrix}$

All variables ≥ 0

Variables → $\begin{pmatrix} x_1 & x_2 & x_3 & x_4 & x_5 & x_6 & x_7 & x_8 & x_9 \end{pmatrix}$

$c \to \begin{pmatrix} -\frac{3}{4} & 20 & -20 & -\frac{1}{2} & 6 & 0 & 0 & 0 & 0 \end{pmatrix}$

6.7 The Revised Simplex Method

$$b \to \begin{pmatrix} 2 \\ 3 \\ 1 \end{pmatrix} \quad A \to \begin{pmatrix} \frac{1}{4} & -8 & 8 & -1 & 9 & -1 & 0 & 1 & 0 \\ \frac{1}{2} & -12 & 12 & -\frac{1}{2} & 3 & 0 & 0 & 0 & 1 \\ 0 & 0 & 0 & 1 & 1 & 0 & 1 & 0 & 0 \end{pmatrix}$$

Artificial Objective Function $\to x_8 + x_9$

***** Iteration 1 (Phase 1) *****

Basic variables $\to (x_7 \quad x_8 \quad x_9)$

Values of basic variables $\to (1 \quad 2 \quad 3)$

Nonbasic variables $\to (x_1 \quad x_2 \quad x_3 \quad x_4 \quad x_5 \quad x_6)$

$f \to 0$ Art. Obj. $\to 5$

$$B \to \begin{pmatrix} 0 & 1 & 0 \\ 0 & 0 & 1 \\ 1 & 0 & 0 \end{pmatrix} \quad c_B \to \begin{pmatrix} 0 \\ 1 \\ 1 \end{pmatrix} \quad w \to \begin{pmatrix} 1 \\ 1 \\ 0 \end{pmatrix}$$

Reduced cost coefficients, $r \to \left(-\frac{3}{4} \quad 20 \quad -20 \quad \frac{3}{2} \quad -12 \quad 1 \right)$

$$x_B \to \begin{pmatrix} 1 \\ 2 \\ 3 \end{pmatrix} \quad d \to \begin{pmatrix} 0 \\ -\frac{1}{4} \\ -\frac{1}{2} \end{pmatrix} \quad \text{Ratios, } -x/d \to \begin{pmatrix} \infty \\ 8 \\ 6 \end{pmatrix}$$

New basic variable $\to (x_1)$

Variable going out of basic set $\to (x_9)$

Step length, $\alpha \to 6$

New variable values $\to \left(6 \quad 0 \quad 0 \quad 0 \quad 0 \quad 0 \quad 1 \quad \frac{1}{2} \quad 0 \right)$

***** Iteration 2 (Phase 1) *****

Basic variables $\to (x_1 \quad x_7 \quad x_8)$

Values of basic variables $\to \left(6 \quad 1 \quad \frac{1}{2} \right)$

Nonbasic variables $\to (x_2 \quad x_3 \quad x_4 \quad x_5 \quad x_6 \quad x_9)$

$f \to -\frac{9}{2}$ Art. Obj. $\to \frac{1}{2}$

$$B \to \begin{pmatrix} \frac{1}{4} & 0 & 1 \\ \frac{1}{2} & 0 & 0 \\ 0 & 1 & 0 \end{pmatrix} \quad c_B \to \begin{pmatrix} 0 \\ 0 \\ 1 \end{pmatrix} \quad w \to \begin{pmatrix} 1 \\ -\frac{1}{2} \\ 0 \end{pmatrix}$$

Reduced cost coefficients, $r \to \left(2 \quad -2 \quad \frac{3}{4} \quad -\frac{15}{2} \quad 1 \quad \frac{3}{2} \right)$

$$x_B \to \begin{pmatrix} 6 \\ 1 \\ \frac{1}{2} \end{pmatrix} \quad d \to \begin{pmatrix} -24 \\ 0 \\ -2 \end{pmatrix} \quad \text{Ratios, } -x/d \to \begin{pmatrix} \frac{1}{4} \\ \infty \\ \frac{1}{4} \end{pmatrix}$$

New basic variable $\to (x_3)$

Variable going out of basic set $\to (x_1)$

Step length, $\alpha \to \dfrac{1}{4}$

New variable values $\to \begin{pmatrix} 0 & 0 & \dfrac{1}{4} & 0 & 0 & 0 & 1 & 0 & 0 \end{pmatrix}$

***** Iteration 3 (Phase 1) *****

Basic variables $\to (x_3 \ x_7 \ x_8)$

Values of basic variables $\to \begin{pmatrix} \dfrac{1}{4} & 1 & 0 \end{pmatrix}$

Nonbasic variables $\to (x_1 \ x_2 \ x_4 \ x_5 \ x_6 \ x_9)$

$f \to -5$ Art. Obj. $\to 0$

$B \to \begin{pmatrix} 8 & 0 & 1 \\ 12 & 0 & 0 \\ 0 & 1 & 0 \end{pmatrix} \quad c_B \to \begin{pmatrix} 0 \\ 0 \\ 1 \end{pmatrix} \quad w \to \begin{pmatrix} 1 \\ -\dfrac{2}{3} \\ 0 \end{pmatrix}$

Reduced cost coefficients, $r \to \begin{pmatrix} \dfrac{1}{12} & 0 & \dfrac{2}{3} & -7 & 1 & \dfrac{5}{3} \end{pmatrix}$

$x_B \to \begin{pmatrix} \dfrac{1}{4} \\ 1 \\ 0 \end{pmatrix} \quad d \to \begin{pmatrix} -\dfrac{1}{4} \\ -1 \\ -7 \end{pmatrix} \quad \text{Ratios, } -x/d \to \begin{pmatrix} 1 \\ 1 \\ 0 \end{pmatrix}$

New basic variable $\to (x_5)$

Variable going out of basic set $\to (x_8)$

Step length, $\alpha \to 0$

New variable values $\to \begin{pmatrix} 0 & 0 & \dfrac{1}{4} & 0 & 0 & 0 & 1 & 0 & 0 \end{pmatrix}$

***** Iteration 4 (Phase 1) *****

Basic variables $\to (x_3 \ x_5 \ x_7)$

Values of basic variables $\to \begin{pmatrix} \dfrac{1}{4} & 0 & 1 \end{pmatrix}$

Nonbasic variables $\to (x_1 \ x_2 \ x_4 \ x_6 \ x_8 \ x_9)$

$f \to -5$ Art. Obj. $\to 0$

$B \to \begin{pmatrix} 8 & 9 & 0 \\ 12 & 3 & 0 \\ 0 & 1 & 1 \end{pmatrix} \quad c_B \to \begin{pmatrix} 0 \\ 0 \\ 0 \end{pmatrix} \quad w \to \begin{pmatrix} 0 \\ 0 \\ 0 \end{pmatrix}$

Reduced cost coefficients, $r \to \begin{pmatrix} 0 & 0 & 0 & 0 & 1 & 1 \end{pmatrix}$

***** Iteration 5 (Phase 2) *****

Basic variables $\to (x_3 \ x_5 \ x_7)$

Values of basic variables $\to \begin{pmatrix} \dfrac{1}{4} & 0 & 1 \end{pmatrix}$

Nonbasic variables $\to (x_1 \ x_2 \ x_4 \ x_6)$

$f \to -5$

$$B \to \begin{pmatrix} 8 & 9 & 0 \\ 12 & 3 & 0 \\ 0 & 1 & 1 \end{pmatrix} \quad c_B \to \begin{pmatrix} -20 \\ 6 \\ 0 \end{pmatrix} \quad w \to \begin{pmatrix} \frac{11}{7} \\ -\frac{19}{7} \\ 0 \end{pmatrix}$$

Reduced cost coefficients, $r \to \begin{pmatrix} \frac{3}{14} & 0 & -\frac{2}{7} & \frac{11}{7} \end{pmatrix}$

$$x_B \to \begin{pmatrix} \frac{1}{4} \\ 0 \\ 1 \end{pmatrix} \quad d \to \begin{pmatrix} \frac{1}{56} \\ \frac{2}{21} \\ -\frac{23}{21} \end{pmatrix} \quad \text{Ratios, } -x/d \to \begin{pmatrix} \infty \\ \infty \\ \frac{21}{23} \end{pmatrix}$$

New basic variable $\to (x_4)$

Variable going out of basic set $\to (x_7)$

Step length, $\alpha \to \frac{21}{23}$

New variable values $\to \begin{pmatrix} 0 & 0 & \frac{49}{184} & \frac{21}{23} & \frac{2}{23} & 0 & 0 \end{pmatrix}$

***** Iteration 6 (Phase 2) *****

Basic variables $\to (x_3 \ x_4 \ x_5)$

Values of basic variables $\to \begin{pmatrix} \frac{49}{184} & \frac{21}{23} & \frac{2}{23} \end{pmatrix}$

Nonbasic variables $\to (x_1 \ x_2 \ x_6 \ x_7)$

$f \to -\frac{121}{23}$

$$B \to \begin{pmatrix} 8 & -1 & 9 \\ 12 & -\frac{1}{2} & 3 \\ 0 & 1 & 1 \end{pmatrix} \quad c_B \to \begin{pmatrix} -20 \\ -\frac{1}{2} \\ 6 \end{pmatrix} \quad w \to \begin{pmatrix} \frac{37}{23} \\ -\frac{63}{23} \\ -\frac{6}{23} \end{pmatrix}$$

Reduced cost coefficients, $r \to \begin{pmatrix} \frac{5}{23} & 0 & \frac{37}{23} & \frac{6}{23} \end{pmatrix}$

***** Optimum solution after 6 iterations *****

Basis $\to \{\{\{x_3, x_4, x_5\}\}\}$

Variable values $\to \left\{\left\{\left\{ x1 \to 0, x2 \to -\frac{49}{184}, x3 \to \frac{21}{23}, x4 \to \frac{2}{23} \right\}\right\}\right\}$

Objective function $\to -\frac{121}{23}$

Example 6.24 *Stock Cutting* This example demonstrates formulating and solving a typical stock-cutting problem to minimize waste. A carpenter is working on a job that needs 25 boards that are 3.5 ft long and 35 boards that are 7.5 ft long. The local lumber yard sells boards only in the lengths of 8 ft and 12 ft. How many boards of each length should he buy, and how should he cut them, to meet his needs while minimizing waste?

The optimization variables are the different ways in which the given boards can be cut to yield boards of required length. Thus, we have the following set of variables:

From the 8 ft board

x_{81} One 3.5 ft board with 4.5 ft waste
x_{82} One 7.5 ft board with 0.5 ft waste
x_{83} Two 3.5 ft boards with 1 ft waste

From the 12 ft board

x_{121} One 3.5 ft board with 8.5 ft waste
x_{122} Two 3.5 ft boards with 5 ft waste
x_{123} Three 3.5 ft boards with 1.5 ft waste
x_{124} One 7.5 ft board with 4.5 ft waste
x_{125} One each 7.5 ft and 3.5 ft long boards with 1 ft waste

Using these cutting patterns, the total waste to be minimized is as follows:

$$f = 4.5 x_{81} + 0.5 x_{82} + x_{83} + 8.5 x_{121} + 5 x_{122} + 1.5 x_{123} + 4.5 x_{124} + x_{125}$$

The constraints are the number of boards of required length.

3.5 ft boards: $x_{81} + 2 x_{83} + x_{121} + 2 x_{122} + 3 x_{123} + x_{125} \geq 25$

7.5 ft boards: $x_{82} + x_{124} + x_{125} \geq 35$

In terms of *Mathematica* expressions, the problem is formulated as follows:

```
vars = {x81, x82, x83, x121, x122, x123, x124, x125};
f = 9/2 x81 + 1/2 x82 + x83 + 17/2 x121 + 5 x122 + 3/2 x123 + 9/2 x124 + x125;
g = {x81 + 2 x83 + x121 + 2 x122 + 3 x123 + x125 ≥ 25, x82 + x124 + x125 ≥ 35};
```

Using RevisedSimplex, the optimum is obtained as follows:

```
RevisedSimplex[f, g, vars, PrintLevel → 2];
```

**** Problem in Starting Simplex Form ****

New variables are defined as →
$\{x_{81} \to x_1, x_{82} \to x_2, x_{83} \to x_3, x_{121} \to x_4, x_{122} \to x_5, x_{123} \to x_6, x_{124} \to x_7, x_{125} \to x_8\}$

Slack/surplus variables → $\{x_9, x_{10}\}$

Artificial variables → $\{x_{11}, x_{12}\}$

Minimize $\dfrac{9 x_1}{2} + \dfrac{x_2}{2} + x_3 + \dfrac{17 x_4}{2} + 5 x_5 + \dfrac{3 x_6}{2} + \dfrac{9 x_7}{2} + x_8$

Subject to $\begin{pmatrix} x_1 + 2 x_3 + x_4 + 2 x_5 + 3 x_6 + x_8 - x_9 + x_{11} == 25 \\ x_2 + x_7 + x_8 - x_{10} + x_{12} == 35 \end{pmatrix}$

6.7 The Revised Simplex Method

```
All variables ≥ 0
Variables → (x₁  x₂  x₃  x₄  x₅  x₆  x₇  x₈  x₉  x₁₀  x₁₁  x₁₂)
```

$$c \to \begin{pmatrix} \frac{9}{2} & \frac{1}{2} & 1 & \frac{17}{2} & 5 & \frac{3}{2} & \frac{9}{2} & 1 & 0 & 0 & 0 & 0 \end{pmatrix}$$

$$b \to \begin{pmatrix} 25 \\ 35 \end{pmatrix} \quad A \to \begin{pmatrix} 1 & 0 & 2 & 1 & 2 & 3 & 0 & 1 & -1 & 0 & 1 & 0 \\ 0 & 1 & 0 & 0 & 0 & 0 & 1 & 1 & 0 & -1 & 0 & 1 \end{pmatrix}$$

Artificial Objective Function → $x_{11} + x_{12}$

***** Iteration 1 (Phase 1) *****

Basic variables → $(x_{11} \quad x_{12})$

Values of basic variables → $(25 \quad 35)$

Nonbasic variables → $(x_1 \quad x_2 \quad x_3 \quad x_4 \quad x_5 \quad x_6 \quad x_7 \quad x_8 \quad x_9 \quad x_{10})$

f → 0   Art. Obj. → 60

$$B \to \begin{pmatrix} 1 & 0 \\ 0 & 1 \end{pmatrix} \quad c_B \to \begin{pmatrix} 1 \\ 1 \end{pmatrix} \quad w \to \begin{pmatrix} 1 \\ 1 \end{pmatrix}$$

Reduced cost coefficients, r → $(-1 \quad -1 \quad -2 \quad -1 \quad -2 \quad -3 \quad -1 \quad -2 \quad 1 \quad 1)$

$$x_B \to \begin{pmatrix} 25 \\ 35 \end{pmatrix} \quad d \to \begin{pmatrix} -1 \\ 0 \end{pmatrix} \quad \text{Ratios, } -x/d \to \begin{pmatrix} 25 \\ \infty \end{pmatrix}$$

New basic variable → (x_1)

Variable going out of basic set → (x_{11})

Step length, $\alpha \to 25$

New variable values → $(25 \quad 0 \quad 0 \quad 0 \quad 0 \quad 0 \quad 0 \quad 0 \quad 0 \quad 0 \quad 0 \quad 35)$

***** Iteration 2 (Phase 1) *****

Basic variables → $(x_1 \quad x_{12})$

Values of basic variables → $(25 \quad 35)$

Nonbasic variables → $(x_2 \quad x_3 \quad x_4 \quad x_5 \quad x_6 \quad x_7 \quad x_8 \quad x_9 \quad x_{10} \quad x_{11})$

$f \to \frac{225}{2}$ Art. Obj. → 35

$$B \to \begin{pmatrix} 1 & 0 \\ 0 & 1 \end{pmatrix} \quad c_B \to \begin{pmatrix} 0 \\ 1 \end{pmatrix} \quad w \to \begin{pmatrix} 0 \\ 1 \end{pmatrix}$$

Reduced cost coefficients, r → $(-1 \quad 0 \quad 0 \quad 0 \quad 0 \quad -1 \quad -1 \quad 0 \quad 1 \quad 1)$

$$x_B \to \begin{pmatrix} 25 \\ 35 \end{pmatrix} \quad d \to \begin{pmatrix} 0 \\ -1 \end{pmatrix} \quad \text{Ratios, } -x/d \to \begin{pmatrix} \infty \\ 35 \end{pmatrix}$$

New basic variable → (x_2)

Variable going out of basic set → (x_{12})

Step length, $\alpha \to 35$

New variable values → $(25 \quad 35 \quad 0 \quad 0 \quad 0 \quad 0 \quad 0 \quad 0 \quad 0 \quad 0 \quad 0 \quad 0)$

***** Iteration 3 (Phase 1) *****

Basic variables $\to (x_1 \quad x_2)$

Values of basic variables $\to (25 \quad 35)$

Nonbasic variables $\to (x_3 \quad x_4 \quad x_5 \quad x_6 \quad x_7 \quad x_8 \quad x_9 \quad x_{10} \quad x_{11} \quad x_{12})$

$f \to 130$ Art. Obj. $\to 0$

$B \to \begin{pmatrix} 1 & 0 \\ 0 & 1 \end{pmatrix}$ $c_B \to \begin{pmatrix} 0 \\ 0 \end{pmatrix}$ $w \to \begin{pmatrix} 0 \\ 0 \end{pmatrix}$

Reduced cost coefficients, $r \to (0 \quad 0 \quad 0 \quad 0 \quad 0 \quad 0 \quad 0 \quad 0 \quad 1 \quad 1)$

***** Iteration 4 (Phase 2) *****

Basic variables $\to (x_1 \quad x_2)$

Values of basic variables $\to (25 \quad 35)$

Nonbasic variables $\to (x_3 \quad x_4 \quad x_5 \quad x_6 \quad x_7 \quad x_8 \quad x_9 \quad x_{10})$

$f \to 130$

$B \to \begin{pmatrix} 1 & 0 \\ 0 & 1 \end{pmatrix}$ $c_B \to \begin{pmatrix} \frac{9}{2} \\ \frac{1}{2} \end{pmatrix}$ $w \to \begin{pmatrix} \frac{9}{2} \\ \frac{1}{2} \end{pmatrix}$

Reduced cost coefficients, $r \to \left(-8 \quad 4 \quad -4 \quad -12 \quad 4 \quad -4 \quad \frac{9}{2} \quad \frac{1}{2}\right)$

$x_B \to \begin{pmatrix} 25 \\ 35 \end{pmatrix}$ $d \to \begin{pmatrix} -2 \\ 0 \end{pmatrix}$ Ratios, $-x/d \to \begin{pmatrix} \frac{25}{2} \\ \infty \end{pmatrix}$

New basic variable $\to (x_3)$

Variable going out of basic set $\to (x_1)$

Step length, $\alpha \to \dfrac{25}{2}$

New variable values $\to \left(0 \quad 35 \quad \dfrac{25}{2} \quad 0 \quad 0 \quad 0 \quad 0 \quad 0 \quad 0 \quad 0\right)$

***** Iteration 5 (Phase 2) *****

Basic variables $\to (x_2 \quad x_3)$

Values of basic variables $\to \left(35 \quad \dfrac{25}{2}\right)$

Nonbasic variables $\to (x_1 \quad x_4 \quad x_5 \quad x_6 \quad x_7 \quad x_8 \quad x_9 \quad x_{10})$

$f \to 30$

$B \to \begin{pmatrix} 0 & 2 \\ 1 & 0 \end{pmatrix}$ $c_B \to \begin{pmatrix} \frac{1}{2} \\ 1 \end{pmatrix}$ $w \to \begin{pmatrix} \frac{1}{2} \\ \frac{1}{2} \end{pmatrix}$

Reduced cost coefficients, $r \to \left(4 \quad 8 \quad 4 \quad 0 \quad 4 \quad 0 \quad \frac{1}{2} \quad \frac{1}{2}\right)$

***** Optimum solution after 5 iterations *****

Basis $\to \{\{\{x_2, x_3\}\}\}$

```
Variable values →
   {{{x₈₁ → 0, x₈₂ → 35, x₈₃ → 25/2, x₁₂₁ → 0, x₁₂₂ → 0, x₁₂₃ → 0, x₁₂₄ → 0, x₁₂₅ → 0}}}
Objective function → 30
```

The problem indeed is a discrete optimization problem. Since it was solved with the assumption that the variables are continuous, we got an optimum that is not an integer. Based on this solution, we need to buy and cut boards as follows:

35 boards each 8 ft $\Longrightarrow$ 35 boards each 7.5 ft   Waste = 17.5 ft

13 boards each 8 ft $\Longrightarrow$ Each board yields two 3.5 ft boards

Waste = 12 + 4.5 ft

Actual waste = 34 ft.

The problem also has multiple optimum solutions. Solving the problem using the BasicSimplex, we obtain a different optimum solution with the same value of the objective function.

BasicSimplex[f, g, vars];

```
Problem variables redefined as:
  {x₈₁ → x₁, x₈₂ → x₂, x₈₃ → x₃, x₁₂₁ → x₄, x₁₂₂ → x₅, x₁₂₃ → x₆, x₁₂₄ → x₇, x₁₂₅ → x₈}
Minimize  9x₁/2 + x₂/2 + x₃ + 17x₄/2 + 5x₅ + 3x₆/2 + 9x₇/2 + x₈
Subject to  (x₁ + 2x₃ + x₄ + 2x₅ + 3x₆ + x₈ ≥ 25)
            (x₂ + x₇ + x₈ ≥ 35)
All variables ≥ 0
********** Initial simplex tableau **********
New problem variables: {x₁, x₂, x₃, x₄, x₅, x₆, x₇, x₈, s₁, s₂, a₁, a₂}
```

$$\begin{pmatrix}
\text{Basis} & 1 & 2 & 3 & 4 & 5 & 6 \\
--- & --- & --- & --- & --- & --- & --- \\
11 & 1 & 0 & 2 & 1 & 2 & 3 \\
12 & 0 & 1 & 0 & 0 & 0 & 0 \\
\text{Obj.} & \frac{9}{2} & \frac{1}{2} & 1 & \frac{17}{2} & 5 & \frac{3}{2} \\
\text{ArtObj.} & -1 & -1 & -2 & -1 & -2 & -3 \\
& 7 & 8 & 9 & 10 & 11 & 12 & \text{RHS} \\
& --- & --- & --- & --- & --- & --- & --- \\
& 0 & 1 & -1 & 0 & 1 & 0 & 25 \\
& 1 & 1 & 0 & -1 & 0 & 1 & 35 \\
& \frac{9}{2} & 1 & 0 & 0 & 0 & 0 & f \\
& -1 & -2 & 1 & 1 & 0 & 0 & -60 + \phi
\end{pmatrix}$$

```
********** Final simplex tableau **********
```

$$\begin{pmatrix}
\text{Basis} & 1 & 2 & 3 & 4 & 5 & 6 \\
\text{---} & \text{---} & \text{---} & \text{---} & \text{---} & \text{---} & \text{---} \\
6 & \frac{1}{3} & 0 & \frac{2}{3} & \frac{1}{3} & \frac{2}{3} & 1 \\
2 & 0 & 1 & 0 & 0 & 0 & 0 \\
\text{Obj.} & 4 & 0 & 0 & 8 & 4 & 0
\end{pmatrix}$$

$$\begin{pmatrix}
 & 7 & 8 & 9 & 10 & 11 & 12 & \text{RHS} \\
 & \text{---} & \text{---} & \text{---} & \text{---} & \text{---} & \text{---} & \text{---} \\
 & 0 & \frac{1}{3} & -\frac{1}{3} & 0 & \frac{1}{3} & 0 & \frac{25}{3} \\
 & 1 & 1 & 0 & -1 & 0 & 1 & 35 \\
 & 4 & 0 & \frac{1}{2} & \frac{1}{2} & -\frac{1}{2} & -\frac{1}{2} & -30+f
\end{pmatrix}$$

Optimum solution →

$$\{\{\{x_{81} \to 0, x_{82} \to 35, x_{83} \to 0, x_{121} \to 0, x_{122} \to 0, x_{123} \to \frac{25}{3}, x_{124} \to 0, x_{125} \to 0\}\}\}$$

Optimum objective function value → 30
** The problem may have multiple solutions.

6.8 Sensitivity Analysis Using the Revised Simplex Method

The revised simplex formulation is also advantageous for deriving the sensitivity analysis procedures. For the following discussion, we assume that we know $\mathbf{x}^*$ to be an optimum solution of the following LP problem:

$$\text{Minimize } f = \mathbf{c}^T\mathbf{x} \equiv (\mathbf{c}_B^T \quad \mathbf{c}_N^T) \begin{pmatrix} \mathbf{x}_B \\ \mathbf{x}_N \end{pmatrix}$$

$$\text{Subject to } \mathbf{A}\mathbf{x} \equiv (\mathbf{B} \quad \mathbf{N}) \begin{pmatrix} \mathbf{x}_B \\ \mathbf{x}_N \end{pmatrix} = \mathbf{b}$$

where $\mathbf{B}$ refers to quantities associated with basic variables and $\mathbf{N}$ to those associated with nonbasic variables.

6.8.1 Lagrange Multipliers

The negative of the multipliers $\mathbf{w}^T = \mathbf{c}_B^T \mathbf{B}^{-1}$ computed in step 3 of the revised simplex algorithm turn out to be the Lagrange multipliers. As mentioned before, in actual computations, they are obtained by solving the following system of linear equations.

$$\mathbf{B}^T \mathbf{w} = \mathbf{c}_B$$

6.8.2 Changes in Objective Function Coefficients

In this section, we consider the effect of changes in the objective function coefficients on the optimum solution. Assume that the modified coefficients are written as follows:

$$\mathbf{c} + \alpha \Delta \mathbf{c} \equiv (\mathbf{c}_B + \alpha \Delta \mathbf{c}_B) + (\mathbf{c}_N + \alpha \Delta \mathbf{c}_N)$$

where $\Delta \mathbf{c}$ is a vector of changes in the coefficients and α is a scale factor. Our goal is to find a suitable range for α for which we can determine a new optimum solution without actually solving the modified problem. Since the constraints are not changed, the feasible domain remains the same. For $\mathbf{x}^*$ to remain optimum, the reduced cost coefficients for the modified objective function should all still be positive. Thus,

$$(\mathbf{c}_N + \alpha \Delta \mathbf{c}_N)^T - (\mathbf{c}_B + \alpha \Delta \mathbf{c}_B)^T \mathbf{B}^{-1} \mathbf{N} \geq \mathbf{0}$$

or

$$\left(\mathbf{c}_N^T - \mathbf{c}_B^T \mathbf{B}^{-1} \mathbf{N}\right) + \alpha \left(\Delta \mathbf{c}_N^T - \Delta \mathbf{c}_B^T \mathbf{B}^{-1} \mathbf{N}\right) \equiv \mathbf{r}^T + \alpha \bar{\mathbf{r}}^T \geq 0$$

The vector $\mathbf{r}$ is the reduced cost of the original problem. The second term represents the reduced cost due to changes in the objective function coefficients. Thus, the following condition must be satisfied for $\mathbf{x}^*$ to remain optimum.

$$\alpha \bar{\mathbf{r}} \geq -\mathbf{r}$$

All components of $\mathbf{r}$ should be positive but the components of $\bar{\mathbf{r}}$ can either be positive or negative. Therefore, we get the following lower and upper limits for α that will satisfy this condition:

$$\alpha_{\min} = \text{Max}\left[\left\{\frac{-r_i}{\bar{r}_i}, \bar{r}_i > 0, i \in \text{nonbasic}\right\}, -\infty\right]$$

$$\alpha_{\max} = \text{Min}\left[\left\{\frac{-r_i}{\bar{r}_i}, \bar{r}_i < 0, i \in \text{nonbasic}\right\}, \infty\right]$$

The above discussion applies to simultaneous changes in as many coefficients as desired. In practice, however, it is more common to study the effect of changes in one of the coefficients at a time. The following two cases arise naturally. Note that the same rules were given with the basic simplex procedure but without their derivations.

Special Case: Changes in Coefficient of the ith Nonbasic Variable

In this case, $\Delta \mathbf{c}_B = \mathbf{0}$ and $\Delta \mathbf{c}_N = \mathbf{e}^i$, where $\mathbf{e}^i$ is a vector with 1 corresponding to the ith nonbasic variable whose coefficient is being changed, and 0 everywhere else. Then we have

$$\left(\mathbf{c}_N^T - \mathbf{c}_B^T \mathbf{B}^{-1} \mathbf{N}\right) + \alpha \left(\mathbf{e}^i\right)^T \geq 0$$

or

$$\mathbf{r}^T + \alpha \left(\mathbf{e}^i\right)^T \geq 0 \implies \alpha \geq -r_i$$

Thus, the change should be larger than the negative of the reduced cost coefficient corresponding to the ith variable. There is no upper limit. Therefore, the allowable range for the ith cost coefficient that corresponds to a nonbasic variable is

$$-r_i \leq \Delta c_i < \infty$$

Special Case: Changes in Coefficient of the ith Basic Variable

In this case, $\Delta \mathbf{c}_N = \mathbf{0}$ and $\Delta \mathbf{c}_B = \mathbf{e}^i$, where $\mathbf{e}^i$ is a vector with 1 corresponding to the ith basic variable whose coefficient is being changed, and 0 everywhere else. Then we have

$$\left(\mathbf{c}_N^T - \mathbf{c}_B^T \mathbf{B}^{-1} \mathbf{N}\right) - \alpha \left(\mathbf{e}^i\right)^T \mathbf{B}^{-1} \mathbf{N} \geq 0$$

or

$$\mathbf{r}^T - \alpha \left(\mathbf{B}^{-1} \mathbf{N}\right)^{(i)} \geq 0$$

Referring to the matrix form of the simplex tableau, we see that $(\mathbf{B}^{-1}\mathbf{N})^{(i)}$ represents entries in the ith row of the constraint matrix corresponding to the nonbasic variables. Therefore the allowable range for the ith cost coefficient that corresponds to a basic variable is

$$(\Delta c_i)_{\min} = \text{Max} \left[\left\{ \frac{-r_j}{(a_N^{(i)})_j}, (a_N^{(i)})_j > 0, j \in \text{nonbasic} \right\}, -\infty \right]$$

$$(\Delta c_i)_{\max} = \text{Min} \left[\left\{ \frac{-r_j}{(a_N^{(i)})_j}, (a_N^{(i)})_j < 0, j \in \text{nonbasic} \right\}, \infty \right]$$

where $(a_N^{(i)})_j$ refers to the jth element in the ith row of the $(\mathbf{B}^{-1}\mathbf{N})$ matrix.

6.8.3 Changes in the Right-hand Sides of Constraints

In this section, we consider the effect of changes in the right-hand sides of constraints on the optimum solution. Assume that the modified constraints are written as follows:

$$\mathbf{Ax} = \mathbf{b} + \alpha \Delta \mathbf{b}$$

where $\Delta \mathbf{b}$ is a vector of changes in the right-hand sides of constraints and α is a scale factor. The reduced cost coefficients do not change and are all positive. Since the feasible domain is now changed, the optimum solution $\mathbf{x}^*$ may change. If the basic variables remain the same, then for optimality we must have

$$\overline{\mathbf{x}}_B = \mathbf{B}^{-1}(\mathbf{b} + \alpha \Delta \mathbf{b}) \geq \mathbf{0}$$

or

$$\mathbf{B}^{-1}\mathbf{b} + \alpha \mathbf{B}^{-1} \Delta \mathbf{b} \geq \mathbf{0} \implies \alpha \overline{\Delta \mathbf{b}} \geq -\overline{\mathbf{b}}$$

where $\overline{\mathbf{b}} = \mathbf{B}^{-1}\mathbf{b}$ and $\overline{\Delta \mathbf{b}} = \mathbf{B}^{-1}\Delta \mathbf{b}$. Note that all entries in $\overline{\mathbf{b}}$ must be positive because they represent the solution for basic variables for the original problem. Entries in $\overline{\Delta \mathbf{b}}$ may be either positive or negative. Thus, we get the following lower limit and upper limit for α in order to meet the feasibility condition.

$$\alpha_{\min} = \text{Max}\left[\text{Max}\left\{\frac{-\overline{b}_i}{\overline{\Delta b}_i}, \overline{\Delta b}_i > 0, i = 1, \ldots m\right\}, -\infty\right]$$

$$\alpha_{\max} = \text{Min}\left[\text{Min}\left\{\frac{-\overline{b}_i}{\overline{\Delta b}_i}, \overline{\Delta b}_i < 0, i = 1, \ldots m\right\}, \infty\right]$$

For α values within the allowable range, the new optimum solution can be obtained as follows:

$$\mathbf{x}_B^* = \alpha \overline{\Delta \mathbf{b}} + \overline{\mathbf{b}} \quad \text{and} \quad f^* = \mathbf{c}_B^T \mathbf{x}_B^*$$

Note that, in contrast with the rules given with the basic simplex procedure, here it is possible to get both the optimum value of the objective function as well as the optimum values of the optimization variables.

Special Case: Changes in the Right-hand Side of the ith Constraint

The above discussion applies to simultaneous changes in as many constraints as desired. As a special case, we consider the effect of changes in one of the constraint right-hand sides at a time. In this case, we will get the same rules that were given with the basic simplex method.

The change in the right-hand side of the ith constraint can be written as $\Delta \mathbf{b} = \mathbf{e}^i$, where $\mathbf{e}^i$ is a vector with 1 in the ith location, and 0 everywhere else. Then we have

$$\mathbf{B}^{-1}\mathbf{b} + \alpha \mathbf{B}^{-1}\Delta \mathbf{b} = \overline{\mathbf{b}} + \alpha \mathbf{B}^{-1}\mathbf{e}^i \geq 0 \implies \alpha \overline{\mathbf{B}_i} \geq -\overline{\mathbf{b}}$$

where $\overline{\mathbf{B}_i}$ is the ith column of the inverse of the $\mathbf{B}$ matrix. Thus, we get the following lower limit and upper limit for α in order to meet the feasibility condition.

$$\alpha_{\min} = \text{Max}\left[\left\{\frac{-\overline{b}_k}{\overline{B}_{ik}}, \overline{B}_{ik} > 0, k = 1, \ldots m\right\}, -\infty\right]$$

$$\alpha_{\max} = \text{Min}\left[\left\{\frac{-\overline{b}_k}{\overline{B}_{ik}}, \overline{B}_{ik} < 0, k = 1, \ldots m\right\}, \infty\right]$$

It can be seen from the following numerical examples that the ith column of the inverse of the $\mathbf{B}$ matrix is the same as the slack/artificial variable column used in the sensitivity analysis with the basic simplex procedure. Note a slight change in notation in the numerical examples. Symbol bb is used to refer to $\overline{b}$ for convenience in implementation.

6.8.4 The Sensitivity Analysis Option of the RevisedSimplex Function

The SensitivityAnalysis option of the RevisedSimplex function returns Lagrange multipliers and allowable ranges for right-hand sides and cost coefficients.

```
Options[RevisedSimplex]
```
{UnrestrictedVariables → {} , MaxIterations → 10,
  ProblemType → Min, StandardVariableName → x,
  PrintLevel → 1, SensitivityAnalysis → False}

```
?SensitivityAnalysis
```

6.8 Sensitivity Analysis Using the Revised Simplex Method

SensitivityAnalysis is an option of simplex method. It controls whether a post-optimality (sensitivity) analysis is performed after obtaining an optimum solution. Default is False.

Example 6.25

```
f = -3x1 - 2x2; g = {x1 + x2 ≤ 40, 2x1 + x2 ≤ 60}; vars = {x1, x2};
RevisedSimplex[f, g, vars, SensitivityAnalysis → True, PrintLevel → 2];
```
**** Problem in Starting Simplex Form ****
New variables are defined as $\rightarrow \{x1 \rightarrow x_1, x2 \rightarrow x_2\}$
Slack/surplus variables $\rightarrow \{x_3, x_4\}$
Minimize $-3x_1 - 2x_2$
Subject to $\begin{pmatrix} x_1 + x_2 + x_3 == 40 \\ 2x_1 + x_2 + x_4 == 60 \end{pmatrix}$
All variables ≥ 0
Variables $\rightarrow (x_1 \ x_2 \ x_3 \ x_4)$
$c \rightarrow (-3 \ -2 \ 0 \ 0)$
$b \rightarrow \begin{pmatrix} 40 \\ 60 \end{pmatrix} \quad A \rightarrow \begin{pmatrix} 1 & 1 & 1 & 0 \\ 2 & 1 & 0 & 1 \end{pmatrix}$

***** Iteration 1 (Phase 2) *****
Basic variables $\rightarrow (x_3 \ x_4)$
Values of basic variables $\rightarrow (40 \ 60)$
Nonbasic variables $\rightarrow (x_1 \ x_2)$
$f \rightarrow 0$
$B \rightarrow \begin{pmatrix} 1 & 0 \\ 0 & 1 \end{pmatrix} \quad c_B \rightarrow \begin{pmatrix} 0 \\ 0 \end{pmatrix} \quad w \rightarrow \begin{pmatrix} 0 \\ 0 \end{pmatrix}$
Reduced cost coefficients, $r \rightarrow (-3 \ -2)$
$x_B \rightarrow \begin{pmatrix} 40 \\ 60 \end{pmatrix} \quad d \rightarrow \begin{pmatrix} -1 \\ -2 \end{pmatrix} \quad$ Ratios, $-x/d \rightarrow \begin{pmatrix} 40 \\ 30 \end{pmatrix}$
New basic variable $\rightarrow (x_1)$
Variable going out of basic set $\rightarrow (x_4)$
Step length, $\alpha \rightarrow 30$
New variable values $\rightarrow (30 \ 0 \ 10 \ 0)$

***** Iteration 2 (Phase 2) *****
Basic variables $\rightarrow (x_1 \ x_3)$
Values of basic variables $\rightarrow (30 \ 10)$
Nonbasic variables $\rightarrow (x_2 \ x_4)$
$f \rightarrow -90$

$$B \to \begin{pmatrix} 1 & 1 \\ 2 & 0 \end{pmatrix} \quad c_B \to \begin{pmatrix} -3 \\ 0 \end{pmatrix} \quad w \to \begin{pmatrix} 0 \\ -\frac{3}{2} \end{pmatrix}$$

Reduced cost coefficients, $r \to \begin{pmatrix} -\frac{1}{2} & \frac{3}{2} \end{pmatrix}$

$$x_B \to \begin{pmatrix} 30 \\ 10 \end{pmatrix} \quad d \to \begin{pmatrix} -\frac{1}{2} \\ -\frac{1}{2} \end{pmatrix} \quad \text{Ratios, } -x/d \to \begin{pmatrix} 60 \\ 20 \end{pmatrix}$$

New basic variable $\to (x_2)$

Variable going out of basic set $\to (x_3)$

Step length, $\alpha \to 20$

New variable values $\to (20 \quad 20 \quad 0 \quad 0)$

***** Iteration 3 (Phase 2) *****

Basic variables $\to (x_1 \quad x_2)$

Values of basic variables $\to (20 \quad 20)$

Nonbasic variables $\to (x_3 \quad x_4)$

$f \to -100$

$$B \to \begin{pmatrix} 1 & 1 \\ 2 & 1 \end{pmatrix} \quad c_B \to \begin{pmatrix} -3 \\ -2 \end{pmatrix} \quad w \to \begin{pmatrix} -1 \\ -1 \end{pmatrix}$$

Reduced cost coefficients, $r \to (1 \quad 1)$

***** Optimum solution after 3 iterations *****

Basis $\to \{\{\{x_1, x_2\}\}\}$

Variable values $\to \{\{\{x1 \to 20, x2 \to 20\}\}\}$

Objective function $\to -100$

Lagrange multipliers

$$B \to \begin{pmatrix} 1 & 1 \\ 2 & 1 \end{pmatrix} \quad c_B \to \begin{pmatrix} -3 \\ -2 \end{pmatrix}$$

Lagrange multipliers $\to (1 \quad 1)$

Sensitivity to constraint constants

$$\Delta b \to \begin{pmatrix} 1 \\ 0 \end{pmatrix} \quad bb \to \begin{pmatrix} 20 \\ 20 \end{pmatrix} \quad \Delta bb \to \begin{pmatrix} -1 \\ 2 \end{pmatrix} \quad -bb/\Delta bb \to \begin{pmatrix} 20 \\ -10 \end{pmatrix}$$

$$\Delta b \to \begin{pmatrix} 0 \\ 1 \end{pmatrix} \quad bb \to \begin{pmatrix} 20 \\ 20 \end{pmatrix} \quad \Delta bb \to \begin{pmatrix} 1 \\ -1 \end{pmatrix} \quad -bb/\Delta bb \to \begin{pmatrix} -20 \\ 20 \end{pmatrix}$$

Allowable constraint RHS changes and ranges

$-10 \le \Delta b_1 \le 20 \quad 30 \le b_1 \le 60$

$-20 \le \Delta b_2 \le 20 \quad 40 \le b_2 \le 80$

Sensitivity to objective function coefficients

$$\Delta c \to \begin{pmatrix} 1 \\ 0 \\ 0 \\ 0 \end{pmatrix} \quad r \to \begin{pmatrix} 1 \\ 1 \end{pmatrix} \quad rb \to \begin{pmatrix} 1 \\ -1 \end{pmatrix} \quad -r/rb \to \begin{pmatrix} -1 \\ 1 \end{pmatrix}$$

6.8 Sensitivity Analysis Using the Revised Simplex Method

$$\Delta c \to \begin{pmatrix} 0 \\ 1 \\ 0 \\ 0 \end{pmatrix} \quad r \to \begin{pmatrix} 1 \\ 1 \end{pmatrix} \quad rb \to \begin{pmatrix} 2 \\ 1 \end{pmatrix} \quad -r/rb \to \begin{pmatrix} \frac{1}{2} \\ -1 \end{pmatrix}$$

Allowable cost coefficient changes and ranges

$-1 \leq \Delta c_1 \leq 1 \qquad -4 \leq c_1 \leq -2$
$-1 \leq \Delta c_2 \leq \frac{1}{2} \qquad -3 \leq c_2 \leq -\frac{3}{2}$

Figure 6.8 gives a graphical interpretation of the changes in the objective function coefficients. The current minimum point is labelled as b. With changes in the objective function coefficients alone, the feasible region remains the same. Only the slope of the objective function contours change. If the change in slope is small, the set of current basic variables and their values will not change. The limiting changes in the cost coefficients are those that will make the vertex before or after the current optimum (points a and c) as the new optimum.

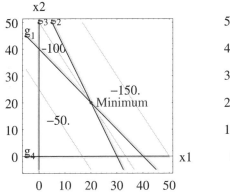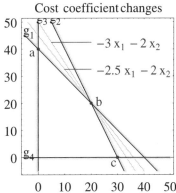

FIGURE 6.8 Solution with the original and the modified objective functions.

The graphs shown in Figure 6.9 illustrate the effect of changes in the right-hand sides of the constraints. If the changes are within the allowable limits, the basis do not change and therefore, the new optimum can be determined simply by computing a new basic solution. If the changes are outside the limits, the basic variables change and a complete new solution is required.

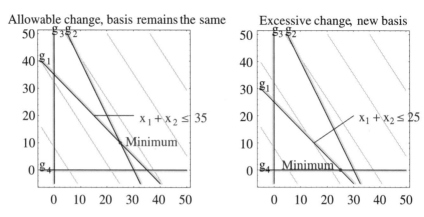

FIGURE 6.9 A graph illustrating solutions with two different right-hand sides for g_1.

Example 6.26

```
f = -3x1/4 + 20x2 - x3/2 + 6x4;
g = {x1/4 - 8x2 - x3 + 9x4 ≥ 2, x1/2 - 12x2 - x3/2 + 3x4 == 3, x3 + x4 ≤ 1};
vars = {x1, x2, x3, x4};
RevisedSimplex[f, g, vars, UnrestrictedVariables → {x2},
   SensitivityAnalysis → True, PrintLevel → 2];
```

**** Problem in Starting Simplex Form ****

New variables are defined as $\to \{x1 \to x_1, x2 \to x_2 - x_3, x3 \to x_4, x4 \to x_5\}$

Slack/surplus variables $\to \{x_6, x_7\}$

Artificial variables $\to \{x_8, x_9\}$

Minimize $\dfrac{-3x_1}{4} + 20x_2 - 20x_3 - \dfrac{x_4}{2} + 6x_5$

Subject to $\begin{pmatrix} \dfrac{x_1}{4} - 8x_2 + 8x_3 - x_4 + 9x_5 - x_6 + x_8 == 2 \\ \dfrac{x_1}{2} - 12x_2 + 12x_3 - \dfrac{x_4}{2} + 3x_5 + x_9 == 3 \\ x_4 + x_5 + x_7 == 1 \end{pmatrix}$

All variables ≥ 0

Variables $\to \begin{pmatrix} x_1 & x_2 & x_3 & x_4 & x_5 & x_6 & x_7 & x_8 & x_9 \end{pmatrix}$

$c \to \begin{pmatrix} -\dfrac{3}{4} & 20 & -20 & -\dfrac{1}{2} & 6 & 0 & 0 & 0 & 0 \end{pmatrix}$

$b \to \begin{pmatrix} 2 \\ 3 \\ 1 \end{pmatrix}$ $A \to \begin{pmatrix} \dfrac{1}{4} & -8 & 8 & -1 & 9 & -1 & 0 & 1 & 0 \\ \dfrac{1}{2} & -12 & 12 & -\dfrac{1}{2} & 3 & 0 & 0 & 0 & 1 \\ 0 & 0 & 0 & 1 & 1 & 0 & 1 & 0 & 0 \end{pmatrix}$

Artificial Objective Function $\to x_8 + x_9$

***** Iteration 1 (Phase 1) *****

Basic variables $\to (x_7 \ x_8 \ x_9)$

Values of basic variables $\to (1 \ 2 \ 3)$

Nonbasic variables $\to (x_1 \ x_2 \ x_3 \ x_4 \ x_5 \ x_6)$

$f \to 0$ Art. Obj. $\to 5$

$B \to \begin{pmatrix} 0 & 1 & 0 \\ 0 & 0 & 1 \\ 1 & 0 & 0 \end{pmatrix}$ $c_B \to \begin{pmatrix} 0 \\ 1 \\ 1 \end{pmatrix}$ $w \to \begin{pmatrix} 1 \\ 1 \\ 0 \end{pmatrix}$

Reduced cost coefficients, $r \to \left(-\dfrac{3}{4} \quad 20 \quad -20 \quad \dfrac{3}{2} \quad -12 \quad 1\right)$

$x_B \to \begin{pmatrix} 1 \\ 2 \\ 3 \end{pmatrix}$ $d \to \begin{pmatrix} 0 \\ -\frac{1}{4} \\ -\frac{1}{2} \end{pmatrix}$ Ratios, $-x/d \to \begin{pmatrix} \infty \\ 8 \\ 6 \end{pmatrix}$

New basic variable $\to (x_1)$

Variable going out of basic set $\to (x_9)$

Step length, $\alpha \to 6$

New variable values $\to \left(6 \ 0 \ 0 \ 0 \ 0 \ 0 \ 1 \ \dfrac{1}{2} \ 0\right)$

***** Iteration 2 (Phase 1) *****

Basic variables $\to (x_1 \ x_7 \ x_8)$

Values of basic variables $\to \left(6 \ 1 \ \dfrac{1}{2}\right)$

Nonbasic variables $\to (x_2 \ x_3 \ x_4 \ x_5 \ x_6 \ x_9)$

$f \to -\dfrac{9}{2}$ Art. Obj. $\to \dfrac{1}{2}$

$B \to \begin{pmatrix} \frac{1}{4} & 0 & 1 \\ \frac{1}{2} & 0 & 0 \\ 0 & 1 & 0 \end{pmatrix}$ $c_B \to \begin{pmatrix} 0 \\ 0 \\ 1 \end{pmatrix}$ $w \to \begin{pmatrix} 1 \\ -\frac{1}{2} \\ 0 \end{pmatrix}$

Reduced cost coefficients, $r \to \left(2 \quad -2 \quad \dfrac{3}{4} \quad -\dfrac{15}{2} \quad 1 \quad \dfrac{3}{2}\right)$

$x_B \to \begin{pmatrix} 6 \\ 1 \\ \frac{1}{2} \end{pmatrix}$ $d \to \begin{pmatrix} -24 \\ 0 \\ -2 \end{pmatrix}$ Ratios, $-x/d \to \begin{pmatrix} \frac{1}{4} \\ \infty \\ \frac{1}{4} \end{pmatrix}$

New basic variable $\to (x_3)$

Variable going out of basic set $\to (x_1)$

Step length, $\alpha \to \dfrac{1}{4}$

New variable values $\to \left(0 \ 0 \ \dfrac{1}{4} \ 0 \ 0 \ 0 \ 1 \ 0 \ 0\right)$

***** Iteration 3 (Phase 1) *****
Basic variables $\to (x_3 \ x_7 \ x_8)$

Values of basic variables $\to \left(\dfrac{1}{4} \ 1 \ 0\right)$

Nonbasic variables $\to (x_1 \ x_2 \ x_4 \ x_5 \ x_6 \ x_9)$

$f \to -5$ Art. Obj. $\to 0$

$$B \to \begin{pmatrix} 8 & 0 & 1 \\ 12 & 0 & 0 \\ 0 & 1 & 0 \end{pmatrix} \quad c_B \to \begin{pmatrix} 0 \\ 0 \\ 1 \end{pmatrix} \quad w \to \begin{pmatrix} 1 \\ -\frac{2}{3} \\ 0 \end{pmatrix}$$

Reduced cost coefficients, $r \to \left(\dfrac{1}{12} \ 0 \ \dfrac{2}{3} \ -7 \ 1 \ \dfrac{5}{3}\right)$

$$x_B \to \begin{pmatrix} \frac{1}{4} \\ 1 \\ 0 \end{pmatrix} \quad d \to \begin{pmatrix} -\frac{1}{4} \\ -1 \\ -7 \end{pmatrix} \quad \text{Ratios, } -x/d \to \begin{pmatrix} 1 \\ 1 \\ 0 \end{pmatrix}$$

New basic variable $\to (x_5)$

Variable going out of basic set $\to (x_8)$

Step length, $\alpha \to 0$

New variable values $\to \left(0 \ 0 \ \dfrac{1}{4} \ 0 \ 0 \ 0 \ 1 \ 0 \ 0\right)$

***** Iteration 4 (Phase 1) *****
Basic variables $\to (x_3 \ x_5 \ x_7)$

Values of basic variables $\to \left(\dfrac{1}{4} \ 0 \ 1\right)$

Nonbasic variables $\to (x_1 \ x_2 \ x_4 \ x_6 \ x_8 \ x_9)$

$f \to -5$ Art. Obj. $\to 0$

$$B \to \begin{pmatrix} 8 & 9 & 0 \\ 12 & 3 & 0 \\ 0 & 1 & 1 \end{pmatrix} \quad c_B \to \begin{pmatrix} 0 \\ 0 \\ 0 \end{pmatrix} \quad w \to \begin{pmatrix} 0 \\ 0 \\ 0 \end{pmatrix}$$

Reduced cost coefficients, $r \to (0 \ 0 \ 0 \ 0 \ 1 \ 1)$

***** Iteration 5 (Phase 2) *****
Basic variables $\to (x_3 \ x_5 \ x_7)$

Values of basic variables $\to \left(\dfrac{1}{4} \ 0 \ 1\right)$

Nonbasic variables $\to (x_1 \ x_2 \ x_4 \ x_6)$

$f \to -5$

$$B \to \begin{pmatrix} 8 & 9 & 0 \\ 12 & 3 & 0 \\ 0 & 1 & 1 \end{pmatrix} \quad c_B \to \begin{pmatrix} -20 \\ 6 \\ 0 \end{pmatrix} \quad w \to \begin{pmatrix} \frac{11}{7} \\ -\frac{19}{7} \\ 0 \end{pmatrix}$$

Reduced cost coefficients, $r \to \left(\dfrac{3}{14} \ 0 \ -\dfrac{2}{7} \ \dfrac{11}{7}\right)$

6.8 Sensitivity Analysis Using the Revised Simplex Method

$$x_B \to \begin{pmatrix} \frac{1}{4} \\ 0 \\ 1 \end{pmatrix} \quad d \to \begin{pmatrix} \frac{1}{56} \\ \frac{2}{21} \\ -\frac{23}{21} \end{pmatrix} \quad \text{Ratios, } -x/d \to \begin{pmatrix} \infty \\ \infty \\ \frac{21}{23} \end{pmatrix}.$$

New basic variable $\to (x_4)$

Variable going out of basic set $\to (x_7)$

Step length, $\alpha \to \frac{21}{23}$

New variable values $\to \begin{pmatrix} 0 & 0 & \frac{49}{184} & \frac{21}{23} & \frac{2}{23} & 0 & 0 \end{pmatrix}$

***** Iteration 6 (Phase 2) *****

Basic variables $\to (x_3 \ x_4 \ x_5)$

Values of basic variables $\to \begin{pmatrix} \frac{49}{184} & \frac{21}{23} & \frac{2}{23} \end{pmatrix}$

Nonbasic variables $\to (x_1 \ x_2 \ x_6 \ x_7)$

$f \to -\frac{121}{23}$

$$B \to \begin{pmatrix} 8 & -1 & 9 \\ 12 & -\frac{1}{2} & 3 \\ 0 & 1 & 1 \end{pmatrix} \quad c_B \to \begin{pmatrix} -20 \\ -\frac{1}{2} \\ 6 \end{pmatrix} \quad w \to \begin{pmatrix} \frac{37}{23} \\ -\frac{63}{23} \\ -\frac{6}{23} \end{pmatrix}$$

Reduced cost coefficients, $r \to \begin{pmatrix} \frac{5}{23} & 0 & \frac{37}{23} & \frac{6}{23} \end{pmatrix}$

***** Optimum solution after 6 iterations *****

Basis $\to \{\{\{x_3, x_4, x_5\}\}\}$

Variable values $\to \left\{\left\{\left\{x1 \to 0, x2 \to -\frac{49}{184}, x3 \to \frac{21}{23}, x4 \to \frac{2}{23}\right\}\right\}\right\}$

Objective function $\to -\frac{121}{23}$

Lagrange multipliers

$$B \to \begin{pmatrix} 8 & -1 & 9 \\ 12 & -\frac{1}{2} & 3 \\ 0 & 1 & 1 \end{pmatrix} \quad c_B \to \begin{pmatrix} -20 \\ -\frac{1}{2} \\ 6 \end{pmatrix}$$

Lagrange multipliers $\to \begin{pmatrix} -\frac{37}{23} & \frac{63}{23} & \frac{6}{23} \end{pmatrix}$

Sensitivity to constraint constants

$$\Delta b \to \begin{pmatrix} 1 \\ 0 \\ 0 \end{pmatrix} \quad bb \to \begin{pmatrix} \frac{49}{184} \\ \frac{21}{23} \\ \frac{2}{23} \end{pmatrix} \quad \Delta bb \to \begin{pmatrix} -\frac{7}{184} \\ -\frac{3}{23} \\ \frac{3}{23} \end{pmatrix} \quad -bb/\Delta bb \to \begin{pmatrix} 7 \\ 7 \\ -\frac{2}{3} \end{pmatrix}$$

$$\Delta b \to \begin{pmatrix} 0 \\ 1 \\ 0 \end{pmatrix} \quad bb \to \begin{pmatrix} \frac{49}{184} \\ \frac{21}{23} \\ \frac{2}{23} \end{pmatrix} \quad \Delta bb \to \begin{pmatrix} \frac{5}{46} \\ \frac{2}{23} \\ -\frac{2}{23} \end{pmatrix} \quad -bb/\Delta bb \to \begin{pmatrix} -\frac{49}{20} \\ -\frac{21}{2} \\ 1 \end{pmatrix}$$

$$\Delta b \to \begin{pmatrix} 0 \\ 0 \\ 1 \end{pmatrix} \quad bb \to \begin{pmatrix} \frac{49}{184} \\ \frac{21}{23} \\ \frac{2}{23} \end{pmatrix} \quad \Delta bb \to \begin{pmatrix} \frac{3}{184} \\ \frac{21}{23} \\ \frac{2}{23} \end{pmatrix} \quad -bb/\Delta bb \to \begin{pmatrix} -\frac{49}{3} \\ -1 \\ -1 \end{pmatrix}$$

Allowable constraint RHS changes and ranges

$-\frac{2}{3} \le \Delta b_1 \le 7 \qquad \frac{4}{3} \le b_1 \le 9$
$-\frac{49}{20} \le \Delta b_2 \le 1 \qquad \frac{11}{20} \le b_2 \le 4$
$-1 \le \Delta b_3 \le \infty \qquad 0 \le b_3 \le \infty$

Sensitivity to objective function coefficients

$$\Delta c \to \begin{pmatrix} 1 \\ 0 \\ 0 \\ 0 \\ 0 \\ 0 \\ 0 \end{pmatrix} \quad r \to \begin{pmatrix} \frac{5}{23} \\ 0 \\ \frac{37}{23} \\ \frac{6}{23} \end{pmatrix} \quad rb \to \begin{pmatrix} 1 \\ 0 \\ 0 \\ 0 \end{pmatrix} \quad -r/rb \to \begin{pmatrix} -\frac{5}{23} \\ - \\ - \\ - \end{pmatrix}$$

$$\Delta c \to \begin{pmatrix} 0 \\ 1 \\ 0 \\ 0 \\ 0 \\ 0 \\ 0 \end{pmatrix} \quad r \to \begin{pmatrix} \frac{5}{23} \\ 0 \\ \frac{37}{23} \\ \frac{6}{23} \end{pmatrix} \quad rb \to \begin{pmatrix} 0 \\ 1 \\ 0 \\ 0 \end{pmatrix} \quad -r/rb \to \begin{pmatrix} - \\ 0 \\ - \\ - \end{pmatrix}$$

$$\Delta c \to \begin{pmatrix} 0 \\ 0 \\ 1 \\ 0 \\ 0 \\ 0 \\ 0 \end{pmatrix} \quad r \to \begin{pmatrix} \frac{5}{23} \\ 0 \\ \frac{37}{23} \\ \frac{6}{23} \end{pmatrix} \quad rb \to \begin{pmatrix} -\frac{33}{736} \\ 1 \\ -\frac{7}{184} \\ -\frac{3}{184} \end{pmatrix} \quad -r/rb \to \begin{pmatrix} \frac{160}{33} \\ 0 \\ \frac{296}{7} \\ 16 \end{pmatrix}$$

$$\Delta c \to \begin{pmatrix} 0 \\ 0 \\ 0 \\ 1 \\ 0 \\ 0 \\ 0 \end{pmatrix} \quad r \to \begin{pmatrix} \frac{5}{23} \\ 0 \\ \frac{37}{23} \\ \frac{6}{23} \end{pmatrix} \quad rb \to \begin{pmatrix} -\frac{1}{92} \\ 0 \\ -\frac{3}{23} \\ -\frac{21}{23} \end{pmatrix} \quad -r/rb \to \begin{pmatrix} 20 \\ - \\ \frac{37}{3} \\ \frac{2}{7} \end{pmatrix}$$

$$\Delta c \to \begin{pmatrix} 0 \\ 0 \\ 0 \\ 0 \\ 1 \\ 0 \\ 0 \end{pmatrix} \quad r \to \begin{pmatrix} \frac{5}{23} \\ 0 \\ \frac{37}{23} \\ \frac{6}{23} \end{pmatrix} \quad rb \to \begin{pmatrix} \frac{1}{92} \\ 0 \\ \frac{3}{23} \\ -\frac{2}{23} \end{pmatrix} \quad -r/rb \to \begin{pmatrix} -20 \\ - \\ -\frac{37}{3} \\ 3 \end{pmatrix}$$

Allowable cost coefficient changes and ranges

6.8 Sensitivity Analysis Using the Revised Simplex Method

$$-\tfrac{5}{23} \le \Delta c_1 \le \infty \qquad -\tfrac{89}{92} \le c_1 \le \infty$$
$$0 \le \Delta c_2 \le \infty \qquad 20 \le c_2 \le \infty$$
$$0 \le \Delta c_3 \le \tfrac{160}{33} \qquad -20 \le c_3 \le -\tfrac{500}{33}$$
$$-\infty \le \Delta c_4 \le \tfrac{2}{7} \qquad -\infty \le c_4 \le -\tfrac{3}{14}$$
$$-\tfrac{37}{3} \le \Delta c_5 \le 3 \qquad -\tfrac{19}{3} \le c_5 \le 9$$

Example 6.27 *Plant operation* In this example, we consider the solution of the tire manufacturing plant operations problem presented in Chapter 1. The problem statement is as follows:

A tire manufacturing plant has the ability to produce both radial and bias-ply automobile tires. During the upcoming summer months, they have contracts to deliver tires as follows.

Date	Radial tires	Bias-ply tires
June 30	5,000	3,000
July 31	6,000	3,000
August 31	4,000	5,000
Total	15,000	11,000

The plant has two types of machines, gold machines and black machines, with appropriate molds to produce these tires. The following production hours are available during the summer months.

Month	On gold machines	On black machines
June	700	1,500
July	300	400
August	1,000	300

The production rates for each machine type and tire combination, in terms of hours per tire, are as follows.

Type	On gold machines	On black machines
Radial	0.15	0.16
Bias-Ply	0.12	0.14

The labor cost of producing tires is $10.00 per operating hour, regardless of which machine type is being used or which tire is being produced. The material cost for radial tires is $5.25 per tire and for bias-ply tires is $4.15 per tire. Finishing, packing and shipping cost is $0.40 per tire. The excess tires are carried over into the next month but are subject to an inventory-carrying charge of $0.15 per tire. Wholesale prices have been set at $20 per tire for radials and $15 per tire for bias-ply.

How should the production be scheduled in order to meet the delivery requirements while maximizing profit for the company during the three-month period?

The optimization variables are as follows:

x_1	Number of radial tires produced in June on the gold machines
x_2	Number of radial tires produced in July on the gold machines
x_3	Number of fadial tires produced in August on the gold machines
x_4	Number of bias-ply tires produced in June on the gold machines
x_5	Number of bias-ply tires produced in July on the gold machines
x_6	Number of bias-ply tires produced in August on the gold machines
x_7	Number of radial tires produced in June on the black machines
x_8	Number of radial tires produced in July on the black machines
x_9	Number of radial tires produced in August on the black machines
x_{10}	Number of bias-ply tires produced in June on the black machines
x_{11}	Number of bias-ply tires produced in July on the black machines
x_{12}	Number of bias-ply tires produced in August on the black machines

The objective of the company is to maximize profit. The following expressions used in defining the objective function were presented in Chapter 1.

```
sales = 20(x₁ + x₂ + x₃ + x₇ + x₈ + x₉) + 15(x₄ + x₅ + x₆ + x₁₀ + x₁₁ + x₁₂);

materialsCost = 5.25(x₁ + x₂ + x₃ + x₇ + x₈ + x₉) + 4.15(x₄ + x₅ + x₆ + x₁₀ + x₁₁ + x₁₂);

laborCost = 10(0.15(x₁ + x₂ + x₃) + 0.16(x₇ + x₈ + x₉) + 0.12(x₄ + x₅ + x₆) + 0.14(x₁₀ +
x₁₁ + x₁₂));

handlingCost = 0.40(x₁ + x₂ + x₃ + x₄ + x₅ + x₆ + x₇ + x₈ + x₉ + x₁₀ + x₁₁ + x₁₂);

inventoryCost = 0.15((x₁ + x₇ - 5000) + (x₄ + x₁₀ - 5000) + (x₁ + x₂ + x₇ + x₈ - 11000) +
(x₄ + x₅ + x₁₀ + x₁₁ - 6000));
```

6.8 Sensitivity Analysis Using the Revised Simplex Method

The production hour limitations are expressed as follows:

```
productionLimitations = {0.15x₁ + 0.12x₄ ≤ 700, 0.15x₂ + 0.12x₅ ≤ 300, 0.15x₃ +
  0.12x₆ ≤ 1000, 0.16x₇ + 0.14x₁₀ ≤ 1500, 0.16x₈ + 0.14x₁₁ ≤ 400, 0.16x₉ + 0.14x₁₂ ≤
  300};
```

Delivery contract constraints are written as follows:

```
deliveryConstraints = {x₁+x₇ ≥ 5000, x₄+x₁₀ ≥ 3000, x₁+x₂+x₇+x₈ ≥ 11000, x₄+x₅+
  x₁₀+x₁₁ ≥ 6000, x₁+x₂+x₃+x₇+x₈+x₉ == 15000, x₄+x₅+x₆+x₁₀+x₁₁+x₁₂ == 11000};
```

Thus, the problem is stated as follows. Note that by multiplying the profit with a negative sign, the problem is defined as a minimization problem.

```
vars = Table[xᵢ, {i, 1, 12}];
f = -(sales - (materialsCost + laborCost + handlingCost + inventoryCost));
g = Join[productionLimitations, deliveryConstraints];
```

The solution is obtained by using the RevisedSimplex function as follows:

```
RevisedSimplex[f, g, vars, SensitivityAnalysis → True, PrintLevel →
  1, MaxIterations → 30];
```

Slack/surplus variables → $\{x_{13}, x_{14}, x_{15}, x_{16}, x_{17}, x_{18}, x_{19}, x_{20}, x_{21}, x_{22}\}$
Artificial variables → $\{x_{23}, x_{24}, x_{25}, x_{26}, x_{27}, x_{28}\}$
Minimize $-13.25x_1 - 13.25x_2 - 13.25x_3 - 9.65x_4 - 9.65x_5 - 9.65x_6 - 13.15x_7 - 13.15x_8 - 13.15x_9 - 9.45x_{10} - 9.45x_{11} - 9.45x_{12}$

Subject to
$$\begin{cases} 0.15x_1 + 0.12x_4 + x_{13} == 700 \\ 0.15x_2 + 0.12x_5 + x_{14} == 300 \\ 0.15x_3 + 0.12x_6 + x_{15} == 1000 \\ 0.16x_7 + 0.14x_{10} + x_{16} == 1500 \\ 0.16x_8 + 0.14x_{11} + x_{17} == 400 \\ 0.16x_9 + 0.14x_{12} + x_{18} == 300 \\ x_1 + x_7 - x_{19} + x_{23} == 5000 \\ x_4 + x_{10} - x_{20} + x_{24} == 3000 \\ x_1 + x_2 + x_7 + x_8 - x_{21} + x_{25} == 11000 \\ x_4 + x_5 + x_{10} + x_{11} - x_{22} + x_{26} == 6000 \\ x_1 + x_2 + x_3 + x_7 + x_8 + x_9 + x_{27} == 15000 \\ x_4 + x_5 + x_6 + x_{10} + x_{11} + x_{12} + x_{28} == 11000 \end{cases}$$

All variables ≥ 0
***** Optimum solution after 18 iterations *****
Basis → $\{\{\{x_1, x_3, x_4, x_5, x_6, x_7, x_8, x_9, x_{16}, x_{18}, x_{19}, x_{20}\}\}\}$
Variable values →
  $\{\{\{x_1 \to 1866.67, x_2 \to 0., x_3 \to 2666.67, x_4 \to 3500., x_5 \to 2500., x_6 \to 5000.,$
  $x_7 \to 6633.33, x_8 \to 2500., x_9 \to 1333.33, x_{10} \to 0, x_{11} \to 0., x_{12} \to 0.\}\}\}$
$f \to -303853.$

Lagrange multipliers

Lagrange multipliers →
$$(0.666667\ 0.666667\ 0.666667\ 0.\ 0.\ 0.\ 0.\ 0.\ 0.\ 0.\ 13.15\ 9.57)$$

Sensitivity to objective function coefficients
Allowable cost coefficient changes and ranges

$-0.15 \leq \Delta c_1 \leq 0.$ $-13.4 \leq c_1 \leq -13.25$
$0. \leq \Delta c_2 \leq \infty$ $-13.25 \leq c_2 \leq \infty$
$0. \leq \Delta c_3 \leq 0.01$ $-13.25 \leq c_3 \leq -13.15$
$0. \leq \Delta c_4 \leq 0.12$ $-9.65 \leq c_4 \leq -9.53$
$-\infty \leq \Delta c_5 \leq 0.$ $-\infty \leq c_5 \leq -9.65$
$-\infty \leq \Delta c_6 \leq 0.$ $-\infty \leq c_6 \leq -9.65$
$0. \leq \Delta c_7 \leq 0.15$ $-13.15 \leq c_7 \leq -13.$
$-\infty \leq \Delta c_8 \leq 0.$ $-\infty \leq c_8 \leq -13.15$
$-0.1 \leq \Delta c_9 \leq 0.$ $-13.25 \leq c_9 \leq -13.15$
$-0.12 \leq \Delta c_{10} \leq \infty$ $-9.57 \leq c_{10} \leq \infty$
$-0.12 \leq \Delta c_{11} \leq \infty$ $-9.57 \leq c_{11} \leq \infty$
$-0.12 \leq \Delta c_{12} \leq \infty$ $-9.57 \leq c_{12} \leq \infty$

Sensitivity to constraint constants
Allowable constraint RHS changes and ranges

$-280. \leq \Delta b_1 \leq 995.$ $420. \leq b_1 \leq 1695.$
$-280. \leq \Delta b_2 \leq 60.$ $20. \leq b_2 \leq 360.$
$-81.25 \leq \Delta b_3 \leq 200.$ $918.75 \leq b_3 \leq 1200.$
$-438.667 \leq \Delta b_4 \leq \infty$ $1061.33 \leq b_4 \leq \infty$
$-400. \leq \Delta b_5 \leq 960.$ $0. \leq b_5 \leq 960.$
$-86.6667 \leq \Delta b_6 \leq \infty$ $213.333 \leq b_6 \leq \infty$
$-\infty \leq \Delta b_7 \leq 3500.$ $-\infty \leq b_7 \leq 8500.$
$-\infty \leq \Delta b_8 \leq 500.$ $-\infty \leq b_8 \leq 3500.$
$-541.667 \leq \Delta b_9 \leq 1333.33$ $10458.3 \leq b_9 \leq 12333.3$
$-500. \leq \Delta b_{10} \leq 1666.67$ $5500. \leq b_{10} \leq 7666.67$
$-1333.33 \leq \Delta b_{11} \leq 541.666$ $13666.7 \leq b_{11} \leq 15541.7$
$-1666.67 \leq \Delta b_{12} \leq 677.083$ $9333.33 \leq b_{12} \leq 11677.1$

Using the sensitivity analysis, we can answer the following types of questions, without actually solving the modified problem.

(a) What would be the company's profit if the labor cost goes up from $10 to $10.50 per hour?

With this change, the new labor cost and the objective function are as follows:

newLaborCost = 10.5(0.15(x$_1$ + x$_2$ + x$_3$) + 0.16(x$_7$ + x$_8$ + x$_9$) + 0.12(x$_4$ + x$_5$ + x$_6$) + 0.14(x$_{10}$ + x$_{11}$ + x$_{12}$));

newf = -(sales - (materialsCost + newLaborCost + handlingCost + inventoryCost));
Expand[newf]

$-13.175 x_1 - 13.175 x_2 - 13.175 x_3 - 9.59 x_4 - 9.59 x_5 - 9.59 x_6 - 13.07 x_7 - 13.07 x_8 - 13.07 x_9 - 9.38 x_{10} - 9.38 x_{11} - 9.38 x_{12}$

6.8 Sensitivity Analysis Using the Revised Simplex Method 419

Comparing the coefficients in the modified objective function with their allowable ranges, we see that all are within the limits. Thus, the revised profit can simply be computed by substituting the optimum values of variables into the new objective function.

```
newf/.{x₁ → 1866.67, x₂ → 0., x₃ → 2666.67, x₄ → 3500., x₅ → 2500., x₆ →
5000., x₇ → 6633.33, x₈ → 2500., x₉ → 1333.33, x₁₀ → 0, x₁₁ → 0, x₁₂ → 0}
-302016.
```

(b) Based on the Lagrange multiplier values, comment on what effect the changes in the production hours available will have on the company's profit.

From the Lagrange multiplier values, we see that the first three multipliers are positive while the next three are 0. The first three multipliers correspond to constraints based on production hours available on the gold machine. The other three correspond to the production hours available on the black machine. This means that as long as the changes are within the allowable limits, the company's profit will not change as a result of changes in the production hours of the black machine. Furthermore, since the first three Lagrange multipliers are the same, it means that changes in the production hours of the gold machine in any month will have the same effect on the profit.

(c) Assume that during June, the gold machine breaks down, and thus the number of production hours available reduces to 600. What effect would this have on the company's profit?

The situation represents reducing the right-hand side of g_1 by 100. From the right-hand side ranges, we see that the allowable range is $420 \leq b_1 \leq 1695$. The change is allowable. Since the Lagrange multipliers for this constraint is 2/3, the new objective function value as a result of this change would be as follows:

$$\text{New } f^* = -303853. - 0.666667(-100) = -303787.$$

Thus, the profit will decrease by about $67.

(d) Based on the Lagrange multiplier values, comment on what effect the changes in the delivery schedule will have on the company's profit.

From the Lagrange multiplier values, we see that the only two delivery constraints with positive Lagrange multipliers are those that represent the total demand that must be met. Thus, reasonable changes in delivery within each month will have no effect on the profit. However, if for some reason enough tires cannot be produced, then reduction in radial tire delivery will hurt the company more than that in the bias-ply tires. This conclusion is based on the fact that the former constraint has a larger Lagrange multiplier then the latter.

(e) Assume that because of a strike at the plant, total production of Radial tires drops by 1000. What effect would this have on the company's profit?

The situation represents reducing the right-hand side of g_{11} by 1000. From the right-hand side ranges, we see that the allowable range is $13666.7 \leq b_{11} \leq 15541.7$. The change is allowable. Thus, the new objective function value as a result of this change would be as follows.

$$\text{New } f^* = -303853. - 13.15\,(-1000) = -290703.$$

where 13.15 is the Lagrange multiplier associated with this constraint.

6.9 Concluding Remarks

Since its introduction in the late 1940s, the simplex method has been one of the most widely used methods for solving large scale linear programming problems. The BasicSimplex and the RevisedSimplex functions discussed in this chapter are very useful for educational purposes because they provide intermediate computational details. They also perform sensitivity analysis that the built-in *Mathematica* functions LinearProgramming, ConstrainedMin, and ConstrainedMax do not. Several public domain and commercial linear programming packages are also available. The Optimization Technology Center, a joint enterprise of Argonne National Laboratory and Northwestern University, maintains a World Wide Web page that contains a wealth of information on the available software. Currently, the Web page for linear programming software is located at http://www-c.mcs.anl.gov/home/otc/Guide/SoftwareGuide/Categories/linearprog.html.

A lot of books are available that cover the subject from a variety of different points of view. Readers interested in more details should consult a recent book coauthored by one of the original developers of the method, Dantzig and Thapa [1997]. Other excellent sources for additional information are Avriel and Golany [1996], Bazarra, Jarvis, and Sherali [1990], Luenberger [1984], and Nash and Sofer [1996]. Those interested in practical applications and guidelines for forming a wide variety of linear programming problems should consult Kolman and Beck [1980], Nazareth [1987], Pannell [1997], and Rardin [1998].

6.10 Problems

Standard LP Problem

For the following problems, write the problem statement in the standard LP form. Unless stated otherwise, all variables are restricted to be positive.

6.1. Minimize $f = x_1 + 3x_2 - 2x_3$
Subject to $x_1 - 2x_2 - 2x_3 \geq -2$, and $2x_1 - 3x_2 - x_3 \leq -2$.

6.2. Maximize $z = -x_1 + 2x_2$
Subject to $x_1 - 2x_2 + 2 \geq 0$, and $2x_1 - 3x_2 \leq 3$.

6.3. Maximize $z = x_1 + 3x_2 - 2x_3 + x_4$
Subject to $3x_1 + 2x_2 + x_3 + x_4 \leq 20$, $2x_1 + x_2 + x_4 = 10$, and $5x_1 - 2x_2 - x_3 + 2x_4 \geq 3$.

6.4. Minimize $f = 3x_1 + 4x_2 + 5x_3 + 6x_4$
Subject to $3x_1 + 4x_2 + 5x_3 + 6x_4 \geq 20$, $2x_1 + 3x_2 + 4x_4 \leq 10$, and $5x_1 - 6x_2 - 7x_3 + 8x_4 \geq 3$. Variables x_1 and x_4 are unrestricted in sign.

6.5. Maximize $z = -13x_1 + 3x_2 + 5$
Subject to $3x_1 + 5x_2 \leq 20$, $2x_1 + x_2 \geq 10$, $5x_1 + 2x_2 \geq 3$, and $x_1 + 2x_2 \geq 3$. x_1 is unrestricted in sign.

Solution of Linear System of Equations

Find general solutions of the following system of equations using the Gauss-Jordan form.

6.6. Two equations in five unknowns: $x_1 - 2x_2 - 2x_3 + x_4 = -2$, and $2x_1 - 3x_2 - x_3 + x_5 = -2$.

6.7. Two equations in four unknowns: $x_1 - 2x_2 + 2 = x_3$, and $2x_1 - 3x_2 + x_4 = 3$.

6.8. Three equations in six unknowns: $3x_1 + 2x_2 + x_3 + x_4 + x_5 = 20$, $2x_1 + x_2 + x_4 = 10$, and $5x_1 - 2x_2 - x_3 + 2x_4 - x_6 = 3$.

6.9. Three equations in seven unknowns: $3x_1 + 4x_2 + 5x_3 + 6x_4 - x_5 = 20$, $2x_1 + 3x_2 + 4x_4 + x_6 = 10$, and $5x_1 - 6x_2 - 7x_3 + 8x_4 - x_7 = 3$.

6.10. Four equations in six unknowns: $3x_1 + 5x_2 + x_3 = 20$, $2x_1 + x_2 - x_4 = 10$, $5x_1 + 2x_2 - x_5 = 3$, and $x_1 + 2x_2 - x_6 = 3$.

Basic Solutions of an LP Problem

For the following problems, find all basic feasible solutions. Draw the feasible region and mark points on the feasible region that correspond to these solutions. Unless stated otherwise, all variables are restricted to be positive.

6.11. Minimize $f = -100x_1 - 80x_2$
Subject to $5x_1 + 3x_2 \leq 15$ and $x_1 + x_2 \leq 4$.

6.12. Maximize $z = -x_1 + 2x_2$
Subject to $x_1 - 2x_2 + 2 \geq 0$ and $2x_1 - 3x_2 \leq 3$.

6.13. Maximize $z = x_1 + 3x_2$
Subject to $x_1 + 3x_2 \leq 10$ and $x_1 + 2x_2 \leq 10$.

6.14. Maximize $z = -13x_1 + 3x_2 + 5$
Subject to $3x_1 + 5x_2 \leq 20$, $2x_1 + x_2 \geq 10$, $5x_1 + 2x_2 \geq 3$, and $x_1 + 2x_2 \geq 3$. x_1 is unrestricted in sign.

6.15. Minimize $f = -3x_1 - 4x_2$
Subject to $x_1 + 2x_2 \leq 10$, $x_1 + x_2 \leq 10$, and $3x_1 + 5x_2 \leq 20$.

The Simplex Method

Find the optimum solution of the following problems using the simplex method. Unless stated otherwise, all variables are restricted to be positive. Verify solutions graphically for two variable problems.

6.16. Maximize $z = x_1 + 3x_2$
Subject to $x_1 + 4x_2 \leq 10$ and $x_1 + 2x_2 \leq 10$.

6.17. Maximize $z = -x_1 + 2x_2$
Subject to $x_1 - 4x_2 + 2 \geq 0$ and $2x_1 - 3x_2 \leq 3$.

6.18. Minimize $f = -2x_1 + 2x_2 + x_3 - 3x_4$
Subject to $x_1 + x_2 + x_3 + x_4 \leq 18$, $x_1 - 2x_3 + 4x_4 \leq 12$, $x_1 + x_2 \leq 18$, and $x_3 + 2x_4 \leq 16$.

6.19. Minimize $f = -3x_1 - 4x_2$
Subject to $x_1 + 2x_2 \leq 10$, $x_1 + x_2 \leq 10$, and $3x_1 + 5x_2 \leq 20$.

6.20. Minimize $f = -100x_1 - 80x_2$

Subject to $5x_1 + 3x_2 \leq 15$ and $x_1 + x_2 \leq 4$.

6.21. Maximize $z = -x_1 + 2x_2$

Subject to $x_1 - 2x_2 + 2 \geq 0$ and $2x_1 - 3x_2 \leq 3$.

6.22. Minimize $f = x_1 + 3x_2 - 2x_3$

Subject to $x_1 - 2x_2 - 2x_3 \geq -2$ and $2x_1 - 3x_2 - x_3 \leq -2$.

6.23. Maximize $z = x_1 + 3x_2 - 2x_3 + x_4$

Subject to $3x_1 + 2x_2 + x_3 + x_4 \leq 20$, $2x_1 + x_2 + x_4 = 10$, and $5x_1 - 2x_2 - x_3 + 2x_4 \geq 3$.

6.24. Minimize $f = 3x_1 + 4x_2 + 5x_3 + 6x_4$

Subject to $3x_1 + 4x_2 + 5x_3 + 6x_4 \geq 20$, $2x_1 + 3x_2 + 4x_4 \leq 10$, and $5x_1 - 6x_2 - 7x_3 + 8x_4 \geq 3$. Variables x_1 and x_4 are unrestricted in sign.

6.25. Minimize $f = 13x_1 - 3x_2 - 5$

Subject to $3x_1 + 5x_2 \leq 20$, $2x_1 + x_2 \geq 10$, $5x_1 + 2x_2 \geq 3$, and $x_1 + 2x_2 \geq 3$. x_1 is unrestricted in sign.

6.26. Minimize $f = 5x_1 + 2x_2 + 3x_3 + 5x_4$

Subject to $x_1 - x_2 + 7x_3 + 3x_4 \geq 4$, $x_1 + 2x_2 + 2x_3 + x_4 = 9$, and $2x_1 + 3x_2 + x_3 - 4x_4 \leq 5$.

6.27. Minimize $f = -3x_1 + 8x_2 - 2x_3 + 4x_4$

Subject to $x_1 - 2x_2 + 4x_3 + 6x_4 \leq 0$, $x_1 - 4x_2 - x_3 + 6x_4 \leq 2$, $x_3 \leq 3$, and $x_4 \geq 3$.

6.28. Minimize $f = 3x_1 + 2x_2$

Subject to $2x_1 + 2x_2 + x_3 + x_4 = 10$, $2x_1 - 3x_2 + 2x_3 = 10$. The variables should not be greater than 10.

6.29. Maximize $z = x_1 + x_2 + x_3$

Subject to $x_1 + 2x_2 \leq 2$, $x_2 - x_3 \leq 3$, and $x_1 + x_2 + x_3 = 3$.

6.30. Minimize $f = -2x_1 + 5x_2 + 3x_3$

Subject to $x_1 - x_2 - x_3 \leq -3$ and $2x_1 + x_2 \geq 1$.

6.31. Hawkeye foods owns two types of trucks. Truck type I has a refrigerated capacity of 15 m³ and a nonrefrigerated capacity of 25 m³. Truck type II has a refrigerated capacity of 15 m³ and nonrefrigerated capacity of 10 m³. One of their stores in Gofer City needs products that require 150 m³ of refrigerated capacity and 130 m³ of nonrefrigerated capacity. For the round trip from the distribution center to Gofer City, Truck type I uses 300 liters of fuel, while Truck type II uses 200 liters. Determine the number of trucks of each type that the company must use in order to meet the store's needs while minimizing the fuel consumption.

6.32. A manufacturer requires an alloy consisting of 50% tin, 30% lead, and 20% zinc. This alloy can be made by mixing a number of available alloys, the properties and costs of which are tabulated. The goal is to find the cheapest blend. Formulate the problem as an optimization problem. Use the basic simplex method to find an optimum.

	Available alloys				
Properties	A	B	C	D	E
Lead (%)	10	10	40	60	30
Zinc (%)	10	30	50	30	30
Tin (%)	80	60	10	10	40
Cost: ($/lb alloy)	8.2	9.3	11.2	13	17

6.33. A company can produce three different types of concrete blocks, identified as A, B, and C. The production process is constrained by facilities available for mixing, vibration, and inspection/drying. Using the data given in the following table, formulate the production problem in order to maximize the profit. Use the basic simplex method to find an optimum.

	Blocks			
	A	B	C	Available
Mixing (hours/batch)	1	3	9	900
Vibration (hours/batch)	2	3	6	1,200
Inspection/Drying (hours/batch)	0.7	0.8	1	400
Profit: ($/batch)	7	17	30	

6.34. A mining company operates two mines, identified as A and B. Each mine can produce high-, medium-, and low-grade iron ores. The weekly demand for different ores and the daily production rates and operating costs are given in the following table. Formulate an optimization problem to determine the production schedule for the two mines in order to meet the weekly demand at the lowest cost to the company. Use the basic simplex method to find an optimum.

Ore grade	Weekly demand (tons)	Daily production Mine A (tons)	Daily production Mine B (tons)
High	12,000	2,000	1,000
Medium	8,000	1,000	1,000
Low	24,000	5,000	2,000
Operations cost ($/day)		210,000	170,000

6.35. Assignment of parking spaces for its employees has become an issue for an automobile company located in an area with harsh climate. There are enough parking spaces available for all employees; however, some employees must be assigned spaces in lots that are not adjacent to the buildings in which they work. The following table shows the distances in meters between parking lots (identified as 1, 2, and 3) and office buildings (identified as A, B, C, and D). The number of spaces in each lot, and the number of employees who need spaces are also tabulated. Formulate the parking assignment problem to minimize the distances walked by the employees from their parking spaces to their offices. Use the basic simplex method to find an optimum.

Parking lot	Distances from parking lot (m) Building A	Building B	Building C	Building D	Spaces available
1	290	410	260	410	80
2	430	350	330	370	100
3	310	260	290	380	40
# of employees	40	40	60	60	

6.36. Hawkeye Pharmaceuticals can manufacture a new drug using any one of the three processes identified as A, B, and C. The costs and quantities of ingredients used in *one batch* of these processes are given in the following table. The quantity of new drug produced during each batch of different processes is also given in the table.

Process	Cost ($ per batch)	Ingredients used per batch (tons)		Quantity of drug produced
		Ingredient I	Ingredient II	
A	$12,000	3	2	2
B	$25,000	2	6	5
C	$9,000	7	2	1

The company has a supply of 80 tons of ingredient I and 70 tons of ingredient II at hand and would like to produce 60 tons of new drug at a minimum cost.

Formulate the problem as an optimization problem. Use the basic simplex method to find an optimum.

6.37. A major auto manufacturer in Detroit, Michigan, needs two types of seat assemblies during 1998 on the following quarterly schedule:

	Type 1	Type 2
First Quarter	25,000	25,000
Second Quarter	35,000	30,000
Third Quarter	35,000	25,000
Fourth Quarter	25,000	30,000
Total	120,000	110,000

The excess seats from each quarter are carried over to the next quarter but are subject to an inventory-carrying charge of $20 per thousand seats. However, assume no inventory is carried over to 1999.

The company has contracted with an auto seat manufacturer that has two plants: one in Detroit and the other in Waterloo, Iowa. Each plant can manufacture either type of seat; however, their maximum capacities and production costs are different. The production costs per seat and the annual capacity at each of the two plants in terms of number of seat assemblies is given as follows:

	Quarterly capacity (either type)	Production cost	
		Type 1	Type 2
Detroit Plant	30,000	$225	$240
Waterloo Plant	35,000	$165	$180

The packing and shipping costs from the two plants to the auto manufacturer are as follows:

	Cost/100 seats
Detroit Plant	$10
Waterloo Plant	$80

Formulate the problem to determine a seat acquisition schedule from the two plants to minimize the overall cost of this operation to the auto manufacturer for the year. Use the basic simplex method to find an optimum.

6.38. A small company needs pipes in the following lengths:

0.5 m   100 pieces

0.6 m   300 pieces

1.2 m   200 pieces

The local supplier sells pipes only in the following three lengths:

4 m

6 m

8 m

After cutting the necessary lengths, the excess pipe must be thrown away. The company obviously wants to minimize this waste. Formulate the problem as a linear programming problem. Use the basic simplex method to find an optimum.

6.39. Consider the problem of finding the shortest route between two cities while traveling on a given network of available roads. The network is shown in Figure 6.10. The nodes represent cities and the links are the roads that connect these cities. The distances in kilometers along each road are noted in the figure. Use the basic simplex method to find the shortest route.

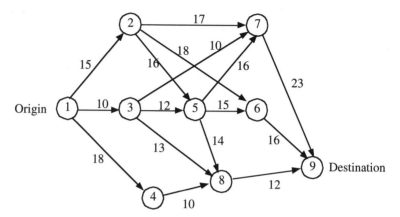

FIGURE 6.10 A network diagram showing distances and direction of travel between cities.

Post-Optimality Analysis

For the following problems, determine Lagrange multipliers and allowable ranges for constraint right-hand sides and objective function coefficients for which the basis do not change. Using this sensitivity analysis, determine the effect on the optimum of the following changes.

(a) A 10% increase (in absolute value) of the coefficient of the first variable in the objective function.

(b) A 10% increase (in absolute value) of the constant term in the first constraint.

6.40. Maximize $z = x_1 + 3x_2$

Subject to $x_1 + 4x_2 \leq 10$ and $x_1 + 2x_2 \leq 10$.

6.41. Maximize $z = -x_1 + 2x_2$

Subject to $x_1 - 4x_2 + 2 \geq 0$ and $2x_1 - 3x_2 \leq 3$.

6.42. Minimize $f = -2x_1 + 2x_2 + x_3 - 3x_4$

Subject to $x_1 + x_2 + x_3 + x_4 \leq 18$, $x_1 - 2x_3 + 4x_4 \leq 12$, $x_1 + x_2 \leq 18$, and $x_3 + 2x_4 \leq 16$.

6.43. Minimize $f = -3x_1 - 4x_2$

Subject to $x_1 + 2x_2 \leq 10$, $x_1 + x_2 \leq 10$ and $3x_1 + 5x_2 \leq 20$.

6.44. Minimize $f = -100x_1 - 80x_2$

Subject to $5x_1 + 3x_2 \leq 15$ and $x_1 + x_2 \leq 4$.

6.45. Maximize $z = -x_1 + 2x_2$

Subject to $x_1 - 2x_2 + 2 \geq 0$ and $2x_1 - 3x_2 \leq 3$.

6.46. Minimize $f = x_1 + 3x_2 - 2x_3$

Subject to $x_1 - 2x_2 - 2x_3 \geq -2$ and $2x_1 - 3x_2 - x_3 \leq -2$.

6.47. Maximize $z = x_1 + 3x_2 - 2x_3 + x_4$

Subject to $3x_1 + 2x_2 + x_3 + x_4 \leq 20$, $2x_1 + x_2 + x_4 = 10$, and $5x_1 - 2x_2 - x_3 + 2x_4 \geq 3$.

6.48. Minimize $f = 3x_1 + 4x_2 + 5x_3 + 6x_4$

Subject to $3x_1 + 4x_2 + 5x_3 + 6x_4 \geq 20$, $2x_1 + 3x_2 + 4x_4 \leq 10$, and $5x_1 - 6x_2 - 7x_3 + 8x_4 \geq 3$. Variables x_1 and x_4 is are unrestricted in sign.

6.49. Minimize $f = 13x_1 - 3x_2 - 5$

Subject to $3x_1 + 5x_2 \leq 20$, $2x_1 + x_2 \geq 10$, $5x_1 + 2x_2 \geq 3$, and $x_1 + 2x_2 \geq 3$. x_1 is unrestricted in sign.

6.50. Minimize $f = 5x_1 + 2x_2 + 3x_3 + 5x_4$

Subject to $x_1 - x_2 + 7x_3 + 3x_4 \geq 4$, $x_1 + 2x_2 + 2x_3 + x_4 = 9$, and $2x_1 + 3x_2 + x_3 - 4x_4 \leq 5$.

6.51. Minimize $f = -3x_1 + 8x_2 - 2x_3 + 4x_4$

Subject to $x_1 - 2x_2 + 4x_3 + 6x_4 \leq 0$, $x_1 - 4x_2 - x_3 + 6x_4 \leq 2$, $x_3 \leq 3$, and $x_4 \geq 3$.

6.52. Minimize $f = 3x_1 + 2x_2$

Subject to $2x_1 + 2x_2 + x_3 + x_4 = 10$, $2x_1 - 3x_2 + 2x_3 = 10$. The variables should not be greater than 10.

6.53. Maximize $z = x_1 + x_2 + x_3$

Subject to $x_1 + 2x_2 \leq 2$, $x_2 - x_3 \leq 3$, and $x_1 + x_2 + x_3 = 3$.

6.54. Minimize $f = -2x_1 + 5x_2 + 3x_3$

Subject to $x_1 - x_2 - x_3 \leq -3$ and $2x_1 + x_2 \geq 1$.

6.55. Using sensitivity analysis, determine the effect of following changes in the Hawkeye food problem 6.31.

(a) The required capacity for refrigerated goods increases to 160 m^3.

(b) The required capacity for nonrefrigerated goods increases to 140 m^3.

(c) Gas consumption of truck type I increases to 320 gallons.

(d) Gas consumption of truck type II decreases to 180 gallons.

6.56. Using sensitivity analysis, determine the effect of following changes in the alloy manufacturing problem 6.32.

(a) The cost of alloy B increases to $10/lb.

(b) The tin content in the alloy produced is decreased to 48% and the zinc content is increased to 22%.

6.57. Using sensitivity analysis, determine the effect of following changes in the concrete block manufacturing problem 6.33.

(a) The profit from block type A increases to $8 and that from block type C decreases to $28.

(b) The number of available mixing hours increases to 1,000.

(c) The number of available inspection/drying hours increases to 450.

6.58. Using sensitivity analysis, determine the effect of following changes in the mining company problem 6.34.

(a) The operations cost at the mine B increase to $180,000 per day.

(b) The weekly demand for medium grade ore increases to 9,000 tons.

(c) The weekly demand for low grade ore decreases to 22,000 tons.

6.59. Using sensitivity analysis, determine the effect of following changes in the parking spaces assignment problem 6.35.

(a) The number of employees working in building B increases to 45.

(b) The number of spaces available in parking lot 3 increases to 45.

(c) The number of spaces available in parking lot 2 decreases to 90.

6.60. Using sensitivity analysis, determine the effect of following changes in the drug manufacturing problem 6.36.

(a) The cost per batch of process C increases to $10,000.

(b) The supply of ingredient II increases to 75 tons.

6.61. Using sensitivity analysis, determine the effect of following changes in the auto seat problem 6.37.

(a) The production costs at the Waterloo plant increase to $175 and $185, respectively, for the two seat types.

(b) In the third quarter, the demand of type II seats increases to 27,000.

(c) The shipping cost from the Detroit plant increases to $15.

Revised Simplex Method

Find the optimum solution of the following problems using the revised simplex method. Unless stated otherwise, all variables are restricted to be positive. Verify solutions graphically for two-variable problems.

6.62. Maximize $z = x_1 + 3x_2$

Subject to $x_1 + 4x_2 \leq 10$ and $x_1 + 2x_2 \leq 10$.

6.63. Maximize $z = -x_1 + 2x_2$

Subject to $x_1 - 4x_2 + 2 \geq 0$ and $2x_1 - 3x_2 \leq 3$.

6.64. Minimize $f = -2x_1 + 2x_2 + x_3 - 3x_4$

Subject to $x_1 + x_2 + x_3 + x_4 \leq 18$, $x_1 - 2x_3 + 4x_4 \leq 12$, $x_1 + x_2 \leq 18$, and $x_3 + 2x_4 \leq 16$.

6.65. Minimize $f = -3x_1 - 4x_2$

Subject to $x_1 + 2x_2 \leq 10$, $x_1 + x_2 \leq 10$, and $3x_1 + 5x_2 \leq 20$.

6.66. Minimize $f = -100x_1 - 80x_2$

Subject to $5x_1 + 3x_2 \leq 15$ and $x_1 + x_2 \leq 4$.

6.67. Maximize $z = -x_1 + 2x_2$

Subject to $x_1 - 2x_2 + 2 \geq 0$ and $2x_1 - 3x_2 \leq 3$.

6.68. Minimize $f = x_1 + 3x_2 - 2x_3$

Subject to $x_1 - 2x_2 - 2x_3 \geq -2$ and $2x_1 - 3x_2 - x_3 \leq -2$.

6.69. Maximize $z = x_1 + 3x_2 - 2x_3 + x_4$

Subject to $3x_1 + 2x_2 + x_3 + x_4 \leq 20$, $2x_1 + x_2 + x_4 = 10$, and $5x_1 - 2x_2 - x_3 + 2x_4 \geq 3$.

6.70. Minimize $f = 3x_1 + 4x_2 + 5x_3 + 6x_4$

Subject to $3x_1 + 4x_2 + 5x_3 + 6x_4 \geq 20$, $2x_1 + 3x_2 + 4x_4 \leq 10$, and $5x_1 - 6x_2 - 7x_3 + 8x_4 \geq 3$. Variables x_1 and x_4 are unrestricted in sign.

6.71. Minimize $f = 13x_1 - 3x_2 - 5$

Subject to $3x_1 + 5x_2 \leq 20$, $2x_1 + x_2 \geq 10$, $5x_1 + 2x_2 \geq 3$, and $x_1 + 2x_2 \geq 3$. x_1 is unrestricted in sign.

6.72. Minimize $f = 5x_1 + 2x_2 + 3x_3 + 5x_4$

Subject to $x_1 - x_2 + 7x_3 + 3x_4 \geq 4$, $x_1 + 2x_2 + 2x_3 + x_4 = 9$, and $2x_1 + 3x_2 + x_3 - 4x_4 \leq 5$.

6.73. Minimize $f = -3x_1 + 8x_2 - 2x_3 + 4x_4$

Subject to $x_1 - 2x_2 + 4x_3 + 6x_4 \leq 0$, $x_1 - 4x_2 - x_3 + 6x_4 \leq 2$, $x_3 \leq 3$, and $x_4 \geq 3$.

6.74. Minimize $f = 3x_1 + 2x_2$

Subject to $2x_1 + 2x_2 + x_3 + x_4 = 10$ and $2x_1 - 3x_2 + 2x_3 = 10$. The variables should not be greater than 10.

6.75. Maximize $z = x_1 + x_2 + x_3$

Subject to $x_1 + 2x_2 \leq 2$, $x_2 - x_3 \leq 3$, and $x_1 + x_2 + x_3 = 3$.

6.76. Minimize $f = -2x_1 + 5x_2 + 3x_3$

Subject to $x_1 - x_2 - x_3 \leq -3$ and $2x_1 + x_2 \geq 1$.

6.77. Use the revised simplex method to solve the Hawkeye food problem 6.31.

6.78. Use the revised simplex method to solve the alloy manufacturing problem 6.32.

6.79. Use the revised simplex method to solve the concrete block manufacturing problem 6.33.

6.80. Use the revised simplex method to solve the mining company problem 6.34.

6.81. Use the revised simplex method to solve the parking spaces assignment problem 6.35.

6.82. Use the revised simplex method to solve the drug manufacturing problem 6.36.

6.83. Use the revised simplex method to solve the auto seat problem 6.37.

6.84. Use the revised simplex method to solve the stock-cutting problem 6.38.

6.85. Use the revised simplex method to solve the network problem 6.39.

Sensitivity Analysis Using the Revised Simplex Method

For the following problems, use the revised simplex method to determine Lagrange multipliers and allowable ranges for constraint right-hand sides and objective function coefficients for which the basis does not change. Using this sensitivity analysis, determine the effect on the optimum of the following changes.

(a) A 10% increase (in absolute value) of the coefficient of the first variable in the objective function.

(b) A 10% increase (in absolute value) of the constant term in the first constraint.

6.86. Maximize $z = x_1 + 3x_2$

Subject to $x_1 + 4x_2 \leq 10$ and $x_1 + 2x_2 \leq 10$.

6.87. Maximize $z = -x_1 + 2x_2$

Subject to $x_1 - 4x_2 + 2 \geq 0$ and $2x_1 - 3x_2 \leq 3$.

6.88. Minimize $f = -2x_1 + 2x_2 + x_3 - 3x_4$

Subject to $x_1 + x_2 + x_3 + x_4 \leq 18$, $x_1 - 2x_3 + 4x_4 \leq 12$, $x_1 + x_2 \leq 18$, and $x_3 + 2x_4 \leq 16$.

6.89. Minimize $f = -3x_1 - 4x_2$

Subject to $x_1 + 2x_2 \leq 10$, $x_1 + x_2 \leq 10$ and $3x_1 + 5x_2 \leq 20$.

6.90. Minimize $f = -100x_1 - 80x_2$

Subject to $5x_1 + 3x_2 \leq 15$ and $x_1 + x_2 \leq 4$.

6.91. Maximize $z = -x_1 + 2x_2$

Subject to $x_1 - 2x_2 + 2 \geq 0$ and $2x_1 - 3x_2 \leq 3$.

6.92. Minimize $f = x_1 + 3x_2 - 2x_3$

Subject to $x_1 - 2x_2 - 2x_3 \geq -2$ and $2x_1 - 3x_2 - x_3 \leq -2$.

6.93. Maximize $z = x_1 + 3x_2 - 2x_3 + x_4$

Subject to $3x_1 + 2x_2 + x_3 + x_4 \leq 20$, $2x_1 + x_2 + x_4 = 10$, and $5x_1 - 2x_2 - x_3 + 2x_4 \geq 3$.

6.94. Minimize $f = 3x_1 + 4x_2 + 5x_3 + 6x_4$

Subject to $3x_1 + 4x_2 + 5x_3 + 6x_4 \geq 20$, $2x_1 + 3x_2 + 4x_4 \leq 10$, and $5x_1 - 6x_2 - 7x_3 + 8x_4 \geq 3$. Variables x_1 and x_4 are unrestricted in sign.

6.95. Minimize $f = 13x_1 - 3x_2 - 5$

Subject to $3x_1 + 5x_2 \leq 20$, $2x_1 + x_2 \geq 10$, $5x_1 + 2x_2 \geq 3$, and $x_1 + 2x_2 \geq 3$. x_1 is unrestricted in sign.

6.96. Minimize $f = 5x_1 + 2x_2 + 3x_3 + 5x_4$

Subject to $x_1 - x_2 + 7x_3 + 3x_4 \geq 4$, $x_1 + 2x_2 + 2x_3 + x_4 = 9$, and $2x_1 + 3x_2 + x_3 - 4x_4 \leq 5$.

6.97. Minimize $f = -3x_1 + 8x_2 - 2x_3 + 4x_4$

Subject to $x_1 - 2x_2 + 4x_3 + 6x_4 \leq 0$, $x_1 - 4x_2 - x_3 + 6x_4 \leq 2$, $x_3 \leq 3$, and $x_4 \geq 3$.

6.98. Minimize $f = 3x_1 + 2x_2$

Subject to $2x_1 + 2x_2 + x_3 + x_4 = 10$ and $2x_1 - 3x_2 + 2x_3 = 10$. The variables should not be greater than 10.

6.99. Maximize $z = x_1 + x_2 + x_3$

Subject to $x_1 + 2x_2 \leq 2$, $x_2 - x_3 \leq 3$, and $x_1 + x_2 + x_3 = 3$.

6.100. Minimize $f = -2x_1 + 5x_2 + 3x_3$

Subject to $x_1 - x_2 - x_3 \leq -3$ and $2x_1 + x_2 \geq 1$.

6.101. Using sensitivity analysis based on the revised simplex method, determine the effect of following changes in the Hawkeye food problem 6.31.

(a) The required capacity for refrigerated goods increases to 160 m^3.

(b) The required capacity for nonrefrigerated goods increases to 140 m^3.

(c) Gas consumption of truck type I increases to 320 gallons.

(d) Gas consumption of truck type II decreases to 180 gallons.

6.102. Using sensitivity analysis based on the revised simplex method, determine the effect of following changes in the alloy manufacturing problem 6.32.

(a) The cost of alloy B increases to $10/lb.

(b) The tin content in the alloy produced is decreased to 48% and the zinc content is increased to 22%.

6.103. Using sensitivity analysis based on the revised simplex method, determine the effect of following changes in the concrete block manufacturing problem 6.33.

(a) The profit from block type A increases to $8 and that from block type C decreases to $28.

(b) The number of available mixing hours increases to 1,000.

(c) The number of available inspection/drying hours increases to 450.

6.104. Using sensitivity analysis based on the revised simplex method, determine the effect of following changes in the mining company problem 6.34.

(a) The operations cost at the mine B increase to $180,000 per day.

(b) The weekly demand for medium grade ore increases to 9,000 tons.

(c) The weekly demand for low-grade ore decreases to 22,000 tons.

6.105. Using sensitivity analysis based on the revised simplex method, determine the effect of following changes in the parking spaces assignment problem 6.35.

(a) The number of employees working in building B increases to 45.

(b) The number of spaces available in parking lot 3 increases to 45.

(c) The number of spaces available in parking lot 2 decreases to 90.

6.106. Using sensitivity analysis based on the revised simplex method, determine the effect of following changes in the drug manufacturing problem 6.36.

(a) The cost per batch of process C increases to $10,000.

(b) The supply of ingredient II increases to 75 tons.

6.107. Using sensitivity analysis based on the revised simplex method, determine the effect of following changes in the auto seat problem 6.37.

(a) The production costs at the Waterloo plant increase to $175 and $185, respectively, for the two seat types.

(b) In the third quarter, the demand of type II seats increases to 27,000.

(c) The shipping cost from the Detroit plant increase to $15.

CHAPTER SEVEN

Interior Point Methods

The simplex method starts from a basic feasible solution and moves along the boundary of the feasible region until an optimum is reached. At each step, the algorithm brings only one new variable into the basic set, regardless of the total number of variables. Thus, for problems with a large number of variables, the method may take many steps before terminating. In fact, relatively simple examples exist in which the simplex method visits all vertices of the feasible region before finding the optimum.

This behavior of the simplex method motivated researchers to develop another class of methods known as interior point methods for solving linear programming (LP) problems. As the name implies, in these methods, one starts from an interior feasible point and takes appropriate steps along descent directions until an optimum is found. The following figure contrasts the approach used in the simplex method with the one used in the interior point methods.

Since most interior point methods make direct use of Karush-Kuhn-Tucker optimality conditions, the first section presents a special form taken by these conditions for LP problems. This section also presents the dual of an LP problem and examines the relationship between the primal and the dual variables. The remaining sections present two relatively simple but computationally effective Interior point methods. The first method is known as the primal affine scaling method. The second method is known as the primal-dual method and

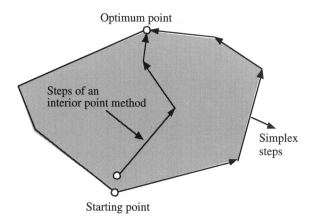

FIGURE 7.1 Typical search paths in the simplex and interior point methods.

is emerging as an effective alternative to the simplex method for large-scale problems.

7.1 Optimality Conditions for Standard LP

Consider a linear programming problem written in the standard LP form as follows:

Minimize $c^T x$

Subject to $\begin{pmatrix} Ax = b \\ x \geq 0 \end{pmatrix}$

where x is an $n \times 1$ vector of optimization variables, c is an $n \times 1$ vector containing coefficients of the objective function, and A is an $m \times n$ matrix of constraint coefficients.

7.1.1 The KT Condition for a Primal LP Problem

Following the presentation in Chapter 4, the Lagrangian for the standard LP problem is written as follows:

$$L(x, u, v, s) = c^T x + u^T(-x + s^2) + v^T(-Ax + b)$$

7.1 Optimality Conditions for Standard LP

where $\mathbf{u}$ is an $n \times 1$ vector of Lagrange multipliers associated with positivity constraints, $\mathbf{s}$ is a vector of slack variables, and $\mathbf{v}$ is an $m \times 1$ vector of Lagrange multipliers associated with equality constraints. The necessary conditions for the minimum of the Lagrangian results in the following system of equations:

$$\begin{pmatrix} \mathbf{c} - \mathbf{u} - \mathbf{A}^T \mathbf{v} = \mathbf{0} \\ \mathbf{A}\mathbf{x} - \mathbf{b} = \mathbf{0} \\ -\mathbf{x} + \mathbf{s}^2 = \mathbf{0} \\ s_i u_i = 0, i = 1, \ldots, n \\ u_i \geq 0, i = 1, \ldots, n \end{pmatrix}$$

The first two sets of equations are linear, while the other two are nonlinear. However, we can eliminate the slack variables $\mathbf{s}$ by using the complementary slackness conditions $u_i s_i = 0$, as follows.

(a) If $u_i = 0$, then $s_i \geq 0$. The third condition then implies that $-x_i \leq 0$ or $x_i \geq 0$. Thus when $u_i = 0$, $x_i \geq 0$.

(b) If $s_i = 0$, then $u_i \geq 0$. The third condition implies that $-x_i = 0$. Thus when $u_i \geq 0$, $x_i = 0$.

Thus, the variable s_i in the complementary slackness conditions can be replaced by x_i. The optimum of the LP problem can therefore be obtained by solving the following system of equations:

$$\begin{pmatrix} \mathbf{c} - \mathbf{u} - \mathbf{A}^T \mathbf{v} = \mathbf{0} \\ \mathbf{A}\mathbf{x} - \mathbf{b} = \mathbf{0} \\ x_i u_i = 0, i = 1, \ldots, n \\ x_i \geq 0, u_i \geq 0, i = 1, \ldots, n \end{pmatrix}$$

It is convenient to express the complementary slackness conditions in the matrix form as well. For this purpose, we define $n \times n$ diagonal matrices:

$$\mathbf{U} = \text{diag}[u_i] \qquad \mathbf{X} = \text{diag}[x_i]$$

Further defining an $n \times 1$ vector $\mathbf{e}$, whose entries are all equal to 1, i.e.,

$$\mathbf{e}^T = (1, 1, \ldots, 1)$$

The complementary slackness conditions are written as follows:

$$x_i u_i = 0, i = 1, \ldots, n \quad \Longrightarrow \quad \mathbf{X}\mathbf{U}\mathbf{e} = \mathbf{0}$$

7.1.2 Dual LP Problem

Following the general Lagrangian duality discussion in Chapter 4, it is possible to define an explicit dual function for an LP problem as follows:

$$M(\mathbf{u}, \mathbf{v}) = \underset{\mathbf{x}}{\text{Min}}[\mathbf{c}^T\mathbf{x} - \mathbf{u}^T\mathbf{x} + \mathbf{v}^T(-\mathbf{A}\mathbf{x} + \mathbf{b})] \quad u_i \geq 0, i = 1, \ldots, n$$

The minimum over $\mathbf{x}$ can easily be computed by differentiating with respect to $\mathbf{x}$ and setting it equal to zero. That is,

$$\underset{\mathbf{x}}{\text{Min}}[\mathbf{c}^T\mathbf{x} - \mathbf{u}^T\mathbf{x} + \mathbf{v}^T(-\mathbf{A}\mathbf{x} + \mathbf{b})] \Longrightarrow \mathbf{c} - \mathbf{u} - \mathbf{A}^T\mathbf{v} = \mathbf{0}$$

Thus, the dual LP problem can be stated as follows:

Maximize $\mathbf{c}^T\mathbf{x} - \mathbf{u}^T\mathbf{x} + \mathbf{v}^T(-\mathbf{A}\mathbf{x} + \mathbf{b})$

Subject to $\begin{pmatrix} \mathbf{c} - \mathbf{u} - \mathbf{A}^T\mathbf{v} = \mathbf{0} \\ u_i \geq 0, i = 1, \ldots, n \end{pmatrix}$

Transposing the first three terms of the dual objective function, we can group the terms as follows:

$$\mathbf{c}^T\mathbf{x} - \mathbf{u}^T\mathbf{x} + \mathbf{v}^T(-\mathbf{A}\mathbf{x} + \mathbf{b}) \quad \Longrightarrow \quad \mathbf{x}^T(\mathbf{c} - \mathbf{u} - \mathbf{A}^T\mathbf{v}) + \mathbf{v}^T\mathbf{b}$$

Using the constraint, the term in the parentheses vanishes; thus, the dual LP problem can be stated as follows:

Maximize $\mathbf{v}^T\mathbf{b}$

Subject to $u_i \geq 0, i = 1, \ldots, n$

Furthermore, using the constraint equation, the Lagrange multipliers $\mathbf{u}$ can be written in terms of $\mathbf{v}$ as follows:

$$\mathbf{c} - \mathbf{u} - \mathbf{A}^T\mathbf{v} = \mathbf{0} \quad \Longrightarrow \quad \mathbf{u} = \mathbf{c} - \mathbf{A}^T\mathbf{v}$$

Thus, the dual LP problem can be written entirely in terms of Lagrange multipliers for the equality constraints as follows:

Maximize $\mathbf{v}^T\mathbf{b}$

Subject to $\mathbf{c} - \mathbf{A}^T\mathbf{v} \geq \mathbf{0}$

Note that the variables in the dual problem are the Lagrange multipliers of the primal problem. Also, as mentioned in Chapter 4, the maximum of the dual problem is the same as the minimum of the primal problem only if feasible solutions exist for both the primal and the dual problems. The relationship does not hold if either the primal or the dual problem is infeasible.

Example 7.1 Construct the dual of the following LP problem. By solving the primal and the dual problems, demonstrate the relationship between the variables in the two formulations.

Minimize $f = 6x_1 - 9x_2$

Subject to $\begin{pmatrix} 3x_1 + 7x_2 \leq 15 \\ x_1 + x_2 \geq 3 \\ x_i \geq 0, i = 1, 2 \end{pmatrix}$

Introducing slack and surplus variables, the LP in the standard form is as follows:

```
Minimize 6x₁ - 9x₂
Subject to (3x₁ + 7x₂ + x₃ == 15)
           ( x₁ + x₂ - x₄ == 3 )
All variables ≥ 0
```

For the matrix form, we can identify the following matrices and vectors:

$$A \to \begin{pmatrix} 3 & 7 & 1 & 0 \\ 1 & 1 & 0 & -1 \end{pmatrix} \quad b \to \begin{pmatrix} 15 \\ 3 \end{pmatrix} \quad c \to \begin{pmatrix} 6 \\ -9 \\ 0 \\ 0 \end{pmatrix}$$

Substituting these into the general dual LP form, we get the following dual problem:

```
Variables → (v₁  v₂)
Maximize → 15v₁ + 3v₂
                ( 6 - 3v₁ - v₂ ≥ 0 )
                ( -9 - 7v₁ - v₂ ≥ 0 )
Subject to  →   (     -v₁ ≥ 0      )
                (      v₂ ≥ 0      )
```

Graphical solutions of the primal and the dual problems are shown in Figure 7.2. As seen from the first graph, the primal problem is feasible, and has the minimum value of -4.5 at $x_1 = 1.5$ and $x_2 = 1.5$. The graph of the dual problem shows that the dual is also feasible and has a maximum value of -4.5 at $v_1 = -3.75$ and $v_2 = 17.25$. Thus, both the primal and the dual give the same optimum value.

The solutions can be verified by solving the primal problem directly by using the KT conditions as follows.

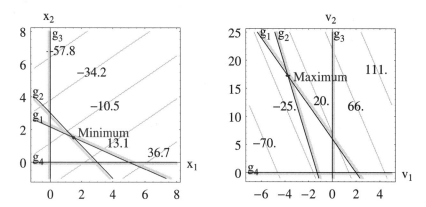

FIGURE 7.2 Graphical solutions of the primal and the dual problems.

```
KTSolution[f, {15 - 3x₁ - 7x₂ - x₃ == 0, 3 - x₁ - x₂ + x₄ == 0,
    Table[xᵢ ≥ 0, {i, 1, 4}]}, Table[xᵢ, {i, 1, 4}]];
```

***** Lagrangian → $u_1 (s_1^2 - x_1) + 6x_1 + u_2 (s_2^2 - x_2) - 9x_2 + u_3 (s_3^2 - x_3)$
$+ v_1 (15 - 3x_1 - 7x_2 - x_3) + u_4 (s_4^2 - x_4) + v_2 (3 - x_1 - x_2 + x_4)$

***** Valid KT Point(s) *****

$f \to -4.5$
$x_1 \to 1.5$
$x_2 \to 1.5$
$x_3 \to 0$
$x_4 \to 0$
$u_1 \to 0$
$u_2 \to 0$
$u_3 \to 3.75$
$u_4 \to 17.25$
$s_1^2 \to 1.5$
$s_2^2 \to 1.5$
$s_3^2 \to 0$
$s_4^2 \to 0$
$v_1 \to -3.75$
$v_2 \to 17.25$

The Lagrange multipliers of this problem correspond to the optimum solution of the dual variables. Note that there is no significance to the sign of **v** multipliers. In this example, the constraint expressions were written in such a way that we came up with the same signs. If we had written the equality constraints by multiplying both sides by a negative sign, we would have obtained same numerical values, but the signs would have been opposite.

The FormDualLP Function

A *Mathematica* function called FormDualLP has been created to automate the process of creating a dual LP.

```
Needs["OptimizationToolbox`InteriorPoint`"];
?FormDualLP
```

FormDualLP[f, g, vars, options]. Forms dual of the given LP. f is the
  objective function, g is a list of constraints, and vars is a list of
  variables. See Options[FormDualLP] to find out about a list of valid
  options for this function.

```
OptionsUsage[FormDualLP]
```
{UnrestrictedVariables → {}, ProblemType → Min,
 StandardVariableName → x, DualLPVariableName → v}

UnrestrictedVariables is an option for LP and several QP problems.
  A list of variables that are not restricted to be positive can be
  specified with this option. Default is {}.

ProblemType is an option for most optimization methods. It can either be
  Min (default) or Max.

StandardVariableName is an option for LP and QP methods. It specifies
  the symbol to use when creating variable names during conversion to
  the standard form. Default is x.

DualLPVariableName is an option for dual interior point methods. It
  defines the symbol to use when creating dual variable names (Lagrange
  multipliers for equality constraints). Default is v.

Example 7.2 Construct the dual of the following LP problem. Demonstrate graphically that the primal is feasible but the solution is unbounded. The corresponding dual problem has no solution.

$$\text{Minimize } f = -3x_1 - x_2$$
$$\text{Subject to } \begin{pmatrix} x_1 - 2x_2 \leq 5 \\ x_i \geq 0, i = 1, 2 \end{pmatrix}$$

```
f = -3x₁ - x₂;
g = {x₁ - 2x₂ ≤ 5}; vars = {x₁, x₂};
```

Using the FormDualLP, the dual problem can easily be written as follows:

```
{df, dg, dv} = FormDualLP[f, g, vars];
```

Primal problem
Minimize $-3x_1 - x_2$
Subject to $\left(x_1 - 2x_2 + x_3 == 5\right)$
All variables ≥ 0

$$A \to \begin{pmatrix} 1 & -2 & 1 \end{pmatrix} \quad b \to \begin{pmatrix} 5 \end{pmatrix} \quad c \to \begin{pmatrix} -3 \\ -1 \\ 0 \end{pmatrix}$$

Dual LP problem

Variables $\to (v_1)$

Maximize $\to 5v_1$

Subject to $\to \begin{pmatrix} -3 - v_1 \geq 0 \\ -1 + 2v_1 \geq 0 \\ -v_1 \geq 0 \end{pmatrix}$

The graphical solution of the primal problem is shown in Figure 7.3. As seen from the graph, the primal problem is feasible, but the minimum is unbounded. The dual, however, has no solution because the constraints are contradictory. The last constraint says $v_1 \leq 0$, while the second constraint says that $v_1 \geq 1/2$. This example indicates that when the primal is unbounded, the dual has no solution.

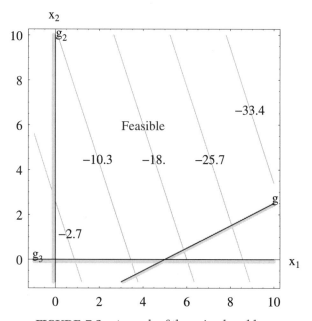

FIGURE 7.3 A graph of the primal problem.

7.1.3 Summary of Optimality Conditions for Standard LP

The previous discussion can be summarized into the following set of optimality conditions. Usual names associated with these conditions are indicated as well.

$$\mathbf{Ax} - \mathbf{b} = \mathbf{0} \quad \text{Primal feasibility}$$
$$\mathbf{A}^T\mathbf{v} + \mathbf{u} = \mathbf{c} \quad \text{Dual feasibility}$$
$$\mathbf{XUe} = \mathbf{0} \quad \text{Complementary slackness conditions}$$
$$x_i \geq 0, u_i \geq 0, i = 1, \ldots, n$$

7.2 The Primal Affine Scaling Method

The Primal Affine Scaling (PAS) method is one of the simplest interior point methods. The basic idea is to start with a point in the interior of the feasible region. Scale variables so that the point is near the center of the transformed domain. Determine a descent direction based on the maximum possible reduction in the objective function. Compute an appropriate step length along this direction so that the next point maintains feasibility of variables (i.e., none of the variables becomes negative). Repeat the process until the optimality conditions are satisfied.

The following example is used to illustrate these ideas:

Minimize $f = -5x_1 - x_2$

Subject to $\begin{pmatrix} 2x_1 + 3x_2 \leq 12 \\ 2x_1 + x_2 \leq 8 \\ x_i \geq 0, i = 1, 2 \end{pmatrix}$

Introducing slack variables x_3 and x_4, the problem is written in the standard LP form as follows:

Minimize $f = -5x_1 - x_2$

Subject to $\begin{pmatrix} 2x_1 + 3x_2 + x_3 = 12 \\ 2x_1 + x_2 + x_4 = 8 \\ x_i \geq 0, i = 1, \ldots, 4 \end{pmatrix}$

In matrix notation, the problem is written as follows:

Minimize $\mathbf{c}^T\mathbf{x}$ subject to $\mathbf{Ax} = \mathbf{b}$ and $\mathbf{x} \geq 0$

$$\mathbf{c}^T = \{-5, -1, 0, 0\} \qquad \mathbf{A} = \begin{pmatrix} 2 & 3 & 1 & 0 \\ 2 & 1 & 0 & 1 \end{pmatrix} \qquad \mathbf{b} = \begin{pmatrix} 12 \\ 8 \end{pmatrix}$$

To start the process, we need an initial point that satisfies all constraint conditions. A procedure to determine the initial interior point will be presented later. For the time being, we can use trial-and-error to determine a point that satisfies the constraint equations. For the example problem, we use the following point:

$$\mathbf{x}^0 = \{1/2, 2, 5, 5\}^T$$

Since all values are positive and $\mathbf{Ax}^0 - \mathbf{b} = \mathbf{0}$, this is a feasible solution for the problem.

7.2.1 Scaling Transformation

The first step in the PAS method is to scale the problem so that the current point is near the center of the feasible domain. A simple scaling that moves all points a unit distance away from the coordinates is to divide the variables by their current values. By defining an $n \times n$ transformation matrix $\mathbf{T}^k$ whose diagonal elements are equal to the current point, this scaling transformation is written as follows:

$$\mathbf{y}^k = (\mathbf{T}^k)^{-1}\mathbf{x} \qquad \text{or} \qquad \mathbf{x} = \mathbf{T}^k\mathbf{y}^k$$

where

$$\mathbf{T}^k = \begin{pmatrix} x_1^k & 0 & 0 & 0 \\ 0 & x_2^k & 0 & 0 \\ \vdots & \vdots & \ddots & \vdots \\ 0 & 0 & 0 & x_n^k \end{pmatrix} \qquad (\mathbf{T}^k)^{-1} = \begin{pmatrix} 1/x_1^k & 0 & 0 & 0 \\ 0 & 1/x_2^k & 0 & 0 \\ \vdots & \vdots & \ddots & \vdots \\ 0 & 0 & 0 & 1/x_n^k \end{pmatrix}$$

The superscript k refers to the iteration number. The scaled problem can then be written as follows:

Minimize $\mathbf{c}^T\mathbf{T}^k\mathbf{y}^k$

Subject to $\mathbf{AT}^k\mathbf{y}^k = \mathbf{b}$ and $\mathbf{y} \geq 0$

Introducing the notation $\mathbf{c}^k = \mathbf{T}^k\mathbf{c}$ and $\mathbf{A}^k = \mathbf{AT}^k$, the scaled LP problem is as follows:

7.2 The Primal Affine Scaling Method

Minimize $(\mathbf{c}^k)^T \mathbf{y}^k$

Subject to $\mathbf{A}^k \mathbf{y}^k = \mathbf{b}$ and $\mathbf{y} \geq 0$

For the example problem, the scaled problem is developed as follows:

$$\mathbf{T}^k = \begin{pmatrix} \frac{1}{2} & 0 & 0 & 0 \\ 0 & 2 & 0 & 0 \\ 0 & 0 & 5 & 0 \\ 0 & 0 & 0 & 5 \end{pmatrix} \quad (\mathbf{T}^k)^{-1} = \begin{pmatrix} 2 & 0 & 0 & 0 \\ 0 & \frac{1}{2} & 0 & 0 \\ 0 & 0 & \frac{1}{5} & 0 \\ 0 & 0 & 0 & \frac{1}{5} \end{pmatrix} \quad \mathbf{A}^k = \mathbf{A}\mathbf{T}^k = \begin{pmatrix} 1 & 6 & 5 & 0 \\ 1 & 2 & 0 & 5 \end{pmatrix}$$

$$\mathbf{c}^k = \mathbf{T}^k \mathbf{c} = \left\{ -\frac{5}{2}, -2, 0, 0 \right\}^T \quad \text{and} \quad \mathbf{y}^k = \left(\mathbf{T}^k\right)^{-1} \mathbf{x} = \left\{ 2x_1, \frac{x_2}{2}, \frac{x_3}{5}, \frac{x_4}{5} \right\}^T$$

The scaled problem is

Minimize $\{-\frac{5}{2}, -2, 0, 0\} \begin{pmatrix} y_1 \\ y_2 \\ y_3 \\ y_4 \end{pmatrix}$

Subject to $\begin{pmatrix} 1 & 6 & 5 & 0 \\ 1 & 2 & 0 & 5 \end{pmatrix} \begin{pmatrix} y_1 \\ y_2 \\ y_3 \\ y_4 \end{pmatrix} = \begin{pmatrix} 12 \\ 8 \end{pmatrix}$ and $y_i \geq 0, i = 1, \ldots, 4$

Since the problem has four variables, we obviously cannot directly plot a graph of this scaled problem. However, if we ignore the slack variables, we can see that the original problem is now transformed as follows:

Minimize $-\frac{5}{2} y_1 - 2 y_2$

Subject to $\begin{pmatrix} y_1 + 6y_2 \leq 12 \\ y_1 + 2y_2 \leq 8 \\ y_i \geq 0, i = 1, 2 \end{pmatrix}$

The graphs of the original and scaled problems are shown in Figure 7.4. The current point in the original space ($x_1 = 1/2$ and $x_2 = 2$) is mapped to $y_1 = 2x_1 = 1$ and $y_2 = x_2/2 = 1$. It is clear from this graph that the scaling changes the shape of the feasible region and the current point is a unit distance away from the coordinates.

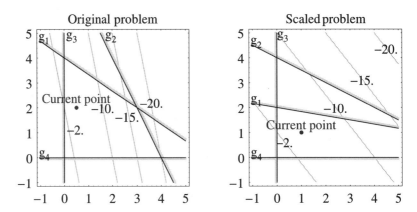

FIGURE 7.4 Graphs of the original and the scaled problems.

7.2.2 Direction of Descent

After scaling, the next step is to determine a descent direction. A suitable direction is the one along which the objective function reduces rapidly. At the same time, it must be possible to move in this direction by a reasonable amount (step length) and still remain feasible. The first goal is achieved by moving in the negative gradient direction of the objective function, since this represents the steepest descent direction. The feasibility will be maintained if we move in the null space of matrix $\mathbf{A}^k$. (See the appendix of this chapter for a review of concepts of null space and range space from linear algebra.) Thus, a feasible descent direction is the projection of the negative gradient of the objective function on the null space.

$$\mathbf{d}^k = \mathbf{P}^k(-\mathbf{c}^k)$$

where $\mathbf{c}^k$ is the gradient of the objective function, and $\mathbf{P}^k$ is an $n \times n$ projection matrix. The following formula for this projection matrix is derived in the appendix.

$$\mathbf{P}^k = \mathbf{I} - \mathbf{A}^{kT}(\mathbf{A}^k \mathbf{A}^{kT})^{-1} \mathbf{A}^k$$

where $\mathbf{I}$ is an $n \times n$ identity matrix. Thus,

$$\mathbf{d}^k = 1\left[\mathbf{I} - \mathbf{A}^{kT}(\mathbf{A}^k \mathbf{A}^{kT})^{-1} \mathbf{A}^k\right] = -\left[\mathbf{c}^k - \mathbf{A}^{kT}(\mathbf{A}^k (\mathbf{A}^k \mathbf{A}^{kT})^{-1} \mathbf{A}^k \mathbf{c}^k\right]$$

7.2 The Primal Affine Scaling Method

Defining $\mathbf{w}^k = (\mathbf{A}^k \mathbf{A}^{kT})^{-1}\mathbf{A}^k \mathbf{c}^k$, we can express the direction as

$$\mathbf{d}^k = -[\mathbf{c}^k - \mathbf{A}^{kT} \mathbf{w}^k]$$

Substituting the scaling transformation $\mathbf{c}^k = \mathbf{T}^k \mathbf{c}$ and $\mathbf{A}^k = \mathbf{A}\mathbf{T}^k$, we can express $\mathbf{w}^k$ and $\mathbf{d}^k$ in terms of the original problem variables as follows:

$$\mathbf{w}^k = (\mathbf{A}\,\mathbf{T}^k\,\mathbf{T}^k\,\mathbf{A}^T)^{-1}\mathbf{A}\,\mathbf{T}^k\,\mathbf{T}^k\,\mathbf{c}$$

$$\mathbf{d}^k = -[\mathbf{T}^k \mathbf{c} - \mathbf{T}^k\,\mathbf{A}^T\,\mathbf{w}^k] = -\mathbf{T}^k[\mathbf{c} - \mathbf{A}^T \mathbf{w}^k] \equiv -\mathbf{T}^k\,\mathbf{r}^k$$

where $\mathbf{r}^k = \mathbf{c} - \mathbf{A}^T \mathbf{w}^k$, $\mathbf{A}$ is the original constraint coefficient matrix, and $\mathbf{c}$ is the cost coefficient vector. The definition of vector $\mathbf{w}^k$ indicates matrix inversion. In actual implementations, however, it is more efficient to compute the vector $\mathbf{w}^k$ as follows.

$$\mathbf{w}^k = (\mathbf{A}^k \mathbf{A}^{kT})^{-1}\mathbf{A}^k \mathbf{c}^k \quad \text{or} \quad (\mathbf{A}^k \mathbf{A}^{kT})\mathbf{w}^k = \mathbf{A}^k \mathbf{c}^k$$

Thus, $\mathbf{w}^k$ can be obtained by solving the following linear system of equations:

$$(\mathbf{A}\,\mathbf{T}^k\,\mathbf{T}^k\,\mathbf{A}^T)\mathbf{w}^k = \mathbf{A}\,\mathbf{T}^k\,\mathbf{T}^k\,\mathbf{c}$$

For the example problem, the direction is computed as follows:

$$\mathbf{A}^k = \begin{pmatrix} 1 & 6 & 5 & 0 \\ 1 & 2 & 0 & 5 \end{pmatrix} \quad \mathbf{c}^k = \left\{-\frac{5}{2}, -2, 0, 0\right\}^T$$

$$\mathbf{A}^k \mathbf{A}^{kT} = \begin{pmatrix} 62 & 13 \\ 13 & 30 \end{pmatrix} \quad \mathbf{A}^k \mathbf{c}^k = \begin{pmatrix} -\frac{29}{2} \\ -\frac{13}{2} \end{pmatrix}$$

Solving the system of equations $(\mathbf{A}^k \mathbf{A}^{kT})\mathbf{w}^k = \mathbf{A}^k \mathbf{c}^k$, we get

$$\mathbf{w}^k = \left\{-\frac{701}{3{,}382}, -\frac{429}{3{,}382}\right\}^T$$

The descent direction is then as follows:

$$\mathbf{d}^k = -[\mathbf{c}^k - \mathbf{A}^{kT} \mathbf{w}^k] = -\begin{pmatrix} -5/2 \\ -2 \\ 0 \\ 0 \end{pmatrix} + \begin{pmatrix} 1 & 1 \\ 6 & 2 \\ 5 & 0 \\ 0 & 5 \end{pmatrix} \begin{pmatrix} -\frac{701}{3{,}382} \\ -\frac{429}{3{,}382} \end{pmatrix} = \begin{pmatrix} 2.16587 \\ 0.5026611 \\ -1.03636 \\ -0.6342400 \end{pmatrix}$$

7.2.3 Step Length and the Next Point

After determining the feasible descent direction, the next step is to determine the largest possible step in this direction. The equality constraints have already been accounted for during the direction computations. The restriction that the variables remain positive is taken into account in determining the step length. Thus, we need to determine α such that the following condition is satisfied:

$$\mathbf{y}^{k+1} = \mathbf{y}^k + \alpha \mathbf{d}^k \geq \mathbf{0}$$

Since at the current point, the scaled variables y_i are all equal to 1, the step length is restricted as follows:

$$\alpha d_i^k \geq -1, \ i = 1, \ldots, n$$

If all elements of the direction vector are positive, then obviously, we can take any step in that direction without becoming infeasible. This situation means that the problem is unbounded and there is no need to proceed any further. Thus, the actual step length is determined by the negative entries in the $\mathbf{d}^k$ vector and therefore, step length is given by

$$\alpha \leq \{-1/d_i^k, d_i^k < 0, i = 1, \ldots, n\}$$

This step length will make at least one of the variables go to zero. In order to stay inside the feasible region, the actual step taken should be slightly smaller than this maximum. Thus, α is set to $\beta \alpha_{\max}$ with $0 < \beta < 1$. Usually, $\beta = 0.99$ is chosen in order to go as far as possible but without actually being on the boundary of a constraint.

For the example problem, using the descent direction computed in the previous step, the step length is computed as follows:

$$1/\mathbf{d}^k = \{0.461706, 1.98941, -0.964907, -1.57669\}^T$$

$$\alpha_{\max} = \text{Min}[-1/d_i^k, d_i^k < 0, i = 1, \ldots, n] = \min\{0.964907, 1.57669\}^T$$

$$= 0.964907$$

With $\beta = 0.99$, the step length is

$$\alpha = 0.99 \times 0.964907 = 0.955258$$

With this step length, in terms of scaled variables, the next point is as follows:

$$\mathbf{y}^{k+1} = \mathbf{y}^k + \alpha \mathbf{d}^k$$

7.2 The Primal Affine Scaling Method

Using the scaling transformation, we can express the next point in terms of actual problem variables as follows:

$$\mathbf{x}^{k+1} = \mathbf{T}^k(\mathbf{y}^k + \alpha \mathbf{d}^k) = \mathbf{x}^k + \alpha \mathbf{T}^k \mathbf{d}^k$$

For the example problem, the next point therefore is as follows:

$$\mathbf{x}^{(1)} = \begin{pmatrix} 1/2 \\ 2 \\ 5 \\ 5 \end{pmatrix} + 0.955258 \begin{pmatrix} \frac{1}{2} & 0 & 0 & 0 \\ 0 & 2 & 0 & 0 \\ 0 & 0 & 5 & 0 \\ 0 & 0 & 0 & 5 \end{pmatrix} \begin{pmatrix} 2.16588 \\ 0.502661 \\ -1.03637 \\ -0.63424 \end{pmatrix} = \begin{pmatrix} 1.53449 \\ 2.96034 \\ 0.05 \\ 1.97068 \end{pmatrix}$$

7.2.4 Convergence Criteria

Starting with this new point, the previous series of steps is repeated until an optimum is found. Theoretically, the optimum is reached when $\mathbf{d}^k = \mathbf{0}$. Thus, we can define the first convergence criteria as follows:

$$\sigma_1 \equiv \text{Norm}[\mathbf{d}^k] \leq \epsilon_1$$

where ϵ_1 is a small positive number.

In addition to this, because of the presence of round-off errors, the numerical implementations of the algorithm also check the following conditions derived from the KT optimality conditions.

Feasibility

The constraints must be satisfied at the optimum, i.e., $\mathbf{A}\mathbf{x}^k - \mathbf{b} = \mathbf{0}$. To use as convergence criteria, this requirement is expressed in a normalized form, as follows:

$$\sigma_2 \equiv \frac{||\mathbf{A}\mathbf{x}^k - \mathbf{b}||}{||\mathbf{b}|| + 1} \leq \epsilon_2$$

where ϵ_2 is a small positive number. The 1 is added to the denominator to avoid division by small numbers.

Reduced Cost Coefficients

From the revised simplex method, we know that the vector $\mathbf{r}^k \equiv \mathbf{c} - \mathbf{A}^T \mathbf{w}^k$ represents the reduced cost coefficients appearing in the last row of the simplex tableau. Therefore, the nonnegativity of the reduced cost coefficients gives the following convergence criteria:

$$\sigma_3 \equiv \frac{||\mathbf{r}^k||}{||\mathbf{c}||+1} \leq \epsilon_3$$

where ϵ_3 is a small positive number. Again, a 1 is added to the denominator to avoid division by a small number.

Primal-Dual Solutions

The vector $\mathbf{w}^k$ is related to the Lagrange multipliers that are the variables in the dual LP problem. Thus, another convergence criterion is the difference between the primal solution and the dual solution, as follows.

$$\sigma_4 \equiv \text{Abs}[\mathbf{c}^k \mathbf{x}^k - \mathbf{b}^T \mathbf{w}^k] \leq \epsilon_4$$

where ϵ_4 is a small positive number.

7.2.5 Finding the Initial Interior Point

In order to use the Primal Affine Scaling algorithm, we need to start from an interior feasible point. To achieve this, similar to the simplex method, we define a Phase I problem during which the goal is to find an initial interior point. Choose an arbitrary starting point $\mathbf{x}^0 > \mathbf{0}$, say $\mathbf{x}^0 = \{1, 1, \ldots, 1\}$. Then from the constraint equations, we have

$$\mathbf{z}^0 \equiv \mathbf{b} - \mathbf{A}\mathbf{x}^0$$

If $\mathbf{z}^0 = \mathbf{0}$, we have a starting interior point. If not, we introduce an artificial variable and define a Phase I LP problem as follows:

Minimize a

Subject to $\begin{pmatrix} \mathbf{A}\mathbf{x} + a\mathbf{z} = \mathbf{b} \\ \mathbf{x} \geq 0 \\ a \geq 0 \end{pmatrix}$

The minimum of this problem is reached when $a = 0$, and at that point, $\mathbf{A}\mathbf{x}^k = \mathbf{b}$, which makes $\mathbf{x}^k$ an interior point for the original problem. Furthermore, if we set $a = 1$, then any arbitrary $\mathbf{x}^0$ becomes a starting interior point for the

Phase I problem. Thus, we apply the PAS algorithm to the Phase I problem until $a = 0$ and then switch over to the actual problem for Phase II.

Note that in the PAS algorithm, Phase I needs only one artificial variable. In contrast, recall from Chapter 6 that Phase I of the simplex method needs as many artificial variables as the number of equality or greater than type constraints.

7.2.6 Finding the Exact Optimum Solution

The exact optimum of an LP problem lies on the constraint boundary. However, by design, all points generated by an interior point algorithm are always slightly inside the feasible domain. Therefore, even after convergence, we only have an approximate optimum solution. An exact optimum solution can be obtained if we can identify the basic variables for the vertex that is closest to the interior optimum point. The optimum is then computed by setting the nonbasic variables to zero and solving for the basic variables from the constraint equations. This is known as the *purification* procedure.

It can be shown that the magnitude of diagonal elements of matrix

$$\mathbf{P} = \mathbf{T}\mathbf{A}^T(\mathbf{A}\mathbf{T}^2\mathbf{A}^T)^{-1}\mathbf{A}\mathbf{T}$$

serve as an indicator of the basic variables. Using this matrix, a procedure for identifying basic variables from the approximate interior point optimum is as follows:

1. Define a diagonal scaling matrix $\mathbf{T}$ with the diagonal elements set to the converged optimum solution from the PAS method.

2. Compute matrix $\mathbf{P} = \mathbf{T}\mathbf{A}^T(\mathbf{A}\mathbf{T}^2\mathbf{A}\mathbf{T})^{-1}\mathbf{A}\mathbf{T}$. Set vector $\mathbf{p}$ to diagonal elements of this matrix.

3. Elements of vector $\mathbf{p}$ that are close to 1 correspond to basic variables and those that are close to zero define nonbasic variables. In case the elements of vector $\mathbf{p}$ do not give clear indication, repeat calculations of step 2 by defining a new diagonal scaling matrix $\mathbf{T}$ with diagonal elements set to vector $\mathbf{p}$. The procedure is stopped when the correct number of basic variables has been identified. A solution of constraint equations in terms of these basic variables then gives the exact vertex optimum.

7.2.7 The Complete PAS Algorithm

The complete Primal Affine Scaling (PAS) algorithm can be summarized in the following steps.

Phase II—Known Interior Starting Point

Given: Constraint coefficient matrix $\mathbf{A}$, constraint right-hand side vector $\mathbf{b}$, objective function coefficient vector $\mathbf{c}$, current interior point $\mathbf{x}^k$, step-length parameter β, and convergence tolerance parameters.

The next point $\mathbf{x}^{k+1}$ is computed as follows:

1. Form scaling matrix.

$$\mathbf{T}^k = \begin{pmatrix} x_1^k & 0 & 0 & 0 \\ 0 & x_2^k & 0 & 0 \\ \vdots & \vdots & \ddots & \vdots \\ 0 & 0 & 0 & x_n^k \end{pmatrix}$$

2. Solve the system of linear equations for $\mathbf{w}^k$.

$$(\mathbf{A}\mathbf{T}^k\mathbf{T}^k\mathbf{A}^T)\mathbf{w}^k = \mathbf{A}\mathbf{T}^k\mathbf{T}^k\mathbf{c}$$

3. Compute $\mathbf{r}^k = \mathbf{c} - \mathbf{A}^T\mathbf{w}^k$.
4. Check for convergence.

    If $\left[\text{Norm}[\mathbf{d}^k] \leq \epsilon_1, \frac{\|\mathbf{A}\mathbf{x}^k - \mathbf{b}\|}{\|\mathbf{b}\|+1} \leq \epsilon_2, \frac{\|\mathbf{r}^k\|}{\|\mathbf{c}\|+1} \leq \epsilon_3, \text{Abs}[\mathbf{c}^T\mathbf{x}^k - \mathbf{b}^T\mathbf{w}^k] \leq \epsilon_4\right]$, we have the optimum. Go to the purification phase. Otherwise, continue.

5. Compute direction $\mathbf{d}^k = -\mathbf{T}^k\mathbf{r}^k$.
6. Compute step length $\alpha = \beta \, \text{Min}[-1/d_i^k, d_i^k < 0, i = 1, \ldots, n]$.
7. Compute the next point $\mathbf{x}^{k+1} = \mathbf{x}^k + \alpha\mathbf{T}^k\mathbf{d}^k$.

Phase I—To Find Initial Interior Point

With the addition of an artificial variable, the length of the solution vector in this phase is $n + 1$. Initially, all entries in the solution vector are arbitrarily set to 1. At each iteration, the artificial objective function vector is defined as $\mathbf{c} = \{0, 0, \ldots, 0, x_{n+1}^k\}$, a column $\mathbf{z}^k = \mathbf{A}\mathbf{x}^k - \mathbf{b}$ is appended to the matrix $\mathbf{A}$, and the steps of the Phase II algorithm are repeated until $\frac{\|\mathbf{A}\mathbf{x}^k - \mathbf{b}\|}{\|\mathbf{b}\|+1} \leq \text{tol}$.

Purification Phase

After obtaining the interior point optimum from Phase II, use the following steps to get an exact vertex solution:

1. Set diagonal scaling matrix **T** to the elements of the converged interior optimum solution.
2. Compute matrix $\mathbf{P} \equiv T A^T (\mathbf{A}\,\mathbf{T}^2\,\mathbf{A}^T)^{-1} \mathbf{A}\,\mathbf{T}$. Set vector **p** to diagonal elements of this matrix.
3. Elements of vector **p** that are close to 1 correspond to basic variables, and those that are close to zero define nonbasic variables. In case the elements of vector **p** do not give clear indication, repeat calculations of step 2 by defining a new scaling matrix **T** with diagonal elements set to vector **p**.
4. Set the nonbasic variables to zero and solve the constraint equations $\mathbf{Ax} = \mathbf{b}$ for the optimum values of basic variables.

The PrimalAffineLP Function

The following PrimalAffineLP function implements the Primal Affine algorithm for solving LP problems. The function usage and its options are explained first. The function is intended to be used for educational purposes. Several intermediate results can be printed to gain understanding of the process.

```
Needs["OptimizationToolbox`InteriorPoint`"];
?PrimalAffineLP
```

PrimalAffineLP[f, g, vars, options]. Solves an LP problem using
  Primal Affine algorithm. f is the objective function, g is a list of
  constraints, and vars is a list of variables. See
  Options[PrimalAffineLP] to find out about a list of valid options for
  this function.

```
OptionsUsage[PrimalAffineLP]
```
{UnrestrictedVariables → {}, MaxIterations → 20, ProblemType → Min,
  StandardVariableName → x, PrintLevel → 1, StepLengthFactor → 0.99,
  ConvergenceTolerance → {0.001, 0.2, 2, 0.5}, StartingVector → {}}

UnrestrictedVariables is an option for LP and several QP problems.
  A list of variables that are not restricted to be positive can be
  specified with this option. Default is {}.

MaxIterations is an option for several optimization methods. It
  specifies maximum number of iterations allowed.

ProblemType is an option for most optimization methods. It can either be
  Min (default) or Max.

StandardVariableName is an option for LP and QP methods. It specifies the symbol to use when creating variable names during conversion to the standard form. Default is x.

PrintLevel is an option for most functions in the OptimizationToolbox. It is specified as an integer. The value of the integer indicates how much intermediate information is to be printed. A PrintLevel→0 suppresses all printing. Default for most functions is set to 1 in which case they print only the initial problem setup. Higher integers print more intermediate results.

StepLengthFactor is an option for interior point methods. It is the reduction factor applied to the computed step length to maintain feasibility. Default is 0.99

ConvergenceTolerance is an option for most optimization methods. Most methods require only a single zero tolerance value. Some interior point methods require a list of convergence tolerance values.

StartingVector is an option for several interior point methods. Default is $\{1,\ldots,1\}$.

Example 7.3 The complete solution of the example problem used in the previous section is obtained here using the PrimalAffineLP function.

```
f = -5x1 - x2;
g = {2x1 + 3x2 ≤ 12, 2x1 + x2 ≤ 8};
vars = {x1, x2};
```

All calculations for the first two iterations are as follows:

PrimalAffineLP[f, g, vars, PrintLevel → 2,
  StartingVector → {0.5, 2., 5., 5.}, MaxIterations → 2];

Minimize $-5x_1 - x_2$

Subject to $\begin{pmatrix} 2x_1 + 3x_2 + x_3 == 12 \\ 2x_1 + x_2 + x_4 == 8 \end{pmatrix}$

All variables ≥ 0

Problem variables redefined as: $\{x1 \to x_1, x2 \to x_2\}$

$b \to \begin{pmatrix} -5 \\ -1 \\ 0 \\ 0 \end{pmatrix} \quad c \to \begin{pmatrix} 12 \\ 8 \end{pmatrix}$

$A \to \begin{pmatrix} 2 & 3 & 1 & 0 \\ 2 & 1 & 0 & 1 \end{pmatrix}$

Starting point → {0.5, 2., 5., 5.}
  Objective function → -4.5 Status → NonOptimum

***** **Iteration 1 (Phase 2)** *****
Tk[diagonal] → {0.5, 2., 5., 5.}

$$A.Tk.Tk.AT \to \begin{pmatrix} 62. & 13. \\ 13. & 30. \end{pmatrix}$$

$$A.Tk.Tk.c \to \begin{pmatrix} -14.5 \\ -6.5 \end{pmatrix} \quad w \to \begin{pmatrix} -0.207274 \\ -0.126848 \end{pmatrix}$$

$$r \to \begin{pmatrix} -4.33176 \\ -0.251331 \\ 0.207274 \\ 0.126848 \end{pmatrix} \quad d \to \begin{pmatrix} 2.16588 \\ 0.502661 \\ -1.03637 \\ -0.63424 \end{pmatrix}$$

Convergence parameters $\to \{2.53378, 0., 0.712547, 0.99793\}$ β (-1/d) $\to$
$\{-0.457089, -1.96952, 0.955258, 1.56092\}$ Step length, $\alpha \to 0.955258$
New point $\to \{1.53449, 2.96034, 0.05, 1.97068\}$ Objective function $\to$
-10.6328 Status $\to$ NonOptimum

***** Iteration 2 (Phase 2) *****

Tk[diagonal] $\to \{1.53449, 2.96034, 0.05, 1.97068\}$

$$A.Tk.Tk.AT \to \begin{pmatrix} 88.2937 & 35.7095 \\ 35.7095 & 22.0658 \end{pmatrix}$$

$$A.Tk.Tk.c \to \begin{pmatrix} -49.8374 \\ -32.3101 \end{pmatrix} \quad w \to \begin{pmatrix} 0.0803362 \\ -1.59427 \end{pmatrix}$$

$$r \to \begin{pmatrix} -1.97213 \\ 0.353262 \\ -0.0803362 \\ 1.59427 \end{pmatrix} \quad d \to \begin{pmatrix} 3.02621 \\ -1.04578 \\ 0.00401681 \\ -3.1418 \end{pmatrix}$$

Convergence parameters $\to \{4.48582, 0., 0.420017, 1.15735\}$ β (-1/d) $\to$
$\{-0.327142, 0.946666, -246.464, 0.315106\}$ Step length, $\alpha \to 0.315106$
New point $\to \{2.99774, 1.98482, 0.0500633, 0.0197068\}$ Objective function $\to$
-16.9735 Status $\to$ NonOptimum

***** NonOptimum solution after 2 iterations *****

Interior solution $\to \{x1 \to 2.99774, x2 \to 1.98482\}$ Objective function $\to$
-16.9735
Convergence parameters $\to \{4.48582, 0., 0.420017, 1.15735\}$

The function is allowed to run until convergence.

```
{sol, history} = PrimalAffineLP[f, g, vars,
    StartingVector → {0.5, 2., 5., 5.}, MaxIterations → 20];
```

***** Optimum solution after 6 iterations *****

Interior solution $\to \{x1 \to 3.99986, x2 \to 0.000196054\}$ Objective function $\to$
-19.9995
Convergence parameters $\to$
$\{0.000373245, 9.06322 \times 10^{-13}, 0.478024, 0.000523892\}$

Nearest vertex solution $\to \{x1 \to 4, x2 \to 0\}$ Objective function $\to -20$ Status $\to$
Feasible

The history of points computed by the PAS method is extracted. Using GraphicalSolution, this search path is shown on the graph.

```
xhist = Transpose[Transpose[history][[{1, 2}]]]; TableForm[xhist]
```

0.5	2.
1.53449	2.96034
2.99774	1.98482
3.01963	1.96054
3.99015	0.0196054
3.99986	0.000196054
3.99986	0.000196054

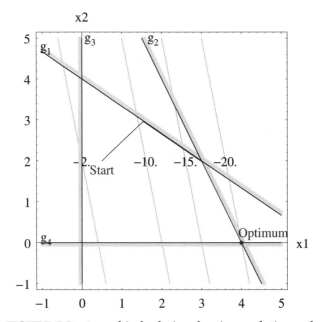

FIGURE 7.5 A graphical solution showing a solution path.

Example 7.4 Solve the following LP problem using the Primal Affine Scaling method.

```
f = x1 + x2 + x3 + x4;
g = {x1 + 2x2 - x3 + 3x4 ≤ 12,
     x1 + 3x2 + x3 + 2x4 ≤ 8,
     2x1 - 3x2 - x3 + 2x4 ≤ 7};
vars = {x1, x2, x3, x4};
```

7.2 The Primal Affine Scaling Method

Intermediate calculations are shown for the first two iterations. The Phase I procedure is started with an arbitrary starting point.

```
PrimalAffineLP[f, g, vars, ProblemType → Max, PrintLevel → 2,
   MaxIterations → 2];
```
Minimize $-x_1 - x_2 - x_3 - x_4$

Subject to $\begin{pmatrix} x_1 + 2x_2 - x_3 + 3x_4 + x_5 == 12 \\ x_1 + 3x_2 + x_3 + 2x_4 + x_6 == 8 \\ 2x_1 - 3x_2 - x_3 + 2x_4 + x_7 == 7 \end{pmatrix}$

All variables ≥ 0

Problem variables redefined as: $\{x1 \to x_1, x2 \to x_2, x3 \to x_3, x4 \to x_4\}$

$b \to \begin{pmatrix} -1 \\ -1 \\ -1 \\ -1 \\ 0 \\ 0 \\ 0 \end{pmatrix} \quad c \to \begin{pmatrix} 12 \\ 8 \\ 7 \end{pmatrix}$

$A \to \begin{pmatrix} 1 & 2 & -1 & 3 & 1 & 0 & 0 \\ 1 & 3 & 1 & 2 & 0 & 1 & 0 \\ 2 & -3 & -1 & 2 & 0 & 0 & 1 \end{pmatrix}$

Starting point → {1., 1., 1., 1., 1., 1., 1.} Objective function → −4. Status → NonOptimum

***** Iteration 1 (Phase 1) *****

Tk[diagonal] → {1., 1., 1., 1., 1., 1., 1., 1}

$A.Tk.Tk.A^T \to \begin{pmatrix} 52. & 12. & 39. \\ 12. & 16. & -4. \\ 39. & -4. & 55. \end{pmatrix}$

$A.Tk.Tk.c \to \begin{pmatrix} 6. \\ 0. \\ 6. \end{pmatrix} \quad w \to \begin{pmatrix} 0.129032 \\ -0.094086 \\ 0.0107527 \end{pmatrix}$

$r \to \begin{pmatrix} -0.0564516 \\ 0.0564516 \\ 0.233871 \\ -0.22043 \\ -0.129032 \\ 0.094086 \\ -0.0107527 \\ 0.16129 \end{pmatrix} \quad d \to \begin{pmatrix} 0.0564516 \\ -0.0564516 \\ -0.233871 \\ 0.22043 \\ 0.129032 \\ -0.094086 \\ 0.0107527 \\ -0.16129 \end{pmatrix}$

Convergence parameters → {0.40161, 0., 0.200805, 0.129032}
β (−1/d) → {−17.5371, 17.5371, 4.2331, −4.49122,
            −7.6725, 10.5223, −92.07, 6.138}
Step length, α → 4.2331

New point → {1.23897, 0.761034, 0.01,
             1.9331, 1.54621, 0.601724, 1.04552}
Artificial objective function → 0.317241

Objective function → -3.9431
Status → NonOptimum

***** Iteration 2 (Phase 1) *****

Tk[diagonal] → {1.23897, 0.761034, 0.01, 1.9331,
              1.54621, 0.601724, 1.04552, 0.317241}

$$A.Tk.Tk.AT \to \begin{pmatrix} 40.2392 & 27.4313 & 22.3811 \\ 27.4313 & 22.0573 & 12.805 \\ 22.3811 & 12.805 & 27.7581 \end{pmatrix}$$

$$A.Tk.Tk.c \to \begin{pmatrix} 0.060773 \\ 2.83576 \times 10^{-17} \\ 0.060773 \end{pmatrix} \quad w \to \begin{pmatrix} 0.0101794 \\ -0.0125181 \\ -0.000243507 \end{pmatrix}$$

$$r \to \begin{pmatrix} 0.00282574 \\ 0.0164651 \\ 0.0224541 \\ -0.00501496 \\ -0.0101794 \\ 0.0125181 \\ 0.000243507 \\ 0.298329 \end{pmatrix} \quad d \to \begin{pmatrix} -0.00350099 \\ -0.0125305 \\ -0.000224541 \\ 0.00969444 \\ 0.0157395 \\ -0.00753247 \\ -0.000254591 \\ -0.0946423 \end{pmatrix}$$

Convergence parameters → {0.0975961, 0.107914, 0.227836, 0.0803387}
$\beta(-1/d) \to$ {282.777, 79.0073, 4409., -102.12,
              -62.8991, 131.431, 3888.59, 10.4604}
Step length, $\alpha \to 10.4604$
New point → {1.19359, 0.661282, 0.00997651,
              2.12914, 1.80078, 0.554312, 1.04273}
Artificial objective function → 0.00317241
Objective function → -3.99399
Status → NonOptimum

***** NonOptimum solution after 2 iterations *****

Interior solution → {x1 → 1.19359, x2 → 0.661282,
              x3 → 0.00997651, x4 → 2.12914}
Objective function → 3.99399
Convergence parameters → {0.09759608, 0.107914, 0.227836, 0.0803387}

Calculations are allowed to run until the optimum is found.

{sol, history} = PrimalAffineLP[f, g, vars, ProblemType → Max];

***** Optimum solution after 11 iterations *****

Interior solution → {x1 → 4.5641, x2 → 0.000104029,
              x3 → 3.43536, x4 → 0.000102124}
Objective function → 7.99966
Convergence parameters → {0.000233334, 0.108415, 0.816497, 0.000337118}

***** Finding nearest vertex solution *****

Nearest vertex solution → {x1 → 5, x2 → 0, x3 → 3, x4 → 0} Objective function →
8 Status → Feasible

7.2 The Primal Affine Scaling Method

The history of points computed is as follows:

```
xhist = Transpose[Transpose[history][[{1, 2, 3, 4}]]]; TableForm[xhist]
```

1.	1.	1.	1.
1.2389	0.761034	0.01	1.9331
1.19359	0.661282	0.00997651	2.12914
1.19359	0.661282	0.00997651	2.12914
2.85016	0.926912	0.0102189	1.17667
4.43891	1.1739	0.0103702	0.0117667
4.54767	1.1381	0.0113538	0.0107107
4.56279	1.13621	0.025652	0.000107107
4.56409	1.04029	0.314812	0.000102906
4.5641	0.0104029	3.40446	0.000102625
4.5641	0.000104029	3.43536	0.000102124
4.5641	0.000104029	3.43536	0.000102124

Example 7.5 *Unbounded solution*

```
f = x1 - 2x2;
g = {2x1 - x2 ≥ 0, -2x1 + 3x2 ≤ 6}; vars = {x1, x2};

{sol, history} = PrimalAffineLP[f, g, vars, MaxIterations → 10];
```

Minimize $x_1 - 2x_2$

Subject to $\begin{pmatrix} 2x_1 - x_2 - x_3 == 0 \\ -2x_1 + 3x_2 + x_4 == 6 \end{pmatrix}$

All variables ≥ 0

Problem variables redefined as: $\{x1 \to x_1, x2 \to x_2\}$

***** Unbounded solution after 9 iterations *****

Interior solution $\to \{x1 \to 1.89411 \times 10^{68}, x2 \to 1.26274 \times 10^{68}\}$
Objective function $\to -6.3137 \times 10^{67}$
Convergence parameters $\to \{3.64522 \times 10^{67}, 1.41745 \times 10^{53},$
$\qquad\qquad\qquad\qquad\quad 0.2085, 6.3137 \times 10^{67}\}$

The history of points computed by the PAS method is extracted, and the search path is shown on the graph.

```
xhist = Transpose[Transpose[history][[{1, 2}]]]; TableForm[xhist]
```

1.	1.
1.22629	2.24457
1.22629	2.24457
1.64044	3.0747
2.2659	3.49721
1412.64	943.749
2.04709×10^{11}	1.36472×10^{11}
4.2976×10^{29}	2.86507×10^{29}
1.89411×10^{68}	1.26274×10^{68}
1.89411×10^{68}	1.26274×10^{68}

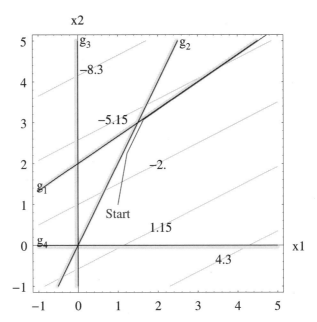

FIGURE 7.6 A graphical solution showing a solution path.

Example 7.6 *No feasible solution*   Consider the following LP problem:

```
f = 4x1 + x2 + x3 + 3x4;
g = {2x1 + x2 + 3x3 + x4 ≥ 12,
     3x1 + 2x2 + 4x3 == 5,
     2x1 - x2 + 2x3 + 3x4 == 8,
     3x1 + 4x2 + 3x3 + x4 ≥ 16};
vars = {x1, x2, x3, x4};
PrimalAffineLP[f, g, vars, ProblemType → Max];
```

Minimize $-4x_1 - x_2 - x_3 - 3x_4$

Subject to $\begin{pmatrix} 2x_1 + x_2 + 3x_3 + x_4 - x_5 == 12 \\ 3x_1 + 2x_2 + 4x_3 == 5 \\ 2x_1 - x_2 + 2x_3 + 3x_4 == 8 \\ 3x_1 + 4x_2 + 3x_3 + x_4 - x_6 == 16 \end{pmatrix}$

All variables ≥ 0

Problem variables redefined as: $\{x1 \to x_1, x2 \to x_2, x3 \to x_3, x4 \to x_4\}$

Phase I ended with no feasible solution
New point →
          {0.00459891, 3.98552, 0.00436602, 3.48881, 0.0000959518, 7.96127}
Convergence parameters → {0.0004121, 0.0844611, 0.0875652, 0.197076}
Constraint values norm → 0.311477
$Aborted

7.2 The Primal Affine Scaling Method

The problem has no feasible solution. The built-in ConstrainedMin function returns the same conclusion.

`ConstrainedMin[-f, g, {x1, x2, x3, x4}]`

ConstrainedMin::nsat: The specified constraints cannot be satisfied.
ConstrainedMin[
  -4x1 - x2 - x3 - 3x4,
  {2x1 + x2 + 3x3 + x4 ≥ 12, 3x1 + 2x2 + 4x3 == 5,
  2x1 - x2 + 2x3 + 3x4 == 8, 3x1 + 4x2 + 3x3 + x4 ≥ 16}, {x1, x2, x3, x4}]

Example 7.7 *The method gets trapped at a nonoptimum point* This example demonstrates that, in some situations, the PAS method gets trapped at a nonoptimum point.

```
f = -x1 - x2 + 2x4;
g = {3x1 + x2 + 3x3 + 2x4 == 10, x1 - 3x2 + 2x3 ≤ 7, x1 + 2x2 + 3x3 + x4 ≥ 4};
vars = {x1, x2, x3, x4};
```

`PrimalAffineLP[f, g, vars, MaxIterations → 20];`

Minimize $-x_1 - x_2 + 2x_4$

Subject to $\begin{pmatrix} 3x_1 + x_2 + 3x_3 + 2x_4 == 10 \\ x_1 - 3x_2 + 2x_3 + x_5 == 7 \\ x_1 + 2x_2 + 3x_3 + x_4 - x_6 == 4 \end{pmatrix}$

All variables ≥ 0
Problem variables redefined as: $\{x1 \to x_1, x2 \to x_2, x3 \to x_3, x4 \to x_4\}$

***** **Optimum solution after 10 iterations** *****
Interior solution → {x1 → 3.01693, x2 → 0.720208,
                    x3 → 0.0000470621, x4 → 0.0000687285}
Objective function → -3.737
Convergence parameters → {0.000212405, 0.105786, 0.971915, 0.136996}

***** **Finding nearest vertex solution** *****
Nearest vertex solution → $\left\{x1 \to \frac{16}{5}, x2 \to \frac{2}{5}, x3 \to 0, x4 \to 0\right\}$
Objective function → $-\frac{18}{5}$
Status → Feasible

Using the default step-length parameter $\beta = 0.99$, the optimum returned by the PAS method has an f value of $-18/5$. A simplex solution returns $f = -10$, which is the correct solution. Several techniques have been proposed to overcome this difficulty. Refer to recent books and articles on advanced linear programming to learn about these refinements. For this example, though, we can get the correct solution simply by reducing the step length parameter β to 0.95.

```
PrimalAffineLP[f,g,vars,MaxIterations → 20, StepLengthFactor → .95];
***** Optimum solution after 14 iterations *****
Interior solution → {x1 → 0.000338359, x2 → 9.75706,
            x3 → 0.0000194353, x4 → 0.0000445125}
Objective function → -9.75731
Convergence parameters → {0.000702174, 0.111818, 1.56115, 0.242692}

***** Finding nearest vertex solution *****
Nearest vertex solution → {x1 → 0, x2 → 10, x3 → 0, x4 → 0}
Objective function → -10  Status → Feasible
```

7.3 The Primal-Dual Interior Point Method

As seen from section 7.1, the optimum of the LP problem can be obtained by solving the following system of equations:

$$\mathbf{Ax} - \mathbf{b} = \mathbf{0} \quad \text{Primal feasibility}$$

$$\mathbf{A}^T\mathbf{v} + \mathbf{u} = \mathbf{c} \quad \text{Dual feasibility}$$

$$\mathbf{XUe} = \mathbf{0} \quad \text{Complementary slackness conditions}$$

$$x_i \geq 0 \quad \text{and} \quad u_i \geq 0, i = 1, \ldots, n$$

The primal-dual interior point method is based on trying to solve the above system of equations directly. Because of the complementary slackness conditions, the complete system of equations is nonlinear. We can use an iterative method, such as the Newton-Raphson method, to solve this system. The linear equations will be satisfied exactly at each iteration. The error in satisfying the complementary slackness condition should get smaller as the iterations progress. Using $\mu > 0$ as an indicator of this error, at a given iteration, the complementary slackness conditions are of the following form:

$$\mathbf{XUe} = \mu \mathbf{e}$$

7.3.1 Direction Using the Newton-Raphson Method

Recall from Chapter 3 that each iteration of the Newton-Raphson method for solving a system of nonlinear equations $\mathbf{F}(\mathbf{x}) = \mathbf{0}$ is written as follows:

7.3 The Primal-Dual Interior Point Method

$$\mathbf{x}^{k+1} = \mathbf{x}^k + \Delta\mathbf{x}, k = 0, 1, \ldots$$

where the change $\Delta\mathbf{x}$ is determined from solving the following equation:

$$\mathbf{J}(\mathbf{x}^k)\Delta\mathbf{x}^k = -\mathbf{F}(\mathbf{x}^k)$$

The Jacobian matrix $\mathbf{J}$ consists of partial derivatives of the functions with respect to the solution variables. For the solution of equations resulting from KT conditions for an LP problem, at each iteration we need to compute new values of $2n + m$ variables ($\mathbf{x}$, $\mathbf{u}$, and $\mathbf{v}$). Therefore, the Jacobian matrix is obtained by differentiating equations with respect to $\mathbf{x}$, $\mathbf{u}$, and $\mathbf{v}$ as follows:

$$\mathbf{J} = \begin{pmatrix} \mathbf{A} & \mathbf{0} & \mathbf{0} \\ \mathbf{0} & \mathbf{I} & \mathbf{A}^T \\ \mathbf{U} & \mathbf{X} & \mathbf{0} \end{pmatrix} \begin{matrix} \Leftarrow \text{From primal feasibility} \\ \Leftarrow \text{From dual feasibility} \\ \Leftarrow \text{From complementary slackness} \end{matrix}$$

Denoting the changes in variables as $\mathbf{d}_x$, $\mathbf{d}_u$, and $\mathbf{d}_v$, the Newton-Raphson method gives the following system of equations:

$$\begin{pmatrix} \mathbf{A} & \mathbf{0} & \mathbf{0} \\ \mathbf{0} & \mathbf{I} & \mathbf{A}^T \\ \mathbf{U} & \mathbf{X} & \mathbf{0} \end{pmatrix} \begin{pmatrix} \mathbf{d}_x \\ \mathbf{d}_u \\ \mathbf{d}_v \end{pmatrix} = - \begin{pmatrix} \mathbf{A}\mathbf{x}^k - \mathbf{b} \\ \mathbf{A}^T\mathbf{v}^k + \mathbf{u}^k - \mathbf{c} \\ \mathbf{X}\mathbf{U}\mathbf{e} - \mu^k \mathbf{e} \end{pmatrix}$$

Note that the matrices $\mathbf{U}$ and $\mathbf{X}$ are defined by using the known values at the current iteration. However, to simplify notation, superscript k is not used on these terms. We can perform the computations more efficiently by writing the three sets of equations explicitly as follows:

(a) $\mathbf{A}\mathbf{d}_x = -\mathbf{A}\mathbf{x}^k + \mathbf{b} \equiv \mathbf{r}_p$

(b) $\mathbf{d}_u + \mathbf{A}^T\mathbf{d}_v = -\mathbf{A}^T\mathbf{v}^k - \mathbf{u}^k + \mathbf{c} \equiv \mathbf{r}_d$

(c) $\mathbf{U}\mathbf{d}_x + \mathbf{X}\mathbf{d}_u = -\mathbf{X}\mathbf{U}\mathbf{e} + \mu^k \mathbf{e}$

From the third equation, we get

$$\mathbf{d}_u = -\mathbf{X}^{-1}\mathbf{U}\mathbf{d}_x - \mathbf{X}^{-1}\mathbf{X}\mathbf{U}\mathbf{e} + \mu^k \mathbf{X}^{-1}\mathbf{e}$$

or

$$\mathbf{d}_u = -\mathbf{X}^{-1}\mathbf{U}\mathbf{d}_x - \mathbf{U}\mathbf{e} + \mu^k \mathbf{X}^{-1}\mathbf{e}$$

Introducing the notation $\mathbf{r}_c = -\mathbf{U}\mathbf{e} + \mu^k \mathbf{X}^{-1}\mathbf{e}$

$$\mathbf{d}_u = -\mathbf{X}^{-1}\mathbf{U}\mathbf{d}_x + \mathbf{r}_c$$

Multiplying both sides of equation (b) by $\mathbf{AXU}^{-1}$, we get

$$\mathbf{AXU}^{-1}\mathbf{d}_u + \mathbf{AXU}^{-1}\mathbf{A}^T\mathbf{d}_v = \mathbf{AXU}^{-1}\mathbf{r}_d$$

Substituting for $\mathbf{d}_u$, we get

$$-\mathbf{AXU}^{-1}\mathbf{X}^{-1}\mathbf{U}\mathbf{d}_x + \mathbf{AXU}^{-1}\mathbf{r}_c + \mathbf{AXU}^{-1}\mathbf{A}^T\mathbf{d}_v = \mathbf{AXU}^{-1}\mathbf{r}_d$$

From the form of $\mathbf{X}$ and $\mathbf{U}$ matrices, it is easy to see that

$$\mathbf{XU}^{-1}\mathbf{X}^{-1}\mathbf{U} = \mathbf{I}$$

Therefore, using equation (a), we have

$$-\mathbf{r}_p + \mathbf{AXU}^{-1}\mathbf{r}_c + \mathbf{AXU}^{-1}\mathbf{A}^T\mathbf{d}_v = \mathbf{AXU}^{-1}\mathbf{r}_d$$

or

$$\mathbf{AXU}^{-1}\mathbf{A}^T\mathbf{d}_v = \mathbf{r}_p - \mathbf{AXU}^{-1}\mathbf{r}_c + \mathbf{AXU}^{-1}\mathbf{r}_d$$

Introducing the notation $\mathbf{D} \equiv \mathbf{XU}^{-1}$, we have

$$\mathbf{ADA}^T\mathbf{d}_v = \mathbf{r}_p + \mathbf{AD}(-\mathbf{r}_c + \mathbf{r}_d)$$

This system of equations can be solved for $\mathbf{d}_v$. Using this solution, the other two increments can be calculated as follows:

From (b) $\mathbf{d}_u = -\mathbf{A}^T\mathbf{d}_v + \mathbf{r}_d$
From (c) $\mathbf{d}_x = \mathbf{U}^{-1}\mathbf{X}(-\mathbf{d}_u + \mathbf{r}_c)$ or $\mathbf{d}_x = \mathbf{D}(-\mathbf{d}_u + \mathbf{r}_c)$

7.3.2 Step-Length Calculations

The previous derivation did not take into account the requirement that $x_i \geq 0$ and $u_i \geq 0$. Assuming that we start from positive initial values, we take care of these requirements by introducing step-length parameters α_p and α_d.

$$\mathbf{x}^{k+1} = \mathbf{x}^k + \alpha_p \mathbf{d}_x$$
$$\mathbf{u}^{k+1} = \mathbf{u}^k + \alpha_d \mathbf{d}_u$$
$$\mathbf{v}^{k+1} = \mathbf{v}^k + \alpha_d \mathbf{d}_v$$

The maximum value of the step length is the one that will make one of the x_i or u_i values go to zero.

$$x_i + \alpha_p d_{xi} \geq 0 \quad \text{and} \quad u_i + \alpha_d d_{ui} \geq 0 \quad i = 1, \ldots, n$$

Variables with positive increments will obviously remain positive regardless of the step length. Also, in order to strictly maintain feasibility, the actual step length should be slightly smaller than the above maximum. The actual step length is chosen as follows:

$$\alpha = \beta \alpha_{\max} \quad \text{where } \beta = 0.999\ldots$$

Thus, the maximum step lengths are determined as follows:

$$\alpha_p = \text{Min}\,[1, -\beta x_i/d_{xi}, d_{xi} < 0] \quad \text{and} \quad \alpha_d = \text{Min}\,[1, -\beta u_i/d_{ui}, d_{ui} < 0]$$

The variables are therefore updated as follows:

$$\mathbf{x}^{k+1} = \mathbf{x}^k + \alpha_p d_x$$
$$\mathbf{u}^{k+1} = \mathbf{u}^k + \alpha_d d_u$$
$$\mathbf{v}^{k+1} = \mathbf{v}^k + \alpha_d d_v$$

7.3.3 Convergence Criteria

Primal Feasibility

The constraints must be satisfied at the optimum, i.e., $\mathbf{A}\mathbf{x}^k - \mathbf{b} = \mathbf{0}$. To use as convergence criteria, this requirement is expressed as follows:

$$\sigma_p = \frac{\|\mathbf{A}\mathbf{x}^k - \mathbf{b}\|}{\|\mathbf{b}\| + 1} \leq \epsilon_1$$

where ϵ_1 is a small positive number. The 1 is added to the denominator to avoid division by small numbers.

Dual Feasibility

We also have the requirement that

$$\mathbf{A}^T \mathbf{v}^k + \mathbf{u}^k - \mathbf{c} = \mathbf{0}$$

This gives the following convergence criteria:

$$\sigma_d = \frac{||\mathbf{r}_d||}{||\mathbf{c}|| + 1} \leq \epsilon_2$$

where ϵ_2 is a small positive number.

Complementary Slackness

The value of parameter μ determines how well complementary slackness conditions are satisfied. Numerical experiments suggest defining an average value of μ as follows:

$$\mu^k = \frac{(\mathbf{x}^k)^T \mathbf{u}}{n}$$

where n = number of optimization variables. This parameter should be zero at the optimum. Thus, for convergence

$$\mu \leq \epsilon_3$$

where ϵ_3 is a small positive number.

7.3.4 Complete Primal-Dual Algorithm

The complete primal-dual algorithm can be summarized in the following steps:

Algorithm

Given: Constraint coefficient matrix $\mathbf{A}$, constraint right-hand side vector $\mathbf{b}$, objective function coefficient vector $\mathbf{c}$, step-length parameter β, and convergence tolerance parameters.

Initialization: $k = 0$, arbitrary initial values (≥ 0), say $\mathbf{x}^k = \mathbf{u}^k = \mathbf{e}$ (vector with all entries 1) and $\mathbf{v}^k = \mathbf{0}$.

The next point $\mathbf{x}^{k+1}$ is computed as follows:

1. Set $\mu^k = \left[\frac{(\mathbf{x}^k)^T \mathbf{u}}{n}\right]/(k+1)$. If $\left[\frac{||\mathbf{A}\mathbf{x}^k - \mathbf{b}||}{||\mathbf{b}||+1} \leq \epsilon_1, \frac{||\mathbf{r}_d||}{||\mathbf{c}||+1} \leq \epsilon_2, \mu^k \leq \epsilon_3\right]$, we have the optimum. Otherwise, continue.

2. Form:
$$\mathbf{D} = \mathbf{X}\mathbf{U}^{-1} = \text{diag}[x_i/u_i]$$
$$\mathbf{r}_p = -\mathbf{A}\mathbf{x}^k + \mathbf{b}$$
$$\mathbf{r}_d = -\mathbf{A}^T\mathbf{v}^k - \mathbf{u}^k + \mathbf{c}$$
$$\mathbf{r}_c = -\mathbf{u}^k + \mu^k \mathbf{X}^{-1}\mathbf{e}$$

3. Solve the system of linear equations for $\mathbf{d}_v$:
$$\mathbf{A}\mathbf{D}\mathbf{A}^T \mathbf{d}_v = \mathbf{r}_p + \mathbf{A}\mathbf{D}(-\mathbf{r}_c + \mathbf{r}_d)$$

4. Compute increments:
$$\mathbf{d}_u = -\mathbf{A}^T \mathbf{d}_v + \mathbf{r}_d$$
$$\mathbf{d}_x = \mathbf{D}(-\mathbf{d}_u + \mathbf{r}_c)$$

5. Check for unboundedness:

   Primal is unbounded if $\mathbf{r}_p = 0$, $\mathbf{d}_x > 0$, and $\mathbf{c}^T\mathbf{d}_x < 0$

   Dual is unbounded if $\mathbf{r}_d = 0$, $\mathbf{d}_u > 0$, and $\mathbf{b}^T\mathbf{d}_v > 0$

   If either of these conditions is true, *stop*. Otherwise, continue.

6. Compute step lengths:
$$\alpha_p = \text{Min}[1, -\beta x_i/d_{xi}, d_{xi} < 0] \quad \text{and} \quad \alpha_d = \text{Min}[1, -\beta u_i/d_{ui}, d_{ui} < 0]$$

7. Compute the next point:
$$\mathbf{x}^{k+1} = \mathbf{x}^k + \alpha_p \mathbf{d}_x$$
$$\mathbf{u}^{k+1} = \mathbf{u}^k + \alpha_d \mathbf{d}_u$$
$$\mathbf{v}^{k+1} = \mathbf{v}^k + \alpha_d \mathbf{d}_v$$

Purification Phase

After obtaining the interior point optimum, we use the same steps as those for the PAS method to get an exact vertex solution.

1. Set the diagonal of scaling matrix $\mathbf{T}$ to the interior point optimum.
2. Compute matrix $\mathbf{P} \equiv \mathbf{T}\mathbf{A}^T(\mathbf{A}\mathbf{T}^2\mathbf{A}^T)^{-1}\mathbf{A}\mathbf{T}$. Set vector $\mathbf{p}$ to diagonal elements of this matrix.

3. Elements of vector **p** that are close to 1 correspond to basic variables, and those that are close to zero define nonbasic variables. In case the elements of vector **p** do not give clear indication, repeat calculations of step 2 by defining a new scaling matrix **T** with diagonal elements set to vector **p**.
4. Set the nonbasic variables to zero and solve the constraint equations $\mathbf{Ax} = \mathbf{b}$ for the optimum values of basic variables.

The PrimalDualLP Function

The following PrimalDualLP function implements the above algorithm for solving LP problems. The function usage and its options are explained first. Several intermediate results can be printed to gain understanding of the process.

```
Needs["OptimizationToolbox`InteriorPoint`"];
?PrimalDualLP
```

PrimalDualLP[f, g, vars, options]. Solves an LP problem using
  Interior Point algorithm based on solving KT conditions using
  the Newton-Raphson method. f is the objective function, g is
  a list of constraints, and vars is a list of variables. See
  Options[PrimalDualLP] to find out about a list of valid options
  for this function.

OptionsUsage[PrimalDualLP]
{UnrestrictedVariables → {}, MaxIterations → 20,
 ProblemType → Min, StandardVariableName → x,
 PrintLevel → 1, StepLengthFactor → 0.99,
 ConvergenceTolerance → 0.0001, StartingVector → {}}

UnrestrictedVariables is an option for LP and several QP problems.
  A list of variables that are not restricted to be positive can be
  specified with this option. Default is {}.

MaxIterations is an option for several optimization methods. It
  specifies maximum number of iterations allowed.

ProblemType is an option for most optimization methods. It can either be
  Min (default) or Max.

StandardVariableName is an option for LP and QP methods. It specifies
  the symbol to use when creating variable names during conversion to
  the standard form. Default is x.

PrintLevel is an option for most functions in the OptimizationToolbox.
  It is specified as an integer. The value of the integer indicates
  how much intermediate information is to be printed. A PrintLevel→0
  suppresses all printing. Default for most functions is set to 1 in
  which case they print only the initial problem setup. Higher integers
  print more intermediate results.

StepLengthFactor is an option for interior point methods. It is the reduction factor applied to the computed step length to maintain feasibility. Default is 0.99

ConvergenceTolerance is an option for most optimization methods. Most methods require only a single zero tolerance value. Some interior point methods require a list of convergence tolerance values.

StartingVector is an option for several interior point methods. Default is $\{1, \ldots, 1\}$.

Example 7.8 Solve the following LP problem using the primal-dual interior point method:

```
f = -5x1 - x2;
g = {2x1 + 3x2 ≤ 12, 2x1 + x2 ≤ 8};
vars = {x1, x2};
```

All calculations for the first two iterations are as follows:

PrimalDualLP[f, g, vars, PrintLevel → 2, MaxIterations → 2];

Minimize $-5x_1 - x_2$

Subject to $\begin{pmatrix} 2x_1 + 3x_2 + x_3 == 12 \\ 2x_1 + x_2 + x_4 == 8 \end{pmatrix}$

All variables ≥ 0

Problem variables redefined as: $\{x1 \to x_1, x2 \to x_2\}$

$b \to \begin{pmatrix} 12 \\ 8 \end{pmatrix} \quad c \to \begin{pmatrix} -5 \\ -1 \\ 0 \\ 0 \end{pmatrix}$

$A \to \begin{pmatrix} 2 & 3 & 1 & 0 \\ 2 & 1 & 0 & 1 \end{pmatrix}$

**** Starting vectors ****

Primal vars (x) → {1., 1., 1., 1.}
Dual vars (u) → {1., 1., 1., 1.}
Multipliers (v) → {0., 0.}
Objective function → -6.   Status → NonOptimum

***** **Iteration 1** *****

D[diagonal] → {1., 1., 1., 1.}

$r_p \to \begin{pmatrix} 6. \\ 4. \end{pmatrix} \quad r_d \to \begin{pmatrix} -6. \\ -2. \\ -1. \\ -1. \end{pmatrix} \quad r_c \to \begin{pmatrix} 0. \\ 0. \\ 0. \\ 0. \end{pmatrix}$

Parameters: $\sigma p \to 0.467579$ $\sigma d \to 1.06259$ $\mu \to 1$.

$\text{A.D.AT} \to \begin{pmatrix} 14. & 7. \\ 7. & 6. \end{pmatrix}$

rp+A.D(-rc+rd) → {-13., -11.}
dv → {-0.0285714, -1.8}
du → {-2.34286, -0.114286, -0.971429, 0.8}
dx → {2.34286, 0.114286, 0.971429, -0.8}
-β x/dx → {∞, ∞, ∞, 1.2375}
-β u/du → {0.422561, 8.6625, 1.01912, ∞}
$\alpha_p \to 1$ $\alpha_d \to 0.422561$
New primal vars (x) → {3.34286, 1.11429, 1.97143, 0.2}
New dual vars (u) → {0.01, 0.951707, 0.589512, 1.33805}
New multipliers (v) → {-0.0120732, -0.76061}
Objective function → -17.8286 Status → NonOptimum

***** **Iteration 2** *****

D[diagonal] → {334.286, 1.17083, 3.34417, 0.149471}

$r_p \to \begin{pmatrix} 0. \\ 0. \end{pmatrix}$ $r_d \to \begin{pmatrix} -3.46463 \\ -1.15488 \\ -0.577439 \\ -0.577439 \end{pmatrix}$ $r_c \to \begin{pmatrix} 0.0843689 \\ -0.668601 \\ -0.429495 \\ 0.23926 \end{pmatrix}$

Parameters: $\sigma p \to 0$. $\sigma d \to 0.613579$ $\mu \to 0.315462$

$\text{A.D.AT} \to \begin{pmatrix} 1351.02 & 1340.66 \\ 1340.66 & 1338.46 \end{pmatrix}$

rp+A.D(-rc+rd) → {-2374.96, -2373.45}
dv → {0.290696, -2.06444}
du → {0.0828534, 0.0374745, -0.868135, 1.487}
dx → {0.506597, -0.826693, 1.46688, -0.186502}
-β x/dx → {∞, 1.3344, ∞, 1.06165}
-β u/du → {∞, ∞, 0.672266, ∞}
$\alpha_p \to 1$ $\alpha_d \to 0.672266$
New primal vars (x) → {3.84945, 0.287593, 3.43831, 0.0134985}
New dual vars (u) → {0.0656995, 0.9769, 0.00589512, 2.33771}
New multipliers (v) → {-0.183351, -2.14846}
Objective function → -19.5349 Status → NonOptimum

***** **NonOptimum solution after 2 iterations** *****
Interior solution → {x1 → 3.84945, x2 → 0.287593}
Objective function → -19.5349

The complete solution is obtained using the PrimalDualLP function.

{sol, history} = PrimalDualLP[f, g, vars];

7.3 The Primal-Dual Interior Point Method

***** Optimum solution after 7 iterations *****
Interior solution → {x1 → 3.99985, x2 → 0.000192359}
Objective function → -19.9994

***** Finding nearest vertex solution *****
Nearest vertex solution → {x1 → 4, x2 → 0}
Objective function → -20
Status → Feasible

The history of points computed by the method are extracted and are shown on the graph in Figure 7.7. Notice that the method takes a little more direct path to the optimum as compared to the PAS algorithm.

xhist = Transpose[Transpose[history][[{1, 2}]]]; TableForm[xhist]

1.	1.
3.34286	1.11429
3.84945	0.287593
3.98951	0.00287593
3.99545	0.00568681
3.99907	0.00116336
3.99985	0.000192359
3.99985	0.000192359

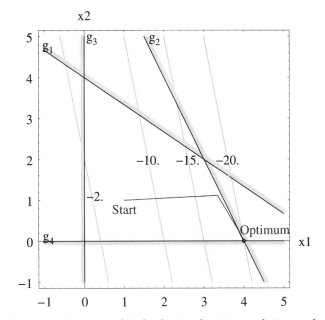

FIGURE 7.7 A graphical solution showing a solution path.

Example 7.9 Solve the following LP problem using the primal-dual method:

```
f = x1 + x2 + x3 + x4;
g = {x1 + 2x2 - x3 + 3x4 ≤ 12,
     x1 + 3x2 + x3 + 2x4 ≤ 8,
     2x1 - 3x2 - x3 + 2x4 ≤ 7};
vars = {x1, x2, x3, x4};
```

Intermediate calculations are shown for the first two iterations.

```
PrimalDualLP[f, g, vars, ProblemType → Max, PrintLevel → 2,
  MaxIterations → 2];
```

Minimize $-x_1 - x_2 - x_3 - x_4$

Subject to $\begin{pmatrix} x_1 + 2x_2 - x_3 + 3x_4 + x_5 == 12 \\ x_1 + 3x_2 + x_3 + 2x_4 + x_6 == 8 \\ 2x_1 - 3x_2 - x_3 + 2x_4 + x_7 == 7 \end{pmatrix}$

All variables ≥ 0

Problem variables redefined as: $\{x1 \to x_1, x2 \to x_2, x3 \to x_3, x4 \to x_4\}$

$b \to \begin{pmatrix} 12 \\ 8 \\ 7 \end{pmatrix} \quad c \to \begin{pmatrix} -1 \\ -1 \\ -1 \\ -1 \\ 0 \\ 0 \\ 0 \end{pmatrix}$

$A \to \begin{pmatrix} 1 & 2 & -1 & 3 & 1 & 0 & 0 \\ 1 & 3 & 1 & 2 & 0 & 1 & 0 \\ 2 & -3 & -1 & 2 & 0 & 0 & 1 \end{pmatrix}$

**** Starting vectors ****
Primal vars (x) → {1., 1., 1., 1., 1., 1., 1.}
Dual vars (u) → {1., 1., 1., 1., 1., 1., 1.}
Multipliers (v) → {0., 0., 0.}
Objective function → -4. Status → NonOptimum

***** Iteration 1 *****
D[diagonal] → {1., 1., 1., 1., 1., 1., 1.}

$r_p \to \begin{pmatrix} 6. \\ 0. \\ 6. \end{pmatrix} \quad r_d \to \begin{pmatrix} -2. \\ -2. \\ -2. \\ -2. \\ -1. \\ -1. \\ -1. \end{pmatrix} \quad r_c \to \begin{pmatrix} 0. \\ 0. \\ 0. \\ 0. \\ 0. \\ 0. \\ 0. \end{pmatrix}$

Parameters: σ p → 0.498219   σ d → 1.45297   μ → 1.

$A.D.A^T \to \begin{pmatrix} 16. & 12. & 3. \\ 12. & 16. & -4. \\ 3. & -4. & 19. \end{pmatrix}$

7.3 The Primal-Dual Interior Point Method

rp+A.D(-rc+rd) → {-5., -15., 5.}
dv → {1.16667, -1.89236, -0.319444}
du → {-0.635417, 0.385417, 0.739583,
    -1.07639, -2.16667, 0.892361, -0.680556}
dx → {0.635417, -0.385417, -0.739583,
    1.07639, 2.16667, -0.892361, 0.680556}
$-\beta$ x/dx → {∞, 2.56865, 1.33859, ∞, ∞, 1.10942, ∞}
$-\beta$ u/du → {1.55803, ∞, ∞, 0.919742, 0.456923, ∞, 1.45469}
α_p → 1 α_d → 0.456923
New primal vars (x) → {1.63542, 0.614583, 0.260417, 2.07639,
                3.16667, 0.107639, 1.68056}
New dual vars (u) → {0.709663, 1.1761, 1.33793, 0.508173,
                0.01, 1.40774, 0.689038}
New multipliers (v) → {0.533077, -0.864663, -0.145962}
Objective function → -4.58681   Status → NonOptimum

***** **Iteration 2** *****

D[diagonal] → {2.3045, 0.522558, 0.194641,
        4.08599, 316.667, 0.0764622, 2.43899}

$r_p \to \begin{pmatrix} 1.77636 \times 10^{-15} \\ 0. \\ 0. \end{pmatrix}$ $r_d \to \begin{pmatrix} -1.08615 \\ -1.08615 \\ -1.08615 \\ -1.08615 \\ -0.543077 \\ -0.543077 \\ -0.543077 \end{pmatrix}$ $r_c \to \begin{pmatrix} -0.507524 \\ -0.638209 \\ -0.0684953 \\ -0.348963 \\ 0.0943945 \\ 1.66348 \\ -0.492328 \end{pmatrix}$

Parameters: σ p → 1.043 × 10^{-16} σ d → 0.789072 μ → 0.330583

$A.D.A^T \to \begin{pmatrix} 358.03 & 29.7611 & 26.1842 \\ 29.7611 & 23.6226 & 16.0553 \\ 26.1842 & 16.0553 & 32.8986 \end{pmatrix}$

rp+A.D(-rc+rd) → {-212.506, -8.42679, -7.91467}
dv → {-0.63118, 0.389863, 0.0715204}
du → {-0.987878, -0.778821, -2.03568, -0.11538,
    0.0881033, -0.93294, -0.614597}
dx → {1.10697, 0.0734783, 0.382894, -0.9544153,
    1.99221, 0.198528, 0.298213}
$-\beta$ x/dx → {∞, ∞, ∞, 2.153801, ∞, ∞, ∞}
$-\beta$ u/du → {0.711188, 1.49501, 0.65067, 4.36029, ∞, 1.49384, 1.10991}
α_p → 1 α_d → 0.65067
New primal vars (x) → {2.74239, 0.688062, 0.643311, 1.12197, 5.15888,
                0.306167, 1.97877}
New dual vars (u) → {0.0668814, 0.66935, 0.0133793, 0.433099, 0.0673262,
                0.800704, 0.289139}
New multipliers (v) → {0.122387, -0.610991, -0.0994254}
Objective function → -5.19574   Status → NonOptimum

```
***** NonOptimum solution after 2 iterations *****
Interior solution → {x1 → 2.74239, x2 → 0.688062,
                    x3 → 0.643311, x4 → 1.12197}
Objective function → 5.19574
```

Calculations are allowed to run until the optimum is found.

```
{sol, hist} = PrimalDualLP[f, g, vars, ProblemType → Max];

***** Optimum solution after 7 iterations *****
Interior solution → {x1 → 3.25621, x2 → 0.000351426,
                    x3 → 4.7406, x4 → 0.000713895}
Objective function → 7.99788
***** Finding nearest vertex solution *****
Nearest vertex solution → {x1 → 5, x2 → 0, x3 → 3, x4 → 0}
Objective function → 8
Status → Feasible

xhist = Transpose[Transpose[hist][[{1, 2, 3, 4}]]]; TableForm[xhist]
```

1.	1.	1.	1.
1.63542	0.614583	0.260417	2.07639
2.74239	0.688067	0.643311	1.12197
3.82294	0.00688062	3.27992	0.342731
4.26952	0.0200827	3.57756	0.00342731
3.53977	0.00305149	4.43303	0.00596975
3.25621	0.000351426	4.7406	0.000713895
3.25621	0.000351426	4.7406	0.000713895

Example 7.10 *Unbounded solution*

```
f = x1 - 2x2;
g = {2x1 - x2 ≥ 0, -2x1 + 3x2 ≤ 6}; vars = {x1, x2};

{sol, history} = PrimalDualLP[f, g, vars];
```
Minimize $x_1 - 2x_2$

Subject to $\begin{pmatrix} 2x_1 - x_2 - x_3 == 0 \\ -2x_1 + 3x_2 + x_4 == 6 \end{pmatrix}$

All variables ≥ 0

Problem variables redefined as: $\{x1 \to x_1, x2 \to x_2\}$

$b \to \begin{pmatrix} 0 \\ 6 \end{pmatrix} \quad c \to \begin{pmatrix} 1 \\ -2 \\ 0 \\ 0 \end{pmatrix}$

$A \to \begin{pmatrix} 2 & -1 & -1 & 0 \\ -2 & 3 & 0 & 1 \end{pmatrix}$

***** Unbounded solution after 3 iterations *****

Interior solution → {x1 → 14.8277, x2 → 11.85752}

Objective function → -8.88663

The history of points computed by the method are extracted and are shown on the graph in Figure 7.8.

xhist = Transpose[Transpose[history][[{1, 2}]]]; TableForm[xhist]

1.	1.
2.25714	2.91429
3.58721	4.38119
14.8277	11.8572

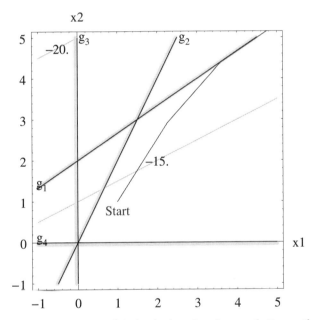

FIGURE 7.8 A graphical solution showing a solution path.

Example 7.11 *Plant operation* In this example, we consider the solution of the tire manufacturing plant operations problem presented in Chapter 1. The problem statement is as follows:

A tire manufacturing plant has the ability to produce both radial and bias-ply automobile tires. During the upcoming summer months, they have contracts to deliver tires as follows.

Date	Radial tires	Bias-ply tires
June 30	5,000	3,000
July 31	6,000	3,000
August 31	4,000	5,000
Total	15,000	11,000

The plant has two types of machines, gold machines and black machines, with appropriate molds to produce these tires. The following production hours are available during the summer months:

Month	On gold machines	On black machines
June	700	1,500
July	300	400
August	1,000	300

The production rates for each machine type and tire combination, in terms of hours per tire, are as follows:

Type	On gold machines	On black machines
Radial	0.15	0.16
Bias-Ply	0.12	0.14

The labor cost of producing tires is $10.00 per operating hour, regardless of which machine type is being used or which tire is being produced. The material cost for radial tires is $5.25 per tire and for bias-ply tires is $4.15 per tire. Finishing, packing, and shipping cost is $0.40 per tire. The excess tires are carried over into the next month but are subjected to an inventory-carrying charge of $0.15 per tire. Wholesale prices have been set at $20 per tire for radials and $15 per tire for bias-ply.

How should the production be scheduled in order to meet the delivery requirements while maximizing profit for the company during the three-month period?

7.3 The Primal-Dual Interior Point Method

The optimization variables are as follows:

x_1	Number of radial tires produced in June on the gold machines
x_2	Number of radial tires produced in July on the gold machines
x_3	Number of radial tires produced in August on the gold machines
x_4	Number of bias-ply tires produced in June on the gold machines
x_5	Number of bias-ply tires produced in July on the gold machines
x_6	Number of bias-ply tires produced in August on the gold machines
x_7	Number of radial tires produced in June on the black machines
x_8	Number of radial tires produced in July on the black machines
x_9	Number of radial tires produced in August on the black machines
x_{10}	Number of bias-ply tires produced in June on the black machines
x_{11}	Number of bias-ply tires produced in July on the black machines
x_{12}	Number of bias-ply tires produced in August on the black machines

The objective of the company is to maximize profit. The following expressions used in defining the objective function were presented in Chapter 1.

```
sales = 20(x₁ + x₂ + x₃ + x₇ + x₈ + x₉) + 15(x₄ + x₅ + x₆ + x₁₀ + x₁₁ + x₁₂);

materialsCost = 5.25(x₁ + x₂ + x₃ + x₇ + x₈ + x₉) + 4.15(x₄ + x₅ + x₆ + x₁₀ + x₁₁ + x₁₂);

laborCost = 10(0.15(x₁ + x₂ + x₃) + 0.16(x₇ + x₈ + x₉) + 0.12(x₄ + x₅ + x₆)
    + 0.14(x₁₀ + x₁₁ + x₁₂));

handlingCost = 0.40(x₁ + x₂ + x₃ + x₄ + x₅ + x₆ + x₇ + x₈ + x₉ + x₁₀ + x₁₁ + x₁₂);

inventoryCost = 0.15((x₁ + x₇ - 5000) + (x₄ + x₁₀ - 5000) + (x₁ + x₂ + x₇ + x₈ - 11000)
    + (x₄ + x₅ + x₁₀ + x₁₁ - 6000));
```

The production hour limitations are expressed as follows:

```
productionLimitations =
    {0.15x₁ + 0.12x₄ ≤ 700, 0.15x₂ + 0.12x₅ ≤ 300, 0.15x₃ + 0.12x₆ ≤ 1000,
     0.16x₇ + 0.14x₁₀ ≤ 1500, 0.16x₈ + 0.14x₁₁ ≤ 400, 0.16x₉ + 0.14x₁₂ ≤ 300};
```

Delivery contract constraints are written as follows:

```
deliveryConstraints =
    {x₁ + x₇ ≥ 5000, x₄ + x₁₀ ≥ 3000, x₁ + x₂ + x₇ + x₈ ≥ 11000, x₄ + x₅ + x₁₀ + x₁₁ ≥ 6000,
     x₁ + x₂ + x₃ + x₇ + x₈ + x₉ == 15000, x₄ + x₅ + x₆ + x₁₀ + x₁₁ + x₁₂ == 11000};
```

Chapter 7 Interior Point Methods

Thus, the problem is stated as follows. Note by multiplying the profit with a negative sign, the problem is defined as a minimization problem.

```
vars = Table[x_i, {i, 1, 12}];
f = -(sales - (materialsCost + laborCost + handlingCost + inventoryCost));
g = Join[productionLimitations, deliveryConstraints];
```

The solution is obtained using the PrimalDualLP, as follows:

PrimalDualLP[f, g, vars, MaxIterations → 30];

Minimize $-13.25x_1 - 13.25x_2 - 13.25x_3 - 9.65x_4 - 9.65x_5 - 9.65x_6 - 13.15x_7 - 13.15x_8 - 13.15x_9 - 9.45x_{10} - 9.45x_{11} - 9.45x_{12}$

Subject to
$$\begin{pmatrix} 0.15x_1 + 0.2x_4 + x_{13} == 700 \\ 0.15x_2 + 0.2x_5 + x_{14} == 300 \\ 0.15x_3 + 0.2x_6 + x_{15} == 1,000 \\ 0.16x_7 + 0.14x_{10} + x_{16} == 1,500 \\ 0.16x_8 + 0.14x_{11} + x_{17} == 400 \\ 0.16x_9 + 0.14x_{12} + x_{18} == 300 \\ x_1 + x_7 - x_{19} == 5,000 \\ x_4 + x_{10} - x_{20} == 3,000 \\ x_1 + x_2 + x_7 + x_8 - x_{21} == 11,000 \\ x_4 + x_5 + x_{10} + x_{11} - x_{22} == 6,000 \\ x_1 + x_2 + x_3 + x_7 + x_8 + x_9 == 15,000 \\ x_4 + x_5 + x_6 + x_{10} + x_{11} + x_{12} == 11,000 \end{pmatrix}$$

All variables ≥ 0

Problem variables redefined as:
$\{x_1 \to x_1, x_2 \to x_2, x_3 \to x_3, x_4 \to x_4, x_5 \to x_5, x_6 \to x_6,$
$x_7 \to x_7, x_8 \to x_8, x_9 \to x_9, x_{10} \to x_{10}, x_{11} \to x_{11}, x_{12} \to x_{12}\}$

***** **Optimum solution after 13 iterations** *****

Interior solution →
$\{x_1 \to 1,759.51, x_2 \to 99.7596, x_3 \to 2,674.05, x_4 \to 3,633.93,$
$x_5 \to 2,375.29, x_6 \to 4,990.75, x_7 \to 6,664.86, x_8 \to 2,484.5,$
$x_9 \to 1,317.3, x_{10} \to 0.00491385, x_{11} \to 0.0049176, x_{12} \to 0.00491642\}$

Objective function → -303853.

***** **Finding nearest vertex solution** *****

Nearest vertex solution →
$\{x_1 \to 1,866.67, x_2 \to 0, x_3 \to 2,666.67, x_4 \to 3,500.,$
$x_5 \to 2,500., x_6 \to 5,000., x_7 \to 6,633.33, x_8 \to 2,500.,$
$x_9 \to 1,333.33, x_{10} \to 0, x_{11} \to 0, x_{12} \to 0\}$

Objective function → -303,853.

Status → Feasible

The same solution was obtained in Chapter 6 using the revised simplex method.

7.4 Concluding Remarks

Since the publication of L.G. Khachiyan's original paper in 1979 [Fang and Puthenpura, 1993], a wide variety of interior point methods have been proposed. The main motivation for this development has been to devise a method that has superior convergence properties over the simplex method. Examples exist for which the simplex method visits every vertex before finding the optimum. However, for most practical problems, the convergence is quite rapid. On the other hand, most interior point methods have been proven to have good theoretical convergence rates, but realizing this performance on actual large-scale problems is still an open area. Commercial implementations of interior point methods are still relatively rare.

The goal of this chapter was to provide a practical introduction to the interior point methods for solving linear programming problems. Two relatively simple-to-understand methods were presented. The *Mathematica* implementations, PrimalAffineLP and PrimalDualLP, should be useful for solving small-scale problems and for developing an understanding of how these methods work. For a comprehensive treatment of this area, refer to the books by Fang and Puthenpura [1993], Hertog [1994], Megiddo [1989], and Padberg [1995].

7.5 Appendix—Null and Range Spaces

The concepts of null and range spaces from linear algebra are used in the derivation of Primal Affine scaling algorithm. These concepts are reviewed briefly in this appendix.

Null and Range spaces

Consider an $m \times n$ matrix **A** with $m \leq n$. Assume that all rows are linearly independent and thus, rank of matrix **A** is m. The *null space* of matrix **A** (denoted by Null(**A**)) is defined by the set of $n - m$ linearly independent vectors $\mathbf{p}_i$ that satisfy the following relationship:

$$\mathbf{A}\mathbf{p}_i = 0$$

It is clear from this definition that each vector $\mathbf{p}_i$ has n elements and forms the basis vectors for the null space of **A**. The basis vectors defining null space are not unique and can be determined in a number of different ways. The following example illustrates a simple procedure based on solving systems of equations.

The *range space* of matrix **A** (denoted by Range(**A**)) is the set of linearly independent vectors obtained from the columns of matrix **A**. The range space of matrix $\mathbf{A}^T$, rather than of **A**, is of more interest because of its important relationship with the null space of matrix **A**. Since rank of matrix **A** is m, all m columns of matrix $\mathbf{A}^T$ are linearly independent and form the range space. Any vector **q** in the range space of $\mathbf{A}^T$ can be written as a linear combination of columns of matrix $\mathbf{A}^T$. That is,

$$\mathbf{q} = \mathbf{A}^T \alpha \text{ for some } m \times 1 \text{ vector } \alpha$$

Since the dimension of range space of $\mathbf{A}^T$ is m and that of null space of **A** is $n - m$, together the two spaces span the entire n dimensional space. Furthermore, by considering the dot product of vector **q** in the range space of $\mathbf{A}^T$ and a vector **p** in the null space of **A**, it is easy to see that the two subspaces are orthogonal to each other.

$$\mathbf{q}^T \mathbf{p} = \alpha^T \mathbf{A} \mathbf{p} = 0$$

Example 7.12 Consider the following 2×4 matrix:

```
A = {{1, 2, 1, 3}, {2, 1, 1, -4}};
MatrixForm[A]
```

$$\begin{pmatrix} 1 & 2 & 1 & 3 \\ 2 & 1 & 1 & -4 \end{pmatrix}$$

The range space of $\mathbf{A}^T$ is simply the columns of the transpose of **A** (or rows of matrix **A**). Thus, the basis for the range space of $\mathbf{A}^T$ is

```
{q₁, q₂} = A
{{1, 2, 1, 3}, {2, 1, 1, -4}}
```

The basis vectors for the null space of matrix **A** are determined by considering solutions of the system of equations $\mathbf{Ap} = 0$. Since we have only two equations, we can solve for two unknowns in terms of the other two and thus, there are two basis vectors for the null space.

p = {a₁, a₂, a₃, a₄};
eqns = A.p == {0, 0}

$\{a_1 + 2a_2 + a_3 + 3a_4, 2a_1 + a_2 + a_3 - 4a_4\} == \{0, 0\}$

If we choose $a_1 = 1$ and $a_2 = 0$ (arbitrarily), we get the following solution:

Solve[eqns/.{a₁ → 1, a₂ → 0}]

$\left\{\left\{a_3 \to -\dfrac{10}{7},\ a_4 \to \dfrac{1}{7}\right\}\right\}$

If we choose $a_1 = 0$ and $a_2 = 1$, we get the following solution:

Solve[eqns/.{a₁ → 0, a₂ → 1}]

$\left\{\left\{a_3 \to -\dfrac{11}{7},\ a_4 \to -\dfrac{1}{7}\right\}\right\}$

Thus, we have following two basis vectors for the null space of **A**:

p₁ = {1, 0, -10/7, 1/7};
p₂ = {0, 1, -11/7, -1/7};

There are obviously many other possibilities for the basis vectors. However, we can only have two linearly independent basis vectors. Any other vector in the null space can be written in terms of the basis vector. As an illustration consider another vector in the null space as follows:

Solve[eqns/.{a₁ → -3, a₂ → 2}]

$\left\{\left\{a_3 \to \dfrac{8}{7},\ a_4 \to -\dfrac{5}{7}\right\}\right\}$

p = {-3, 3, 8/7, -5/7};

Clearly, this vector can be obtained by the linear combination $-3\mathbf{p}_1 + 2\mathbf{p}_2$.

-3p₁ + 2p₂

$\left\{-3,\ 2,\ \dfrac{8}{7},\ -\dfrac{5}{7}\right\}$

Next, we numerically demonstrate that the two subspaces (range space of $\mathbf{A}^T$ and null space of **A**) are orthogonal to each other.

```
{q₁.p₁, q₁.p₂, q₂.p₁, q₂.p₂}
{0,0,0,0}
```

There is a built-in *Mathematica* function **NullSpace** to produce null space basis vectors. The following basis vectors are obtained using this function:

```
NullSpace[A]
{{11,-10,0,3},{-1,-1,3,0}}
```

Null Space Projection Matrix

Any vector $\mathbf{x}$ can be represented by two orthogonal components, one in the range space of $\mathbf{A}^T$, and the other in the null space of $\mathbf{A}$:

$$\mathbf{x} = \mathbf{p} + \mathbf{q}$$

The null space component $\mathbf{p}$ can be determined as follows:

$$\mathbf{p} = \mathbf{x} - \mathbf{q} \quad \text{or} \quad \mathbf{p} = \mathbf{x} - \mathbf{A}^T \alpha$$

Multiplying both sides by $\mathbf{A}$, we get

$$\mathbf{A}\mathbf{p} = \mathbf{A}\mathbf{x} - \mathbf{A}\mathbf{A}^T \alpha$$

Since $\mathbf{A}\mathbf{p} = 0$, we get

$$\mathbf{A}\mathbf{x} = \mathbf{A}\mathbf{A}^T \alpha \quad \text{giving} \quad \alpha = (\mathbf{A}\mathbf{A}^T)^{-1} \mathbf{A}\mathbf{x}$$

Therefore,

$$\mathbf{p} = \mathbf{x} - \mathbf{A}^T(\mathbf{A}\mathbf{A}^T)^{-1}\mathbf{A}\mathbf{x} = \left[\mathbf{I} - \mathbf{A}^T(\mathbf{A}\mathbf{A}^T)^{-1}\mathbf{A}\right]\mathbf{x} \equiv \mathbf{P}\mathbf{x}$$

The $n \times n$ matrix $\mathbf{P}$ is called the null space projection matrix.

$$\mathbf{P} = \mathbf{I} - \mathbf{A}^T(\mathbf{A}\mathbf{A}^T)^{-1}\mathbf{A}$$

The range space component $\mathbf{q}$ is obtained as follows:

$$\mathbf{q} = \mathbf{A}^T\alpha = \mathbf{A}^T(\mathbf{A}\mathbf{A}^T)^{-1}\mathbf{A}\mathbf{x} \equiv \mathbf{R}\mathbf{x}$$

where the range space projection matrix is as follows:

$$\mathbf{R} = \mathbf{A}^T(\mathbf{A}\mathbf{A}^T)^{-1}\mathbf{A}$$

Example 7.13 Consider the following 1 × 2 matrix:

```
A = {{1, 2}};
```

The basis vector for the range space of A^T is

```
q = {1, 2};
```

The null space basis vector is computed using the NullSpace function:

```
p = First[NullSpace[A]]
{-2, 1}
```

It is easy to see that the two spaces are orthogonal to each other.

```
q.p
0
```

The orthogonal projection matrices are

```
R = Transpose[A].Inverse[A.Transpose[A]].A;
MatrixForm[R]
```

$$\begin{pmatrix} \frac{1}{5} & \frac{2}{5} \\ \frac{2}{5} & \frac{4}{5} \end{pmatrix}$$

```
P = IdentityMatrix[2] - R; MatrixForm[P]
```

$$\begin{pmatrix} \frac{4}{5} & -\frac{2}{5} \\ -\frac{2}{5} & \frac{1}{5} \end{pmatrix}$$

Given a two-dimensional vector, we can resolve it into two orthogonal components using these projection matrices:

```
vec = {3, -4};
pv = P.vec; qv = R.vec;
{pv, qv}
```

{{4, -2}, {-1, -2}}

These vectors are plotted in Figure 7.9. From the figure, it is clear that the components p and q of the given vector are in the two subspaces.

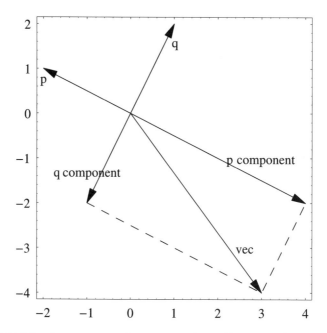

FIGURE 7.9   Components of a vector in the null space and range space.

7.6 Problems

Dual LP

Construct the duals for the following LP problems. If possible, use KT conditions or graphical methods to demonstrate the relationship between the primal and the dual problems.

7.1. Maximize $z = x_1 + 3x_2$

Subject to $x_1 + 4x_2 \leq 10$ and $x_1 + 2x_2 \leq 10$.

7.2. Maximize $z = -x_1 + 2x_2$

Subject to $x_1 - 4x_2 + 2 \geq 0$ and $2x_1 - 3x_2 \leq 3$.

7.3. Minimize $f = -2x_1 + 2x_2 + x_3 - 3x_4$

Subject to $x_1 + x_2 + x_3 + x_4 \leq 18$, $x_1 - 2x_3 + 4x_4 \leq 12$, $x_1 + x_2 \leq 18$, and $x_3 + 2x_4 \leq 16$.

7.4. Minimize $f = -3x_1 - 4x_2$

Subject to $x_1 + 2x_2 \leq 10$, $x_1 + x_2 \leq 10$, and $3x_1 + 5x_2 \leq 20$.

7.5. Minimize $f = -100x_1 - 80x_2$

Subject to $5x_1 + 3x_2 \leq 15$ and $x_1 + x_2 \leq 4$.

Primal Affine Scaling Method

Find the optimum solution of the following problems using the primal affine scaling method. Unless stated otherwise, all variables are restricted to be positive. Verify solutions graphically for two variable problems.

7.6. Maximize $z = x_1 + 3x_2$

Subject to $x_1 + 4x_2 \leq 10$ and $x_1 + 2x_2 \leq 10$.

7.7. Maximize $z = -x_1 + 2x_2$

Subject to $x_1 - 4x_2 + 2 \geq 0$ and $2x_1 - 3x_2 \leq 3$.

7.8. Minimize $f = -2x_1 + 2x_2 + x_3 - 3x_4$

Subject to $x_1 + x_2 + x_3 + x_4 \leq 18$, $x_1 - 2x_3 + 4x_4 \leq 12$, $x_1 + x_2 \leq 18$, and $x_3 + 2x_4 \leq 16$.

7.9. Minimize $f = -3x_1 - 4x_2$

Subject to $x_1 + 2x_2 \leq 10$, $x_1 + x_2 \leq 10$, and $3x_1 + 5x_2 \leq 20$.

7.10. Minimize $f = -100x_1 - 80x_2$

Subject to $5x_1 + 3x_2 \leq 15$ and $x_1 + x_2 \leq 4$.

7.11. Maximize $z = -x_1 + 2x_2$

Subject to $x_1 - 2x_2 + 2 \geq 0$ and $2x_1 - 3x_2 \leq 3$.

7.12. Minimize $f = x_1 + 3x_2 - 2x_3$

Subject to $x_1 - 2x_2 - 2x_3 \geq -2$ and $2x_1 - 3x_2 - x_3 \leq -2$.

7.13. Maximize $z = x_1 + 3x_2 - 2x_3 + x_4$

Subject to $3x_1 + 2x_2 + x_3 + x_4 \leq 20$, $2x_1 + x_2 + x_4 = 10$, and $5x_1 - 2x_2 - x_3 + 2x_4 \geq 3$.

7.14. Minimize $f = 3x_1 + 4x_2 + 5x_3 + 6x_4$

Subject to $3x_1 + 4x_2 + 5x_3 + 6x_4 \geq 20$, $2x_1 + 3x_2 + 4x_4 \leq 10$, and $5x_1 - 6x_2 - 7x_3 + 8x_4 \geq 3$. Variables x_1 and x_4 are unrestricted in sign.

7.15. Minimize $f = 13x_1 - 3x_2 - 5$

Subject to $3x_1 + 5x_2 \leq 20$, $2x_1 + x_2 \geq 10$, $5x_1 + 2x_2 \geq 3$, and $x_1 + 2x_2 \geq 3$. x_1 is unrestricted in sign.

7.16. Minimize $f = 5x_1 + 2x_2 + 3x_3 + 5x_4$

Subject to $x_1 - x_2 + 7x_3 + 3x_4 \geq 4$, $x_1 + 2x_2 + 2x_3 + x_4 = 9$, and $2x_1 + 3x_2 + x_3 - 4x_4 \leq 5$.

7.17. Minimize $f = -3x_1 + 8x_2 - 2x_3 + 4x_4$

Subject to $x_1 - 2x_2 + 4x_3 + 6x_4 \leq 0$, $x_1 - 4x_2 - x_3 + 6x_4 \leq 2$, $x_3 \leq 3$, and $x_4 \geq 3$.

7.18. Minimize $f = 3x_1 + 2x_2$

Subject to $2x_1 + 2x_2 + x_3 + x_4 = 10$, and $2x_1 - 3x_2 + 2x_3 = 10$. The variables should not be greater than 10.

7.19. Minimize $f = -2x_1 + 5x_2 + 3x_3$

Subject to $x_1 - x_2 - x_3 \leq -3$ and $2x_1 + x_2 \geq 1$.

7.20. Hawkeye foods owns two types of trucks. Truck type I has a refrigerated capacity of 15 m³ and a nonrefrigerated capacity of 25 m³. Truck type II has a refrigerated capacity of 15 m³ and non-refrigerated capacity of 10 m³. One of their stores in Gofer City needs products that require 150 m³ of refrigerated capacity and 130 m³ of nonrefrigerated capacity. For the round trip from the distribution center to Gofer City, truck type I uses 300 gallons of gasoline while truck type II uses 200 gallons. Formulate the problem of determining the number of trucks of each type that the company must use in order to meet the store's needs while minimizing gas consumption. Use the PAS method to find an optimum.

7.21. A manufacturer requires an alloy consisting of 50% tin, 30% lead, and 20% zinc. This alloy can be made by mixing a number of available alloys, the properties and costs of which are tabulated. The goal is to find the cheapest blend. Formulate the problem as an optimization problem. Use the PAS method to find an optimum.

	Available alloys				
Properties	A	B	C	D	E
Lead (%)	10	10	40	60	30
Zinc (%)	10	30	50	30	30
Tin (%)	80	60	10	10	40
Cost: ($/lb alloy)	8.2	9.3	11.2	13	17

7.22. A company can produce three different types of concrete blocks, identified as A, B, and C. The production process is constrained by facilities available for mixing, vibration, and inspection/drying. Using the data given in the following table, formulate the production problem in order to maximize the profit. Use the PAS method to find an optimum.

	Blocks			
	A	B	C	Available
Mixing (hours/batch)	1	3	9	900
Vibration (hours/batch)	2	3	6	1200
Inspection/drying (hours/batch)	0.7	0.8	1	400
Profit: ($/batch)	7	17	30	

7.23. A mining company operates two mines, identified as A and B. Each mine can produce high-, medium-, and low-grade iron ores. The weekly demand for different ores and the daily production rates and operating costs are given in the following table. Formulate an optimization problem to determine the production schedule for the two mines in order to meet the weekly demand at lowest cost to the company. Use the PAS method to find an optimum.

Ore grade	Weekly demand (tons)	Daily production	
		Mine A (tons)	Mine B (tons)
High	12,000	2,000	1,000
Medium	8,000	1,000	1,000
Low	24,000	5,000	2,000
Operations cost ($/day)		210,000	170,000

7.24. Assignment of parking spaces for its employees has become an issue for an automobile company located in an area with harsh climate. There are enough parking spaces available for all employees; however, some employees must be assigned spaces in lots that are not adjacent to the buildings in which they work. The following table shows the distances in meters between parking lots (identified as 1, 2, and 3) and office buildings (identified as A, B, C, and D). The number of spaces in the lots and the number of employees who need spaces are also tabulated. Formulate the parking assignment problem to minimize the distances walked by the employees from their parking spaces to their offices. Use the PAS method to find an optimum.

	Distances from parking lot (m)				Spaces available
Parking Lot	Building A	Building B	Building C	Building D	
1	290	410	260	410	80
2	430	350	330	370	100
3	310	260	290	380	40
# of employees	40	40	60	60	

7.25. Hawkeye Pharmaceuticals can manufacture a new drug using any one of the three processes identified as A, B, and C. The costs and quantities of ingredients used in *one batch* of these processes are given in the following table. The quantity of new drug produced during each batch of different processes is also given in the table.

		Ingredients used per batch (tons)		
Process	Cost ($ per batch)	Ingredient I	Ingredient II	Quantity of drug produced
A	$12,000	3	2	2
B	$25,000	2	6	5
C	$9,000	7	2	1

The company has a supply of 80 tons of ingredient I and 70 tons of ingredient II at hand and would like to produce 60 tons of new drug at a minimum cost.

Formulate the problem as an optimization problem. Use the PAS method to find an optimum.

7.26. A major auto manufacturer in Detroit, Michigan needs two types of seat assemblies during 1998 on the following quarterly schedule:

	Type 1	Type 2
First Quarter	25,000	25,000
Second Quarter	35,000	30,000
Third Quarter	35,000	25,000
Fourth Quarter	25,000	30,000
Total	120,000	110,000

The excess seats from each quarter are carried over to the next quarter but are subjected to an inventory-carrying charge of $20 per thousand seats. However, assume no inventory is carried over to 1999.

The company has contracted with an auto seat manufacturer that has two plants: one in Detroit and the other in Waterloo, Iowa. Each plant can manufacture both types of seats; however, their maximum capacities and production costs are different. The production costs per seat and the annual capacity at each of the two plants in terms of number of seat assemblies is given as follows:

	Quarterly capacity (either type)	Production cost Type 1	Production cost Type 2
Detroit plant	30,000	$225	$240
Waterloo plant	35,000	$165	$180

The packing and shipping costs from the two plants to the auto manufacturer are as follows:

	Cost/100 seats
Detroit plant	$10
Waterloo plant	$80

Formulate the problem to determine a seat acquisition schedule from the two plants to minimize the overall cost of this operation to the auto manufacturer for the year. Use the PAS method to find an optimum.

7.27. A small company needs pipes in the following lengths:

    0.5 m   100 pieces

    0.6 m   300 pieces

    1.2 m   200 pieces

The local supplier sells pipes only in the following three lengths:

    4 m

    6 m

    8 m

After cutting the necessary lengths, the excess pipe must be thrown away. The company obviously wants to minimize this waste. Formulate the problem as a linear programming problem. Use the PAS method to find an optimum.

7.28. Consider the problem of finding the shortest route between two cities while traveling on a given network of available roads. The network is shown in Figure 7.10. The nodes represent cities, and the links are the roads that connect these cities. The distances in kilometers along each road are noted in the figure. Use the PAS method to find the shortest route.

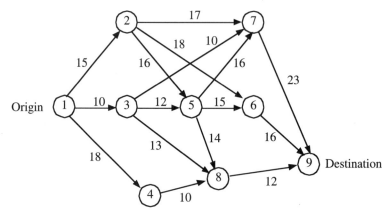

FIGURE 7.10 A network diagram showing distances and direction of travel between cities.

7.6 Problems

Primal-Dual Method

Find the optimum solution of the following problems using the primal-dual method. Unless stated otherwise, all variables are restricted to be positive. Verify solutions graphically for two variable problems.

7.29. Maximize $z = x_1 + 3x_2$

Subject to $x_1 + 4x_2 \leq 10$ and $x_1 + 2x_2 \leq 1$.

7.30. Maximize $z = -x_1 + 2x_2$

Subject to $x_1 - 4x_2 + 2 \geq 0$ and $2x_1 - 3x_2 \leq 3$.

7.31. Minimize $f = -2x_1 + 2x_2 + x_3 - 3x_4$

Subject to $x_1 + x_2 + x_3 + x_4 \leq 18$, $x_1 - 2x_3 + 4x_4 \leq 12$, $x_1 + x_2 \leq 18$, and $x_3 + 2x_4 \leq 1$.

7.32. Minimize $f = -3x_1 - 4x_2$

Subject to $x_1 + 2x_2 \leq 10$, $x_1 + x_2 \leq 10$, $and 3x_1 + 5x_2 \leq 20$.

7.33. Minimize $f = -100x_1 - 80x_2$

Subject to $5x_1 + 3x_2 \leq 15$ and $x_1 + x_2 \leq 4$.

7.34. Maximize $z = -x_1 + 2x_2$

Subject to $x_1 - 2x_2 + 2 \geq 0$ and $2x_1 - 3x_2 \leq 3$.

7.35. Minimize $f = x_1 + 3x_2 - 2x_3$

Subject to $x_1 - 2x_2 - 2x_3 \geq -2$ and $2x_1 - 3x_2 - x_3 \leq -2$.

7.36. Maximize $z = x_1 + 3x_2 - 2x_3 + x_4$

Subject to $3x_1 + 2x_2 + x_3 + x_4 \leq 20$, $2x_1 + x_2 + x_4 = 10$, and $5x_1 - 2x_2 - x_3 + 2x_4 \geq 3$.

7.37. Minimize $f = 3x_1 + 4x_2 + 5x_3 + 6x_4$

Subject to $3x_1 + 4x_2 + 5x_3 + 6x_4 \geq 20$, $2x_1 + 3x_2 + 4x_4 \leq 10$ and $5x_1 - 6x_2 - 7x_3 + 8x_4 \geq 3$. Variables x_1 and x_4 are unrestricted in sign.

7.38. Minimize $f = 13x_1 - 3x_2 - 5$

Subject to $3x_1 + 5x_2 \leq 20$, $2x_1 + x_2 \geq 10$, $5x_1 + 2x_2 \geq 3$, and $x_1 + 2x_2 \geq 3$. x_1 is unrestricted in sign.

7.39. Minimize $f = 5x_1 + 2x_2 + 3x_3 + 5x_4$

Subject to $x_1 - x_2 + 7x_3 + 3x_4 \geq 4$, $x_1 + 2x_2 + 2x_3 + x_4 = 9$, and $2x_1 + 3x_2 + x_3 - 4x_4 \leq 5$.

7.40. Minimize $f = -3x_1 + 8x_2 - 2x_3 + 4x_4$

Subject to $x_1 - 2x_2 + 4x_3 + 6x_4 \leq 0$, $x_1 - 4x_2 - x_3 + 6x_4 \leq 2$, $x_3 \leq 3$, and $x_4 \geq 3$.

7.41. Minimize $f = 3x_1 + 2x_2$

Subject to $2x_1 + 2x_2 + x_3 + x_4 = 10$ and $2x_1 - 3x_2 + 2x_3 = 10$. The variables should not be greater than 10.

7.42. Minimize $f = -2x_1 + 5x_2 + 3x_3$

Subject to $x_1 - x_2 - x_3 \leq -3$ and $2x_1 + x_2 \geq 1$.

7.43. Use the primal-dual method to solve the Hawkeye food problem 7.20.

7.44. Use the primal-dual method to solve the alloy manufacturing problem 7.21.

7.45. Use the primal-dual method to solve the concrete block manufacturing problem 7.22.

7.46. Use the primal-dual method to solve the mining company problem 7.23.

7.47. Use the primal-dual method to solve the parking spaces assignment problem 7.24.

7.48. Use the primal-dual method to solve the drug manufacturing problem 7.25.

7.49. Use the primal-dual method to solve the auto seat problem 7.26.

7.50. Use the primal-dual method to solve the stock-cutting problem 7.27.

7.51. Use the primal-dual method to solve the network problem 7.28.

CHAPTER EIGHT

Quadratic Programming

When the objective function is a quadratic function of the optimization variables and all constraints are linear, the problem is called a quadratic programming (QP) problem. Several important practical problems may directly be formulated as QP. The portfolio management problem, presented in Chapter 1, is such an example. QP problems also arise as a result of approximating more general nonlinear problems by linear and quadratic functions.

Since QP problems have linear constraints, their solution methods are relatively simple extensions of LP problems. The first section introduces a standard form for QP problems and presents KT conditions for its optimality. Sections 2 and 3 present extensions of two interior point methods (Primal Affine Scaling and primal-dual) to solving QP problems. Active set methods for primal and dual QP, using the conjugate gradient method, are presented in the last two sections. These methods exploit the special structure of the QP problems typically encountered as subproblems during the solution of general nonlinear programming problems discussed in the following chapter.

8.1 KT Conditions for Standard QP

Quadratic programming (QP) problems have quadratic objective functions and linear constraints. To simplify discussion, unless stated otherwise, it will be assumed that a QP is written in the following standard form:

Minimize $\mathbf{c}^T\mathbf{x} + \frac{1}{2}\mathbf{x}^T\mathbf{Q}\mathbf{x}$

Subject to $\begin{pmatrix} \mathbf{A}\mathbf{x} = \mathbf{b} \\ \mathbf{x} \geq \mathbf{0} \end{pmatrix}$

where $\mathbf{x}$ is an $n \times 1$ vector of optimization variables, $\mathbf{A}$ is an $m \times n$ constraint coefficient matrix, $\mathbf{b}$ is an $m \times 1$ vector of constraint right-hand sides, $\mathbf{c}$ is an $n \times 1$ vector containing coefficients of linear terms, and $n \times n$ matrix $\mathbf{Q}$ contains coefficients of square and mixed terms in the objective function. Conversion of constraints to the standard form is the same as that discussed for LP in Chapter 6. For the methods presented in this chapter, it is not necessary to have positive right-hand sides of constraints as was the case for the simplex method. Writing a quadratic function in the above matrix form was discussed in Chapter 3. Thus, all QP problems can be written in the above standard form with little effort. Furthermore, it is easy to see that a QP problem is convex if the matrix $\mathbf{Q}$ is at least positive semidefinite.

Following the presentation in Chapter 4, the Lagrangian for the standard QP problem is as follows:

$$\mathbf{L}(\mathbf{x}, \mathbf{u}, \mathbf{v}, \mathbf{s}) = \mathbf{c}^T\mathbf{x} + \frac{1}{2}\mathbf{x}^T\mathbf{Q}\mathbf{x} + \mathbf{u}^T\left(-\mathbf{x} + \mathbf{s}^2\right) + \mathbf{v}^T\left(-\mathbf{A}\mathbf{x} + \mathbf{b}\right)$$

where $\mathbf{u} \geq \mathbf{0}$ is an $n \times 1$ vector of Lagrange multipliers associated with positivity constraints, $\mathbf{s}$ is a vector of slack variables, and $\mathbf{v}$ is an $m \times 1$ vector of Lagrange multipliers associated with equality constraints. Differentiating the Lagrangian with respect to all variables results in the following system of equations:

$\frac{\partial L}{\partial \mathbf{x}} = 0 \Longrightarrow \mathbf{c} + \mathbf{Q}\mathbf{x} - \mathbf{u} - \mathbf{A}^T\mathbf{v} = \mathbf{0}$ or $-\mathbf{Q}\mathbf{x} + \mathbf{A}^T\mathbf{v} + \mathbf{u} = \mathbf{c}$

$\frac{\partial L}{\partial \mathbf{v}} = 0 \Longrightarrow \mathbf{A}\mathbf{x} - \mathbf{b} = \mathbf{0}$

$\frac{\partial L}{\partial \mathbf{u}} = 0 \Longrightarrow -\mathbf{x} + \mathbf{s}^2 = \mathbf{0}$

$\frac{\partial L}{\partial \mathbf{s}} = 0 \Longrightarrow u_i s_i = 0, i = 1, \ldots, n$

$u_i \geq 0, i = 1, \ldots, n$

The first two sets of equations are linear, while the other two are nonlinear. We can eliminate $\mathbf{s}$ by noting that the conditions $u_i s_i = 0$, say that either $u_i = 0$ (in which case, $-x_i < 0$) or the corresponding $s_i = 0$ (in which case, $x_i = 0$). Thus, by maintaining the positivity of x_i explicitly, the optimum of the QP problem can be obtained by solving the following system of equations:

$$\mathbf{Ax - b = 0}$$

$$-\mathbf{Qx} + \mathbf{A}^T\mathbf{v} + \mathbf{u} = \mathbf{c}$$

$$u_i x_i = 0, i = 1, \ldots, n$$

$$x_i \geq 0, u_i \geq 0, i = 1, \ldots, n$$

It is convenient to express the complementary slackness conditions ($u_i x_i = 0$) in the matrix form as well. For this purpose, we define $n \times n$ diagonal matrices:

$$\mathbf{U} = \text{diag}[u_i] \quad \mathbf{X} = \text{diag}[x_i]$$

Further, by defining an $n \times 1$ vector $\mathbf{e}$, all of whose entries are 1

$$\mathbf{e}^T = (1, 1, \ldots, 1)$$

the complementary slackness conditions are written as follows:

$$\mathbf{XUe = 0}$$

8.1.1 Dual QP Problem

Using the concept of Lagrangian duality presented in the last section of Chapter 4, it is possible to define an explicit dual QP problem. The dual function is as follows:

$$M(\mathbf{u}, \mathbf{v}) = \underset{\mathbf{x}}{\text{Min}} \left[\mathbf{c}^T\mathbf{x} + \frac{1}{2}\mathbf{x}^T\mathbf{Q}\mathbf{x} - \mathbf{u}^T\mathbf{x} + \mathbf{v}^T(-\mathbf{Ax} + \mathbf{b}) \right] \quad u_i \geq 0, i = 1, \ldots, n$$

The minimum can easily be computed by differentiating with respect to $\mathbf{x}$ and solving the resulting system of equations:

$$\mathbf{c} + \mathbf{Qx} - \mathbf{u} - \mathbf{A}^T\mathbf{v} = \mathbf{0}$$

Taking the transpose and multiplying by $\mathbf{x}$

$$\mathbf{c}^T\mathbf{x} + \mathbf{x}^T\mathbf{Qx} - \mathbf{u}^T\mathbf{x} - \mathbf{v}^T\mathbf{Ax} = 0$$

or

$$\mathbf{c}^T\mathbf{x} - \mathbf{u}^T\mathbf{x} - \mathbf{v}^T\mathbf{Ax} = -\mathbf{x}^T\mathbf{Qx}$$

Substituting into the dual function, we get

$$M(\mathbf{u}, \mathbf{v}) = -\frac{1}{2}\mathbf{x}^T\mathbf{Q}\mathbf{x} + \mathbf{v}^T\mathbf{b} \quad u_i \geq 0, i = 1, \ldots, n$$

The complete dual QP problem can be stated as follows:

Maximize $-\frac{1}{2}\mathbf{x}^T\mathbf{Q}\mathbf{x} + \mathbf{v}^T\mathbf{b}$

Subject to $\begin{pmatrix} \mathbf{Q}\mathbf{x} + \mathbf{c} - \mathbf{u} - \mathbf{A}^T\mathbf{v} = \mathbf{0} \\ u_i \geq 0, i = 1, \ldots, n \end{pmatrix}$

For the special case when the matrix $\mathbf{Q}$ is positive definite, and therefore the inverse of $\mathbf{Q}$ exists, the primal variables $\mathbf{x}$ can be eliminated from the dual by using the equality constraint.

$$\mathbf{x} = \mathbf{Q}^{-1}\left[-\mathbf{c} + \mathbf{u} + \mathbf{A}^T\mathbf{v}\right]$$

The dual QP is then written as follows:

Maximize $-\frac{1}{2}[-\mathbf{c} + \mathbf{u} + \mathbf{A}^T\mathbf{v}]^T\mathbf{Q}^{-1}[-\mathbf{c} + \mathbf{u} + \mathbf{A}^T\mathbf{v}] + \mathbf{v}^T\mathbf{b}$

Subject to $u_i \geq 0, i = 1, \ldots, n$

Computationally, it may be advantageous to solve the dual problem since it has simple bound constraints.

The FormDualQP Function

A *Mathematica* function called FormDualQP has been created to automate the process of creating a dual QP.

```
Needs["OptimizationToolbox`QuadraticProgramming`"];
?FormDualQP
```

FormDualQP[f, g, vars, options]. Forms dual of the given QP. f is the
   objective function, g is a list of constraints, and vars is a list of
   variables. See Options[FormDualQP] to find out about a list of valid
   options for this function.

```
OptionsUsage[FormDualQP]
```
{UnrestrictedVariables → {} , ProblemType → Min,
  StandardVariableName → x, DualQPVariableNames → {u, v}}

UnrestrictedVariables is an option for LP and several QP problems.
   A list of variables that are not restricted to be positive can be
   specified with this option. Default is {}.

ProblemType is an option for most optimization methods. It can either be
   Min (default) or Max.

StandardVariableName is an option for LP and QP methods. It specifies
the symbol to use when creating variable names during conversion to
the standard form. Default is x.

DualQPVariableNames→ symbols to use when creating dual variable names.
Default is {u, v}.

Example 8.1 Construct the dual of the following QP problem. Solve the primal and the dual problems using KT conditions to demonstrate that they both give the same solution.

$$\text{Minimize } f = -6x_1x_2 + 2x_1^2 + 9x_2 - 18x_1 + 9x_2^2$$

$$\text{Subject to } \begin{pmatrix} x_1 + 2x_2 = 15 \\ x_i \geq 0, i = 1, 2 \end{pmatrix}$$

```
f = -6x₁x₂ + 2x₁² + 9x₂ - 18x₁ + 9x₂²;
g = {x₁ + 2x₂ == 15}; vars = {x₁, x₂};
```

The dual is written as follows:

```
{df, dg, dvars, dv} = FormDualQP[f, g, vars];
```

Primal problem

Minimize → $-18x_1 + 2x_1^2 + 9x_2 - 6x_1x_2 + 9x_2^2$

Subject to → $(x_1 + 2x_2 == 15)$

and → $(x_1 \geq 0 \quad x_2 \geq 0)$

A → $(1 \quad 2)$ b → (15)

c → $\begin{pmatrix} -18 \\ 9 \end{pmatrix}$ Q → $\begin{pmatrix} 4 & -6 \\ -6 & 18 \end{pmatrix}$

It is easy to see that **Q** is a positive definite matrix and therefore, using its inverse, we can write the dual in terms of Lagrange multipliers alone.

$Q^{-1} \to \begin{pmatrix} \frac{1}{2} & \frac{1}{6} \\ \frac{1}{6} & \frac{1}{9} \end{pmatrix}$ $-c+u+A^T.v \to \begin{pmatrix} 18 + u_1 + v_1 \\ -9 + u_2 + 2v_1 \end{pmatrix}$

Solution → $\begin{pmatrix} x_1 \to \frac{15}{2} + \frac{u_1}{2} + \frac{u_2}{6} + \frac{5v_1}{6} \\ x_2 \to 2 + \frac{u_1}{6} + \frac{u_2}{9} + \frac{7v_1}{18} \end{pmatrix}$

Dual QP problem

Variables → $(u_1 \quad u_2 \quad v_1)$

Maximize → $-\frac{117}{2} - \frac{15u_1}{2} - \frac{u_1^2}{4} - 2u_2 - \frac{u_1u_2}{6} - \frac{u_2^2}{18} + \frac{7v_1}{2} - \frac{5u_1v_1}{6} - \frac{7u_2v_1}{18} - \frac{29v_1^2}{36}$

Subject to → $\begin{pmatrix} u_1 \geq 0 \\ u_2 \geq 0 \end{pmatrix}$

The primal and the dual problems are solved using the KT conditions as follows:

KTSolution[f, {g, Thread[vars ≥ 0]}, vars];

Minimize $f \to -18x_1 + 2x_1^2 + 9x_2 - 6x_1 x_2 + 9x_2^2$

$\nabla f \to \begin{pmatrix} -18 + 4x_1 - 6x_2 \\ 9 - 6x_1 + 18x_2 \end{pmatrix}$

***** LE constraints & their gradients

$g_1 \to -x_1 \leq 0 \quad g_2 \to -x_2 \leq 0$

$\nabla g_1 \to \begin{pmatrix} -1 \\ 0 \end{pmatrix} \quad \nabla g_2 \to \begin{pmatrix} 0 \\ -1 \end{pmatrix}$

***** EQ constraints & their gradients

$h_1 \to -15 + x_1 + 2x_2 == 0 \quad \nabla h_1 \to \begin{pmatrix} 1 \\ 2 \end{pmatrix}$

***** Lagrangian $\to u_1 \left(s_1^2 - x_1 \right) - 18x_1 + 2x_1^2 + u_2 \left(s_2^2 - x_2 \right) + 9x_2 - 6x_1 x_2 + 9x_2^2 + v_1 \left(-15 + x_1 + 2x_2 \right)$

$\nabla L = 0 \to \begin{pmatrix} -18 - u_1 + v_1 + 4x_1 - 6x_2 == 0 \\ 9 - u_2 + 2v_1 - 6x_1 + 18x_2 == 0 \\ s_1^2 - x_1 == 0 \\ s_2^2 - x_2 == 0 \\ -15 + x_1 + 2x_2 == 0 \\ 2s_1 u_1 == 0 \\ 2s_2 u_2 == 0 \end{pmatrix}$

***** Valid KT Point(s) *****

$f \to -54.6983$
$x_1 \to 9.31034$
$x_2 \to 2.84483$
$u_1 \to 0$
$u_2 \to 0$
$s_1^2 \to 9.31034$
$s_2^2 \to 2.84483$
$v_1 \to -2.17241$

KTSolution[-df, dg, dvars, KTVarNames $\to$ {U, S, V}];

Minimize $f \to \frac{117}{2} + \frac{15u_1}{2} + \frac{u_1^2}{4} + 2u_2 + \frac{u_1 u_2}{6} + \frac{u_2^2}{18} - \frac{7v_1}{2} + \frac{5u_1 v_1}{6} + \frac{7u_2 v_1}{18} + \frac{29v_1^2}{36}$

$\nabla f \to \begin{pmatrix} \frac{15}{2} + \frac{u_1}{2} + \frac{u_2}{6} + \frac{5v_1}{6} \\ 2 + \frac{u_1}{6} + \frac{u_2}{9} + \frac{7v_1}{18} \\ -\frac{7}{2} + \frac{5u_1}{6} + \frac{7u_2}{18} + \frac{29v_1}{18} \end{pmatrix}$

***** LE constraints & their gradients

$g_1 \to -u_1 \leq 0 \quad g_2 \to -u_2 \leq 0$

$$\nabla g_1 \to \begin{pmatrix} -1 \\ 0 \\ 0 \end{pmatrix} \quad \nabla g_2 \to \begin{pmatrix} 0 \\ -1 \\ 0 \end{pmatrix}$$

***** Lagrangian $\to \dfrac{117}{2} + \dfrac{15 u_1}{2} + \dfrac{u_1^2}{4} + 2 u_2 + \dfrac{u_1 u_2}{6} + \dfrac{u_2^2}{18} - \dfrac{7 v_1}{2} + \dfrac{5 u_1 v_1}{6} + \dfrac{7 u_2 v_1}{18} + \dfrac{29 v_1^2}{36} +$
$\left(-u_1 + S_1^2\right) U_1 + \left(-u_2 + S_2^2\right) U_2$

$$\nabla L = 0 \to \begin{pmatrix} \dfrac{15}{2} + \dfrac{u_1}{2} + \dfrac{u_2}{6} + \dfrac{5 v_1}{6} - U_1 == 0 \\ 2 + \dfrac{u_1}{6} + \dfrac{u_2}{9} + \dfrac{7 v_1}{18} - U_2 == 0 \\ -\dfrac{7}{2} + \dfrac{5 u_1}{6} + \dfrac{7 u_2}{18} + \dfrac{29 v_1}{18} == 0 \\ -u_1 + S_1^2 == 0 \\ -u_2 + S_2^2 == 0 \\ 2 S_1 U_1 == 0 \\ 2 S_2 U_2 == 0 \end{pmatrix}$$

***** Valid KT Point(s) *****

$f \to 54.6983$
$u_1 \to 0$
$u_2 \to 0$
$v_1 \to 2.17241$
$U_1 \to 9.31034$
$U_2 \to 2.84483$
$S_1^2 \to 0$
$S_2^2 \to 0$

Both solutions are identical except for the sign of v_1. As pointed out in Chapter 6, the sign of multipliers for the equality constraints is arbitrary. The multiplier will have a negative sign if both sides of the equality constraint are multiplied by a negative sign.

Example 8.2 Construct the dual of the following QP problem. Solve the primal and the dual problems using KT conditions to demonstrate that they both give the same solution.

$$\text{Minimize } f = -2 x_1 + \dfrac{x_1^2}{2} - 6 x_2 - x_1 x_2 + x_2^2$$
$$\text{Subject to } \begin{pmatrix} 3 x_1 + x_2 \leq 25 \\ -x_1 + 2 x_2 \leq 10 \\ x_i \geq 0, i = 1, 2 \end{pmatrix}$$

```
f = -2x₁ + x₁²/2 - 6x₂ - x₁x₂ + x₂²;
g = {3x₁ + x₂ ≤ 25, -x₁ + 2x₂ ≤ 10}; vars = {x₁, x₂};
```

The dual is written as follows:

`{df, dg, dvars, dv} = FormDualQP[f, g, vars];`

Primal problem

$$\text{Minimize} \to -2x_1 + \frac{x_1^2}{2} - 6x_2 - x_1 x_2 + x_2^2$$

$$\text{Subject to} \to \begin{pmatrix} 3x_1 + x_2 + x_3 == 25 \\ -x_1 + 2x_2 + x_4 == 10 \end{pmatrix}$$

$$\text{and} \to \begin{pmatrix} x_1 \geq 0 & x_2 \geq 0 & x_3 \geq 0 & x_4 \geq 0 \end{pmatrix}$$

$$A \to \begin{pmatrix} 3 & 1 & 1 & 0 \\ -1 & 2 & 0 & 1 \end{pmatrix} \quad b \to \begin{pmatrix} 25 \\ 10 \end{pmatrix}$$

$$c \to \begin{pmatrix} -2 \\ -6 \\ 0 \\ 0 \end{pmatrix} \quad Q \to \begin{pmatrix} 1 & -1 & 0 & 0 \\ -1 & 2 & 0 & 0 \\ 0 & 0 & 0 & 0 \\ 0 & 0 & 0 & 0 \end{pmatrix}$$

The matrix **Q** is clearly not a positive definite matrix and therefore, the dual cannot be written in terms of Lagrange multipliers alone.

Dual QP problem

$$\text{Variables} \to \begin{pmatrix} u_1 & u_2 & u_3 & u_4 & v_1 & v_2 & x_1 & x_2 & x_3 & x_4 \end{pmatrix}$$

$$\text{Maximize} \to 25v_1 + 10v_2 - \frac{x_1^2}{2} + x_1 x_2 - x_2^2$$

$$\text{Subject to} \to \begin{pmatrix} -2 - u_1 - 3v_1 + v_2 + x_1 - x_2 == 0 \\ -6 - u_2 - v_1 - 2v_2 - x_1 + 2x_2 == 0 \\ -u_3 - v_1 == 0 \\ -u_4 - v_2 == 0 \\ u_1 \geq 0 \\ u_2 \geq 0 \\ u_3 \geq 0 \\ u_4 \geq 0 \end{pmatrix}$$

Using KT conditions, it can easily be verified that both the primal and the dual problems give the same solution:

$$x_1 = 6.36 \quad x_2 = 5.92 \quad f = -30.62$$

8.2 The Primal Affine Scaling Method for Convex QP

The primal affine scaling (PAS) algorithm for convex QP problems is based on exactly the same ideas as those used for the PAS method for LP problems. Starting from an interior feasible point, the key steps to compute the next point

8.2 The Primal Affine Scaling Method for Convex QP

are scaling, descent direction, and step length. Two scaling transformations are used to derive the descent direction. The step-length computations are a little more complicated because the optimum of a QP problem can either be at the boundary or in the interior of the feasible region.

The following example illustrates the ideas:

Minimize $f = -2x_1 + \frac{x_1^2}{2} - 6x_2 - x_1 x_2 + x_2^2$

Subject to $\begin{pmatrix} 3x_1 + x_2 \leq 25 \\ -x_1 + 2x_2 \leq 10 \\ x_1 + 2x_2 \leq 15 \\ x_i \geq 0, i = 1, 2 \end{pmatrix}$

Introducing slack variables x_3, x_4, and x_5, the problem is written in the standard form as follows:

Minimize $f = -2x_1 + \frac{x_1^2}{2} - 6x_2 - x_1 x_2 + x_2^2$

Subject to $\begin{pmatrix} 3x_1 + x_2 + x_3 = 25 \\ -x_1 + 2x_2 + x_4 = 10 \\ x_1 + 2x_2 + x_5 = 15 \\ x_i \geq 0, i = 1, \ldots, 5 \end{pmatrix}$

The problem is now in the standard QP form with the following vectors and matrices:

$$\mathbf{c} = \begin{pmatrix} -2 \\ -6 \\ 0 \\ 0 \\ 0 \end{pmatrix} \quad \mathbf{Q} = \begin{pmatrix} 1 & -1 & 0 & 0 & 0 \\ -1 & 2 & 0 & 0 & 0 \\ 0 & 0 & 0 & 0 & 0 \\ 0 & 0 & 0 & 0 & 0 \\ 0 & 0 & 0 & 0 & 0 \end{pmatrix}$$

$$\mathbf{A} = \begin{pmatrix} 3 & 1 & 1 & 0 & 0 \\ -1 & 2 & 0 & 1 & 0 \\ 1 & 2 & 0 & 0 & 1 \end{pmatrix} \quad \mathbf{b} = \begin{pmatrix} 25 \\ 10 \\ 15 \end{pmatrix}$$

8.2.1 Finding an Initial Interior Point

An initial interior point must have all values greater than zero and satisfy the constraint equations. Since the actual objective function does not enter into these considerations, the Phase I procedure of the PAS algorithm for linear problems is applicable. Thus, choose an arbitrary starting point $\mathbf{x}^0 > \mathbf{0}$, say

$\mathbf{x}^0 = [1, 1, \ldots, 1]$, and then from the constraint equations, we have

$$\mathbf{z}^0 \equiv \mathbf{b} - \mathbf{A}\mathbf{x}^0$$

If $\mathbf{z}^0 = \mathbf{0}$, we have a starting interior point. If not, we introduce an artificial variable and define a Phase I LP problem as follows.

Minimize a

Subject to $\begin{pmatrix} \mathbf{A}\mathbf{x} + a\mathbf{z} = \mathbf{b} \\ \mathbf{x} \geq 0 \\ a \geq 0 \end{pmatrix}$

The minimum of this problem is reached when $a = 0$ and at that point, $\mathbf{A}\mathbf{x}^k = \mathbf{b}$, which makes $\mathbf{x}^k$ an interior point for the original problem. Furthermore, if we set $a = 1$, then any arbitrary $\mathbf{x}^0$ becomes a starting interior point for the Phase I problem. We apply the PAS algorithm for LP to this problem until $a = 0$ and then switch over to the QP algorithm for the actual problem (Phase II).

The following starting point for the example problem, however, was not determined by using the above procedure. It was written directly, after few trials, to satisfy the constraints. The goal was to have a starting solution that did not involve many decimal places. This avoided having to carry many significant figures in the illustrative example. Later examples use the above Phase I procedure to get an initial interior point.

$$\mathbf{x}^0 = \{5/2, 17/4, 53/4, 4, 4\}^T$$

It is easy to see $\mathbf{A}\mathbf{x}^0 - \mathbf{b} = \mathbf{0}$ and therefore, this is a feasible solution for the problem:

```
A = {{3, 1, 1, 0, 0}, {-1, 2, 0, 1, 0}, {1, 2, 0, 0, 1}};
b = {25, 10, 15};
c = {-2, -6, 0, 0, 0};
Q = {{1, -1, 0, 0, 0}, {-1, 2, 0, 0, 0}, {0, 0, 0, 0, 0},
     {0, 0, 0, 0, 0}, {0, 0, 0, 0, 0}};
xk = {5/2, 17/4, 53/4, 4, 4};
A.xk - b
{0, 0, 0}
```

8.2.2 Determining a Descent Direction

Similar to the LP case, a feasible descent direction is the projection of the negative gradient of the objective function on the nullspace of the constraint

8.2 The Primal Affine Scaling Method for Convex QP

coefficient matrix. After introducing two transformations and a lengthy series of manipulations (see the appendix to this chapter for detailed derivation), the direction vector is expressed as follows:

$$\mathbf{d}^k = -[\mathbf{I} - \mathbf{H}^k \mathbf{A}^T (\mathbf{A} \mathbf{H}^k \mathbf{A}^T)^{-1} \mathbf{A}] \mathbf{H}^k (\mathbf{Q} \mathbf{x}^k + \mathbf{c})$$

where $\mathbf{Q}$ and $\mathbf{c}$ define the objective function, $\mathbf{A}$ is the constraint coefficient matrix, $\mathbf{x}^k$ is current feasible interior point, and

$$\mathbf{H}^k = [\mathbf{Q} + (\mathbf{T}^k)^{-2}]^{-1}$$

$$(\mathbf{T}^k)^{-2} = \begin{pmatrix} 1/(x_1^k)^2 & 0 & 0 & 0 \\ 0 & 1/(x_2^k)^2 & 0 & 0 \\ \vdots & \vdots & \ddots & \vdots \\ 0 & 0 & 0 & 1/(x_n^k)^2 \end{pmatrix}$$

By introducing the following additional notation, the direction vector can be written in a very simple form:

Define $\mathbf{w}^k = (\mathbf{A} \mathbf{H}^k \mathbf{A}^T)^{-1} \mathbf{A} \mathbf{H}^k (\mathbf{Q} \mathbf{x}^k + \mathbf{c})$
Then $\mathbf{d}^k = -\mathbf{H}^k (\mathbf{Q} \mathbf{x}^k + \mathbf{c}) + \mathbf{H}^k \mathbf{A}^T \mathbf{w}^k = -\mathbf{H}^k (\mathbf{Q} \mathbf{x}^k + \mathbf{c} - \mathbf{A}^T \mathbf{w}^k)$
Further defining $\mathbf{s}^k = \mathbf{Q} \mathbf{x}^k + \mathbf{c} - \mathbf{A}^T \mathbf{w}^k$

we have

$$\mathbf{d}^k = -\mathbf{H}^k \mathbf{s}^k$$

Note that instead of inverting the matrix as indicated in its definition, $\mathbf{w}^k$ is computed more efficiently by solving a linear system of equations as follows:

$$(\mathbf{A} \mathbf{H}^k \mathbf{A}^T) \mathbf{w}^k = \mathbf{A} \mathbf{H}^k (\mathbf{Q} \mathbf{x}^k + \mathbf{c})$$

It is interesting to observe that a QP problem reduces to an LP problem if matrix $\mathbf{Q}$ is a zero matrix. The descent direction formula given here also reduces to the one given for an LP problem if matrix $\mathbf{Q}$ is a zero matrix.

For the example problem, the descent direction is computed as follows:

```
Hk = Inverse[Q + DiagonalMatrix[1./(xk)^2]]; MatrixForm[Hk]
```

$$\begin{pmatrix} 1.48485 & 0.722428 & 0. & 0. & 0. \\ 0.722428 & 0.838016 & 0. & 0. & 0. \\ 0. & 0. & 175.562 & 0. & 0. \\ 0. & 0. & 0. & 16. & 0. \\ 0. & 0. & 0. & 0. & 16. \end{pmatrix}$$

AHkA = A.Hk.Transpose[A]; MatrixForm[AHkA]

$$\begin{pmatrix} 194.099 & 0.833617 & 11.1876 \\ 0.833617 & 17.9472 & 1.86721 \\ 11.1876 & 1.86721 & 23.7266 \end{pmatrix}$$

Qxc = Q.xk + c

$$\left\{ -\frac{15}{4}, 0, 0, 0, 0 \right\}$$

w = LinearSolve[AHkA, A.Hk.Qxc]
{-0.0753651, 0.0567999, -0.431975}

sk = Qxc - Transpose[A].w
{-3.03513, 0.825715, 0.0753651, -0.0567999, 0.431975}

dk = -Hk.sk
{3.9102, 1.5007, -13.2313, 0.908799, -6.9116}

8.2.3 Step Length and the Next Point

After knowing the feasible descent direction, the next step is to determine the largest possible step in this direction. Only the equality constraints are used during direction computations. The requirement that variables be positive is taken into consideration during the step-length calculations. Thus, we need to find the largest α such that the following conditions are satisfied in terms of scaled and original variables:

$$\mathbf{x}^{k+1} = \mathbf{x}^k + \alpha \mathbf{d}^k > \mathbf{0}$$

Since the optimum of a QP can lie either inside the feasible domain or on the constraint boundary, we must examine two possibilities. If the optimum is on the constraint boundary, we have a situation similar to the LP case. Thus, the step length is determined by the negative entries in the $\mathbf{d}^k$ vector.

$$\alpha_1 = \beta \text{Min}[-x_i/d_i^k, d_i^k < 0, i = 1, \ldots, n]$$

As for the LP case, the parameter β (with $0 < \beta < 1$) is introduced to ensure that the next point is inside the feasible region. Usually $\beta = 0.99$ is chosen in order to go as far as possible without a variable becoming negative.

If the optimum of QP is inside the feasible region, then the problem is essentially unconstrained. The step length is now determined by minimizing the objective function along the given direction. Thus, we solve the following one-dimensional problem:

Find α to minimize $\quad \mathbf{c}^T(\mathbf{x}^k + \alpha \mathbf{d}^k) + \frac{1}{2}(\mathbf{x}^k + \alpha \mathbf{d}^k)^T \mathbf{Q}(\mathbf{x}^k + \alpha \mathbf{d}^k)$

The necessary condition for the minimum is that the first derivative with respect to α be zero. Thus, we get the following equation:

$$\mathbf{c}^T \mathbf{d}^k + \frac{1}{2}(\mathbf{d}^k)^T \mathbf{Q}(\mathbf{x}^k + \alpha \mathbf{d}^k) + \frac{1}{2}(\mathbf{x}^k + \alpha \mathbf{d}^k)^T \mathbf{Q}(\mathbf{d}^k) = 0$$

or

$$(\mathbf{x}^k + \alpha \mathbf{d}^k)^T \mathbf{Q}(\mathbf{d}^k) = -\mathbf{c}^T \mathbf{d}^k$$

or

$$\mathbf{x}^{kT} \mathbf{Q} \mathbf{d}^k + \alpha \mathbf{d}^{kT} \mathbf{Q} \mathbf{d}^k = -\mathbf{c}^T \mathbf{d}^k$$

Giving

$$\alpha_2 = -\frac{\mathbf{c}^T \mathbf{d}^k + \mathbf{x}^{kT} \mathbf{Q} \mathbf{d}^k}{\mathbf{d}^{kT} \mathbf{Q} \mathbf{d}^k} = -\frac{\mathbf{d}^{kT}(\mathbf{Q}\mathbf{x}^k + \mathbf{c})}{\mathbf{d}^{kT} \mathbf{Q} \mathbf{d}^k}$$

In practice, since we don't know whether the optimum is on the boundary or inside the feasible domain, we determine both α_1 and α_2 and select the smaller of the two. The next interior feasible point is then given by

$$\mathbf{x}^{k+1} = \mathbf{x}^k + \alpha \mathbf{d}^k$$

where $\alpha = \text{Min}[\alpha_1, \alpha_2]$

For the example problem, the step length and the next point are computed as follows:

N[xk]
{2.5, 4.25, 13.25, 4., 4.}

dk
{3.9102, 1.5007, -13.2313, 0.908799, -6.9116}

xk/dk
{0.639354, 2.83201, -1.00141, 4.40142, -0.578737}

α1 = 0.99Min[Select[-xk/dk, Positive]]
0.57295

α2 = -dk.(Q.xk + c)/(dk.Q.dk)
1.81976

α = Min[α1, α2]
0.57295

The new point is therefore as follows:

xk1 = xk + αdk
{4.74035, 5.10983, 5.66913, 4.5207, 0.04}

8.2.4 Convergence Criteria

Starting with this new point, the previous series of steps is repeated until an optimum is found. Theoretically, the optimum is reached when $\mathbf{d}^k = \mathbf{0}$. Thus, we can define the first convergence criteria as follows:

$$\sigma_1 \equiv \text{Norm}[\mathbf{d}^k] \leq \epsilon_1$$

where ϵ_1 is a small positive number.

In addition to this, because of the presence of round-off errors, the numerical implementations of the algorithm also check the following conditions derived from the KT optimality conditions.

Feasibility

The constraints must be satisfied at the optimum, i.e., $\mathbf{A}\mathbf{x}^k - \mathbf{b} = \mathbf{0}$. To use as convergence criteria, this requirement is expressed as follows:

$$\sigma_2 \equiv \frac{||\mathbf{A}\mathbf{x}^k - \mathbf{b}||}{||\mathbf{b}|| + 1} \leq \epsilon_2$$

where ϵ_2 is a small positive number. The 1 is added to the denominator to improve numerical performance.

Dual Feasibility

It can be seen from section 1 that the following vector is related to the Lagrange multipliers $\mathbf{u}$ of the dual QP problem.

$$\mathbf{s}^k \equiv \mathbf{Q}\mathbf{x}^k + \mathbf{c} - \mathbf{A}^T \mathbf{w}^k$$

At the optimum, these dual variables should be positive; therefore, we have the following convergence criteria:

$$\sigma_3 \equiv \frac{||\mathbf{s}^k||}{||\mathbf{Q}\mathbf{x}^k + \mathbf{c}|| + 1} \leq \epsilon_3$$

where ϵ_3 is a small positive number.

Complementary Slackness

The primal and dual variables must satisfy the following complementary slackness condition:

$$\mathbf{x}^{kT}\mathbf{s}^k = 0$$

Thus, we can define another convergence criteria as follows:

$$\sigma_4 \equiv \text{Abs}[\mathbf{x}^{kT}\mathbf{s}^k] \leq \epsilon_4$$

where ϵ_4 is a small positive number.

8.2.5 Complete PAS Algorithm for QP Problems

Phase I of the PAS algorithm for a QP problem is identical to that of an LP problem. The computations in Phase II are as follows.

Phase II

Given: Constraint coefficient matrix $\mathbf{A}$, constraint right-hand side vector $\mathbf{b}$, objective function coefficient vector $\mathbf{c}$, quadratic form matrix $\mathbf{Q}$, current interior point $\mathbf{x}^k$, step-length parameter β, and convergence tolerance parameters.

The next point $\mathbf{x}^{k+1}$ is computed as follows:

1. Form scaling matrix:

$$(\mathbf{T}^k)^{-2} = \begin{pmatrix} 1/(x_1^k)^2 & 0 & 0 & 0 \\ 0 & 1/(x_2^k)^2 & 0 & 0 \\ \vdots & \vdots & \ddots & \vdots \\ 0 & 0 & 0 & 1/(x_n^k)^2 \end{pmatrix}$$

$$\mathbf{H}^k = [\mathbf{Q} + (\mathbf{T}^k)^{-2}]^{-1}$$

2. Solve system of linear equations for $\mathbf{w}^k$:

$$(\mathbf{A}\mathbf{H}^k\mathbf{A}^T)\mathbf{w}^k = \mathbf{A}\mathbf{H}^k(\mathbf{Q}\mathbf{x}^k + \mathbf{c})$$

3. Compute $\mathbf{s}^k = \mathbf{Q}\mathbf{x}^k + \mathbf{c} - \mathbf{A}^T\mathbf{w}^k$.
4. Compute direction $\mathbf{d}^k = -\mathbf{H}^k\mathbf{s}^k$.

5. Check for convergence. If

$$\left[\text{Norm}[\mathbf{d}^k] \leq \epsilon_1, \frac{\|\mathbf{A}\mathbf{x}^k - \mathbf{b}\|}{\|\mathbf{b}\| + 1} \leq \epsilon_2, \frac{\|\mathbf{s}^k\|}{\|\mathbf{Q}\mathbf{x}^k + \mathbf{c}\| + 1} \leq \epsilon_3, \text{Abs}[\mathbf{x}^{k\text{T}}\mathbf{s}^k] \leq \epsilon_4 \right],$$

we have the optimum. Otherwise, continue.

6. Compute step length $\alpha = \text{Min}[\alpha_1, \alpha_2]$:

$$\alpha_1 = \beta \text{Min}\left[-x_i/d_i^k, d_i^k < 0, i = 1, \ldots, n \right]$$

$$\alpha_2 = -\frac{\mathbf{d}^{k\text{T}}(Q\mathbf{x}^k + c)}{\mathbf{d}^{k\text{T}} \mathbf{Q} \mathbf{d}^k}$$

7. Compute the next point $\mathbf{x}^{k+1} = \mathbf{x}^k + \alpha \mathbf{d}^k$.

The PrimalAffineQP Function

The following PrimalAffineQP function implements the Primal Affine algorithm for solving QP problems. The function usage and its options are explained first. Several intermediate results can be printed to gain understanding of the process.

```
Needs["OptimizationToolbox`QuadraticProgramming`"];
?PrimalAffineQP
```

PrimalAffineQP[f, g, vars, options]. Solves a convex QP problem using
  Primal Affine algorithm. f is the objective function, g is a list of
  constraints, and vars is a list of variables. See
  Options[PrimalAffineQP] to find out about a list of valid options for
  this function.

OptionsUsage[PrimalAffineQP]

{UnrestrictedVariables → {}, MaxIterations → 20,
 ProblemType → Min, StandardVariableName → x, PrintLevel → 1,
 OptimizationToolbox`QuadraticProgramming`StepLengthFactor → 0.99,
 ConvergenceTolerance → {0.001, 0.2, 2, 0.5},
 OptimizationToolbox`QuadraticProgramming`StartingVector → {}}

UnrestrictedVariables is an option for LP and several QP problems.
  A list of variables that are not restricted to be positive can be
  specified with this option. Default is {}.

MaxIterations is an option for several optimization methods. It
  specifies maximum number of iterations allowed.

8.2 The Primal Affine Scaling Method for Convex QP

ProblemType is an option for most optimization methods. It can either be Min (default) or Max.

StandardVariableName is an option for LP and QP methods. It specifies the symbol to use when creating variable names during conversion to the standard form. Default is x.

PrintLevel is an option for most functions in the OptimizationToolbox. It is specified as an integer. The value of the integer indicates how much intermediate information is to be printed. A PrintLevel→ 0 suppresses all printing. Default for most functions is set to 1 in which case they print only the initial problem setup. Higher integers print more intermediate results.

StepLengthFactor is an option for interior point methods. It is the reduction factor applied to the computed step length to maintain feasibility. Default is 0.99

ConvergenceTolerance is an option for most optimization methods. Most methods require only a single zero tolerance value. Some interior point methods require a list of convergence tolerance values.

StartingVector is an option for several interior point methods. Default is {1,...,1}.

Example 8.3 The complete solution of the example problem used in the previous section is obtained in this example using the PrimalAffineQP function.

$$\text{Minimize } f = -2x_1 + \frac{x_1^2}{2} - 6x_2 - x_1 x_2 + x_2^2$$

$$\text{Subject to } \begin{pmatrix} 3x_1 + x_2 \leq 25 \\ -x_1 + 2x_2 \leq 10 \\ x_1 + 2x_2 \leq 15 \\ x_i \geq 0, i = 1, 2 \end{pmatrix}$$

```
Clear[x1, x2];
f = -2x1 + x1²/2 - 6x2 - x1x2 + x2²;
g = {3x1 + x2 ≤ 25, -x1 + 2x2 ≤ 10, x1 + 2x2 ≤ 15}; vars = {x1, x2};
```

All intermediate calculations are shown for the first two iterations.

```
PrimalAffineQP[f, g, vars, PrintLevel → 2, MaxIterations → 2,
    StartingVector → N[{5/2, 17/4, 53/4, 4, 4}]];
```

$$\text{Minimize} \to -2x_1 + \frac{x_1^2}{2} - 6x_2 - x_1 x_2 + x_2^2$$

$$\text{Subject to} \to \begin{pmatrix} 3x_1 + x_2 + x_3 == 25 \\ -x_1 + 2x_2 + x_4 == 10 \\ x_1 + 2x_2 + x_5 == 15 \end{pmatrix}$$

Chapter 8 Quadratic Programming

and $\rightarrow \begin{pmatrix} x_1 \geq 0 & x_2 \geq 0 & x_3 \geq 0 & x_4 \geq 0 & x_5 \geq 0 \end{pmatrix}$

Problem variables redefined as: $\{x1 \rightarrow x_1, x2 \rightarrow x_2\}$

$$A \rightarrow \begin{pmatrix} 3 & 1 & 1 & 0 & 0 \\ -1 & 2 & 0 & 1 & 0 \\ 1 & 2 & 0 & 0 & 1 \end{pmatrix} \quad b \rightarrow \begin{pmatrix} 25 \\ 10 \\ 15 \end{pmatrix}$$

$$c \rightarrow \begin{pmatrix} -2 \\ -6 \\ 0 \\ 0 \\ 0 \end{pmatrix} \quad Q \rightarrow \begin{pmatrix} 1 & -1 & 0 & 0 & 0 \\ -1 & 2 & 0 & 0 & 0 \\ 0 & 0 & 0 & 0 & 0 \\ 0 & 0 & 0 & 0 & 0 \\ 0 & 0 & 0 & 0 & 0 \end{pmatrix}$$

Convex problem. Principal minors of $Q \rightarrow \{1, 1, 0, 0, 0\}$
Starting point $\rightarrow \{2.5, 4.25, 13.25, 4., 4.\}$
Objective function $\rightarrow -19.9375$ Status $\rightarrow$ NonOptimum

***** Iteration 1 (Phase 2) *****

Tk2[diagonal] $\rightarrow \{0.16, 0.0553633, 0.00569598, 0.0625, 0.0625\}$

$$Q+Tk2 \rightarrow \begin{pmatrix} 1.16 & -1 & 0 & 0 & 0 \\ -1 & 2.05536 & 0 & 0 & 0 \\ 0 & 0 & 0.00569598 & 0 & 0 \\ 0 & 0 & 0 & 0.0625 & 0 \\ 0 & 0 & 0 & 0 & 0.0625 \end{pmatrix}$$

$$H \rightarrow \begin{pmatrix} 1.48485 & 0.722428 & 0. & 0. & 0. \\ 0.722428 & 0.838016 & 0. & 0. & 0. \\ 0. & 0. & 175.562 & 0. & 0. \\ 0. & 0. & 0. & 16. & 0. \\ 0. & 0. & 0. & 0. & 16. \end{pmatrix}$$

$$A.H \rightarrow \begin{pmatrix} 5.17698 & 3.0053 & 175.562 & 0. & 0. \\ -0.039996 & 0.953605 & 0. & 16. & 0. \\ 2.92971 & 2.39846 & 0. & 0. & 16. \end{pmatrix}$$

$$A.H.A^T \rightarrow \begin{pmatrix} 194.099 & 0.833617 & 11.1876 \\ 0.833617 & 17.9472 & 1.86721 \\ 11.1876 & 1.86721 & 23.7266 \end{pmatrix}$$

$$Qx+c \rightarrow \begin{pmatrix} -3.75 \\ 0. \\ 0. \\ 0. \\ 0. \end{pmatrix} \quad A.H.(Qx+c) \rightarrow \begin{pmatrix} -19.4137 \\ 0.149985 \\ -10.9864 \end{pmatrix}$$

$$w \rightarrow \begin{pmatrix} -0.0753651 \\ 0.0567999 \\ -0.431975 \end{pmatrix} \quad s \rightarrow \begin{pmatrix} -3.03513 \\ 0.825715 \\ 0.0753651 \\ -0.0567999 \\ 0.431975 \end{pmatrix} \quad d \rightarrow \begin{pmatrix} 3.9102 \\ 1.5007 \\ -13.2313 \\ 0.908799 \\ -6.9116 \end{pmatrix}$$

Convergence parameters $\rightarrow \{15.5308, 0., 0.668709, 1.57925\}$
d.(Qx+c) $\rightarrow -14.6632$ d.Q.d $\rightarrow 8.05778$ $\alpha 2 \rightarrow 1.81976$

$-\beta\ x/d \to \begin{pmatrix} -0.63296 & -2.80369 & 0.9914 & -4.3574 & 0.57295 \end{pmatrix}$

$\alpha \to 0.57295$

New point $\to \{4.74035, 5.10983, 5.66913, 4.5207, 0.04\}$

Objective function $\to -27.0162$ Status $\to$ NonOptimum

***** Iteration 2 (Phase 2) *****

Tk2[diagonal] $\to \{0.044502, 0.038299, 0.0311148, 0.0489316, 625.\}$

$Q+Tk2 \to \begin{pmatrix} 1.0445 & -1 & 0 & 0 & 0 \\ -1 & 2.0383 & 0 & 0 & 0 \\ 0 & 0 & 0.0311148 & 0 & 0 \\ 0 & 0 & 0 & 0.0489316 & 0 \\ 0 & 0 & 0 & 0 & 625. \end{pmatrix}$

$H \to \begin{pmatrix} 1.80539 & 0.885734 & 0. & 0. & 0. \\ 0.885734 & 0.925151 & 0. & 0. & 0. \\ 0. & 0. & 32.139 & 0. & 0. \\ 0. & 0. & 0. & 20.4367 & 0. \\ 0. & 0. & 0. & 0. & 0.0016 \end{pmatrix}$

$A.H \to \begin{pmatrix} 6.3019 & 3.58235 & 32.139 & 0. & 0. \\ -0.0339227 & 0.964568 & 0. & 20.4367 & 0. \\ 3.57686 & 2.73604 & 0. & 0. & 0.0016 \end{pmatrix}$

$A.H.AT \to \begin{pmatrix} 54.6271 & 0.862799 & 13.4666 \\ 0.862799 & 22.3998 & 1.89521 \\ 13.4666 & 1.89521 & 9.05053 \end{pmatrix}$

$Qx+c \to \begin{pmatrix} -2.36948 \\ -0.520696 \\ 0. \\ 0. \\ 0. \end{pmatrix}$ $A.H.(Qx+c) \to \begin{pmatrix} -16.7975 \\ -0.421868 \\ -9.89993 \end{pmatrix}$

$w \to \begin{pmatrix} -0.0557946 \\ 0.0700823 \\ -1.02551 \end{pmatrix}$ $s \to \begin{pmatrix} -1.1065 \\ 1.44595 \\ 0.0557946 \\ -0.0700823 \\ 1.02551 \end{pmatrix}$ $d \to \begin{pmatrix} 0.716946 \\ -0.357653 \\ -1.79319 \\ 1.43225 \\ -0.00164081 \end{pmatrix}$

Convergence parameters $\to \{2.4308, 0., 0.610507, 2.18384\}$

$d.(Qx+c) \to -1.51256$ $d.Q.d \to 1.28268$ $\alpha 2 \to 1.17922$

$-\beta\ x/d \to \begin{pmatrix} -6.54574 & 14.1443 & 3.12987 & -3.12479 & 24.1344 \end{pmatrix}$

$\alpha \to 1.17922$

New point $\to \{5.58579, 4.68807, 3.55457, 6.20964, 0.0380651\}$

Objective function $\to -27.9081$ Status $\to$ NonOptimum

***** NonOptimum solution after 2 iterations *****

Interior solution $\to \{x1 \to 5.58579, x2 \to 4.68807\}$

Objective function $\to -27.9081$

Convergence parameters $\to \{2.4308, 0., 0.610507, 2.18384\}$

Chapter 8 Quadratic Programming

The procedure is allowed to run until the optimum point is obtained.

```
{sol, history} =
PrimalAffineQP[f, g, vars, StartingVector → N[{5/2, 17/4, 53/4, 4, 4}]];
```

***** Optimum solution after 5 iterations *****

Interior solution → {x1 → 5.59985, x2 → 4.69988}
 Objective function → -27.9496

Convergence parameters →
 {0.0000108694, 6.81389 × 10^{-16}, 0.317952, 0.000414031}

The graph in Figure 8.1 confirms the solution obtained by the PAS method. The plot also shows the solution history.

```
xhist = Transpose[Transpose[history][[{1, 2}]]]; TableForm[xhist]
```

2.5	4.25
4.74035	5.10983
5.58579	4.68807
5.57839	4.71062
5.59985	4.69988
5.59985	4.69988

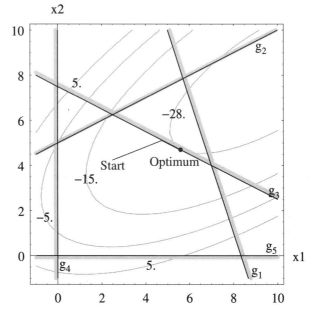

FIGURE 8.1 A graphical solution showing the search path.

8.2 The Primal Affine Scaling Method for Convex QP

Example 8.4 Solve the following QP problem using the PrimalAffineQP method.

```
f = -6x1x2 + 2x1^2 + 9x2^2 - 18x1 + 9x2;
g = {x1 + 2x2 ≤ 15}; vars = {x1, x2};

PrimalAffineQP[f, g, vars, PrintLevel → 2, MaxIterations → 2];
```

Minimize $\to -18x_1 + 2x_1^2 + 9x_2 - 6x_1x_2 + 9x_2^2$

Subject to $\to (x_1 + 2x_2 + x_3 == 15)$

and $\to (x_1 \geq 0 \quad x_2 \geq 0 \quad x_3 \geq 0)$

Problem variables redefined as: $\{x1 \to x_1, x2 \to x_2\}$

$A \to \begin{pmatrix} 1 & 2 & 1 \end{pmatrix} \quad b \to (15)$

$c \to \begin{pmatrix} -18 \\ 9 \\ 0 \end{pmatrix} \quad Q \to \begin{pmatrix} 4 & -6 & 0 \\ -6 & 18 & 0 \\ 0 & 0 & 0 \end{pmatrix}$

Convex problem. Principal minors of $Q \to \{4, 36, 0\}$

Starting point $\to \{1, 1, 1\}$ Objective function $\to -4$ Status $\to$ NonOptimum

***** **Iteration 1 (Phase 1)** *****

Tk[diagonal] $\to \{1, 1, 1, 1\}$

$A.Tk.Tk.A^T \to (127)$

$A.Tk.Tk.c \to (11) \quad w \to \left(\frac{11}{127}\right)$

$r \to \begin{pmatrix} -\frac{11}{127} \\ -\frac{22}{127} \\ -\frac{11}{127} \\ \frac{6}{127} \end{pmatrix} \quad d \to \begin{pmatrix} \frac{11}{127} \\ \frac{22}{127} \\ \frac{11}{127} \\ -\frac{6}{127} \end{pmatrix}$

Convergence parameters $\to \left\{0.217357, 0, 0.108679, \frac{38}{127}\right\}$

β (-1/d) $\to \{-11.43, -5.715, -11.43, 20.955\}$ Step length, $\alpha \to 20.955$

New point $\to \{2.815, 4.63, 2.815\}$ Artificial objective function $\to$ 0.01 Objective function $\to 121.58$ Status $\to$ NonOptimum

***** **Iteration 2 (Phase 1)** *****

Tk[diagonal] $\to \{2.815, 4.63, 2.815, 0.01\}$

$A.Tk.Tk.A^T \to (101.596)$

$A.Tk.Tk.c \to (1.1 \times 10^{-7}) \quad w \to (1.08272 \times 10^{-9})$

$r \to \begin{pmatrix} -1.08272 \times 10^{-9} \\ -2.16544 \times 10^{-9} \\ -1.08272 \times 10^{-9} \\ 0.01 \end{pmatrix} \quad d \to \begin{pmatrix} 3.04785 \times 10^{-9} \\ 1.0026 \times 10^{-8} \\ 3.04785 \times 10^{-9} \\ -0.0001 \end{pmatrix}$

Convergence parameters →
{0.0001, 0.00680625, 0.00990099, 0.0000999838}
β (-1/d) → {-3.24819 × 10^8, -9.87435 × 10^7, -3.24819 × 10^8, 9900.}
Step length, α → 0

New point → {2.815, 4.63, 2.815} Artificial objective function → 0.01
Objective function → 121.58 Status → Optimum

***** **NonOptimum solution after 2 iterations** *****

Interior solution → {x1 → 2.815, x2 → 4.63} Objective function → 121.58

Convergence parameters →
{0.0001, 0.00680625, 0.00990099, 0.0000999838}

The procedure is allowed to run until the optimum point is obtained.

```
{sol, history} = PrimalAffineQP[f, g, vars];
```

***** **Optimum solution after 5 iterations** *****

Interior solution → {x1 → 7.5006, x2 → 1.99962} Objective function → -58.5

Convergence parameters →
{0.000732458, 0.006875, 0.0113432, 0.0144618}

The graph in Figure 8.2 confirms the solution obtained by the PAS method. The optimum solution is inside the feasible domain; hence, the problem is essentially unconstrained.

```
xhist = Transpose[Transpose[history][[{1, 2}]]]; TableForm[xhist]
```

1	1
2.815	4.63
2.815	4.63
7.36173	1.93731
7.5006	1.99962
7.5006	1.99962

8.2 The Primal Affine Scaling Method for Convex QP

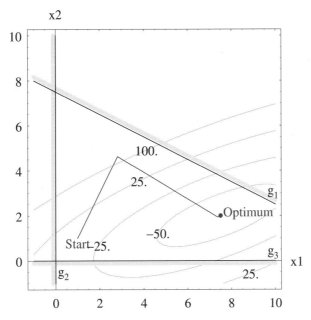

FIGURE 8.2 A graphical solution showing the search path.

Example 8.5 Solve the following QP problem using the PrimalAffineQP method. The variable x2 is unrestricted in sign.

```
f = x1^2 + x1x2 + 2x2^2 + 2x3^2 + 2x2x3 + 4x1 + 6x2 + 12x3;
g = {x1 + x2 + x3 ≥ 6, -x1 - x2 + 2x3 ≥ 2}; vars = {x1, x2, x3};
```

Intermediate results are printed for the first two iterations.

```
PrimalAffineQP[f, g, vars, UnrestrictedVariables → {x2}, PrintLevel →
2, MaxIterations → 2];
```

$\text{Minimize} \to 4x_1 + x_1^2 + 6x_2 + x_1 x_2 + 2x_2^2 - 6x_3 - x_1 x_3 - 4x_2 x_3 + 2x_3^2 + 12x_4 + 2x_2 x_4 - 2x_3 x_4 + 2x_4^2$

$\text{Subject to} \to \begin{pmatrix} x_1 + x_2 - x_3 + x_4 - x_5 == 6 \\ -x_1 - x_2 + x_3 + 2x_4 - x_6 == 2 \end{pmatrix}$

$\text{and} \to \begin{pmatrix} x_1 \geq 0 & x_2 \geq 0 & x_3 \geq 0 & x_4 \geq 0 & x_5 \geq 0 & x_6 \geq 0 \end{pmatrix}$

Problem variables redefined as: $\{x1 \to x_1, x2 \to x_2 - x_3, x3 \to x_4\}$

$A \to \begin{pmatrix} 1 & 1 & -1 & 1 & -1 & 0 \\ -1 & -1 & 1 & 2 & 0 & -1 \end{pmatrix} \quad b \to \begin{pmatrix} 6 \\ 2 \end{pmatrix}$

$$c \to \begin{pmatrix} 4 \\ 6 \\ -6 \\ 12 \\ 0 \\ 0 \end{pmatrix} \quad Q \to \begin{pmatrix} 2 & 1 & -1 & 0 & 0 & 0 \\ 1 & 4 & -4 & 2 & 0 & 0 \\ -1 & -4 & 4 & -2 & 0 & 0 \\ 0 & 2 & -2 & 4 & 0 & 0 \\ 0 & 0 & 0 & 0 & 0 & 0 \\ 0 & 0 & 0 & 0 & 0 & 0 \end{pmatrix}$$

Convex problem. Principal minors of $Q \to \{2, 7, 0, 0, 0, 0\}$
Starting point $\to \{1, 1, 1, 1, 1, 1\}$
Objective function $\to 19$ Status $\to$ NonOptimum

***** **Iteration 1 (Phase 1)** *****

Tk[diagonal] $\to \{1, 1, 1, 1, 1, 1, 1\}$

$$\text{A.Tk.Tk.AT} \to \begin{pmatrix} 30 & 9 \\ 9 & 12 \end{pmatrix}$$

$$\text{A.Tk.Tk.c} \to \begin{pmatrix} 5 \\ 2 \end{pmatrix} \quad w \to \begin{pmatrix} \frac{14}{93} \\ \frac{5}{93} \end{pmatrix}$$

$$r \to \begin{pmatrix} -\frac{3}{31} \\ -\frac{3}{31} \\ \frac{3}{31} \\ -\frac{8}{31} \\ \frac{14}{93} \\ \frac{5}{93} \\ \frac{13}{93} \end{pmatrix} \quad d \to \begin{pmatrix} \frac{3}{31} \\ \frac{3}{31} \\ -\frac{3}{31} \\ \frac{8}{31} \\ -\frac{14}{93} \\ -\frac{5}{93} \\ -\frac{13}{93} \end{pmatrix}$$

Convergence parameters $\to \{0.373878, 0, 0.186939, \frac{1}{93}\}$
 $\beta(-1/d) \to \{-10.23, -10.23, 10.23, -3.83625, 6.57643, 18.414, 7.08231\}$
 Step length, $\alpha \to 6.57643$
New point $\to \{1.63643, 1.63643, 0.363571, 2.69714, 0.01, 0.646429\}$
 Artificial objective function $\to 0.0807143$
 Objective function $\to 75.9651$
Status $\to$ NonOptimum

***** **Iteration 2 (Phase 1)** *****

Tk[diagonal] $\to$
$\{1.63643, 1.63643, 0.363571, 2.69714, 0.01, 0.646429, 0.0807143\}$

$$\text{A.Tk.Tk.AT} \to \begin{pmatrix} 12.7637 & 9.0616 \\ 9.0616 & 35.0043 \end{pmatrix}$$

$$\text{A.Tk.Tk.c} \to \begin{pmatrix} 0.000212213 \\ 0.0000848851 \end{pmatrix} \quad w \to \begin{pmatrix} 0.0000182607 \\ -2.30217 \times 10^{-6} \end{pmatrix}$$

$$r \to \begin{pmatrix} -0.0000205628 \\ -0.0000205628 \\ 0.0000205628 \\ -0.0000136563 \\ 0.0000182607 \\ -2.30217 \times 10^{-6} \\ 0.0807073 \end{pmatrix} \quad d \to \begin{pmatrix} 0.0000336496 \\ 0.0000336496 \\ -7.47606 \times 10^{-6} \\ 0.0000368331 \\ -1.82607 \times 10^{-7} \\ 1.48819 \times 10^{-6} \\ -0.00651423 \end{pmatrix}$$

8.2 The Primal Affine Scaling Method for Convex QP

```
Convergence parameters → {0.00651451, 0.054553, 0.0746796, 0.00640984}
  β(-1/d) → {-29420.8, -29420.8, 132423., -26878.,
             5.42149×10⁶, -665239., 151.975}
Step length, α → 151.975
New point → {1.6448, 1.6448, 0.363158, 2.71224, 0.00999972, 0.646575}
  Artificial objective function → 0.000807143  Objective function → 76.5792
  Status → NonOptimum
```

***** **NonOptimum solution after 2 iterations** *****

```
Interior solution → {x1 → 1.6448, x2 → 1.28164, x3 → 2.71224}
  Objective function → 76.5792
Convergence parameters → {0.00651451, 0.054553, 0.0746796, 0.00640984}
```

The optimum solution for the problem is computed as follows:

PrimalAffineQP[f, g, vars, UnrestrictedVariables → {x2}];

***** **Optimum solution after 9 iterations** *****

```
Interior solution → {x1 → 4.0984, x2 → -0.964673, x3 → 2.49501}
  Objective function → 62.8866
Convergence parameters → {0.000669959, 0.0546009, 0.50019, 0.0166818}
```

The solution is fairly close to the one obtained by using KT conditions. Note that we must add the positivity constraints before solving the problem using KT conditions. In the PAS method, these are always implicit but not in the case of the KT solution.

KTSolution[f, Join[g, {x1 ≥ 0, x3 ≥ 0}], vars];

Lagrangian → $4x1 + x1^2 + 6x2 + x1x2 + 2x2^2 + 12x3 + 2x2x3 + 2x3^2 + \left(6 - x1 - x2 - x3 + s_1^2\right)u_1 + \left(2 + x1 + x2 - 2x3 + s_2^2\right)u_2 + \left(-x1 + s_3^2\right)u_3 + \left(-x3 + s_4^2\right)u_4$

***** **Valid KT Point(s)** *****

$f \to 68.6667$
$x1 \to 4.33333$
$x2 \to -1.$
$x3 \to 2.66667$
$u_1 \to 14.6667$
$u_2 \to 3.$
$u_3 \to 0$
$u_4 \to 0$
$s_1^2 \to 0$
$s_2^2 \to 0$
$s_3^2 \to 4.33333$
$s_4^2 \to 2.66667$

8.3 The Primal-Dual Method for Convex QP

The necessary conditions for the optimum of QP problems are represented by the following system of equations:

$$\begin{aligned}
\mathbf{Ax} - \mathbf{b} &= \mathbf{0} & &\text{Primal feasibility} \\
-\mathbf{Qx} + \mathbf{A}^T\mathbf{v} + \mathbf{u} &= \mathbf{c} & &\text{Dual feasibility} \\
\mathbf{XUe} &= \mathbf{0} & &\text{Complementary slackness conditions} \\
x_i \geq 0, \; u_i &\geq 0, \quad i = 1, \ldots, n
\end{aligned}$$

The primal-dual interior point method is based on trying to solve the above system of equations directly. Because of the complementary slackness conditions, the complete system of equations is nonlinear. We can use an iterative method, such as the Newton-Raphson method, to solve this system. The linear equations will be satisfied exactly at each iteration. The error in satisfying the complementary slackness condition will be reduced at each iteration. Using $\mu > 0$ as an indicator of this error, at a given iteration, the complementary slackness conditions are of the following form:

$$\mathbf{XUe} = \mu \mathbf{e}$$

8.3.1 Direction Using the Newton-Raphson Method

Similar to the LP case, at each iteration we need to compute new values of $2n+m$ variables ($\mathbf{x}$, $\mathbf{u}$, and $\mathbf{v}$). Using the Newton-Raphson method, the changes in these variables are obtained from solving the following system of equations:

$$\begin{pmatrix} \mathbf{A} & \mathbf{0} & \mathbf{0} \\ -\mathbf{Q} & \mathbf{I} & \mathbf{A}^T \\ \mathbf{U} & \mathbf{X} & \mathbf{0} \end{pmatrix} \begin{pmatrix} \mathbf{d}_x \\ \mathbf{d}_u \\ \mathbf{d}_v \end{pmatrix} = - \begin{pmatrix} \mathbf{Ax}^k - \mathbf{b} \\ -\mathbf{Qx} + \mathbf{A}^T\mathbf{v}^k + \mathbf{u}^k - \mathbf{c} \\ \mathbf{XUe} - \mu^k \mathbf{e} \end{pmatrix}$$

Note that the matrices $\mathbf{U}$ and $\mathbf{X}$ are defined by using the known values at the current iteration. However, to simplify notation, superscript k is not used on these terms. We can perform the computations more efficiently by first writing the three sets of equations explicitly, as follows:

(a) $\mathbf{A}\mathbf{d}_x = -\mathbf{A}\mathbf{x}^k + \mathbf{b} \equiv \mathbf{r}_p$

(b) $-\mathbf{Q}\mathbf{d}_x + \mathbf{d}_u + \mathbf{A}^T \mathbf{d}_v = \mathbf{Q}\mathbf{x} - \mathbf{A}^T \mathbf{v}^k - \mathbf{u}^k + \mathbf{c} \equiv \mathbf{r}_d$

(c) $\mathbf{U}\mathbf{d}_x + \mathbf{X}\mathbf{d}_u = -\mathbf{X}\mathbf{U}\mathbf{e} + \mu^k \mathbf{e} \equiv \mathbf{r}_c$

Multiplying equation (b) by $\mathbf{X}$ and substituting for $\mathbf{X}\mathbf{d}_u$ from equation (c), we get

$$-\mathbf{X}\mathbf{Q}\mathbf{d}_x + \mathbf{r}_c - \mathbf{U}\mathbf{d}_x + \mathbf{X}\mathbf{A}^T \mathbf{d}_v = \mathbf{X}\mathbf{r}_d$$

or

$$-[\mathbf{X}\mathbf{Q} + \mathbf{U}]\mathbf{d}_x + \mathbf{X}\mathbf{A}^T \mathbf{d}_v = \mathbf{X}\mathbf{r}_d - \mathbf{r}_c$$

(d) $-\mathbf{d}_x + [\mathbf{X}\mathbf{Q} + \mathbf{U}]^{-1} \mathbf{X}\mathbf{A}^T \mathbf{d}_v = [\mathbf{X}\mathbf{Q} + \mathbf{U}]^{-1} (\mathbf{X}\mathbf{r}_d - \mathbf{r}_c)$

Now multiplying by $\mathbf{A}$ and using equation (a), we get the following system of equations for solving for $\mathbf{d}_v$:

$$\mathbf{A}[\mathbf{X}\mathbf{Q} + \mathbf{U}]^{-1} \mathbf{X}\mathbf{A}^T \mathbf{d}_v = \mathbf{r}_p + \mathbf{A}[\mathbf{X}\mathbf{Q} + \mathbf{U}]^{-1} (\mathbf{X}\mathbf{r}_d - \mathbf{r}_c)$$

Once $\mathbf{d}_v$ is known, we can compute $\mathbf{d}_x$ from (d) and $\mathbf{d}_u$ from (c) as follows:

From (d) $\quad \mathbf{d}_x = [\mathbf{X}\mathbf{Q} + \mathbf{U}]^{-1} \mathbf{X}\mathbf{A}^T \mathbf{d}_v - [\mathbf{X}\mathbf{Q} + \mathbf{U}]^{-1} (\mathbf{X}\mathbf{r}_d - \mathbf{r}_c)$
or $\quad \mathbf{d}_x = [\mathbf{X}\mathbf{Q} + \mathbf{U}]^{-1} (\mathbf{X}\mathbf{A}^T \mathbf{d}_v - \mathbf{X}\mathbf{r}_d + \mathbf{r}_c)$
From (c) $\quad \mathbf{X}\mathbf{d}_u = \mathbf{r}_c - \mathbf{U}\mathbf{d}_x$
or $\quad \mathbf{d}_u = \mathbf{X}^{-1} (\mathbf{r}_c - \mathbf{U}\mathbf{d}_x)$

8.3.2 Step-Length Calculations

The previous derivation did not take into account the positivity requirements $x_i \geq 0$ and $u_i \geq 0$. Assuming that we start from positive initial values, we take care of these requirements by introducing a step-length parameter α.

$$\mathbf{x}^{k+1} = \mathbf{x}^k + \alpha_p \mathbf{d}_x$$

$$\mathbf{u}^{k+1} = \mathbf{u}^k + \alpha_d \mathbf{d}_u$$

$$\mathbf{v}^{k+1} = \mathbf{v}^k + \alpha_d \mathbf{d}_v$$

The maximum value of the step length is the one that will make one of the x_i or u_i values go to zero.

$$x_i + \alpha_p d_{xi} \geq 0 \quad u_i + \alpha_d d_{ui} \geq 0 \quad i = 1, \ldots, n$$

Variables with positive increments obviously will remain positive regardless of the step length. Similar to the LP case, in order to strictly maintain feasibility, the actual step length should be slightly smaller than the maximum. The actual step length is chosen as follows:

$$\alpha_p = \beta \operatorname{Min}[1, -x_i/d_{xi}, d_{xi} < 0] \quad \alpha_d = \beta \operatorname{Min}[1, -u_i/d_{ui}, d_{ui} < 0]$$

The parameter β is between 0 and 1 with a usual value of $\beta = 0.999$. The variables are then updated as follows:

$$\mathbf{x}^{k+1} = \mathbf{x}^k + \alpha_p d_x$$
$$\mathbf{u}^{k+1} = \mathbf{u}^k + \alpha_d d_u$$
$$\mathbf{v}^{k+1} = \mathbf{v}^k + \alpha_d d_v$$

8.3.3 Convergence Criteria

Feasibility

The constraints must be satisfied at the optimum, i.e., $\mathbf{Ax}^k - \mathbf{b} = \mathbf{0}$. To use as convergence criteria, this requirement is expressed as follows:

$$\sigma_p = \frac{\|\mathbf{Ax}^k - \mathbf{b}\|}{\|\mathbf{b}\| + 1} \leq \epsilon_1$$

where ϵ_1 is a small positive number. The 1 is added to the denominator to avoid division by small numbers.

Dual Feasibility

We also have the requirement that

$$-\mathbf{Qx} + \mathbf{A}^T \mathbf{v}^k + \mathbf{u}^k = \mathbf{c}$$

This gives the following convergence criteria:

$$\sigma_d = \frac{||\mathbf{r}_d||}{||\mathbf{Q}\mathbf{x}^k + \mathbf{c}|| + 1} \leq \epsilon_2$$

where ϵ_2 is a small positive number.

Complementary Slackness

The value of parameter μ determines how well complementary slackness conditions are satisfied. Numerical experiments suggests defining an average value of μ as follows:

$$\mu^k = \frac{(\mathbf{x}^k)^T \mathbf{u}}{n}$$

where n = number of optimization variables. This parameter should be zero at the optimum. Thus, for convergence,

$$\mu \leq \epsilon_3$$

where ϵ_3 is a small positive number.

8.3.4 Complete Primal-Dual QP Algorithm

Given: Constraint coefficient matrix $\mathbf{A}$, constraint right-hand side vector $\mathbf{b}$, objective function coefficient vector $\mathbf{c}$, step-length parameter β, and convergence tolerance parameters.

Initialization: $k = 0$, arbitrary initial values (≥ 0), say $\mathbf{x}^k = \mathbf{u}^k = \mathbf{e}$ (vector with all entries 1) and $\mathbf{v}^k = \mathbf{0}$.

The next point $\mathbf{x}^{k+1}$ is computed as follows.

1. Set $\mu^k = [\frac{(\mathbf{x}^k)^T \mathbf{u}}{n}]/(k+1)$. Check for convergence. If

$$\left[\frac{||\mathbf{A}\mathbf{x}^k - \mathbf{b}||}{||\mathbf{b}|| + 1} \leq \epsilon_1, \frac{||\mathbf{r}_d||}{||\mathbf{Q}\mathbf{x}^k + \mathbf{c}|| + 1} \leq \epsilon_2, \mu^k \leq \epsilon_2 \right],$$

we have the optimum. Otherwise, continue.

2. Form:
$$r_p = -Ax^k + b$$
$$r_d = Qx^k - A^T v^k - u^k + c$$
$$r_c = -XUe + \mu^k e$$

3. Solve the system of linear equations for d_v:
$$A[XQ+U]^{-1} X A^T d_v = r_p + A[XQ+U]^{-1}(Xr_d - r_c)$$

4. Compute increments:
$$d_x = [XQ+U]^{-1}(XA^T d_v - Xr_d + r_c)$$
$$d_u = X^{-1}(r_c - U d_x)$$

5. Compute step lengths:
$$\alpha_p = \beta \operatorname{Min}[1, -x_i/d_{xi}, d_{xi} < 0] \quad \alpha_d = \beta \operatorname{Min}[1, -u_i/d_{ui}, d_{ui} < 0]$$

6. Compute the next point:
$$x^{k+1} = x^k + \alpha_p d_x$$
$$u^{k+1} = u^k + \alpha_d d_u$$
$$v^{k+1} = v^k + \alpha_d d_v$$

The PrimalDualQP Function

The following PrimalDualQP function implements the PrimalDual algorithm for solving QP problems. The function usage and its options are explained first. Several intermediate results can be printed to gain understanding of the process.

```
Needs["OptimizationToolbox`QuadraticProgramming`"];
?PrimalDualQP
```

```
PrimalDualQP[f, g, vars, options]. Solves an QP problem using
   Interior Point algorithm based on solving KT conditions using
   the Newton-Raphson method. f is the objective function, g is
   a list of constraints, and vars is a list of variables. See
   Options[PrimalDualQP] to find out about a list of valid options
   for this function.
```

8.3 The Primal-Dual Method for Convex QP

```
OptionsUsage[PrimalDualQP]
{UnrestrictedVariables → {}, MaxIterations → 20, ProblemType → Min,
 StandardVariableName → x, PrintLevel → 1, StepLengthFactor → 0.99,
 ConvergenceTolerance → 0.0001, StartingVector → {}}
```

UnrestrictedVariables is an option for LP and several QP problems.
  A list of variables that are not restricted to be positive can be
  specified with this option. Default is {}.

MaxIterations is an option for several optimization methods. It
  specifies maximum number of iterations allowed.

ProblemType is an option for most optimization methods. It can either be
  Min (default) or Max.

StandardVariableName is an option for LP and QP methods. It specifies
  the symbol to use when creating variable names during conversion to
  the standard form. Default is x.

PrintLevel is an option for most functions in the OptimizationToolbox.
  It is specified as an integer. The value of the integer indicates
  how much intermediate information is to be printed. A PrintLevel→ 0
  suppresses all printing. Default for most functions is set to 1 in
  which case they print only the initial problem setup. Higher integers
  print more intermediate results.

StepLengthFactor is an option for interior point methods. It is the
  reduction factor applied to the computed step length to maintain
  feasibility. Default is 0.99

ConvergenceTolerance is an option for most optimization methods. Most
  methods require only a single zero tolerance value. Some interior
  point methods require a list of convergence tolerance values.

StartingVector is an option for several interior point methods. Default
  is {1,...,1}.

Example 8.6 Solve the following QP problem using the PrimalDualQP method.

Minimize $f = -2x_1 + \frac{x_1^2}{2} - 6x_2 - x_1 x_2 + x_2^2$

Subject to $\begin{pmatrix} 3x_1 + x_2 \leq 25 \\ -x_1 + 2x_2 \leq 10 \\ x_1 + 2x_2 \leq 15 \\ x_i \geq 0, i = 1, 2 \end{pmatrix}$

```
f = -2x1 + x1²/2 - 6x2 - x1x2 + x2²;
g = {3x1 + x2 ≤ 25, -x1 + 2x2 ≤ 10, x1 + 2x2 ≤ 15}; vars = {x1, x2};
```

All intermediate calculations are shown for the first two iterations.

PrimalDualQP[f, g, vars, PrintLevel → 2, MaxIterations → 2];

Minimize → $-2x_1 + \dfrac{x_1^2}{2} - 6x_2 - x_1 x_2 + x_2^2$

Subject to → $\begin{pmatrix} 3x_1 + x_2 + x_3 == 25 \\ -x_1 + 2x_2 + x_4 == 10 \\ x_1 + 2x_2 + x_5 == 15 \end{pmatrix}$

and → $(x_1 \geq 0 \quad x_2 \geq 0 \quad x_3 \geq 0 \quad x_4 \geq 0 \quad x_5 \geq 0)$

Problem variables redefined as: $\{x1 \to x_1, x2 \to x_2\}$

$A \to \begin{pmatrix} 3 & 1 & 1 & 0 & 0 \\ -1 & 2 & 0 & 1 & 0 \\ 1 & 2 & 0 & 0 & 1 \end{pmatrix} \quad b \to \begin{pmatrix} 25 \\ 10 \\ 15 \end{pmatrix}$

$c \to \begin{pmatrix} -2 \\ -6 \\ 0 \\ 0 \\ 0 \end{pmatrix} \quad Q \to \begin{pmatrix} 1 & -1 & 0 & 0 & 0 \\ -1 & 2 & 0 & 0 & 0 \\ 0 & 0 & 0 & 0 & 0 \\ 0 & 0 & 0 & 0 & 0 \\ 0 & 0 & 0 & 0 & 0 \end{pmatrix}$

Convex problem. Principal minors of $Q \to \{1, 1, 0, 0, 0\}$

**** Starting vectors ****

Primal vars (x) → {1., 1., 1., 1., 1.}

Dual vars (u) → {1., 1., 1., 1., 1.}

Multipliers (v) → {0., 0., 0.}

Objective function → -7.5   Status → NonOptimum

***** Iteration 1 *****

$r_p \to \begin{pmatrix} 20. \\ 8. \\ 11. \end{pmatrix} \quad r_d \to \begin{pmatrix} -3. \\ -6. \\ -1. \\ -1. \\ -1. \end{pmatrix} \quad r_c \to \begin{pmatrix} 0. \\ 0. \\ 0. \\ 0. \\ 0. \end{pmatrix}$

Parameters: $\sigma p \to 0.760063 \quad \sigma d \to 1.08505 \quad \mu \to 1.$

$X[\text{diagonal}] \to \begin{pmatrix} 1. \\ 1. \\ 1. \\ 1. \\ 1. \end{pmatrix} \quad U[\text{diagonal}] \to \begin{pmatrix} 1. \\ 1. \\ 1. \\ 1. \\ 1. \end{pmatrix}$

$X.Q + U \to \begin{pmatrix} 2. & -1. & 0. & 0. & 0. \\ -1. & 3. & 0. & 0. & 0. \\ 0. & 0. & 1. & 0. & 0. \\ 0. & 0. & 0. & 1. & 0. \\ 0. & 0. & 0. & 0. & 1. \end{pmatrix}$

8.3 The Primal-Dual Method for Convex QP

$$\text{Inverse}[X.Q + U] \to \begin{pmatrix} 0.6 & 0.2 & 0. & 0. & 0. \\ 0.2 & 0.4 & 0. & 0. & 0. \\ 0. & 0. & 1. & 0. & 0. \\ 0. & 0. & 0. & 1. & 0. \\ 0. & 0. & 0. & 0. & 1. \end{pmatrix}$$

$$\text{A.Inverse}(X.Q + U).X.AT \to \begin{pmatrix} 8. & 2.22045 \times 10^{-16} & 4. \\ 2.22045 \times 10^{-16} & 2.4 & 1. \\ 4. & 1. & 4. \end{pmatrix}$$

$$\text{rp+A.Inverse}[X.Q + U](X.\text{rd-rc}) \to \begin{pmatrix} 7. \\ 4. \\ 1. \end{pmatrix} \quad dv \to \begin{pmatrix} 2.19079 \\ 2.76316 \\ -2.63158 \end{pmatrix}$$

$$dx \to \begin{pmatrix} 4.19737 \\ 4.21711 \\ 3.19079 \\ 3.76316 \\ -1.63158 \end{pmatrix} \quad du \to \begin{pmatrix} -4.19737 \\ -4.21711 \\ -3.19079 \\ -3.76316 \\ 1.63158 \end{pmatrix}$$

$-x/dx \to (\infty \quad \infty \quad \infty \quad \infty \quad 0.612903) \quad \alpha_p \to 0.606774$

$-u/du \to (0.238245 \quad 0.237129 \quad 0.313402 \quad 0.265734 \quad \infty) \quad \alpha_d \to 0.234758$

New primal vars $\to$ (3.54685 3.55883 2.93609 3.28339 0.01)

New dual vars $\to$ (0.0146334 0.01 0.250936 0.116568 1.38303)

New multipliers $\to$ (0.514306 0.648674 -0.617785)

Objective function $\to$ -22.114   Status $\to$ NonOptimum

***** Iteration 2 *****

$$r_p \to \begin{pmatrix} 7.86452 \\ 3.14581 \\ 4.32548 \end{pmatrix} \quad r_d \to \begin{pmatrix} -2.30307 \\ -3.01528 \\ -0.765242 \\ -0.765242 \\ -0.765242 \end{pmatrix} \quad r_c \to \begin{pmatrix} 0.0701804 \\ 0.0864946 \\ -0.614688 \\ -0.260655 \\ 0.108253 \end{pmatrix}$$

Parameters: $\sigma p \to 0.298876 \quad \sigma d \to 0.967466 \quad \mu \to 0.1220823$

$$X[\text{diagonal}] \to \begin{pmatrix} 3.54685 \\ 3.55883 \\ 2.93609 \\ 3.28339 \\ 0.01 \end{pmatrix} \quad U[\text{diagonal}] \to \begin{pmatrix} 0.0146334 \\ 0.01 \\ 0.250936 \\ 0.116568 \\ 1.38303 \end{pmatrix}$$

$$X.Q + U \to \begin{pmatrix} 3.56149 & -3.54685 & 0. & 0. & 0. \\ -3.55883 & 7.12766 & 0. & 0. & 0. \\ 0. & 0. & 0.250936 & 0. & 0. \\ 0. & 0. & 0. & 0.116568 & 0. \\ 0. & 0. & 0. & 0. & 1.38303 \end{pmatrix}$$

$$\text{Inverse}[X.Q + U] \to \begin{pmatrix} 0.558488 & 0.277914 & 0. & 0. & 0. \\ 0.278852 & 0.27906 & 0. & 0. & 0. \\ 0. & 0. & 3.98508 & 0. & 0. \\ 0. & 0. & 0. & 8.57869 & 0. \\ 0. & 0. & 0. & 0. & 0.723052 \end{pmatrix}$$

$$\text{A.Inverse}(X.Q + U).X.A^T \to \begin{pmatrix} 36.4558 & 0.988872 & 14.8522 \\ 0.988872 & 30.1644 & 1.99164 \\ 14.8522 & 1.99164 & 9.91681 \end{pmatrix}$$

$$rp + \text{A.Inverse}[X.Q + U](X.rd - rc) \to \begin{pmatrix} -26.7785 \\ -19.1975 \\ -13.9981 \end{pmatrix} \quad dv \to \begin{pmatrix} -0.490931 \\ -0.583417 \\ -0.559126 \end{pmatrix}$$

$$dx \to \begin{pmatrix} 1.99266 \\ 1.12653 \\ 0.760003 \\ 2.88541 \\ 0.0797626 \end{pmatrix} \quad du \to \begin{pmatrix} 0.0115655 \\ 0.0211388 \\ -0.27431 \\ -0.181825 \\ -0.206116 \end{pmatrix}$$

$-x/dx \to (\infty \quad \infty \quad \infty \quad \infty \quad \infty) \quad \alpha_p \to 1$

$-u/du \to (\infty \quad \infty \quad 0.914789 \quad 0.641101 \quad 6.70994) \quad \alpha_d \to 0.63469$

New primal vars $\to$ (5.53952 4.68536 3.69609 6.16879 0.0897626)

New dual vars $\to$ (0.0219739 0.0234166 0.076834 0.00116568 1.25221)

New multipliers $\to$ (0.202717 0.278385 $-$0.972656)

Objective function $\to$ $-$27.8501 Status $\to$ NonOptimum

***** NonOptimum solution after 2 iterations *****

Interior solution $\to$ {x1 $\to$ 5.53952, x2 $\to$ 4.68536}

Objective function $\to$ $-$27.8501

The procedure is allowed to run until the optimum point is obtained.

{sol, history} = PrimalDualQP[f, g, vars];

***** Optimum solution after 7 iterations *****

Interior solution $\to$ {x1 $\to$ 5.59981, x2 $\to$ 4.69992}

Objective function $\to$ $-$27.9496

The graph in Figure 8.3 confirms the solution obtained and shows the path taken by the method.

xhist = Transpose[Transpose[history][[{1, 2}]]]; TableForm[xhist]

1.	1.
3.54685	3.55883
5.53952	4.68536
5.57472	4.68986
5.59438	4.69782
5.59888	4.69954
5.59981	4.69992
5.59981	4.69992

8.3 The Primal-Dual Method for Convex QP

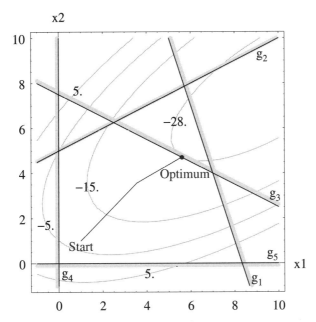

FIGURE 8.3 A graphical solution showing the search path.

Example 8.7 Solve the following QP problem using the PrimalDualQP method. The variable x2 is unrestricted in sign.

```
f = x1^2 + x1x2 + 2x2^2 + 2x3^2 + 2x2x3 + 4x1 + 6x2 + 12x3;
g = {x1 + x2 + x3 ≥ 6, -x1 - x2 + 2x3 ≥ 2}; vars = {x1, x2, x3};
```

Intermediate results are printed for the first two iterations.

```
PrimalDualQP[f, g, vars, UnrestrictedVariables → {x2},
   PrintLevel → 2, MaxIterations → 2];
```

Minimize → $4x_1 + x_1^2 + 6x_2 + x_1 x_2 + 2x_2^2 - 6x_3 - x_1 x_3 - 4x_2 x_3 + 2x_3^2 + 12x_4 + 2x_2 x_4 - 2x_3 x_4 + 2x_4^2$

Subject to → $\begin{pmatrix} x_1 + x_2 - x_3 + x_4 - x_5 == 6 \\ -x_1 - x_2 + x_3 + 2x_4 - x_6 == 2 \end{pmatrix}$

and → $\begin{pmatrix} x_1 \geq 0 & x_2 \geq 0 & x_3 \geq 0 & x_4 \geq 0 & x_5 \geq 0 & x_6 \geq 0 \end{pmatrix}$

Problem variables redefined as: $\{x1 \to x_1, x2 \to x_2 - x_3, x3 \to x_4\}$

$A \to \begin{pmatrix} 1 & 1 & -1 & 1 & -1 & 0 \\ -1 & -1 & 1 & 2 & 0 & -1 \end{pmatrix} \quad b \to \begin{pmatrix} 6 \\ 2 \end{pmatrix}$

Chapter 8 Quadratic Programming

$$c \to \begin{pmatrix} 4 \\ 6 \\ -6 \\ 12 \\ 0 \\ 0 \end{pmatrix} \quad Q \to \begin{pmatrix} 2 & 1 & -1 & 0 & 0 & 0 \\ 1 & 4 & -4 & 2 & 0 & 0 \\ -1 & -4 & 4 & -2 & 0 & 0 \\ 0 & 2 & -2 & 4 & 0 & 0 \\ 0 & 0 & 0 & 0 & 0 & 0 \\ 0 & 0 & 0 & 0 & 0 & 0 \end{pmatrix}$$

Convex problem. Principal minors of $Q \to \{2, 7, 0, 0, 0, 0\}$

**** Starting vectors ****

Primal vars (x) $\to \{1., 1., 1., 1., 1., 1.\}$

Dual vars (u) $\to \{1., 1., 1., 1., 1., 1.\}$

Multipliers (v) $\to \{0., 0.\}$

Objective function $\to 19.$ Status $\to$ NonOptimum

***** Iteration 1 *****

$$r_p \to \begin{pmatrix} 5. \\ 2. \end{pmatrix} \quad r_d \to \begin{pmatrix} 5. \\ 8. \\ -10. \\ 15. \\ -1. \\ -1. \end{pmatrix} \quad r_c \to \begin{pmatrix} 0. \\ 0. \\ 0. \\ 0. \\ 0. \\ 0. \end{pmatrix}$$

Parameters: $\sigma p \to 0.735221 \quad \sigma d \to 0.914324 \quad \mu \to 1.$

$$X[\text{diagonal}] \to \begin{pmatrix} 1. \\ 1. \\ 1. \\ 1. \\ 1. \\ 1. \end{pmatrix} \quad U[\text{diagonal}] \to \begin{pmatrix} 1. \\ 1. \\ 1. \\ 1. \\ 1. \\ 1. \end{pmatrix}$$

$$X.Q + U \to \begin{pmatrix} 3. & 1. & -1. & 0. & 0. & 0. \\ 1. & 5. & -4. & 2. & 0. & 0. \\ -1. & -4. & 5. & -2. & 0. & 0. \\ 0. & 2. & -2. & 5. & 0. & 0. \\ 0. & 0. & 0. & 0. & 1. & 0. \\ 0. & 0. & 0. & 0. & 0. & 1. \end{pmatrix}$$

Inverse[X.Q + U] $\to$

$$\begin{pmatrix} 0.366337 & -0.049505 & 0.049505 & 0.039604 & 0. & 0. \\ -0.049505 & 0.574257 & 0.425743 & -0.0594059 & 0. & 0. \\ 0.049505 & 0.425743 & 0.574257 & 0.0594059 & 0. & 0. \\ 0.039604 & -0.0594059 & 0.0594059 & 0.247525 & 0. & 0. \\ 0. & 0. & 0. & 0. & 1. & 0. \\ 0. & 0. & 0. & 0. & 0. & 1. \end{pmatrix}$$

$$A.\text{Inverse}(X.Q + U).X.A^T \to \begin{pmatrix} 1.55446 & -0.049505 \\ -0.049505 & 2.77228 \end{pmatrix}$$

$$rp + A.\text{Inverse}[X.Q + U](X.rd - rc) \to \begin{pmatrix} 10.7723 \\ 6.75247 \end{pmatrix} \quad dv \to \begin{pmatrix} 7.01149 \\ 2.56092 \end{pmatrix}$$

$$dx \to \begin{pmatrix} 0.135632 \\ 0.521839 \\ 1.47816 \\ -0.190805 \\ -6.01149 \\ -1.56092 \end{pmatrix} \quad du \to \begin{pmatrix} -0.135632 \\ -0.521839 \\ -1.47816 \\ 0.190805 \\ 6.01149 \\ 1.56092 \end{pmatrix}$$

$-x/dx \to \begin{pmatrix} \infty & \infty & \infty & 5.24096 & 0.166348 & 0.640648 \end{pmatrix} \quad \alpha_p \to 0.164685$

$-u/du \to \begin{pmatrix} 7.37288 & 1.9163 & 0.676516 & \infty & \infty & \infty \end{pmatrix} \quad \alpha_d \to 0.669751$

New primal vars $\to$
$\begin{pmatrix} 1.02234 & 1.08594 & 1.24343 & 0.968577 & 0.01 & 0.742941 \end{pmatrix}$

New dual vars $\to$
$\begin{pmatrix} 0.90916 & 0.650498 & 0.01 & 1.12779 & 5.02621 & 2.04543 \end{pmatrix}$

New multipliers $\to \begin{pmatrix} 4.69596 & 1.71518 \end{pmatrix}$

Objective function $\to 17.2723$ Status $\to$ NonOptimum

***** **Iteration 2** *****

$$r_p \to \begin{pmatrix} 4.17658 \\ 1.67063 \end{pmatrix} \quad r_d \to \begin{pmatrix} 1.99724 \\ 4.69825 \\ -5.35875 \\ 6.30522 \\ -0.330249 \\ -0.330249 \end{pmatrix} \quad r_c \to \begin{pmatrix} -0.570255 \\ -0.347188 \\ 0.346778 \\ -0.733141 \\ 0.30895 \\ -1.16042 \end{pmatrix}$$

Parameters: $\sigma p \to 0.614141 \quad \sigma d \to 0.455202 \quad \mu \to 0.359212$

$$X[\text{diagonal}] \to \begin{pmatrix} 1.02234 \\ 1.08594 \\ 1.24343 \\ 0.968577 \\ 0.01 \\ 0.742941 \end{pmatrix} \quad U[\text{diagonal}] \to \begin{pmatrix} 0.90916 \\ 0.650498 \\ 0.01 \\ 1.12779 \\ 5.02621 \\ 2.04543 \end{pmatrix}$$

$$X.Q + U \to \begin{pmatrix} 2.95383 & 1.02234 & -1.02234 & 0. & 0. & 0. \\ 1.08594 & 4.99425 & -4.34376 & 2.17188 & 0. & 0. \\ -1.24343 & -4.97372 & 4.98372 & -2.48686 & 0. & 0. \\ 0. & 1.93715 & -1.93715 & 5.0021 & 0. & 0. \\ 0. & 0. & 0. & 0. & 5.02621 & 0. \\ 0. & 0. & 0. & 0. & 0. & 2.04543 \end{pmatrix}$$

Inverse$[X.Q + U] \to$
$$\begin{pmatrix} 0.379125 & -0.00146237 & 0.0951269 & 0.0479286 & 0. & 0. \\ -0.00155335 & 1.51697 & 1.32115 & -0.00183456 & 0. & 0. \\ 0.115699 & 1.51275 & 1.596 & 0.136645 & 0. & 0. \\ 0.0454083 & -0.0016363 & 0.106441 & 0.253545 & 0. & 0. \\ 0. & 0. & 0. & 0. & 0.198957 & 0. \\ 0. & 0. & 0. & 0. & 0. & 0.488895 \end{pmatrix}$$

A.Inverse$(X.Q + U).X.AT \to \begin{pmatrix} 0.56635 & -0.0907451 \\ -0.0907451 & 2.19055 \end{pmatrix}$

$$\text{rp+A.Inverse}[X.Q + U](X.\text{rd-rc}) \to \begin{pmatrix} 6.67612 \\ 2.0794 \end{pmatrix} \quad dv \to \begin{pmatrix} 12.0198 \\ 1.44719 \end{pmatrix}$$

$$dx \to \begin{pmatrix} 2.87953 \\ 1.01677 \\ 1.31898 \\ 1.63747 \\ 0.0382106 \\ -0.973019 \end{pmatrix} \quad du \to \begin{pmatrix} -3.11855 \\ -0.928778 \\ 0.268281 \\ -2.66356 \\ 11.6896 \\ 1.11694 \end{pmatrix}$$

$-x/dx \to \begin{pmatrix} \infty & \infty & \infty & \infty & \infty & 0.763542 \end{pmatrix} \quad \alpha_p \to 0.755906$

$-u/du \to \begin{pmatrix} 0.291533 & 0.70038 & \infty & 0.423415 & \infty & \infty \end{pmatrix} \quad \alpha_d \to 0.288617$

New primal vars $\to$
$\begin{pmatrix} 3.19899 & 1.85452 & 2.24046 & 2.20635 & 0.0388837 & 0.00742941 \end{pmatrix}$

New dual vars $\to$
$\begin{pmatrix} 0.0090916 & 0.382436 & 0.0874305 & 0.359043 & 8.40002 & 2.3678 \end{pmatrix}$

New multipliers $\to \begin{pmatrix} 8.16509 & 2.13286 \end{pmatrix}$

Objective function $\to 54.2863$ Status $\to$ NonOptimum

***** NonOptimum solution after 2 iterations *****

Interior solution $\to \{x1 \to 3.19899, x2 \to -0.385934, x3 \to 2.20635\}$
Objective function $\to 54.2863$

The optimum solution for the problem is computed as follows:

{sol, hist} = PrimalDualQP[f, g, vars, UnrestrictedVariables → {x2}];

***** Optimum solution after 7 iterations *****

Interior solution $\to \{x1 \to 4.33334, x2 \to -1.00004, x3 \to 2.66672\}$
Objective function $\to 68.6674$

The complete search history, in terms of original variables, is as follows:

TableForm[Map[{#[[1]], #[[2]] - #[[3]], #[[4]]}&, hist]]

1.	0.	1.
1.02234	-0.157491	0.968577
3.19899	-0.385934	2.20635
4.0051	-0.839674	2.54169
4.33364	-1.00113	2.66845
4.33358	-1.00037	2.66692
4.33334	-1.00004	2.66672
4.33334	-1.00004	2.66672

Example 8.8 *Portfolio management* Consider the optimum solution of the following portfolio management problem, first discussed in Chapter 1. The problem statement is as follows.

8.3 The Primal-Dual Method for Convex QP

A portfolio manager for an investment company is looking to make investment decisions such that the investors will get at least a 10% rate of return while minimizing the risk of major losses. For the past six years, the rates of return in four major investment types are as follows:

Type	Annual rates of return						Average
Blue-chip stocks	18.24	12.12	15.23	5.26	2.62	10.42	10.6483
Technology stocks	12.24	19.16	35.07	23.46	−10.62	−7.43	11.98
Real estate	8.23	8.96	8.35	9.16	8.05	7.29	8.34
Bonds	8.12	8.26	8.34	9.01	9.11	8.95	8.6317

The optimization variables are as follows:

x_1	Portion of capital invested in blue chip stocks
x_2	Portion of capital invested in technology stocks
x_3	Portion of capital invested in real estate
x_4	Portion of capital invested in bonds

The objective is to minimize risk of losses. Considering *covariance* between investments as a measure of risk, the following objective function was derived in Chapter 1:

Minimize $f = 29.0552 x_1^2 + 80.7818 x_2 x_1 - 0.575767 x_3 x_1 - 3.90639 x_4 x_1 + 267.344 x_2^2 + 0.375933 x_3^2 + 0.159714 x_4^2 + 13.6673 x_2 x_3 - 7.39403 x_2 x_4 - 0.113267 x_3 x_4$

The constraints are as follows:

Total investment $x_1 + x_2 + x_3 + x_4 = 1$

Desired rate of return $10.6483 x_1 + 11.98 x_2 + 8.34 x_3 + 8.6317 x_4 \geq 10$

All optimization variables must be positive.

The following *Mathematica* expressions generate the above objective function and constraints directly from the given data.

```
blueChipStocks = {18.24, 12.12, 15.23, 5.26, 2.62, 10.42};
techStocks = {12.24, 19.16, 35.07, 23.46, -10.62, -7.43};
realEstate = {8.23, 8.96, 8.35, 9.16, 8.05, 7.29};
bonds = {8.12, 8.26, 8.34, 9.01, 9.11, 8.95};
returns = {blueChipStocks, techStocks, realEstate, bonds};
averageReturns = Map[Apply[Plus, #]/Length[#] returns]
```
{10.6483, 11.98, 8.34, 8.63167}

```
coVariance[x_, y_] := Module[{xb, yb, n = Length[x]},
  xb = Apply[Plus, x]/n; yb = Apply[Plus, y]/n;
  Apply[Plus, (x - xb) (y - yb)]/n];

Q = Outer[coVariance, returns, returns, 1]; MatrixForm[Q]
```

$$\begin{pmatrix} 29.0552 & 40.3909 & -0.2878833 & -1.9532 \\ 40.3909 & 267.344 & 6.83367 & -3.69702 \\ -0.2878833 & 6.83367 & 0.375933 & -0.0566333 \\ -1.9532 & -3.69702 & -0.0566333 & 0.159714 \end{pmatrix}$$

```
Clear[x];
  vars = Table[x_i, {i, 1, Length[averageReturns]}];
f = Expand[vars.Q.vars]
```

$29.0552 x_1^2 + 80.7818 x_1 x_2 + 267.344 x_2^2 - 0.575767 x_1 x_3 + 13.6673 x_2 x_3 + 0.375933 x_3^2 - 3.90639 x_1 x_4 - 7.39403 x_2 x_4 - 0.113267 x_3 x_4 + 0.159714 x_4^2$

```
g = {Apply[Plus, vars] == 1, averageReturns.vars ≥ 10}
```

$\{x_1 + x_2 + x_3 + x_4 == 1, 10.6483 x_1 + 11.98 x_2 + 8.34 x_3 + 8.63167 x_4 \geq 10\}$

```
{sol, hist} = PrimalDualQP[f, g, vars];
```

Minimize → $29.0552 x_1^2 + 80.7818 x_1 x_2 + 267.344 x_2^2 - 0.575767 x_1 x_3 + 13.6673 x_2 x_3 + 0.375933 x_3^2 - 3.90639 x_1 x_4 - 7.39403 x_2 x_4 - 0.113267 x_3 x_4 + 0.159714 x_4^2$

Subject to → $\begin{pmatrix} x_1 + x_2 + x_3 + x_4 == 1 \\ 10.6483 x_1 + 11.98 x_2 + 8.34 x_3 + 8.63167 x_4 - x_5 == 10 \end{pmatrix}$

and → $(x_1 \geq 0 \quad x_2 \geq 0 \quad x_3 \geq 0 \quad x_4 \geq 0 \quad x_5 \geq 0)$

Problem variables redefined as: $\{x_1 \to x_1, x_2 \to x_2, x_3 \to x_3, x_4 \to x_4\}$

A → $\begin{pmatrix} 1 & 1 & 1 & 1 & 0 \\ 10.6483 & 11.98 & 8.34 & 8.63167 & -1 \end{pmatrix}$ b → $\begin{pmatrix} 1 \\ 10 \end{pmatrix}$

c → $\begin{pmatrix} 0 \\ 0 \\ 0 \\ 0 \\ 0 \end{pmatrix}$ Q → $\begin{pmatrix} 58.1104 & 80.7818 & -0.575767 & -3.90639 & 0 \\ 80.7818 & 534.689 & 13.6673 & -7.39403 & 0 \\ -0.575767 & 13.6673 & 0.751867 & -0.113267 & 0 \\ -3.90639 & -7.39403 & -0.113267 & 0.319428 & 0 \\ 0 & 0 & 0 & 0 & 0 \end{pmatrix}$

Convex problem.
Principal minors of Q → {58.1104, 24545.3, 6151.37, 117.253, 0.}

***** Optimum solution after 8 iterations *****
Interior solution →
$\{x_1 \to 0.629234, x_2 \to 0.0296837, x_3 \to 0.000018564, x_4 \to 0.341064\}$
Objective function → 12.3538

The complete history of the actual four variables during iteration follows:

```
TableForm[Transpose[Drop[Transpose[hist], -1]]]
```

1.	1.	1.	1.
0.575346	0.503383	0.424391	0.987077
0.557122	0.33987	0.00424391	1.04404
0.621348	0.0417323	0.0207309	0.336163
0.625833	0.0324388	0.00372627	0.338002
0.628512	0.030269	0.000742737	0.340477
0.629146	0.0297603	0.000128444	0.340966
0.629234	0.0296837	0.000018564	0.341064
0.629234	0.0296837	0.000018564	0.341064

The optimum solution indicates that, under the given conditions, the portfolio manager should invest 63% of capital in blue chip stocks, 3% in technology stocks, 0% in real estate, and 34% in bonds.

8.4 Active Set Method

In this section, we consider solving QP problems of the following form:

Minimize $\mathbf{c}^T\mathbf{x} + \frac{1}{2}\mathbf{x}^T\mathbf{Q}\mathbf{x}$

Subject to $\begin{pmatrix} \mathbf{a}_i^T\mathbf{x} = b_i & i = 1, \ldots, p \\ \mathbf{a}_i^T\mathbf{x} \leq b_i & i = p+1, \ldots, m \end{pmatrix}$

Note that the optimization variables are not restricted to be positive. The first p constraints are linear equalities, and the remaining are less than (LE) type inequalities. Furthermore, it will be assumed that $\mathbf{Q}$ is a positive definite matrix. Problems of this form arise during the direction-finding phase when solving general nonlinear optimization problems using linearization techniques, discussed in the next chapter.

Obviously, a QP of this type can be solved by first converting to standard QP form and then using the methods discussed in the previous sections. However, since the above form is similar to a general nonlinear problem, it is more convenient to use it directly for direction finding, as will be seen in Chapter 9.

As before, starting from an arbitrary point $\mathbf{x}^0$, our basic iteration is of the following form:

$$\mathbf{x}^{k+1} = \mathbf{x}^k + \alpha\mathbf{d}^k$$

where α is a step length and $\mathbf{d}$ is a suitable descent direction.

8.4.1 Direction Finding

At the current point $\mathbf{x}^k$, some of the inequality constraints may be satisfied as equalities. These constraints together with the equality constraints constitute what we call an active set. The set of active constraint indices can therefore be defined as follows:

$$\mathbf{w}^k = \{1, \ldots, p\} \cup \left\{i \,|\, \mathbf{a}_i^T \mathbf{x}^k = b_i, i = p+1, \ldots, m\right\}$$

where the symbol $\cup$ stands for the union of two sets. The direction-finding problem can therefore be stated as follows:

Minimize $\mathbf{c}^T(\mathbf{x}^k + \mathbf{d}^k) + \frac{1}{2}(\mathbf{x}^k + \mathbf{d}^k)^T \mathbf{Q}(\mathbf{x}^k + \mathbf{d}^k)$
Subject to $\mathbf{a}_i^T(\mathbf{x}^k + \mathbf{d}^k) = b_i \quad i \in \mathbf{w}^k$

Expanding and ignoring the constant terms in the objective function, we have

Minimize $[\mathbf{Q}\mathbf{x}^k + \mathbf{c}]^T \mathbf{d}^k + \frac{1}{2}\mathbf{d}^{kT} \mathbf{Q} \mathbf{d}^k$
Subject to $\mathbf{a}_i^T \mathbf{d}^k = b_i - \mathbf{a}_i^T \mathbf{x}^k = 0 \quad i \in \mathbf{w}^k$

Introducing the notation

$$\mathbf{g}^k = \mathbf{Q}\mathbf{x}^k + \mathbf{c} \quad \text{and} \quad \mathbf{A} = \begin{pmatrix} \mathbf{a}_1^T \\ \mathbf{a}_2^T \\ \vdots \end{pmatrix} \quad i \in \mathbf{w}^k$$

We have the following equality-constrained QP problem for direction $\mathbf{d}^k$:

Minimize $\mathbf{g}^{kT} \mathbf{d}^k + \frac{1}{2}\mathbf{d}^{kT} \mathbf{Q} \mathbf{d}^k$
Subject to $\mathbf{A} \mathbf{d}^k = \mathbf{0}$

The KT conditions for the minimum of this problem give the following system of equations:

$$\mathbf{Q} \mathbf{d}^k + \mathbf{g}^k + \mathbf{A}^T \mathbf{v} = \mathbf{0}$$

$$\mathbf{A} \mathbf{d}^k = \mathbf{0}$$

where $\mathbf{v}$ is a vector of Lagrange multipliers associated with active constraints. Both equations can be written in a matrix form, as follows:

$$\begin{pmatrix} \mathbf{Q} & \mathbf{A}^T \\ \mathbf{A} & \mathbf{0} \end{pmatrix} \begin{pmatrix} \mathbf{d}^k \\ \mathbf{v} \end{pmatrix} = \begin{pmatrix} -\mathbf{g}^k \\ \mathbf{0} \end{pmatrix}$$

The direction, together with the Lagrange multipliers, can be calculated by solving the above linear system of equations.

8.4.2 Step-Length Calculations

After computing the direction, the step length α is selected to be as large as possible to maintain feasibility with respect to inactive inequality constraints.

$$\mathbf{a}_i^T(\mathbf{x}^k + \alpha \mathbf{d}^k) \leq b_i \quad i \notin \mathbf{w}^k$$

Clearly, $\mathbf{a}_i^T \mathbf{d}^k < 0$ cannot cause any constraint violation; therefore, the step length is selected as follows:

$$\alpha \mathbf{a}_i^T \mathbf{d}^k \leq b_i - \mathbf{a}_i^T \mathbf{x}^k \quad i \notin \mathbf{w}^k$$

$$\alpha = \text{Min}\left[1, \frac{b_i - \mathbf{a}_i^T \mathbf{x}^k}{\mathbf{a}_i^T \mathbf{d}^k} \text{ for } \mathbf{a}_i^T \mathbf{d}^k > 0 \text{ and } i \notin \mathbf{w}^k \right]$$

8.4.3 Adjustments to the Active Set

From the step-length calculations, it is clear that whenever $\alpha < 1$, a new inequality constraint becomes active and must be included into the active set for the next iteration. From Chapter 4, we know that Lagrange multipliers for all LE constraints must be positive. Thus, if an element in the $\mathbf{v}$ vector corresponding to an inequality constraint in the active set is negative, this means that the constraint cannot be active and must be dropped from the active set. If there are several active inequalities with negative multiplier, usually the constraint with the most negative multiplier is dropped from the active set.

Because of the above rule, all active constraints at a given point may not be included in the active set. In this regard, strictly speaking, the term *active set* is not appropriate. For this reason, some authors prefer to call the set of constraints used for direction finding a *working set*. However, active set terminology is more common and is used here.

8.4.4 Finding a Starting Feasible Solution

In the presentation so far, it has been assumed that we have a starting feasible point $\mathbf{x}^0$. Usually, we will be starting from an arbitrary point; therefore, we need to find a starting feasible solution. We can use the standard Phase I simplex procedure to find a starting feasible solution. By introducing one artificial variable for each constraint, $x_{n+1}, x_{n+2}, \ldots, x_{n+m}$, the Phase I problem is stated as follows:

Minimize $\phi = x_{n+1} + x_{n+2} + \cdots$

Subject to $\begin{pmatrix} \mathbf{a}_i^T \mathbf{x} + x_{n+i} = b_i & i = 1, \ldots, p \\ \mathbf{a}_i^T \mathbf{x} + x_{n+i} \leq b_i & i = p+1, \ldots, m \end{pmatrix}$

The solution of this problem is when $\phi = 0$, giving us a point that is feasible for the original problem.

This Phase I LP can be solved using any of the methods discussed in the previous chapters.

8.4.5 Complete ActiveSet QP Algorithm

Initialization: Set $k = 0$. If there are equality constraints, start by setting $\mathbf{x}^0$ to the solution of these constraints; otherwise, set $\mathbf{x}^0 = \mathbf{0}$. If this $\mathbf{x}^0$ satisfies inequality constraints as well, we have a starting feasible solution. If not, determine the starting feasible solution $\mathbf{x}^0$ by using the Phase I procedure discussed above. Set $\mathbf{w}^0$ to the indices of the constraints active at $\mathbf{x}^0$.

1. Form:

$$\mathbf{g}^k = \mathbf{Q}\mathbf{x}^k + \mathbf{c}$$

2. Solve the following system of equations:

$$\begin{pmatrix} \mathbf{Q} & \mathbf{A}^T \\ \mathbf{A} & \mathbf{0} \end{pmatrix} \begin{pmatrix} \mathbf{d}^k \\ \mathbf{v} \end{pmatrix} = \begin{pmatrix} -\mathbf{g}^k \\ \mathbf{0} \end{pmatrix}$$

3. If $\mathbf{d}^k = \mathbf{0}$, go to step 6 to check for optimality. Otherwise, continue.

4. Compute step length:

$$\alpha = \text{Min}\left[1, \frac{b_i - \mathbf{a}_i^T \mathbf{x}^k}{\mathbf{a}_i^T \mathbf{d}^k} \text{ for } \mathbf{a}_i^T \mathbf{d}^k > 0 \text{ and } i \notin \mathbf{w}^k\right]$$

If $\alpha < 1$, add the constraint controlling the step length to the active set.

5. Update:

$$\mathbf{x}^{k+1} = \mathbf{x}^k + \alpha \mathbf{d}^k \quad k = k+1 \text{ and go to step 1.}$$

6. Check the sign of Lagrange multipliers corresponding to inequality constraints. If they are all positive, stop. We have the optimum. Otherwise, remove the constraint that corresponds to the most negative multiplier and go to step 2.

The ActiveSetQP Function

The following ActiveSetQP function implements the Active Set algorithm for solving QP problems. The function usage and its options are explained first. Several intermediate results can be printed to gain understanding of the process.

```
Needs["OptimizationToolbox`QuadraticProgramming`"];
?ActiveSetQP
```

ActiveSetQP[f, g, vars, options]. Solves an QP problem using Active set
   algorithm. f is the objective function, g is a list of constraints,
   and vars is a list of variables. See Options[ActiveSetQP] to find out
   about a list of valid options for this function.

OptionsUsage[ActiveSetQP]
{MaxIterations → 20, ProblemType → Min,
  PrintLevel → 1, ConvergenceTolerance → 0.0001}

MaxIterations is an option for several optimization methods. It
   specifies maximum number of iterations allowed.

ProblemType is an option for most optimization methods. It can either be
   Min (default) or Max.

PrintLevel is an option for most functions in the OptimizationToolbox.
   It is specified as an integer. The value of the integer indicates
   how much intermediate information is to be printed. A PrintLevel→ 0
   suppresses all printing. Default for most functions is set to 1 in
   which case they print only the initial problem setup. Higher integers
   print more intermediate results.

ConvergenceTolerance is an option for most optimization methods. Most
   methods require only a single zero tolerance value. Some interior
   point methods require a list of convergence tolerance values.

Example 8.9 Solve the following QP problem using the ActiveSetQP method:

Minimize $f = -2x_1 + \frac{x_1^2}{2} - 6x_2 - x_1 x_2 + x_2^2$

Subject to $\begin{pmatrix} 3x_1 + x_2 \leq 25 \\ -x_1 + 2x_2 \leq 10 \\ x_1 + 2x_2 \leq 15 \\ x_i \geq 0, i = 1, 2 \end{pmatrix}$

```
f = -2x_1 + 1/2x_1^2 - 6x_2 - x_1x_2 + x_2^2;
g = {3x_1 + x_2 ≤ 25, -x_1 + 2x_2 ≤ 10, x_1 + 2x_2 ≤ 15, x_1 ≥ 0, x_2 ≥ 0}; vars = {x_1, x_2};

{sol, xhist} = ActiveSetQP[f, g, vars, PrintLevel → 2];
```

Minimize $\to -2x_1 + \dfrac{x_1^2}{2} - 6x_2 - x_1 x_2 + x_2^2$

LE Constraints $\to \begin{pmatrix} 3x_1 + x_2 \le 25 \\ -x_1 + 2x_2 \le 10 \\ x_1 + 2x_2 \le 15 \\ -x_1 \le 0 \\ -x_2 \le 0 \end{pmatrix}$

$c \to \begin{pmatrix} -2 \\ -6 \end{pmatrix} \quad Q \to \begin{pmatrix} 1 & -1 \\ -1 & 2 \end{pmatrix}$

LE constraints: $A \to \begin{pmatrix} 3 & 1 \\ -1 & 2 \\ 1 & 2 \\ -1 & 0 \\ 0 & -1 \end{pmatrix} \quad b \to \begin{pmatrix} 25 \\ 10 \\ 15 \\ 0 \\ 0 \end{pmatrix}$

EQ constraints: $A \to \{\} \quad b \to \{\}$

Convex problem. Principal minors of $Q \to \{1, 1\}$

**** **Iteration 1** ****

Current point $\to \begin{pmatrix} 1 & 1 \end{pmatrix}$

Active set $\to \{\{\}\}$

$A \to \{\} \quad gk \to \begin{pmatrix} -2 \\ -5 \end{pmatrix}$

LE Constraint values $\to \begin{pmatrix} -21 & -9 & -12 & -1 & -1 \end{pmatrix}$

Equations: lhs $\to \begin{pmatrix} 1 & -1 \\ -1 & 2 \end{pmatrix}$ rhs $\to \begin{pmatrix} 2 \\ 5 \end{pmatrix}$

Solution: dk $\to \begin{pmatrix} 9 \\ 7 \end{pmatrix}$ uk $\to \{\}$ vk $\to \{\}$

Inactive set $\to \begin{pmatrix} 1 & 2 & 3 & 4 & 5 \end{pmatrix}$

bi-Ai.x ($i \in$ Inactive) $\to \begin{pmatrix} 21 & 9 & 12 & 1 & 1 \end{pmatrix}$

Ai.d ($i \in$ Inactive) $\to \begin{pmatrix} 34 & 5 & 23 & -9 & -7 \end{pmatrix}$

bi-Ai.x/Ai.d $\to \begin{pmatrix} \dfrac{21}{34} & \dfrac{9}{5} & \dfrac{12}{23} & \infty & \infty \end{pmatrix}$

Step length, $\alpha \to \dfrac{12}{23}$ New active set $\to \begin{pmatrix} 3 \end{pmatrix}$

New point $\to \begin{pmatrix} \dfrac{131}{23} & \dfrac{107}{23} \end{pmatrix}$

**** Iteration 2 ****

Current point $\to \begin{pmatrix} \frac{131}{23} & \frac{107}{23} \end{pmatrix}$

Active set $\to (3)$

$A \to \begin{pmatrix} 1 & 2 \end{pmatrix}$ gk $\to \begin{pmatrix} -\frac{22}{23} \\ -\frac{55}{23} \end{pmatrix}$

LE Constraint values $\to \begin{pmatrix} -\frac{75}{23} & -\frac{147}{23} & 0 & -\frac{131}{23} & -\frac{107}{23} \end{pmatrix}$

Equations: lhs $\to \begin{pmatrix} 1 & -1 & 1 \\ -1 & 2 & 2 \\ 1 & 2 & 0 \end{pmatrix}$ rhs $\to \begin{pmatrix} \frac{22}{23} \\ \frac{55}{23} \\ 0 \end{pmatrix}$

Solution: dk $\to \begin{pmatrix} -\frac{11}{115} \\ \frac{11}{230} \end{pmatrix}$ uk $\to \begin{pmatrix} \frac{11}{10} \end{pmatrix}$ vk $\to \{\}$

Inactive set $\to \begin{pmatrix} 1 & 2 & 4 & 5 \end{pmatrix}$

bi - Ai.x(i ∈ Inactive) $\to \begin{pmatrix} \frac{75}{23} & \frac{147}{23} & \frac{131}{23} & \frac{107}{23} \end{pmatrix}$

Ai.d(i ∈ Inactive) $\to \begin{pmatrix} -\frac{11}{46} & \frac{22}{115} & \frac{11}{115} & -\frac{11}{230} \end{pmatrix}$

bi-Ai.x/Ai.d $\to \begin{pmatrix} \infty & \frac{735}{22} & \frac{655}{11} & \infty \end{pmatrix}$

Step length, $\alpha \to 1$ New active set $\to (3)$

New point $\to \begin{pmatrix} \frac{28}{5} & \frac{47}{10} \end{pmatrix}$

**** Iteration 3 ****

Current point $\to \begin{pmatrix} \frac{28}{5} & \frac{47}{10} \end{pmatrix}$

Active set $\to (3)$

$A \to \begin{pmatrix} 1 & 2 \end{pmatrix}$ gk $\to \begin{pmatrix} -\frac{11}{10} \\ -\frac{11}{5} \end{pmatrix}$

LE Constraint values $\to \begin{pmatrix} -\frac{7}{2} & -\frac{31}{5} & 0 & -\frac{28}{5} & -\frac{47}{10} \end{pmatrix}$

Equations: lhs $\to \begin{pmatrix} 1 & -1 & 1 \\ -1 & 2 & 2 \\ 1 & 2 & 0 \end{pmatrix}$ rhs $\to \begin{pmatrix} \frac{11}{10} \\ \frac{11}{5} \\ 0 \end{pmatrix}$

Solution: dk $\to \begin{pmatrix} 0 \\ 0 \end{pmatrix}$ uk $\to \begin{pmatrix} \frac{11}{10} \end{pmatrix}$ vk $\to \{\}$

***** Optimum solution after 3 iterations *****

Solution $\to \begin{pmatrix} x_1 \to \frac{28}{5} & x_2 \to \frac{47}{10} \end{pmatrix}$

Lagrange multipliers for LE constraints (u) → $\begin{pmatrix} 0 & 0 & \frac{11}{10} & 0 & 0 \end{pmatrix}$

Lagrange multipliers for EQ constraints (v) → {{}}

Objective function → $-\frac{559}{20}$

TableForm[xhist]

1	1
$\frac{131}{23}$	$\frac{107}{23}$
$\frac{28}{5}$	$\frac{47}{10}$

The graph in Figure 8.4 shows the search path taken by the algorithm:

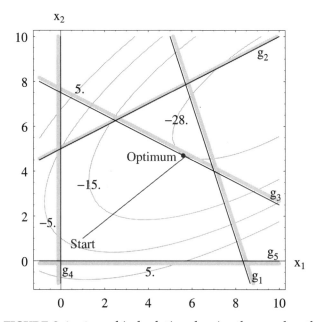

FIGURE 8.4 A graphical solution showing the search path.

Example 8.10 Solve the following QP problem using the ActiveSetQP method.

```
f = -6x₁x₂ + 2x₁² + 9x₂² - 18x₁ + 9x₂;
g = {x₁ + 2x₂ ≤ 15, x₁ ≥ 0, x₂ ≥ 0}; vars = {x₁, x₂};

{sol, xhist} = ActiveSetQP[f, g, vars, PrintLevel → 2];
```

8.4 Active Set Method

Minimize $\rightarrow -18x_1 + 2x_1^2 + 9x_2 - 6x_1x_2 + 9x_2^2$

LE Constraints $\rightarrow \begin{pmatrix} x_1 + 2x_2 \le 15 \\ -x_1 \le 0 \\ -x_2 \le 0 \end{pmatrix}$

$c \rightarrow \begin{pmatrix} -18 \\ 9 \end{pmatrix}$ $Q \rightarrow \begin{pmatrix} 4 & -6 \\ -6 & 18 \end{pmatrix}$

LE constraints: $A \rightarrow \begin{pmatrix} 1 & 2 \\ -1 & 0 \\ 0 & -1 \end{pmatrix}$ $b \rightarrow \begin{pmatrix} 15 \\ 0 \\ 0 \end{pmatrix}$

EQ constraints: $A \rightarrow \{\}$ $b \rightarrow \{\}$

Convex problem. Principal minors of $Q \rightarrow \{4, 36\}$

**** Iteration 1 ****

Current point $\rightarrow (1 \quad 1)$

Active set $\rightarrow \{\{\}\}$

$A \rightarrow \{\}$ $gk \rightarrow \begin{pmatrix} -20 \\ 21 \end{pmatrix}$

LE Constraint values $\rightarrow (-12 \quad -1 \quad -1)$

Equations: lhs $\rightarrow \begin{pmatrix} 4 & -6 \\ -6 & 18 \end{pmatrix}$ rhs $\rightarrow \begin{pmatrix} 20 \\ -21 \end{pmatrix}$

Solution: dk $\rightarrow \begin{pmatrix} \frac{13}{2} \\ 1 \end{pmatrix}$ uk $\rightarrow \{\}$ vk $\rightarrow \{\}$

Inactive set $\rightarrow (1 \quad 2 \quad 3)$

bi-Ai.x (i$\in$ Inactive) $\rightarrow (12 \quad 1 \quad 1)$

Ai.d (i$\in$ Inactive) $\rightarrow \left(\frac{17}{2} \quad -\frac{13}{2} \quad -1 \right)$

bi-Ai.x/Ai.d $\rightarrow \left(\frac{24}{17} \quad \infty \quad \infty \right)$

Step length, $\alpha \rightarrow 1$ New active set $\rightarrow \{\{\}\}$

New point $\rightarrow \left(\frac{15}{2} \quad 2 \right)$

**** Iteration 2 ****

Current point $\rightarrow \left(\frac{15}{2} \quad 2 \right)$

Active set $\rightarrow \{\{\}\}$

$A \rightarrow \{\}$ $gk \rightarrow \begin{pmatrix} 0 \\ 0 \end{pmatrix}$

LE Constraint values $\rightarrow \left(-\frac{7}{2} \quad -\frac{15}{2} \quad -2 \right)$

Equations: lhs $\rightarrow \begin{pmatrix} 4 & -6 \\ -6 & 18 \end{pmatrix}$ rhs $\rightarrow \begin{pmatrix} 0 \\ 0 \end{pmatrix}$

Solution: $dk \to \begin{pmatrix} 0 \\ 0 \end{pmatrix}$ $uk \to \{\}$ $vk \to \{\}$

***** **Optimum solution after 2 iterations** *****

Solution $\to \left(x_1 \to \dfrac{15}{2} \quad x_2 \to 2\right)$

Lagrange multipliers for LE constraints (u) $\to \begin{pmatrix} 0 & 0 & 0 \end{pmatrix}$

Lagrange multipliers for EQ constraints (v) $\to \{\{\}\}$

Objective function $\to -\dfrac{117}{2}$

The graph in Figure 8.5 confirms the solution:

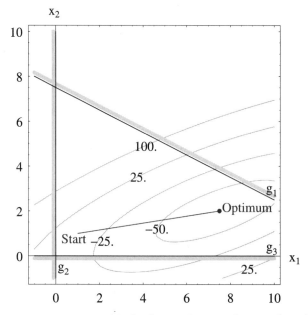

FIGURE 8.5 A graphical solution showing the search path.

Example 8.11 Solve the following QP problem using the ActiveSetQP method:

```
f = x₁² + x₁x₂ + 2x₂² + 2x₃² + 2x₂x₃ + 4x₁ + 6x₂ + 12x₃;
g = {x₁ + x₂ + x₃ ≥ 6, -x₁ - x₂ + 2x₃ ≥ 2, x₁ ≥ 0, x₃ ≥ 0}; vars = {x₁, x₂, x₃};
```

8.4 Active Set Method

Intermediate results are printed for the first iteration:

{sol, xhist} = ActiveSetQP[f, g, vars, PrintLevel → 2];

Minimize → $4x_1 + x_1^2 + 6x_2 + x_1 x_2 + 2x_2^2 + 12x_3 + 2x_2 x_3 + 2x_3^2$

LE Constraints → $\begin{pmatrix} -x_1 - x_2 - x_3 \le -6 \\ x_1 + x_2 - 2x_3 \le -2 \\ -x_1 \le 0 \\ -x_3 \le 0 \end{pmatrix}$

$c \to \begin{pmatrix} 4 \\ 6 \\ 12 \end{pmatrix} \quad Q \to \begin{pmatrix} 2 & 1 & 0 \\ 1 & 4 & 2 \\ 0 & 2 & 4 \end{pmatrix}$

LE constraints: $A \to \begin{pmatrix} -1 & -1 & -1 \\ 1 & 1 & -2 \\ -1 & 0 & 0 \\ 0 & 0 & -1 \end{pmatrix} \quad b \to \begin{pmatrix} -6 \\ -2 \\ 0 \\ 0 \end{pmatrix}$

EQ constraints: $A \to \{\} \quad b \to \{\}$

Convex problem. Principal minors of $Q \to \{2, 7, 20\}$

Phase I problem: Art. Obj. coefficients → $\{0, 0, 0, 0, 0, 0, 1, 1, 1, 1\}$

Constraints expressed as GE type

Constraint coefficients → $\begin{pmatrix} 1 & 1 & 1 & -1 & -1 & -1 & 1 & 0 & 0 & 0 \\ -1 & -1 & 2 & 1 & 1 & -2 & 0 & 1 & 0 & 0 \\ 1 & 0 & 0 & -1 & 0 & 0 & 0 & 0 & 1 & 0 \\ 0 & 0 & 1 & 0 & 0 & -1 & 0 & 0 & 0 & 1 \end{pmatrix}$

Constraint rhs → $\begin{pmatrix} 6 & 2 & 0 & 0 \end{pmatrix}$

Phase I-LP solution → $\left\{ \frac{10}{3}, 0, \frac{8}{3}, 0, 0, 0, 0, 0, 0, 0 \right\}$

**** Iteration 1 ****

Current point → $\begin{pmatrix} \frac{10}{3} & 0 & \frac{8}{3} \end{pmatrix}$

Active set → $\begin{pmatrix} 1 & 2 \end{pmatrix}$

$A \to \begin{pmatrix} -1 & -1 & -1 \\ 1 & 1 & -2 \end{pmatrix} \quad gk \to \begin{pmatrix} \frac{32}{3} \\ \frac{44}{3} \\ \frac{68}{3} \end{pmatrix}$

LE Constraint values → $\begin{pmatrix} 0 & 0 & -\frac{10}{3} & -\frac{8}{3} \end{pmatrix}$

Equations: lhs → $\begin{pmatrix} 2 & 1 & 0 & -1 & 1 \\ 1 & 4 & 2 & -1 & 1 \\ 0 & 2 & 4 & -1 & -2 \\ -1 & -1 & -1 & 0 & 0 \\ 1 & 1 & -2 & 0 & 0 \end{pmatrix} \quad \text{rhs} \to \begin{pmatrix} -\frac{32}{3} \\ -\frac{44}{3} \\ -\frac{68}{3} \\ 0 \\ 0 \end{pmatrix}$

Solution: $dk \to \begin{pmatrix} 1 \\ -1 \\ 0 \end{pmatrix}$ $uk \to \begin{pmatrix} \frac{44}{3} \\ 3 \\ 3 \end{pmatrix}$ $vk \to \{\}$

Inactive set $\to (3 \quad 4)$

$bi - Ai.x(i \in Inactive) \to \left(\frac{10}{3} \quad \frac{8}{3}\right)$

$Ai.d(i \in Inactive) \to (-1 \quad 0)$

$bi-Ai.x/Ai.d \to (\infty \quad \infty)$

Step length, $\alpha \to 1$ New active set $\to (1 \quad 2)$

New point $\to \left(\frac{13}{3} \quad -1 \quad \frac{8}{3}\right)$

**** Iteration 2 ****

Current point $\to \left(\frac{13}{3} \quad -1 \quad \frac{8}{3}\right)$

Active set $\to (1 \quad 2)$

$A \to \begin{pmatrix} -1 & -1 & -1 \\ 1 & 1 & -2 \end{pmatrix}$ $gk \to \begin{pmatrix} \frac{35}{3} \\ \frac{35}{3} \\ \frac{62}{3} \end{pmatrix}$

LE Constraint values $\to \left(0 \quad 0 \quad -\frac{13}{3} \quad -\frac{8}{3}\right)$

Equations: lhs $\to \begin{pmatrix} 2 & 1 & 0 & -1 & 1 \\ 1 & 4 & 2 & -1 & 1 \\ 0 & 2 & 4 & -1 & -2 \\ -1 & -1 & -1 & 0 & 0 \\ 1 & 1 & -2 & 0 & 0 \end{pmatrix}$ rhs $\to \begin{pmatrix} -\frac{35}{3} \\ -\frac{35}{3} \\ -\frac{62}{3} \\ 0 \\ 0 \end{pmatrix}$

Solution: $dk \to \begin{pmatrix} 0 \\ 0 \\ 0 \end{pmatrix}$ $uk \to \begin{pmatrix} \frac{44}{3} \\ 3 \\ 3 \end{pmatrix}$ $vk \to \{\}$

***** Optimum solution after 2 iterations *****

Solution $\to \left(x_1 \to \frac{13}{3} \quad x_2 \to -1 \quad x_3 \to \frac{8}{3}\right)$

Lagrange multipliers for LE constraints (u) $\to \left(\frac{44}{3} \quad 3 \quad 0 \quad 0\right)$

Lagrange multipliers for EQ constraints (v) $\to \{\{\}\}$

Objective function $\to \frac{206}{3}$

Example 8.12 *Portfolio management* Find the optimum solution of the portfolio management problem, considered in Example 8.8 and first presented in Chapter 1, using the ActiveSetQP method.

8.4 Active Set Method

The objective and the constraint functions for the problem are generated using *Mathematica* statements (see Chapter 1 for details).

```
blueChipStocks = {18.24, 12.12, 15.23, 5.26, 2.62, 10.42};
techStocks = {12.24, 19.16, 35.07, 23.46, -10.62, -7.43};
realEstate = {8.23, 8.96, 8.35, 9.16, 8.05, 7.29};
bonds = {8.12, 8.26, 8.34, 9.01, 9.11, 8.95};
returns = {blueChipStocks, techStocks, realEstate, bonds};
averageReturns = Map[Apply[Plus, #]/Length[#] returns];
coVariance[x_, y_] := Module[{xb, yb, n = Length[x]},
    xb = Apply[Plus, x]/n; yb = Apply[Plus, y]/n;
    Apply[Plus, (x - xb) (y - yb)]/n];
Q = Outer[coVariance, returns, returns, 1];
Clear[x];
    vars = Table[x_i, {i, 1, Length[averageReturns]}];
f = Expand[vars.Q.vars];
g = Flatten[{Apply[Plus, vars] == 1,
    averageReturns.vars ≥ 10, Thread[vars ≥ 0]}];

{sol, hist} = ActiveSetQP[f, g, vars, PrintLevel → 2];
```

Minimize → $29.0552x_1^2 + 80.7818x_1x_2 + 267.344x_2^2 - 0.575767x_1x_3 + 13.6673x_2x_3 + 0.375933x_3^2 - 3.90639x_1x_4 - 7.39403x_2x_4 - 0.113267x_3x_4 + 0.159714x_4^2$

LE Constraints → $\begin{pmatrix} -10.6483x_1 - 11.98x_2 - 8.34x_3 - 8.63167x_4 \le -10 \\ -1.x_1 \le 0 \\ -1.x_2 \le 0 \\ -1.x_3 \le 0 \\ -1.x_4 \le 0 \end{pmatrix}$

EQ Constraints → $(x_1 + x_2 + x_3 + x_4 == 1)$

$c \to \begin{pmatrix} 0 \\ 0 \\ 0 \\ 0 \end{pmatrix}$ $Q \to \begin{pmatrix} 58.1104 & 80.7818 & -0.575767 & -3.90639 \\ 80.7818 & 534.689 & 13.6673 & -7.39403 \\ -0.575767 & 13.6673 & 0.751867 & -0.113267 \\ -3.90639 & -7.39403 & -0.113267 & 0.319428 \end{pmatrix}$

LE constraints: $A \to \begin{pmatrix} -10.6483 & -11.98 & -8.34 & -8.63167 \\ -1 & 0 & 0 & 0 \\ 0 & -1 & 0 & 0 \\ 0 & 0 & -1 & 0 \\ 0 & 0 & 0 & -1 \end{pmatrix}$ $b \to \begin{pmatrix} -10 \\ 0 \\ 0 \\ 0 \\ 0 \end{pmatrix}$

EQ constraints: $A \to (1 \quad 1 \quad 1 \quad 1)$ $b \to (1)$

Convex problem.
    Principal minors of $Q \to \{58.1104, 24545.3, 6151.37, 117.253\}$

**** Iteration 1 ****

Current point → $(1 \quad 0 \quad 0 \quad 0)$

Active set → $(3 \quad 4 \quad 5)$

Chapter 8 Quadratic Programming

$$A \to \begin{pmatrix} 1 & 1 & 1 & 1 \\ 0 & -1 & 0 & 0 \\ 0 & 0 & -1 & 0 \\ 0 & 0 & 0 & -1 \end{pmatrix} \quad gk \to \begin{pmatrix} 58.1104 \\ 80.7818 \\ -0.575767 \\ -3.90639 \end{pmatrix}$$

LE Constraint values $\to (-0.648333 \quad -1. \quad 0. \quad 0. \quad 0.)$

EQ Constraint values $\to (0)$

Equations:

$$lhs \to \begin{pmatrix} 58.1104 & 80.7818 & -0.575767 & -3.90639 & 1 & 0 & 0 & 0 \\ 80.7818 & 534.689 & 13.6673 & -7.39403 & 1 & -1 & 0 & 0 \\ -0.575767 & 13.6673 & 0.751867 & -0.113267 & 1 & 0 & -1 & 0 \\ -3.90639 & -7.39403 & -0.113267 & 0.319428 & 1 & 0 & 0 & -1 \\ 1 & 1 & 1 & 1 & 0 & 0 & 0 & 0 \\ 0 & -1 & 0 & 0 & 0 & 0 & 0 & 0 \\ 0 & 0 & -1 & 0 & 0 & 0 & 0 & 0 \\ 0 & 0 & 0 & -1 & 0 & 0 & 0 & 0 \end{pmatrix}$$

$$rhs \to \begin{pmatrix} -58.1104 \\ -80.7818 \\ 0.575767 \\ 3.90639 \\ 0 \\ 0 \\ 0 \\ 0 \end{pmatrix}$$

Solution: $dk \to \begin{pmatrix} 4.63309 \times 10^{-16} \\ -3.05583 \times 10^{-16} \\ -5.7534 \times 10^{-17} \\ 1.55068 \times 10^{-16} \end{pmatrix}$ $uk \to \begin{pmatrix} 22.6713 \\ -58.6862 \\ -62.0168 \end{pmatrix}$ $vk \to (-58.1104)$

**** Iteration 2 ****

Current point $\to (1 \quad 0 \quad 0 \quad 0)$

Active set $\to (3 \quad 4)$

$$A \to \begin{pmatrix} 1 & 1 & 1 & 1 \\ 0 & -1 & 0 & 0 \\ 0 & 0 & -1 & 0 \end{pmatrix} \quad gk \to \begin{pmatrix} 58.1104 \\ 80.7818 \\ -0.575767 \\ -3.90639 \end{pmatrix}$$

LE Constraint values $\to (-0.648333 \quad -1. \quad 0. \quad 0. \quad 0.)$

EQ Constraint values $\to (0)$

Equations:

$$lhs \to \begin{pmatrix} 58.1104 & 80.7818 & -0.575767 & -3.90639 & 1 & 0 & 0 \\ 80.7818 & 534.689 & 13.6673 & -7.39403 & 1 & -1 & 0 \\ -0.575767 & 13.6673 & 0.751867 & -0.113267 & 1 & 0 & -1 \\ -3.90639 & -7.39403 & -0.113267 & 0.319428 & 1 & 0 & 0 \\ 1 & 1 & 1 & 1 & 0 & 0 & 0 \\ 0 & -1 & 0 & 0 & 0 & 0 & 0 \\ 0 & 0 & -1 & 0 & 0 & 0 & 0 \end{pmatrix}$$

$$\text{rhs} \to \begin{pmatrix} -58.1104 \\ -80.7818 \\ 0.575767 \\ 3.90639 \\ 0 \\ 0 \\ 0 \end{pmatrix}$$

$$\text{Solution: } dk \to \begin{pmatrix} -0.936207 \\ 4.81697 \times 10^{-18} \\ -7.94992 \times 10^{-17} \\ 0.936207 \end{pmatrix} \quad uk \to \begin{pmatrix} -1.81888 \\ -0.192621 \end{pmatrix} \quad vk \to (-0.0498496)$$

Inactive set $\to \begin{pmatrix} 1 & 2 & 5 \end{pmatrix}$

bi - Ai.x(i $\in$ Inactive) $\to \begin{pmatrix} 0.648333 & 1. & 0. \end{pmatrix}$

Ai.d(i $\in$ Inactive) $\to \begin{pmatrix} 1.88802 & 0.936207 & -0.936207 \end{pmatrix}$

bi-Ai.x/Ai.d $\to \begin{pmatrix} 0.343394 & 1.06814 & \infty \end{pmatrix}$

Step length, $\alpha \to 0.343394$ New active set $\to \begin{pmatrix} 1 & 3 & 4 \end{pmatrix}$

New point $\to \begin{pmatrix} 0.678512 & 1.65412 \times 10^{-18} & -2.72995 \times 10^{-17} & 0.321488 \end{pmatrix}$

**** Iteration 3 ****

Current point $\to \begin{pmatrix} 0.678512 & 1.65412 \times 10^{-18} & -2.72995 \times 10^{-17} & 0.321488 \end{pmatrix}$

Active set $\to \begin{pmatrix} 1 & 3 & 4 \end{pmatrix}$

$$A \to \begin{pmatrix} 1 & 1 & 1 & 1 \\ -10.6483 & -11.98 & -8.34 & -8.63167 \\ 0 & -1 & 0 & 0 \\ 0 & 0 & -1 & 0 \end{pmatrix} \quad gk \to \begin{pmatrix} 38.1728 \\ 52.4343 \\ -0.427079 \\ -2.54784 \end{pmatrix}$$

LE Constraint values $\to$
$\begin{pmatrix} 0. & -0.678512 & -1.65412 \times 10^{-18} & 2.72995 \times 10^{-17} & -0.321488 \end{pmatrix}$

EQ Constraint values $\to (0.)$

Equations:

$$\text{lhs} \to \begin{pmatrix} 58.1104 & 80.7818 & -0.575767 & -3.90639 & 1 & -10.6483 & 0 & 0 \\ 80.7818 & 534.689 & 13.6673 & -7.39403 & 1 & -11.98 & -1 & 0 \\ -0.575767 & 13.6673 & 0.751867 & -0.113267 & 1 & -8.34 & 0 & -1 \\ -3.90639 & -7.39403 & -0.113267 & 0.319428 & 1 & -8.63167 & 0 & 0 \\ 1 & 1 & 1 & 1 & 0 & 0 & 0 & 0 \\ -10.6483 & -11.98 & -8.34 & -8.63167 & 0 & 0 & 0 & 0 \\ 0 & -1 & 0 & 0 & 0 & 0 & 0 & 0 \\ 0 & 0 & -1 & 0 & 0 & 0 & 0 & 0 \end{pmatrix}$$

$$\text{rhs} \to \begin{pmatrix} -38.1728 \\ -52.4343 \\ 0.42707 \\ 2.54784 \\ 0 \\ 0 \\ 0 \\ 0 \end{pmatrix}$$

Solution: $dk \to \begin{pmatrix} 3.55436 \times 10^{-15} \\ -1.09967 \times 10^{-17} \\ -1.85425 \times 10^{-16} \\ -4.58817 \times 10^{-15} \end{pmatrix}$ $uk \to \begin{pmatrix} 20.192 \\ -12.6275 \\ 8.01011 \end{pmatrix}$ $vk \to (176.839)$

**** Iteration 4 ****

Current point $\to (0.678512 \quad 1.65412 \times 10^{-18} \quad -2.72995 \times 10^{-17} \quad 0.321488)$

Active set $\to (1 \quad 4)$

$A \to \begin{pmatrix} 1 & 1 & 1 & 1 \\ -10.6483 & -11.98 & -8.34 & -8.63167 \\ 0 & 0 & -1 & 0 \end{pmatrix}$ $gk \to \begin{pmatrix} 38.1728 \\ 52.4343 \\ -0.427079 \\ -2.54784 \end{pmatrix}$

LE Constraint values $\to$
$(0. \quad -0.678512 \quad -1.65412 \times 10^{-18} \quad 2.72995 \times 10^{-17} \quad -0.321488)$

EQ Constraint values $\to (0.)$

Equations:

$lhs \to \begin{pmatrix} 58.1104 & 80.7818 & -0.575767 & -3.90639 & 1 & -10.6483 & 0 \\ 80.7818 & 534.689 & 13.6673 & -7.39403 & 1 & -11.98 & 0 \\ -0.575767 & 13.6673 & 0.751867 & -0.113267 & 1 & -8.34 & -1 \\ -3.90639 & -7.39403 & -0.113267 & 0.319428 & 1 & -8.63167 & 0 \\ 1 & 1 & 1 & 1 & 0 & 0 & 0 \\ -10.6483 & -11.98 & -8.34 & -8.63167 & 0 & 0 & 0 \\ 0 & 0 & -1 & 0 & 0 & 0 & 0 \end{pmatrix}$

$rhs \to \begin{pmatrix} -38.1728 \\ -52.4343 \\ 0.42707 \\ 2.54784 \\ 0 \\ 0 \\ 0 \end{pmatrix}$

Solution: $dk \to \begin{pmatrix} -0.0492651 \\ 0.0296718 \\ 4.01429 \times 10^{-17} \\ 0.0195932 \end{pmatrix}$ $uk \to \begin{pmatrix} 19.9333 \\ 8.38703 \end{pmatrix}$ $vk \to (174.627)$

Inactive set $\to (2 \quad 3 \quad 5)$

bi-Ai.x (i$\in$ Inactive) $\to (0.678512 \quad 1.65412 \times 10^{-18} \quad 0.321488)$

Ai.d (i$\in$ Inactive) $\to (0.0492651 \quad -0.0296718 \quad -0.0195932)$

bi-Ai.x/Ai.d $\to (13.7727 \quad \infty \quad \infty)$

Step length, $\alpha \to 1$ New active set $\to (1 \quad 4)$

New point $\to (0.629247 \quad 0.0296718 \quad 1.28434 \times 10^{-17} \quad 0.341081)$

**** Iteration 5 ****

Current point $\to (0.629247 \quad 0.0296718 \quad 1.28434 \times 10^{-17} \quad 0.341081)$

Active set $\to (1 \quad 4)$

$$A \to \begin{pmatrix} 1 & 1 & 1 & 1 \\ -10.6483 & -11.98 & -8.34 & -8.63167 \\ 0 & 0 & -1 & 0 \end{pmatrix} \quad gk \to \begin{pmatrix} 37.6304 \\ 64.1749 \\ 0.00460216 \\ -2.56853 \end{pmatrix}$$

LE Constraint values →
$$\begin{pmatrix} 1.77636 \times 10^{-15} & -0.629247 & -0.0296718 & -1.28434 \times 10^{-17} & -0.341081 \end{pmatrix}$$

EQ Constraint values → $\begin{pmatrix} -9.99201 \times 10^{-16} \end{pmatrix}$

Equations:

$$\text{lhs} \to \begin{pmatrix} 58.1104 & 80.7818 & -0.575767 & -3.90639 & 1 & -10.6483 & 0 \\ 80.7818 & 534.689 & 13.6673 & -7.39403 & 1 & -11.98 & 0 \\ -0.575767 & 13.6673 & 0.751867 & -0.113267 & 1 & -8.34 & -1 \\ -3.90639 & -7.39403 & -0.113267 & 0.319428 & 1 & -8.63167 & 0 \\ 1 & 1 & 1 & 1 & 0 & 0 & 0 \\ -10.6483 & -11.98 & -8.34 & -8.63167 & 0 & 0 & 0 \\ 0 & 0 & -1 & 0 & 0 & 0 & 0 \end{pmatrix}$$

$$\text{rhs} \to \begin{pmatrix} -37.6304 \\ -64.1749 \\ -0.00460216 \\ 2.56853 \\ 0 \\ 0 \\ 0 \end{pmatrix}$$

Solution: $dk \to \begin{pmatrix} 2.54508 \times 10^{-15} \\ 1.07114 \times 10^{-16} \\ -3.54914 \times 10^{-16} \\ -3.54326 \times 10^{-15} \end{pmatrix}$ $uk \to \begin{pmatrix} 19.9333 \\ 8.38703 \end{pmatrix}$ $vk \to \begin{pmatrix} 174.627 \end{pmatrix}$

***** Optimum solution after 5 iterations *****

Solution →
$\begin{pmatrix} x_1 \to 0.629247 & x_2 \to 0.0296718 & x_3 \to 1.28434 \times 10^{-17} & x_4 \to 0.341081 \end{pmatrix}$

Lagrange multipliers for LE constraints (u) → $\begin{pmatrix} 19.9333 & 0 & 0 & 8.38703 & 0 \end{pmatrix}$

Lagrange multipliers for EQ constraints (v) → $\begin{pmatrix} 174.627 \end{pmatrix}$

Objective function → 12.3535

The complete history of the actual four variables during iteration is as follows:

TableForm[hist]

1	0	0	0
1	0	0	0
0.678512	1.65412×10^{-18}	-2.72995×10^{-17}	0.321488
0.678512	1.65412×10^{-18}	-2.72995×10^{-17}	0.321488
0.629247	0.0296718	1.28434×10^{-17}	0.341081

The optimum solution is the same as that obtained in Example 8.8.

8.5 Active Set Method for the Dual QP Problem

In this section, we consider a special form of QP problem that arises by taking the dual of a problem in which $\mathbf{Q}$ is invertible. The problem involves a quadratic objective function and simple nonnegativity constraints on problem variables. Using the standard notation for variables, the problem can be expressed in the following form:

Minimize $f(\mathbf{x}) = \mathbf{c}^T\mathbf{x} + \frac{1}{2}\mathbf{x}^T\mathbf{Q}\mathbf{x}$
Subject to $x_i \geq 0, i = 1, \ldots, m \leq n$

where $\mathbf{x}$ is an $n \times 1$ vector of optimization variables, $\mathbf{c}$ is an $n \times 1$ vector containing coefficients of linear terms, and $n \times n$ symmetric matrix $\mathbf{Q}$ contains coefficients of square and mixed terms. The first $m \leq n$ variables are constrained to be nonnegative.

This problem can obviously be solved by using any of the QP algorithms discussed so far. However, because of the simple nature of constraints, a more efficient method can be developed by a simple extension of any of the methods for unconstrained minimization. In this section, the conjugate gradient method is extended to solve this special QP problem.

8.5.1 Optimality Conditions

By introducing a vector of slack variables, $\mathbf{s}$, and the Lagrange multipliers, $u_i \geq 0, i = 1\ldots, m$ and $u_i = 0, i = m+1\ldots, n$, the Lagrangian for the problem is as follows:

$$L(\mathbf{x}, \mathbf{u}, \mathbf{s}) = f(\mathbf{x}) + \mathbf{u}^T(-\mathbf{x} + \mathbf{s}^2)$$

The necessary optimality conditions are:

$$\nabla f(\mathbf{x}) - \mathbf{u} = \mathbf{0}$$

$$u_i \geq 0, i = 1\ldots, m, \quad \text{and} \quad u_i = 0, i = m+1\ldots, n$$

$$u_i s_i = 0, i = 1, \ldots, m \quad \text{(switching conditions)}$$

From the switching conditions, either $u_i = 0$ or $s_i = 0$. If $u_i = 0$, then the gradient condition says that the corresponding partial derivative $\partial f/\partial x_i$ must be zero. If $s_i = 0$, then there is no slack and therefore, $x_i = 0$. Furthermore,

8.5 Active Set Method for the Dual QP Problem

since now $u_i \neq 0$, the gradient condition says that the partial derivative $\partial f/\partial x_i$ must be greater than or equal to zero. Thus, the optimality conditions for the problem can be written entirely in terms of optimization variables x_i, without explicitly involving the Lagrange multipliers, as follows:

$$\begin{pmatrix} \partial f/\partial x_i = 0 \text{ if } x_i > 0 \\ \partial f/\partial x_i \geq 0 \text{ if } x_i = 0 \end{pmatrix} \quad \text{for } i = 1, \ldots, m$$
$$\partial f/\partial x_i = 0 \quad \text{for } i = m+1, \ldots, n$$

8.5.2 Conjugate Gradient Algorithm for Quadratic Function with Nonnegativity Constraints

A conjugate gradient algorithm for minimizing an unconstrained function was presented in Chapter 5. Two simple extensions are introduced into this algorithm to make it suitable for the present situation. First, an active set idea is introduced. Active variables are defined as those variables that are either greater than zero or that violate the optimality condition. The remaining variables are called passive variables. By setting passive variables to zero, the objective function is expressed in terms of active variables only. The standard conjugate gradient algorithm for unconstrained problems is used to find direction with respect to active variables. The second modification is in the step-length calculations. The step length given by the standard conjugate gradient algorithm for a quadratic function is first computed. If this step length does not make any of the restricted variables take on a negative value, then it is accepted and we proceed to the next iteration. If the standard step length is too large, a smaller value is computed that makes one of the active variables take a zero value. This variable is removed from the active set for the subsequent iteration, and the process is repeated until the optimality conditions are satisfied. The complete algorithm is as follows:

Initialization: Set $k = 0$. Choose an arbitrary starting point $\mathbf{x}^0$.

A. Test for optimality: If $\partial f/\partial x_i = 0$ for $x_i^k > 0$ and $\partial f/\partial x_i \geq 0$ for $x_i^k = 0$ for $i = 1, \ldots, m$ and $\partial f/\partial x_i = 0$ for $i = m+1, \ldots, n$, then stop. We have an optimum. Otherwise, continue.

B. Define the set of active variables consisting of those indices for which either $x_i > 0$ or $x_i = 0$ but corresponding $\partial f/\partial x_i < 0$. The remaining variables are passive variables. All unrestricted variables are considered active. Thus,

$$\mathbf{I}_a^k = \{i | (x_i^k > 0) \text{ or} (x_i^k = 0 \text{ and } \partial f / \partial x_i < 0) \text{ or } (i > m)\} \quad \mathbf{I}_p^k = \{i | i \notin \mathbf{I}_a^k\}$$

C. Set passive variables to zero and use the conjugate gradient algorithm to minimize $f(\mathbf{x})$ with respect to active variables.

1. Set the iteration counter $i = 0$.
2. Compute $\nabla f(\mathbf{x}^i)$ with respect to active variables.
3. Compute $\beta = \begin{pmatrix} 0 & i = 0 \\ \frac{[\nabla f(\mathbf{x}^i)]^T \nabla f(\mathbf{x}^i)}{[\nabla f(\mathbf{x}^{i-1})]^T \nabla f(\mathbf{x}^{i-1})} & i > 0 \end{pmatrix}$
4. Compute direction $\mathbf{d}^{i+1} = \begin{pmatrix} \nabla f(\mathbf{x}^i) & i = 0 \\ \nabla f(\mathbf{x}^i) + \beta \mathbf{d}^i & i > 0 \end{pmatrix}$
5. If $||\mathbf{d}^{i+1}|| \leq$ tol, we have found the minimum of $f(\mathbf{x})$ with respect to active variables. Go to step (A) to check for optimality. Otherwise, continue.
6. Compute step length, keeping in mind that none of the variables should become negative.

$$\alpha = \text{Min}\left[-\mathbf{d}^{i+1^T} \nabla f(\mathbf{x}^i) / \mathbf{d}^{i+1^T} \mathbf{Q} \mathbf{d}^{i+1}, \text{Min}[-x_j^i / d^j, d^j > 0, j = 1, \ldots, m]\right]$$

7. Update $\mathbf{x}^{i+1} = \mathbf{x}^i + \alpha \mathbf{d}^{i+1}$.
8. If a new variable with an index less than or equal to m has reached a zero value, move that to the passive variable. Set $i = i + 1$ and go to step (2). The algorithm is implemented in a *Mathematica* function called ActiveSetDualQP. The function should be loaded using the following needs command prior to its use.

`Needs["OptimizationToolbox`QuadraticProgramming`"];`

Example 8.13 Find the minimum of the following QP problem:

$$f = \frac{1}{2}x_1^2 + 12x_2^2 + 2x_3^2 + 4x_4^2 - x_1 x_2 - x_2 x_3 - 2x_4 + 3x_1 \text{ subject to } x_3 \text{ and } x_4 \geq 0$$

```
f = x1^2/2 + 12x2^2 + 2x3^2 + 4x4^2 - x1 x2 - x2 x3 - 2x4 + 3x1;
vars = {x1, x2, x3, x4};
freeVars = {x1, x2};
```

Using ActiveSetDualQP, the solution is as follows. All intermediate results are printed using notation used in describing the algorithm.

8.5 Active Set Method for the Dual QP Problem

```
ActiveSetDualQP[f, vars, PrintLevel → 2, UnrestrictedVariables → freeVars];
```

Minimize → $3x_1 + \frac{x_1^2}{2} - x_1 x_2 + 12x_2^2 - x_2 x_3 + 2x_3^2 - 2x_4 + 4x_4^2$

Subject to → $(x_3 \geq 0 \quad x_4 \geq 0)$

Unrestricted variables → $(x_1 \quad x_2)$

Order of variables → $\begin{pmatrix} 1 & 2 & 3 & 4 \\ x_3 & x_4 & x_1 & x_2 \end{pmatrix}$

QP Solution Optimality Check 1

$\begin{pmatrix} x_3 \\ x_4 \\ x_1 \\ x_2 \end{pmatrix} \rightarrow \begin{pmatrix} 1. \\ 1. \\ 1. \\ 1. \end{pmatrix} \quad \nabla f \rightarrow \begin{pmatrix} 3. \\ 6. \\ 3. \\ 22. \end{pmatrix}$ Optimality status → $\begin{pmatrix} \text{False} \\ \text{False} \\ \text{False} \\ \text{False} \end{pmatrix}$

**** QP Solution Pass 1 ****

Active variables → $(x_3 \quad x_4 \quad x_1 \quad x_2)$

Current point → $(1. \quad 1. \quad 1. \quad 1.)$

$c \rightarrow \begin{pmatrix} 0 \\ -2 \\ 3 \\ 0 \end{pmatrix} \quad Q \rightarrow \begin{pmatrix} 4 & 0 & 0 & -1 \\ 0 & 8 & 0 & 0 \\ 0 & 0 & 1 & -1 \\ -1 & 0 & -1 & 24 \end{pmatrix} \quad \nabla f \rightarrow \begin{pmatrix} 3. \\ 6. \\ 3. \\ 22. \end{pmatrix}$

>> Conjugate Gradient Iteration 1

$d \rightarrow \begin{pmatrix} -3. \\ -6. \\ -3. \\ -22. \end{pmatrix} \quad \nabla f \rightarrow \begin{pmatrix} 3. \\ 6. \\ 3. \\ 22. \end{pmatrix}$

$||\nabla f|| \rightarrow 23.1948 \quad \beta \rightarrow 0.$

$\alpha 1 \rightarrow 0.0460419$

-pt/d → $(0.333333 \quad 0.166667 \quad 0.333333 \quad 0.0454545)$

$\alpha 2 \rightarrow 0.166667$

$\alpha \rightarrow 0.0460419 \quad$ CG-Status → NonOptimum

New point → $(0.861874 \quad 0.723748 \quad 0.861874 \quad -0.0129226)$

Objective function → 5.11472

>> Conjugate Gradient Iteration 2

$d \rightarrow \begin{pmatrix} -3.71408 \\ -4.2973 \\ -4.12845 \\ 0.173734 \end{pmatrix} \quad \nabla f \rightarrow \begin{pmatrix} 3.46042 \\ 3.78999 \\ 3.8748 \\ -2.03389 \end{pmatrix}$

$||\nabla f|| \rightarrow 6.74457 \quad \beta \rightarrow 0.08455253$

$\alpha 1 \rightarrow 0.203617$

-pt/d → $(0.232056 \quad 0.168419 \quad 0.208764 \quad 0.0743813)$

$\alpha 2 \rightarrow 0.168419$

$\alpha \to 0.168419$ CG-Status $\to$ NonOptimum

New point $\to \begin{pmatrix} 0.236352 & 0. & 0.166563 & 0.0163376 \end{pmatrix}$

Objective function $\to 0.621906$

New var to be made passive $\to (x_4)$

**** QP Solution Pass 2 ****

Active variables $\to \begin{pmatrix} x_3 & x_1 & x_2 \end{pmatrix}$

Current point $\to \begin{pmatrix} 0.236352 & 0.166563 & 0.0163376 \end{pmatrix}$

$c \to \begin{pmatrix} 0 \\ 3 \\ 0 \end{pmatrix}$ $Q \to \begin{pmatrix} 4 & 0 & -1 \\ 0 & 1 & -1 \\ -1 & -1 & 24 \end{pmatrix}$ $\nabla f \to \begin{pmatrix} 0.929071 \\ 3.15023 \\ -0.0108132 \end{pmatrix}$

>> Conjugate Gradient Iteration 1

$d \to \begin{pmatrix} -0.929071 \\ -3.15023 \\ 0.0108132 \end{pmatrix}$ $\nabla f \to \begin{pmatrix} 0.929071 \\ 3.15023 \\ -0.0108132 \end{pmatrix}$

$||\nabla f|| \to 3.28439$ $\beta \to 0.$

$\alpha 1 \to 0.800973$

$-pt/d \to \begin{pmatrix} 0.254396 & 0.0528734 & -1.5109 \end{pmatrix}$

$\alpha 2 \to 0.254396$

$\alpha \to 0.254396$ CG-Status $\to$ NonOptimum

New point $\to \begin{pmatrix} -2.77556 \times 10^{-17} & -0.634842 & 0.0190884 \end{pmatrix}$

Objective function $\to -1.68652$

New var to be made passive $\to (x_3)$

**** QP Solution Pass 3 ****

Active variables $\to \begin{pmatrix} x_1 & x_2 \end{pmatrix}$

Current point $\to \begin{pmatrix} -0.634842 & 0.0190884 \end{pmatrix}$

$c \to \begin{pmatrix} 3 \\ 0 \end{pmatrix}$ $Q \to \begin{pmatrix} 1 & -1 \\ -1 & 24 \end{pmatrix}$ $\nabla f \to \begin{pmatrix} 2.34607 \\ 1.09296 \end{pmatrix}$

>> Conjugate Gradient Iteration 1

$d \to \begin{pmatrix} -2.34607 \\ -1.09296 \end{pmatrix}$ $\nabla f \to \begin{pmatrix} 2.34607 \\ 1.09296 \end{pmatrix}$

$||\nabla f|| \to 2.58817$ $\beta \to 0.$

$\alpha 1 \to 0.230626$

$\alpha \to 0.230626$ CG-Status $\to$ NonOptimum

New point $\to \begin{pmatrix} -1.17591 & -0.232977 \end{pmatrix}$

Objective function $\to -2.45896$

>> Conjugate Gradient Iteration 2

$d \to \begin{pmatrix} -10.3676 \\ 0.543924 \end{pmatrix}$ $\nabla f \to \begin{pmatrix} 2.05707 \\ -4.41554 \end{pmatrix}$

8.5 Active Set Method for the Dual QP Problem 557

$||\nabla f|| \to 4.8712 \quad \beta \to 3.54231$
$\alpha 1 \to 0.188523$
$\alpha \to 0.188523 \quad$ CG-Status $\to$ NonOptimum
New point $\to (-3.13043 \quad -0.130435)$
Objective function $\to -4.69565$
>> Conjugate Gradient Iteration 3
$d \to \begin{pmatrix} -4.71219 \times 10^{-15} \\ -2.19527 \times 10^{-15} \end{pmatrix} \quad \nabla f \to \begin{pmatrix} 0. \\ 2.44249 \times 10^{-15} \end{pmatrix}$
$||\nabla f|| \to 2.44249 \times 10^{-15} \quad \beta \to 4.54512 \times 10^{-16}$
$\alpha \to 0 \quad$ CG-Status $\to$ Optimum
New point $\to (-3.13043 \quad -0.130435)$
Objective function $\to -4.69565$

QP Solution Optimality Check 2

$\begin{pmatrix} x_3 \\ x_4 \\ x_1 \\ x_2 \end{pmatrix} \to \begin{pmatrix} 0 \\ 0 \\ -3.13043 \\ -0.130435 \end{pmatrix} \quad \nabla f \to \begin{pmatrix} 0.130435 \\ -2. \\ 0 \\ 0 \end{pmatrix} \quad$ Optimality status $\to \begin{pmatrix} \text{True} \\ \text{False} \\ \text{True} \\ \text{True} \end{pmatrix}$

**** QP Solution Pass 1 ****

Active variables $\to (x_4 \quad x_1 \quad x_2)$
Current point $\to (0 \quad -3.13043 \quad -0.130435)$
$c \to \begin{pmatrix} -2 \\ 3 \\ 0 \end{pmatrix} \quad Q \to \begin{pmatrix} 8 & 0 & 0 \\ 0 & 1 & -1 \\ 0 & -1 & 24 \end{pmatrix} \quad \nabla f \to \begin{pmatrix} -2. \\ 0. \\ 2.44249 \times 10^{-15} \end{pmatrix}$
>> Conjugate Gradient Iteration 1
$d \to \begin{pmatrix} 2. \\ 0. \\ -2.44249 \times 10^{-15} \end{pmatrix} \quad \nabla f \to \begin{pmatrix} -2. \\ 0. \\ 2.44249 \times 10^{-15} \end{pmatrix}$
$||\nabla f|| \to 2. \quad \beta \to 0.$
$\alpha 1 \to 0.125$
-pt/d $\to (0 \quad$ ComplexInfinity $\quad -5.34024 \times 10^{13})$
$\alpha 2 \to \infty$
$\alpha \to 0.125 \quad$ CG-Status $\to$ NonOptimum
New point $\to (0.25 \quad -3.13043 \quad -0.130435)$
Objective function $\to -4.94565$
>> Conjugate Gradient Iteration 2
$d \to \begin{pmatrix} 1.79958 \times 10^{-29} \\ -4.44089 \times 10^{-16} \\ 4.88498 \times 10^{-15} \end{pmatrix} \quad \nabla f \to \begin{pmatrix} 0. \\ 4.44089 \times 10^{-16} \\ -4.88498 \times 10^{-15} \end{pmatrix}$
$||\nabla f|| \to 4.90513 \times 10^{-15} \quad \beta \to 8.99794 \times 10^{-30}$
$\alpha \to 0 \quad$ CG-Status $\to$ Optimum

Chapter 8 Quadratic Programming

New point $\to \begin{pmatrix} 0.25 & -3.13043 & -0.130435 \end{pmatrix}$

Objective function $\to -4.94565$

QP Solution Optimality Check 3

$$\begin{pmatrix} x_3 \\ x_4 \\ x_1 \\ x_2 \end{pmatrix} \to \begin{pmatrix} 0 \\ 0.25 \\ -3.13043 \\ -0.130435 \end{pmatrix} \quad \nabla f \to \begin{pmatrix} 0.130435 \\ 0 \\ 0 \\ 0 \end{pmatrix} \quad \text{Optimality status} \to \begin{pmatrix} \text{True} \\ \text{True} \\ \text{True} \\ \text{True} \end{pmatrix}$$

The solution can easily be confirmed by solving the problem directly by using KT conditions.

```
KTSolution[f, {x₃ ≥ 0, x₄ ≥ 0}, vars];
```

***** Lagrangian →

$$3x_1 + \frac{x_1^2}{2} - x_1 x_2 + 12x_2^2 + u_1 \left(s_1^2 - x_3 \right) - x_2 x_3 + 2x_3^2 + u_2 \left(s_2^2 - x_4 \right) - 2x_4 + 4x_4^2$$

***** Valid KT Point(s) *****

$f \to -4.94565$
$x_1 \to -3.13043$
$x_2 \to -0.130435$
$x_3 \to 0$
$x_4 \to 0.25$
$u_1 \to 0.130435$
$u_2 \to 0$
$s_1^2 \to 0$
$s_2^2 \to 0.25$

Example 8.14 Construct the dual of the following QP problem and then solve the dual using the ActiveSetDualQP method.

$$f = x_1^2 + x_1 x_2 + 2x_2^2 + 2x_3^2 + 2x_2 x_3 + 4x_1 + 6x_2 + 12x_3$$

$$\text{Subject to} \begin{pmatrix} x_1 + x_2 + x_3 = 6 \\ -x_1 - x_2 + 2x_3 = 2 \\ x_i \geq 0, i = 1, \ldots, 3 \end{pmatrix}$$

```
f = x₁² + x₁x₂ + 2x₂² + 2x₃² + 2x₂x₃ + 4x₁ + 6x₂ + 12x₃;
g = {x₁ + x₂ + x₃ == 6, -x₁ - x₂ + 2x₃ == 2};
vars = {x₁, x₂, x₃};
```

The dual QP is constructed as follows:

```
{df, dg, dvars, dv} = FormDualQP[f, g, vars];
```

Primal problem

8.5 Active Set Method for the Dual QP Problem

Minimize $\to 4x_1 + x_1^2 + 6x_2 + x_1 x_2 + 2x_2^2 + 12x_3 + 2x_2 x_3 + 2x_3^2$

Subject to $\to \begin{pmatrix} x_1 + x_2 + x_3 == 6 \\ -x_1 - x_2 + 2x_3 == 2 \end{pmatrix}$

and $\to (x_1 \geq 0 \quad x_2 \geq 0 \quad x_3 \geq 0)$

$A \to \begin{pmatrix} 1 & 1 & 1 \\ -1 & -1 & 2 \end{pmatrix} \quad b \to \begin{pmatrix} 6 \\ 2 \end{pmatrix}$

$c \to \begin{pmatrix} 4 \\ 6 \\ 12 \end{pmatrix} \quad Q \to \begin{pmatrix} 2 & 1 & 0 \\ 1 & 4 & 2 \\ 0 & 2 & 4 \end{pmatrix}$

$Q^{-1} \to \begin{pmatrix} \frac{3}{5} & -\frac{1}{5} & \frac{1}{10} \\ -\frac{1}{5} & \frac{2}{5} & -\frac{1}{5} \\ \frac{1}{10} & -\frac{1}{5} & \frac{7}{20} \end{pmatrix} \quad -c + u + A^T . v \to \begin{pmatrix} -4 + u_1 + v_1 - v_2 \\ -6 + u_2 + v_1 - v_2 \\ -12 + u_3 + v_1 + 2v_2 \end{pmatrix}$

Solution $\to \begin{cases} x_1 \to -\frac{12}{5} + \frac{3u_1}{5} - \frac{u_2}{5} + \frac{u_3}{10} + \frac{v_1}{2} - \frac{v_2}{5} \\ x_2 \to \frac{4}{5} - \frac{u_1}{5} + \frac{2u_2}{5} - \frac{u_3}{5} - \frac{3v_2}{5} \\ x_3 \to -\frac{17}{5} + \frac{u_1}{10} - \frac{u_2}{5} + \frac{7u_3}{20} + \frac{v_1}{4} + \frac{4v_2}{5} \end{cases}$

Dual QP problem

Variables $\to (u_1 \quad u_2 \quad u_3 \quad v_1 \quad v_2)$

Maximize $\to -\frac{114}{5} + \frac{12u_1}{5} - \frac{3u_1^2}{10} - \frac{4u_2}{5} + \frac{u_1 u_2}{5} - \frac{u_2^2}{5} + \frac{17u_3}{5} - \frac{u_1 u_3}{10} + \frac{u_2 u_3}{5} - \frac{7u_3^2}{40} + 11v_1 - \frac{u_1 v_1}{2} - \frac{u_3 v_1}{4} - \frac{3v_1^2}{8} + \frac{36v_2}{5} + \frac{u_1 v_2}{5} + \frac{3u_2 v_2}{5} - \frac{4u_3 v_2}{5} - \frac{6v_2^2}{5}$

Subject to $\to \begin{pmatrix} u_1 \geq 0 \\ u_2 \geq 0 \\ u_3 \geq 0 \end{pmatrix}$

The dual QP is solved using ActiveSetDualQP, showing all intermediate calculations.

ActiveSetDualQP[-df, dvars, PrintLevel → 2, UnrestrictedVariables → dv];

Minimize $\to \frac{114}{5} - \frac{12u_1}{5} + \frac{3u_1^2}{10} + \frac{4u_2}{5} - \frac{u_1 u_2}{5} + \frac{u_2^2}{5} - \frac{17u_3}{5} + \frac{u_1 u_3}{10} - \frac{u_2 u_3}{5} + \frac{7u_3^2}{40} - 11v_1 + \frac{u_1 v_1}{2} + \frac{u_3 v_1}{4} + \frac{3v_1^2}{8} - \frac{36v_2}{5} - \frac{u_1 v_2}{5} - \frac{3u_2 v_2}{5} + \frac{4u_3 v_2}{5} + \frac{6v_2^2}{5}$

Subject to $\to (u_1 \geq 0 \quad u_2 \geq 0 \quad u_3 \geq 0)$

Unrestricted variables $\to (v_1 \quad v_2)$

Order of variables $\to \begin{pmatrix} 1 & 2 & 3 & 4 & 5 \\ u_1 & u_2 & u_3 & v_1 & v_2 \end{pmatrix}$

QP Solution Optimality Check 1

$$\begin{pmatrix} u_1 \\ u_2 \\ u_3 \\ v_1 \\ v_2 \end{pmatrix} \to \begin{pmatrix} 1. \\ 1. \\ 1. \\ 1. \\ 1. \end{pmatrix} \quad \nabla f \to \begin{pmatrix} -1.6 \\ 0.2 \\ -2.1 \\ -9.5 \\ -4.8 \end{pmatrix} \quad \text{Optimality status} \to \begin{pmatrix} \text{False} \\ \text{False} \\ \text{False} \\ \text{False} \\ \text{False} \end{pmatrix}$$

**** QP Solution Pass 1 ****

Active variables $\to \begin{pmatrix} u_1 & u_2 & u_3 & v_1 & v_2 \end{pmatrix}$

Current point $\to \begin{pmatrix} 1. & 1. & 1. & 1. & 1. \end{pmatrix}$

$$c \to \begin{pmatrix} -\frac{12}{5} \\ \frac{4}{5} \\ -\frac{17}{5} \\ -11 \\ -\frac{36}{5} \end{pmatrix} \quad Q \to \begin{pmatrix} \frac{3}{5} & -\frac{1}{5} & \frac{1}{10} & \frac{1}{2} & -\frac{1}{5} \\ -\frac{1}{5} & \frac{2}{5} & -\frac{1}{5} & 0 & -\frac{3}{5} \\ \frac{1}{10} & -\frac{1}{5} & \frac{7}{20} & \frac{1}{4} & \frac{4}{5} \\ \frac{1}{2} & 0 & \frac{1}{4} & \frac{3}{4} & 0 \\ -\frac{1}{5} & -\frac{3}{5} & \frac{4}{5} & 0 & \frac{12}{5} \end{pmatrix} \quad \nabla f \to \begin{pmatrix} -1.6 \\ 0.2 \\ -2.1 \\ -9.5 \\ -4.8 \end{pmatrix}$$

\>\> Conjugate Gradient Iteration 1

$$d \to \begin{pmatrix} 1.6 \\ -0.2 \\ 2.1 \\ 9.5 \\ 4.8 \end{pmatrix} \quad \nabla f \to \begin{pmatrix} -1.6 \\ 0.2 \\ -2.1 \\ -9.5 \\ -4.8 \end{pmatrix}$$

$||\nabla f|| \to 10.9681 \quad \beta \to 0.$

$\alpha 1 \to 0.722826$

$-pt/d \to \begin{pmatrix} -0.625 & 5. & -0.47619 & -0.105263 & -0.208333 \end{pmatrix}$

$\alpha 2 \to 5.$

$\alpha \to 0.722826 \quad$ CG-Status $\to$ NonOptimum

New point $\to \begin{pmatrix} 2.15652 & 0.855435 & 2.51794 & 7.86685 & 4.46957 \end{pmatrix}$

Objective function $\to -41.178$

\>\> Conjugate Gradient Iteration 2

$$d \to \begin{pmatrix} -1.31947 \\ 2.38762 \\ -2.15646 \\ 7.51669 \\ -2.51275 \end{pmatrix} \quad \nabla f \to \begin{pmatrix} 2.01413 \\ -2.47446 \\ 3.06821 \\ -3.39212 \\ 4.59674 \end{pmatrix}$$

$||\nabla f|| \to 7.22704 \quad \beta \to 0.434165$

$\alpha 1 \to 0.830494$

$-pt/d \to \begin{pmatrix} 1.63439 & -0.358279 & 1.16762 & -1.04658 & 1.77875 \end{pmatrix}$

$\alpha 2 \to 1.16762$

$\alpha \to 0.830494 \quad$ CG-Status $\to$ NonOptimum

New point $\to \begin{pmatrix} 1.06071 & 2.83834 & 0.727008 & 14.1094 & 2.38274 \end{pmatrix}$

```
Objective function → -62.8664
>> Conjugate Gradient Iteration 3
```

$$d \to \begin{pmatrix} -5.07819 \\ 1.22451 \\ -3.06616 \\ 4.02721 \\ 1.37037 \end{pmatrix} \quad \nabla f \to \begin{pmatrix} 4.31962 \\ 0.148147 \\ 1.8264 \\ 0.294167 \\ -2.81496 \end{pmatrix}$$

```
||∇f|| → 5.47972    β → 0.574904
α1 → 2.84371
-pt/d → (0.208876  -2.31794  0.237107  -3.50352  -1.73876)
α2 → 0.208876
α → 0.208876    CG-Status → NonOptimum
New point → (0.  3.09411  0.0865586  14.9506  2.66898)
Objective function → -68.908
New var to be made passive → (u₁)
```

**** QP Solution Pass 2 ****

```
Active variables → (u₂  u₃  v₁  v₂)
Current point → (3.09411  0.0865586  14.9506  2.66898)
```

$$c \to \begin{pmatrix} \frac{4}{5} \\ -\frac{17}{5} \\ -11 \\ -\frac{36}{5} \end{pmatrix} \quad Q \to \begin{pmatrix} \frac{2}{5} & -\frac{1}{5} & 0 & -\frac{3}{5} \\ -\frac{1}{5} & \frac{7}{20} & \frac{1}{4} & \frac{4}{5} \\ 0 & \frac{1}{4} & \frac{3}{4} & 0 \\ -\frac{3}{5} & \frac{4}{5} & 0 & \frac{12}{5} \end{pmatrix} \quad \nabla f \to \begin{pmatrix} 0.418945 \\ 1.88431 \\ 0.23459 \\ -2.58167 \end{pmatrix}$$

```
>> Conjugate Gradient Iteration 1
```

$$d \to \begin{pmatrix} -0.418945 \\ -1.88431 \\ -0.23459 \\ 2.58167 \end{pmatrix} \quad \nabla f \to \begin{pmatrix} 0.418945 \\ 1.88431 \\ 0.23459 \\ -2.58167 \end{pmatrix}$$

```
||∇f|| → 3.23205    β → 0.
α1 → 0.969941
-pt/d → (7.38549  0.0459366  63.7307  -1.03382)
α2 → 0.0459366
α → 0.0459366    CG-Status → NonOptimum
New point → (3.07487  1.38778 × 10⁻¹⁷  14.9398  2.78757)
Objective function → -69.3765
New var to be made passive → (u₂)
```

**** QP Solution Pass 3 ****

```
Active variables → (u₃  v₁  v₂)
Current point → (3.07487  14.9398  2.78757)
```

$$c \to \begin{pmatrix} \frac{4}{5} \\ -11 \\ -\frac{36}{5} \end{pmatrix} \quad Q \to \begin{pmatrix} \frac{2}{5} & 0 & -\frac{3}{5} \\ 0 & \frac{3}{4} & 0 \\ -\frac{3}{5} & 0 & \frac{12}{5} \end{pmatrix} \quad \nabla f \to \begin{pmatrix} 0.357403 \\ 0.204868 \\ -2.35474 \end{pmatrix}$$

\>\> Conjugate Gradient Iteration 1

$$d \to \begin{pmatrix} -0.357403 \\ -0.204868 \\ 2.35474 \end{pmatrix} \quad \nabla f \to \begin{pmatrix} 0.357403 \\ 0.204868 \\ -2.35474 \end{pmatrix}$$

$||\nabla f|| \to 2.39051 \quad \beta \to 0.$

$\alpha 1 \to 0.396841$

$-pt/d \to \begin{pmatrix} 8.60336 & 72.924 & -1.18381 \end{pmatrix}$

$\alpha 2 \to 8.60336$

$\alpha \to 0.396841 \quad$ CG-Status $\to$ NonOptimum

New point $\to \begin{pmatrix} 2.93304 & 14.8585 & 3.72203 \end{pmatrix}$

Objective function $\to -70.5104$

\>\> Conjugate Gradient Iteration 2

$$d \to \begin{pmatrix} 0.254436 \\ -0.147085 \\ 0.063632 \end{pmatrix} \quad \nabla f \to \begin{pmatrix} -0.260005 \\ 0.143893 \\ -0.0269445 \end{pmatrix}$$

$||\nabla f|| \to 0.298385 \quad \beta \to 0.0155803$

$\alpha 1 \to 2.74711$

$\alpha \to 2.74711 \quad$ CG-Status $\to$ NonOptimum

New point $\to \begin{pmatrix} 3.632 & 14.4545 & 3.89684 \end{pmatrix}$

Objective function $\to -70.6327$

\>\> Conjugate Gradient Iteration 3

$$d \to \begin{pmatrix} 0.180531 \\ 0.1041 \\ 0.0506095 \end{pmatrix} \quad \nabla f \to \begin{pmatrix} -0.0853015 \\ -0.159151 \\ -0.0267936 \end{pmatrix}$$

$||\nabla f|| \to 0.182547 \quad \beta \to 0.374276$

$\alpha 1 \to 2.03843$

$\alpha \to 2.03843 \quad$ CG-Status $\to$ NonOptimum

New point $\to \begin{pmatrix} 4. & 14.6667 & 4. \end{pmatrix}$

Objective function $\to -70.6667$

\>\> Conjugate Gradient Iteration 4

$$d \to \begin{pmatrix} -2.21622 \times 10^{-15} \\ 1.30475 \times 10^{-16} \\ 6.28068 \times 10^{-15} \end{pmatrix} \quad \nabla f \to \begin{pmatrix} 2.44249 \times 10^{-15} \\ 0. \\ -6.21725 \times 10^{-15} \end{pmatrix}$$

$||\nabla f|| \to 6.67982 \times 10^{-15} \quad \beta \to 1.25335 \times 10^{-15}$

$\alpha \to 0 \quad$ CG-Status $\to$ Optimum

New point $\to \begin{pmatrix} 4. & 14.6667 & 4. \end{pmatrix}$

Objective function $\to -70.6667$

QP Solution Optimality Check 2

$$\begin{pmatrix} u_1 \\ u_2 \\ u_3 \\ v_1 \\ v_2 \end{pmatrix} \to \begin{pmatrix} 0 \\ 4. \\ 0 \\ 14.6667 \\ 4. \end{pmatrix} \quad \nabla f \to \begin{pmatrix} 3.33333 \\ 0 \\ 2.66667 \\ 0 \\ 0 \end{pmatrix} \quad \text{Optimality status} \to \begin{pmatrix} \text{True} \\ \text{True} \\ \text{True} \\ \text{True} \\ \text{True} \end{pmatrix}$$

Once the Lagrange multipliers are known, the primal solution can be obtained by using the relationship between the **x**, **v**, and **u**.

$$\text{Solution} \to \begin{pmatrix} x_1 \to -\frac{12}{5} + \frac{3u_1}{5} - \frac{u_2}{5} + \frac{u_3}{10} + \frac{v_1}{2} - \frac{v_2}{5} \\ x_2 \to \frac{4}{5} - \frac{u_1}{5} + \frac{2u_2}{5} - \frac{u_3}{5} - \frac{3v_2}{5} \\ x_3 \to -\frac{17}{5} + \frac{u_1}{10} - \frac{u_2}{5} + \frac{7u_3}{20} + \frac{v_1}{4} + \frac{4v_2}{5} \end{pmatrix} \to \begin{pmatrix} x_1 \to 3.33333 \\ x_2 \to 0 \\ x_3 \to 2.66667 \end{pmatrix}$$

The solution is verified by solving the primal problem directly using the KT conditions.

`KTSolution[f, {g, Thread[vars ≥ 0]}, vars];`

Lagrangian $\to u_1 \left(s_1^2 - x_1\right) + 4x_1 + x_1^2 + u_2 \left(s_2^2 - x_2\right) + 6x_2 + x_1 x_2 + 2x_2^2 + u_3 \left(s_3^2 - x_3\right) + 12 x_3 + 2 x_2 x_3 + 2 x_3^2 + v_1 \left(-6 + x_1 + x_2 + x_3\right) + v_2 \left(-2 - x_1 - x_2 + 2 x_3\right)$

***** Valid KT Point(s) *****

$f \to 70.6667$
$x_1 \to 3.33333$
$x_2 \to 0$
$x_3 \to 2.66667$
$u_1 \to 0$
$u_2 \to 4.$
$u_3 \to 0$
$s_1^2 \to 3.33333$
$s_2^2 \to 0$
$s_3^2 \to 2.66667$
$v_1 \to -14.6667$
$v_2 \to -4.$

8.6 Appendix — Derivation of the Descent Direction Formula for the PAS Method

The detailed derivation of the descent direction formula for the PAS method is presented in this section. The basic idea is not very different from that used in

Transformed QP Problem

It is well known that gradient-based methods tend to *zigzag* if the objective function contours are elliptic. The performance is much better if the contours are spherical. Therefore, before proceeding with the affine scaling algorithm, it is advantageous to introduce the following transformation designed to replace $\mathbf{Q}$ by an identity matrix in the objective function.

Using Choleski decomposition, any symmetric positive semidefinite matrix can be decomposed into the product of a Lower ($\mathbf{L}$) and an Upper triangular ($\mathbf{L}^T$) matrix. Thus, $\mathbf{Q}$ can be written as

$$\mathbf{Q} = \mathbf{L}\mathbf{L}^T$$

Introducing a transformation

$$\mathbf{x}' = \mathbf{L}^T\mathbf{x} \quad \text{or} \quad \mathbf{x} = \left(\mathbf{L}^T\right)^{-1}\mathbf{x}' = \left(\mathbf{L}^{-1}\right)^T\mathbf{x}'$$

the QP objective function is written as follows:

$$f = \mathbf{c}^T \left(\mathbf{L}^{-1}\right)^T \mathbf{x}' + \frac{1}{2}\left[\left(\mathbf{L}^{-1}\right)^T \mathbf{x}'\right]^T \mathbf{Q} \left[\left(\mathbf{L}^{-1}\right)^T \mathbf{x}'\right]$$

Denoting $\mathbf{c}'^T = \mathbf{c}^T(\mathbf{L}^{-1})^T$ or $\mathbf{c}' = \mathbf{L}^{-1}\mathbf{c}$, we have

$$f = \mathbf{c}'^T\mathbf{x}' + \frac{1}{2}\mathbf{x}'^T\mathbf{L}^{-1}\mathbf{L}\mathbf{L}^T\left(\mathbf{L}^{-1}\right)^T \mathbf{x}' = \mathbf{c}'^T\mathbf{x}' + \frac{1}{2}\mathbf{x}'^T\mathbf{x}'$$

Thus, with this transformation, we have achieved the goal of transforming matrix $\mathbf{Q}$ to an identity matrix. We can write the transformed objective function into an even simpler form if we note that $\frac{1}{2}\mathbf{c}'^T\mathbf{c}'$ is a constant, and adding a constant to the objective function does not change the optimum solution. Therefore,

$$f = \mathbf{c}'^T\mathbf{x}' + \frac{1}{2}\mathbf{x}'^T\mathbf{x}' + \frac{1}{2}\mathbf{c}'^T\mathbf{c}' \equiv \frac{1}{2}\left(\mathbf{c}' + \mathbf{x}'\right)^T\left(\mathbf{c}' + \mathbf{x}'\right)$$

Thus, the transformed problem can be written as follows:

Minimize $\frac{1}{2}(\mathbf{c}' + \mathbf{x}')^T(\mathbf{c}' + \mathbf{x}')$

8.6 Derivation of the Descent Direction Formula for the PAS Method

$$\text{Subject to} \begin{pmatrix} \mathbf{Ax} = \mathbf{b} \\ \mathbf{x} = (\mathbf{L}^{-1})^T \mathbf{x}' \\ \mathbf{x} \geq 0 \end{pmatrix}$$

Note that $\mathbf{x}'$ is not restricted to be positive. The problem can be written in a more compact form by writing the constraints into a matrix form as follows:

Minimize $\frac{1}{2}(\mathbf{c}' + \mathbf{x}')^T(\mathbf{c}' + \mathbf{x}')$

Subject to $\begin{pmatrix} \mathbf{A} & \mathbf{0} \\ \mathbf{I} & -(\mathbf{L}^T)^{-1} \end{pmatrix} \begin{pmatrix} \mathbf{x} \\ \mathbf{x}' \end{pmatrix} = \begin{pmatrix} \mathbf{b} \\ \mathbf{0} \end{pmatrix}$ and $\mathbf{x} \geq 0$

where $\mathbf{I}$ is an $n \times n$ identity matrix. The constraint coefficient matrix for the transformed QP is

$$\mathbf{U} \equiv \begin{pmatrix} \mathbf{A} & \mathbf{0} \\ \mathbf{I} & -(\mathbf{L}^T)^{-1} \end{pmatrix}$$

Example 8.15 The following example illustrates the ideas:

Minimize $f = -2x_1 + \frac{x_1^2}{2} - 6x_2 - x_1 x_2 + x_2^2$

Subject to $\begin{pmatrix} 3x_1 + x_2 \leq 25 \\ -x_1 + 2x_2 \leq 10 \\ x_1 + 2x_2 \leq 15 \\ x_i \geq 0, i = 1, 2 \end{pmatrix}$

Introducing slack variables x_3, x_4, and x_5, the problem is written in the standard form as follows:

Minimize $f = -2x_1 + \frac{x_1^2}{2} - 6x_2 - x_1 x_2 + x_2^2$

Subject to $\begin{pmatrix} 3x_1 + x_2 + x_3 = 25 \\ -x_1 + 2x_2 + x_4 = 10 \\ x_1 + 2x_2 + x_5 = 15 \\ x_i \geq 0, i = 1, \ldots, 5 \end{pmatrix}$

The problem is now in the standard QP form with the following vectors and matrices:

$$\mathbf{c} = \begin{pmatrix} -2 \\ -6 \end{pmatrix} \quad \mathbf{Q} = \begin{pmatrix} 1 & -1 \\ -1 & 2 \end{pmatrix}$$

$$\mathbf{A} = \begin{pmatrix} 3 & 1 & 1 & 0 & 0 \\ -1 & 2 & 0 & 1 & 0 \\ 1 & 2 & 0 & 0 & 1 \end{pmatrix} \quad \mathbf{b} = \begin{pmatrix} 25 \\ 10 \\ 15 \end{pmatrix}$$

```
c = {-2, -6}; Q = {{1, -1}, {-1, 2}};
A = {{3, 1, 1, 0, 0}, {-1, 2, 0, 1, 0}, {1, 2, 0, 0, 1}};
b = {25, 10, 15};
```

The Choleski factors are computed by using the CholeskyDecomposition function that is part of the standard *Mathematica* LinearAlgebra package.

```
<< LinearAlgebra`Cholesky`;
```

```
LT = CholeskyDecomposition[Q]
```

{{1, -1}, {0, 1}}

```
L = Transpose[LT];
L.LT
```

{{1, -1}, {-1, 2}}

We can easily see that L.LT is equal to the original matrix. The first two entries in the transformed objective function coefficient vector $\mathbf{c}'$ can now be computed as follows:

```
cp = Inverse[L].c
```

{-2, -8}

The remaining entries are all zero. Thus, the transformed problem for the example is

$$\text{Minimize } \tfrac{1}{2}(-2 + x_1' \quad -8 + x_2' \quad x_3' \quad x_4' \quad x_5') \begin{pmatrix} -2 + x_1' \\ -8 + x_2' \\ x_3' \\ x_4' \\ x_5' \end{pmatrix}$$

$$\text{Subject to } \begin{pmatrix} 3 & 1 & 1 & 0 & 0 & 0 & 0 & 0 & 0 & 0 \\ -1 & 2 & 0 & 1 & 0 & 0 & 0 & 0 & 0 & 0 \\ 1 & 2 & 0 & 0 & 1 & 0 & 0 & 0 & 0 & 0 \\ 1 & 0 & 0 & 0 & 0 & -1 & -1 & 0 & 0 & 0 \\ 0 & 1 & 0 & 0 & 0 & 0 & -1 & 0 & 0 & 0 \\ 0 & 0 & 1 & 0 & 0 & 0 & 0 & 0 & 0 & 0 \\ 0 & 0 & 0 & 1 & 0 & 0 & 0 & 0 & 0 & 0 \\ 0 & 0 & 0 & 0 & 1 & 0 & 0 & 0 & 0 & 0 \end{pmatrix} \begin{pmatrix} x_1 \\ x_2 \\ x_3 \\ x_4 \\ x_5 \\ x_1' \\ x_2' \\ x_3' \\ x_4' \\ x_5' \end{pmatrix} = \begin{pmatrix} 25 \\ 10 \\ 15 \\ 0 \\ 0 \\ 0 \\ 0 \\ 0 \end{pmatrix}$$

8.6 Derivation of the Descent Direction Formula for the PAS Method

Scaling Transformation

The actual problem variables are the only ones that are scaled by the current interior point. Thus, the scaling transformation is as follows:

$$\mathbf{y}^k = (\mathbf{T}^k)^{-1}\mathbf{x} \quad \text{or} \quad \mathbf{x} = \mathbf{T}^k\mathbf{y}^k$$

where

$$\mathbf{T}^k = \begin{pmatrix} x_1^k & 0 & 0 & 0 \\ 0 & x_2^k & 0 & 0 \\ \vdots & \vdots & \ddots & \vdots \\ 0 & 0 & 0 & x_n^k \end{pmatrix} \quad (\mathbf{T}^k)^{-1} = \begin{pmatrix} 1/x_1^k & 0 & 0 & 0 \\ 0 & 1/x_2^k & 0 & 0 \\ \vdots & \vdots & \ddots & \vdots \\ 0 & 0 & 0 & 1/x_n^k \end{pmatrix}$$

Since the objective function involves the transformed variables ($\mathbf{x}'$) only, it does not change. Only the constraints get scaled, as follows:

$$\begin{pmatrix} \mathbf{A} & \mathbf{0} \\ \mathbf{I} & -(\mathbf{L}^T)^{-1} \end{pmatrix} \begin{pmatrix} \mathbf{T}^k \mathbf{y}^k \\ \mathbf{x}' \end{pmatrix} = \begin{pmatrix} \mathbf{b} \\ \mathbf{0} \end{pmatrix} \quad \text{or} \quad \begin{pmatrix} \mathbf{A}\mathbf{T}^k & \mathbf{0} \\ \mathbf{T}^k & -(\mathbf{L}^T)^{-1} \end{pmatrix} \begin{pmatrix} \mathbf{y}^k \\ \mathbf{x}' \end{pmatrix} = \begin{pmatrix} \mathbf{b} \\ \mathbf{0} \end{pmatrix}$$

Thus, the scaled constraint coefficient matrix is

$$\mathbf{U}^k \equiv \begin{pmatrix} \mathbf{A}\mathbf{T}^k & \mathbf{0} \\ \mathbf{T}^k & -(\mathbf{L}^T)^{-1} \end{pmatrix}$$

Projection

Similar to the LP case, a feasible descent direction is the projection of the negative gradient of the objective function on the nullspace of the constraint coefficient matrix $\mathbf{U}^k$.

$$\mathbf{d}^k = \mathbf{P}^k(-\nabla f^k)$$

The ∇f^k is the gradient of the objective function at the current point

$$\nabla f^k = \begin{pmatrix} \mathbf{0} \\ \mathbf{c}' + \mathbf{x}'^k \end{pmatrix}$$

The projection matrix $\mathbf{P}^k$ is given by

$$\mathbf{P}^k = \mathbf{I} - \mathbf{U}^{kT}(\mathbf{U}^k\mathbf{U}^{kT})^{-1}\mathbf{U}^k$$

where **I** is an $n \times n$ identity matrix. Thus,

$$\mathbf{d}^k = -\nabla f^k + \mathbf{U}^{kT}(\mathbf{U}^k\mathbf{U}^{kT})^{-1}\mathbf{U}^k\nabla f^k$$

Defining $\mathbf{w}^k = (\mathbf{U}^k\mathbf{U}^{kT})^{-1}\mathbf{U}^k\nabla f^k$, we can express the direction as

$$\mathbf{d}^k = -[\nabla f^k - \mathbf{U}^{kT}\mathbf{w}^k]$$

(a) Manipulation of w terms

The starting point of the manipulations is the expansion of the terms in the $\mathbf{w}^k$ vector:

$$\mathbf{w}^k = (\mathbf{U}^k\mathbf{U}^{kT})^{-1}\mathbf{U}^k\nabla f^k$$

or

$$(\mathbf{U}^k\mathbf{U}^{kT})\mathbf{w}^k = \mathbf{U}^k\nabla f^k$$

Substituting for the $\mathbf{U}^k$ term, we have

$$\begin{pmatrix} \mathbf{A}\mathbf{T}^k & \mathbf{0} \\ \mathbf{T}^k & -(\mathbf{L}^T)^{-1} \end{pmatrix} \begin{pmatrix} \mathbf{T}^k\mathbf{A}^T & \mathbf{T}^k \\ \mathbf{0} & -\mathbf{L}^{-1} \end{pmatrix} \begin{pmatrix} \mathbf{w}_1^k \\ \mathbf{w}_2^k \end{pmatrix} = \begin{pmatrix} \mathbf{A}\mathbf{T}^k & \mathbf{0} \\ \mathbf{T}^k & -(\mathbf{L}^T)^{-1} \end{pmatrix} \begin{pmatrix} \mathbf{0} \\ \mathbf{c}' + \mathbf{x}'^k \end{pmatrix}$$

or

$$\begin{pmatrix} \mathbf{A}\mathbf{T}^{k2}\mathbf{A}^T & \mathbf{A}\,\mathbf{T}^{k2} \\ \mathbf{T}^{k2}\,\mathbf{A}^T & \mathbf{T}^{k2} + \mathbf{Q}^{-1} \end{pmatrix} \begin{pmatrix} \mathbf{w}_1^k \\ \mathbf{w}_2^k \end{pmatrix} = \begin{pmatrix} \mathbf{0} \\ -(\mathbf{L}^T)^{-1}(\mathbf{c}' + \mathbf{x}'^k) \end{pmatrix}$$

Our next task is to solve these two matrix equations for $\mathbf{w}_1^k$ and $\mathbf{w}_2^k$. The first equation is

$$\mathbf{A}\mathbf{T}^{k2}\mathbf{A}^T\mathbf{w}_1^k + \mathbf{A}\mathbf{T}^{k2}\mathbf{w}_2^k = \mathbf{0}$$

The second equation is

$$\mathbf{T}^{k2}\mathbf{A}^T\mathbf{w}_1^k + (\mathbf{T}^{k2} + \mathbf{Q}^{-1})\mathbf{w}_2^k = -(\mathbf{L}^T)^{-1}(\mathbf{c}' + \mathbf{x}'^k)$$

or

$$\mathbf{w}_2^k = (\mathbf{T}^{k2} + \mathbf{Q}^{-1})^{-1}[-\mathbf{T}^{k2}\mathbf{A}^T\mathbf{w}_1^k - (\mathbf{L}^T)^{-1}(\mathbf{c}' + \mathbf{x}'^k)]$$

8.6 Derivation of the Descent Direction Formula for the PAS Method

The term $(\mathbf{L}^T)^{-1}(\mathbf{c}' + \mathbf{x}'^k)$ can be written as

$$(\mathbf{L}^T)^{-1}(\mathbf{c}' + \mathbf{x}'^k) = (\mathbf{L}^T)^{-1}\mathbf{L}^{-1}\mathbf{c} + (\mathbf{L}^T)^{-1}\mathbf{L}^T\mathbf{x}^k$$

Since $\mathbf{Q} = \mathbf{L}\mathbf{L}^T$, we have $\mathbf{Q}^{-1} = (\mathbf{L}^T)^{-1}\mathbf{L}^{-1}$ and thus,

$$(\mathbf{L}^T)^{-1}(\mathbf{c}' + \mathbf{x}'^k) = \mathbf{Q}^{-1}\mathbf{c} + \mathbf{x}^k$$

Substituting into the $\mathbf{w}_2^k$ expression, we get

$$\mathbf{w}_2^k = (\mathbf{T}^{k2} + \mathbf{Q}^{-1})^{-1}\left[-\mathbf{T}^{k2}\mathbf{A}^T\mathbf{w}_1^k - (\mathbf{Q}^{-1}\mathbf{c} + \mathbf{x}^k)\right]$$

Substituting this into the first equation gives $\mathbf{w}_1^k$; however, it takes considerable manipulation to get into a simple form given in section 8.22 (without the subscript 1).

$$\mathbf{A}\mathbf{T}^{k2}\mathbf{A}^T\mathbf{w}_1^k = \mathbf{A}\mathbf{T}^{k2}(\mathbf{T}^{k2} + \mathbf{Q}^{-1})^{-1}\left[\mathbf{T}^{k2}\mathbf{A}^T\mathbf{w}_1^k + (\mathbf{Q}^{-1}\mathbf{c} + \mathbf{x}^k)\right]$$

or

$$\mathbf{A}\mathbf{T}^k\mathbf{T}^k\mathbf{A}^T\mathbf{w}_1^k - \mathbf{A}\mathbf{T}^k\mathbf{T}^k(\mathbf{T}^{k2} + \mathbf{Q}^{-1})^{-1}\mathbf{T}^k\mathbf{T}^k\mathbf{A}^T\mathbf{w}_1^k$$
$$= \mathbf{A}\mathbf{T}^{k2}(\mathbf{T}^{k2} + \mathbf{Q}^{-1})^{-1}(\mathbf{Q}^{-1}\mathbf{c} + \mathbf{x}^k)$$

Grouping together terms involving $\mathbf{w}_1^k$, we have

$$\mathbf{A}\mathbf{T}^k\left[\mathbf{I} - \mathbf{T}^k(\mathbf{T}^{k2} + \mathbf{Q}^{-1})^{-1}\mathbf{T}^k\right]\mathbf{T}^k\mathbf{A}^T\mathbf{w}_1^k$$
$$= \mathbf{A}\mathbf{T}^{k2}(\mathbf{T}^{k2} + \mathbf{Q}^{-1})^{-1}(\mathbf{x}^k + \mathbf{Q}^{-1}\mathbf{c})$$

The term $\mathbf{I} - \mathbf{T}^k(\mathbf{T}^{k2} + \mathbf{Q}^{-1})^{-1}\mathbf{T}^k$ is manipulated into a symmetric form as follows. Note that the matrix inversion rule $[\mathbf{AB}]^{-1} = \mathbf{B}^{-1}\mathbf{A}^{-1}$ is used several times in the following manipulations.

First, since $\mathbf{T}^k = [(\mathbf{T}^k)^{-1}]^{-1}$, we have

$$\mathbf{I} - \mathbf{T}^k(\mathbf{T}^{k2} + \mathbf{Q}^{-1})^{-1}\mathbf{T}^k$$
$$= \mathbf{I} - \left[(\mathbf{T}^k)^{-1}\right]^{-1}(\mathbf{T}^{k2} + \mathbf{Q}^{-1})^{-1}\left[(\mathbf{T}^k)^{-1}\right]^{-1}$$

Using the rule for the inverse of a product of matrices:

$$= \mathbf{I} - \left[(\mathbf{T}^{k2}+\mathbf{Q}^{-1})(\mathbf{T}^k)^{-1}\right]^{-1}\left[(\mathbf{T}^k)^{-1}\right]^{-1}$$
$$= \mathbf{I} - \left[\mathbf{T}^{k2}(\mathbf{T}^k)^{-1}+\mathbf{Q}^{-1}(\mathbf{T}^k)^{-1}\right]^{-1}\left[(\mathbf{T}^k)^{-1}\right]^{-1}$$
$$= \mathbf{I} - \left[\mathbf{T}^k+\mathbf{Q}^{-1}(\mathbf{T}^k)^{-1}\right]^{-1}\left[(\mathbf{T}^k)^{-1}\right]^{-1}$$
$$= \mathbf{I} - \left[(\mathbf{T}^k)^{-1}\mathbf{T}^k+(\mathbf{T}^k)^{-1}\mathbf{Q}^{-1}(\mathbf{T}^k)^{-1}\right]^{-1}$$
$$= \mathbf{I} - \left[\mathbf{I}+(\mathbf{T}^k)^{-1}\mathbf{Q}^{-1}(\mathbf{T}^k)^{-1}\right]^{-1}$$

Since $\left[\mathbf{I}+(\mathbf{T}^k)^{-1}\mathbf{Q}^{-1}(\mathbf{T}^k)^{-1}\right]\left[\mathbf{I}+(\mathbf{T}^k)^{-1}\mathbf{Q}^{-1}(\mathbf{T}^k)^{-1}\right]^{-1} \equiv \mathbf{I}$, we have

$$= \left[\mathbf{I}+(\mathbf{T}^k)^{-1}\mathbf{Q}^{-1}(\mathbf{T}^k)^{-1}\right]\left[\mathbf{I}+(\mathbf{T}^k)^{-1}\mathbf{Q}^{-1}(\mathbf{T}^k)^{-1}\right]^{-1}$$
$$\quad - \left[\mathbf{I}+(\mathbf{T}^k)^{-1}\mathbf{Q}^{-1}(\mathbf{T}^k)^{-1}\right]^{-1}$$
$$= \left[\mathbf{I}+(\mathbf{T}^k)^{-1}\mathbf{Q}^{-1}(\mathbf{T}^k)^{-1}-\mathbf{I}\right]\left[\mathbf{I}+(\mathbf{T}^k)^{-1}\mathbf{Q}^{-1}(\mathbf{T}^k)^{-1}\right]^{-1}$$
$$= (\mathbf{T}^k)^{-1}\mathbf{Q}^{-1}(\mathbf{T}^k)^{-1}\left[\mathbf{I}+(\mathbf{T}^k)^{-1}\mathbf{Q}^{-1}(\mathbf{T}^k)^{-1}\right]^{-1}$$

Since $\mathbf{I} = \mathbf{T}^k(\mathbf{T}^k)^{-1} = (\mathbf{T}^k)^{-1}\mathbf{T}^k$

$$= (\mathbf{T}^k)^{-1}\mathbf{Q}^{-1}(\mathbf{T}^k)^{-1}\left[\mathbf{T}^k(\mathbf{T}^k)^{-1}+(\mathbf{T}^k)^{-1}\mathbf{Q}^{-1}(\mathbf{T}^k)^{-1}\right]^{-1}$$
$$= (\mathbf{T}^k)^{-1}\mathbf{Q}^{-1}(\mathbf{T}^k)^{-1}\left[\{\mathbf{T}^k+(\mathbf{T}^k)^{-1}\mathbf{Q}^{-1}\}(\mathbf{T}^k)^{-1}\right]^{-1}$$
$$= (\mathbf{T}^k)^{-1}\mathbf{Q}^{-1}(\mathbf{T}^k)^{-1}\mathbf{T}^k\left[\mathbf{T}^k+(\mathbf{T}^k)^{-1}\mathbf{Q}^{-1}\right]^{-1}$$
$$= (\mathbf{T}^k)^{-1}\mathbf{Q}^{-1}\left[\mathbf{T}^k+(\mathbf{T}^k)^{-1}\mathbf{Q}^{-1}\right]^{-1}$$
$$= (\mathbf{T}^k)^{-1}\mathbf{Q}^{-1}\left[(\mathbf{T}^k)^{-1}\mathbf{T}^k\mathbf{T}^k+(\mathbf{T}^k)^{-1}\mathbf{Q}^{-1}\right]^{-1}$$
$$= (\mathbf{T}^k)^{-1}\mathbf{Q}^{-1}\left[(\mathbf{T}^k)^{-1}\{\mathbf{T}^{k2}+\mathbf{Q}^{-1}\}\right]^{-1}$$
$$= (\mathbf{T}^k)^{-1}\mathbf{Q}^{-1}\left[\mathbf{T}^{k2}+\mathbf{Q}^{-1}\right]^{-1}\mathbf{T}^k$$
$$= (\mathbf{T}^k)^{-1}\left[\mathbf{T}^{k2}\mathbf{Q}+\mathbf{I}\right]^{-1}\mathbf{T}^k$$

Since $\mathbf{I} = \mathbf{T}^{k2}(\mathbf{T}^{k2})^{-1}$

$$= (\mathbf{T}^k)^{-1}\left[\mathbf{T}^{k2}\mathbf{Q}+\mathbf{T}^{k2}(\mathbf{T}^{k2})^{-1}\right]^{-1}\mathbf{T}^k$$
$$= (\mathbf{T}^k)^{-1}\left[\mathbf{T}^{k2}\{\mathbf{Q}+(\mathbf{T}^k)^{-2}\}\right]^{-1}\mathbf{T}^k$$
$$= (\mathbf{T}^k)^{-1}\left[\mathbf{Q}+(\mathbf{T}^k)^{-2}\right]^{-1}(\mathbf{T}^k)^{-1}$$

8.6 Derivation of the Descent Direction Formula for the PAS Method

Thus, we have

$$\mathbf{I} - \mathbf{T}^k(\mathbf{T}^{k2} + \mathbf{Q}^{-1})^{-1}\mathbf{T}^k = (\mathbf{T}^k)^{-1}\left[\mathbf{Q} + (\mathbf{T}^k)^{-2}\right]^{-1}(\mathbf{T}^k)^{-1}$$

Denoting

$$\mathbf{H}^k \equiv \left[\mathbf{Q} + (\mathbf{T}^k)^{-2}\right]^{-1}$$

Therefore, $\mathbf{I} - \mathbf{T}^k(\mathbf{T}^{k2} + \mathbf{Q}^{-1})^{-1}\mathbf{T}^k = (\mathbf{T}^k)^{-1}\mathbf{H}^k(\mathbf{T}^k)^{-1}$

The term $\mathbf{T}^{k2}(\mathbf{T}^{k2} + \mathbf{Q}^{-1})^{-1}$ can be written as follows:

$$\mathbf{T}^{k2}(\mathbf{T}^{k2} + \mathbf{Q}^{-1})^{-1} = \left[(\mathbf{T}^k)^{-2}\right]^{-1}(\mathbf{T}^{k2} + \mathbf{Q}^{-1})^{-1} = \left[\mathbf{I} + \mathbf{Q}^{-1}(\mathbf{T}^k)^{-2}\right]^{-1}$$

Since $\mathbf{I} = \mathbf{Q}^{-1}\mathbf{Q}$

$$= \left[\mathbf{Q}^{-1}\mathbf{Q} + \mathbf{Q}^{-1}(\mathbf{T}^k)^{-2}\right]^{-1} = \left[\mathbf{Q}^{-1}\{\mathbf{Q} + (\mathbf{T}^k)^{-2}\}\right]^{-1}$$

$$= \left[\mathbf{Q} + (\mathbf{T}^k)^{-2}\right]^{-1}\mathbf{Q}$$

Therefore, $\mathbf{T}^{k2}(\mathbf{T}^{k2} + \mathbf{Q}^{-1})^{-1} = \mathbf{H}^k\mathbf{Q}$

Now we return back to the $\mathbf{w}_1^k$ equation

$$\mathbf{A}\mathbf{T}^k\left[\mathbf{I} - \mathbf{T}^k(\mathbf{T}^{k2} + \mathbf{Q}^{-1})^{-1}\mathbf{T}^k\right]\mathbf{T}^k\mathbf{A}^T\mathbf{w}_1^k = \mathbf{A}\mathbf{T}^{k2}(\mathbf{T}^{k2} + \mathbf{Q}^{-1})^{-1}(\mathbf{x}^k + \mathbf{Q}^{-1}\mathbf{c})$$

$$\mathbf{A}\mathbf{T}^k\left[(\mathbf{T}^k)^{-1}\mathbf{H}^k(\mathbf{T}^k)^{-1}\right]\mathbf{T}^k\mathbf{A}^T\mathbf{w}_1^k = \mathbf{A}\mathbf{H}^k\mathbf{Q}(\mathbf{x}^k + \mathbf{Q}^{-1}\mathbf{c})$$

we have

$$\mathbf{A}\mathbf{H}^k\mathbf{A}^T\mathbf{w}_1^k = \mathbf{A}\mathbf{H}^k(\mathbf{Q}\mathbf{x}^k + \mathbf{c})$$

Therefore,

$$\mathbf{w}_1^k = (\mathbf{A}\mathbf{H}^k\mathbf{A}^T)^{-1}\mathbf{A}\mathbf{H}^k(\mathbf{Q}\mathbf{x}^k + \mathbf{c})$$

(b) Expression for direction

Returning now to the direction vector, we have

$$\mathbf{d}^k = -\left[\nabla f^k - \mathbf{U}^{kT}\mathbf{w}^k\right] = -\begin{pmatrix}\mathbf{0}\\\mathbf{c}' + \mathbf{x}'^k\end{pmatrix} + \mathbf{U}^{kT}\begin{pmatrix}\mathbf{w}_1^k\\\mathbf{w}_2^k\end{pmatrix}$$

or

$$\begin{pmatrix}\mathbf{d}_y^k\\\mathbf{d}_{x'}^k\end{pmatrix} = -\begin{pmatrix}\mathbf{0}\\\mathbf{c}' + \mathbf{x}'^k\end{pmatrix} + \begin{pmatrix}\mathbf{T}^k\mathbf{A}^T & \mathbf{T}^k\\ \mathbf{0} & -\mathbf{L}^{-1}\end{pmatrix}\begin{pmatrix}\mathbf{w}_1^k\\\mathbf{w}_2^k\end{pmatrix}$$

where the subscripts y and x' are used on components of vector $\mathbf{d}$ corresponding to scaled and transformed variables. Thus,

$$\mathbf{d}_y^k = \mathbf{T}^k \mathbf{A}^T \mathbf{w}_1^k + \mathbf{T}^k \mathbf{w}_2^k = \mathbf{T}^k (\mathbf{A}^T \mathbf{w}_1^k + \mathbf{w}_2^k)$$

Using the expressions for $\mathbf{w}$, the term $\mathbf{A}^T \mathbf{w}_1^k + \mathbf{w}_2^k$ is written as follows:

$$\begin{aligned}
\mathbf{A}^T \mathbf{w}_1^k + \mathbf{w}_2^k &= \mathbf{A}^T \mathbf{w}_1^k + (\mathbf{T}^{k2} + \mathbf{Q}^{-1})^{-1} \left[-\mathbf{T}^{k2} \mathbf{A}^T \mathbf{w}_1^k - (\mathbf{x}^k + \mathbf{Q}^{-1} \mathbf{c}) \right] \\
&= \left[\mathbf{I} - (\mathbf{T}^{k2} + \mathbf{Q}^{-1})^{-1} \mathbf{T}^{k2} \right] \mathbf{A}^T \mathbf{w}_1^k - (\mathbf{T}^{k2} + \mathbf{Q}^{-1})^{-1} (\mathbf{x}^k + \mathbf{Q}^{-1} \mathbf{c})
\end{aligned}$$

The term in the square brackets can be written as

$$\begin{aligned}
\mathbf{I} - (\mathbf{T}^{k2} + \mathbf{Q}^{-1})^{-1} \mathbf{T}^{k2} &= \mathbf{I} - (\mathbf{T}^{k2} + \mathbf{Q}^{-1})^{-1} \left[(\mathbf{T}^k)^{-2} \right]^{-1} \\
&= \mathbf{I} - \left[(\mathbf{T}^k)^{-2} \mathbf{T}^{k2} + (\mathbf{T}^k)^{-2} \mathbf{Q}^{-1} \right]^{-1} \\
&= \mathbf{I} - \left[\mathbf{I} + (\mathbf{T}^k)^{-2} \mathbf{Q}^{-1} \right]^{-1} \\
&= \left[\mathbf{I} + (\mathbf{T}^k)^{-2} \mathbf{Q}^{-1} \right] \left[\mathbf{I} + (\mathbf{T}^k)^{-2} \mathbf{Q}^{-1} \right]^{-1} \\
&\quad - \left[\mathbf{I} + (\mathbf{T}^k)^{-2} \mathbf{Q}^{-1} \right]^{-1} \\
&= \left[\mathbf{I} + (\mathbf{T}^k)^{-2} \mathbf{Q}^{-1} - \mathbf{I} \right] \left[\mathbf{I} + (\mathbf{T}^k)^{-2} \mathbf{Q}^{-1} \right]^{-1} \\
&= (\mathbf{T}^k)^{-2} \mathbf{Q}^{-1} \left[\mathbf{I} + (\mathbf{T}^k)^{-2} \mathbf{Q}^{-1} \right]^{-1}
\end{aligned}$$

Also,

$$\begin{aligned}
\mathbf{T}^{k2} (\mathbf{T}^{k2} + \mathbf{Q}^{-1})^{-1} &= \left[(\mathbf{T}^k)^{-2} \right]^{-1} (\mathbf{T}^{k2} + \mathbf{Q}^{-1})^{-1} \\
&= \left[\mathbf{T}^{k2} (\mathbf{T}^k)^{-2} + \mathbf{Q}^{-1} (\mathbf{T}^k)^{-2} \right]^{-1} = \left[\mathbf{I} + \mathbf{Q}^{-1} (\mathbf{T}^k)^{-2} \right]^{-1}
\end{aligned}$$

or

$$(\mathbf{T}^{k2} + \mathbf{Q}^{-1})^{-1} = (\mathbf{T}^k)^{-2} \left[\mathbf{I} + \mathbf{Q}^{-1} (\mathbf{T}^k)^{-2} \right]^{-1}$$

Therefore,

$$\begin{aligned}
\mathbf{A}^T \mathbf{w}_1^k + \mathbf{w}_2^k &= (\mathbf{T}^k)^{-2} \mathbf{Q}^{-1} \left[\mathbf{I} + (\mathbf{T}^k)^{-2} \mathbf{Q}^{-1} \right]^{-1} \mathbf{A}^T \mathbf{w}_1^k \\
&\quad - (\mathbf{T}^k)^{-2} \left[\mathbf{I} + \mathbf{Q}^{-1} (\mathbf{T}^k)^{-2} \right]^{-1} (\mathbf{x}^k + \mathbf{Q}^{-1} \mathbf{c}) \\
&= (\mathbf{T}^k)^{-2} \left[\mathbf{Q} + (\mathbf{T}^k)^{-2} \right]^{-1} \mathbf{A}^T \mathbf{w}_1^k \\
&\quad - (\mathbf{T}^k)^{-2} \left[\mathbf{Q} + (\mathbf{T}^k)^{-2} \right]^{-1} \mathbf{Q} (\mathbf{x}^k + \mathbf{Q}^{-1} \mathbf{c}) \\
&= (\mathbf{T}^k)^{-2} \left[\mathbf{H}^k \mathbf{A}^T \mathbf{w}_1^k - \mathbf{H}^k \mathbf{Q} (\mathbf{x}^k + \mathbf{Q}^{-1} \mathbf{c}) \right]
\end{aligned}$$

Substituting for $\mathbf{w}_1^k$ the term $\mathbf{A}^T\mathbf{w}_1^k + \mathbf{w}_2^k$ is written as follows:

$$\mathbf{A}^T\mathbf{w}_1^k + \mathbf{w}_2^k = (\mathbf{T}^k)^{-2}\left[\mathbf{H}^k\mathbf{A}^T(\mathbf{A}\mathbf{H}^k\mathbf{A}^T)^{-1}\mathbf{A}\mathbf{H}^k(\mathbf{Q}\mathbf{x}^k + \mathbf{c}) - \mathbf{H}^k(\mathbf{Q}\mathbf{x}^k + \mathbf{c})\right]$$

or

$$\mathbf{A}^T\mathbf{w}_1^k + \mathbf{w}_2^k = (\mathbf{T}^k)^{-2}\left[\mathbf{H}^k\mathbf{A}^T(\mathbf{A}\mathbf{H}^k\mathbf{A}^T)^{-1}\mathbf{A} - \mathbf{I}\right]\mathbf{H}^k(\mathbf{Q}\mathbf{x}^k + \mathbf{c})$$

or

$$\mathbf{A}^T\mathbf{w}_1^k + \mathbf{w}_2^k = -(\mathbf{T}^k)^{-2}\left[\mathbf{I} - \mathbf{H}^k\mathbf{A}^T(\mathbf{A}\mathbf{H}^k\mathbf{A}^T)^{-1}\mathbf{A}\right]\mathbf{H}^k(\mathbf{Q}\mathbf{x}^k + \mathbf{c})$$

Therefore, the direction in terms of the scaled variables is

$$\mathbf{d}_y^k = \mathbf{T}^k(\mathbf{A}^T\mathbf{w}_1^k + \mathbf{w}_2^k) = -(\mathbf{T}^k)^{-1}\left[\mathbf{I} - \mathbf{H}^k\mathbf{A}^T(\mathbf{A}\mathbf{H}^k\mathbf{A}^T)^{-1}\mathbf{A}\right]\mathbf{H}^k(\mathbf{Q}\mathbf{x}^k + \mathbf{c})$$

Finally, in terms of original variables, the direction vector is

$$\mathbf{d}_x^k = \mathbf{T}^k\mathbf{d}_y^k = -\left[\mathbf{I} - \mathbf{H}^k\mathbf{A}^T(\mathbf{A}\mathbf{H}^k\mathbf{A}^T)^{-1}\mathbf{A}\right]\mathbf{H}^k(\mathbf{Q}\mathbf{x}^k + \mathbf{c})$$

where $\mathbf{H}^k = [\mathbf{Q} + (\mathbf{T}^k)^{-2}]^{-1}$

This is the formula given in section 8.2.2 for direction. For convenience, the subscripts $\mathbf{x}$ from $\mathbf{d}$ and 1 from $\mathbf{w}$ are dropped.

8.7 Problems

Dual QP

Construct the dual QP problem for the following primal QP problems. If possible, use KT conditions to verify that both the primal and dual have the same solution.

8.1. Minimize $f(x, y) = x^2 + 2y^2$

Subject to $\begin{pmatrix} x + y = 1 \\ x, y \geq 0 \end{pmatrix}$

8.2. Minimize $f(x, y) = 2x^2 + y^2 - 2xy - 5x - 2y$

Subject to $\begin{pmatrix} 5x - 3y = -4 \\ x, y \geq 0 \end{pmatrix}$

8.3. Minimize $f(x, y) = 2x^2 + y^2 - 20x - 10y$

Subject to $\begin{pmatrix} 8 \geq -x + 3y \\ 12 = -x + y \\ x, y \text{ unrestricted in sign} \end{pmatrix}$

8.4. Minimize $f(x, y) = x^2 + 2y^2 - 24x - 20y$

Subject to $\begin{pmatrix} x + 2y \geq 0 \\ x + 2y \leq 9 \\ x + y \leq 8 \\ x + y \geq 0 \end{pmatrix}$

8.5. Minimize $f(x, y) = -x + \frac{x^2}{2} + 3y - xy + \frac{y^2}{2}$

Subject to $\begin{pmatrix} x + 4y \geq 7 \\ 3x + 2y \geq 8 \\ x, y \geq 0 \end{pmatrix}$

Primal Affine Scaling Method

Solve the following QP problems using the Primal Affine Scaling method for QP. For problems with two variables, verify the solution graphically. Unless stated otherwise, all variables are assumed to be positive.

8.6. Minimize $f(x, y) = x^2 + 2y^2$

Subject to $\begin{pmatrix} x + y \geq 1 \\ x, y \geq 0 \end{pmatrix}$

8.7. Minimize $f(x, y) = 2x^2 + y^2 - 2xy - 5x - 2y$

Subject to $\begin{pmatrix} 3x + 2y \leq 20 \\ 5x - 3y = -4 \\ x, y \geq 0 \end{pmatrix}$

8.8. Minimize $f(x, y) = 2x^2 + y^2 - 20x - 10y$

Subject to $\begin{pmatrix} 8 \geq -x + 3y \\ 12 = -x + y \\ x, y \text{ unrestricted in sign} \end{pmatrix}$

8.9. Minimize $f(x, y) = x^2 + 2y^2 - 24x - 20y$

Subject to $\begin{pmatrix} x + 2y \geq 0 \\ x + 2y \leq 9 \\ x + y \leq 8 \\ x + y \geq 0 \end{pmatrix}$

8.10. Minimize $f(x, y) = -x + \frac{x^2}{2} + 3y - xy + \frac{y^2}{2}$

Subject to $\begin{pmatrix} x + 4y \geq 7 \\ 3x + 2y \geq 8 \\ x, y \geq 0 \end{pmatrix}$

8.11. Minimize $f(x, y) = 2x + x^2 + 2y - xy + \frac{y^2}{2}$

Subject to $\begin{pmatrix} 3x + 5y \geq 12 \\ x + 6y \geq 9 \\ x, y \geq 0 \end{pmatrix}$

8.12. Minimize $f(x, y, z) = x + x^2 + 2xy + 3y^2 + 2z + 2yz + z^2$

Subject to $\begin{pmatrix} x - y + 2z \geq 1 \\ 5y + 4z \geq 2 \\ 3x - 2y \geq 1 \\ x, y, z \geq 0 \end{pmatrix}$

8.13. Minimize $f = x_1^2 + 8x_2^2 + 12x_3^2 - 10x_2 x_3 + 8x_1 - 14x_2 + 20x_3$

Subject to $\begin{pmatrix} x_1 + 4x_2 \leq 2 \\ -x_2 + 3x_3 \geq 1 \\ x_i \geq 0, i = 1, \ldots, 3 \end{pmatrix}$

8.14. Minimize $f = x_1^2 + 2x_2^2 + 3x_1 + 4x_3 + 5x_4$

Subject to $\begin{pmatrix} x_1 + 2x_2 + 3x_3 = 4 \\ x_1 + 2x_2 + 3x_4 = 4 \\ x_i \geq 0, i = 1, \ldots, 4 \end{pmatrix}$

8.15. Minimize $f = x_1^2 + 20x_1 x_2 + 100x_2^2 + 5x_3^2 + 10x_3 x_4 + 5x_4^2$

Subject to $\begin{pmatrix} x_1 + x_2 + x_3 + x_4 \geq 1 \\ x_i \geq 0, i = 1, \ldots, 4 \end{pmatrix}$

8.16. An investor is looking to make investment decisions such that she will get at least a 10% rate of return while minimizing the risk of major losses. For the past six years, the rates of return in three major investment types that she is considering are as follows:

Type	Annual rates of return					
Stocks	18.24	17.12	22.23	15.26	12.62	15.42
Mutual funds	12.24	11.16	10.07	8.46	6.62	8.43
Bonds	5.12	6.26	6.34	7.01	6.11	5.95

Formulate the problem as an optimization problem. Find an optimum using the PAS method.

Primal-Dual Method

Solve the following QP problems using the primal-dual method for QP. For problems with two variables, verify the solution graphically. Unless stated otherwise, all variables are assumed to be positive.

8.17. Minimize $f(x, y) = x^2 + 2y^2$

Subject to $\begin{pmatrix} x + y \geq 1 \\ x, y \geq 0 \end{pmatrix}$

8.18. Minimize $f(x, y) = 2x^2 + y^2 - 2xy - 5x - 2y$

Subject to $\begin{pmatrix} 3x + 2y \leq 20 \\ 5x - 3y = -4 \\ x, y \geq 0 \end{pmatrix}$

8.19. Minimize $f(x, y) = 2x^2 + y^2 - 20x - 10y$

Subject to $\begin{pmatrix} 8 \geq -x + 3y \\ 12 = -x + y \\ x, y \text{ unrestricted in sign} \end{pmatrix}$

8.20. Minimize $f(x, y) = x^2 + 2y^2 - 24x - 20y$

Subject to $\begin{pmatrix} x + 2y \geq 0 \\ x + 2y \leq 9 \\ x + y \leq 8 \\ x + y \geq 0 \end{pmatrix}$

8.21. Minimize $f(x, y) = -x + \frac{x^2}{2} + 3y - xy + \frac{y^2}{2}$

Subject to $\begin{pmatrix} x + 4y \geq 7 \\ 3x + 2y \geq 8 \\ x, y \geq 0 \end{pmatrix}$

8.22. Minimize $f(x, y) = 2x + x^2 + 2y - xy + \frac{y^2}{2}$

Subject to $\begin{pmatrix} 3x + 5y \geq 12 \\ x + 6y \geq 9 \\ x, y \geq 0 \end{pmatrix}$

8.23. Minimize $f(x, y, z) = x + x^2 + 2xy + 3y^2 + 2z + 2yz + z^2$

Subject to $\begin{pmatrix} x - y + 2z \geq 1 \\ 5y + 4z \geq 2 \\ 3x - 2y \geq 1 \\ x, y, z \geq 0 \end{pmatrix}$

8.24. Minimize $f = x_1^2 + 8x_2^2 + 12x_3^2 - 10x_2x_3 + 8x_1 - 14x_2 + 20x_3$

Subject to $\begin{pmatrix} x_1 + 4x_2 \leq 2 \\ -x_2 + 3x_3 \geq 1 \\ x_i \geq 0, i = 1, \ldots, 3 \end{pmatrix}$

8.25. Minimize $f = x_1^2 + 2x_2^2 + 3x_1 + 4x_3 + 5x_4$

Subject to $\begin{pmatrix} x_1 + 2x_2 + 3x_3 = 4 \\ x_1 + 2x_2 + 3x_4 = 4 \\ x_i \geq 0, i = 1, \ldots, 4 \end{pmatrix}$

8.26. Minimize $f = x_1^2 + 20x_1x_2 + 100x_2^2 + 5x_3^2 + 10x_3x_4 + 5x_4^2$

Subject to $\begin{pmatrix} x_1 + x_2 + x_3 + x_4 \geq 1 \\ x_i \geq 0, i = 1, \ldots, 4 \end{pmatrix}$

8.27. Use the primal-dual method to solve the investment problem 8.16.

Active Set Method

Solve the following QP problems using the active set method for QP. For problems with two variables, verify the solution graphically.

8.28. Minimize $f(x, y) = x^2 + 2y^2$

Subject to $\begin{pmatrix} x + y \geq 1 \\ x, y \geq 0 \end{pmatrix}$

8.29. Minimize $f(x, y) = 2x^2 + y^2 - 2xy - 5x - 2y$

Subject to $\begin{pmatrix} 3x + 2y \leq 20 \\ 5x - 3y = -4 \\ x, y \geq 0 \end{pmatrix}$

8.30. Minimize $f(x, y) = 2x^2 + y^2 - 20x - 10y$

Subject to $\begin{pmatrix} 8 \geq -x + 3y \\ 12 = -x + y \end{pmatrix}$

8.31. Minimize $f(x, y) = x^2 + 2y^2 - 24x - 20y$

Subject to $\begin{pmatrix} x + 2y \geq 0 \\ x + 2y \leq 9 \\ x + y \leq 8 \\ x + y \geq 0 \end{pmatrix}$

8.32. Minimize $f(x, y) = -x + \frac{x^2}{2} + 3y - xy + \frac{y^2}{2}$

Subject to $\begin{pmatrix} x + 4y \geq 7 \\ 3x + 2y \geq 8 \\ x, y \geq 0 \end{pmatrix}$

8.33. Minimize $f(x, y) = 2x + x^2 + 2y - xy + \frac{y^2}{2}$

Subject to $\begin{pmatrix} 3x + 5y \geq 12 \\ x + 6y \geq 9 \\ x, y \geq 0 \end{pmatrix}$

8.34. Minimize $f(x, y, z) = x + x^2 + 2xy + 3y^2 + 2z + 2yz + z^2$

Subject to $\begin{pmatrix} x - y + 2z \geq 1 \\ 5y + 4z \geq 2 \\ 3x - 2y \geq 1 \\ x, y, z \geq 0 \end{pmatrix}$

8.35. Minimize $f = x_1^2 + 8x_2^2 + 12x_3^2 - 10x_2x_3 + 8x_1 - 14x_2 + 20x_3$

Subject to $\begin{pmatrix} x_1 + 4x_2 \leq 2 \\ -x_2 + 3x_3 \geq 1 \\ x_i \geq 0, i = 1, \ldots, 3 \end{pmatrix}$

8.36. Minimize $f = x_1^2 + 2x_2^2 + 3x_1 + 4x_3 + 5x_4$

Subject to $\begin{pmatrix} x_1 + 2x_2 + 3x_3 = 4 \\ x_1 + 2x_2 + 3x_4 = 4 \\ x_i \geq 0, i = 1, \ldots, 4 \end{pmatrix}$

8.37. Minimize $f = x_1^2 + 20x_1x_2 + 100x_2^2 + 5x_3^2 + 10x_3x_4 + 5x_4^2$

Subject to $\begin{pmatrix} x_1 + x_2 + x_3 + x_4 \geq 1 \\ x_i \geq 0, i = 1, \ldots, 4 \end{pmatrix}$

8.38. Use the active set QP method to solve the investment problem 8.16.

Active Set Dual QP Method

Solve the following quadratic problems using the Active set dual QP method.

8.39. Minimize $f(x, y) = 2x + x^2 + 2y - xy + \frac{y^2}{2}$

Subject to $x, y \geq 0$

8.40. Minimize $f(x, y, z) = x + x^2 + 2xy + 3y^2 + 2z + 2yz + z^2$

8.41. Minimize $f = x_1^2 + 20x_1x_2 + 100x_2^2 + 5x_3^2 + 10x_3x_4 + 5x_4^2$

Subject to x_2 and $x_3 \geq 0$

Construct the dual QP problem for the following primal problems and solve the dual using the active set dual QP algorithm. Find the primal solution using the relationship between the dual and the primal variables.

8.42. Minimize $f(x, y) = x^2 + 2y^2$

Subject to $\begin{pmatrix} x + y = 1 \\ x, y \geq 0 \end{pmatrix}$

8.43. Minimize $f(x, y) = 2x^2 + y^2 - 2xy - 5x - 2y$

Subject to $\begin{pmatrix} 5x - 3y = -4 \\ x, y \geq 0 \end{pmatrix}$

8.44. Construct the dual of the investment problem 8.16 and solve the dual using the active set dual QP method.

CHAPTER NINE

Constrained Nonlinear Problems

Numerical methods for solving general nonlinear constrained optimization problems are discussed in this chapter. A large number of methods and their variations are available in the literature for solving these problems. As is frequently the case with nonlinear problems, there is no single method that is clearly better than the others. Each method has its own strengths and weaknesses. The quest for a general method that works effectively for all types of problems continues. Most current journals and conferences on optimization contain new methods or refinements of existing methods for solving constrained nonlinear problems. A thorough review of all these developments is beyond the scope of this chapter. Instead, the main purpose of this chapter is to present the development of two methods that are generally considered among the best in their class, the ALPF (Augmented Lagrangian Penalty Function) method and the SQP (Sequential Quadratic Programming) method. For additional details refer to Fiacco [1983], Fiacco and McCormick [1968], Fletcher [1987], Gill, Murray, and Wright [1991], Gomez and Hennart [1994], McCormick [1983], Scales [1985], and Shapiro [1979].

Most methods for solving nonlinear programming problems benefit from some type of scaling or normalization. A discussion of constraint normalization is contained in the first section. Penalty methods that are classified as indirect methods are discussed in the second section. After presenting the basic ideas of interior and exterior penalty functions, the Augmented Lagrangian Penalty Function method is discussed in detail. The remaining sections in this chapter discuss the so-called direct methods. Most direct methods for solving a nonlinear optimization problem involve some type of linearization. Therefore, the third section in this chapter is devoted to a discussion of linearization of a nonlinear problem using the Taylor series. One of the simplest methods based on the linearization idea is the Sequential Linear Programming method. This method is discussed in section 4. Section 5 presents the Sequential Quadratic Programming method, which is currently considered the best direct method for solving general nonlinear programming problems. Some of the relatively recent refinements to the SQP method are discussed in the last section.

9.1 Normalization

Performance of numerical algorithms, discussed in later sections, usually improves if the problems are normalized before proceeding with their solution. Some simple normalization schemes are discussed in this section.

9.1.1 Constraint Normalization

The idea behind constraint normalization is to try to rewrite constraints so that all constraints in a problem have similar magnitudes when they are near a constraint boundary. If all constraints can be normalized in this manner, it becomes easier to determine the best search directions and step lengths numerically.

Without normalization, from the numerical values alone, one cannot determine how close one is to the boundary of a constraint. Since the choice of direction and step length in most numerical algorithms is based on the knowledge of the constraint boundaries, constraint normalization is usually beneficial.

The following example illustrates these ideas:

Example 9.1 *Constraints in original form* Consider the following problem with two constraints:

$$\text{Minimize } f = 230x_1 + 20x_2$$

$$\text{Subject to } \begin{pmatrix} 0.1x_1^4 - \text{Sin}[x_2^2] \geq 0.001 \\ 230x_1 + 20x_2^3 \leq 1{,}000 \end{pmatrix}$$

9.1 Normalization

In the standard LE ($\leq$) form, the constraints are written as follows:

$$g_1 = -0.1x_1^4 + \text{Sin}\left[x_2^2\right] + 0.001 \leq 0$$
$$g_2 = 230x_1 + 20x_2^3 - 1{,}000 \leq 0$$

At an arbitrarily selected point, say $x_1 = 1$ and $x_2 = 3.4$, the constraint functions come out to have the following values:

$$g_1 = -0.943896 \quad \text{and} \quad g_2 = 16.08$$

From these values, we can see that the second constraint is violated at the selected point but we cannot determine how bad this violation is. Similarly, we can see that the first constraint is satisfied but cannot tell how much margin we have before this constraint becomes violated. In fact, as the following graph shows, the chosen point (1, 3.4) is very close to the boundary of constraint g_2 and far away from that of g_1. However, based on the numerical values alone, we will probably draw the opposite conclusion.

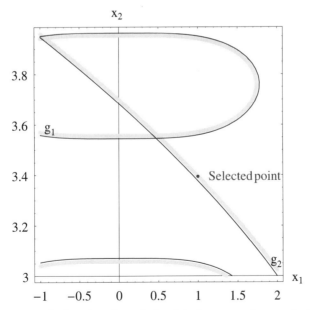

FIGURE 9.1 A graph showing the relationship between the constraint boundaries and the selected point.

Normalization by Dividing by Constant Upper Bounds

In many practical problems, a constraint is naturally expressed as some function of optimization variables that is less than or equal to a specified value. For such constraints the simplest, and very effective, normalization scheme is to divide both sides of constraints by their upper bounds. In this form, all constraints have an upper bound of 1. A point at which a constraint value is close to zero, such as 0.01, would then clearly be very close to the boundary of that constraint.

Example 9.2 *Constraints in the normalized form* Both constraints in example 9.1 can be normalized by dividing by their respective upper bounds. Thus, dividing g_1 by 0.001 and g_2 by 1,000, the normalized constraints are as follows:

$$\text{Normalized constraints:} \begin{pmatrix} \tilde{g}_1 = 1 - 100x_1^4 + 1{,}000\,\text{Sin}[x_2^2] \leq 0 \\ \tilde{g}_2 = -1 + \dfrac{23x_1}{100} + \dfrac{x_2^3}{50} \leq 0 \end{pmatrix}$$

Evaluating these normalized constraints at the same point (1, 3.4) gives the following values:

$$\tilde{g}_1 = -943.896 \quad \text{and} \quad \tilde{g}_2 = 0.01608$$

These values clearly tell us that we are near the boundary of g_2 and far from that of g_1. Looking at these numerical values, and depending upon the desired numerical accuracy, we may treat both constraints as satisfied.

Normalization by Dividing by Optimization Variables

In light of the previous example, it may be tempting to normalize constraints by dividing them by variable upper bounds. Consider a simple example of a constraint in the following form:

$$x_1 \leq x_2 \quad \text{or} \quad x_1 - x_2 \leq 0$$

Normalizing this constraint as

$$\frac{x_1}{x_2} - 1 \leq 0$$

has two problems. The first is that the normalized expression is valid only if x_2 is restricted to be positive. If this is not the case, dividing by a negative number will obviously change the direction of inequality. The normalization can be made valid for all cases if we take the absolute value of the variable, i.e., we write the normalized form as

$$\frac{x_1}{\text{Abs}\,[x_2]} - 1 \leq 0$$

The second problem is that a simple linear constraint has become a nonlinear constraint in the normalized form. Since it is usually simpler to handle a linear constraint than a nonlinear, this normalization is not preferable. Having the absolute values in the constraint expressions further complicate things because of introducing discontinuities in the derivatives. For these reasons, in general, it is not a good idea to normalize constraints by dividing them by optimization variables.

9.1.2 Scaling Optimization Variables

Gradient-based numerical methods usually have difficulty solving problems in which optimization variables have vastly different magnitudes. If suitable upper limits for optimization variables are available, reformulating the problem in terms of scaled variables may improve the performance of numerical methods.

Note that scaling of optimization variables usually changes the shape of the feasible domain. Thus, great care is needed in selecting appropriate scale factors. When in doubt, it is perhaps better not to use any scaling for optimization variables. For a detailed discussion on scaling of variables, and other numerical issues, see the book by Gill, Murray, and Wright [1981].

9.2 Penalty Methods

A large class of indirect methods exist in which the original problem is transformed into an equivalent unconstrained problem which is then solved using the methods discussed for unconstrained problems. Penalty methods belong to this class of indirect methods. The basic idea of a penalty method is to define an unconstrained problem by adding constraints, multiplied by a large penalty term, to the objective function. The additional terms are defined in such a way that if the constraints are satisfied, there is no penalty, but if one or more constraints are violated, a large penalty term is added. Since the objective function is being minimized, the penalty term will indirectly force the constraints to be satisfied.

Several possibilities exist for defining the penalty terms. Exterior, interior, and the so-called exact penalty functions are presented in the following subsections. In the presentation, it is assumed that the optimization problem is written in the following standard form:

Find $\mathbf{x}$ to   Minimize $f(\mathbf{x})$

Subject to $h_i(\mathbf{x}) = 0, i = 1, 2, \ldots, p$ and $g_i(\mathbf{x}) \leq 0, i = 1, 2, \ldots, m$.

9.2.1 The Exterior Penalty Function

One of the simplest techniques, known as the *exterior penalty function*, is to define an equivalent unconstrained problem, as follows:

$$\phi(\mathbf{x}, \mu) = f(\mathbf{x}) + \mu \left[\sum_{i=1}^{m} (\text{Max}[0, g_i(\mathbf{x})])^2 + \sum_{i=1}^{p} (h_i(\mathbf{x}))^2 \right]$$

where μ is a large positive number, known as the penalty number. If an inequality constraint is satisfied, then $g_i < 0$ and $\text{Max}[0, g_i]$ will return 0 and therefore, that constraint will not contribute anything to the function ϕ. If a constraint is violated, i.e., $g_i > 0$ or $h_i \neq 0$, a large term will get added to ϕ. Minimizing ϕ should force the violation to go away. For the method to work, the penalty number μ must be very large. In fact, theoretically, the minimum of ϕ corresponds to the solution of the original problem only as $\mu \to \infty$.

Example 9.3   Consider the following optimization problem:

Minimize $f = \frac{1}{2}x_1^2 + x_1 x_2^2$

Subject to $\begin{pmatrix} x_1 x_2^2 = 10 \\ -2x_1^2 + x_2/3 \leq 0 \end{pmatrix}$

Using KT conditions, it can easily be verified that the optimum solution of the problem is given by ($x_1 = 0.774$, $x_2 = 3.594$).

```
KTSolution [½x₁² + x₁x₂², {x₁x₂² == 10, -2x₁² + x₂/3 ≤ 0}, {x₁, x₂}];

***** Lagrangian → x₁²/2 + u₁(s₁² - 2x₁² + x₂/3) + x₁x₂² + v₁(-10 + x₁x₂²)
```

9.2 Penalty Methods

```
***** Valid KT Point(s) *****
f → 10.2995
x₁ → 0.773997
x₂ → 3.59443
u₁ → 0.2
s₁² → 0
v₁ → -1.01198
```

The unconstrained function ϕ for the problem is as follows:

$\phi = \frac{1}{2}x_1^2 + x_1 x_2^2 + \mu((x_1 x_2^2 - 10)^2 + (\text{Max}[0, (-2x_1^2 + x_2/3)])^2);$

Figure 9.2 shows contour plots of this function for two different μ values. The plots demonstrate that the minimum of the unconstrained problem is close to the optimum solution. The figure also illustrates the numerical difficulty anticipated in minimizing ϕ for large μ values. As the value of μ is increased, the contours become closer together with a long narrow shape. A method, like the steepest descent, would have great difficulty in finding the minimum of such functions. Even more advanced methods usually have difficulty in locating the minimum for such functions.

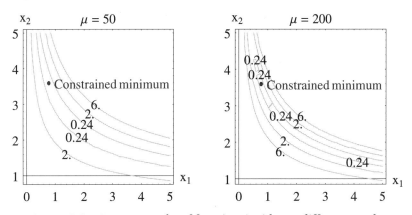

FIGURE 9.2 Contour graphs of function ϕ with two different μ values.

9.2.2 The Interior Penalty Function

In the *interior penalty function* method, the penalty term is defined to keep the solution from leaving the feasible region. It is not possible to directly handle equality constraints in this method. Thus, the original problem for this method is defined as follows:

Find **x** to   Minimize $f(\mathbf{x})$
Subject to $g_i(\mathbf{x}) \leq 0, i = 1, 2, \ldots, m.$

Two commonly used interior penalty functions are defined as follows:

$$\phi(\mathbf{x}, \mu) = f(\mathbf{x}) - \mu \left[\sum_{i=1}^{m} \text{Log}[-g_i(\mathbf{x})] \right] \quad \text{—Logarithmic barrier function}$$

$$\phi(\mathbf{x}, \mu) = f(\mathbf{x}) + \mu \left[\sum_{i=1}^{m} \frac{-1}{g_i(\mathbf{x})} \right] \quad \text{—Inverse barrier function}$$

where μ is a positive number, known as the penalty number. It is assumed that the inequality constraints are satisfied at all times and therefore, $g_i < 0$ and a positive term is added to the objective function. As we move closer to a constraint boundary, $g_i \to 0$ causing a large term to be added to the objective function. Thus, the method keeps the solution away from the constraint boundaries and hence, is also known as the *barrier function* method. Minimizing ϕ should force the violation to go away. If the solution is at a constraint boundary, for the method to work, the penalty number μ must be very small. In fact, theoretically, the minimum of ϕ corresponds to the solution of the original problem only as $\mu \to 0$.

Example 9.4   Consider the following optimization problem:

Minimize $f = xy + y^2$

Subject to $\begin{pmatrix} \frac{1}{4}y^2 - 1 \leq 0 \\ x/5 + y/2 - 1 \leq 0 \end{pmatrix}$

Using KT conditions, it can easily be verified that the optimum solution of the problem is at $(x = 10, y = -2)$.

KTSolution $\left[xy + y^2, \left\{ \frac{1}{4}y^2 - 1 \leq 0, x/5 + y/2 - 1 \leq 0 \right\}, \{x, y\} \right]$;

***** Lagrangian $\to xy + y^2 + \left(-1 + \frac{y^2}{4} + s_1^2\right) u_1 + \left(-1 + \frac{x}{5} + \frac{y}{2} + s_2^2\right) u_2$

***** Valid KT Point(s) *****

$f \to 0$ $f \to -16.$
$x \to 0$ $x \to 10.$
$y \to 0$ $y \to -2.$
$u_1 \to 0$ $u_1 \to 11.$
$u_2 \to 0$ $u_2 \to 10.$
$s_1^2 \to 1.$ $s_1^2 \to 0$
$s_2^2 \to 1.$ $s_2^2 \to 0$

(a) Using the inverse barrier function, the unconstrained function ϕ for the problem is as follows:

$$\phi = xy + y^2 + \mu \left(\frac{-1}{\frac{1}{4}y^2 - 1} + \frac{-1}{x/5 + y/2 - 1} \right);$$

The necessary condition for the minimum of this unconstrained function is that the gradient must be zero. This gives the following two equations:

```
eqns = Thread[Grad[ϕ, {x, y}] == 0]; MatrixForm[eqns]
```

$$\begin{pmatrix} y + \dfrac{\mu}{5(-1 + \frac{x}{5} + \frac{y}{2})^2} == 0 \\ x + 2y + \dfrac{\mu}{2(-1 + \frac{x}{5} + \frac{y}{2})^2} + \dfrac{y\mu}{2(-1 + \frac{y^2}{4})^2} == 0 \end{pmatrix}$$

Several solutions of this system of equations, for different values of penalty parameter μ, are as follows:

$\mu = 1$	$\{x \to 7.3926, y \to -1.65277\}, \{x \to 9.29953, y \to -2.30849\},$ $\{x \to 10.9855, y \to -1.71027\}, \{x \to 12.1669, y \to -2.27360\}$
$\mu = 0.1$	$\{x \to 9.23956, y \to -1.90097\}, \{x \to 9.7521, y \to -2.0962\},$ $\{x \to 10.2764, y \to -1.90566\}, \{x \to 10.7192, y \to -2.09216\}$
$\mu = 0.0001$	$\{x \to 9.99188, y \to -1.99698\}, \{x \to 9.99304, y \to -1.99698\},$ $\{x \to 10.0072, y \to -2.00301\}, \{x \to 10.0079, y \to -2.00301\}$

Clearly, as the value of μ is approaching 0, the solutions of the unconstrained problem are all approaching the solution of the original constrained problem. It is also interesting to note that setting $\mu = 0$ gives $x = y = 0$, which is obviously not the correct solution. Thus, the correct solution can only be obtained through the limiting process as $\mu \to 0$.

(b) Using the logarithmic barrier function, the unconstrained function ϕ for the problem is as follows:

$$\phi = xy + y^2 - \mu \left(\text{Log}\left[-\left(\tfrac{1}{4}y^2 - 1\right) \right] + \text{Log}[-(x/5 + y/2 - 1)] \right);$$

The necessary conditions for the minimum of this unconstrained function are that the gradient must be zero. This gives the following two equations:

```
eqns = Thread[Grad[ϕ, {x, y}] == 0]; MatrixForm[eqns]
```

$$\begin{pmatrix} y + \dfrac{\mu}{5\left(1 - \dfrac{x}{5} - \dfrac{y}{2}\right)} == 0 \\ x + 2y + \dfrac{\mu}{2\left(1 - \dfrac{x}{5} - \dfrac{y}{2}\right)} + \dfrac{y\mu}{2\left(1 - \dfrac{y^2}{4}\right)} == 0 \end{pmatrix}$$

Solutions of this system of equations for different values of penalty parameter μ are as follows:

$\mu = 1$	$\{x \to 0.000775589, y \to -0.183234\}$,
	$\{x \to 9.23548, y \to -1.90425\}$
$\mu = 0.1$	$\{x \to -0.00889614, y \to -0.0197694\}$,
	$\{x \to 9.92693, y \to -1.99087\}$
$\mu = 0.0001$	$\{x \to -9.99888 \times 10^{-6}, y \to -0.0000199998\}$,
	$\{x \to 0.000300084, y \to 1.9999\}$,
	$\{x \to 0.833116, y \to 1.66678\}$,
	$\{x \to 9.99993, y \to -1.99999\}$

Clearly, as the value of μ is approaching 0, the last solution of the unconstrained problem is approaching the solution of the original constrained problem. By checking the second-order conditions, it is easy to see that the other solutions are simply inflection points for function ϕ.

9.2.3 Augmented Lagrangian Penalty Function

Both the interior and the exterior penalty function methods considered briefly in the previous two sections use simple ideas to convert a constrained problem to an equivalent unconstrained problem. Both methods suffer from numerical difficulties in solving the unconstrained problem. Furthermore, the solution of the unconstrained problem approaches the solution of the original problem in the limit, but is never actually equal to the exact solution. To overcome these shortcomings, the so-called exact penalty functions have been developed. As the name indicates, these methods define unconstrained problems in such a way that the exact solution of the original problem is recovered for a *finite* value of the penalty parameter.

The augmented Lagrangian penalty function is an example of such an exact penalty function. In this approach, the unconstrained function is defined by adding the exterior penalty term to the *Lagrangian* (instead of just the objective function) of the original problem. The inequality constraints are converted to equality constraints through the addition of positive slack variables. Thus, if the original problem is stated as follows,

Find **x** to   Minimize $f(\mathbf{x})$

Subject to $h_i(\mathbf{x}) = 0, i = 1, 2, \ldots, p$ and $g_i(\mathbf{x}) \leq 0, i = 1, 2, \ldots, m$.

then the unconstrained augmented Lagrangian penalty function is defined as follows:

$$\phi(\mathbf{x}, \mathbf{u}, \mathbf{v}, \mathbf{s}, \mu) = f(\mathbf{x}) + \sum_{i=1}^{p} v_i h_i(\mathbf{x}) + \sum_{i=1}^{m} u_i \left(g_i(\mathbf{x}) + s_i^2\right)$$

$$+ \mu \left[\sum_{i=1}^{m} \left(g_i(\mathbf{x}) + s_i^2\right)^2 + \sum_{i=1}^{p} (h_i(\mathbf{x}))^2 \right]$$

It is demonstrated in the following example that if the optimum Lagrange multipliers are known, then the solution of this unconstrained problem corresponds to the solution of the original problem regardless of the value of the penalty parameter.

Example 9.5   Consider the following optimization problem:

Minimize $f = \frac{1}{2}x^2 + xy^2$

Subject to $(x + y^2 = 10)$

Using KT conditions, it is easy to compute the optimum solution as follows:

```
KTSolution[½x² + xy², {x + y² - 10 == 0}, {x, y}];

***** Lagrangian → x²/2 + xy² + (-10 + x + y²) v₁
***** Valid KT Point(s) *****
f → 50.
x → 10.
y → 0
v₁ → -10.
```

Optimum: $x = 10$ $y = 0$ Lagrange multiplier, $v = -10$

In the augmented Lagrangian approach, the unconstrained function is defined by adding the exterior penalty term to the Lagrangian of the original problem. For this example, thus, we have the following unconstrained function:

$\phi = \frac{1}{2}x^2 + xy^2 + v(x + y^2 - 10) + \mu(x + y^2 - 10)^2;$

The necessary conditions for the minimum of this function give the following equations:

eqns = Thread[Grad[ϕ, {x, y}] == 0]; MatrixForm[eqns]

$$\begin{pmatrix} v + x + y^2 - 20\mu + 2x\mu + 2y^2\mu == 0 \\ 2vy + 2xy - 40y\mu + 4xy\mu + 4y^3\mu == 0 \end{pmatrix}$$

If v is set to the optimum value of the Lagrange multiplier, we get the following equations:

eqns2 = eqns/.{v → -10}; MatrixForm[eqns2]

$$\begin{pmatrix} -10 + x + y^2 - 20\mu + 2x\mu + 2y^2\mu == 0 \\ -20y + 2xy - 40y\mu + 4xy\mu + 4y^3\mu == 0 \end{pmatrix}$$

The second equation can be written as follows:

$$y\left(-20 + 2x - 40\mu + 4x\mu + 4y^2\mu\right) == 0$$

Thus, $y = 0$ satisfies this equation for any value of μ. Substituting $y = 0$ in the first equation, we get

$$-10 + x - 20\mu + 2x\mu == 0 \quad \text{or} \quad (-10 + x)(1 + 2\mu) == 0$$

Thus, $x = 10$ satisfies this equation. Thus, the Lagrangian penalty function has the property that the optimum solution of the original problem is recovered, if we know the optimum values of the Lagrange multipliers. Therefore, in this sense it is an exact penalty function.

Obviously, when we are solving a problem we don't know the optimum Lagrange multipliers. (If they were known we wouldn't need to spend time in developing new algorithms. We could simply use them with the KT conditions to get a solution). However, the presence of Lagrange multipliers makes the choice of penalty parameter less critical. In a computational procedure based on using the augmented Lagrangian penalty function, we start with arbitrary values of Lagrange multipliers and develop a procedure that moves these multipliers closer to their optimum values. Thus, near the optimum, the function ϕ is not as sensitive to values of μ, and the procedure converges to the true optimum solution.

Elimination of Slack Variables

The augmented Lagrangian penalty function defined so far includes slack variables for inequality constraints. Their presence increases the number of variables in the problem. It is possible to remove these variables by writing the necessary conditions for the minimum with respect to **s**. Before proceeding, it is convenient to combine the two terms involving slack variables by noting that

$$\mu \left(g_i(\mathbf{x}) + s_i^2 + \frac{u_i}{2\mu} \right)^2 = \mu \left[\left(g_i(\mathbf{x}) + s_i^2 \right)^2 + \left(\frac{u_i}{2\mu} \right)^2 + 2 \left(g_i(\mathbf{x}) + s_i^2 \right) \left(\frac{u_i}{2\mu} \right) \right]$$

Rearranging the terms gives

$$\mu \left(g_i(\mathbf{x}) + s_i^2 \right)^2 + u_i \left(g_i(\mathbf{x}) + s_i^2 \right) = \mu \left(g_i(\mathbf{x}) + s_i^2 + \frac{u_i}{2\mu} \right)^2 - \frac{u_i^2}{4\mu}$$

Using this, we can write $\phi(\mathbf{x}, \mathbf{u}, \mathbf{v}, \mathbf{s}, \mu)$, as follows:

$$\phi(\mathbf{x}, \mathbf{u}, \mathbf{v}, \mathbf{s}, \mu) = f(\mathbf{x}) + \sum_{i=1}^{p} v_i h_i(\mathbf{x}) + \mu \sum_{i=1}^{p} (h_i(\mathbf{x}))^2$$

$$+ \mu \sum_{i=1}^{m} \left(g_i(\mathbf{x}) + s_i^2 + \frac{u_i}{2\mu} \right)^2 - \sum_{i=1}^{m} \frac{u_i^2}{4\mu}$$

Writing the necessary conditions for the minimum of ϕ with respect to slack variables, we get

$$\partial \phi / \partial s_i = 0 \implies 2\mu \left(g_i(\mathbf{x}) + s_i^2 + \frac{u_i}{2\mu} \right) 2s_i = 0$$

$$\implies s_i \left(g_i(\mathbf{x}) + s_i^2 + \frac{u_i}{2\mu} \right) = 0, i = 1, \ldots, m$$

These conditions state that

Either $s_i = 0$ Or $g_i(\mathbf{x}) + s_i^2 + \dfrac{u_i}{2\mu} = 0 \implies s_i^2 = -\left[g_i(\mathbf{x}) + \dfrac{u_i}{2\mu} \right]$

These two conditions can be combined together and expressed in the form of the following function:

$$r_i \equiv \text{Max} \left[g_i(\mathbf{x}) + \frac{u_i}{2\mu}, 0 \right], i = 1, \ldots, m$$

Using this function r_i, the unconstrained ϕ function can be written as follows:

$$\phi(\mathbf{x}, \mathbf{u}, \mathbf{v}, \mu) = f(\mathbf{x}) + \sum_{i=1}^{p} v_i h_i(\mathbf{x}) + \mu \sum_{i=1}^{p} (h_i(\mathbf{x}))^2 + \mu \sum_{i=1}^{m} r_i^2 - \sum_{i=1}^{m} \frac{u_i^2}{4\mu}$$

Note that because of the presence of Max, the function r_i is not differentiable. However, r_i^2 is continuous and has well-defined first derivatives, as follows:

$$\frac{\partial (r_i^2)}{\partial x_j} = 2r_i \frac{\partial g_i}{\partial x_j} \quad j = 1, \ldots, n$$

The following example numerically illustrates this fact.

Example 9.6 For the following function, compute the first derivative as indicated above and graphically compare it with its finite difference approximation.

$$f(x) = (\text{Max}[0, \text{Sin}[x]])^2$$

The first derivative of the function is as follows:

$$\frac{df}{dx} = 2\text{Max}[0, \text{Sin}[x]]\text{Cos}[x]$$

The plots of the function and its first derivative are shown in Figure 9.3.

```
Clear[f]; f[x_] := Max[0, Sin[x]]^2;
gr1 = Plot[f[x], {x, 0, 10}, PlotLabel → f[x]];

Clear[df]; (df[x_] := 2Cos[x]Max[0, Sin[x]];)
gr2 = Plot[df[x], {x, 0, 10}, PlotLabel → df[x]];
```

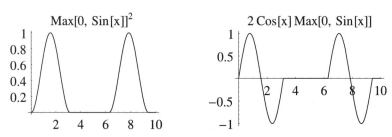

FIGURE 9.3 Graphs of $f(x) = (\text{Max}[0, \text{Sin}[x]])^2$ and its first derivative.

A finite difference approximation of the derivative of $f(x)$ can be generated by taking the difference of the function values close to each other, as follows:

```
finiteDiff[x_] := (f[x + 0.00001] - f[x])/0.00001
```

A plot of this function, shown in Figure 9.4, is identical to the one for analytical df/dx, and therefore demonstrates that the function is differentiable and the analytical expression for the derivative of the given function is correct. The second derivative obtained from the first derivative using finite difference shows discontinuity and thus, we cannot determine the second derivative of the function analytically.

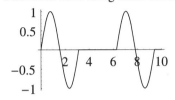
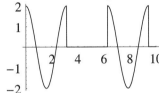

FIGURE 9.4 Plots of first and second derivatives of $f(x) = (\text{Max}[0, \text{Sin}[x]])^2$ using finite differences.

Updating Lagrange Multipliers

In order to use the augmented Lagrangian penalty function to solve constrained optimization problems, we need to determine a procedure that, starting from arbitrary values, leads to near optimum values of Lagrange multipliers.

A simple procedure is based on comparing the necessary conditions for the minimum of the Lagrangian function and the augmented Lagrangian penalty function. The Lagrangian for the problem with equality constraints only is as follows:

$$L(\mathbf{x}, \mathbf{v}) = f(\mathbf{x}) + \sum_{i=1}^{p} v_i h_i(\mathbf{x})$$

(a) $\partial L/\partial x_j = 0 \implies \frac{\partial f}{\partial x_j} + \sum_{i=1}^{p} v_i \frac{\partial h_i}{\partial x_j} = 0$

The corresponding augmented Lagrangian penalty function is

$$\phi(\mathbf{x}, \mathbf{v}, \mu) = f(\mathbf{x}) + \sum_{i=1}^{p} v_i h_i(\mathbf{x}) + \mu \sum_{i=1}^{p} (h_i(\mathbf{x}))^2$$

(b) $\partial \phi / \partial x_j = 0 \implies \frac{\partial f}{\partial x_j} + \sum_{i=1}^{p}(v_i + 2\mu h_i)\frac{\partial h_i}{\partial x_j} = 0$

Comparing the two equations (a) and (b), we get the following rule for updating the Lagrange multipliers at the kth iteration:

$$v_i^{(k+1)} = v_i^{(k)} + 2\mu h_i(\mathbf{x}^{(k)}) \quad i = 1, \ldots, p$$

In the presence of inequality constraints, the above analysis does not work out as cleanly as for the equality case. In practice, the following rule based on similarity with the equality case is adopted:

$$u_i^{(k+1)} = u_i^{(k)} + 2\mu r_i(\mathbf{x}^{(k)}) \quad i = 1, \ldots, m$$

To start the process, arbitrary values, usually zero, are assigned to all multipliers. Also, the multiplier updating is done only after a substantial decrease in constraint violations is achieved.

Complete Augmented Lagrangian Penalty Function Method

(i) Set iteration counter $k = 0$. Set multipliers $u_i = 0, i = 1, \ldots, m$ and $v_i = 0, i = 1, \ldots, p$. Set multiplier update counter $\ell = 0$. Choose a penalty parameter μ and a factor $c > 1$ to increase penalty parameter value during iterations. Typical values are $\mu = 10$ and $c = 2$.

(ii) Set up the unconstrained minimization problem.

$$\phi(\mathbf{x}, \mathbf{u}, \mathbf{v}, \mu) = f(\mathbf{x}) + \sum_{i=1}^{p} v_i h_i(\mathbf{x}) + \mu \sum_{i=1}^{p}(h_i(\mathbf{x}))^2 + \mu \sum_{i=1}^{m} r_i^2 - \sum_{i=1}^{m} \frac{u_i^2}{4\mu}$$

where $r_i = \text{Max}[g_i(\mathbf{x}) + \frac{u_i}{2\mu}, 0]$

Use a suitable unconstrained minimization method to find the minimum point $\mathbf{x}^{(k)}$ of this problem. The derivatives of ϕ with respect to x_j are evaluated using the following equation:

$$\frac{\partial \phi}{\partial x_j} = \frac{\partial f}{\partial x_j} + \sum_{i=1}^{p} v_i \frac{\partial h_i}{\partial x_j} + 2\mu \sum_{i=1}^{p} h_i \frac{\partial h_i}{\partial x_j} + 2\mu \sum_{i=1}^{m} r_i \frac{\partial g_i}{\partial x_j} \quad j = 1, \ldots, n$$

(iii) Check for convergence.

A simple convergence criteria is to stop when all constraints are satisfied and the objective function is not changing much between successive iterations. Thus, stop if the following two conditions are met. Otherwise, continue to step (iv).

$Abs[f(\mathbf{x}^{(k)}) - f(\mathbf{x}^{(k-1)})/f(\mathbf{x}^{(k)})] < \epsilon$

$V(\mathbf{x}^{(k)}) < \epsilon$

where $V(\mathbf{x}^{(k)}) \equiv \text{Max}[g_i(\mathbf{x}^{(k)}), i = 1, \ldots, m, \text{Abs}[h_i(\mathbf{x}^{(k)}), i = 1, \ldots, p]]$ is the maximum constraint violation.

(iv) Update the multipliers and the penalty parameter:

If $V(\mathbf{x}^{(k)}) \leq \frac{1}{4}V(x^{(\ell)})$ then update multipliers

$$u_i^{(\ell+1)} = u_i^{(\ell)} + 2\mu r_i(\mathbf{x}^{(k)}) \quad i = 1, \ldots, m$$

$$v_i^{(\ell 1)} = v_i^{(\ell)} + 2\mu h_i(\mathbf{x}^{(k)}) \quad i = 1, \ldots, p$$

Set $\ell = \ell + 1$

Else update the penalty parameter

$$\mu^{(k+1)} = c\mu^{(k)}$$

(v) Update the iteration counter $k = k + 1$ and go back to step (ii).

The algorithm is implemented in a *Mathematica* function called ALPF.

Needs["OptimizationToolbox`ConstrainedNLP`"];
?ALPF

ALPF[f, cons, vars, pt, opts] solves a constrained optimization
   problem using the augmented Lagrangian penalty function method. f
   = objective function. cons = list of constraints. vars = list of
   problem variables. pt = starting point. opts = options allowed by the
   function. See Options[ALPF] to see all valid options.

OptionsUsage[ALPF]
{ProblemType → Min, PrintLevel → 1, MaxIterations → 20,
 ConvergenceTolerance → 0.001, SolveUnconstrainedUsing → FindMinimum,
 MaxUnconstrainedIterations → 50, PenaltyParameters → {10, 2}}

ProblemType is an option for most optimization methods. It can either be
   Min (default) or Max.

PrintLevel is an option for most functions in the OptimizationToolbox.
   It is specified as an integer. The value of the integer indicates
   how much intermediate information is to be printed. A PrintLevel→ 0
   suppresses all printing. Default for most functions is set to 1 in
   which case they print only the initial problem setup. Higher integers
   print more intermediate results.

MaxIterations is an option for several optimization methods. It
   specifies maximum number of iterations allowed.

Chapter 9 Constrained Nonlinear Problems

```
ConvergenceTolerance is an option for most optimization methods. Most
  methods require only a single zero tolerance value. Some interior
  point methods require a list of convergence tolerance values.

SolveUnconstrainedUsing is an option for ALPF method. It defines the
  solution method used for solving unconstrained subproblem. Currently
  the only choice is the built-in FindMinimum function.

MaxUnconstrainedIterations is an option for ALPF method. It defines
  the maximum number of iterations allowed in solving the unconstrained
  problem. Default is MaxUnconstrainedIterations→ 50.

PenaltyParameters is an option for ALPF method. It defines the starting
  value for parameter μ and scale factor for changing it during
  iterations. Default values are PenaltyParameters→ {10,2}
```

Example 9.7 Solve the following optimization problem using the augmented Lagrangian penalty function method. Use the starting point as {3, 1}.

Minimize $f = \frac{1}{2}x_1^2 + x_1 x_2^2$

Subject to $\begin{pmatrix} x_1 x_2^2 = 10 \\ -2x_1^2 + x_2/3 \leq 0 \\ x_2 \geq -4 \end{pmatrix}$

```
vars = {x₁, x₂}; (pt = {3.0, 1.0};)
f = x₁²/2 + x₁x₂²;
cons = {x₁x₂² == 10, -2x₁² + x₂/3 ≤ 0, x₂ ≥ -4};
```

The solution is obtained by using the ALPF method. All intermediate calculations are shown for the first two iterations.

```
ALPF[f, cons, vars, pt, MaxIterations → 2, PrintLevel → 3,
    PenaltyParameters → {10, 2}];
```

Minimize $f \to \frac{x_1^2}{2} + x_1 x_2^2$

$\nabla f \to \begin{pmatrix} x_1 + x_2^2 & 2x_1 x_2 \end{pmatrix}$

LE Constraints: $\begin{pmatrix} g_1 \\ g_2 \end{pmatrix} \to \begin{pmatrix} -2x_1^2 + \frac{x_2}{3} \leq 0 \\ -4 - x_2 \leq 0 \end{pmatrix}$

$\begin{pmatrix} \nabla g_1 \\ \nabla g_2 \end{pmatrix} \to \begin{pmatrix} -4x_1 & \frac{1}{3} \\ 0 & -1 \end{pmatrix}$

EQ Constraints: $(h_1) \to (-10 + x_1 x_2^2 == 0)$

$(\nabla h_1) \to \begin{pmatrix} x_2^2 & 2x_1 x_2 \end{pmatrix}$

9.2 Penalty Methods

------------------ **ALPF Iteration 1** ------------------

Current point $\to \begin{pmatrix} 3. & 1. \end{pmatrix}$

Function values $\begin{pmatrix} f \\ g \\ h \end{pmatrix} \to \begin{pmatrix} 7.5 \\ \{-17.6667, -5.\} \\ \{-7.\} \end{pmatrix}$

Max Violation $\to 7.$

$\mu \to 10 \quad v \to \begin{pmatrix} 0 \end{pmatrix}$

$r \to \begin{pmatrix} 0 \\ 0 \end{pmatrix} \quad u \to \begin{pmatrix} 0 \\ 0 \end{pmatrix}$

$\phi \to 1000 + 10\text{Max}\left[0, -4 - x_2\right]^2 + 10\text{Max}\left[0, -2x_1^2 + \frac{x_2}{3}\right]^2 + \frac{x_1^2}{2} - 199 x_1 x_2^2 + 10 x_1^2 x_2^4$

$\nabla \phi \to \begin{pmatrix} x_1 - 80\text{Max}\left[0, -2x_1^2 + \frac{x_2}{3}\right] x_1 - 199 x_2^2 + 20 x_1 x_2^4 \\ -20\text{Max}\left[0, -4 - x_2\right] + \frac{20}{3}\text{Max}\left[0, -2x_1^2 + \frac{x_2}{3}\right] - 398 x_1 x_2 + 40 x_1^2 x_2^3 \end{pmatrix}$

Solution $\to \{10.9915, \{x_1 \to 1.35001, x_2 \to 2.7008\}\}$

------------------ **ALPF Iteration 2** ------------------

Current point $\to \begin{pmatrix} 1.35001 & 2.7008 \end{pmatrix}$

Function values $\begin{pmatrix} f \\ g \\ h \end{pmatrix} \to \begin{pmatrix} 10.7587 \\ \{-2.74478, -6.7008\} \\ \{-0.15257\} \end{pmatrix}$

Max Violation $\to 0.15257$

$\mu \to 10 \quad v \to \begin{pmatrix} -3.0514 \end{pmatrix}$

$r \to \begin{pmatrix} 0 \\ 0 \end{pmatrix} \quad u \to \begin{pmatrix} 0 \\ 0 \end{pmatrix}$

$\phi \to 1030.51 + 10\text{Max}\left[0, -4 - x_2\right]^2 + 10\text{Max}\left[0, -2x_1^2 + \frac{x_2}{3}\right]^2 + \frac{x_1^2}{2} - 202.051 x_1 x_2^2 + 10 x_1^2 x_2^4$

$\nabla \phi \to \begin{pmatrix} x_1 - 80\text{Max}\left[0, -2x_1^2 + \frac{x_2}{3}\right] x_1 - 202.051 x_2^2 + 20 x_1 x_2^4 \\ -20\text{Max}\left[0, -4 - x_2\right] + \frac{20}{3}\text{Max}\left[0, -2x_1^2 + \frac{x_2}{3}\right] - 404.103 x_1 x_2 + 40 x_1^2 x_2^3 \end{pmatrix}$

Solution $\to \{10.6337, \{x_1 \to 1.15703, x_2 \to 2.94269\}\}$

NonOptimum point: Iterations performed $\to 2$

Point $\to \begin{pmatrix} 1.15703 & 2.94269 \end{pmatrix}$

Function values $\begin{pmatrix} f \\ g \\ h \end{pmatrix} \to \begin{pmatrix} 10.6885 \\ \{-1.69653, -6.94269\} \\ \{0.0191806\} \end{pmatrix}$

$\mu \to 10 \quad$ Max Violation $\to 0.0191806$

$v \to \begin{pmatrix} -2.66779 \end{pmatrix} \quad u \to \begin{pmatrix} 0 \\ 0 \end{pmatrix}$

The procedure is allowed to run until an optimum is found.

```
{sol, hist} = ALPF[f, cons, vars, pt, PrintLevel → 0, PenaltyParameters →
  {10, 2}];
```

Optimum point: Iterations performed → 11

Point → $(0.773997 \quad 3.59443)$

Function values $\begin{pmatrix} f \\ g \\ h \end{pmatrix} \rightarrow \begin{pmatrix} 10.2995 \\ \{8.4953 \times 10^{-8}, -7.59443\} \\ \{1.48753 \times 10^{-8}\} \end{pmatrix}$

$\mu \rightarrow 640$ Max Violation → 8.4953×10^{-8}

$v \rightarrow (-1.01199)$ $u \rightarrow \begin{pmatrix} 0.199888 \\ 0 \end{pmatrix}$

Using KT conditions, it can easily be verified that the optimum solution and the Lagrange multipliers are exact. The complete history of the iterations is as follows:

TableForm[hist]

Point	f	maxViolation
3. 1.	7.5	7.
1.35001 2.7008	10.7587	0.15257
1.15703 2.94269	10.6885	0.0191806
1.0017 3.16195	10.5166	0.0149226
0.902785 3.32897	10.4122	0.00472714
0.819396 3.49424	10.3403	0.00457097
0.773646 3.59603	10.3036	0.00436978
0.773816 3.5965	10.3086	0.00916707
0.773907 3.59547	10.304	0.00458353
0.773952 3.59495	10.3018	0.00229178
0.773975 3.59469	10.3007	0.00114587
0.773997 3.59443	10.2995	8.4953×10^{-8}

The graph in Figure 9.5 confirms the solution and shows the path taken by the ALPF method.

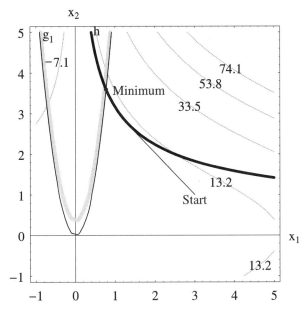

FIGURE 9.5 A graphical solution showing the search path.

Example 9.8 Solve the following optimization problem using the ALPF method.

$$\text{Minimize } f = e^{-x_1-x_2} + x_1 x_2 + x_2^2$$

$$\text{Subject to } \begin{pmatrix} \frac{1}{4}e^{-x_1} + \frac{1}{4}x_2^2 - 1 \leq 0 \\ x_1/5 + x_2/2 - 1 \leq 0 \end{pmatrix}$$

```
vars = {x₁, x₂};
f = Exp[-x₁ - x₂] + x₁x₂ + x₂²;
g = {¼Exp[-x₁] + x₂²/4 - 1 ≤ 0, x₁/5 + x₂/2 - 1 ≤ 0};
pt = {1.0, 1.0};
```

The solution is obtained by using the ALPF method starting from $(1, 1)$. All intermediate calculations are shown for the first two iterations.

```
ALPF[f, g, vars, pt, MaxIterations → 2, PrintLevel → 3, PenaltyParameters →
  {10, 2}];
```

Minimize $f \to E^{-x_1-x_2} + x_1 x_2 + x_2^2$

$\nabla f \to \left(-E^{-x_1-x_2} + x_2 \quad -E^{-x_1-x_2} + x_1 + 2x_2 \right)$

LE Constraints: $\begin{pmatrix} g_1 \\ g_2 \end{pmatrix} \to \begin{pmatrix} -1 + \frac{E^{-x_1}}{4} + \frac{x_2^2}{4} \le 0 \\ -1 + \frac{x_1}{5} + \frac{x_2}{2} \le 0 \end{pmatrix}$

$\begin{pmatrix} \nabla g_1 \\ \nabla g_2 \end{pmatrix} \to \begin{pmatrix} -\frac{E^{-x_1}}{4} & \frac{x_2}{2} \\ \frac{1}{5} & \frac{1}{2} \end{pmatrix}$

------------------ **ALPF Iteration 1** ------------------

Current point $\to (1.\quad 1.)$

Function values $\begin{pmatrix} f \\ g \\ h \end{pmatrix} \to \begin{pmatrix} 2.13534 \\ \{-0.65803, -0.3\} \\ \{\} \end{pmatrix}$

Max Violation $\to 0$

$\mu \to 10 \quad v \to \{\}$

$r \to \begin{pmatrix} 0 \\ 0 \end{pmatrix} \quad u \to \begin{pmatrix} 0 \\ 0 \end{pmatrix}$

$\phi \to E^{-x_1-x_2} + 10\text{Max}\left[0, -1 + \frac{x_1}{5} + \frac{x_2}{2}\right]^2 + 10\text{Max}\left[0, -1 + \frac{E^{-x_1}}{4} + \frac{x_2^2}{4}\right]^2 + x_1 x_2 + x_2^2$

$\nabla \phi \to \begin{pmatrix} -E^{-x_1-x_2} + 4\text{Max}\left[0, -1 + \frac{x_1}{5} + \frac{x_2}{2}\right] - 5E^{-x_1}\text{Max}\left[0, -1 + \frac{E^{-x_1}}{4} + \frac{x_2^2}{4}\right] + x_2 \\ -E^{-x_1-x_2} + 10\text{Max}\left[0, -1 + \frac{x_1}{5} + \frac{x_2}{2}\right] + x_1 + 2x_2 + 10\text{Max}\left[0, -1 + \frac{E^{-x_1}}{4} + \frac{x_2^2}{4}\right] x_2 \end{pmatrix}$

Solution $\to \{-5.61293, \{x_1 \to 5.14616, x_2 \to -2.27336\}\}$

------------------ **ALPF Iteration 2** ------------------

Current point $\to (5.14616 \quad -2.27336)$

Function values $\begin{pmatrix} f \\ g \\ h \end{pmatrix} \to \begin{pmatrix} -6.47436 \\ \{0.293501, -1.10745\} \\ \{\} \end{pmatrix}$

Max Violation $\to 0.293501$

$\mu \to 10 \quad v \to \{\}$

$r \to \begin{pmatrix} 0.293501 \\ 0 \end{pmatrix} \quad u \to \begin{pmatrix} 5.87003 \\ 0 \end{pmatrix}$

$\phi \to -0.86143 + E^{-x_1-x_2} + 10\text{Max}\left[0, -1 + \frac{x_1}{5} + \frac{x_2}{2}\right]^2$
$\quad + 10\text{Max}\left[0, -0.706499 + \frac{E^{-x_1}}{4} + \frac{x_2^2}{4}\right]^2 + x_1 x_2 + x_2^2$

$\nabla \phi \to$
$\begin{pmatrix} -E^{-x_1-x_2} + 4\text{Max}\left[0, -1 + \frac{x_1}{5} + \frac{x_2}{2}\right] - 5E^{-x_1}\text{Max}\left[0, -0.706499 + \frac{E^{-x_1}}{4} + \frac{x_2^2}{4}\right] + x_2 \\ -E^{-x_1-x_2} + 10\text{Max}\left[0, -1 + \frac{x_1}{5} + \frac{x_2}{2}\right] + x_1 + 2x_2 + 10\text{Max}\left[0, -0.706499 + \frac{E^{-x_1}}{4} + \frac{x_2^2}{4}\right] x_2 \end{pmatrix}$

Solution $\to \{-19.8145, \{x_1 \to 13.7027, x_2 \to -2.32061\}\}$

9.2 Penalty Methods

NonOptimum point: Iterations performed → 2
Point → $(13.7027 \quad -2.32061)$

Function values $\begin{pmatrix} f \\ g \\ h \end{pmatrix} \rightarrow \begin{pmatrix} -26.4134 \\ \{0.346306, 0.58024\} \\ \{\} \end{pmatrix}$

$\mu \rightarrow 20$ Max Violation → 0.58024

$v \rightarrow \{\} \quad u \rightarrow \begin{pmatrix} 5.87003 \\ 0 \end{pmatrix}$

The complete solution is as follows:

```
{sol, hist} = ALPF[f, g, vars, pt, PrintLevel → 0, PenaltyParameters → {10, 2}];
```
Optimum point: Iterations performed → 11
Point → $(9.99683 \quad -1.9988)$

Function values $\begin{pmatrix} f \\ g \\ h \end{pmatrix} \rightarrow \begin{pmatrix} -15.9861 \\ \{-0.00119097, -0.0000334746\} \\ \{\} \end{pmatrix}$

$\mu \rightarrow 2560$ Max Violation → 0

$v \rightarrow \{\} \quad u \rightarrow \begin{pmatrix} 17.1014 \\ 10.1677 \end{pmatrix}$

TableForm[hist]

Point	f	maxViolation
1. 1.	2.13534	0
5.14616 -2.27336	-6.47436	0.293501
13.7027 -2.32061	-26.4134	0.58024
11.7066 -2.14577	-20.5153	0.268441
10.8183 -2.06868	-18.0999	0.129309
10.4009 -2.03323	-17.0129	0.063548
9.89409 -1.96204	-15.5626	0
9.94836 -1.98097	-15.7828	0
9.97451 -1.99047	-15.8916	0
9.98732 -1.99523	-15.9457	0
9.99367 -1.99761	-15.9726	0
9.99683 -1.9988	-15.9861	0

The graph in Figure 9.6 confirms the solution and shows the path taken by the ALPF method.

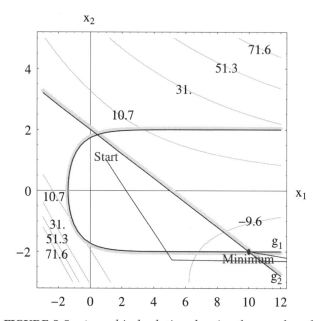

FIGURE 9.6 A graphical solution showing the search path.

Example 9.9 *Building design* Consider an optimum solution of a building design problem, formulated in Chapter 1, using the ALPF method. The problem statement is as follows:

To save energy costs for heating and cooling, an architect is considering designing a partially buried rectangular building. The total floor space needed is 20,000 m². Lot size limits the longer building dimensions to 50 m in plan. The ratio between the plan dimensions must be equal to the *golden ratio* (1.618), and each story must be 3.5 m high. The heating and cooling costs are estimated at $100 m² of the exposed surface area of the building. The owner has specified that the annual energy costs should not exceed $225,000. The objective is to determine building dimensions such that the cost of excavation is minimized.

A formulation for this problem involving following three design variables is presented in Chapter 2.

$$d = \text{Depth of building below ground}$$
$$h = \text{Height of building above ground}$$
$$w = \text{Width of building in plan}$$

Find $(d, h, \text{and } w)$ in order to

Minimize $f = 1.618\, dw^2$

Subject to
$$\begin{pmatrix} 100(5.236\, hw + 1.618 w^2) \leq 225{,}000 \\ 1.618 w \leq 50 \\ 0.462286 (d+h) w^2 \geq 20{,}000 \\ d \geq 0, h \geq 0, \text{ and } w \geq 0 \end{pmatrix}$$

By dividing the constraints by their right-hand sides, the normalized form of the constraints is as follows:

$$100 \left(5.236 hw + 1.618 w^2\right)/225{,}000 \leq 1$$
$$1.618 w / 50 \leq 1$$
$$0.462286 (d+h) w^2 / 20{,}000 \geq 1$$

In terms of *Mathematica* expressions, the problem is stated as follows:

```
Clear[d, h, w];
vars = {d, h, w};
f = 1.618 dw²;
cons = { 100(5.236hw + 1.618w²)/225000 ≤ 1, 1.618w/50 ≤ 1,
         0.462286(d+h)w²/20000 ≥ 1, d ≥ 0, h ≥ 0, w ≥ 0};
```

Starting from the following point, the solution is obtained using the ALPF method. All calculations are shown for the first iteration.

```
pt = {40, 10, 20}; ALPF[f, cons, vars, pt, MaxIterations → 1, PrintLevel → 3];
```

Minimize $f \to 1.618 dw^2$

$\nabla f \to \begin{pmatrix} 1.618 w^2 & 0 & 3.236 dw \end{pmatrix}$

LE Constraints: $\begin{pmatrix} g_1 \\ g_2 \\ g_3 \\ g_4 \\ g_5 \\ g_6 \end{pmatrix} \to \begin{pmatrix} -1 + \dfrac{5.236 hw + 1.618 w^2}{2250} \leq 0 \\ -1 + 0.03236 w \leq 0 \\ 1 - 0.0000231143\,(d+h)\, w^2 \leq 0 \\ -d \leq 0 \\ -h \leq 0 \\ -w \leq 0 \end{pmatrix}$

------------------ **ALPF Iteration 1** ------------------

Current point → $(40 \quad 10 \quad 20)$

Function values $\begin{pmatrix} f \\ g \\ h \end{pmatrix} \to \begin{pmatrix} 25888. \\ \{-0.246933, -0.3528, 0.537714, -40, -10, -20\} \\ \{\} \end{pmatrix}$

Max Violation → 0.537714

$\mu \to 10 \quad v \to \{\}$

$r \to \begin{pmatrix} 0 \\ 0 \\ 0 \\ 0 \\ 0 \\ 0 \end{pmatrix} \quad u \to \begin{pmatrix} 0 \\ 0 \\ 0 \\ 0 \\ 0 \\ 0 \end{pmatrix}$

$\phi \to 1.618 dw^2 + 10\text{Max}[0, -d]^2 + 10\text{Max}[0, -h]^2 + 10\text{Max}[0, -1 + 0.03236w]^2$
$\quad + 10\text{Max}[0, -w]^2 + 10\text{Max}\left[0, 1 - 0.0000231143\,(d+h)\,w^2\right]^2$
$\quad + 10\text{Max}\left[0, -1 + \dfrac{5.236hw + 1.618w^2}{2250}\right]^2$

Solution → $\{10., \{d \to 34.9998, h \to 10.0008, w \to 4.84836 \times 10^{-12}\}\}$

NonOptimum point: Iterations performed → 1

Point → $(34.9998 \quad 10.0008 \quad 4.84836 \times 10^{-12})$

Function values $\begin{pmatrix} f \\ g \\ h \end{pmatrix} \to$

$\begin{pmatrix} 1.33117 \times 10^{-21} \\ \{-1., -1., 1., -34.9998, -10.0008, -4.84836 \times 10^{-12}\} \\ \{\} \end{pmatrix}$

$\mu \to 10 \quad$ Max Violation → 1.

$v \to \{\} \quad u \to \begin{pmatrix} 0 \\ 0 \\ 20. \\ 0 \\ 0 \\ 0 \end{pmatrix}$

The function is allowed to iterate until the solution converges.

```
{sol, hist} = ALPF[f, cons, vars, pt, MaxIterations → 20, PrintLevel → 0,
    MaxUnconstrainedIterations → 100, PenaltyParameters → {1000, 2}];
```

Optimum point: Iterations performed → 14

Point → $(80.217 \quad 13.322 \quad 21.5132)$

9.2 Penalty Methods

Function values $\begin{pmatrix} f \\ g \\ h \end{pmatrix} \to$

$$\begin{pmatrix} 60069.9 \\ \{-0.000232023, -0.303831, -0.000657117, -80.217, -13.322, -21.5132\} \\ \{\} \end{pmatrix}$$

$\mu \to 2048000 \quad$ Max Violation $\to 0$

$v \to \{\} \quad u \to \begin{pmatrix} 15962.8 \\ 0 \\ 72726.1 \\ 0 \\ 0 \\ 0 \end{pmatrix}$

The same solution was obtained in Chapter 4 using KT conditions. The complete iteration history is shown in the following table. It is interesting to note that, initially when the μ value is relatively small, the method made the objective function very small, simply by reducing the building width w. After several iterations, and when μ became large, the method actually started paying attention to the constraints. Once that happened, the convergence to the optimum solution was quite rapid. It should be clear from the solution history that using a larger initial value of μ would make the process converge in fewer iterations.

TableForm[hist]

Point	f	maxViolation
40		
10		
20	25888.	0.537714
34.9804		
10.0783	1.32218×10^{-21}	1.
4.8333×10^{-12}		
34.9804		
10.0783	7.60247×10^{-19}	1.
1.15898×10^{-10}		
34.9804		
10.0783	3.74477×10^{-19}	1.
8.13413×10^{-11}		
34.9804		
10.0783	1.23883×10^{-19}	1.
4.67847×10^{-11}		
34.9804		
10.0783	8.46289×10^{-21}	1.
1.22281×10^{-11}		
34.9804		
10.0783	3.18685×10^{-15}	1.
7.50376×10^{-9}		

```
-0.000934418
38.2354          -0.230548      0.865235
12.3487
25.6515
14.3893          21159.8        0.52815
22.5793
50.8907
13.6888          40507.9        0.265658
22.18
61.1964
12.7303          50167.8        0.134231
22.5092
75.5497
12.448           60388.7        0
22.2266
80.5104
13.3331          60204.2        0
21.498
80.2311
13.3098          60116.8        0
21.5197
80.217
13.322           60069.9        0
21.5132
```

9.3 Linearization of a Nonlinear Problem

Since methods for linear programming are well established, a simple approach to solve a nonlinear constrained optimization problem is to linearize it and then use a linear programming method for the solution. The linearization can easily be achieved by approximating the objective and constraint functions around the current point using the Taylor series. This section concentrates on developing linear approximations of nonlinear problems. For two-variable problems, the original and the linearized problems are compared graphically. A complete algorithm for solving nonlinear programming problems based on linearization is presented in the next section.

For the discussion in this section, we assume that a problem is given in the following form:

Find $\mathbf{x}$ to   Minimize $f(\mathbf{x})$

Subject to $h_i(\mathbf{x}) = 0, i = 1, 2, \ldots, p$ and $g_i(\mathbf{x}) \leq 0, i = 1, 2, \ldots, m$.

The current point is denoted by $\mathbf{x}^k$, and the next point to be determined is denoted by $\mathbf{x}^{k+1}$.

9.3.1 Linearized Problem

Retaining only the first-order terms in the Taylor series, the objective function and the constraints are approximated around the current point, as follows:

$$f(\mathbf{x}^k + \mathbf{d}) \approx f(\mathbf{x}^k) + \nabla f(\mathbf{x}^k)^T \mathbf{d} \quad \text{where} \quad \mathbf{d} \equiv \mathbf{x}^{k+1} - \mathbf{x}^k$$
$$h_i(\mathbf{x}^k + \mathbf{d}) \approx h_i(\mathbf{x}^k) + \nabla h_i(\mathbf{x}^k)^T \mathbf{d} = 0 \quad i = 1, 2, \ldots, p$$
$$g_i(\mathbf{x}^k + \mathbf{d}) \approx g_i(\mathbf{x}^k) + \nabla g_i(\mathbf{x}^k)^T \mathbf{d} \leq 0 \quad i = 1, 2, \ldots, m$$

The complete linearized problem can then be written as follows:

Minimize $f(\mathbf{x}^k) + \nabla f(\mathbf{x}^k)^T \mathbf{d}$

Subject to $\begin{pmatrix} h_i(\mathbf{x}^k) + \nabla h_i(\mathbf{x}^k)^T \mathbf{d} = 0 & i = 1, 2, \ldots, p \\ g_i(\mathbf{x}^k) + \nabla g_i(\mathbf{x}^k)^T \mathbf{d} \leq 0 & i = 1, 2, \ldots, m \end{pmatrix}$

Note that the constant term in the linearized objective function ($f(\mathbf{x}^k)$) can be dropped since it does not influence the minimum. Also, the variables in the linearized problem are the changes in the values of the optimization variables from the current point.

9.3.2 Graphical Comparison of Linearized and Original Problems

The linearized problem is written in terms of variables identified as $\mathbf{d}$ that represent changes from the current point. If we want to graphically compare the original and the linearized problem, we must express the linearized problem in terms of original variables using

$$\mathbf{x}^{k+1} = \mathbf{x}^k + \mathbf{d}$$

Thus, a linearized problem can be written in terms of original variables by setting $d_i = x_i - x_i^k, i = 1, \ldots, n$.

A function called Linearize, included in the OptimizationToolbox package, creates a linear approximation for a given set of functions. It returns linearized functions both in terms of $\mathbf{d}$ variables and the original variables.

```
Needs["OptimizationToolbox`ConstrainedNLP`"];
?Linearize
```

Linearize[f,fLabels,vars,pt,dvars], Linearize function f(vars) at given
   point (pt). fLabels are used to identify the linearized functions. The
   results are returned in terms of dvars and original vars.

Example 9.10 Linearize the following optimization problem around the point $(1, 2)$. Compare graphically the linearized problem with the original problem.

Minimize $f = \frac{1}{2}x_1^2 + x_1 x_2^2$

Subject to $\begin{pmatrix} h = x_1 x_2^2 - 1 = 0 \\ g = -2x_1^2 + x_2/3 \leq 0 \end{pmatrix}$

```
f = x₁²/2 + x₁x₂²; h = x₁x₂² - 1; g = -2x₁² + x₂/3;
vars = {x₁,x₂}; dvars = {d₁,d₂}; pt = {1,2};

{lpd,lpx} = Linearize[{f,g,h},{"f","g","h"},vars,pt, dvars];
***** Gradient vectors
```

$\nabla f \to \begin{pmatrix} x_1 + x_2^2 \\ 2x_1 x_2 \end{pmatrix} \quad \nabla g \to \begin{pmatrix} -4x_1 \\ \frac{1}{3} \end{pmatrix} \quad \nabla h \to \begin{pmatrix} x_2^2 \\ 2x_1 x_2 \end{pmatrix}$

```
***** Function values at → {1,2}
```

$f \to \frac{9}{2} \quad g \to -\frac{4}{3} \quad h \to 3$

```
***** Gradient vectors at → {1,2}
```

$\nabla f \to \begin{pmatrix} 5 \\ 4 \end{pmatrix} \quad \nabla g \to \begin{pmatrix} -4 \\ \frac{1}{3} \end{pmatrix} \quad \nabla h \to \begin{pmatrix} 4 \\ 4 \end{pmatrix}$

```
***** Linearized functions at → {1,2}
```

$\tilde{f} \to \frac{9}{2} + 5d_1 + 4d_2 \quad \tilde{g} \to -\frac{4}{3} - 4d_1 + \frac{d_2}{3} \quad \tilde{h} \to 3 + 4d_1 + 4d_2$

```
***** In terms of original variables → {d₁ → -1 + x₁, d₂ → -2 + x₂}
```

$\tilde{f} \to -\frac{17}{2} + 5x_1 + 4x_2 \quad \tilde{g} \to 2 - 4x_1 + \frac{x_2}{3} \quad \tilde{h} \to -9 + 4x_1 + 4x_2$

Thus, the linearized problems at $(1, 2)$, in terms of increments and in terms of original variables, are as follows:

Find (d_1, d_2) to minimize $\frac{9}{2} + 5d_1 + 4d_2$

subject to $-\frac{4}{3} - 4d_1 + \frac{d_2}{3} \leq 0$ and $3 + 4d_1 + 4d_2 = 0$.

Find (x_1, x_2) to minimize $-\frac{17}{2} + 5x_1 + 4x_2$

subject to $2 - 4x_1 + \frac{x_2}{3} \leq 0$ and $-9 + 4x_1 + 4x_2 = 0$.

```
{lpf, lpg, lph} = lpx
```

$\left\{-\frac{17}{2} + 5x_1 + 4x_2, \ 2 - 4x_1 + \frac{x_2}{3}, \ -9 + 4x_1 + 4x_2\right\}$

The plots of the original and the linearized problem are shown in Figure 9.7. Note that the solution of the linearized problem is well defined. Furthermore,

9.3 Linearization of a Nonlinear Problem

moving in the direction of the solution of the linearized problem takes us closer to the solution of the original problem. This is easier to see from Figure 9.8 in which the two graphs are superimposed on each other.

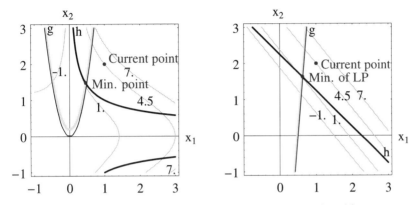

FIGURE 9.7 Plots of the original and the linearized problems.

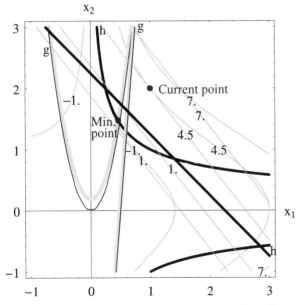

FIGURE 9.8 A plot of the linearized problem superimposed on the original problem.

Example 9.11 *Unbounded linearized problem*

Minimize $f = (x_1 - 4)^2 + (x_2 - 2)^2$

Subject to $\begin{pmatrix} g_1 = 5x_1 + 6x_2 - 15 \leq 0 \\ g_2 = -2x_1^2 - 3x_2^2 + 21 \leq 0 \end{pmatrix}$

```
f = (x₁ - 4)⁴ + (x₂ - 2)²;
g = {5x₁ + 6x₂ - 15, -2x₁² - 3x₂² + 21};
vars = {x₁, x₂}; dvars = {d₁, d₂}; pt = {3, 2};

{lpd, lpx} = Linearize[{f, g}, {"f", "g1", "g2"}, vars, pt, dvars];
```

***** Gradient vectors

$\nabla f \to \begin{pmatrix} -256 + 192x_1 - 48x_1^2 + 4x_1^3 \\ -4 + 2x_2 \end{pmatrix} \quad \nabla g1 \to \begin{pmatrix} 5 \\ 6 \end{pmatrix} \quad \nabla g2 \to \begin{pmatrix} -4x_1 \\ -6x_2 \end{pmatrix}$

***** Function values at → {3, 2}

f → 1   g1 → 12   g2 → -9

***** Gradient vectors at → {3, 2}

$\nabla f \to \begin{pmatrix} -4 \\ 0 \end{pmatrix} \quad \nabla g1 \to \begin{pmatrix} 5 \\ 6 \end{pmatrix} \quad \nabla g2 \to \begin{pmatrix} -12 \\ -12 \end{pmatrix}$

***** Linearized functions at → {3, 2}

$\tilde{f} \to 1 - 4d_1 \quad \tilde{g1} \to 12 + 5d_1 + 6d_2 \quad \tilde{g2} \to -9 - 12d_1 - 12d_2$

***** In terms of original variables → {d₁ → -3 + x₁, d₂ → -2 + x₂}

$\tilde{f} \to 13 - 4x_1 \quad \tilde{g1} \to -15 + 5x_1 + 6x_2 \quad \tilde{g2} \to 51 - 12x_1 - 12x_2$

Thus, the linearized problems at (3, 2), in terms of increments and in terms of original variables, are as follows:

Find (d_1, d_2) to minimize $1 - 4d_1$

subject to $12 + 5d_1 + 6d_2 \leq 0$ and $-9 - 12d_1 - 12d_2 \leq 0$.

Find (x_1, x_2) to minimize $13 - 4x_1$

subject to $-15 + 5x_1 + 6x_2 \leq 0$ and $51 - 12x_1 - 12x_2 \leq 0$.

```
{lpf, lpg} = {First[lpx], Drop[lpx, 1]}
```

$\{13 - 4x_1, \{-15 + 5x_1 + 6x_2, 51 - 12x_1 - 12x_2\}\}$

The plots of the original and the linearized problem are shown in Figure 9.9 individually and are superimposed on each other in Figure 9.10. For this example, the linearized problem is unbounded and therefore cannot be used directly to determine the direction of movement toward solving the original nonlinear problem.

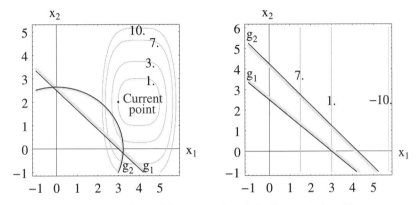

FIGURE 9.9 Plots of the original and the linearized problems.

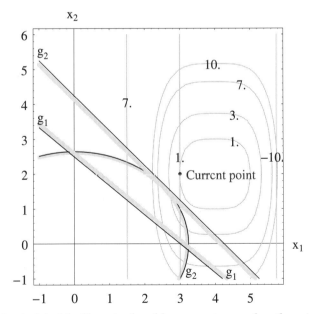

FIGURE 9.10 A plot of the linearized problem superimposed on the original problem.

9.4 Sequential Linear Programming—SLP

In order to transform the linearization idea into an effective solution algorithm for constrained nonlinear problems, it is necessary to define suitable limits on the allowable change from the current point, $\mathbf{x}^k$. There are two reasons for this. The first is that if we move too far away from the current point, the linear approximation based on the Taylor series breaks down. The second is to make the LP subproblem bounded. Since the LP solution is always on the boundary of the linearized constrained region, it is important to make this region bounded. Otherwise, the LP subproblem will not have a solution. One such situation was illustrated in example 9.11, where the nonlinear problem had a well-defined minimum but the linearized problem was unbounded.

In the Sequential Linear Programming algorithm (SLP), simple move limit constraints are added to the linearized problem. These constraints are expressed as some percentage (e.g., 50%) of the current point. If the LP solution gives a large value of $\mathbf{d}$, this percentage is reduced. If there is no feasible solution, the percentage is increased. The complete algorithm is stated as follows:

SLP Algorithm

1. Set $k = 0$. Choose a starting point $\mathbf{x}^0$ and a move limit parameter β (say $\beta = 0.5$).

2. Set up and solve the LP subproblem for $\mathbf{d}$.

    Minimize $\nabla f(\mathbf{x}^k)^T \mathbf{d}$

    Subject to $\begin{pmatrix} h_i(\mathbf{x}^k) + \nabla h_i(\mathbf{x}^k)^T \mathbf{d} = 0 & i = 1, 2, \ldots, p \\ g_i(\mathbf{x}^k) + \nabla g_i(\mathbf{x}^k)^T \mathbf{d} \leq 0 & i = 1, 2, \ldots, m \end{pmatrix}$

    Move limits: $-\beta \mathbf{x}^k \leq \mathbf{d} \leq \beta \mathbf{x}^k$

    Reduce β if the solution results in large values of $\mathbf{d}$. Increase β if there is no feasible solution to the problem. Note that all variables in this problem (d_i's) are unrestricted in sign. This situation must be taken into account when solving this LP problem, say by using the simplex method.

3. Set the new point as

$$\mathbf{x}^{k+1} = \mathbf{x}^k + \mathbf{d}$$

4. If $||\mathbf{d}|| \leq$ tol (a small number), then stop. Otherwise, set $k = k + 1$ and go to step 2.

The algorithm is implemented in a *Mathematica* function SLP.

9.4 Sequential Linear Programming—SLP

```
Needs["OptimizationToolbox`ConstrainedNLP`"];
?SLP
```

SLP[f, con, vars, x0, opts]. Computes minimum of f(vars) subjected
  to specified 'con'straints starting from x0 using SLP method.
  LP subproblem is solved using built-in *Mathematica* function
  ConstrainedMin. See Options[SLP] to see a list of options for the
  function. The function returns {x, hist}. 'x' is either the optimum
  point or the next point after MaxIterations. 'hist' contains history
  of values tried at different iterations.

OptionsUsage[SLP]

{PrintLevel → 1, MaxIterations → 10, ConvergenceTolerance → 0.001,
 DirVarName → d, MoveLimit → 0.5}

PrintLevel is an option for most functions in the OptimizationToolbox.
  It is specified as an integer. The value of the integer indicates
  how much intermediate information is to be printed. A PrintLevel→ 0
  suppresses all printing. Default for most functions is set to 1 in
  which case they print only the initial problem setup. Higher integers
  print more intermediate results.

MaxIterations is an option for several optimization methods. It
  specifies maximum number of iterations allowed.

ConvergenceTolerance is an option for most optimization methods. Most
  methods require only a single zero tolerance value. Some interior
  point methods require a list of convergence tolerance values.

DirVarName is an option for SLP. It defines the variable name to be used
  for defining direction. Default DirVarName→ d.

MoveLimit is an option for SLP. It defines the MoveLimit parameter β.
  Default MoveLimit→ 0.5.

Example 9.12 Starting from the given point, use the SLP method to find the minimum of the following problem.

```
vars = {x1, x2};
f = (x1 - 2)^2 + (x2 - 1)^2;
cons = { x1^2/4 + x2^2 - 1 ≤ 0, x1 - 2 x2 + 1 == 0 };
x0 = {2, 2};
```

Using 70% move limits, all calculations for the first iteration are as follows:

```
SLP[f, cons, vars, x0, MoveLimit → 0.7, PrintLevel → 2, MaxIterations → 1];
```

Minimize $f \to (-2 + x_1)^2 + (-1 + x_2)^2$

$\nabla f \to (-4 + 2x_1 \quad -2 + 2x_2)$

LE Constraints: $(g1) \to \left(-1 + \frac{x_1^2}{4} + x_2^2 \le 0\right)$

$$(\nabla g1) \to \begin{pmatrix} \frac{x_1}{2} & 2x_2 \end{pmatrix}$$

EQ Constraints: $(h1) \to (1 + x_1 - 2x_2 == 0)$

$(\nabla h1) \to (1 \quad -2)$

***** Iteration 1 *****
Function values at $\to \{2, 2\}$
$f \to 1 \quad g1 \to 4 \quad h1 \to -1$
Gradient vectors at $\to \{2, 2\}$

$$\nabla f \to \begin{pmatrix} 0 \\ 2 \end{pmatrix} \quad \nabla g1 \to \begin{pmatrix} 1 \\ 4 \end{pmatrix} \quad \nabla h1 \to \begin{pmatrix} 1 \\ -2 \end{pmatrix}$$

Linearized functions at $\to \{2, 2\}$
$\tilde{f} \to 1 + 2d_2$
$\tilde{g1} \to 4 + d_1 + 4d_2$
$\tilde{h1} \to -1 + d_1 - 2d_2$
-------Complete LP subproblem-------
Minimize $2d_2$

$$\text{Subject to} \to \begin{cases} 4 + d_1 + 4d_2 \le 0 \\ -1.4 + d_1 \le 0 \\ -1.4 + d_2 \le 0 \\ -1.4 - d_1 \le 0 \\ -1.4 - d_2 \le 0 \\ -1 + d_1 - 2d_2 == 0 \end{cases}$$

$$\text{Direction} \to \begin{pmatrix} -1.4 \\ -1.2 \end{pmatrix} \quad \text{New point} \to \begin{pmatrix} 0.6 \\ 0.8 \end{pmatrix}$$

New Point (Non-Optimum): $\{x_1 \to 0.6, x_2 \to 0.8\}$ after 1 iterations

The procedure is allowed to run until convergence. The complete iteration history is shown in the table.

{sol, hist} = SLP[f, cons, vars, x0, MoveLimit $\to$ 0.7, MaxIterations $\to$ 20];
Optimum: $\{0.822876, 0.911438\}$ after 4 iterations

TableForm[hist]

x	$\|\|d\|\|$
2	
2	--
0.6	
0.8	1.84391
0.845455	
0.922727	0.274427
0.823065	
0.911533	0.0250321
0.822876	
0.911438	0.000211802

The graph in Figure 9.11 confirms the solution and shows the path taken by the SLP algorithm.

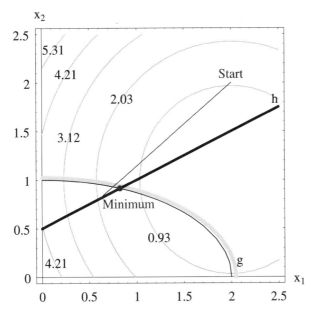

FIGURE 9.11 A graphical solution showing the search path.

Example 9.13 Starting from the given point, use the SLP method to find the minimum of the following problem:

```
vars = {x₁, x₂};
f = (1 - x₁)² - 10 (x₂ - x₁²)² + x₁² - 2x₁x₂ + Exp[-x₁ - x₂];
cons = {x₁² + x₂² - 16 ≤ 0, (x₂ - x₁)² + x₁ - 6 ≤ 0, -x₁ - x₂ + 2 ≤ 0};
x0 = {5.0, 5.0};
```

Using 80% move limits, all calculations for the first iteration are as follows:

SLP[f, cons, vars, x0, MoveLimit → 0.8, PrintLevel → 2, MaxIterations → 1];

Minimize $f \to E^{-x_1-x_2} + (1-x_1)^2 + x_1^2 - 2x_1x_2 - 10\left(-x_1^2 + x_2\right)^2$

$\nabla f \to \left(-2 - E^{-x_1-x_2} + 4x_1 - 40x_1^3 - 2x_2 + 40x_1x_2 \quad -E^{-x_1-x_2} - 2x_1 + 20x_1^2 - 20x_2\right)$

LE Constraints: $\begin{pmatrix} g1 \\ g2 \\ g3 \end{pmatrix} \to \begin{pmatrix} -16 + x_1^2 + x_2^2 \le 0 \\ -6 + x_1 + (-x_1 + x_2)^2 \le 0 \\ 2 - x_1 - x_2 \le 0 \end{pmatrix}$

$$\begin{pmatrix} \nabla g1 \\ \nabla g2 \\ \nabla g3 \end{pmatrix} \to \begin{pmatrix} 2x_1 & 2x_2 \\ 1 + 2x_1 - 2x_2 & -2x_1 + 2x_2 \\ -1 & -1 \end{pmatrix}$$

***** Iteration 1 *****

Function values at $\to \{5., 5.\}$

$f \to -4009.$ $g1 \to 34.$ $g2 \to -1.$

$g3 \to -8.$

Gradient vectors at $\to \{5., 5.\}$

$\nabla f \to \begin{pmatrix} -3992. \\ 390. \end{pmatrix}$ $\nabla g1 \to \begin{pmatrix} 10. \\ 10. \end{pmatrix}$ $\nabla g2 \to \begin{pmatrix} 1. \\ 0. \end{pmatrix}$

$\nabla g3 \to \begin{pmatrix} -1 \\ -1 \end{pmatrix}$

Linearized functions at $\to \{5., 5.\}$

$\tilde{f} \to -4009. - 3992.d_1 + 390.d_2$

$\tilde{g1} \to 34. + 10.d_1 + 10.d_2$

$\tilde{g2} \to -1. + 1.d_1 + 0.d_2$

$\tilde{g3} \to -8. - d_1 - d_2$

-------Complete LP subproblem-------

Minimize $-3992.d_1 + 390.d_2$

Subject to $\to \begin{pmatrix} 34. + 10.d_1 + 10.d_2 \leq 0 \\ -1. + 1.d_1 + 0.d_2 \leq 0 \\ -8. - d_1 - d_2 \leq 0 \\ -4. + d_1 \leq 0 \\ -4. + d_2 \leq 0 \\ -4. - d_1 \leq 0 \\ -4. - d_2 \leq 0 \end{pmatrix}$

Direction $\to \begin{pmatrix} 0.6 \\ -4. \end{pmatrix}$ New point $\to \begin{pmatrix} 5.6 \\ 1. \end{pmatrix}$

New Point (Non-Optimum): $\{x_1 \to 5.6, x_2 \to 1.\}$ after 1 iterations

The procedure is allowed to run until convergence. The complete iteration history is shown in the table:

{sol, hist} = SLP[f, cons, vars, x0, MoveLimit → 0.8, MaxIterations → 20];

Optimum: $\{x_1 \to 3.50563, x_2 \to 1.92628\}$ after 6 iterations

9.4 Sequential Linear Programming—SLP

```
TableForm[hist]
```

x	‖d‖
5. 5.	--
5.6 1.	4.04475
4.04433 1.53176	1.64404
3.58868 1.85254	0.557244
3.50886 1.92335	0.106703
3.50564 1.92627	0.00434652
3.50563 1.92628	7.24474×10^{-6}

The graph in Figure 9.12 confirms the solution and shows the path taken by the SLP algorithm. Note that the objective function contour values are divided by 10,000 for ease of labelling these contours.

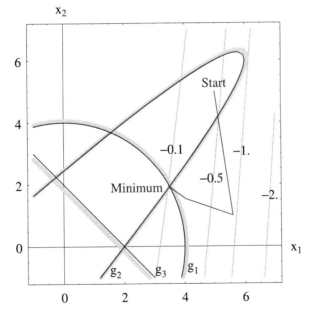

FIGURE 9.12 A graphical solution showing the search path.

9.5 Basic Sequential Quadratic Programming—SQP

The biggest drawback of the Sequential Linear Programming algorithm is the arbitrary nature of the move limits. A small value of move limit parameter β may make the convergence toward the optimum very slow. A large value may cause problems because of poor approximation of the linearized constraints.

Instead of explicitly defining move limits, it is possible to indirectly put restriction on $\mathbf{d}$ by adding the following quadratic term to the linearized objective function:

Quadratic term in the direction-finding subproblem: $\frac{1}{2}\mathbf{d}^T\mathbf{d}$

Since this term is always positive and we are trying to minimize the objective function, $\mathbf{d}$ cannot become very large. The objective function is now a quadratic function. The constraints are still linear. Thus, the direction-finding subproblem is now a quadratic programming (QP) problem, as follows:

Minimize $\nabla f(\mathbf{x}^k)^T\mathbf{d} + \frac{1}{2}\mathbf{d}^T\mathbf{d}$

Subject to $\begin{pmatrix} h_i(\mathbf{x}^k) + \nabla h_i(\mathbf{x}^k)^T\mathbf{d} = 0 & i = 1, 2, \ldots, p \\ g_i(\mathbf{x}^k) + \nabla g_i(\mathbf{x}^k)^T\mathbf{d} \leq 0 & i = 1, 2, \ldots, m \end{pmatrix}$

The optimum of a QP can lie either inside the feasible domain or on the constraint boundary. Thus, a bounded feasible region is not required; therefore, there is no need to define arbitrary move limits. Furthermore, the QP subproblem is convex and therefore, we can use any of the first three methods discussed in Chapter 8 to solve this QP. As discussed in the following subsection, we can also form its dual and solve it using the ActiveSetDualQP presented in Chapter 8.

9.5.1 Dual QP for Direction Finding

Computationally, it may be advantageous to write a dual for the direction-finding problem and actually solve the dual problem to determine the direction. The dual QP can be written by following the standard steps discussed in Chapters 4 and 8.

9.5 Basic Sequential Quadratic Programming—SQP

The Lagrangian function is

$$L(\mathbf{d}, \mathbf{u}, \mathbf{v}) = \nabla f(\mathbf{x}^k)^T \mathbf{d} + \frac{1}{2}\mathbf{d}^T\mathbf{d} + \sum_{i=1}^{m} u_i[g_i(\mathbf{x}^k) + \nabla g_i(\mathbf{x}^k)^T \mathbf{d}]$$

$$+ \sum_{i=1}^{p} v_i[h_i(\mathbf{x}^k) + \nabla h_i(\mathbf{x}^k)^T \mathbf{d}]$$

where the vectors of Lagrange multipliers for inequality and equality constraints are

$$\mathbf{u} = [u_1, \ldots, u_m]^T \geq 0 \quad \mathbf{v} = [v_1, \ldots, v_p]^T$$

Defining rows of matrices $\mathbf{J}_g$ ($m \times n$) and $\mathbf{J}_h$ ($p \times n$) to consist of gradients of constraint functions, the Lagrangian can be written in matrix form, as follows:

$$\mathbf{J}_g(\mathbf{x}^k) = \begin{pmatrix} \nabla g_1(\mathbf{x}^k)^T \\ \nabla g_2(\mathbf{x}^k)^T \\ \vdots \\ \nabla g_m(\mathbf{x}^k)^T \end{pmatrix} \quad \text{and} \quad \mathbf{J}_h(\mathbf{x}^k) = \begin{pmatrix} \nabla h_1(\mathbf{x}^k)^T \\ \nabla h_2(\mathbf{x}^k)^T \\ \vdots \\ \nabla h_p(\mathbf{x}^k)^T \end{pmatrix}$$

or

$$L(\mathbf{d}, \mathbf{u}, \mathbf{v}) = \nabla f(\mathbf{x}^k)^T \mathbf{d} + \frac{1}{2}\mathbf{d}^T\mathbf{d} + \mathbf{u}^T \mathbf{g}(\mathbf{x}^k) + \mathbf{u}^T \mathbf{J}_g(\mathbf{x}^k)\mathbf{d}$$

$$+ \mathbf{v}^T \mathbf{h}(\mathbf{x}^k) + \mathbf{v}^T \mathbf{J}_h(\mathbf{x}^k)\mathbf{d}$$

The dual function is obtained by minimizing with respect to $\mathbf{d}$.

$$M(\mathbf{u}, \mathbf{v}) = \underset{\mathbf{d}}{\text{Min}}[\nabla f(\mathbf{x}^k)^T \mathbf{d} + \frac{1}{2}\mathbf{d}^T\mathbf{d} + \mathbf{u}^T \mathbf{g}(\mathbf{x}^k) + \mathbf{u}^T \mathbf{J}_g(\mathbf{x}^k)\mathbf{d}$$

$$+ \mathbf{v}^T \mathbf{h}(\mathbf{x}^k) + \mathbf{v}^T \mathbf{J}_h(\mathbf{x}^k)\mathbf{d}]$$

Differentiating with respect to $\mathbf{d}$ (necessary conditions for the minimum), we get

$$\nabla f(\mathbf{x}^k) + \mathbf{d} + \mathbf{J}_g(\mathbf{x}^k)^T \mathbf{u} + \mathbf{J}_h(\mathbf{x}^k)^T \mathbf{v} = 0$$

giving

$$\mathbf{d} = -\nabla f(\mathbf{x}^k) - \mathbf{J}_g(\mathbf{x}^k)^T \mathbf{u} - \mathbf{J}_h(\mathbf{x}^k)^T \mathbf{v}$$

Taking the transpose and multiplying by $\mathbf{d}$, we see that

$$-\mathbf{d}^T\mathbf{d} = \nabla f(\mathbf{x}^k)^T \mathbf{d} + \mathbf{u}^T \mathbf{J}_g(\mathbf{x}^k)\mathbf{d} + \mathbf{v}^T \mathbf{J}_h(\mathbf{x}^k)\mathbf{d}$$

Thus, the dual function for the problem is

$$M(\mathbf{u}, \mathbf{v}) = -\frac{1}{2}\mathbf{d}^T\mathbf{d} + \mathbf{u}^T\mathbf{g}(\mathbf{x}^k) + \mathbf{v}^T\mathbf{h}(\mathbf{x}^k)$$

The dual direction-funding problem is therefore as follows:

Maximize $-\frac{1}{2}\mathbf{d}^T\mathbf{d} + \mathbf{u}^T\mathbf{g}(\mathbf{x}^k) + \mathbf{v}^T\mathbf{h}(\mathbf{x}^k)$

Subject to $u_i \geq 0, i = 1, \ldots m$

The problem can be expressed in standard QP form as follows:
Define

Vector of Lagrange multipliers: $\boldsymbol{\lambda} = \begin{pmatrix} \mathbf{u} \\ \mathbf{v} \end{pmatrix}$

Vector of constraint values: $\mathbf{r}(\mathbf{x}^k) = \begin{pmatrix} \mathbf{g}(\mathbf{x}^k) \\ \mathbf{h}(\mathbf{x}^k) \end{pmatrix}$

Jacobian of constraint vector: $\mathbf{J}_r(\mathbf{x}^k) = \begin{pmatrix} \mathbf{J}_g(\mathbf{x}^k) \\ \mathbf{J}_h(\mathbf{x}^k) \end{pmatrix}$

Therefore,

$$\mathbf{d} = -\nabla f(\mathbf{x}^k) - \mathbf{J}_r(\mathbf{x}^k)^T\boldsymbol{\lambda} \quad \text{and} \quad \mathbf{d}^T = -\nabla f(\mathbf{x}^k)^T - \boldsymbol{\lambda}^T\mathbf{J}_r(\mathbf{x}^k)$$

and thus,

$$\mathbf{d}^T\mathbf{d} = -\nabla f(\mathbf{x}^k)^T \left[-\nabla f(\mathbf{x}^k) - \mathbf{J}_r(\mathbf{x}^k)^T\boldsymbol{\lambda}\right]$$
$$- \boldsymbol{\lambda}^T\mathbf{J}_r(\mathbf{x}^k)\left[-\nabla f(\mathbf{x}^k) - \mathbf{J}_r(\mathbf{x}^k)^T\boldsymbol{\lambda}\right]$$

giving

$$\mathbf{d}^T\mathbf{d} = \nabla f(\mathbf{x}^k)^T\nabla f(\mathbf{x}^k) + 2\nabla f(\mathbf{x}^k)^T\mathbf{J}_r(\mathbf{x}^k)^T\boldsymbol{\lambda} + \boldsymbol{\lambda}^T\mathbf{J}_r(\mathbf{x}^k)\mathbf{J}_r(\mathbf{x}^k)^T\boldsymbol{\lambda}$$

The objective function of the dual QP problem can now be written as follows:

$$-\frac{1}{2}\mathbf{d}^T\mathbf{d} + \mathbf{u}^T\mathbf{g}(\mathbf{x}^k) + \mathbf{v}^T\mathbf{h}(\mathbf{x}^k) = -\left[\frac{1}{2}\nabla f(\mathbf{x}^k)^T\nabla f(\mathbf{x}^k)\right.$$
$$\left. + \left\{\nabla f\left(\mathbf{x}^k\right)^T\mathbf{J}_r(\mathbf{x}^k)^T - \mathbf{r}(\mathbf{x}^k)^T\right\}\boldsymbol{\lambda} + \frac{1}{2}\boldsymbol{\lambda}^T\mathbf{J}_r(\mathbf{x}^k)\mathbf{J}_r(\mathbf{x}^k)^T\boldsymbol{\lambda}\right]$$

Changing to a minimization problem and dropping the first term (since it is a constant), we get the following QP subproblem:

9.5 Basic Sequential Quadratic Programming—SQP

Minimize $\mathbf{c}^T\boldsymbol{\lambda} + \frac{1}{2}\boldsymbol{\lambda}^T\mathbf{Q}\boldsymbol{\lambda}$

Subject to $\begin{pmatrix} \lambda_i \geq 0, i = 1, \ldots m \\ \lambda_i \text{ unrestricted}, i = m+1, \ldots, p \end{pmatrix}$

where

$$\mathbf{c} = \mathbf{J}_r(\mathbf{x}^k)\nabla f(\mathbf{x}^k) - \mathbf{r}(\mathbf{x}^k) \quad \text{and} \quad \mathbf{Q} = \mathbf{J}_r(\mathbf{x}^k)\mathbf{J}_r(\mathbf{x}^k)^T$$

This problem can be solved efficiently by using the active set dual method presented in Chapter 8. Once Lagrange multiplier vector $\boldsymbol{\lambda}$ is known, the direction can be computed from

$$\mathbf{d} = -\nabla f(\mathbf{x}^k) - \mathbf{J}_r(\mathbf{x}^k)^T\boldsymbol{\lambda}$$

Example 9.14 For the following minimization problem, construct the direction-finding subproblem at the given point. Find the direction using both the primal form and the dual form.

```
vars = {x₁, x₂}; (xk = {x₁ → 2, x₂ → 2};)
```

$f = \frac{x_1^2}{2} + x_1 x_2^2;$

$h = x_1 x_2^2 - 1 == 0;$

$g = -2x_1^2 + \frac{x_2}{3} \leq 0;$

The function values and the gradients of the functions at the given point are as follows:

```
∇f = Grad[f, vars]; ∇fxk = ∇f/.xk; fx = f/.xk;
PrintLabelledList[{fx, ∇f, ∇fxk}, {"f(xk)", "∇f", "∇f(xk)"}]
```

$f(xk) \to 10 \quad \nabla f \to \begin{pmatrix} x_1 + x_2^2 \\ 2x_1 x_2 \end{pmatrix} \quad \nabla f(xk) \to \begin{pmatrix} 6 \\ 8 \end{pmatrix}$

```
∇h = Grad[h〚1〛, vars]; ∇hxk = ∇h/.xk; hx = h〚1〛/.xk;
PrintLabelledList[{hx, ∇h, ∇hxk}, {"h(xk)", "∇h", "∇h(xk)"}]
```

$h(xk) \to 7 \quad \nabla h \to \begin{pmatrix} x_2^2 \\ 2x_1 x_2 \end{pmatrix} \quad \nabla h(xk) \to \begin{pmatrix} 4 \\ 8 \end{pmatrix}$

```
∇g = Grad[g〚1〛, vars]; ∇gxk = ∇g/.xk; gx = g〚1〛/.xk;
PrintLabelledList[{gx, ∇g, ∇gxk}, {"g(xk)", "∇g", "∇g(xk)"}]
```

$g(xk) \to -\frac{22}{3} \quad \nabla g \to \begin{pmatrix} -4x_1 \\ \frac{1}{3} \end{pmatrix} \quad \nabla g(xk) \to \begin{pmatrix} -8 \\ \frac{1}{3} \end{pmatrix}$

(a) Direction using primal QP

Using the above gradient and function values, we get

$$\mathbf{d} = (d_1, d_2)^T$$

$$\nabla f(\mathbf{x}^k)^T \mathbf{d} + \frac{1}{2}\mathbf{d}^T\mathbf{d} = 6d_1 + 8d_2 + \frac{1}{2}(d_1^2 + d_2^2)$$

$$h(\mathbf{x}^k) + \nabla h(\mathbf{x}^k)^T \mathbf{d} = 7 + 4d_1 + 8d_2 = 0$$

$$g(\mathbf{x}^k) + \nabla g(\mathbf{x}^k)^T \mathbf{d} = -\frac{22}{3} - 8d_1 + \frac{1}{3}d_2 \le 0$$

Thus, the primal QP subproblem for direction finding is as follows:

Minimize $6d_1 + 8d_2 + \frac{1}{2}(d_1^2 + d_2^2)$

Subject to $\begin{pmatrix} 7 + 4d_1 + 8d_2 = 0 \\ -\frac{22}{3} - 8d_1 + \frac{1}{3}d_2 \le 0 \end{pmatrix}$

We can obtain the solution of this problem using the active set QP method discussed in Chapter 8 as follows:

```
ActiveSetQP[6d₁ + 8d₂ + 0.5(d₁² + d₂²),
{7 + 4d₁ + 8d₂ == 0, -22/3 - 8d₁ + d₂/3 ≤ 0}, {d₁, d₂}, PrintLevel → 2];
```

Minimize → $6d_1 + 8d_2 + 0.5\left(d_1^2 + d_2^2\right)$

LE Constraints → $\left(-8d_1 + \frac{d_2}{3} \le \frac{22}{3}\right)$

EQ Constraints → $\left(-4d_1 - 8d_2 == 7\right)$

$c \to \begin{pmatrix} 6 \\ 8 \end{pmatrix} \quad Q \to \begin{pmatrix} 1. & 0 \\ 0 & 1. \end{pmatrix}$

LE constraints: $A \to \left(-8 \quad \frac{1}{3}\right) \quad b \to \left(\frac{22}{3}\right)$

EQ constraints: $A \to \left(-4 \quad -8\right) \quad b \to (7)$

Convex problem. Principal minors of $Q \to \{1., 1.\}$

Phase I problem: Art. Obj. coefficients → $\{0, 0, 0, 0, 1, 1, 1\}$

Constraints expressed as GE type

Constraint coefficients → $\begin{pmatrix} -4 & -8 & 4 & 8 & 1 & 0 & 0 \\ 4 & 8 & -4 & -8 & 0 & 1 & 0 \\ 8 & -\frac{1}{3} & -8 & \frac{1}{3} & 0 & 0 & 1 \end{pmatrix}$

Constraint rhs → $\left(7 \quad -7 \quad -\frac{22}{3}\right)$

Phase I-LP solution → $\left\{0, 0, 0, \frac{7}{8}, 0, 0, 0\right\}$

**** Iteration 1 ****

Current point → $\left(0 \quad -\frac{7}{8}\right)$

Active set → {{}}

$A \to \begin{pmatrix} -4 & -8 \end{pmatrix}$ $gk \to \begin{pmatrix} 6. \\ 7.125 \end{pmatrix}$

LE Constraint values → $\left(-\frac{61}{8} \right)$

EQ Constraint values → (0)

Equations: lhs → $\begin{pmatrix} 1. & 0 & -4 \\ 0 & 1. & -8 \\ -4 & -8 & 0 \end{pmatrix}$ rhs → $\begin{pmatrix} -6. \\ -7.125 \\ 0 \end{pmatrix}$

Solution: dk → $\begin{pmatrix} -1.95 \\ 0.975 \end{pmatrix}$ uk → {} vk → (1.0125)

Inactive set → (1)

bi-Ai.x (i∈ Inactive) → $\left(\frac{61}{8} \right)$

Ai.d (i∈ Inactive) → (15.925)

bi-Ai.x/Ai.d → (0.478807)

Step length, α → 0.478807 New active set → (1)

New point → $(-0.933673 \quad -0.408163)$

**** Iteration 2 ****

Current point → $(-0.933673 \quad -0.408163)$

Active set → (1)

$A \to \begin{pmatrix} -4 & -8 \\ -8 & \frac{1}{3} \end{pmatrix}$ $gk \to \begin{pmatrix} 5.06633 \\ 7.59184 \end{pmatrix}$

LE Constraint values → $(0.)$

EQ Constraint values → $(0.)$

Equations: lhs → $\begin{pmatrix} 1. & 0 & -4 & -8 \\ 0 & 1. & -8 & \frac{1}{3} \\ -4 & -8 & 0 & 0 \\ -8 & \frac{1}{3} & 0 & 0 \end{pmatrix}$ rhs → $\begin{pmatrix} -5.06633 \\ -7.59184 \\ 0 \\ 0 \end{pmatrix}$

Solution: dk → $\begin{pmatrix} 0. \\ 0. \end{pmatrix}$ uk → (0.15556) vk → (0.955461)

***** Optimum solution after 2 iterations *****

Solution → $(d_1 \to -0.933673 \quad d_2 \to -0.408163)$

Lagrange multipliers for LE constraints (u) → (0.15556)

Lagrange multipliers for EQ constraints (v) → (0.955461)

Objective function → -8.34818

(b) Direction using dual QP

The constraint vector and its Jacobian are as follows:

rxk = {gx, hx}

$\left\{-\dfrac{22}{3}, 7\right\}$

Jrxk = {∇gxk, ∇hxk}; MatrixForm[Jrxk]

$\begin{pmatrix} -8 & \frac{1}{3} \\ 4 & 8 \end{pmatrix}$

The dual objective function is as follows:

c = Jrxk.∇fxk - rxk

$\{-38, 81\}$

Q = Jrxk.Transpose[Jrxk]; MatrixForm[Q]

$\begin{pmatrix} \frac{577}{9} & -\frac{88}{3} \\ -\frac{88}{3} & 80 \end{pmatrix}$

dualObj = Expand[c.{u, v} + $\frac{1}{2}${u, v}.Q.{u, v}]

$-38u + \dfrac{577u^2}{18} + 81v - \dfrac{88uv}{3} + 40v^2$

The minimum of this function subject to $u \geq 0$ is found by using the ActiveSet-DualQP method discussed in Chapter 8.

sol = ActiveSetDualQP[-38u + $\dfrac{577u^2}{18}$ + 81v - $\dfrac{88uv}{3}$ + 40v², {u, v},
UnrestrictedVariables → {v}, PrintLevel → 2];

Minimize → $-38u + \dfrac{577u^2}{18} + 81v - \dfrac{88uv}{3} + 40v^2$

Subject to → $(u \geq 0)$

Unrestricted variables → v

Order of variables → $\begin{pmatrix} 1 & 2 \\ u & v \end{pmatrix}$

QP Solution Optimality Check 1

$\begin{pmatrix} u \\ v \end{pmatrix} \to \begin{pmatrix} 1. \\ 1. \end{pmatrix}$ $\nabla f \to \begin{pmatrix} -3.22222 \\ 131.667 \end{pmatrix}$ Optimality status → $\begin{pmatrix} \text{False} \\ \text{False} \end{pmatrix}$

**** QP Solution Pass 1 ****

Active variables → (u v)

Current point → (1. 1.)

9.5 Basic Sequential Quadratic Programming—SQP

$c \to \begin{pmatrix} -38 \\ 81 \end{pmatrix}$ $Q \to \begin{pmatrix} \frac{577}{9} & -\frac{88}{3} \\ -\frac{88}{3} & 80 \end{pmatrix}$ $\nabla f \to \begin{pmatrix} -3.22222 \\ 131.667 \end{pmatrix}$

\>\> Conjugate Gradient Iteration 1

$d \to \begin{pmatrix} 3.22222 \\ -131.667 \end{pmatrix}$ $\nabla f \to \begin{pmatrix} -3.22222 \\ 131.667 \end{pmatrix}$

$||\nabla f|| \to 131.706$ $\beta \to 0.$

$\alpha 1 \to 0.0122812$

$\alpha \to 0.0122812$ CG-Status $\to$ NonOptimum

New point $\to (1.03957 \quad -0.617023)$

Objective function $\to -20.7955$

\>\> Conjugate Gradient Iteration 2

$d \to \begin{pmatrix} -46.3413 \\ -17.7415 \end{pmatrix}$ $\nabla f \to \begin{pmatrix} 46.7475 \\ 1.14403 \end{pmatrix}$

$||\nabla f|| \to 46.7615$ $\beta \to 0.126056$

$\alpha 1 \to 0.0190761$

-pt/d $\to (0.022433 \quad -0.0347786)$

$\alpha 2 \to 0.022433$

$\alpha \to 0.0190761$ CG-Status $\to$ NonOptimum

New point $\to (0.15556 \quad -0.955461)$

Objective function $\to -41.6518$

\>\> Conjugate Gradient Iteration 3

$d \to \begin{pmatrix} -6.89096 \times 10^{-16} \\ 2.81579 \times 10^{-14} \end{pmatrix}$ $\nabla f \to \begin{pmatrix} 0. \\ -2.84217 \times 10^{-14} \end{pmatrix}$

$||\nabla f|| \to 2.84217 \times 10^{-14}$ $\beta \to 1.487 \times 10^{-17}$

$\alpha \to 0$ CG-Status $\to$ Optimum

New point $\to (0.15556 \quad -0.955461)$

Objective function $\to -41.6518$

QP Solution Optimality Check 2

$\begin{pmatrix} u \\ v \end{pmatrix} \to \begin{pmatrix} 0.15556 \\ -0.955461 \end{pmatrix}$ $\nabla f \to \begin{pmatrix} 0 \\ 0 \end{pmatrix}$ Optimality status $\to \begin{pmatrix} \text{True} \\ \text{True} \end{pmatrix}$

sol
{{-41.6518, {u $\to$ 0.15556, v $\to$ -0.955461}}, Optimum}

Knowing the Lagrange multipliers, the direction can now be computed as follows:

d = -∇fxk - Transpose[Jrxk].({u, v}/.sol[[1, 2]])
{-0.933673, -0.408163}

As expected, this solution is the same as the one from the primal form.

9.5.2 Merit Function

The SQP algorithm can be made more robust by adding a step-length control. Thus, the basic iteration of a practical method for solving nonlinear constrained problems is written as follows:

$$\mathbf{x}^{k+1} = \mathbf{x}^k + \alpha \mathbf{d}$$

The direction **d** is computed by solving the QP subproblem. In the case of unconstrained problems, the step length was computed by minimizing the objective function along the computed direction. In the case of interior point LP and QP methods, because the constraints were linear, it was always possible to get an explicit expression for the step length. For a general nonlinear problem, the situation is more complex. We obviously would like to minimize the objective function but not at the expense of violating the constraints. In fact, if the constraints are not satisfied at the current point, our priority perhaps is to first satisfy the constraints before seriously trying to reduce the objective function.

Using ideas from the penalty function methods, a *merit function* that takes into consideration both the constraints and the objective function is used during step-length calculations. The following merit function is commonly used with the SQP method:

$$\phi(\alpha) \equiv f(\mathbf{x}^k + \alpha \mathbf{d}) + \mu V(\mathbf{x}^k + \alpha \mathbf{d})$$

where f is the objective function, μ is a scalar *penalty* parameter, and V is the maximum constraint violation. V is defined as follows:

$$V(\mathbf{x}^k) \equiv \text{Max}\left[0, \{\text{Abs}[h_i(\mathbf{x}^k)], i = 1, \ldots, p\}, \{g_i(\mathbf{x}^k), i = 1, \ldots, m\}\right]$$

An inequality constraint is violated if $g_i > 0$. For an equality constraint, any nonzero value indicates a violation of that constraint. Hence, for equality constraints, an absolute value is used in the above definition. If all constraints are satisfied (all $g_i \leq 0$ and $h_i = 0$), then $V = 0$ and the merit function is the same as the objective function. If a constraint is violated, then $V > 0$ and by minimizing ϕ, we are attempting to minimize the sum of the objective function and the maximum constraint violation. The penalty parameter μ is introduced to allow for an additional control over the definition of ϕ. A large value of μ emphasizes minimizing the constraint violation over minimizing the objective function, and vice versa.

It is possible to show that the SQP method converges if the penalty parameter μ is based on the Lagrange multipliers associated with the constraints, as follows:

$$\mu \geq \text{Max}\left[1, \sum_{i=1}^{p} \text{Abs}\,[v_i] + \sum_{i=1}^{m} u_i\right]$$

where v_i are the Lagrange multipliers associated with equality constraints and u_i are those associated with the inequality constraints. Since v_i can have any sign, their absolute values are used in the above sum. Intuitively, this definition of μ makes sense because of the fact that the magnitude of a Lagrange multiplier determines the sensitivity of the objective function with respect to constraint changes. Thus, for more sensitive constraints, this definition emphasizes a constraint violation in the merit function. If the constraints are not very sensitive, then the objective function is given more importance in the merit function.

We still do not have a procedure for determining μ because the Lagrange multipliers are unknown. The only Lagrange multipliers that we compute are those associated with the direction-finding QP subproblem. Therefore, in actual computations, we adjust μ at each iteration as follows:

$$\mu^{k+1} = \text{Max}\left[\mu^k, 2\left(\sum_{i=1}^{p} \text{Abs}\,[v_i] + \sum_{i=1}^{m} u_i\right)\right]$$

where k is the iteration number, and the Lagrange multipliers are from the solution of the QP problem at the kth iteration.

Since the QP is based on linearization of original functions, near the optimum point, the QP Lagrange multipliers are fairly close to the Lagrange multipliers of the original problem. However, in the early stages of computations, the multipliers from the QP problem may not be very good approximations of the actual multipliers. Therefore, with an adjusted value of μ, the value of merit function may actually increase. That is, we may have the following situation:

$$f(\mathbf{x}^k) + \mu^{k+1} V(\mathbf{x}^k) > f(\mathbf{x}^k) + \mu^k V(\mathbf{x}^k)$$

In this situation, the criteria for minimizing the merit function, presented in the following subsection, may fail, and the method could possibly get stuck. In order to avoid this problem, the new value of μ based on the Lagrange multipliers of the QP problem is accepted if the merit function does not increase. If it does, we use the new value of μ but restart the computations from the beginning to avoid convergence problems. For more details refer to Pshenichnyi [1994].

9.5.3 Minimizing the Merit Function

To determine a suitable step length α along a computed direction **d**, the merit function $\phi(\alpha)$ needs to be minimized. Theoretically, this is the same as the one-dimensional line search presented for unconstrained problems. However, from the definition of ϕ, it should be clear that it is not a differentiable function and therefore, we cannot use the analytical method (i.e., we cannot find α by setting $d\phi/d\alpha = 0$). We must use a numerical search procedure. In principle, we can use a method such as the golden section search, but such a procedure will involve huge computational effort. This is because for each value of α, we must first compute all constraints to determine constraint violation V and then get a value for ϕ. Performing large number of iterations for problems involving a large number of constraints will become prohibitively expensive. Since the goal is not to get an exact minimum of $\phi(\alpha)$ but to solve the original constrained minimization problem, an approximate line-search procedure is usually considered satisfactory.

The following approximate line-search procedure is based on Armijo's rule presented in Chapter 5 for unconstrained problems.

Approximate Line-Search Procedure

The procedure starts by first computing the merit function value with $\alpha = 0$. This value is identified as ϕ_0 and from the definition, we see that

$$\phi_0 \equiv f(\mathbf{x}^k) + \mu V(\mathbf{x}^k)$$

Then $\phi(\alpha) \equiv f(\mathbf{x}^k + \alpha\mathbf{d}) + \mu V(\mathbf{x}^k + \alpha\mathbf{d})$ values are computed successively with $\alpha = 1, 1/2, 1/4, 1/8, \ldots$ until an α is found that satisfies the following criteria.

$$\phi(\alpha) \leq \phi_0 - \alpha\gamma\|\mathbf{d}\|^2$$

where γ is an additional parameter introduced to provide further control over the search process. γ is typically set to 0.5. As can be seen from this criteria, instead of finding the true minimum of $\phi(\alpha)$, in this line search we accept α that reduces the initial value of ϕ by a factor $\alpha\gamma\|\mathbf{d}\|^2$. Since α is included in this factor, as the trial step length is reduced, the expected reduction becomes small as well. Hence, as long as **d** is a descent direction, the test must pass after a finite number of iterations.

9.5.4 Complete SQP Algorithm

Given the penalty parameter μ (say $\mu = 1$) and a step-length parameter γ (say $\gamma = 0.5$)

1. Set an iteration counter $k = 0$. Starting point $\mathbf{x}^0$.
2. Set up and solve the QP subproblem for $\mathbf{d}$. We can use either the primal form or its dual.

    Primal:

    Minimize $\nabla f(\mathbf{x}^k)^T \mathbf{d} + \frac{1}{2} \mathbf{d}^T \mathbf{d}$

    Subject to $\begin{pmatrix} h_i(\mathbf{x}^k) + \nabla h_i(\mathbf{x}^k)^T \mathbf{d} = 0 & i = 1, 2, \ldots, p \\ g_i(\mathbf{x}^k) + \nabla g_i(\mathbf{x}^k)^T \mathbf{d} \leq 0 & i = 1, 2, \ldots, m \end{pmatrix}$

    The primal QP problem can be solved by using the Active set method or the Primal-Dual interior point method discussed in Chapter 8.

    Dual:

    Minimize $\mathbf{c}^T \boldsymbol{\lambda} + \frac{1}{2} \boldsymbol{\lambda}^T \mathbf{Q} \boldsymbol{\lambda}$

    Subject to $\begin{pmatrix} \lambda_i \geq 0, i = 1, \ldots m \\ \lambda_i \text{ unrestricted}, i = m+1, \ldots, p \end{pmatrix}$

    with

    $$\mathbf{c} = \mathbf{J}_r(\mathbf{x}^k) \nabla f(\mathbf{x}^k) - \mathbf{r}(\mathbf{x}^k) \quad \text{and} \quad \mathbf{Q} = \mathbf{J}_r(\mathbf{x}^k) \mathbf{J}_r(\mathbf{x}^k)^T$$

    $$\boldsymbol{\lambda} = \begin{pmatrix} \mathbf{u} \\ \mathbf{v} \end{pmatrix} \quad \mathbf{r}(\mathbf{x}^k) = \begin{pmatrix} \mathbf{g}(\mathbf{x}^k) \\ \mathbf{h}(\mathbf{x}^k) \end{pmatrix} \quad \mathbf{J}_r(\mathbf{x}^k) = \begin{pmatrix} \mathbf{J}_g(\mathbf{x}^k) \\ \mathbf{J}_h(\mathbf{x}^k) \end{pmatrix}$$

    $$\mathbf{J}_g(\mathbf{x}^k) = \begin{pmatrix} \nabla g_1(\mathbf{x}^k)^T \\ \nabla g_2(\mathbf{x}^k)^T \\ \vdots \\ \nabla g_m(\mathbf{x}^k)^T \end{pmatrix} \quad \text{and} \quad \mathbf{J}_h(\mathbf{x}^k) = \begin{pmatrix} \nabla h_1(\mathbf{x}^k)^T \\ \nabla h_2(\mathbf{x}^k)^T \\ \vdots \\ \nabla h_p(\mathbf{x}^k)^T \end{pmatrix}$$

    This problem can be solved by using the active set dual method presented in Chapter 8. Once Lagrange multiplier vector $\boldsymbol{\lambda}$ is known, the direction can be computed from

    $$\mathbf{d} = -\nabla f(\mathbf{x}^k) - \mathbf{J}_r(\mathbf{x}^k)^T \boldsymbol{\lambda}$$

3. Update penalty parameter μ.
    (i) Set $\mu^{k+1} = \text{Max}[\mu^k, 2(\sum_{i=1}^{p} \text{Abs}[v_i] + \sum_{i=1}^{m} u_i)]$.
    (ii) If $f(\mathbf{x}^k) + \mu^{k+1} V(\mathbf{x}^k) > f(\mathbf{x}^k) + \mu^k V(\mathbf{x}^k)$ need to restart, go to step 1. Otherwise, continue to step 4.

4. Compute step length α using approximate line search as follows:
   (i) Compute $\phi_0 \equiv f(\mathbf{x}^k) + \mu^{k+1} V(\mathbf{x}^k)$. Set $i = 0$.
   (ii) Set $\alpha = (1/2)^i$.
   (iii) Compute $\phi(\alpha) \equiv f(\mathbf{x}^k + \alpha \mathbf{d}) + \mu^{k+1} V(\mathbf{x}^k + \alpha \mathbf{d})$.
   (iv) If $\phi(\alpha) \leq \phi_0 - \alpha \gamma ||\mathbf{d}||^2$ then accept α. Otherwise, set $i = i+1$ and go to step (ii).

5. Set the new point as

$$\mathbf{x}^{k+1} = \mathbf{x}^k + \alpha \mathbf{d}$$

6. If $||\mathbf{d}|| \leq$ tol (a small number), then stop. Otherwise, set $k = k+1$ and go to step 2.

The algorithm is implemented in a function called SQP included in the OptimizationToolbox package.

**Needs["OptimizationToolbox`ConstrainedNLP`"];
?SQP**

SQP[f, cons, vars, pt, opts] solves a constrained optimization problem
   using the SQP algorithm. f = objective function. cons = list of
   constraints. vars = list of problem variables. pt = starting point.
   opts = options allowed by the function. See Options[SQP] to see all
   valid options.

OptionsUsage[SQP]
{ProblemType → Min, PrintLevel → 1, MaxIterations → 10,
 ConvergenceTolerance → 0.001, SolveQPUsing → ActiveSetDual,
 MaxQPIterations → 20, StartingMu → 10,
 LineSearchGamma → 0.5, MaxLineSearchIterations → 10}

ProblemType is an option for most optimization methods. It can either be
   Min (default) or Max.

PrintLevel is an option for most functions in the OptimizationToolbox.
   It is specified as an integer. The value of the integer indicates
   how much intermediate information is to be printed. A PrintLevel→ 0
   suppresses all printing. Default for most functions is set to 1 in
   which case they print only the initial problem setup. Higher integers
   print more intermediate results.

MaxIterations is an option for several optimization methods. It
   specifies maximum number of iterations allowed.

ConvergenceTolerance is an option for most optimization methods. Most
   methods require only a single zero tolerance value. Some interior
   point methods require a list of convergence tolerance values.

SolveQPUsing is an option for SQP & RSQP methods. It defines the
   solution method used for solving QP subproblem. Currently the choice
   is between ActiveSet and the ActiveSetDual for SQP and between

ActiveSet and the PrimalDual for RSQP. Default is SolveQPUsing→ ActiveSetDual for SQP and PrimalDual for RSQP.

MaxQPIterations is an option for SQP method. It defines the maximum number of iterations allowed in solving the QP problem. Default is MaxQPIterations→ 20.

StartingMu is an option for SQP method. It defines the starting value of penalty parameter μ used in establishing the merit function. Default is StartingMu→ 10

LineSearchGamma is an option for SQP & RSQP methods. It defines the value of parameter γ used with approximate line search. Default is LineSearchGamma→ 0.5

MaxLineSearchIterations is an option for SQP method. It defines the maximum number of iterations allowed during line search. Default is MaxLineSearchIterations→ 10.

Example 9.15 Solve the following optimization problem using the SQP method. Use the starting point as (3, 1).

Minimize $f = \frac{1}{2}x_1^2 + x_1 x_2^2$

Subject to $\begin{pmatrix} x_1 x_2^2 = 10 \\ -2x_1^2 + x_2/3 \leq 0 \\ x_2 \geq -4 \end{pmatrix}$

```
vars = {x₁, x₂}; (pt = {3.0, 1.0};)
f = x₁²/2 + x₁x₂²;
cons = {x₁x₂² == 10, -2x₁² + x₂/3 ≤ 0, x₂ ≥ -4};
```

Using the SQP function, the solution is as follows. All calculations, except for the solution of the QP subproblem, are shown for the first two iterations. The dual form of QP is used for direction finding. The QP is solved using the ActiveSetDualQP.

```
SQP[f, cons, vars, pt, MaxIterations → 2, PrintLevel → 4];
```

Minimize $f \to \dfrac{x_1^2}{2} + x_1 x_2^2$

$\nabla f \to \begin{pmatrix} x_1 + x_2^2 & 2 x_1 x_2 \end{pmatrix}$

LE Constraints: $\begin{pmatrix} g_1 \\ g_2 \end{pmatrix} \to \begin{pmatrix} -2x_1^2 + \frac{x_2}{3} \leq 0 \\ -4 - x_2 \leq 0 \end{pmatrix}$

$\begin{pmatrix} \nabla g_1 \\ \nabla g_2 \end{pmatrix} \to \begin{pmatrix} -4x_1 & \frac{1}{3} \\ 0 & -1 \end{pmatrix}$

EQ Constraints: $(h_1) \to (-10 + x_1 x_2^2 == 0)$

$(\nabla h_1) \rightarrow \begin{pmatrix} x_2^2 & 2x_1x_2 \end{pmatrix}$

------------------ **SQP Iteration 1** ------------------

Current point $\rightarrow (3. \quad 1.)$

Function values $\begin{pmatrix} f \\ g \\ h \end{pmatrix} \rightarrow \begin{pmatrix} 7.5 \\ \{-17.6667, -5.\} \\ \{-7.\} \end{pmatrix}$

Gradients: $\nabla f \rightarrow \begin{pmatrix} 4. \\ 6. \end{pmatrix}$

$\nabla g \rightarrow \begin{pmatrix} -12. & \frac{1}{3} \\ 0 & -1 \end{pmatrix}$

$\nabla h \rightarrow (1. \quad 6.)$

--------QP subproblem for direction--------

Minimize $4.d_1 + 6.d_2 + \frac{1}{2}(d_1^2 + d_2^2)$

Subject to $\rightarrow \begin{pmatrix} -17.6667 - 12.d_1 + 0.333333d_2 \le 0 \\ -5. - 1.d_2 \le 0 \\ -7. + 1.d_1 + 6.d_2 == 0 \end{pmatrix}$

--------Dual QP subproblem--------

Minimize $-28.3333u_1 - 1.u_2 + 47.v_1 + \frac{1}{2}(u_1(144.111u_1 - 0.333333u_2 - 10.v_1)$
$+ u_2(-0.333333u_1 + 1.u_2 - 6.v_1) + v_1(-10.u_1 - 6.u_2 + 37.v_1))$

Direction $\rightarrow (-1.43318 \quad 1.40553) \quad \|d\| \rightarrow 2.00737$

Lagrange multipliers $\rightarrow (0.110535 \quad 0 \quad -1.2404)$

--------- Step length calculations ---------

$\mu \rightarrow 10 \quad \|d\|^2 \rightarrow 4.02952$

$\phi_0 \rightarrow 77.5 \quad \gamma \rightarrow 0.5$

----- Try $\alpha \rightarrow 1$

Point $\rightarrow (1.56682 \quad 2.40553)$

Function values $\begin{pmatrix} f \\ g \\ h \end{pmatrix} \rightarrow \begin{pmatrix} 10.294 \\ \{-4.10801, -6.40553\} \\ \{-0.933478\} \end{pmatrix}$

$\phi \rightarrow 19.6288 \quad \phi + \alpha\gamma d^2 \rightarrow 21.6435 \quad \phi$ Test $\rightarrow$ Passes

Step length $\rightarrow 1$

New point $\rightarrow \{1.56682, 2.40553\}$

------------------ **SQP Iteration 2** ------------------

Current point $\rightarrow (1.56682 \quad 2.40553)$

Function values $\begin{pmatrix} f \\ g \\ h \end{pmatrix} \rightarrow \begin{pmatrix} 10.294 \\ \{-4.10801, -6.40553\} \\ \{-0.933478\} \end{pmatrix}$

Gradients: $\nabla f \to \begin{pmatrix} 7.35339 \\ 7.53807 \end{pmatrix}$

$\nabla g \to \begin{pmatrix} -6.26728 & \frac{1}{3} \\ 0 & -1 \end{pmatrix}$

$\nabla h \to \begin{pmatrix} 5.78657 & 7.53807 \end{pmatrix}$

-------QP subproblem for direction-------

Minimize $7.35339 d_1 + 7.53807 d_2 + \frac{1}{2}\left(d_1^2 + d_2^2\right)$

Subject to $\to \begin{pmatrix} -4.10801 - 6.26728 d_1 + 0.333333 d_2 \le 0 \\ -6.40553 - 1. d_2 \le 0 \\ -0.933478 + 5.78657 d_1 + 7.53807 d_2 == 0 \end{pmatrix}$

-------Dual QP subproblem-------

Minimize $-39.4651 u_1 - 1.13254 u_2 + 100.307 v_1 +$
$\quad \frac{1}{2} \left(u_1 (39.3899 u_1 - 0.333333 u_2 - 33.7534 v_1) \right.$
$\quad + u_2 (-0.333333 u_1 + 1. u_2 - 7.53807 v_1)$
$\quad \left. + v_1 (-33.7534 u_1 - 7.53807 u_2 + 90.3069 v_1) \right)$

Direction $\to \begin{pmatrix} -0.623429 & 0.602409 \end{pmatrix}$ $\|d\| \to 0.866926$

Lagrange multipliers $\to \begin{pmatrix} 0.0737302 & 0 & -1.08318 \end{pmatrix}$

-------- Step length calculations --------

$\mu \to 10$ $\|d\|^2 \to 0.75156$

$\phi_0 \to 19.6288$ $\gamma \to 0.5$

----- Try $\alpha \to 1$

Point $\to \begin{pmatrix} 0.943391 & 3.00794 \end{pmatrix}$

Function values $\begin{pmatrix} f \\ g \\ h \end{pmatrix} \to \begin{pmatrix} 8.98051 \\ \{-0.777328, -7.00794\} \\ \{-1.46448\} \end{pmatrix}$

$\phi \to 23.6254$ $\phi + \alpha \gamma d^2 \to 24.0011$ ϕ Test $\to$ Fails

----- Try $\alpha \to 0.5$

Point $\to \begin{pmatrix} 1.25511 & 2.70673 \end{pmatrix}$

Function values $\begin{pmatrix} f \\ g \\ h \end{pmatrix} \to \begin{pmatrix} 9.98307 \\ \{-2.24834, -6.70673\} \\ \{-0.80458\} \end{pmatrix}$

$\phi \to 18.0289$ $\phi + \alpha \gamma d^2 \to 18.2168$ ϕ Test $\to$ Passes

Step length $\to 0.5$

New point $\to \{1.25511, 2.70673\}$

NonOptimum point $\to \begin{pmatrix} x_1 \to 1.25511 & x_2 \to 2.70673 \end{pmatrix}$

Iterations performed $\to 2$

The procedure is allowed to run until convergence.

```
{soln, hist} = SQP[f, cons, vars, pt];
```

Optimum point → $(x_1 \to 0.773997 \quad x_2 \to 3.59443)$
Iterations performed → 7

TableForm[hist]

Point	d	‖d‖	φ
3. 1.	−1.43318 1.40553	2.00737	77.5
1.56682 2.40553	−0.623429 0.602409	0.866926	19.6288
1.25511 2.70673	−0.410581 0.561142	0.69531	18.0289
1.04982 2.98731	−0.251395 0.458353	0.522768	16.2342
0.798421 3.44566	−0.0242409 0.146943	0.148929	15.005
0.77418 3.5926	−0.00018242 0.00183101	0.00184008	10.3702
0.773997 3.59443	-2.02155×10^{-8} 1.19016×10^{-8}	2.34587×10^{-8}	10.2995

The graph in Figure 9.13 confirms the solution and shows the path taken by the SQP method.

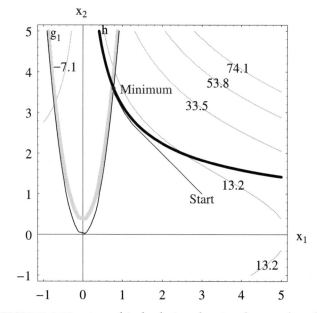

FIGURE 9.13 A graphical solution showing the search path.

9.5 Basic Sequential Quadratic Programming—SQP

Example 9.16 Solve the following optimization problem using the SQP method.

Minimize $f = e^{-x_1-x_2} + x_1 x_2 + x_2^2$

Subject to $\begin{pmatrix} \frac{1}{4}e^{-x_1} + \frac{1}{4}x_2^2 - 1 \leq 0 \\ x_1/5 + x_2/2 - 1 \leq 0 \end{pmatrix}$

```
vars = {x₁, x₂};
f = Exp[-x₁ - x₂] + x₁x₂ + x₂²;
g = {¼Exp[-x₁] + x₂²/4 - 1 ≤ 0, x₁/5 + x₂/2 - 1 ≤ 0};
pt = {1.0, 1.0};
```

Using a SQP function, the solution is as follows. All calculations, except for the solution of the QP subproblem, are shown for the first two iterations. The primal form of QP is used for direction finding. The QP is solved using the ActiveSetQP.

```
SQP[f, g, vars, pt, MaxIterations → 2, PrintLevel → 4,
  SolveQPUsing → ActiveSet];
```

Minimize $f \to E^{-x_1-x_2} + x_1 x_2 + x_2^2$

$\nabla f \to \left(-E^{-x_1-x_2} + x_2 \quad -E^{-x_1-x_2} + x_1 + 2x_2 \right)$

LE Constraints: $\begin{pmatrix} g_1 \\ g_2 \end{pmatrix} \to \begin{pmatrix} -1 + \frac{E^{-x_1}}{4} + \frac{x_2^2}{4} \leq 0 \\ -1 + \frac{x_1}{5} + \frac{x_2}{2} \leq 0 \end{pmatrix}$

$\begin{pmatrix} \nabla g_1 \\ \nabla g_2 \end{pmatrix} \to \begin{pmatrix} -\frac{E^{-x_1}}{4} & \frac{x_2}{2} \\ \frac{1}{5} & \frac{1}{2} \end{pmatrix}$

------------------ **SQP Iteration 1** ------------------

Current point $\to (1. \quad 1.)$

Function values $\begin{pmatrix} f \\ g \\ h \end{pmatrix} \to \begin{pmatrix} 2.13534 \\ \{-0.65803, -0.3\} \\ \{\} \end{pmatrix}$

Gradients: $\nabla f \to \begin{pmatrix} 0.864665 \\ 2.86466 \end{pmatrix}$

$\nabla g \to \begin{pmatrix} -0.0919699 & 0.5 \\ \frac{1}{5} & \frac{1}{2} \end{pmatrix}$

$\nabla h \to \{\}$

-------QP subproblem for direction-------

Minimize $0.864665 d_1 + 2.86466 d_2 + \frac{1}{2}(d_1^2 + d_2^2)$

Subject to $\to \begin{pmatrix} -0.65803 - 0.0919699 d_1 + 0.5 d_2 \leq 0 \\ -0.3 + 0.2 d_1 + 0.5 d_2 \leq 0 \end{pmatrix}$

Chapter 9 Constrained Nonlinear Problems

Direction $\to \begin{pmatrix} -0.864665 & -2.86466 \end{pmatrix}$ $\|d\| \to 2.99231$

Lagrange multipliers $\to \begin{pmatrix} 0 & 0 \end{pmatrix}$

-------- Step length calculations --------

$\mu \to 10$ $\|d\|^2 \to 8.95395$

$\phi_0 \to 2.13534$ $\gamma \to 0.5$

----- Try $\alpha \to 1$

Point $\to \begin{pmatrix} 0.135335 & -1.86466 \end{pmatrix}$

Function values $\begin{pmatrix} f \\ g \\ h \end{pmatrix} \to \begin{pmatrix} 8.86149 \\ \{0.0875994, -1.90527\} \\ \{\} \end{pmatrix}$

$\phi \to 9.73749$ $\phi + \alpha\gamma d^2 \to 14.2145$ ϕ Test $\to$ Fails

----- Try $\alpha \to 0.5$

Point $\to \begin{pmatrix} 0.567668 & -0.432332 \end{pmatrix}$

Function values $\begin{pmatrix} f \\ g \\ h \end{pmatrix} \to \begin{pmatrix} 0.814913 \\ \{-0.811561, -1.10263\} \\ \{\} \end{pmatrix}$

$\phi \to 0.814913$ $\phi + \alpha\gamma d^2 \to 3.0534$ ϕ Test $\to$ Fails

----- Try $\alpha \to 0.25$

Point $\to \begin{pmatrix} 0.783834 & 0.283834 \end{pmatrix}$

Function values $\begin{pmatrix} f \\ g \\ h \end{pmatrix} \to \begin{pmatrix} 0.64685 \\ \{-0.865697, -0.701316\} \\ \{\} \end{pmatrix}$

$\phi \to 0.64685$ $\phi + \alpha\gamma d^2 \to 1.76609$ ϕ Test $\to$ Passes

Step length $\to 0.25$

New point $\to \{0.783834, 0.283834\}$

------------------ **SQP Iteration 2** ------------------

Current point $\to \begin{pmatrix} 0.783834 & 0.283834 \end{pmatrix}$

Function values $\begin{pmatrix} f \\ g \\ h \end{pmatrix} \to \begin{pmatrix} 0.64685 \\ \{-0.865697, -0.701316\} \\ \{\} \end{pmatrix}$

Gradients: $\nabla f \to \begin{pmatrix} -0.0599756 \\ 1.00769 \end{pmatrix}$

$\nabla g \to \begin{pmatrix} -0.114163 & 0.141917 \\ \frac{1}{5} & \frac{1}{2} \end{pmatrix}$

$\nabla h \to \{\}$

-------QP subproblem for direction-------

Minimize $-0.0599756 d_1 + 1.00769 d_2 + \frac{1}{2}\left(d_1^2 + d_2^2\right)$

Subject to $\to \begin{pmatrix} -0.865697 - 0.114163 d_1 + 0.141917 d_2 \leq 0 \\ -0.701316 + 0.2 d_1 + 0.5 d_2 \leq 0 \end{pmatrix}$

9.5 Basic Sequential Quadratic Programming—SQP

Direction → $(0.0599756 \quad -1.00769)$ ‖d‖ → 1.00948
Lagrange multipliers → $(0 \quad 0)$
-------- Step length calculations --------
$\mu \to 10$ ‖d‖² → 1.01904
$\phi_0 \to 0.64685$ $\gamma \to 0.5$
----- Try $\alpha \to 1$
Point → $(0.843809 \quad -0.723858)$
Function values $\begin{pmatrix} f \\ g \\ h \end{pmatrix} \to \begin{pmatrix} 0.800136 \\ \{-0.76149, -1.19317\} \\ \{\} \end{pmatrix}$
$\phi \to 0.800136$ $\phi + \alpha\gamma d^2 \to 1.30966$ ϕ Test → Fails
----- Try $\alpha \to 0.5$
Point → $(0.813822 \quad -0.220012)$
Function values $\begin{pmatrix} f \\ g \\ h \end{pmatrix} \to \begin{pmatrix} 0.421574 \\ \{-0.877108, -0.947242\} \\ \{\} \end{pmatrix}$
$\phi \to 0.421574$ $\phi + \alpha\gamma d^2 \to 0.676334$ ϕ Test → Fails
----- Try $\alpha \to 0.25$
Point → $(0.798828 \quad 0.0319108)$
Function values $\begin{pmatrix} f \\ g \\ h \end{pmatrix} \to \begin{pmatrix} 0.462237 \\ \{-0.887281, -0.824279\} \\ \{\} \end{pmatrix}$
$\phi \to 0.462237$ $\phi + \alpha\gamma d^2 \to 0.589617$ ϕ Test → Passes
Step length → 0.25
New point → $\{0.798828, 0.0319108\}$
NonOptimum point → $(x_1 \to 0.798828 \quad x_2 \to 0.0319108)$
Iterations performed → 2

The procedure is allowed to run until convergence.

```
{soln, hist} = SQP[f, g, vars, pt, MaxIterations → 20,
   SolveQPUsing → ActiveSet];
```

Optimum point → $(x_1 \to 10.0009 \quad x_2 \to -2.00009)$
Iterations performed → 12

TableForm[hist]

Point	d	‖d‖	ϕ
1.	−0.864665	2.99231	2.13534
1.	−2.86466		
0.783834	0.0599756	1.00948	0.64685
0.283834	−1.00769		

0.798828	0.403817	0.587648	0.462237
0.0319108	-0.426922		
1.20264	0.840923	0.841582	0.126888
-0.395011	0.0332904		
2.04357	0.547751	1.25945	-0.422329
-0.361721	-1.1341		
2.59132	1.83019	1.97214	-1.3043
-1.49582	0.734687		
3.50641	1.22121	1.68208	-2.5907
-1.12847	-1.15672		
4.11702	1.79803	1.82666	-4.02399
-1.70684	-0.32215		
5.91505	2.05075	2.05094	-7.56553
-2.02899	0.0280799		
7.9658	2.00382	2.00382	-11.9227
-2.00091	0.000818507		
9.96961	2.0005	2.0005	-15.9384
-2.00009	0.0000755731		
10.0009	-0.000900352	0.000905611	-16.0006
-2.00009	0.0000974477		

The graph in Figure 9.14 confirms the solution and shows the path taken by the SQP method.

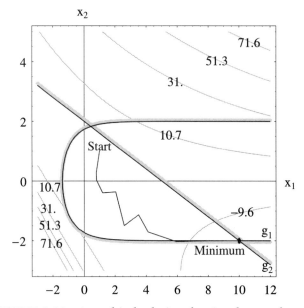

FIGURE 9.14 A graphical solution showing the search path.

The method converges to the same solution even when the solution is started from infeasible points. Two solution paths computed by the method starting from two different points are as follows:

$$\begin{pmatrix} x_1 & x_2 \\ -10 & 10 \\ -9.00382 & 0.990923 \\ 186.022 & -76.443 \\ 100.619 & -38.2477 \\ 52.9403 & -19.1761 \\ 29.2309 & -9.69236 \\ 17.6313 & -5.05253 \\ 12.3053 & -2.92211 \\ 10.3637 & -2.14549 \\ 10.0123 & -2.00492 \\ 9.99999 & -1.99999 \end{pmatrix} \quad \begin{pmatrix} x_1 & x_2 \\ 4 & 4 \\ 2.00017 & -1.99983 \\ 3.32764 & -0.930939 \\ 3.83862 & -1.61831 \\ 5.56592 & -2.04985 \\ 7.64687 & -2.00161 \\ 9.65245 & -2.00012 \\ 9.99997 & -1.99999 \end{pmatrix}$$

Example 9.17 *Building design* Consider the solution of the building design problem, first formulated in Chapter 1, and solved using the ALPF method in Example 9.9. The optimization problem statement is as follows:

d = Depth of building below ground

h = Height of building above ground

w = Width of building in plan

Find (d, h, and w) in order to

Minimize $f = 1.618 dw^2$

Subject to $\begin{pmatrix} 100(5.236hw + 1.618w^2)/225000 \leq 1 \\ 1.618w/50 \leq 1 \\ 0.462286(d+h)w^2/20000 \geq 1 \\ d \geq 0, h \geq 0, \text{ and } w \geq 0 \end{pmatrix}$

In terms of *Mathematica* expressions, the problem is stated as follows:

```
Clear[d, h, w];
vars = {d, h, w};
f = 1.618dw²;
cons = {100(5.236hw + 1.618w²)/225000 ≤ 1, 1.618w/50 ≤ 1,
        0.462286(d+h)w²/20000 ≥ 1, d ≥ 0, h ≥ 0, w ≥ 0};
```

Starting from the following starting point, the solution is obtained using the SQP method. All calculations are shown for the first iteration.

```
pt = {50, 10, 25}; SQP[f, cons, vars, pt, MaxIterations → 1,
  PrintLevel → 4, SolveQPUsing → ActiveSetDual];
```

Minimize $f \to 1.618 dw^2$

$\nabla f \to \begin{pmatrix} 1.618 w^2 & 0 & 3.236 dw \end{pmatrix}$

LE Constraints: $\begin{pmatrix} g_1 \\ g_2 \\ g_3 \\ g_4 \\ g_5 \\ g_6 \end{pmatrix} \to \begin{pmatrix} -1 + \dfrac{5.236 hw + 1.618 w^2}{2250} \le 0 \\ -1 + 0.03236 w \le 0 \\ 1 - 0.0000231143 (d+h) w^2 \le 0 \\ -d \le 0 \\ -h \le 0 \\ -w \le 0 \end{pmatrix}$

$\begin{pmatrix} \nabla g_1 \\ \nabla g_2 \\ \nabla g_3 \\ \nabla g_4 \\ \nabla g_5 \\ \nabla g_6 \end{pmatrix} \to$

$\begin{pmatrix} 0 & 0.00232711 w & 0.00232711 h + 0.00143822 w \\ 0 & 0 & 0.03236 \\ -0.0000231143 w^2 & -0.0000231143 w^2 & -0.0000462286 dw - 0.0000462286 hw \\ -1 & 0 & 0 \\ 0 & -1 & 0 \\ 0 & 0 & -1 \end{pmatrix}$

------------------ SQP Iteration 1 ------------------

Current point $\to \begin{pmatrix} 50 & 10 & 25 \end{pmatrix}$

Function values $\begin{pmatrix} f \\ g \\ h \end{pmatrix} \to \begin{pmatrix} 50562.5 \\ \{0.0312222, -0.191, 0.133214, -50, -10, -25\} \\ \{\} \end{pmatrix}$

Gradients: $\nabla f \to \begin{pmatrix} 1011.25 \\ 0 \\ 4045. \end{pmatrix}$

$\nabla g \to \begin{pmatrix} 0 & 0.0581778 & 0.0592267 \\ 0 & 0 & 0.03236 \\ -0.0144464 & -0.0144464 & -0.0693429 \\ -1 & 0 & 0 \\ 0 & -1 & 0 \\ 0 & 0 & -1 \end{pmatrix}$

$\nabla h \to \{\}$

-------QP subproblem for direction-------

Minimize $1011.25 d_1 + 4045. d_3 + \dfrac{1}{2} \left(d_1^2 + d_2^2 + d_3^2 \right)$

9.5 Basic Sequential Quadratic Programming—SQP

Subject to →
$$\begin{pmatrix} 0.0312222 + 0.0581778d_2 + 0.0592267d_3 \leq 0 \\ -0.191 + 0.03236d_3 \leq 0 \\ 0.133214 - 0.0144464d_1 - 0.0144464d_2 - 0.0693429d_3 \leq 0 \\ -50 - 1.d_1 \leq 0 \\ -10 - 1.d_2 \leq 0 \\ -25 - 1.d_3 \leq 0 \end{pmatrix}$$

--------Dual QP subproblem-------

Minimize $239.541u_1 + 131.087u_2 - 295.234u_3 - 961.25u_4 + 10.u_5 - 4020.u_6$
$+ \frac{1}{2}(u_1 (0.00689245u_1 + 0.00191657u_2 - 0.00494741u_3 + 0.u_4$
$-0.0581778u_5 - 0.0592267u_6)$
$+ u_2 (0.00191657u_1 + 0.001047169u_2 - 0.00224394u_3 + 0.u_4 + 0.u_5 - 0.03236u_6)$
$+ u_4 (0.u_1 + 0.u_2 + 0.0144464u_3 + 1.u_4 + 0.u_5 + 0.u_6)$
$+ u_5 (-0.0581778u_1 + 0.u_2 + 0.0144464u_3 + 0.u_4 + 1.u_5 + 0.u_6)$
$+ u_3 (-0.00494741u_1 - 0.00224394u_2 + 0.005225836u_3$
$+0.0144464u_4 + 0.0144464u_5 + 0.0693429u_6)$
$+ u_6 (-0.0592267u_1 - 0.03236u_2 + 0.0693429u_3 + 0.u_4 + 0.u_5 + 1.u_6))$

Direction → $(52.3751 \quad 10.935 \quad -11.2685)$ $\|d\| \to 54.6782$

Lagrange multipliers → $(18094.4 \quad 0 \quad 73625.4 \quad 0 \quad 0 \quad 0)$

-------- Step length calculations --------

$\mu \to 183440..$ $\|d\|^2 \to 2989.71$
$\phi_0 \to 74999.2$ $\gamma \to 0.5$

----- Try $\alpha \to 1$
Point → $(102.375 \quad 20.935 \quad 13.7315)$

Function values $\begin{pmatrix} f \\ g \\ h \end{pmatrix}$
$\to \begin{pmatrix} 31232.5 \\ \{-0.195438, -0.555649, 0.46258, -102.375, -20.935, -13.7315\} \\ \{\} \end{pmatrix}$

$\phi \to 116088.$ $\phi + \alpha\gamma d^2 \to 117583.$ ϕ Test → Fails

----- Try $\alpha \to 0.5$
Point → $(76.1876 \quad 15.4675 \quad 19.3657)$

Function values $\begin{pmatrix} f \\ g \\ h \end{pmatrix} \to$

$\begin{pmatrix} 46230.7 \\ \{-0.0332483, -0.373325, 0.205479, -76.1876, -15.4675, -19.3657\} \\ \{\} \end{pmatrix}$

$\phi \to 83923.7$ $\phi + \alpha\gamma d^2 \to 84671.1$ ϕ Test → Fails

----- Try $\alpha \to 0.25$
Point → $(63.0938 \quad 12.7338 \quad 22.1829)$

```
Function values (f g h) →
```
$$\begin{pmatrix} 50234.3 \\ \{0.0112018, -0.282162, 0.137532, -63.0938, -12.7338, -22.1829\} \\ \{\} \end{pmatrix}$$

$\phi \to 75463.1 \quad \phi + \alpha\gamma d^2 \to 75836.8 \quad \phi \text{ Test} \to \text{Fails}$
----- Try $\alpha \to 0.125$
Point → (56.5469 11.3669 23.5914)

```
Function values (f g h) →
```
$$\begin{pmatrix} 50920.9 \\ \{0.0242657, -0.236581, 0.12633, -56.5469, -11.3669, -23.5914\} \\ \{\} \end{pmatrix}$$

$\phi \to 74094.9 \quad \phi + \alpha\ \gamma d^2 \to 74281.7 \quad \phi \text{ Test} \to \text{Passes}$
Step length → 0.125
New point → {56.5469, 11.3669, 23.5914}
NonOptimum point → (d → 56.5469 h → 11.3669 w → 23.5914)
Iterations performed → 1

The function is allowed to iterate until solution converges.

pt = {50, 10, 25};
{sol, hist} = SQP[f, cons, vars, pt, MaxIterations → 20];
Optimum point → (d → 80.0277 h → 13.3062 w → 21.5298)
Iterations performed → 9

The same solution was obtained using the ALPF method. The complete iteration history is shown in the following table:

TableForm[hist]

Point	d	‖d‖	ϕ
50	52.3751		
10	10.935	54.6782	50563.8
25	−11.2685		
56.5469	24.9432		
11.3669	3.02476	25.3229	74094.9
23.5914	−3.15203		
81.4901	−6.67753		
14.3916	−1.98643	7.18113	68680.5
20.4394	1.74163		
74.8125	6.6521		
12.4052	1.17885	6.81606	61048.5
22.181	−0.904726		

```
81.4646      -1.85294
13.5841      -0.344218       1.90906         60670.6
21.2763       0.304386
79.6117       0.443132
13.2398       0.0710093      0.452118        60075.2
21.5807      -0.0547943
80.0548      -0.0322867
13.3109      -0.00554176     0.0330793       60023.
21.5259       0.00459353
80.0225       0.00518395
13.3053       0.000861026    0.00530085      60020.4
21.5305      -0.000695909
80.0277      -0.000602412
13.3062      -0.000100187    0.000616043     60020.4
21.5298       0.0000810697
```

9.6 Refined SQP Methods

As mentioned in the introduction to this chapter, the development of methods for solving nonlinear programming problems, particularly for large-scale problems, remains an active area of research. As such, several modifications to the basic SQP algorithm have been introduced and analyzed from a theoretical and computer implementation point of view. Several papers in a recent conference on large-scale optimization, collected in a three-volume series edited by Biegler, Coleman, Conn, and Santosa [1997], present excellent summaries of these developments. The modifications deal both with the direction-finding subproblem and the merit function.

Refinements to the Direction-Finding Phase

To improve direction generated by SQP, several researchers have suggested to add a more general quadratic term to the linearized objective function. Thus, the direction-finding subproblem is defined as follows:

Minimize $\nabla f(\mathbf{x}^k)^T \mathbf{d} + \frac{1}{2}\mathbf{d}^T \mathbf{H}\mathbf{d}$

Subject to $\begin{pmatrix} h_i(\mathbf{x}^k) + \nabla h_i(\mathbf{x}^k)^T \mathbf{d} = 0 & i = 1, 2, \ldots, p \\ g_i(\mathbf{x}^k) + \nabla g_i(\mathbf{x}^k)^T \mathbf{d} \leq 0 & i = 1, 2, \ldots, m \end{pmatrix}$

where $\mathbf{H}$ is an $n \times n$ matrix. Solving KT conditions for the original problem using Newton's method for solving nonlinear equations directly leads to the above form in which matrix **H** represents the Hessian of the Lagrangian

$$\mathbf{H} = \nabla^2 f + \sum_{i=1}^{m} u_i \nabla^2 g_i + \sum_{i=1}^{p} v_i \nabla^2 h_i$$

where u_i and v_i are Lagrange multipliers. There are two difficulties associated with this form. The first problem is that it requires second derivatives of the functions, which may be difficult to obtain. The second problem is that in general, the Hessian of the Lagrangian is not a positive definite matrix and hence, $\mathbf{d}^T \mathbf{H} \mathbf{d}$ may not always be positive.

To overcome these difficulties, a vast majority of researchers advocate using a positive definite **H** matrix generated by using the BFGS method to approximate the Hessian of the Lagrangian function for the problem. Using this approach, we start with an identity matrix for **H** and update it at each iteration, as follows:

$$\mathbf{H}^{(k+1)} = \mathbf{H}^{(k)} + \frac{\gamma \gamma^T}{\mathbf{q}^T \mathbf{s}} - \frac{\mathbf{H}^{(k)} \mathbf{s} \mathbf{s}^T \mathbf{H}^{(k)}}{\mathbf{s}^T \mathbf{H}^{(k)} \mathbf{s}}$$

where

$$\mathbf{s} = \mathbf{x}^{(k+1)} - \mathbf{x}^{(k)} \quad \mathbf{q} = \nabla_x L^{(k+1)} - \nabla_x L^{(k)}$$

$$\nabla_x L = \nabla f + \sum_{i=1}^{m} u_i \nabla g_i + \sum_{i=1}^{p} v_i \nabla h_i$$

$$\gamma = \theta \mathbf{q} + (1 - \theta) \mathbf{H}^{(k)} \mathbf{s}$$

$$\theta = \begin{pmatrix} 1.0 & \text{if } \mathbf{q}^T \mathbf{s} \geq 0.2 \mathbf{s}^T \mathbf{H}^{(k)} \mathbf{s} \\ \frac{0.8 \mathbf{s}^T \mathbf{H}^{(k)} \mathbf{s}}{\mathbf{s}^T \mathbf{H}^{(k)} \mathbf{s} - \mathbf{q}^T \mathbf{s}} & \text{if } \mathbf{q}^T \mathbf{s} < 0.2 \mathbf{s}^T \mathbf{H}^{(k)} \mathbf{s} \end{pmatrix}$$

Note that this BFGS formula approximates the Hessian matrix of a function. This is in contrast to the BFGS formula given in Chapter 5 that approximates the inverse of the Hessian.

The QP subproblem can be infeasible, even when the original problem has a feasible solution. In order to safeguard against this possibility, the constant values in the linearized constraints are reduced slightly, and the direction-finding problem is defined as follows:

Minimize $\nabla f(\mathbf{x}^k)^T \mathbf{d} + \frac{1}{2} \mathbf{d}^T \mathbf{H} \mathbf{d}$

Subject to $\begin{pmatrix} \bar{r} h_i(\mathbf{x}^k) + \nabla h_i(\mathbf{x}^k)^T \mathbf{d} = 0 & i = 1, 2, \ldots, p \\ r_i g_i(\mathbf{x}^k) + \nabla g_i(\mathbf{x}^k)^T \mathbf{d} \leq 0 & i = 1, 2, \ldots, m \end{pmatrix}$

where typically, $\bar{r} = 0.9$ and $r_i = 0.9$ if $g_i(\mathbf{x}) > 0$. Otherwise, it is 1.

References to several other proposals for direction-finding problems can be found in papers contained in Biegler, Coleman, Conn, and Santosa [1997].

Other Merit Functions

It is important to use a merit function that is appropriate for a given direction-finding problem. The convergence proofs are based on several assumptions related to these functions. A merit function used frequently with the QP involving a BFGS update of the Hessian of the Lagrangian is as follows:

$$\phi(\mathbf{x}) = f(\mathbf{x}) + \sum_{i=1}^{m} \bar{u}_i \, \text{Max}[0, g_i] + \sum_{i=1}^{p} \bar{v}_i \, \text{Abs}[h_i]$$

where in the first iteration, $\bar{u}_i$ and $\bar{v}_i$ are the same as the Lagrange multipliers obtained from solving the QP problem (u_i, v_i). For subsequent iterations, these are updated as follows:

$$\bar{u}_i^{(k+1)} = \text{Max}\left[u_i, \frac{1}{2}\left(\bar{u}_i^{(k)} + u_i\right)\right] \quad \bar{v}_i^{(k+1)} = \text{Max}\left[\text{Abs}[v_i], \frac{1}{2}\left(\bar{v}_i^{(k)} + \text{Abs}[v_i]\right)\right]$$

The step length is computed by minimizing this function. Typically, the quadratic interpolation method or the approximate line search is used for this minimization.

References to several other proposals for merit functions can be found in papers contained in Biegler, Coleman, Conn, and Santosa [1997].

These BFGS updates of the Hessian, together with the above merit function, are implemented into a function called RSQP.

```
Needs["OptimizationToolbox`ConstrainedNLP`"];
?RSQP
```

RSQP[f, cons, vars, pt, opts] solves a constrained optimization problem
  using the RSQP algorithm. f = objective function. cons = list of
  constraints. vars = list of problem variables. pt = starting point.
  opts = options allowed by the function. See Options[RSQP] to see all
  valid options.

```
OptionsUsage[RSQP]
```
{ProblemType → Min, PrintLevel → 1, MaxIterations → 10,
 ConvergenceTolerance → 0.01, SolveQPUsing → PrimalDual,
 QPConstraintFactor → 0.9, MaxQPIterations → 20,
 ResetQPHessian → Never, LineSearchDelta → 0.1, LineSearchGamma → 0.5,
 StepLengthUsing → ApproximateLineSearch,
 MaxLineSearchIterations → 20}

ProblemType is an option for most optimization methods. It can either be Min (default) or Max.

PrintLevel is an option for most functions in the OptimizationToolbox. It is specified as an integer. The value of the integer indicates how much intermediate information is to be printed. A PrintLevel→ 0 suppresses all printing. Default for most functions is set to 1 in which case they print only the initial problem setup. Higher integers print more intermediate results.

MaxIterations is an option for several optimization methods. It specifies maximum number of iterations allowed.

ConvergenceTolerance is an option for most optimization methods. Most methods require only a single zero tolerance value. Some interior point methods require a list of convergence tolerance values.

SolveQPUsing is an option for SQP & RSQP methods. It defines the solution method used for solving QP subproblem. Currently the choice is between ActiveSet and the ActiveSetDual for SQP and between ActiveSet and the PrimalDual for RSQP. Default is SolveQPUsing→ ActiveSetDual for SQP and PrimalDual for RSQP.

QPConstraintFactor is an option for RSQP method. It defines the constraint value reduction factor (r) used in the QP subproblem formulation. Default is QPConstraintFactor→ 0.9.

MaxQPIterations is an option for SQP method. It defines the maximum number of iterations allowed in solving the QP problem. Default is MaxQPIterations→ 20.

ResetQPHessian is an option for RSQP method. It defines the number of iterations after which the Hessian should be reset to identity matrix. Default is ResetQPHessian→ Never.

LineSearchDelta is an option for RSQP method. It defines the value of parameter δ used during the line search using Quadratic interpolation. Default is LineSearchDelta→ 0.1.

LineSearchGamma is an option for SQP & RSQP methods. It defines the value of parameter γ used with approximate line search. Default is LineSearchGamma→ 0.5.

StepLengthUsing is an option for RSQP method. It specifies the method used for computing step length. The options are ApproximateLineSearch (default) and QuadraticSearch.

MaxLineSearchIterations is an option for SQP method. It defines the maximum number of iterations allowed during line search. Default is MaxLineSearchIterations→ 10.

Example 9.18 Solve the following optimization problem using the RSQP method. Use the starting point as (3, 1).

9.6 Refined SQP Methods

$$\text{Minimize } f = \tfrac{1}{2}x_1^2 + x_1 x_2^2$$

$$\text{Subject to } \begin{pmatrix} x_1 x_2^2 = 10 \\ -2x_1^2 + x_2/3 \leq 0 \\ x_2 \geq -4 \end{pmatrix}$$

```
vars = {x₁, x₂}; (pt = {3.0, 1.0};)
f = x₁²/2 + x₁x₂²;
cons = {x₁x₂² == 10, -2x₁² + x₂/3 ≤ 0, x₂ ≥ -4};
```

Using the RSQP function, the solution is as follows. The step length is computed by minimizing the merit function using the quadratic interpolation technique discussed in Chapter 5.

```
RSQP[f, cons, vars, pt,
    MaxIterations → 2, PrintLevel → 4,
    QPConstraintFactor → 1];
```

Minimize $f \to \dfrac{x_1^2}{2} + x_1 x_2^2$

$\nabla f \to \begin{pmatrix} x_1 + x_2^2 & 2 x_1 x_2 \end{pmatrix}$

LE Constraints: $\begin{pmatrix} g_1 \\ g_2 \end{pmatrix} \to \begin{pmatrix} -2x_1^2 + \frac{x_2}{3} \leq 0 \\ -4 - x_2 \leq 0 \end{pmatrix}$

$\begin{pmatrix} \nabla g_1 \\ \nabla g_2 \end{pmatrix} \to \begin{pmatrix} -4x_1 & \frac{1}{3} \\ 0 & -1 \end{pmatrix}$

EQ Constraints: $(h_1) \to (-10 + x_1 x_2^2 == 0)$

$(\nabla h_1) \to \begin{pmatrix} x_2^2 & 2x_1 x_2 \end{pmatrix}$

------------------ RSQP Iteration 1 ------------------

Current point → (3. 1.)

Function values $\begin{pmatrix} f \\ g \\ h \end{pmatrix} \to \begin{pmatrix} 7.5 \\ \{-17.6667, -5.\} \\ \{-7.\} \end{pmatrix}$

Gradients: $\nabla f \to \begin{pmatrix} 4. \\ 6. \end{pmatrix}$

$\nabla g \to \begin{pmatrix} -12. & \frac{1}{3} \\ 0 & -1 \end{pmatrix}$

$\nabla h \to (1. \quad 6.)$

-------QP subproblem for direction-------

Minimize $4. d_1 + 6. d_2 + \dfrac{1}{2}\left(d_1^2 + d_2^2\right)$

Subject to → $\begin{pmatrix} -17.6667 - 12.d_1 + 0.333333d_2 \le 0 \\ -5. - 1.d_2 \le 0 \\ -7. + 1.d_1 + 6.d_2 == 0 \end{pmatrix}$

Direction → $(-1.43307 \quad 1.40551)$ ||d|| → 2.00727

Lagrange multipliers → $(0.110568 \quad 0.0000671201 \quad 1.24037)$

$\phi \to \frac{1}{2}(3. - 1.43307\alpha)^2 + (3. - 1.43307\alpha)(1. + 1.40551\alpha)^2$
$\quad + 1.24037\text{Abs}\left[-10 + (3. - 1.43307\alpha)(1. + 1.40551\alpha)^2\right]$
$\quad + 0.0000671201\text{Max}[0, -5. - 1.40551\alpha]$
$\quad + 0.110568\text{Max}\left[0, -2(3. - 1.43307\alpha)^2 + \frac{1}{3}(1. + 1.40551\alpha)\right]$

$\phi_0 \to 16.1826$

----- Try $\alpha \to 1$

$\phi \to 11.4519$ $\phi + \alpha\gamma d^2 \to 13.4665$ ϕ Test → Passes

Step length → 1

$\lambda \to (0.110568 \quad 0.0000671201 \quad 1.24037)$

$H \to \begin{pmatrix} 1 & 0 \\ 0 & 1 \end{pmatrix}$

New point → {1.56693, 2.40551}

------------------ **RSQP Iteration 2** ------------------

Current point → $(1.56693 \quad 2.40551)$

Function values $\begin{pmatrix} f \\ g \\ h \end{pmatrix} \to \begin{pmatrix} 10.2947 \\ \{-4.10872, -6.40551\} \\ \{-0.932966\} \end{pmatrix}$

Gradients: $\nabla f \to \begin{pmatrix} 7.35342 \\ 7.53855 \end{pmatrix}$

$\nabla g \to \begin{pmatrix} -6.26773 & \frac{1}{3} \\ 0 & -1 \end{pmatrix}$

$\nabla h \to (5.78648 \quad 7.53855)$

-------QP subproblem for direction-------

Minimize $7.35342 d_1 + 7.53855 d_2 + \frac{1}{2}(d_1^2 + d_2^2)$

Subject to → $\begin{pmatrix} -4.10872 - 6.26773 d_1 + 0.333333 d_2 \le 0 \\ -6.40551 - 1.d_2 \le 0 \\ -0.932966 + 5.78648 d_1 + 7.53855 d_2 == 0 \end{pmatrix}$

Direction → $(-0.623397 \quad 0.602269)$ ||d|| → 0.866805

Lagrange multipliers → $(0.0738564 \quad 0.0000195637 \quad 1.08312)$

$\phi \to \frac{1}{2}(1.56693 - 0.623397\alpha)^2 + (1.56693 - 0.623397\alpha)(2.40551 + 0.602269\alpha)^2$
$\quad + 1.16174\text{Abs}\left[-10 + (1.56693 - 0.623397\alpha)(2.40551 + 0.602269\alpha)^2\right]$

$$+ 0.0922124 \text{Max}\left[0, -2\left(1.56693 - 0.623397\alpha\right)^2 + \frac{1}{3}\left(2.40551 + 0.602269\alpha\right)\right] +$$
$$0.0000433419 \text{Max}\left[0, -6.40551 - 0.602269\alpha\right]$$
$\phi_0 \to 11.3785$
----- Try $\alpha \to 1$
$\phi \to 10.6819 \quad \phi + \alpha \; \gamma d^2 \to 11.0576 \quad \phi \; \text{Test} \to \text{Passes}$
Step length $\to 1$
$\lambda \to \begin{pmatrix} 0.0922124 & 0.0000433419 & 1.16174 \end{pmatrix}$
$H \to \begin{pmatrix} 1 & 0 \\ 0 & 1 \end{pmatrix}$
New point $\to \{0.943537, 3.00778\}$
NonOptimum point $\to \begin{pmatrix} x_1 \to 0.943537 & x_2 \to 3.00778 \end{pmatrix}$
Function values $\to \begin{pmatrix} 8.98107 \\ \{-0.77793, -7.00778\} \\ \{-1.46406\} \end{pmatrix}$
Iterations performed $\to 2$

The procedure is allowed to run until convergence.

```
{soln, hist} = RSQP[f, cons, vars, pt, MaxIterations → 10,
  QPConstraintFactor → 1];
```
Function values $\to \begin{pmatrix} 10.2938 \\ \{-0.00156928, -7.59263\} \\ \{-0.00595463\} \end{pmatrix}$
Iterations performed $\to 5$

The complete iteration history is as follows. The performance of the RSQP is very similar to the SQP method.

TableForm[hist]

Point	d	$||d||$	ϕ
3. 1.	-1.43307 1.40551	2.00727	16.1826
1.56693 2.40551	-0.623397 0.602269	0.866805	11.4519
0.943537 3.00778	-0.149851 0.496788	0.518897	10.6819
0.793686 3.50457	-0.0193756 0.0880654	0.0901716	10.3464
0.774311 3.59263	-0.000222801 0.00158715	0.00160272	10.3002

Example 9.19 *Open top container* Consider the solution of the open-top rectangular container problem formulated in Chapter 1. The problem statement is as follows.

A company requires open-top rectangular containers to transport material. Using the following data, formulate an optimum design problem to determine the container dimensions for minimum annual cost.

Construction costs	Sides = \$65/m^2 Ends = \$80/m^2 Bottom = \$120/m^2
Useful life	10 years
Salvage value	20% of the initial construction cost
Yearly maintenance cost	\$12/m^2 of the outside surface area
Minimum required volume of the container	1200 m^3
Nominal interest rate	10% (Annual compounding)

The design variables are the dimensions of the box.

b = Width of container ℓ = Length of container h = height of container

Considering time value of money, the annual cost is written as the following function of design variables (see Chapter 1 for details):

$$\text{Annual cost} = 48.0314bh + 30.0236b\ell + 43.5255h\ell$$

The optimization problem can now be stated as follows:

Find b, h, and ℓ to

Minimize annual cost = $48.0314bh + 30.0236b\ell + 43.5255h\ell$

Subject to $bh\ell \geq 1,200$ and $b, h,$ and $\ell \geq 0$

```
Clear[b, h, ℓ];
vars = {b, h, ℓ};
f = 48.0314bh + 30.0236bℓ + 43.5255hℓ;
cons = {bhℓ ≥ 1200, b ≥ 0, h ≥ 0, ℓ ≥ 0};
pt = {5, 5, 5};
```

The first two iterations of the RSQP method are as follows:

```
RSQP[f, cons, vars, pt, PrintLevel → 4, MaxIterations → 2,
   QPConstraintFactor → 0.99, StepLengthUsing → QuadraticSearch];
Minimize f → 48.0314bh + 30.0236bℓ + 43.5255hℓ
∇f → (48.0314h + 30.0236ℓ   48.0314b + 43.5255ℓ   30.0236b + 43.5255h)
```

9.6 Refined SQP Methods

LE Constraints: $\begin{pmatrix} g_1 \\ g_2 \\ g_3 \\ g_4 \end{pmatrix} \to \begin{pmatrix} 1200 - bh\ell \le 0 \\ -b \le 0 \\ -h \le 0 \\ -\ell \le 0 \end{pmatrix}$

$\begin{pmatrix} \nabla g_1 \\ \nabla g_2 \\ \nabla g_3 \\ \nabla g_4 \end{pmatrix} \to \begin{pmatrix} -h\ell & -b\ell & -bh \\ -1 & 0 & 0 \\ 0 & -1 & 0 \\ 0 & 0 & -1 \end{pmatrix}$

------------------ **RSQP Iteration 1** ------------------

Current point $\to \begin{pmatrix} 5. & 5. & 5. \end{pmatrix}$

Function values $\begin{pmatrix} f \\ g \\ h \end{pmatrix} \to \begin{pmatrix} 3039.51 \\ \{1075., -5., -5., -5.\} \\ \{\} \end{pmatrix}$

Gradients: $\nabla f \to \begin{pmatrix} 390.275 \\ 457.784 \\ 367.745 \end{pmatrix}$

$\nabla g \to \begin{pmatrix} -25. & -25. & -25. \\ -1 & 0 & 0 \\ 0 & -1 & 0 \\ 0 & 0 & -1 \end{pmatrix}$

$\nabla h \to \{\}$

-------QP subproblem for direction-------

Minimize $390.275 d_1 + 457.784 d_2 + 367.745 d_3 + \frac{1}{2}\left(d_1^2 + d_2^2 + d_3^2\right)$

Subject to $\to \begin{pmatrix} 1064.25 - 25. d_1 - 25. d_2 - 25. d_3 \le 0 \\ -5. - 1. d_1 \le 0 \\ -5. - 1. d_2 \le 0 \\ -5. - 1. d_3 \le 0 \end{pmatrix}$

Direction $\to \begin{pmatrix} 12.5203 & -4.99999 & 35.0497 \end{pmatrix}$ $\|d\| \to 37.5532$

Lagrange multipliers $\to \begin{pmatrix} 16.1108 & 0.0000217799 & 49.9862 & 0.0000414631 \end{pmatrix}$

$\phi \to 48.0314 \, (5. - 4.99999\alpha) \, (5. + 12.5203\alpha)$
$\quad + 43.5255 \, (5. - 4.99999\alpha) \, (5. + 35.0497\alpha)$
$\quad + 30.0236 \, (5. + 12.5203\alpha) \, (5. + 35.0497\alpha)$
$\quad + 0.0000414631 \text{Max}\,[0, -5. - 35.0497\alpha]$
$\quad + 0.0000217799 \text{Max}\,[0, -5. - 12.5203\alpha]$
$\quad + 49.9862 \text{Max}\,[0, -5. + 4.99999\alpha]$
$\quad + 16.1108 \text{Max}\,[0, 1200 - (5. - 4.99999\alpha) \, (5. + 12.5203\alpha) \, (5. + 35.0497\alpha)]$

**** Iteration 1
$\alpha_\ell \to 0. \quad \phi_\ell \to 20358.7 \quad \alpha_1 \to 0.1 \quad \phi_1 \to 20091.6$
$\alpha_2 \to 0.2 \quad \phi_2 \to 19763.7$

**** Iteration 2
$\alpha_\ell \to 0. \quad \phi_\ell \to 20358.7 \quad \alpha_1 \to 0.2 \quad \phi_1 \to 19763.7$
$\alpha_2 \to 0.4 \quad \phi_2 \to 19773.5$

654   Chapter 9   Constrained Nonlinear Problems

$$\text{Initial three points} \to \begin{pmatrix} 0. \\ 0.2 \\ 0.4 \end{pmatrix} \quad \text{Function values} \to \begin{pmatrix} 20358.7 \\ 19763.7 \\ 19773.5 \end{pmatrix}$$

$$\begin{pmatrix} \alpha_\ell & \alpha_m & \alpha_u & \alpha_q & \phi_\ell & \phi_m & \phi_u & \phi_q & \text{Conv.} \\ 0. & 0.2 & 0.4 & 0.296746 & 20358.7 & 19763.7 & 19773.5 & 19588.1 & 0.00534 \\ 0.2 & 0.296746 & 0.4 & -- & 19763.7 & 19588.1 & 19773.5 & -- & 0.00534 \end{pmatrix}$$

Step length $\to 0.296746$

$\lambda \to \begin{pmatrix} 16.1108 & 0.0000217799 & 49.9862 & 0.0000414631 \end{pmatrix}$

Updating Hessian

$s \to \begin{pmatrix} 3.71534 \\ -1.48373 \\ 10.4009 \end{pmatrix} \quad q \to \begin{pmatrix} -228.684 \\ -1128.53 \\ -43.986 \end{pmatrix}$

$H.s \to \begin{pmatrix} 3.71534 \\ -1.48373 \\ 10.4009 \end{pmatrix} \quad qT.s \to 367.305$

$s.H.sT \to 124.184 \quad \theta \to 1$

$\gamma \to \begin{pmatrix} -228.684 \\ -1128.53 \\ -43.986 \end{pmatrix}$

$\gamma.\gamma\, T \to \begin{pmatrix} 52296.3 & 258077. & 10058.9 \\ 258077. & 1.27359 \times 10^6 & 49639.7 \\ 10058.9 & 49639.7 & 1934.77 \end{pmatrix}$

$H.s.sT.H \to \begin{pmatrix} 13.8038 & -5.51256 & 38.6428 \\ -5.51256 & 2.20145 & -15.4321 \\ 38.6428 & -15.4321 & 108.178 \end{pmatrix}$

$H \to \begin{pmatrix} 143.267 & 702.668 & 27.0745 \\ 702.668 & 3468.36 & 135.27 \\ 27.0745 & 135.27 & 5.39635 \end{pmatrix}$

New point $\to \{8.71534, 3.51627, 15.4009\}$

------------------ **RSQP Iteration 2** ------------------

Current point $\to \begin{pmatrix} 8.71534 & 3.51627 & 15.4009 \end{pmatrix}$

$\text{Function values} \begin{pmatrix} f \\ g \\ h \end{pmatrix} \to \begin{pmatrix} 7858.9 \\ \{728.032, -8.71534, -3.51627, -15.4009\} \\ \{\} \end{pmatrix}$

Gradients: $\nabla f \to \begin{pmatrix} 631.281 \\ 1088.94 \\ 414.713 \end{pmatrix}$

$\nabla g \to \begin{pmatrix} -54.1537 & -134.224 & -30.6455 \\ -1 & 0 & 0 \\ 0 & -1 & 0 \\ 0 & 0 & -1 \end{pmatrix}$

$\nabla h \to \{\}$

```
-------QP subproblem for direction-------
Minimize
```
$$631.281 d_1 + 1088.94 d_2 + 414.713 d_3 + \frac{1}{2} \Big(d_3 \big(27.0745 d_1 + 135.27 d_2 + 5.39635 d_3\big)$$
$$+ d_1 \big(143.267 d_1 + 702.668 d_2 + 27.0745 d_3\big)$$
$$+ d_2 \big(702.668 d_1 + 3468.36 d_2 + 135.27 d_3\big) \Big)$$

$$\text{Subject to} \rightarrow \begin{pmatrix} 720.752 - 54.1537 d_1 - 134.224 d_2 - 30.6455 d_3 \leq 0 \\ -8.71534 - 1.d_1 \leq 0 \\ -3.51627 - 1.d_2 \leq 0 \\ -15.4009 - 1.d_3 \leq 0 \end{pmatrix}$$

Direction $\rightarrow (-1.91279 \quad -0.485211 \quad 29.0243)$ $\|d\| \rightarrow 29.0913$

Lagrange multipliers $\rightarrow (14.8115 \quad 0.0000289631 \quad 0.0000835147 \quad 0.0000127179)$

$\phi \rightarrow 48.0314 \, (8.71534 - 1.91279\alpha) \, (3.51627 - 0.485211\alpha)$
$+ 30.0236 \, (8.71534 - 1.91279\alpha) \, (15.4009 + 29.0243\alpha)$
$+ 43.5255 \, (3.51627 - 0.485211\alpha) \, (15.4009 + 29.0243\alpha)$
$+ 0.0000270905 \text{Max} [0, -15.4009 - 29.0243\alpha]$
$+ 24.9931 \text{Max} [0, -3.51627 + 0.485211\alpha]$
$+ 0.0000289631 \text{Max} [0, -8.71534 + 1.91279\alpha]$
$+ 15.4612 \text{Max} [0,$
$\quad 1200 - (8.71534 - 1.91279\alpha) \, (3.51627 - 0.485211\alpha) \, (15.4009 + 29.0243\alpha)]$

**** Iteration 1
$\alpha_\ell \rightarrow 0. \quad \phi_\ell \rightarrow 19115.1 \quad \alpha_1 \rightarrow 0.1 \quad \phi_1 \rightarrow 19055.$
$\alpha_2 \rightarrow 0.2 \quad \phi_2 \rightarrow 19041.6$

**** Iteration 2
$\alpha_\ell \rightarrow 0. \quad \phi_\ell \rightarrow 19115.1 \quad \alpha_1 \rightarrow 0.2 \quad \phi_1 \rightarrow 19041.6$
$\alpha_2 \rightarrow 0.4 \quad \phi_2 \rightarrow 19144.9$

Initial three points $\rightarrow \begin{pmatrix} 0. \\ 0.2 \\ 0.4 \end{pmatrix}$ Function values $\rightarrow \begin{pmatrix} 19115.1 \\ 19041.6 \\ 19144.9 \end{pmatrix}$

$$\begin{pmatrix} \alpha_\ell & \alpha_m & \alpha_u & \alpha_q & \phi_\ell & \phi_m & \phi_u & \phi_q & \text{Conv.} \\ 0. & 0.2 & 0.4 & 0.183159 & 19115.1 & 19041.6 & 19144.9 & 19040.7 & 0.000014 \\ 0. & 0.183159 & 0.2 & -- & 19115.1 & 19040.7 & 19041.6 & -- & 0.000014 \end{pmatrix}$$

Step length $\rightarrow 0.183159$

$\lambda \rightarrow (15.4612 \quad 0.0000289631 \quad 24.9931 \quad 0.0000270905)$

Updating Hessian

$s \rightarrow \begin{pmatrix} -0.350345 \\ -0.0888708 \\ 5.31606 \end{pmatrix} \quad q \rightarrow \begin{pmatrix} -23.8977 \\ -139.796 \\ 54.6882 \end{pmatrix}$

$H.s \rightarrow \begin{pmatrix} 31.2899 \\ 164.69 \\ 7.18038 \end{pmatrix} \quad q^T.s \rightarrow 311.522$

$s.H.s^T \rightarrow 12.573 \quad \theta \rightarrow 1$

$$\gamma \to \begin{pmatrix} -23.8977 \\ -139.796 \\ 54.6882 \end{pmatrix}$$

$$\gamma . \gamma^T \to \begin{pmatrix} 571.102 & 3340.8 & -1306.92 \\ 3340.8 & 19542.8 & -7645.18 \\ -1306.92 & -7645.18 & 2990.8 \end{pmatrix}$$

$$H.s.sT.H \to \begin{pmatrix} 979.055 & 5153.12 & 224.673 \\ 5153.12 & 27122.8 & 1182.54 \\ 224.673 & 1182.54 & 51.5579 \end{pmatrix}$$

$$H \to \begin{pmatrix} 67.2307 & 303.535 & 5.00965 \\ 303.535 & 1373.87 & 16.6746 \\ 5.00965 & 16.6746 & 10.8963 \end{pmatrix}$$

New point $\to \{8.365, 3.4274, 20.7169\}$

NonOptimum point $\to (b \to 8.365 \quad h \to 3.4274 \quad \ell \to 20.7169)$

Function values $\to \begin{pmatrix} 9670.61 \\ \{606.041, -8.365, -3.4274, -20.7169\} \\ \{\} \end{pmatrix}$

Iterations performed $\to 2$

The complete solution is obtained as follows:

{sol, hist} = RSQP[f, cons, vars, pt, MaxIterations → 30,
  QPConstraintFactor → 0.99, StepLengthUsing → QuadraticSearch];

Optimum point $\to (b \to 11.6188 \quad h \to 8.04337 \quad \ell \to 12.8401)$

Function values $\to \begin{pmatrix} 13463.1 \\ \{0.0351209, -11.6188, -8.04337, -12.8401\} \\ \{\} \end{pmatrix}$

Iterations performed $\to 22$

TableForm[hist]

Point	d	‖d‖	φ
5.	12.5203		
5.	-4.99999	37.5532	20358.7
5.	35.0497		
8.71534	-1.91279		
3.51627	-0.485211	29.0913	19588.1
15.4009	29.0243		
8.365	12.6727		
3.4274	-1.82184	12.8149	19040.7
20.7169	0.553698		
9.81189	7.66351		
3.21939	0.999483	9.57054	18918.9
20.7802	-5.64508		

19.2552	−0.608446		
4.451	1.12082	3.13925	14993.1
13.824	−2.86852		
18.7014	−3.48493		
5.47129	0.534089	3.83334	14495.8
11.2128	1.50482		
13.5769	−1.1555		
6.25665	0.450391	1.51503	13822.4
13.4256	0.870221		
12.3373	−0.850489		
6.73983	0.6351	1.10076	13581.7
14.3591	−0.291526		
11.1902	0.348036		
7.5964	0.191598	0.751008	13504.3
13.9659	−0.637317		
11.5658	0.123022		
7.80318	0.206211	0.531421	13468.8
13.2781	−0.474079		
11.6784	−0.0177588		
7.99193	0.0268227	0.0339148	13464.
12.8442	−0.0107416		
11.6706	−0.142248		
8.00377	0.122383	0.190458	13463.5
12.8394	−0.0325883		
11.6192	−0.0480943		
8.04792	0.0355683	0.0598335	13463.6
12.8277	0.00137353		
11.5931	0.0783625		
8.06729	−0.153633	0.235217	13463.5
12.8284	0.159949		
11.6114	−0.0068949		
8.0314	0.0373916	0.0630097	13463.4
12.8658	−0.050245		
11.6101	−0.000329693		
8.03842	0.0235217	0.0426765	13463.4
12.8564	−0.0356077		
11.61	0.753902		
8.04879	−0.262045	0.899505	13463.4
12.8407	−0.414818		
11.6381	−0.0284982		
8.03901	0.0181158	0.0339467	13463.4
12.8252	0.00347059		
11.636	−0.0249663		
8.04031	0.016168	0.02986	13463.4
12.8255	0.00262668		

11.6335	-0.0337355		
8.04195	-0.00533908	0.0577096	13463.4
12.8257	0.0465167		
11.6218	-0.00920864		
8.0401	0.0100063	0.01458	13463.4
12.8419	-0.00525853		
11.6188	-0.00698019		
8.04337	0.00365139	0.00819461	13463.4
12.8401	0.00225741		

The same solution was obtained in Chapter 4 using the KT conditions and was verified to be the minimum using sufficient conditions. It is interesting to note that the basic SQP method, with the ActiveSetDual method for the QP solution, gets stuck at the first iteration because the QP has no solution. With the active set method for the QP solution, the method performs few iterations but leads to points that are farther away from the optimum and eventually stops again because QP has no solution. The ALPF method, however, has no difficulty with this problem. The method actually converges to the optimum in just two iterations.

SQP[f, cons, vars, pt, PrintLevel → 3, MaxIterations → 1];

Minimize f → 48.0314bh + 30.0236bℓ + 43.5255hℓ

∇f → $(48.0314h + 30.0236\ell \quad 48.0314b + 43.5255\ell \quad 30.0236b + 43.5255h)$

LE Constraints: $\begin{pmatrix} g_1 \\ g_2 \\ g_3 \\ g_4 \end{pmatrix} \to \begin{pmatrix} 1200 - bh\ell \le 0 \\ -b \le 0 \\ -h \le 0 \\ -\ell \le 0 \end{pmatrix}$

$\begin{pmatrix} \nabla g_1 \\ \nabla g_2 \\ \nabla g_3 \\ \nabla g_4 \end{pmatrix} \to \begin{pmatrix} -h\ell & -b\ell & -bh \\ -1 & 0 & 0 \\ 0 & -1 & 0 \\ 0 & 0 & -1 \end{pmatrix}$

------------------ SQP Iteration 1 ------------------

Current point → $(5 \quad 5 \quad 5)$

Function values $\begin{pmatrix} f \\ g \\ h \end{pmatrix} \to \begin{pmatrix} 3039.51 \\ \{1075, -5, -5, -5\} \\ \{\} \end{pmatrix}$

Gradients: ∇f → $\begin{pmatrix} 390.275 \\ 457.784 \\ 367.745 \end{pmatrix}$

∇g → $\begin{pmatrix} -25 & -25 & -25 \\ -1 & 0 & 0 \\ 0 & -1 & 0 \\ 0 & 0 & -1 \end{pmatrix}$

∇h → {}
-------QP subproblem for direction-------
Minimize $390.275d_1 + 457.784d_2 + 367.745d_3 + \frac{1}{2}\left(d_1^2 + d_2^2 + d_3^2\right)$

Subject to → $\begin{pmatrix} 1075 - 25d_1 - 25d_2 - 25d_3 \leq 0 \\ -5 - d_1 \leq 0 \\ -5 - d_2 \leq 0 \\ -5 - d_3 \leq 0 \end{pmatrix}$

-------Dual QP subproblem-------
Minimize $-31470.1u_1 - 385.275u_2 - 452.785u_3 - 362.745u_4$
$+ \frac{1}{2}\left(u_2\left(25u_1 + u_2\right) + u_3\left(25u_1 + u_3\right) + u_4\left(25u_1 + u_4\right)\right.$
$\left. + u_1\left(1875u_1 + 25u_2 + 25u_3 + 25u_4\right)\right)$

SQP::QPFailed: QP failed: status → NonOptimum, current point {5,5,5}.

ALPF[f, cons, vars, pt, PrintLevel → 3];
Minimize f → $48.0314 bh + 30.0236 b\ell + 43.5255 h\ell$
∇f → $(48.0314 h + 30.0236\ell \quad 48.0314 b + 43.5255\ell \quad 30.0236 b + 43.5255 h)$

LE Constraints: $\begin{pmatrix} g1 \\ g2 \\ g3 \\ g4 \end{pmatrix} \to \begin{pmatrix} 1200 - bh\ell \leq 0 \\ -b \leq 0 \\ -h \leq 0 \\ -\ell \leq 0 \end{pmatrix}$

$\begin{pmatrix} \nabla g1 \\ \nabla g2 \\ \nabla g3 \\ \nabla g4 \end{pmatrix} \to \begin{pmatrix} -h\ell & -b\ell & -bh \\ -1 & 0 & 0 \\ 0 & -1 & 0 \\ 0 & 0 & -1 \end{pmatrix}$

------------------ **ALPF Iteration 1** ------------------
Current point → $(5 \quad 5 \quad 5)$

Function values $\begin{pmatrix} f \\ g \\ h \end{pmatrix} \to \begin{pmatrix} 3039.51 \\ \{1075,, -5, -5, -5\} \\ \{\} \end{pmatrix}$

Max Violation → 1075
$\mu \to 10 \quad v \to \{\}$

$r \to \begin{pmatrix} 0 \\ 0 \\ 0 \\ 0 \end{pmatrix} \quad u \to \begin{pmatrix} 0 \\ 0 \\ 0 \\ 0 \end{pmatrix}$

Solution → {13461.9, {b → 11.6357, h → 8.03321, ℓ → 12.8339}}

------------------ **ALPF Iteration 2** ------------------
Current point → $(11.6357 \quad 8.03321 \quad 12.8339)$

Function values $\begin{pmatrix} f \\ g \\ h \end{pmatrix} \to \begin{pmatrix} 13460.5 \\ \{0.383643, -11.6357, -8.03321, -12.8339\} \\ \{\} \end{pmatrix}$

```
Max Violation → 0.383643
μ → 10   v → {}
       ⎛0.383643⎞        ⎛7.67286⎞
       ⎜   0    ⎟        ⎜   0   ⎟
r →    ⎜   0    ⎟   u →  ⎜   0   ⎟
       ⎝   0    ⎠        ⎝   0   ⎠
Solution → {13463.3, {b → 11.6387, h → 8.0278, ℓ → 12.8435}}
Optimum point: Iterations performed → 2
Point → (b → 11.6387   h → 8.0278   ℓ → 12.8435)
                ⎛f⎞        ⎛              13463.4                  ⎞
Function values ⎜g⎟   →    ⎜{-0.00962203, -11.6387, -8.0278, -12.8435}⎟
                ⎝h⎠        ⎝                 {}                    ⎠
μ → 10   Max Violation → 0
                    ⎛7.67286⎞
                    ⎜   0   ⎟
v → {}    u →       ⎜   0   ⎟
                    ⎝   0   ⎠
```

9.7 Problems

Normalization

For the following problems, express the constraints in the standard and normalized form.

9.1. Minimize $f(x_1, x_2) = x_1 + \frac{x_1}{x_2^2} + \frac{x_2}{x_1}$

Subject to $\begin{pmatrix} x_1 + x_2 \geq 2 \\ x_1, x_2 \geq 0 \end{pmatrix}$

9.2. Maximize $f(x_1, x_2) = (x_1 - 2)^2 + (x_2 - 10)^2$

Subject to $\begin{pmatrix} x_1^2 + x_2^2 \leq 50 \\ x_1^2 + x_2^2 + 2x_1x_2 - x_1 - x_2 + 20 \geq 0 \\ x_1, x_2 \geq 0 \end{pmatrix}$

9.3. Maximize $f(x_1, x_2, x_3) = x_1 x_2 + x_3^2$

Subject to $x_1^2 + 2x_2^2 + 3x_3^2 = 4$

9.4. Maximize $f(x_1, x_2, x_3) = 7 \times 10^{-9} x_1^4 x_2 x_3^2$

Subject to $\begin{pmatrix} (x_1^2 x_3^2)/10^7 \leq 0.7 \\ x_1^2 x_2 = 700 \\ x_1 \leq 7 \end{pmatrix}$

9.5. Maximize $f = 5x_1 + e^{-2x_2} - e^{-x_2} + x_1 x_3 + 4x_3 + 6x_4 + \frac{5x_5}{x_5+1} + \frac{6x_6}{x_6+1}$

Subject to $\begin{pmatrix} x_1 + x_2 + x_3 + x_4 + x_5 + x_6 \leq 10 \\ x_1 + x_3 + x_4 \leq 5 \\ x_1 - x_2^2 + x_3 + x_5 + x_6^2 \leq 5 \\ x_2 + 2x_4 + x_5 + 0.8x_6 = 5 \\ x_3^2 + x_5^2 + x_6^2 = 5 \end{pmatrix}$

ALPF Method

Solve the following problems using the augmented Lagrangian penalty function method. Start with default parameter values but adjust if necessary. For two-variable problems, verify solutions graphically.

9.6. Minimize $f(x, y) = x^2 + y^2 - \text{Log}[x^2 y^2]$
Subject to $x \leq \text{Log}[y] \quad x \geq 1 \quad y \geq 1$
Starting point: $\{5, 5\}$

9.7. Minimize $f(x, y, z) = x + y + z$
Subject to $x^{-2} + x^{-2} y^{-2} + x^{-2} y^{-2} z^{-2} \leq 1$
Starting point: $\{1, 2, 3\}$

9.8. Minimize $f(x_1, x_2) = x_1 + \frac{x_1}{x_2^2} + \frac{x_2}{x_1}$
Subject to $\begin{pmatrix} x_1 + x_2 \geq 2 \\ x_1, x_2 \geq 0 \end{pmatrix}$
Starting point: $\{1, 2\}$

9.9. Maximize $f(x_1, x_2) = (x_1 - 2)^2 + (x_2 - 10)^2$
Subject to $\begin{pmatrix} x_1^2 + x_2^2 \leq 50 \\ x_1^2 + x_2^2 + 2x_1 x_2 - x_1 - x_2 + 20 \geq 0 \\ x_1, x_2 \geq 0 \end{pmatrix}$
Starting point: $\{1, 2\}$

9.10. Minimize $f(x, y) = x^2 + 2yx + y^2 - 15x - 20y$
Subject to $\begin{pmatrix} x^2 + y^2 \leq 20 \\ x^2 - y^2 \leq 10 \end{pmatrix}$
Starting point: $\{1, 2\}$

Chapter 9 Constrained Nonlinear Problems

9.11. Minimize $f(x, y) = \frac{1}{xy}$

Subject to $\begin{pmatrix} x + y \leq 5 \\ x, y \geq 1 \end{pmatrix}$

9.12. Minimize $f(x_1, x_2) = \frac{8x_1 + 6x_2 - 5}{-4x_1 + 2x_2 - 40}$

Subject to $\begin{pmatrix} x_1 + x_2 = 10 \\ x_1 \geq 0 \\ 3x_1 - 5x_2 \leq 10 \end{pmatrix}$

Starting point: $\{1, 5\}$

9.13. Minimize $f(x_1, x_2, x_3) = x_1^2 + x_2^2/4 + x_3^2/9 - 1$

Subject to $x_1^2 + x_2^2 + x_3^2 = 1$

Starting point: $\{1, 2, 3\}$

9.14. Minimize $f(x_1, x_2, x_3) = x_1^2 + x_2^2/4 + x_3^2/9 - 1$

Subject to $x_1^2 + x_2^2 + x_3^2 \leq 1$

Starting point: $\{1, 2, 3\}$

9.15. Minimize $f(x_1, x_2, x_3) = 1/(1 + x_1^2 + x_2^2 + x_3^2)$

Subject to $\begin{pmatrix} 2 - 3x_1^2 - 4x_2^2 - 5x_3^2 = 0 \\ 6x_1 + 7x_2 + 8x_3 = 0 \end{pmatrix}$

Starting point: $\{1, 2, 3\}$

9.16. Minimize $f(x_1, x_2, x_3) = 1/(1 + x_1^2 + x_2^2 + x_3^2)$

Subject to $\begin{pmatrix} x_1 + 2x_2 + 3x_3 = 0 \\ 4x_1^2 + 5x_2^2 + 6x_3^2 = 7 \end{pmatrix}$

Starting point: $\{-1, 1, -1\}$

9.17. Minimize $f(x_1, x_2, x_3) = 1/(1 + x_1^2 + x_2^2 + x_3^2)$

Subject to $\begin{pmatrix} x_1 + 2x_2 + 3x_3 = 0 \\ 4x_1^2 + 5x_2^2 + 6x_3^2 \leq 7 \end{pmatrix}$

Starting point: $\{-1, 1, -1\}$

9.18. Maximize $f(x_1, x_2, x_3) = x_1 x_2 + x_3^2$

Subject to $x_1^2 + 2x_2^2 + 3x_3^2 = 4$

Starting point: $\{1, 2, 3\}$

9.19. Minimize $f(x_1, x_2, x_3) = x_1^2 + 9x_2^2 + x_3^2$
Subject to $x_1 x_2 \geq 1$
Starting point: $\{1, 2, 3\}$

9.20. Minimize $f(x_1, x_2, x_3) = x_1^2 + 9x_2^2 + x_3^2$
Subject to $x_1 x_2 x_3 \geq 1$
Starting point: $\{1, 2, 3\}$

9.21. Maximize $f(x_1, x_2, x_3) = 7 \times 10^{-9} x_1^4 x_2 x_3^2$
Subject to $\begin{pmatrix} (x_1^2 x_3^2)/10^7 \leq 0.7 \\ x_1^2 x_2 = 700 \\ x_1 \leq 7 \end{pmatrix}$
Starting point: $\{1, 2, 3\}$

9.22. Maximize $f = 5x_1 + e^{-2x_2} - e^{-x_2} + x_1 x_3 + 4x_3 + 6x_4 + \frac{5x_5}{x_5+1} + \frac{6x_6}{x_6+1}$
Subject to $\begin{pmatrix} x_1 + x_2 + x_3 + x_4 + x_5 + x_6 \leq 10 \\ x_1 + x_3 + x_4 \leq 5 \\ x_1 - x_2^2 + x_3 + x_5 + x_6^2 \leq 5 \\ x_2 + 2x_4 + x_5 + 0.8x_6 = 5 \\ x_3^2 + x_5^2 + x_6^2 = 5 \end{pmatrix}$
Starting point: $\{1, 1, \ldots, 1\}$

9.23. Dust from an older cement manufacturing plant is a major source of dust pollution in a small community. The plant currently emits 2 pounds of dust per barrel of cement produced. The Environmental Protection Agency (EPA) has asked the plant to reduce this pollution by 85% (1.7 lbs/barrel). There are two models of electrostatic dust collectors that the plant can install to control dust emission. The higher-efficiency model would reduce emissions by 1.8 lbs/barrel and would cost $0.70/barrel to operate. The lower-efficiency model would reduce emissions by 1.5 lbs/barrel and would cost $0.50/barrel to operate. Since the higher-efficiency model reduces more than the EPA required amount and the lower-efficiency less than the required, the plant has decided to install one of each. If the plant has a capacity to produce 3 million barrels of cement per year, how many barrels of cement should be produced using each dust control model to meet the EPA requirements at a minimum cost? Formulate the situation as an optimization problem. Use the ALPF method to find an optimum solution.

9.24. A thin steel plate, 10 in. wide and 1/2 in. thick and carrying a tensile force of $T = 150,000$ lbs, is to be connected to a structure through high strength bolts, as shown in Figure 9.15. The plate must extend by a fixed distance $L = 120$ in. from the face of the structure. The material for the steel plate can resist a stress up to $F_u = 58{,}000$ lbs/in^2. The bolts are chosen as A325 type (standard bolts available from a number of steel manufacturers). The bolts are arranged in horizontal and vertical rows. The spacing between horizontal rows is $s = 3d$ inches, where d is the diameter of the bolt and that between vertical rows is $g \geq 3d$ inches. The bolts must be at least 1.5d inches from the edges of the plate or the structure.

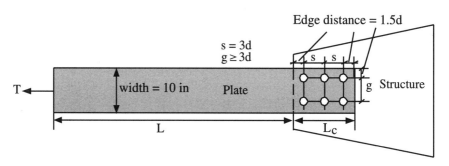

FIGURE 9.15 A steel plate connected to a structure using steel bolts.

The design problem is to determine the diameter of bolts, total number of bolts required, and the arrangement of bolts (i.e., the number of horizontal and vertical rows) in order to minimize total cost of the plate and the bolts. Assuming the cost of a steel plate is $2 per inch and that of a bolt is $5 per bolt, formulate the connection design problem as an optimization problem. Use the ALPF method to find an optimum solution.

The pertinent design requirements are as follows:

(a) Tensile force that one A325 bolt can support is given by the smaller of the following two values, each based on a different failure criteria.

Based on failure of bolt = $12{,}000\pi d^2$ lbs

Based on failure of plate around bolt hole = $2.4 dt F_u$ lbs

where d = diameter of the bolt, t = thickness of the plate, and F_u = ultimate tensile strength of the plate material = 58,000 lbs/in^2.

(b) Vertical rows of bolts (those along the width of the plate) make the plate weaker. Taking these holes into consideration, the maximum tensile force that the plate can support is given by the following equation:

$$\text{Tensile load capacity of plate} = 0.75 F_u (w - n_v d_h)$$

where w = width of plate, n_v = number of bolts in one vertical row across the plate width, and d_h = bolt hole diameter ($= d + 1/8$ in).

(c) Practical requirements. The number of bolts in each row is the same. Each row must have at least two bolts, and there must be at least two rows of bolts. The smallest bolt diameter allowed is $1/2$ in.

9.25. Consider the cantilever beam-mass system shown in Figure 9.16. The beam cross-section is rectangular. The goal is to select cross-sectional dimensions (b and h) to minimize the weight of the beam while keeping the fundamental vibration frequency (ω) larger than 8 rad/sec. Use the ALPF method to find an optimum solution.

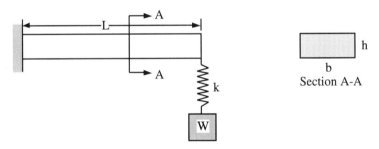

FIGURE 9.16 A rectangular cross-section cantilever beam with a suspended mass.

The numerical data and various equations for the problem are as follows:

Fundamental vibration frequency	$\omega = \sqrt{k_e/m}$ radians/sec
Equivalent spring constant, k_e	$\dfrac{1}{k_e} = \dfrac{1}{k} + \dfrac{L^3}{3EI}$
Mass attached to the spring	$m = W/g$
Gravitational constant	$g = 386$ in/sec^2
Weight attached to the spring	$W = 60$ lbs
Length of beam	$L = 15$ in
Modulus of elasticity	$E = 30 \times 10^6$ lbs/in^2
Spring constant	$k = 10$ lbs/in^2

Moment of inertia	$I = \frac{bh^3}{12}$ in^4
Width of beam cross-section	0.5 in $\leq b \leq$ 1 in
Height of beam cross-section	0.2 in $\leq h \leq$ 2 in
Unit weight of beam material	0.286 lbs/in^3

9.26. Modern aircraft structures and other sheet metal construction require stiffeners that are normally of I-, Z-, or C-shaped cross sections. An I-shaped cross section, shown in Figure 9.17, has been selected for the present situation. Since a large number of these stiffeners are employed in a typical aircraft, it is important to optimally proportion the dimensions, b, t, and h.

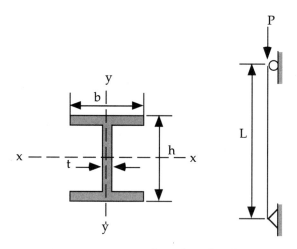

FIGURE 9.17 An I-shaped steel stiffener.

For preliminary investigation, a stiffener is treated as an axially loaded column, as shown in the figure. The goal is to have a least-volume design while meeting the following yield-stress and buckling-load limits.

Yield-stress limit	$P/A \leq \sigma_y$
Overall buckling load	$\dfrac{\pi^2 E I_{min}}{L^2} \geq P$
Flange buckling load	$0.43 \dfrac{\pi^2 E}{12(1-\nu^2)} \left(\dfrac{2t}{b}\right)^2 A \geq P$

Web buckling load	$4\dfrac{\pi^2 E}{12(1-\nu^2)}\left(\dfrac{t}{h}\right)^2 A \geq P$
Area of cross-section, A	$A = 2bt + (h-2t)t$
Minimum moment of inertia, I_{min}	$\text{Min}[I_x, I_y]$ $I_x = \frac{1}{6}bt^3 + \frac{1}{2}bt(h-t)^2 + \frac{1}{12}t(h-2t)^3$ $I_y = \frac{1}{6}tb^3 + \frac{1}{12}t^3(h-2t)$
Applied load, P	2000 lbs
Yield stress, σ_y	25,000 lbs/in^2
Young's modulus E	10^7 lbs/in^2
Poisson's ratio, ν	0.3
Length of stiffener L	25 in

Formulate the problem in the standard optimization problem format. Use the ALPF method to find an optimum solution.

9.27. Consider the optimum design of a rectangular-reinforced concrete beam shown in Figure 9.18. There is steel reinforcement near the bottom. Formwork is required on three sides during construction. The beam must support a given bending moment. A least-cost design is required.

The bending strength of the beam is calculated from the following formula:

$$M_u = 0.90 A_s F_y d \left(1 - 0.59 \left(\dfrac{A_s}{bd}\right)\left(\dfrac{F_y}{f_c'}\right)\right)$$

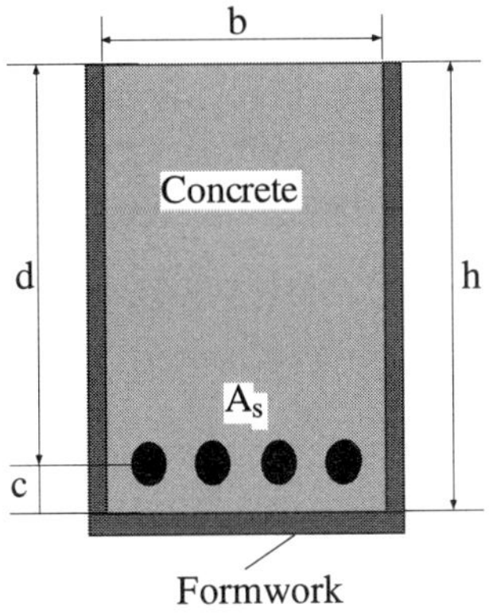

FIGURE 9.18 A reinforced concrete beam.

where F_y = specified yield strength of the steel, and f'_c is the specified compressive strength of concrete. The ductility requirements dictate minimum and maximum limits on steel ratio $\rho = A_s/bd$.

$$\rho_{min} \leq \rho \leq \rho_{max}$$

Use the following numerical data:

Maximum steel ratio	$\rho_{max} = 0.025$
Minimum steel ratio	$\rho_{min} = 0.0033$
Required moment capacity	$M_u \geq 400 \times 10^3$ N-m
Minimum beam width	$b \geq 300$ mm
Concrete cover	$c = 65$ mm
Maximum beam depth	$h \leq 1200$ mm
Concrete cost	$100/m^3$
Formwork cost	$2/m^2$
Steel reinforcement cost	$610/ton (1 ton = 907.18 kg)
Density of steel	7850 kg/m^3
Yield stress of steel, F_y	420 MPa
Ultimate concrete strength, f'_c	35 MPa

Formulate the problem of determining the cross-section variables and amount of steel reinforcement to meet all design requirements at a minimum cost. Assume a unit beam length for cost computations. Use the ALPF method to find an optimum solution.

9.28. A small electronics company is planning to expand two of its manufacturing plants. The additional annual revenue expected from the two plants is as follows:

From plant 1 : $0.00002 x_1^2 - x_2$ From plant 2 : $0.00001 x_2^2$

where x_1 and x_2 are the investments made into upgrading the facilities. Each plant requires a minimum investment of $30,000. The company can borrow a maximum of $100,000 for this upgrade to be paid back in yearly installments in 10 years at an annual interest rate of 12%. The revenue that the company generates can earn interest at an annual rate of 10%. After the 10-year period, the salvage value of the upgrades is expected to be as follows:

$$\text{For plant 1}: 0.1\,x_1 \quad \text{For plant 2}: 0.15\,x_2$$

Formulate an optimization problem to maximize the net present worth of these upgrades. Use the ALPF method to find an optimum solution.

Linearization

Construct linear approximations of the following problem around the given point. For two-variable problems, graphically compare the linearized problem with the original problem.

9.29. Minimize $f(x, y, z) = x + y + z$

Subject to $x^{-2} + x^{-2}y^{-2} + x^{-2}y^{-2}z^{-2} \leq 1$

Starting point: $\{1, 2, 3\}$

9.30. Minimize $f(x_1, x_2) = x_1 + \frac{x_1}{x_2^2} + \frac{x_2}{x_1}$

Subject to $\begin{pmatrix} x_1 + x_2 \geq 2 \\ x_1, x_2 \geq 0 \end{pmatrix}$

Starting point: $\{1, 2\}$

9.31. Maximize $f(x_1, x_2) = (x_1 - 2)^2 + (x_2 - 10)^2$

Subject to $\begin{pmatrix} x_1^2 + x_2^2 \leq 50 \\ x_1^2 + x_2^2 + 2x_1x_2 - x_1 - x_2 + 20 \geq 0 \\ x_1, x_2 \geq 0 \end{pmatrix}$

Starting point: $\{1, 2\}$

9.32. Minimize $f(x_1, x_2, x_3) = x_1^2 + x_2^2/4 + x_3^2/9 - 1$

Subject to $x_1^2 + x_2^2 + x_3^2 = 1$

Starting point: $\{1, 2, 3\}$

9.33. Minimize $f(x_1, x_2, x_3) = x_1^2 + x_2^2/4 + x_3^2/9 - 1$

Subject to $x_1^2 + x_2^2 + x_3^2 \leq 1$

Starting point: $\{1, 2, 3\}$

SLP Method

Solve the following problems using the SLP method. Unless stated otherwise, try 50% move limits and adjust them if necessary. Show calculations for at least two iterations. For two-variable problems, verify the solutions graphically.

9.34. Minimize $f(x, y) = x^2 + y^2 - \text{Log}[x^2 y^2]$
Subject to $x \leq \text{Log}[y] \quad x \geq 1 \quad y \geq 1$
Starting point: $\{5, 5\}$ Move limit $\beta = 0.7$

9.35. Minimize $f(x, y, z) = x + y + z$
Subject to $x^{-2} + x^{-2}y^{-2} + x^{-2}y^{-2}z^{-2} \leq 1$
Starting point: $\{1, 2, 3\}$

9.36. Minimize $f(x_1, x_2) = x_1 + \frac{x_1}{x_2^2} + \frac{x_2}{x_1}$
Subject to $\begin{pmatrix} x_1 + x_2 \geq 2 \\ x_1, x_2 \geq 0 \end{pmatrix}$
Starting point: $\{1, 2\}$

9.37. Maximize $f(x_1, x_2) = (x_1 - 2)^2 + (x_2 - 10)^2$
Subject to $\begin{pmatrix} x_1^2 + x_2^2 \leq 50 \\ x_1^2 + x_2^2 + 2x_1 x_2 - x_1 - x_2 + 20 \geq 0 \\ x_1, x_2 \geq 0 \end{pmatrix}$
Starting point: $\{1, 2\}$

9.38. Minimize $f(x_1, x_2, x_3) = x_1^2 + x_2^2/4 + x_3^2/9 - 1$
Subject to $x_1^2 + x_2^2 + x_3^2 = 1$
Starting point: $\{1, 2, 3\}$

9.39. Minimize $f(x_1, x_2, x_3) = x_1^2 + x_2^2/4 + x_3^2/9 - 1$
Subject to $x_1^2 + x_2^2 + x_3^2 \leq 1$
Starting point: $\{1, 2, 3\}$

9.40. Maximize $f(x_1, x_2, x_3) = x_1 x_2 + x_3^2$
Subject to $x_1^2 + 2x_2^2 + 3x_3^2 = 4$
Starting point: $\{1, 2, 3\}$

9.41. Minimize $f(x_1, x_2, x_3) = x_1^2 + 9x_2^2 + x_3^2$
Subject to $x_1 x_2 \geq 1$
Starting point: $\{1, 2, 3\}$

9.42. Minimize $f(x_1, x_2, x_3) = x_1^2 + 9x_2^2 + x_3^2$

Subject to $x_1 x_2 x_3 \geq 1$

Starting point: $\{1, 2, 3\}$

9.43. Maximize $f = 5x_1 + e^{-2x_2} - e^{-x_2} + x_1 x_3 + 4x_3 + 6x_4 + \frac{5x_5}{x_5+1} + \frac{6x_6}{x_6+1}$

Subject to $\begin{pmatrix} x_1 + x_2 + x_3 + x_4 + x_5 + x_6 \leq 10 \\ x_1 + x_3 + x_4 \leq 5 \\ x_1 - x_2^2 + x_3 + x_5 + x_6^2 \leq 5 \\ x_2 + 2x_4 + x_5 + 0.8x_6 = 5 \\ x_3^2 + x_5^2 + x_6^2 = 5 \end{pmatrix}$

Starting point: $\{1, 1, \ldots, 1\}$

SQP Method

Solve the following problems using the SQP method. Start with default parameter values but adjust if necessary. For two-variable problems, verify the solutions graphically.

9.44. Minimize $f(x, y) = x^2 + y^2 - \text{Log}[x^2 y^2]$

Subject to $x \leq \text{Log}[y] \quad x \geq 1 \quad y \geq 1$

Starting point: $\{5, 5\}$

9.45. Minimize $f(x, y, z) = x + y + z$

Subject to $x^{-2} + x^{-2}y^{-2} + x^{-2}y^{-2}z^{-2} \leq 1$

Starting point: $\{1, 2, 3\}$

9.46. Minimize $f(x_1, x_2) = x_1 + \frac{x_1}{x_2^2} + \frac{x_2}{x_1}$

Subject to $\begin{pmatrix} x_1 + x_2 \geq 2 \\ x_1, x_2 \geq 0 \end{pmatrix}$

Starting point: $\{1, 2\}$

9.47. Maximize $f(x_1, x_2) = (x_1 - 2)^2 + (x_2 - 10)^2$

Subject to $\begin{pmatrix} x_1^2 + x_2^2 \leq 50 \\ x_1^2 + x_2^2 + 2x_1 x_2 - x_1 - x_2 + 20 \geq 0 \\ x_1, x_2 \geq 0 \end{pmatrix}$

Starting point: $\{1, 2\}$

9.48. Minimize $f(x, y) = x^2 + 2yx + y^2 - 15x - 20y$

Subject to $\begin{pmatrix} x^2 + y^2 \le 20 \\ x^2 - y^2 \le 10 \end{pmatrix}$

Starting point: {1, 2}

9.49. Minimize $f(x, y) = \frac{1}{xy}$

Subject to $\begin{pmatrix} x + y \le 5 \\ x, y \ge 1 \end{pmatrix}$

(a) Use the starting point: {5, 5} and solve QP using the ActiveSetQP method.

(b) Use the starting point: {4, 5} and solve QP using the ActiveSetQP method. Why is the convergence slow compared to the first case?

9.50. Minimize $f(x_1, x_2) = \frac{8x_1 + 6x_2 - 5}{-4x_1 + 2x_2 - 40}$

Subject to $\begin{pmatrix} x_1 + x_2 = 10 \\ x_1 \ge 0 \\ 3x_1 - 5x_2 \le 10 \end{pmatrix}$

Starting point: {−1, 5}

9.51. Minimize $f(x_1, x_2, x_3) = x_1^2 + x_2^2/4 + x_3^2/9 - 1$

Subject to $x_1^2 + x_2^2 + x_3^2 = 1$

Starting point: {1, 2, 3}

9.52. Minimize $f(x_1, x_2, x_3) = x_1^2 + x_2^2/4 + x_3^2/9 - 1$

Subject to $x_1^2 + x_2^2 + x_3^2 \le 1$

Starting point: {1, 2, 3}

9.53. Minimize $f(x_1, x_2, x_3) = 1/(1 + x_1^2 + x_2^2 + x_3^2)$

Subject to $\begin{pmatrix} 2 - 3x_1^2 - 4x_2^2 - 5x_3^2 = 0 \\ 6x_1 + 7x_2 + 8x_3 = 0 \end{pmatrix}$

Starting point: {1, 2, 3}

9.54. Minimize $f(x_1, x_2, x_3) = 1/(1 + x_1^2 + x_2^2 + x_3^2)$

Subject to $\begin{pmatrix} x_1 + 2x_2 + 3x_3 = 0 \\ 4x_1^2 + 5x_2^2 + 6x_3^2 = 7 \end{pmatrix}$

Starting point: {−1, 1, −1}

9.55. Minimize $f(x_1, x_2, x_3) = 1/(1 + x_1^2 + x_2^2 + x_3^2)$

Subject to $\begin{pmatrix} x_1 + 2x_2 + 3x_3 = 0 \\ 4x_1^2 + 5x_2^2 + 6x_3^2 \leq 7 \end{pmatrix}$

Starting point: $\{-1, 1, -1\}$

9.56. Maximize $f(x_1, x_2, x_3) = x_1 x_2 + x_3^2$

Subject to $x_1^2 + 2x_2^2 + 3x_3^2 = 4$

Starting point: $\{1, 2, 3\}$

9.57. Minimize $f(x_1, x_2, x_3) = x_1^2 + 9x_2^2 + x_3^2$

Subject to $x_1 x_2 \geq 1$

Starting point: $\{1, 2, 3\}$

9.58. Minimize $f(x_1, x_2, x_3) = x_1^2 + 9x_2^2 + x_3^2$

Subject to $x_1 x_2 x_3 \geq 1$

Starting point: $\{1, 2, 3\}$

9.59. Maximize $f(x_1, x_2, x_3) = 7 \times 10^{-9} x_1^4 x_2 x_3^2$

Subject to $\begin{pmatrix} (x_1^2 x_3^2)/10^7 \leq 0.7 \\ x_1^2 x_2 = 700 \\ x_1 \leq 7 \end{pmatrix}$

Starting point: $\{1, 2, 3\}$

9.60. Maximize $f = 5x_1 + e^{-2x_2} - e^{-x_2} + x_1 x_3 + 4x_3 + 6x_4 + \frac{5x_5}{x_5+1} + \frac{6x_6}{x_6+1}$

Subject to $\begin{pmatrix} x_1 + x_2 + x_3 + x_4 + x_5 + x_6 \leq 10 \\ x_1 + x_3 + x_4 \leq 5 \\ x_1 - x_2^2 + x_3 + x_5 + x_6^2 \leq 5 \\ x_2 + 2x_4 + x_5 + 0.8x_6 = 5 \\ x_3^2 + x_5^2 + x_6^2 = 5 \end{pmatrix}$

Starting point: $\{1, 1, \ldots, 1\}$

9.61. Use the SQP method to solve the dust pollution problem 9.23.

9.62. Use the SQP method to solve the plate connection problem 9.24.

9.63. Use the SQP method to solve the cantilever beam-mass problem 9.25.

9.64. Use the SQP method to solve the stiffener design problem 9.26.

9.65. Use the SQP method to solve the reinforced concrete beam problem 9.27.

9.66. Use the SQP method to solve the electronics company problem 9.28.

RSQP Method

Solve the following problems using the RSQP method. Start with default parameter values but adjust if necessary. For two-variable problems, verify the solutions graphically.

9.67. Minimize $f(x, y, z) = x + y + z$

Subject to $x^{-2} + x^{-2}y^{-2} + x^{-2}y^{-2}z^{-2} \leq 1$

Starting point: $\{1, 2, 3\}$

9.68. Minimize $f(x_1, x_2) = x_1 + \frac{x_1}{x_2^2} + \frac{x_2}{x_1}$

Subject to $\begin{pmatrix} x_1 + x_2 \geq 2 \\ x_1, x_2 \geq 0 \end{pmatrix}$

Starting point: $\{1, 2\}$

9.69. Maximize $f(x_1, x_2) = (x_1 - 2)^2 + (x_2 - 10)^2$

Subject to $\begin{pmatrix} x_1^2 + x_2^2 \leq 50 \\ x_1^2 + x_2^2 + 2x_1x_2 - x_1 - x_2 + 20 \geq 0 \\ x_1, x_2 \geq 0 \end{pmatrix}$

Starting point: $\{1, 2\}$

9.70. Minimize $f(x, y) = x^2 + 2yx + y^2 - 15x - 20y$

Subject to $\begin{pmatrix} x^2 + y^2 \leq 20 \\ x^2 - y^2 \leq 10 \end{pmatrix}$

Starting point: $\{1, 2\}$

9.71. Minimize $f(x, y) = \frac{1}{xy}$

Subject to $\begin{pmatrix} x + y \leq 5 \\ x, y \geq 1 \end{pmatrix}$

9.72. Minimize $f(x_1, x_2) = \frac{8x_1 + 6x_2 - 5}{-4x_1 + 2x_2 - 40}$

Subject to $\begin{pmatrix} x_1 + x_2 = 10 \\ x_1 \geq 0 \\ 3x_1 - 5x_2 \leq 10 \end{pmatrix}$

Starting point: $\{-1, 5\}$

9.73. Minimize $f(x_1, x_2, x_3) = x_1^2 + x_2^2/4 + x_3^2/9 - 1$

Subject to $x_1^2 + x_2^2 + x_3^2 = 1$

Starting point: $\{1, 2, 3\}$

9.74. Minimize $f(x_1, x_2, x_3) = x_1^2 + x_2^2/4 + x_3^2/9 - 1$
Subject to $x_1^2 + x_2^2 + x_3^2 \leq 1$
Starting point: $\{1, 2, 3\}$

9.75. Minimize $f(x_1, x_2, x_3) = 1/(1 + x_1^2 + x_2^2 + x_3^2)$
Subject to $\begin{pmatrix} 2 - 3x_1^2 - 4x_2^2 - 5x_3^2 = 0 \\ 6x_1 + 7x_2 + 8x_3 = 0 \end{pmatrix}$
Starting point: $\{1, 2, 3\}$

9.76. Minimize $f(x_1, x_2, x_3) = 1/(1 + x_1^2 + x_2^2 + x_3^2)$
Subject to $\begin{pmatrix} x_1 + 2x_2 + 3x_3 = 0 \\ 4x_1^2 + 5x_2^2 + 6x_3^2 = 7 \end{pmatrix}$
Starting point: $\{-1, 1, -1\}$

9.77. Minimize $f(x_1, x_2, x_3) = 1/(1 + x_1^2 + x_2^2 + x_3^2)$
Subject to $\begin{pmatrix} x_1 + 2x_2 + 3x_3 = 0 \\ 4x_1^2 + 5x_2^2 + 6x_3^2 \leq 7 \end{pmatrix}$
Starting point: $\{-1, 1, -1\}$

9.78. Maximize $f(x_1, x_2, x_3) = x_1 x_2 + x_3^2$
Subject to $x_1^2 + 2x_2^2 + 3x_3^2 = 4$
Starting point: $\{1, 2, 3\}$

9.79. Minimize $f(x_1, x_2, x_3) = x_1^2 + 9x_2^2 + x_3^2$
Subject to $x_1 x_2 \geq 1$
Starting point: $\{1, 2, 3\}$

9.80. Minimize $f(x_1, x_2, x_3) = x_1^2 + 9x_2^2 + x_3^2$
Subject to $x_1 x_2 x_3 \geq 1$
Starting point: $\{1, 2, 3\}$

9.81. Use the RSQP method to solve the dust pollution problem 9.23.

9.82. Use the RSQP method to solve the plate connection problem 9.24.

9.83. Use the RSQP method to solve the cantilever beam-mass problem 9.25.

9.84. Use the RSQP method to solve the stiffener design problem 9.26.

9.85. Use the RSQP method to solve the reinforced concrete beam problem 9.27.

9.86. Use the RSQP method to solve the electronics company problem 9.28.

APPENDIX

An Introduction to *Mathematica*

Mathematica is a powerful tool for performing both symbolic and numerical calculations. It is available on all major computing platforms, including Macintosh, Windows, and UNIX systems. The *notebook* interface on all systems is very similar. A user normally enters *commands* in *cells* in a notebook window and *executes* these commands to see the results appear in the same notebook. Starting with version 3.0 of *Mathematica,* the notebook interface has become quite sophisticated. In addition to *Mathematica* input and output, the notebooks can now contain graphics and typeset text, including all mathematics symbols and complicated equations. The *style* of a cell determines whether it contains live input and output or just text. Similar to a word processor, several other specialized cell styles are available to create section headings, subheadings, etc. A large number of pull-down menus are available to manipulate information in different notebook cells.

Mathematica is a complex system and requires considerable practice to become proficient in its use. However, getting started with it and using it as an advanced calculator requires understanding of only a few basic concepts and knowledge of a few commands. This brief introduction is intended as a quick-start guide for the users of this text. Serious users should consult many excellent books on *Mathematica* listed at the end of this chapter. *Mathematica* has an excellent online Help system as well. This Help system has usage instructions and examples of all *Mathematica* commands. The entire *Mathematica* book is available online as well, and is accessible through the Help browser.

A.1 Basic Manipulations in *Mathematica*

Mathematica performs a wide variety of numeric, graphic, and symbolic operations. Expressions are entered using a syntax similar to that of common programming languages such as BASIC, FORTRAN, or C. Some of the operators available are: + (add), − (subtract), * (multiply), / (divide), ^ (exponentiate), and Sqrt[ ] (square root). The multiplication operator is optional between two variables.

A command can be split into any number of lines. The *Enter key* (or Option-Return) on the keyboard is used to actually execute the command line containing the insertion point (blinking vertical line). Note on some keyboards the usual Return key (the one used to move to the next line) is labelled as Enter. As far as *Mathematica* is concerned, this is still a Return key. A semicolon [;] is used at the end of a line of *Mathematica* input if no output is desired. This feature should be used freely to suppress printout of long intermediate expressions that no one cares to look at anyway.

1 + 5 + 32 * 43/34^3

$$\frac{29650}{4913}$$

As seen from this example, *Mathematica* normally does rational arithmetic. To get a numerical answer, the function N can be used:

N[1 + 5 + 32 * 43/34^3]

6.03501

The N function (or any other function requiring a single argument) can also be applied using the // operator, as follows:

1 + 5 + 32 * 43/34^3//N

6.03501

A.1.1 Caution on *Mathematica* Input Syntax

It is important to remember that *Mathematica* commands are case-sensitive. The built-in functions all start with an uppercase letter. Also you must be very careful in using brackets. Different brackets have different meanings. Arguments to functions must be enclosed in square brackets ([ ]). Curly braces ({ }) are used to define lists and matrices. Parentheses '( )' do not have any built-in function and therefore, can be used to group terms in expressions, if desired.

To indicate a multiplication operation, you can use either an asterisk (*) or simply one or more blank spaces between the two variables. When a number precedes a variable, even a blank is not necessary. Sometimes, this convention can lead to unexpected problems, as demonstrated in the following examples.

Suppose you want the product of variables *a* and *b*. A space is necessary between the two variables. Without a space, *Mathematica* simply defines a new symbol *ab* and does not give the intended product.

a = 10; b = 3;
2ab
2 ab

No space is needed between 2 and *a*, but there must be a space or an asterisk (*) between *a* and *b* to get the product 2*ab*.

2 a b
60

If we enter 2a^3b without any spaces, it is interpreted as $2a^3 b$ and not as $2a^{3b}$.

2a^3b
6000

In situations like these, it is much better to use parentheses to make the association clear. For example, there is no ambiguity in the following form:

(2a)^(3b)
512000000000

A.1.2 Using Built-In Functions

Mathematica has a huge library of built-in functions. A few common ones are introduced here. The functions *Simplify* and *Expand* are used to get expressions in simpler forms. Note the use of *Clear* to remove any previously defined values for *a*.

Clear[a];
Simplify$\left[a^2 + 3a + (a+b)a\right]$
a (3 + 2a + b)

Expand$\left[a^2 + 3a + (a + b)a\right]$

$3a + 2a^2 + ab$

Just like a programming language, we can assign names to expressions so that they can be referred to later. For example, we define an expression e1 as follows:

e1 = x + Sin[x] Cos[x^3] / Exp[x]

$x + E^{-x} \text{Cos}[x^3] \text{Sin}[x]$

With this definition, the expression on the righthand side of the equal sign can be manipulated easily. For example, it can be differentiated with respect to x as follows. For later reference, the derivative expression is denoted by de1.

de1 = D[e1, x]

$1 + E^{-x} \text{Cos}[x] \text{Cos}[x^3] - E^{-x} \text{Cos}[x^3] \text{Sin}[x] - 3E^{-x}x^2 \text{Sin}[x] \text{Sin}[x^3]$

We can integrate de1 as follows:

Integrate[de1, x]

$\frac{1}{2}\left(2x + E^{-x} \text{Sin}\left[x - x^3\right] + E^{-x} \text{Sin}\left[x + x^3\right]\right)$

This does not look like e1. But the following expansion, using the trigonometric identities, shows that it is indeed equal to e1. Note the use of the % symbol to refer to the result of the previous evaluation as a short cut.

Expand[%, Trig- > True]

$\frac{1}{2}\left(2x + 2E^{-x} \text{Cos}[x^3] \text{Sin}[x]\right)$

Note that % refers to the result of the previous evaluation and not the result appearing in the cell just above the current cell. In a notebook with many input cells, one can execute any cell simply by highlighting it (or placing the insertion point anywhere in the cell). The % will always refer to the most recent evaluation. In fact, the evaluation does not have to be in the current notebook and can be in any other notebook open at the same time. Therefore, the use of % as a shortcut is strongly discouraged. It is better to assign a variable to refer to an evaluation in other places.

Mathematica can also compute definite integrals involving complicated expressions using numerical integration.

NIntegrate[e1, {x, 0, 1}]

0.724231

A.1.3 Substitution Rule

The symbol /. (slash-period) is a useful operator and is used to substitute different variables or numerical values into existing expressions. For example, the expression e1 can be evaluated at $x = 1.5$, as follows:

e1 /. x->1.5

1.28346

The arrow symbol is a combination of a hyphen (-) and a greater than sign (>), and is known as a *rule* in *Mathematica*. Of course, more complicated substitutions are possible. The following example illustrates defining a new expression called e2 by substituting $x = a + \operatorname{Sin}[b]$ in expression e1.

e2 = e1 /. {x->a + Sin[b]}

$10 + \operatorname{Sin}[3] + E^{-10-\operatorname{Sin}[3]} \operatorname{Cos}[(10 + \operatorname{Sin}[3])^3] \operatorname{Sin}[10 + \operatorname{Sin}[3]]$

Substitute $a = 1$ and $b = 2$ in expression e2.

e2 /. {a->1, b->2}

$1 + \operatorname{Sin}[2] + E^{-10-\operatorname{Sin}[2]} \operatorname{Cos}[(1 + \operatorname{Sin}[2])^2] \operatorname{Sin}[1 + \operatorname{Sin}[2]]$

To get the numerical result after substituting $a = 1$ and $b = 2$ in expression e2, we can use the *N* function:

N[e2 /. {a->1, b->2}]

1.90928

Note that the same result can also be obtained as follows:

a = 1;
b = 2;
N[e2]

1.90928

However, the substitution form is much more desirable. This second form defines the symbols *a* and *b* to have the indicated numerical values. *Mathematica* will automatically substitute these values into any subsequent evaluation that involves *a* and *b*.

a + b x

1 + 2x

The only way to get symbolic expressions involving *a* and *b* later would be to explicitly clear these variables, as follows:

```
Clear[a, b];
a + b x
```

a + bx

A.2 Lists and Matrices

Lists are quantities enclosed in curly braces ({}). They are one of the basic data structures used by *Mathematica*. Many *Mathematica* functions expect arguments in the form of lists and return results as lists. When working with these functions, it is important to pay close attention to the nesting of braces. Two expressions that are otherwise identical, except that one is enclosed in a single set of braces and the other in double braces, mean two entirely different things to *Mathematica*. Beginners often don't pay close attention to braces and get frustrated when *Mathematica* does not do what they think it should be doing.

A.2.1 Lists

Here is an example of a one-dimensional list with six elements:

```
s = {1, 3, Sin[bb], 2 * x, 5, (3 * y)/x^3}
```

$\left\{1, 3, \text{Sin}[bb], 2x, 5, \dfrac{3y}{x^3}\right\}$

The Length function gives the number of elements in a list.

Length[s]

6

We can extract elements of a list by enclosing the element indices in double square brackets. For example, to get elements 3, 4, and 6 from *s*, we use

```
ss = s[[{3, 4, 6}]]
```

$\left\{\text{Sin}[bb], 2x, \dfrac{3y}{x^3}\right\}$

Most standard algebraic operations can be applied to all elements of a list. For example, to differentiate all elements of a list by x, we simply need to use the *D* function on the list.

A.2 Lists and Matrices

D[ss,x]

$\left\{0, 2, -\frac{9y}{x^4}\right\}$

Multidimensional lists can be defined in a similar way. Here is an example of a two-dimensional list with the first *row* having one element, the second row three elements, and the third row as the *s* list defined above.

a = {{1}, {2, 3, 4}, s}

$\left\{\{1\}, \{2, 3, 4\}, \left\{1, 3, \text{Sin}[bb], 2x, 5, \frac{3y}{x^3}\right\}\right\}$

Length[a]

3

Elements of a list can be extracted by specifying the row number and the element number within each row in pairs of square brackets. For example, to get the second row of list *a*, we use the following structure:

a[[2]]

{2, 3, 4}

To get the sixth element of the third row of list *a*, we use the following structure:

a[[3,6]]

$\frac{3y}{x^3}$

Obviously, *Mathematica* will generate an error message if you try to get an element that is not defined in a list. For example, trying to get the fourth element from the second row will produce the following result:

a[[2,4]]

Part :: partw : Part {4 of {2, 3, 4} does not exist.}

$\left\{\{1\}, \{2, 3, 4\}, \left\{1, 3, \text{Sin}[bb], 2x, 5, \frac{3y}{x^3}\right\}\right\}[[2, 4]]$

Sometimes it is necessary to change the structure of a list. Flatten and Partition functions are simple ways to achieve this. Flatten removes all internal braces and converts any multidimensional list into a single *flat* list, keeping the order the same as that in the original list. For example, we can define a new one-dimensional list *b* by flattening list *a* as follows:

b = Flatten[a]

$\left\{1, 2, 3, 4, 1, 3, \text{Sin}[bb], 2x, 5, \frac{3y}{x^3}\right\}$

Using Partition, we can create partitions (row structure) in a given list. The second argument of Partition specifies the number of elements in each row. Extra elements that do not define a complete row are ignored.

Partition[b, 3]

{{1, 2, 3}, {4, 1, 3}, {Sin[bb], 2x, 5}}

Partition[b, 4]

{{1, 2, 3, 4}, {1, 3, Sin[bb], 2x}}

A.2.2 Matrices

Matrices are special two-dimensional lists in which each row has exactly the same number of elements. Here we define a 3 × 4 matrix m:

m = {{1,2,3,4},{5,6,7,8},{9,10,11,12}}

{{1, 2, 3, 4}, {5, 6, 7, 8}, {9, 10, 11, 12}}

To see the result displayed in a conventional matrix form, we use MatrixForm.

MatrixForm[m]

$$\begin{pmatrix} 1 & 2 & 3 & 4 \\ 5 & 6 & 7 & 8 \\ 9 & 10 & 11 & 12 \end{pmatrix}$$

TableForm is another way of displaying multidimensional lists.

TableForm[m]

```
1  2   3   4
5  6   7   8
9  10  11  12
```

The following two-dimensional list is not a valid matrix because the number of elements in each row is not the same. However, there is no error message produced because it is still a valid list in *Mathematica*.

a = {{1}, {2, 3}, {4, 5, 6}}

{{1}, {2, 3}, {4, 5, 6}}

MatrixForm[a]

$$\begin{pmatrix} \{1\} \\ \{2,3\} \\ \{4,5,6\} \end{pmatrix}$$

Once a matrix is defined, standard matrix operations can be performed on it. For example, we can define a new matrix *mt* as the Transpose of *m*.

mt = Transpose[m]; MatrixForm[mt]

$$\begin{pmatrix} 1 & 5 & 9 \\ 2 & 6 & 10 \\ 3 & 7 & 11 \\ 4 & 8 & 12 \end{pmatrix}$$

Matrices can be multiplied by using . (period) between their names.

MatrixForm[mt.m]

$$\begin{pmatrix} 107 & 122 & 137 & 152 \\ 122 & 140 & 158 & 176 \\ 137 & 158 & 179 & 200 \\ 152 & 176 & 200 & 224 \end{pmatrix}$$

Using * or a blank will produce an error or simply element-by-element multiplication.

mt * m

Thread :: tdlen : Objects of unequal length in
{{1, 5, 9}, {2, 6, 10}, {3, 7, 11}, {4, 8, 12}}
{{1, 2, 3, 4}, {5, 6, 7, 8}, {9, 10, 11, 12}} cannot be combined.

{{1, 2, 3, 4}, {5, 6, 7, 8}, {9, 10, 11, 12}}
{{1, 5, 9}, {2, 6, 10}, {3, 7, 11}, {4, 8, 12}}

Here we get an element-by-element product and not the matrix product.

mt * mt

{{1, 25, 81}, {4, 36, 100}, {9, 49, 121}, {16, 64, 144}}

When using MatrixForm, it is very important to note that the MatrixForm is used for display purposes only. If a matrix is defined with the MatrixForm in it, that matrix definition cannot be used in any subsequent calculation. For example, consider the following definition of matrix *m* used previously:

m = MatrixForm[{{1, 2, 3, 4}, {5, 6, 7, 8}, {9, 10, 11, 12}}]

$$\begin{pmatrix} 1 & 2 & 3 & 4 \\ 5 & 6 & 7 & 8 \\ 9 & 10 & 11 & 12 \end{pmatrix}$$

The output looks exactly like it did in the earlier case. However, we cannot perform any operations on matrix *m* in this form. For example, using the Transpose function on *m* simply returns the initial matrix *m* wrapped in Transpose.

Transpose[m]

$$\text{Transpose}\left[\begin{pmatrix} 1 & 2 & 3 & 4 \\ 5 & 6 & 7 & 8 \\ 9 & 10 & 11 & 12 \end{pmatrix}\right]$$

The solution obviously is not to use the MatrixForm in the definition of matrices. After the definition, we can use the MatrixForm to get a nice-looking display.

m = {{1, 2, 3, 4}, {5, 6, 7, 8}, {9, 10, 11, 12}}; MatrixForm[m]

$$\begin{pmatrix} 1 & 2 & 3 & 4 \\ 5 & 6 & 7 & 8 \\ 9 & 10 & 11 & 12 \end{pmatrix}$$

Elements of matrices can be extracted using the double square brackets. For example, the second row of m can be extracted as follows:

m2 = m[[2]]

{5, 6, 7, 8}

The element in the second row and the fourth column is

m[[2, 4]]

8

Extracting a column is a little more difficult. The simplest way is to transpose the matrix and then take its desired row (which obviously corresponds to the original column). For example, the second column of matrix m is extracted as follows:

c2 = Transpose[m][[2]]

{2, 6, 10}

Matrices can also be defined in terms of symbolic expressions. For example, here is a 2×2 matrix called m.

m = {{y, 2*x}, {(3*y)/x^3, 4*y}}; MatrixForm[m]

$$\begin{pmatrix} y & 2x \\ \frac{3y}{x^3} & 4y \end{pmatrix}$$

Determinant of m:

Det[m]

$$-\frac{6y}{x^2} + 4y^2$$

Inverse of m:

```
mi = Inverse[m]; MatrixForm[mi]
```
$$\begin{pmatrix} \dfrac{4y}{-\dfrac{6y}{x^2}+4y^2} & -\dfrac{2x}{-\dfrac{6y}{x^2}+4y^2} \\ -\dfrac{3y}{x^3\left(-\dfrac{6y}{x^2}+4y^2\right)} & \dfrac{y}{-\dfrac{6y}{x^2}+4y^2} \end{pmatrix}$$

Check the inverse:

```
MatrixForm[Simplify[mi.m]]
```
$$\begin{pmatrix} 1 & 0 \\ 0 & 1 \end{pmatrix}$$

Operations, such as differentiation and integration, are performed on each element of a matrix.

```
D[m,y]//MatrixForm
```
$$\begin{pmatrix} 1 & 0 \\ \dfrac{3}{x^3} & 4 \end{pmatrix}$$

```
Integrate[m,x]//MatrixForm
```
$$\begin{pmatrix} xy & x^2 \\ -\dfrac{3y}{2x^2} & 4xy \end{pmatrix}$$

A.2.3 Generating Lists with the Table Function

The Table function is a handy tool to generate lists or matrices with a systematic pattern. For example, to generate a 3×5 matrix with all entries as zero, we use the following form:

```
m = Table[0, {3}, {5}]; MatrixForm[m]
```
$$\begin{pmatrix} 0 & 0 & 0 & 0 & 0 \\ 0 & 0 & 0 & 0 & 0 \\ 0 & 0 & 0 & 0 & 0 \end{pmatrix}$$

The first argument to the Table function can be any expression. Here is a 3×4 matrix whose entries are equal to the sum of row and column indices divided by $3^\wedge$ (column index).

```
z = Table[(i+j)/3^j, {i, 1, 3}, {j, 1, 4}]; MatrixForm[z]
```
$$\begin{pmatrix} \dfrac{2}{3} & \dfrac{1}{3} & \dfrac{4}{27} & \dfrac{5}{81} \\ 1 & \dfrac{4}{9} & \dfrac{5}{27} & \dfrac{2}{27} \\ \dfrac{4}{3} & \dfrac{5}{9} & \dfrac{2}{9} & \dfrac{7}{81} \end{pmatrix}$$

Using the If command, we can construct complicated matrices. For example, here is a 4 × 4 upper triangular matrix:

```
z = Table[If[i > j, 0, (i + j)/3^j], {i, 1, 4}, {j, 1, 4}];
MatrixForm[z]
```

$$\begin{pmatrix} \frac{2}{3} & \frac{1}{3} & \frac{4}{27} & \frac{5}{81} \\ 0 & \frac{4}{9} & \frac{5}{27} & \frac{2}{27} \\ 0 & 0 & \frac{2}{9} & \frac{7}{81} \\ 0 & 0 & 0 & \frac{8}{81} \end{pmatrix}$$

Note the syntax of the If command is as follows:

If[test, Statements to be executed if test is True,
Statements to be executed if test is False]

A.2.4 Caution When Dealing with Row and Column Vectors

Since *Mathematica* treats matrices as essentially lists, it does not distinguish between a column or a row vector. The actual form is determined from syntax in which it is used. For example, define two vectors a and b as follows:

```
a = {1, 2, 3}; b = {4, 5, 6};
```

The matrix inner product (Transpose[a] . b) is evaluated simply as a . b, resulting in a scalar. Explicitly evaluating Transpose[a]. b will produce an error.

```
a . b
```

32

```
Transpose[a] . b
```

Transpose :: nmtx : The first two levels of the one-
    dimensional list {1, 2, 3} cannot be transposed.
Transpose[{1, 2, 3}].{4, 5, 6}

If we want to treat *a* as a column vector (3 × 1) and *b* as a row vector (1 × 3) to get a 3 × 3 matrix from the product, we need to use the Outer function of *Mathematica*, as follows:

```
ab = Outer[Times, a, b]; MatrixForm[ab]
```

$$\begin{pmatrix} 4 & 5 & 6 \\ 8 & 10 & 12 \\ 12 & 15 & 18 \end{pmatrix}$$

Another way to explicitly define a 3 × 1 column vector is to enter it as a two-dimensional list. Here, each entry defines a row of a matrix with one column.

a = {{1}, {2}, {3}}; MatrixForm[a]

$\begin{pmatrix} 1 \\ 2 \\ 3 \end{pmatrix}$

Now Transpose[a] . b makes sense. However, the result is a 1 × 1 matrix and not a scalar.

Transpose[a] . b

{32}

Obviously, now a . b will produce an error because the dimensions do not match.

a . b

Dot :: dotsh : Tensors {{1}, {2}, {3}}
    and {4, 5, 6} have incompatible shapes.
{{1}, {2}, {3}} . {4, 5, 6}

To get a 3 × 3 matrix, we need to define b as a two-dimensional matrix with only one row, as follows:

b = {{4, 5, 6}}

{{4, 5, 6}}

Now a (3 × 1) . b (1 × 3) can be evaluated directly, as one would expect.

MatrixForm[a . b]

$\begin{pmatrix} 4 & 5 & 6 \\ 8 & 10 & 12 \\ 12 & 15 & 18 \end{pmatrix}$

A.3 Solving Equations

An equation in *Mathematica* is defined as an expression with two parts separated by two equal signs (==), without any spaces between the two signs. Thus, the following is an equation:

x^3 - 3 == 0

$-3 + x^3 == 0$

Note that you get the following strange-looking error message if you try to define an equation with a single equal sign. (The error message makes sense if you know more details of how *Mathematica* handles expressions internally.)

x^3 - 3 = 0

Set :: write : Tag Plus in $-3+x^3$ is Protected.

0

Equations can also be assigned to variables for later reference. For example, we can call the above equation *eq* so that we can refer to it in other computations.

eq = x^3 - 3 == 0

$-3 + x^3 == 0$

A single equation, or a systems of equations, can be solved using the Solve command.

sol = Solve[eq, x]

$\{\{x \to -(-3)^{1/3}\}, \{x \to 3^{1/3}\}, \{x \to (-1)^{2/3} 3^{1/3}\}\}$

Note that the solution is returned in the form of a two-dimensional list of substitution *rules*. Because of this form, we can substitute any solution into another expression if desired. For example, we can substitute the second solution back into the equation to verify that the solution is correct, as follows:

eq /. sol[[2]]

True

The Solve command tries to find all possible solutions using symbolic manipulations. If a numerical solution is desired, the NSolve command is usually much faster.

NSolve[x^3 - 3 == 0, x]

$\{\{x \to -0.721125 + 1.24902 I\}, \{x \to -0.721125 - 1.24902 I\}, \{x \to 1.44225\}\}$

Both Solve and NSolve commands can be used for systems of linear or nonlinear equations.

eqn1 = x + 2y + 3z == 41;
eqn2 = 5x + 5y + 4z == 20;
eqn3 = 3y + 4z == 125;
sol = Solve[{eqn1, eqn2, eqn3}, {x, y, z}]

$\left\{\left\{x \to -\frac{527}{13}, y \to \frac{635}{13}, z \to -\frac{70}{13}\right\}\right\}$

Notice that even when there is only one solution, the result is a two-dimensional list. If we want to evaluate a given function at this solution point, we still need to extract the first element of this list. For example:

a = N[Sin[xy] Cos[y^z]/.sol[[1]]]

-0.810238

The following substitution produces the same result but in the form of a list whose structure must be kept in mind if it is to be used in other expressions.

b = N[Sin[xy] Cos[y^z]/.sol]

{-0.810238}

For example, since *a* is a scalar as defined by the first form, it can be used in any matrix operation. However, *b* can only be used with appropriate-sized matrices.

a{a1, a2, a3}

{-0.810238a1, -0.810238a2, -0.810238a3}

b{a1, a2, a3}

```
Thread :: tdlen : Objects of unequal length
     in {-0.810238`}{a1, a2, a3} cannot be combined.
```
{-0.810238}{a1, a2, a3}

A linear system of equations, written in matrix form as $Kd = R$, can be solved very efficiently using the LinearSolve[K, R] command, as follows:

K = {{2, 4, 6}, {7, 9, 2}, {1, 2, 13}};
R = {1, 2, 3};
LinearSolve[K, R]

$\left\{\dfrac{21}{20}, -\dfrac{13}{20}, \dfrac{1}{4}\right\}$

A.4 Plotting in *Mathematica*

Mathematica supports both two-dimensional and three-dimensional graphics. The Plot command provides support for two-dimensional graphs of one or more functions specified as expressions or functions. The following shows the plot of an expression in the interval from -2 to 2.

```
p1 = Plot[Sin[x] Cos[x]/(1+x^2),{x,-2,2}];
```

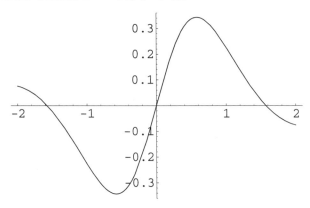

Multiple plots can be shown on the same graph using the Show command. The following example plots an expression and its first derivative. Each plot command automatically displays a resulting graph. For multiple plots, usually the final plot showing all graphs superimposed is of interest. Using Display-Function as Identity suppresses the display of intermediate plots.

```
de1 = D[Sin[x] Cos[x]/(1+x^2),x];
p2 = Plot[de1,{x,-2,2}, PlotStyle->{RGBColor[0,0,1]},
    DisplayFunction -> Identity];
Show[{p1,p2}];
```

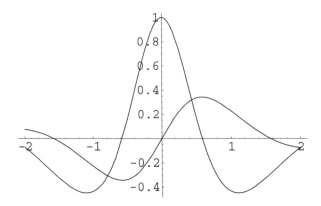

The RGBColor function returns a combination color by mixing specified amounts of red, green and blue colors. In the example, the Blue component is set to 1 and the others to 0 to get a blue color. Use *Mathematica* online Help to explore many other plot options.

Note that before generating the plot for the derivative, we defined a new expression and then used it in the plot function. If we had actually used the derivative operation inside the argument of the Plot function, we would get a series of strange error messages, as follows:

Plot[D[Sin[x] Cos[x]/(1+x^2),x],{x,-2,2}];

General :: ivar : -2. is not a valid variable.

General :: ivar : -2. is not a valid variable.

General :: ivar : -2. is not a valid variable.

General :: stop : Further output of
  General :: ivar will be suppressed during this calculation.

Plot :: plnr :
$\partial_x \frac{\operatorname{Sin}[x]\operatorname{Cos}[x]}{1+x^2}$ is not a machine-size real number at x = -2..

Plot :: plnr :
$\partial_x \frac{\operatorname{Sin}[x]\operatorname{Cos}[x]}{1+x^2}$ is not a machine-size real number at x = -1.83773.

Plot :: plnr :
$\partial_x \frac{\operatorname{Sin}[x]\operatorname{Cos}[x]}{1+x^2}$ is not a machine-size real number at x = -1.66076.

General :: stop : Further output of
  Plot :: plnr will be suppressed during this calculation.

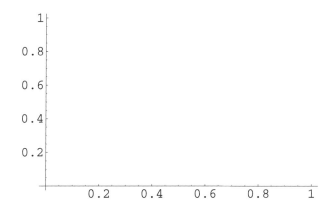

The reason for the error is that Plot and several other built-in functions do not evaluate their arguments (for valid reasons but difficult to explain here). We can force them to evaluate the arguments by enclosing the computation inside Evaluate, as follows:

```
Plot[Evaluate[D[Sin[x] Cos[x]/(1+x^2),x]],{x,-2,2}];
```

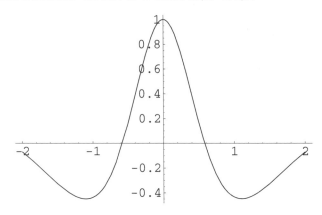

Functions of two variables can be shown either as contour plots or as three-dimensional surface plots.

```
f = Sin[x] Cos[y]
```

Cos[y] Sin[x]

```
ContourPlot[f, {x, -π, π}, {y, -π, π}, PlotPoints- > 50];
```

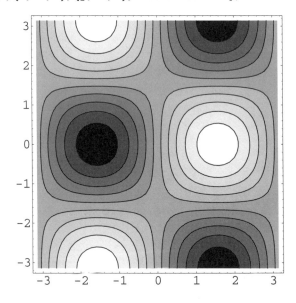

```
Plot3D[f, {x, -π, π}, {y, -π, π}];
```

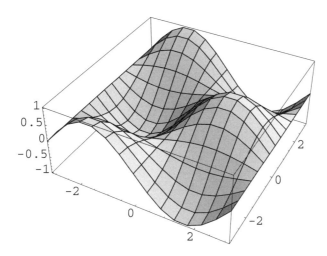

A.5 Programming in *Mathematica*

A.5.1 Defining New Functions

Calculations that involve many steps using an interactive approach may become tedious. *Mathematica* offers a rich and sophisticated programming environment to create functions that perform complex series of computations.

The general syntax of the function definition is as follows:

```
newFunctionName[var1_, var2_, ...] := Module[
           {localVar1, localVar2, ... },
           statement 1;
           statement 2;
           ⋮
           last statement
    ]
```

The newFunctionName can be any name that the user wants. Since all built-in functions start with an uppercase letter, it may be a good idea to start your functions with a lowercase letter to avoid any conflict with built-in functions. All needed variables for the function are enclosed in square brackets. For reasons beyond the scope of this introductory tutorial, an underscore must be appended to all names in the variable list. The definition of the function starts with := followed by the word *Module* and an opening square bracket. The first line in the definition is a list of local variables that are to be used only inside the body of the function. Outside of the function, these variables do not exist. The list of local variables ends with a comma. The remainder of the function body can contain as many statements as needed to achieve the goal of the function. Each statement ends with a semicolon. The last statement does not have a semicolon and is the one that is returned by the function when it is used. The end of the function is indicated by the closing square bracket.

Simple one-line functions that do not need any local variables can be defined simply as follows:

```
oneLineFcn[var1_, var2_, ...] := expression involving vars
```

An example of a one-line function is presented in the next section. As an example of Module, we define the following function to return stresses in thick-walled cylinders. The tangential and radial stresses in an open-ended thick cylinder are given by the following formulas:

$$\sigma_t = \frac{p_i r_i^2}{r_0^2 - r_i^2}\left(1 + \frac{r_0^2}{r^2}\right) \qquad \sigma_r = \frac{p_i r_i^2}{r_0^2 - r_i^2}\left(1 - \frac{r_0^2}{r^2}\right)$$

where p_i is internal pressure on the cylinder, r_i and r_0 are inner and outer radii and r is radius of the point where the stress is desired.

```
thickCylinderStresses[pi_, ri_, r0_, r_] :=
  Module[{c1, c2, σt, σr},
    c1 = pi * ri^2 / (r0^2 - ri^2);
    c2 = r0^2 / r^2;
    σt = c1 (1 + c2);
    σr = c1 (1 - c2);
    {σt, σr}
  ]
```

After entering the function definition, the input must be executed to actually tell *Mathematica* about the new function. Unless there is an error, the execution of the function definition line does not produce any output. After this, the function can be used as any other *Mathematica* function. For example, we can compute stresses in a cylinder with $pi = 20$, $ri = 5$, $r0 = 15$, and $r = 10$ as follows:

A.5 Programming in *Mathematica*

```
thickCylinderStresses[20, 5, 15, 10]
```

$$\left\{\frac{65}{8}, -\frac{25}{8}\right\}$$

Leaving some variables in symbolic form, we can get symbolic results:

```
{st, sr} = thickCylinderStresses[20, 5, 15, r]
```

$$\left\{\frac{5}{2}\left(1 + \frac{225}{r^2}\right), \frac{5}{2}\left(1 - \frac{225}{r^2}\right)\right\}$$

These expressions can be plotted to see the stress distribution through the cylinder wall.

```
Plot[{st, sr}, {r, 5, 15}, AxesLabel-> {"r", "Stress"},
  PlotStyle-> {{GrayLevel[0]}, {GrayLevel[0.5]}}];
```

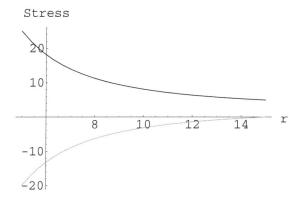

A natural question might be to see what happens to the stresses as the cylinder becomes thin. This question can be answered by evaluating the stresses at the center of wall thickness as a function of inner radius, as follows:

```
{st, sr} = thickCylinderStresses[20, ri, 15, (ri + 15) /2]
```

$$\left\{\frac{20 \text{ri}^2 \left(1 + \frac{900}{(15 + \text{ri})^2}\right)}{225 - \text{ri}^2}, \frac{20 \text{ri}^2 \left(1 - \frac{900}{(15 + \text{ri})^2}\right)}{225 - \text{ri}^2}\right\}$$

A plot of these expressions is as follows:

```
Plot[{st, sr}, {ri, 5, 14}, AxesLabel-> {"ri", "Stress"},
  PlotStyle-> {{GrayLevel[0.5]}, {GrayLevel[0]}}];
```

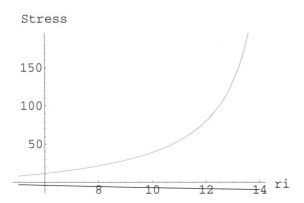

We can see that σ_t becomes the predominant stress as the wall becomes thin. The usual approximation for thin-walled cylinders is $\sigma_t = \frac{p_i r}{t}$. We can graphically see how this approximation compares with the thick cylinder solution, as follows:

```
Plot[{st, (20(ri+15)/2)/(15-ri)},
  {ri, 5, 14}, AxesLabel->{"ri", "Stress"},
  PlotStyle->{{GrayLevel[0]}, {GrayLevel[0.5]}}];
```

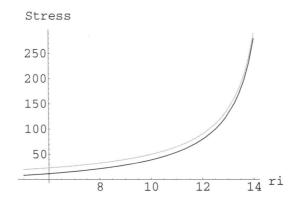

A.5.2 Use of Map, Apply, Thread, and Outer

Frequently, we need to perform operations on a list of functions or variables. Many built-in functions are designed to operate on lists directly. However, there are still many situations where it is necessary to perform some operation on a list of quantities. In traditional programming language, we typically

A.5 Programming in *Mathematica*

perform these operations using a Do loop or an equivalent. *Mathematica* has a Do loop that can be used in a similar way. However, it also offers many other elegant and more convenient ways of doing the same thing. Map, Thread, and Apply are three of these handy functions. These functions have been used in several examples in this text. We illustrate the use of these functions in the portfolio example considered in Chapter 1.

Suppose we are given historical rates of returns from different investments that we have entered as lists in *Mathematica*.

```
blueChipStocks = {18.24, 12.12, 15.23, 5.26, 2.62, 10.42};
techStocks = {12.24, 19.16, 35.07, 23.46, -10.62, -7.43};
realEstate = {8.23, 8.96, 8.35, 9.16, 8.05, 7.29};
bonds = {8.12, 8.26, 8.34, 9.01, 9.11, 8.95};
```

We want to compute the average rate of return. The computation is fairly simple. We just need to add all rates of returns and divide them by the number of returns. Computing the sum or multiplication of all elements of a list is done very conveniently by using Apply. It does not matter how long the list is or whether it consists of numerical or symbolic quantities.

```
Apply[Plus, blueChipStocks]
```

63.89

The first argument of Apply is the function to be applied, and the second is the list of items. Multiplication of all elements can be computed in exactly the same way.

```
Apply[Times, blueChipStocks]
```

483484.

Obviously, the function to be applied must expect a list as its argument. For example, applying Sin function to a list will not produce any useful result.

```
Apply[Sin, blueChipStocks]
```

Sin :: argx : Sin called with 6 arguments; 1 argument is expected.

Sin[18.24, 12.12, 15.23, 5.26, 2.62, 10.42]

The Map function is similar to Apply, but it applies a function to each element of a list. For example, if we want to compute the Sin of each term in the blueChipStocks list, we can do it as follows:

```
Map[Sin, blueChipStocks]
```

{-0.572503, -0.431695, 0.459972, -0.853771, 0.498262, -0.83888}

To compute the average, all we need to do now is divide the sum by the number of entries in the list. The Length function, described earlier, does exactly this. Thus, the following one-line program can compute the average of any list.

```
average[n_] := Apply[Plus, n]/Length[n]
average[blueChipStocks]
```

10.6483

To compute average rates of returns of all investment types, we can use the average function repeatedly on other lists of returns. However, it is much more convenient to define a list of all investments and simply Map the average function to each element of a list. We first define a list of all returns and then use the average function defined above.

```
returns = {blueChipStocks, techStocks, realEstate, bonds}
```

{{18.24, 12.12, 15.23, 5.26, 2.62, 10.42},
 {12.24, 19.16, 35.07, 23.46, -10.62, -7.43},
 {8.23, 8.96, 8.35, 9.16, 8.05, 7.29},
 {8.12, 8.26, 8.34, 9.01, 9.11, 8.95}}

```
averageReturns = Map[average, returns]
```

{10.6483, 11.98, 8.34, 8.63167}

The elegance of these constructs is that we never need to know how long the lists really are. We can keep adding or deleting elements into any of the lists and the process will keep working. Map and Apply functions are used so frequently in *Mathematica* programming that several short cuts have been designed to make their use even more efficient. One such useful technique is combining Map with the function definition itself. In the example of computing averages, we had to define a function (called average) and then apply it to elements of the list using Map. We can do exactly the same thing, without explicitly defining the function, as follows:

```
averageReturns = Map[Apply[Plus, #]/Length[#]&, returns]
```

{10.6483, 11.98, 8.34, 8.63167}

We can see that the first argument is exactly the definition of the function. The pound (#) sign stands for the function argument. The ampersand (&) at the end is very important. It essentially tells *Mathematica* that we are defining a function as the first argument of Map.

The computation of the covariance matrix involves several uses of Apply. We define *covariance* between two investments i and j as follows:

$$v_{ij} = \frac{1}{n}\sum_{k=1}^{n}(r_{ik} - \mu_i)(r_{jk} - \mu_j)$$

where n = total number of observations, r_{jk} = rate of return of investment j for the kth observation, and μ_j is the average value of the investment j. The following function implements this computation. The first two lines compute averages, and the third line returns the above sum.

```
coVariance[x_, y_] := Module[{xb, yb, n = Length[x]},
    xb = Apply[Plus, x]/n;
    yb = Apply[Plus, y]/n;
    Apply[Plus, (x - xb) (y - yb)]/n
];
```

Using this function, we can compute the covariance between, say, blueChip-Stocks and bonds as follows:

coVariance[blueChipStocks, bonds]

-1.9532

Computing the entire covariance matrix this way would be tedious. The Outer function, described earlier, feeds all combinations of investments to the coVariance function and thus, we can generate the entire matrix using the following line of *Mathematica* input.

Q = Outer[coVariance, returns, returns, 1]; MatrixForm[Q]

$$\begin{pmatrix} 29.0552 & 40.3909 & -0.287883 & -1.9532 \\ 40.3909 & 267.344 & 6.83367 & -3.69702 \\ -0.287883 & 6.83367 & 0.375933 & -0.0566333 \\ -1.9532 & -3.69702 & -0.0566333 & 0.159714 \end{pmatrix}$$

The Thread function is similar to Map as it threads a function over its arguments. The most common use of this function in the text has been to define rules for substitution into an expression. Suppose we have a function of four variables that we would like to evaluate at a given point.

```
f = x_1 x_2 + Sin[x_3] Cos[x_4];
pt = {1.1, 2.23, 3.2, 4.556};
vars = {x_1, x_2, x_3, x_4};
```

A tedious way to evaluate f at the given point is as follows:

f/.{x_1 -> 1.1, x_2 -> 2.23, x_3 -> 3.2, x_4 -> 4.556}

2.46209

A more convenient way is to use Thread to define the substitution rule.

Thread[vars->pt]

$\{x_1 \to 1.1, x_2 \to 2.23, x_3 \to 3.2, x_4 \to 4.556\}$

f/.Thread[vars->pt]

2.46209

Again, the advantage of the last form is clear. We don't have to change anything if the number of variables is increased or decreased.

A.6 Packages in *Mathematica*

Mathematica comes with a wide variety of special packages. Among these are the linear algebra and graphics packages. These packages provide additional commands for manipulating matrices and plots. Loading the matrix manipulation commands from the LinearAlgebra Package and Graphics' Legend packages is accomplished as follows:

Needs["LinearAlgebra`MatrixManipulation`"];
Needs["Graphics`Legend`"];

With Graphics'Legend, we can show labels for different graphs on the same plot. For example,

Plot[{st, (20(ri + 15)/2)/(15 - ri)},
 {ri, 5, 14}, AxesLabel-> {"ri", "Stress"},
 PlotStyle-> {{GrayLevel[0]}, {GrayLevel[0.5]}},
 PlotLegend-> {"Thick", "Thin"},
 LegendPosition-> {1, 0}];

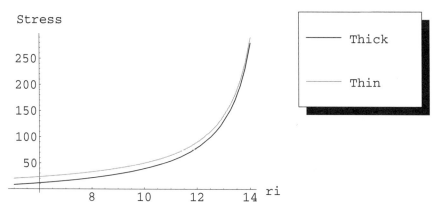

The large number of functions created for this text are included in the OptimizationToolbox package. Specific instructions for loading and using this package are included on the accompanying CD.

A.7 Online Help

Mathematica contains a complete online Help system. The Help menu provides access to all *Mathematica* features. In fact, the entire *Mathematica* book is online. In addition you can obtain information about any command by typing ? followed by the command name. This form also supports the use of wildcards. For example, to get a listing of all commands that start with letter G, type

?G*

Gamma	Generic	GraphicsSpacing
GammaRegularized	Get	GrayLevel
GaussianIntegers	GetContext	Greater
GaussKronrod	GetLinebreakInformationPacket	GreaterEqual
GaussPoints	GoldenRatio	GridBaseline
GCD	Goto	GridBox
Gear	Gradient	GridBoxOptions
GegenbauerC	Graphics	GridFrame
General	Graphics3D	GridLines
GenerateBitmapCaches	GraphicsArray	GroebnerBasis
GenerateConditions	GraphicsData	GroupPageBreakWithin
GeneratedCell		

Detailed instructions about a specific command can be obtained by typing ? followed by the command name. For example,

?GaussPoints

```
GaussPoints is an option for NIntegrate. With
    GaussPoints->n, the Gaussian part of Gauss-Kronrod
    quadrature uses n points. With GaussPoints->Automatic,
    an internal algorithm chooses the number of points.
```

Bibliography

Abel, M.L. and Braselton, J.P., *Mathematica by Example*, Academic Press, San Diego, 1997.

Bahder, T., *Mathematica for Scientists and Engineers*, Addison-Wesley, Reading, MA, 1995.

Gray, J.W., *Mastering Mathematica*, AP Professional, Cambridge, MA, 1994.

Gray, T. and Glynn, J., *A Beginner's Guide to Mathematica*, Version 2, Addison-Wesley, Reading, MA, 1992.

Maeder, R., *Programming in Mathematica*, 3rd Edition, Addison-Wesley, Redwood City, CA, 1997.

Shaw, W.T. and Tigg, J., *Applied Mathematica: Getting Started, Getting It Done*, Addison-Wesley, Reading, MA, 1994.

Wagner, D.B., *Power Programming with Mathematica*, McGraw-Hill, New York, 1996.

Wickham-Jones, T., *Mathematica Graphics: Techniques and Applications*, TELOS (Springer-Verlag), Santa Clara, CA, 1994.

Wolfram, S., *The Mathematica Book*, 3rd Edition, Wolfram Media/Cambridge University Press, Boston, MA, 1996.

Bibliography

Arora, J.S., *Introduction to Optimum Design*, McGraw-Hill, New York, 1989.

Avriel, M. and Golany, B., *Mathematical Programming for Industrial Engineers*, Marcel Dekker, New York, 1996.

Bates, D.M. and Watts, D.G., *Nonlinear Regression Analysis and Its Applications*, Wiley, New York, 1988.

Bazarra, M.S., Jarvis, J.J., and Sherali, H.D., *Linear Programming and Network Flows*, Wiley, New York, 1990.

Bazarra, M.S., Sherali, H.D., and Shetty, C.M., *Nonlinear Programming: Theory and Algorithms*, 2nd Edition, Wiley, New York, 1993.

Bertsekas, D.P., *Nonlinear Programming*, Athena Scientific, Belmont, MA, 1995.

Beveridge, Gordon S.G., and Schechter, R.S., *Optimization: Theory and Practice*, McGraw-Hill, New York, 1970.

Biegler, L.T., Coleman, T.F., Conn, A.R., and Santosa, F.N. (editors), *Large-Scale Optimization with Applications, Part I: Optimization in Inverse Problems and Design*, Springer-Verlag, New York, 1997.

Biegler, L.T., Coleman, T.F., Conn, A.R., and Santosa, F.N. (editors), *Large-Scale Optimization with Applications, Part II: Optimal Design and Control*, Springer-Verlag, New York, 1997.

Biegler, L.T., Coleman, T.F., Conn, A.R., and Santosa, F.N. (editors), *Large-Scale Optimization with Applications, Part III: Molecular Structure and Optimization*, Springer-Verlag, New York, 1997.

Dantzig, G.B. and Thapa, M.N., *Linear Programming 1: Introduction*, Springer-Verlag, New York, 1997.

Degarmo, E.P., Sullivan, W.G., Bontadelli, J.A., and Wicks, E.M., *Engineering Economy*, Prentice-Hall, Englewood Cliffs, NJ, 1997.

Dennis, J.E., Schnabel, R.B., *Numerical Methods for Unconstrained Optimization and Nonlinear Equations*, Prentice-Hall, Englewood Cliffs, NJ, 1983.

Ertas, A. and Jones, J.C., *The Engineering Design Process*, 2nd Edition, Wiley, New York, 1996.

Fang, S.-C. and Puthenpura, S., *Linear Optimization and Extensions*, AT&T, Prentice-Hall, Englewood Cliffs, NJ, 1993.

Fiacco, A.V., *Introduction to Sensitivity and Stability Analysis in Nonlinear Programming*, Academic Press, New York, 1983.

Fiacco, A.V. and McCormick, G.P., *Nonlinear Programming: Sequential Unconstrained Minimization Techniques*, Wiley, New York, 1968.

Fletcher, R., *Practical Methods of Optimizations*, Second Edition, Wiley, New York, 1987.

Gen, M. and Cheng, R., *Genetic Algorithms and Engineering Design*, Wiley, New York, 1997.

Gill, P.E., Murray, W., and Wright, M.H., *Practical Optimization*, Academic Press, New York, 1981.

Gill, P.E., Murray, W., and Wright, M.H., *Numerical Linear Algebra and Optimization*, Addison-Wesley, Reading, MA, 1991.

Goldberg, D.E., *Genetic Algorithms in Search, Optimization, and Machine Learning*, Addison-Wesley, Reading, MA, 1989.

Gomez, S. and Hennart, J.-P. (editors), *Advances in Optimization and Numerical Analysis*, Kluwer Academic Publishers, Dordrecht, 1994.

Hansen, E., *Global Optimization Using Interval Analysis*, Marcel Dekker, New York, 1992.

Hayhurst, G., *Mathematical Programming Applications*, Macmillan, New York, 1987.

Hentenryck, P. Van, Michel, L., and Deville, Y., *Numerica: A Modeling Language for Global Optimization*, MIT Press, Cambridge, MA, 1997.

Hertog, D. den, *Interior Point Approach to Linear, Quadratic and Convex Programming*, Kluwer Academic Publishers, Dordrecht, 1994.

Hestenes, M., *Conjugate Direction Methods in Optimization*, Springer-Verlag, Berlin, 1980.

Himmelblau, D.M., *Applied Nonlinear Programming*, McGraw-Hill, New York, 1972.

Hymann, B., *Fundamentals of Engineering Design*, Prentice-Hall, Englewood Cliffs, NJ, 1998.

Jahn, J., *Introduction to the Theory of Nonlinear Optimization*, 2nd Revised Edition, Springer-Verlag, Berlin, 1996.

Jeter, M.W., *Mathematical Programming: An Introduction to Optimization*, Marcel Dekker, New York, 1986.

Kolman, B. and Beck, R.E., *Elementary Linear Programming with Applications*, Academic Press, New York, 1980.

Luenberger, D.G., *Linear and Nonlinear Programming*, 2nd Edition, Addison-Wesley, Reading, MA, 1984.

McAloon, K. and Tretkoff, C., *Optimization and Computational Logic*, Wiley, New York, 1996.

McCormick, G.P., *Nonlinear Programming: Theory, Algorithms, and Applications*, Wiley, New York, 1983.

Megiddo, N. *Progress in Mathematical Programming: Interior-Point and Related Methods*, Springer-Verlag, New York, 1989.

Nash, S.G. and Sofer, A., *Linear and Nonlinear Programming*, McGraw-Hill, New York, 1996.

Nazareth, J.L., *Computer Solution of Linear Programs*, Oxford University Press, New York, 1987.

Ozan, T., *Applied Mathematical Programming for Engineering and Production Management*, Prentice-Hall, Englewood Cliffs, NJ, 1986.

Padberg, M., *Linear Optimization and Extensions*, Springer-Verlag, Berlin, 1995.

Pannell, D.J., *Introduction to Practical Linear Programming*, Wiley, New York, 1997.

Papalambros, P.Y. and Wilde, D.J., *Principles of Optimal Design: Modeling and Computation*, Cambridge University Press, Cambridge, 1988.

Peressini, A.L., Sullivan, F.E., and Uhl, Jr., J.J., *The Mathematics of Nonlinear Programming*, Springer-Verlag, New York, 1988.

Pierre, D.A. and Lowe, M.J., *Mathematical Programming Via Augmented Lagrangians: An Introduction with Computer Programs*, Addison-Wesley, Reading, MA, 1975.

Pike, R.W., *Optimization for Engineering Systems*, Van Nostrand Reinhold, New York, 1986.

Polak, E., *Computational Methods in Optimization*, Academic Press, New York, 1971.

Polak, E. and Polak, E., *Optimization: Algorithms and Consistent Approximations*, Springer-Verlag, New York, 1997.

Pshenichnyj, B. N., *The Linearization Method for Constrained Optimization*, Springer-Verlag, Berlin, 1994.

Rao, S.S., *Engineering Optimization: Theory and Practice*, 3rd Edition, Wiley, New York, 1996.

Rardin, R.L., *Optimization in Operations Research*, Prentice-Hall, Upper Saddle River, NJ, 1998.

Reeves, C.R. (editor), *Modern Heuristic Techniques for Combinatorial Problems*, Blackwell Scientific Publications, Oxford, 1993.

Scales, L. E., *Introduction to Nonlinear Optimization*, Macmillan, London, 1985.

Schittkowski, K., *Nonlinear Programming Codes: Information, Tests, Performance*, Springer-Verlag, Berlin, 1980.

Schittkowski, K., *More Test Examples for Nonlinear Programming Codes*, Springer-Verlag, Berlin, 1987.

Shapiro, J.F., *Mathematical Programming: Structures and Algorithms*, Wiley, New York, 1979.

Shor, N.Z., *Minimization Methods for Non-Differentiable Functions*, Springer-Verlag, New York, 1985.

Stark, R.M. and Nicholls, R.L., *Mathematical Foundations for Design: Civil Engineering Systems*, McGraw-Hill, New York, 1972.

Starkey, C.V., *Basic Engineering Design*, Edward Arnold, London, 1988.

Suh, N.P., *The Principles of Design*, Oxford University Press, New York, 1990.

Xie, Y. M. and Steven, G.P., *Evolutionary Structural Optimization*, Springer-Verlag, Berlin, 1997.

Index

Abel, M.L., 703
Active constraint, 148
Active set method, 535
   for dual QP, 552
ActiveSetDualQP function, 554
Active set QP algorithm, 538
ActiveSetQP function, 539
   options, 539
Additive property of constraints, 144
Aggregate constraint, 144
ALPF algorithm, 596
ALPF function, 597
   options, 597
ALPF method, 581
Analytical line search, 232
AnalyticalLineSearch function, 234
   options, 234
Angle between vectors, 79
Approximate line search
   constrained, 630
   unconstrained, 251
Approximation using Taylor series, 89
ArmijoLineSearch function, 252
Armijo's rule, 251

Arora, J.S., 31, 705
Artificial objective function, 350
Augmented Lagrangian penalty
   function, 590
Averiel, M., 420, 705

Bahder, T., 703
Barrier function, 588
Basic feasible solutions of LP, 334
Basic idea of simplex method, 339
Basic set, bringing a new variable into, 341
BasicSimplex function, 353
BasicSimplex function, options, 353
Basic solutions of LP, 334
BasicSolutions function, 338
Bates, D.M., 13, 705
Bazarra, M.S., 199, 420, 705
Beck, R.E., 420, 706
Bertsekas, D.P., 199, 705
Beveridge, 131, 705
BFGS formula for Hessian, 646
BFGS formula for inverse Hessian, 289
Biegler, L.T., 645, 647, 705

Bland's rule, 371, 390
Bontadelli, J.A., 18, 706
Braselton, J.P., 703
Building design example
   ALPF solution, 604
   graphical solution, 64
   KT solution, 161
   problem formulation, 2
   sensitivity analysis, 180
   SQP solution, 641

Capital recovery factor (crf), 21
Car payment example, 22
Cash flow diagram, 24
CashFlowDiagram function, 24
Changes in constraint rhs, 378
Changes in objective function coefficients, 382
Cheng, R., 303, 706
Choleski decomposition, 564
CholeskiDecomposition function, 566
Cofactors of a matrix, 82
Coleman, T.F., 645, 647, 705
Combinatorial problems, 303
Complementary slackness, 151, 439, 497
Composite objective function, 14
Compound interest formulas, 19
Concave functions, 113
Conjugate gradient method, 262
   with non-negativity constraints, 553
ConjugateGradient function, 262
   options, 263
Conn, A.R., 645, 647, 705
Constant term in the objective function, 317
Constraint function contours, 50
Constraint normalization, 582
Convergence criteria
   interior point, 451, 508, 522
   unconstrained, 228
Conversion to standard LP form, 317
Converting maximization to minimization, 137
Convex feasible sets, 118

Convex functions, 112
Convex optimization problem, 121, 502
Convex set, 117
Convexity, check for complicated functions, 113, 116
ConvexityCheck function, 122
Covariance, 10
coVariance function, 68
crf function, 23
Cycling in LP, 370

Dantzig, G.B., 420, 705
Data fitting example
   problem formulation, 12
   solution using optimality conditions, 142
   solution using QuasiNewtonMethod, 299
Degarmo, E.P., 18, 706
Degenerate solution, 370
Dennis, J.E., 302, 706
Descent direction, 229, 504, 520, 536, 563
DescentDirectionCheck function, 230
Design variables, 2
Determinant of a matrix, 80
Deville, Y., 303, 706
DFP formula, 288
Disjoint feasible region, 61
Dual feasibility, 467
Dual function, 621
Dual LP problem, 440
Dual QP for direction finding, 620
Dual QP problem, 497, 620

Economic factors, 24
Eigenvalues, 84
Elementary row operations, 320
Equal interval search, 236
EqualIntervalSearch function, 237
Equality (EQ) constraints, 14
Ertas, A., 31, 706
Exact penalty function, 590
Exterior penalty function, 586

Index

Fang, S.-C., 481, 706
Feasibility condition, 151
Feasible region, 50
Fiacco, A.V., 581, 706
Fletcher, R., 581, 706
Fletcher-Reeves formula, 262
FormDualLP function, 443
   options, 443
FormDualQP function, 443
   options, 443

Gauss-Jordan form, 320
GaussJordanForm function, 324
Gen, M., 303, 706
Genetic algorithm, 303
Gill, P.E., 581, 585, 706
Global optimization, 303
Glynn, J., 704
Golany, B., 420, 705
Goldberg, D.E., 303, 706
Golden ratio, 3, 243
Golden section search, 243
GoldenSectionSearch function, 244
Gomez, S., 581, 706
Gordon, S.G., 131, 705
Gradient condition, 151
Gradient vector, 92, 95
Graphical optimization, 16, 47
Graphical solution, appropriate range for variables, 49
GraphicalSolution function, 57
   options, 58
Gray, J.W., 704
Gray, T., 704
Greater-than type (GE) constraints, 14

Hansen, E., 303, 706
Hayhurst, G., 31, 706
Hennart, J.-P., 581, 706
Hentenryck, P.V., 303, 706
Hertog, D., 481, 706
Hessian matrix, 92, 95
Hessian of Lagrangian, 646
Hestenes, M., 302, 706
Himmelblau, D.M., 302, 706

http://www-c.mcs.anl.gov/home/otc/Guide/SoftwareGuide, 420
Hymann, B., 31, 706

Identity matrix, 76, 564
Inactive constraint, 148
Inconsistent system of equations, 325
Inequality constraint, 14
Inflection point, 133
Initial interior point, 452, 503
Interest functions, 19
Interior penalty function, 587
Interior point methods for LP, 17, 437
Interior point methods for QP, 17, 495, 503
Inventory-carrying charges, 7
Inverse barrier function, 588, 589
Inverse Hessian update, 288
Inverse of a matrix, 81
Investment risk, 10

Jacobian matrix, 100, 622
Jahn, J., 131, 706
Jeter, M.W., 131, 706
Jones, J.C., 31, 706

Karush-Kuhn-Tucker (KT) conditions, 147
Khachiyan, L.G., 481
Kolman, B., 420, 706
KT conditions, 147
   abnormal case, 173
   geometric interpretation of, 165
   for LP, 438
   for QP, 495, 536
KT point, 151
KTSolution function, 158
   options, 159

Lagrange multipliers, 150, 498, 536, 552, 591, 621
   recovering from LP, 378
Lagrangian duality in NLP, 199
Lagrangian function, 150, 552, 591, 621
Lagrangian for standard QP, 496

Length of a vector, 79
Less-than type (LE) constraints, 14
Life cycle cost, 18
Line search techniques, 231
Linear independence of vectors, 87
Linear programming (LP), 15, 315, 437
Linear system of equations, 319
LinearAlgebra'Choleski' package, 566
Linearization, 608
Linearize function, 609
Linearized problem, 609
Local duality, 199
Local maximum, 133
Local minimum, 133
Logarithmic barrier function, 588, 589
Lowe, M.J., 131, 707
LU decomposition algorithm, 330
LU decomposition, solving equations, 328
LUDecompositionSteps function, 332
Luenberger, D.G., 420, 706

Maeder, R., 704
Main diagonal of a matrix, 76
Matrix, 76
Matrix inverse identities, 82
Matrix inversion using Gauss-Jordan form, 327
Matrix operations, 76
Matrix relationships, 77
Maximization problem, 14
Maximum constraint violation, 628
McAloon, K., 131, 707
McCormick, G.P., 581, 706, 707
Megiddo, N., 481, 707
Merit function, 628, 647
    minimum of, 630
Michel, L., 303, 706
Minimization problem, 14
ModifiedNewton function, 273
    options, 273
Modified Newton method, 272
Multiple objective functions, 14
Multiple solutions, 368
Murray, W., 581, 585, 706

Nash, S.G., 199, 420, 707
Nazareth, J.L., 420, 707
Necessary conditions, 507, 589
    unconstrained problem, 132
Negative values on constraint right-hand sides, 317
Network problem, 359
NewFUsingSensitivity function, 177
NewtonRaphson function, 101
Newton-Raphson method, 100, 465, 520
Nicholls, R.L., 31, 708
No feasible solution, 365
Nonconvex functions, 113
Nonconvex set, 117
Nondegenerate solution, 370
Non-differentiable functions, 302
Nonlinear programming (NLP), 16, 581
Normal to a surface, 97
Null space, 448, 481
    projection matrix, 484, 567
NullSpace function, 484
Numerical methods
    for general NLP, 581
    for unconstrained, 17, 227

Objective function
    adding a constant, 136
    change using Lagrange multipliers, 176, 380
    contours, 52
    rate of change of, 176
Open top container example
    KT solution and sufficient conditions, 196
    RSQP solution, 651
Open-top rectangular container, problem formulation, 29
Optimality conditions, 552
    for convex problems, 181
    finding an optimum using, 134
    for standard LP, 438
    for unconstrained problems, 132
Optimality criteria methods, 16, 131
Optimization problem formulation, 2
    involving annual cost, 29

Index

Optimization problems
  classification of, 15
  solution methods, 16
OptimizationToolbox'Chap3Tools'
  package, 90, 101
OptimizationToolbox'CommonFunctions'
  package, 87, 96, 110, 122
OptimizationToolbox'ConstrainedNLP'
  package, 597, 609, 615, 632
OptimizationToolbox'EconomicFactors'
  package, 23
OptimizationToolbox'GraphicalSolution'
  package, 57
OptimizationToolbox'InteriorPoint'
  package, 443, 455, 470
OptimizationToolbox'LPSimplex'
  package, 324, 332, 338, 353, 393
OptimizationToolbox'Optimality
  Conditions' package, 137, 158, 177
OptimizationToolbox'Quadratic
  Programming' package, 498, 510,
  524, 539, 554
OptimizationToolbox'Unconstrained'
  package, 230, 234, 237, 241, 244,
  249, 252, 254, 262, 273, 289
Optimization variables, 2
Optimum of LP problems, 319
Ozan, T., 31, 707

Padberg, M., 481, 707
Pannell, D.J., 31, 420, 707
Papalambros, P.Y., 31, 707
PAS algorithm for LP, 454
PAS algorithm for QP, 509
Penalty methods, 585
Penalty number, 586
Penalty parameter, 628
  effect on convergence, 607
Peressini, A.L., 131, 707
Performance of different unconstrained
  methods, 281
Phase I & II simplex method, 350
Physical interpretation of Lagrange
  multipliers, 176
Pierre, D.A., 131, 707

Pike, R.W., 31, 707
PlotSearchPath function, 254
Polak, E., 302, 707
Polak-Ribiere formula, 262
Portfolio management example
  ActiveSetQP solution, 546
  Graphical solution, 67
  PrimalDualQP solution, 532
  problem formulation, 9
Positive definite matrix, 498, 646
Positive semidefinite matrix, 564
Possible cases using KTconditions, 153
Post-optimality analysis, 376
Primal affine scaling method for LP, 445
Primal affine scaling method for QP, 502
PrimalAffineLP function, 455
  options, 455
PrimalAffineQP function, 510
  options, 510
Primal-dual interior point for LP, 464
Primal-dual LP algorithm, 468
PrimalDualLP function, 470
  options, 470
PrimalDualQP function, 524
  options, 525
Primal-dual QP algorithm, 523
Primal dual QP method, 520
Primal feasibility, 467
Primal LP problem, 438
Principal minors, 85
PrincipalMinors function, 110
Procedure for graphical solution, 48
Projection matrix, 448
Pshenichnyj, B.N., 629, 707
Purchase versus rent example, 24
Purification procedure, 453
Puthenpura, S., 481, 706

Quadratic form, 106
  differentiation of, 107
  matrix form of, 107
  sign of, 108
QuadraticForm function, 110
Quadratic interpolation, 246
Quadratic programming (QP), 15, 495

QuadraticSearch function, 249
Quasi-Newton methods, 288
QuasiNewtonMethod function, 289
  options, 289

Range space, 481
Range space projection matrix, 484
Rank function, 87
Rank of a matrix, 86
Rao, S.S., 31, 707
Rardin, R.L., 31, 420, 707
Rectangular matrix, 76
Reeves, C.R., 303, 707
Refined sequential quadratic
  programming, 645
Regularity condition, 149
RevisedSimplex function, 393
  options, 393
Revised simplex algorithm, 390
Revised simplex method, 387
Row exchanges, 322
RSQP function, 647
  options, 647

Salvage value, 24, 29
Santosa, F.N., 645, 647, 705
Scales, L.E., 581, 707
Scaling objective function, 137
Scaling optimization variables, 585
Scaling transformation for LP, 446
Scaling transformation for QP, 564, 567
Schechter, R.S., 131, 705
Schnabel, R.B., 302, 706
Search methods, 302
Section search, 240
SectionSearch function, 241
Sensitivity analysis
  in NLP, 175
  using revised simplex, 402
SensitivityAnalysis option of the
  BasicSimplex function, 385
SensitivityAnalysis option of the
  RevisedSimplex function, 406
Sequential linear programming, 614
Sequential quadratic programming, 620

Series function, 90
sfdf function, 23
Shapiro, J.F., 581, 707
Shaw, W.T., 704
Sherali, H.D., 199, 420, 705
Shetty, C.M., 199, 420, 705
Shittkowski, K., 302, 707
Shor, N.Z., 131, 707
Shortest route problem, 359
Sign of a quadratic form
  eigenvalue test, 109
  principal minors test, 109
Simplex method, 17, 340
  matrix form of, 387
  unusual situations, 365
Simplex tableau, 344
Single payment compound amount
  factor (spcaf), 19
Single payment present worth factor
  (sppwf), 19
Sinking fund deposit factor (sfdf), 21
Slack variables, 591
  elimination of, 592
  for nonlinear problems, 150
SLP algorithm, 614
SLP function, 615
  options, 615
Sofer, A., 199, 420, 707
Solution of nonlinear equations, 100
Solving equations, 608
spcaf function, 23
sppwf function, 23
SQP algorithm, 631
SQP function, 632
  options, 632
SQP method, 581
SQP, refinements to, 645
Square matrix, 76
Standard form of an optimization
  problem, 13
Standard LP problem, 316
Stark, R.M., 31, 708
Starkey, C.V., 31, 708
Starting feasible solution, 340, 537
Stationary point, 133

Status of constraints, 377
Steepest descent method, 254
SteepestDescent function, 254
  options, 254
Step length, 506, 521, 537
Step length calculations, 231
Steven, G.P., 303, 708
Stock cutting example, 397
Sufficient conditions for NLP, 187
Sufficient conditions for unconstrained
  problems, 133
Suh, N.P., 31, 708
Sullivan, F.E., 131, 707
Sullivan, W.G., 18, 706
Surfaces, 96
Switching conditions, 151, 552
Symmetric matrix, 76

Tangent plane, 97
TaylorSeriesApprox function, 89
Thapa, M.N., 420, 705
Tigg, J., 704
Time value of money, 18
Tire manufacturing plant
  PrimalDualLP solution, 477
  problem formulation, 5
  RevisedSimplex solution, 415
Transpose of a matrix, 76
Tretkoff, C., 131, 707

Uhl, J.J., Jr., 131, 707
Unbounded solution, 366

Unconstrained minimization
  techniques, 15, 253
UnconstrainedOptimality function, 137
  options, 137
Uniform series compound amount
  factor (uscaf), 20
Uniform series present worth factor
  (uspwf), 21
Unrestricted variables, 318
Updating Lagrange multipliers, 595, 647
uscaf function, 23
uspwf function, 23

Vector, 76
Vector norm, 79
Vertex optimum for LP, 453

Wagner, D.B., 704
Water treatment facilities, 26
Watts, D.G., 13, 705
Web page for LP software, 420
Weighted sum of functions, 14
Wickham-Jones, T., 704
Wicks, E.M., 18, 706
Wilde, D.J., 31, 707
Wolfram, S., 704
Wright, M.H., 581, 585, 706

Xie, Y.M., 303, 708

Zigzagging in steepest descent method,
  259

Practical Optimization Methods

REGISTRATION CARD

Since this field is fast-moving, we expect updates and changes to occur that might necessitate sending you the most current pertinent information by paper, electronic media, or both, regarding *Practical Optimization Methods*. Therefore, in order to not miss out on receiving your important update information, please fill out this card and return it to us promptly. Thank you.

Name: _____

Title: _____

Company: _____

Address: _____

City: _____ State: _____ Zip: _____

Country: _____ Phone: _____

E-mail: _____

Areas of Interest/Technical Expertise: _____

Comments on this Publication: _____

_____

_____

❏ Please check this box to indicate that we may use your comments in our promotion and advertising for this publication.

Purchased from: _____
Date of Purchase: _____

❏ Please add me to your mailing list to receive updated information on *Practical Optimizaton Methods* and other TELOS publications.

❏ I have a(n)   ❏ IBM compatible   ❏ Macintosh   ❏ UNIX   ❏ other

Designate specific model _____

THE ELECTRONIC LIBRARY OF SCIENCE

Return your postage-paid registration card today!

PLEASE TAPE HERE

FOLD HERE

BUSINESS REPLY MAIL
FIRST-CLASS MAIL   PERMIT NO. 5863   NEW YORK, NY

POSTAGE WILL BE PAID BY ADDRESSEE

NO POSTAGE
NECESSARY
IF MAILED
IN THE
UNITED STATES

TELOS PROMOTION
SPRINGER-VERLAG NEW YORK, INC.
ATTN: K. QUINN
175 FIFTH AVENUE
NEW YORK NY 10160-0266

ROWAN COLLEGE OF NEW JERSEY

3 3001 00831 0483

QA 402.5 .B49 2000
Bhatti, M. Asghar.
Practical optimization
methods

DATE DUE

APR 01 2004		
APR 10 2005		